Physical Geography

A Landscape Appreciation

Jana Rosenkoetter
631-6642

Physical Geography

A Landscape Appreciation

Tom L. McKnight
University of California, Los Angeles

With cartographic collaboration of
Patricia Caldwell, Ph.D., University of California, Los Angeles
Caldwell & Associates

PRENTICE-HALL, INC., Englewood Cliffs, N.J. 07632

Library of Congress Cataloging in Publication Data

McKnight, Tom L. (Tom Lee), (date)
Physical geography.
Bibliography: p.
Includes index.
1. Physical geography. I. Title.
GB54.5.M39 1984 551 83-11199
ISBN 0-13-669101-3

Editorial/production supervision: Karen J. Clemments
Interior design and page layout: Caliber Design Planning, Inc.
Cover design: Linda Conway and Jayne Conte
Art Direction: Linda Conway
Photo researcher: Teri Stratford and Anita Duncan
Manufacturing buyer: John Hall

Cover photograph: Mt. Rainier, Washington, by Hans Wendler, The Image Bank

Photo credits: Chapter 1 (page 1) NASA; Chapter 2 (page 22) NASA; Chapter 3 (page 39) NOAA; Chapter 4 (page 50) U.S. Public Health Service; Chapter 5 (page 76) NASA; Chapter 6 (page 96) TLM Photo; Chapter 7 (page 117) NASA; Chapter 8 (page 140) Arizona-Sonora Desert Museum; Chapter 9 (page 169) USDA-Soil Conservation Service; Chapter 10 (page 186) TLM Photo; Chapter 11 (page 201) TLM Photo; Chapter 12 (page 238) TLM Photo; Chapter 13 (page 272) TLM Photo; Chapter 14 (page 289) TLM Photo; Chapter 15 (page 318) National Park Service; Chapter 16 (page 334) NASA; Chapter 17 (page 369) TLM Photo; Chapter 18 (page 393) TLM Photo; and Chapter 19 (page 414) TLM Photo.

Printed in the United States of America

10 9 8 7 6 5 4 3 2 1

ISBN 0-13-669101-3

Prentice-Hall International, Inc., *London*
Prentice-Hall of Australia Pty. Limited, *Sydney*
Editora Prentice-Hall do Brasil, Ltda., *Rio de Janeiro*
Prentice-Hall Canada Inc., *Toronto*
Prentice-Hall of India Private Limited, *New Delhi*
Prentice-Hall of Japan, Inc., *Tokyo*
Prentice-Hall of Southeast Asia Pte. Ltd., *Singapore*
Whitehall Books Limited, *Wellington*, *New Zealand*

This book is dedicated with respect and love to my mentors—*Edwin J. Foscue, Harold H. Hoffmeister, John W. Alexander, Glen T. Trewartha, C. Langdon White.* I only wish I had paid closer attention to the tutelage they offered and the examples they set.

Contents

11 Terrestrial Flora and Fauna 201

12 Soils 238

Preface

Despite having a highly literate and educated society, Americans are perhaps the world's most geographically ignorant people. In many other countries, geography is a basic field of study in both primary and secondary schools, as well as being a firmly established university subject. Over much of the world, schoolchildren are exposed to geographic training for most of their school years. This applies not only to such developed countries as England, Sweden, and New Zealand, but also to such less-developed lands as India, Tanzania, and Ecuador. Not so in the United States, where a pupil last hears the word *geography* in about the third grade, and the rare introduction of geographic course content at any higher level is usually muffled under the heading of "social studies."

Most American students, then, are surprised to find geography courses offered in colleges and universities. This discovery can evoke memories of third-grade recitations of people, places, and products, and raises questions as to the intellectual validity of such a discipline. What can a grade-school subject like geography provide to the questing college student?

Geography as a Field of Learning

The fact of the matter is that geography is an ancient and honorable field of learning, with its roots firmly set in classical antiquity. The Greek derivation of the word *geo-graphy* is "earth description." During the Hellenistic period, when Greek civilization, culture, and intellectual achievement dominated the "known world," many notable Greek scholars were more "earth describers" than anything else. Some of the prominent names that shine through the ages—such as Anaximander, Aristotle, Erathosthenes, Hecataeus, Herodotus, Hipparchus, Plato, Ptolemy—are famed as philosophers or historians or mathematicians or something else, but their contribution as geographers is at least as great.

An important corollary of this historical vignette is the realization that geography was (and is) a generalized—as opposed to a specialized—discipline. Its viewpoint is one of broad understanding. During the first centuries of the Christian era, there was a trend away from generalized earth description and toward more specific scholarly specializations. This led to a variety of more specialized disciplines—such as geology, meteorology, economics, botany, etc.—along with a concomitant eclipsing of geography. If "Mother geography gave birth to many offshoot sciences," as some geographers have said, it is clear that these developing disciplines soon became better known than their progenitor.

It was not until the 1600s that there began a rekindling of interest in geography in the European world, and it was another two centuries before there was a strong impetus given to the discipline by geographers from various countries, most notably the Germans Alexander von Humboldt and Karl Ritter in the middle of the nineteenth century. A prominent theme that has persisted through the centuries in the history of the discipline, and one that was clearly enunciated by both Humboldt and Ritter, is that geographers study how things differ from place to place over the earth—the areal differentiation of the Earth's surface. There is a multiplicity of "things" to be found on the Earth, and nearly all of them are distributed unevenly, thus providing a spatial, or geographic aspect.

The following table provides an abbreviated tabulation of the kinds of "things" to which we are referring; in this context they can be thought of as elements of geographic study. They are divided into two lists: one of physical or environmental elements that are natural in origin, and the other of cultural or human elements that are the products of some sort of human endeavor. The list of physical elements is short and essentially complete as a broad tabulation. The list of cultural elements, on the other hand, is quite incomplete and is merely suggestive of a much longer inventory of both material and nonmaterial features.

It is readily apparent that this tabulation of geographic elements contains no new words or concepts. The listed items are familiar to us all. This familiarity highlights another basic characteristic of geography as a field

The Elements of Geography

Physical	Cultural
Rocks	Population
Landforms	Settlements
Soil	Economic activities
Flora	Transportation
Fauna	Recreation activities
Climate	Languages
Water	Religion
Minerals	Political systems
	Traditions
	And many others

of learning: Geography has no peculiar body of facts or objects of study that it can call its own. The particular focus of geology is rocks, the primary interest of botany is plants, the attention of economics is fastened on economic systems, and almost all other academic disciplines have some distinctive assemblage of objects, facts, or concepts as a center of attention. Geography, however, like history, is a very broad field of inquiry and "borrows" its objects of study from related disciplines. Geographers, too, are interested in rocks, plants, and economic systems, but only in certain aspects of these elements. In simplest terms, geographers are concerned with the spatial, or distributional, aspects of these various elements. Thus geography is neither a physical science nor a social science; rather it combines characteristics of both and can be conceptualized as bridging the gap between the two. A geographer is not interested in the environment for the environment's sake, nor in humans for humans' sake. Indeed, geography is concerned with the environment as it provides a home for humanity, and the way that humans utilize that environmental home.

Another basic characteristic of geography is its interest in interrelationships. One cannot understand the distribution of soils, for example, without knowing something of the rocks from which the soils were derived, the slopes on which the soils developed, and the climate and vegetation under which the soils evolved. Similarly, it is impossible to comprehend the distribution of agriculture without an understanding of climate, topography, soil, drainage, population, economic conditions, technology, historical development, and a host of other factors, both physical and cultural. Thus the elements tabulated in the table are enmeshed in an intricate web of interrelationships, all of which are encompassed in the field of geographic study. To summarize: Geographers are concerned with the distribution of phenomena, how that distribution came about, and the significance of that distribution (to an understanding of the geography of the world). Thus the fundamental question of geographic inquiry is, "Why What is Where, and so What?"

Physical and Human Geography

Geography is a broad and integrating discipline that brings together important aspects of both physical and social sciences as well as impinging on the humanities. As a field of learning, then, geography can be viewed as a dichotomous subject—one that lends itself to subdivision into two broad and separate subfields: physical geography and human geography. The subject matter of this book is physical geography. Thus an attempt is made to summarize the various components of the "natural" environment, considering the nature and characteristics of the physical elements, the processes involved in their development, their distribution over the earth, and their basic interrelationships. Relationships with the various elements of human geography are largely ignored except where they help to explain development or contemporary distribution patterns of the physical elements.

Contents and Approach

This book presents the physical or environmental portion of geography in terms of its essentials. Written and designed for the introductory survey of physical geography, the text assumes that the student has little or no prior formal training in geography. The level of presentation doubles as both a survey for the liberal arts undergraduate and as a broad and systematic overview for the student who plans to major in the subject. An informal writing style, devoid of excessive jargon, an abundance of exceptionally accurate high-quality 4-color illustrations, and the use of "focus boxes" all work together to provide students with a solid background on which to build their future studies.

Written from a geographer's point of view, the text continually explores the processes of physical geography, carefully explaining the interrelationships of phenomena. There is a preliminary assessment of the earth's planetary characteristics, and a systematic presentation of the major components of physical geography with special emphasis on climate and topography. Within each chapter "focus boxes" provide useful vignettes that expand on topics that are presented more briefly in the text.

The first two chapters are preliminary in concept. Chapter 1 provides an introductory look at the Earth, discussing particularly its planetary setting and the environmental effects of its relationship with the Sun. Chapter 2 deals with the concepts and practicalities of portraying and representing the Earth's surface in ways understandable and meaningful to interested humans. The traditional roles of globes and maps are presented, and the exciting new techniques of remote sensing are broadly introduced.

Chapters 3 through 8 are focussed on the atmosphere. Emphasis is on the nature of the weather elements and the dynamics of atmospheric processes, leading to a consideration of the interaction of air masses, fronts, and storms. This section concludes with a discussion of the general subject of climatic classification and a detailed presentation of the widely used Köppen system for the world.

Chapter 9 is an introduction to the hydrosphere, which surveys the principal components of the aqueous part of our environment.

Chapters 10 and 11 comprise a more systematic and coherent presentation of the biotic part of the environment than is normally found in a physical geography textbook. The increasing ecological awareness of our populace calls for this extended treatment of the biosphere.

Chapter 12 is an expansive consideration of soils, dealing at some length with soil-forming factors, soil components, and soil-generating regimes. The latter half of the chapter considers soil classification in general and the United States Comprehensive Soil Classification in particular, outlining the world distribution pattern.

Chapters 13 through 19 are focussed on terrain analysis. These chapters proceed systematically from a conceptual introduction through a broad treatment of the internal crust-shaping processes (plate tectonics, diastrophism, vulcanism) to a more detailed consideration of the external processes of weathering, mass wasting, erosion, and deposition. The principal external land-shaping systems—fluvial, aeolian, coastal, solution, and glacial—are discussed in detail.

The writer believes that a useful definition of geography is "landscape appreciation," and has prepared this book with that theme in mind. "Landscape" is considered to include everything that one sees when looking out a window, everything that one hears upon listening at a window, and everything that one smells upon smelling at a window; referring to every window, actual and theoretical, in the world. "Appreciation" is used in the sense of understanding. Any proper exposition of geography should serve to heighten one's understanding of all that is seen, heard, and smelled at the window, whether an actual experience at a nearby window or a vicarious experience on the other side of the Earth. Thus it is the purpose of this book to make the environmental landscape of the world more understandable to the reader, at least at an introductory level.

What do you see when you cross the Mohave Desert from Los Angeles to Las Vegas? Three hundred miles of Not Much? A geographer sees three hundred miles of Quite A Lot. It is hoped that this book will help the reader expand his/her capacity for landscape appreciation from the former to the latter.

Acknowledgments

My heartfelt appreciation is extended to the following individuals. Arnold Court of California State University, Northridge provided a large measure of both substantive help and critical acumen to the entire manuscript; I hope he is not groaning too much over some of the things I didn't delete. It is not completely true that he suffers fools lightly. Norman Thrower of University of California, Los Angeles contributed significantly to both the text and the illustrative material of Chapter Two. I am also grateful to the following reviewers who afforded thoughtful and valuable suggestions for the manuscript's improvement: David E. Greenland, University of Colorado; John G. Hehr, University of Arkansas; David M. Helgren, University of Miami; David S. McArthur, San Diego State University; Derek Milton, University of Manitoba; Rodney Steiner, California State University–Long Beach; Thomas A. Terich, Western Washington University; Colin E. Thorn, University of Illinois; Thomas R. Vale, University of Wisconsin; Jean Wheeler, California State University–Long Beach; T. H. Lee Williams, University of Kansas; Craig ZumBrunnen, University of Washington.

Patricia Caldwell of Caldwell & Associates took my half-baked notions and transformed them into an attractive package of visuals. She supervised and designed all, and executed some, of the cartographic work. Her ability and diligence were beacons for me. The splendid drawings and diagrams flowed from the talented pen of Clint McKnight, which is a rousing affirmation of the concept of recessive genes.

Kathy Porter and Julia Witz were conscientious and thorough research assistants. Cheerful and efficient typing was supplied by Jeanne Douglas and Susan Murray. Geography editor, Betsy Perry, of Prentice-Hall inherited this project and saw it to fruition with persistence, tenacity, and kindness.

T. L. M.

Physical Geography

A Landscape Appreciation

1 Introduction to the Earth

The Earth is the ancestral, and so far the only, home of mankind. Thus it is the only planet of interest for geographic study. See Figure 1-1. People live on the surface of the Earth in a physical environment that is extraordinarily complex, extremely diverse, infinitely renewing, and yet ultimately fragile.

This habitable environment exists over almost the entire face of the Earth, which means that its horizontal dimensions are exceedingly vast. Its vertical extent, however, is very limited; the vast majority of all earthly life inhabits a zone less than 3 miles (5 km) thick, and the total vertical extent of the life zone is less than 20 miles (32 km).

Figure 1-2. The physical landscape of the Earth's surface can be considered to be composed of four overlapping and interacting "spheres": atmosphere, hydrosphere, biosphere, and lithosphere.

The Environmental "Spheres"

From the standpoint of physical geography, the surface of the Earth comprises a complex interface where the four principal components of the environment meet and to some degree overlap. See Figure 1-2. The solid, inorganic portion of the earthly fundament is the *lithosphere*, comprised of the rocks of the Earth's crust as well as the broken and unconsolidated particles of mineral matter that overlie the unfragmented bedrock. The vast gaseous envelope of air that surrounds the Earth is the *atmosphere*. The waters of the Earth—mostly consisting of oceans, lakes, rivers, and glaciers—compose the *hydrosphere*. All living things, plant and animal, collectively comprise the *biosphere*.

These four "spheres" are not discrete and separated but have considerable mixing. This is readily apparent when considering an ocean or other large body of water, a body that is clearly a major component of the hydrosphere and yet may contain a vast quantity of fish and other organic life that are part of the biosphere. An even better example is soil, which is composed largely of bits of mineral matter (lithospheric components), but it also contains life forms (biosphere), soil moisture (hydrosphere), and air (atmosphere) in pore spaces.

Figure 1-1. The spaceship Earth, as seen from the moon. [*Apollo 8* photo, courtesy NASA.]

Thus lithosphere, atmosphere, hydrosphere, and biosphere are not separate subsystems within a single earthly system. They are more properly considered as overlapping components of the environment of the Earth's surface. Moreover, they can serve as important organizing concepts for the systematic study of the Earth's physical geography and will be used in that sense as the basic units of the structural framework of this book.

Before focussing our attention on these four spheres, however, we must first set the stage by considering the planetary characteristics of the Earth and noting the salient relationships between our planet and its basic source of energy, the sun. Our Earth is merely an infinitesimal speck in the vastness of space. Yet it is the only speck on which we are sure that life exists. Its locational relationship with the totality of the universe is still imperfectly understood, but at least we can put it into perspective.

The Earth as a Planet

Before we consider the nature of the Earth itself, it is useful to try to visualize its setting in space.

Location in the Universe

The geographer's concern with spatial relationships properly begins with the location of our spaceship Earth within the universe. The Earth is one of the nine planets of

Focus on Distance Measurement in Space

On an historic day in 1969, a human first walked on the moon. One of the minor sidelights to that event was the communication delay between Houston Mission Control and the moon-walking astronaut. When Mission Control asked a question, almost three seconds elapsed before the astronaut's reply was heard. The reason for the delay was that it took 1.28 seconds for the radio message to reach the moon and a similar time for the return message. Radio waves travel at the speed of light, which is 186,282 miles (299,793 km) per second.

Similarly, the sunlight that we see departed from the sun $8\frac{1}{3}$ minutes before it reached us, after traversing more than 93 million miles (150,000,000 km) of space. Beyond our solar system, the staggering immensity of space can be visualized by considering the travel time for starlight. Light from the nearest star requires more than 4 years to reach the Earth, whereas starlight from the prominent Andromeda galaxy needs 2,200,000 years to travel to the Earth. If at some time we were able to receive a meaningful radio signal from Andromeda, and reply immediately to that message, the sender on Andromeda would have to wait nearly $4\frac{1}{2}$ million years for the response!

Distances in space are so enormous that a unit of measure larger than the familiar miles or kilometers is needed to avoid numbers so large that they would be difficult to write or to imagine. Since the speed of light is believed to be unvarying throughout the universe, the distance that light can travel in a year can be calculated to be almost 6 trillion miles (10,000,000,000,000 km). This unit of length, known as *a light-year*, is used to measure distances in the universe.

The distance that light travels in one year thus serves as an indication of the vast extent of the universe. The equivalent of a light-year in miles can be determined by multiplying the speed of light (officially 186,282.3976 mi/sec) by the 60 seconds in a minute, then by the 60 minutes in an hour, then by the 24 hours in a day, and finally by the 365.2422 days in a year

186,282.3976	mi/sec
× 60	
11,176,943.85	mi/min
× 60	
670,616,631.3	mi/hr
× 24	
16,094,799,150	mi/day
× 365.2422	
5,878,499,648,919	mi/yr

So, one light-year is equal to almost 6 trillion miles. The distance to the Andromeda galaxy—2,200,000 light-years—would be equivalent to 12,-932,700,000,000,000,000 miles (12 quintillion, 932 quadrillion, 700 trillion mi). It is easy to see why the use of light-years as a unit of measurement greatly increases our ability to express and comprehend such vast distances.

For the relatively shorter distances within our own solar system, however, the light-year is too long, whereas miles or kilometers are still awkwardly tiny. For these "intermediate" measurements, we can use the average distance from the Earth to the sun, called the *astronomical unit* (AU), which is 92,955,806 miles ± 3 miles (149,597,892 kilometers ± 5 km).

our solar system. The solar system also contains at least 31 other moons or satellites revolving around the planets, scores of comets, thousands of asteroids, and millions of meteors (most of them tiny specks of dust). The sun is the central star of our solar system and makes up more than 90 percent of its total mass. Our sun is also a star in the Milky Way galaxy—one of perhaps 100,000,-000,000 stars in a disk-shaped galaxy that is some 100,000 light-years in diameter and 10,000 light-years thick at the center. The Milky Way galaxy is only one of at least one billion galaxies in the universe!

The vastness of the universe is clearly beyond normal comprehension. As a small sample of astronomical distances, we might consider a model at a reduced scale. If the distance between the Earth and sun could be reduced from its actual figure of 93,000,000 miles (150,-000,000 km) to just 1 inch (2.5 cm), the distance to the nearest star would be a mere 4.5 miles (7.25 km), the distance to the nearest galaxy would become about 150,000 miles (240,000 km), and the distance to the Andromeda galaxy would be close to 2,000,000 miles (3,-200,000 km).

The schematic representation of our galaxy (Figure 1-3) shows that our sun is on the Orion arm, an outer spiral arm of this thin, disk-shaped outer edge of our galaxy. Our sun is some 30,000 light-years from the center of our galaxy, and it moves in a nearly circular path around the center at a speed of 124 to 186 miles (200 to 300 km) per second.

As we look up into the nighttime sky, we can see a few of the nearby stars that are also on Orion's arm. For example, our nearest neighbors, Proxima Centauri and Alpha Centauri, only 4.2 light-years away from the Earth, are the brightest stars in the sky of the Southern Hemisphere. But when we look back from the edge of our galaxy into the densest part of its disk, we see what appears as a long foggy pathway of stars across the sky; thus, the name of our galaxy—the Milky Way.

The Solar System

Most astronomers believe that the celestial patterns in the universe began to develop some 20 billion years ago with a massive cosmic explosion, or "big bang," after which all the material began dispersing at enormous speeds. Our solar system apparently originated when a *nebula*, a cloud of interstellar gas and dust perhaps one light-year in diameter, began to contract inward, forming a sphere of hot gas that became our sun. This hot center was surrounded by a cold, revolving disk that formed

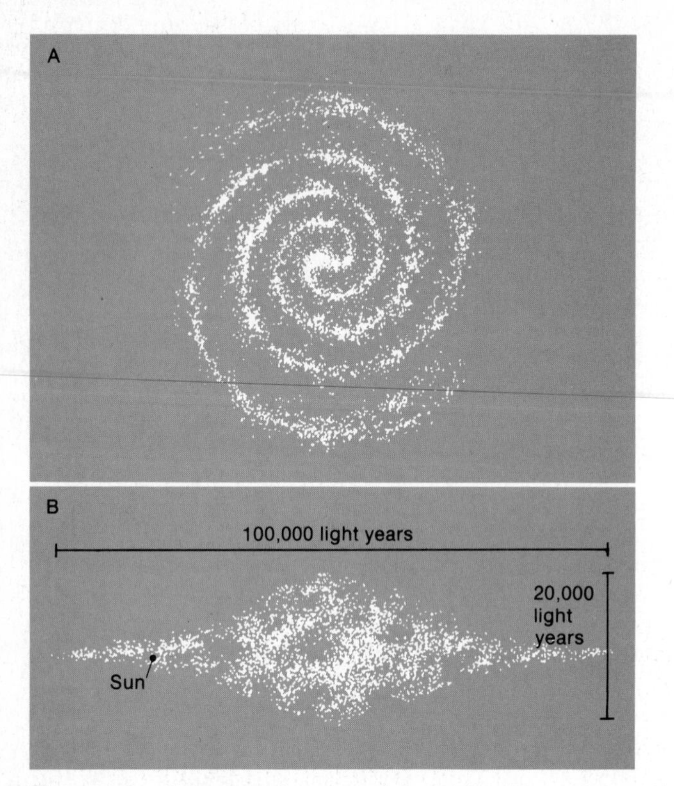

Figure 1-3. Schematic drawing of the Milky Way galaxy. (A) The spiral arrangement of the larger stars is conspicuous. (B) A cross section through the galaxy shows a disk-shaped pattern with thinning of the clusters outward from the center. Our solar system is positioned toward the outer end of the disk.

the planets. It is thought that the *outer* (or *Jovian*) planets—Jupiter, Saturn, Uranus, Neptune, Pluto—formed as subconcentrations in this revolving disk, whereas the *inner* (or *terrestrial*) planets—Mercury, Venus, Earth, Mars—formed by the gathering of smaller particles. See Figure 1-4. The planetary satellites, such as our moon, may have been formed in this manner elsewhere in the nebula. Their orbits around the sun then brought them close enough to the planets to be captured by mutual gravitational influence.

The members of our solar system have a strikingly orderly organization. The orbits of all nine planets are in nearly the same plane, perhaps revealing their relationship to the formative nebular disk. The four inner planets are smaller, denser, slower in rotation on their own axes, and less oblate (that is, more nearly spherical). Also, the inner planets (except Earth) are composed principally of mineral matter and have either no atmospheric covering or only a shallow outer covering rich in carbon dioxide and water vapor, released from the crustal rocks of the planets.

By contrast, the outer planets are much larger, at least 100 times heavier, less dense, rotate more rapidly, and are much more oblate. The compression of their axes of rotation is probably the result of their gaseous composition. The Jovian planets seem to be composed entirely of gases, principally hydrogen, which becomes liquid and then frozen toward the interior of the planets. Their chemical composition is similar to that of the sun, indicat-

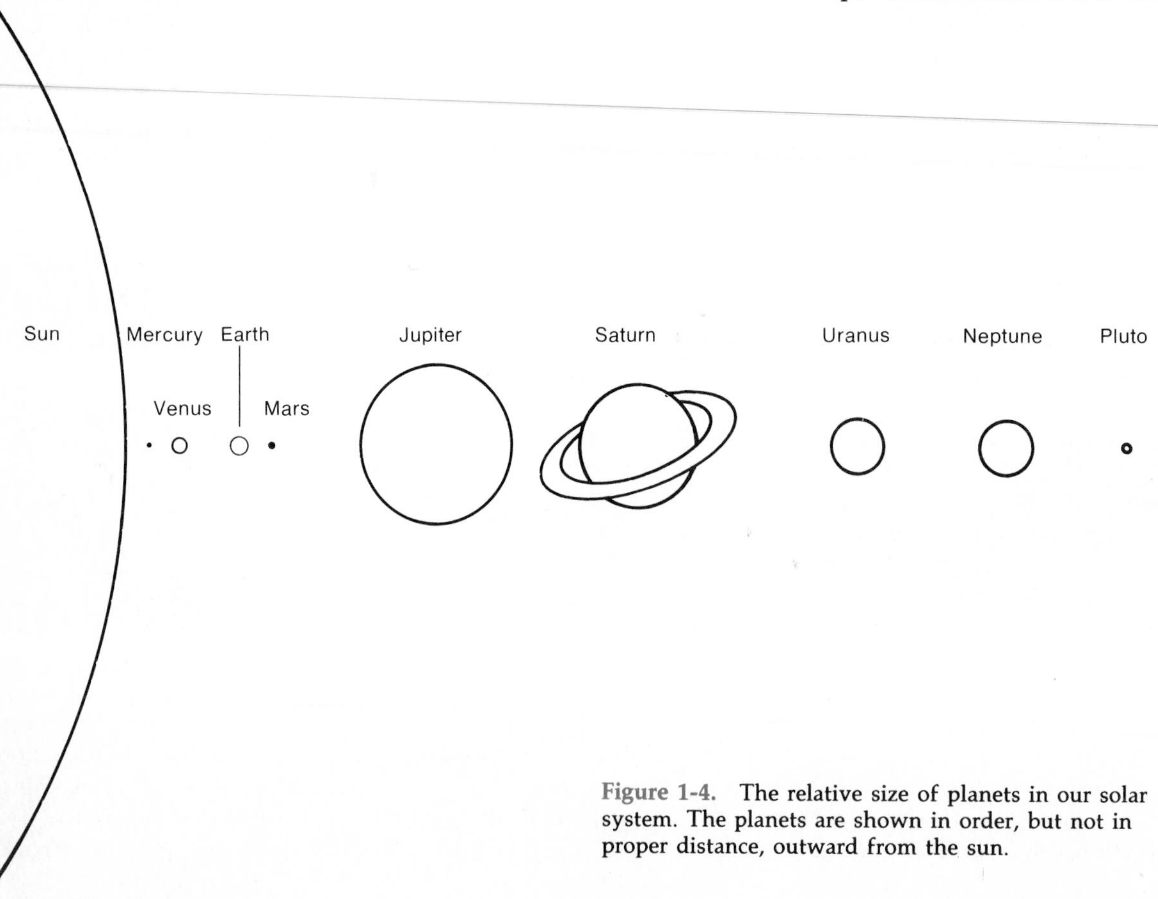

Figure 1-4. The relative size of planets in our solar system. The planets are shown in order, but not in proper distance, outward from the sun.

ing that they were formed in a similar fashion. Jupiter and Saturn, with their rings and satellites, can be viewed as smaller models of our solar system. Indeed, Jupiter and Saturn each emit more energy than they receive from the sun.

The sun itself rotates from west to east, and each of the planets revolves around the sun from west to east. Moreover, each of the planets, except Venus and Uranus, rotates from west to east on its own axis. Generally, planets move more slowly and have a lower effective temperature in succession away from the sun.

Even at the "local" scale of our own solar system, the vast distances involved are still extraordinarily impressive. If the sun could be represented by a large beach ball two feet (0.6 m) in diameter, the Earth would become a body about the size of a pea located at a distance of 215 feet (66 m). Jupiter would be the size of a large orange about 2000 feet (600 m) away, and the nearest star, Proxima Centauri, would be about as far away as Honolulu is from New York City. See Figure 1-5.

Figure 1-5. The universe contains an enormous number of celestial bodies. The planet Saturn is a prominent member of our solar system. [*Voyager 1* photo, courtesy NASA.]

Influence of Stars and Planets on the Earth

Various celestial bodies are significant to our lives because of the effects they exert on the physical characteristics of our home planet or on our understanding of some of these characteristics. Prominent among such influences are the following:

Location and Time The heavenly bodies visible to us appear to make a daily trip across the sky as the Earth continually rotates. The precision of these apparent movements is such that the stars can serve both as beacons to provide exact reference points for measuring location and as clocks to mark the passage of time. Mankind still uses the positions of the stars and planets to verify precise locations on Earth and to judge the accuracy of clocks and calendars.

Intellectual Curiosity Perhaps the greatest importance of the stars and planets is as objects of intellectual curiosity. By looking into space, we can look into the past. When energy in the form of light reaches us from a galaxy that is 100 million light-years away, it tells us of events that took place 100 million years ago. Similarly, analysis of data from other planets gives us clues from the past and maybe hints for the future of our own planet. Since our ultimate survival may well depend on an understanding of the nature of our planet, the value of understanding outer space and other planets is inestimable.

Source of Energy and Gravitational Attraction amount of energy coming to us directly from all stars, except our sun, or reflected to Earth by the planets, is extremely small. However, energy in the form of light, radio waves, and other fractions of the electromagnetic spectrum does reach the Earth from much of the universe. Nearly all the data we have about other celestial bodies have been obtained from careful measurements of the quantity and quality of the energy that they generate or reflect.

Newton's *Law of Universal Gravitation* states:

> Any two objects in space are attracted by a force proportional to the product of their masses and inversely proportional to the square of the distance that separates them.

Expressed algebraically, the force (F) attracting two masses (m_1 and m_2), which are separated by a distance (d), is:

$$F = k\frac{m_1 m_2}{d^2}$$

(k is the constant of gravitation)

Most heavenly bodies are so far from the Earth that their distance, especially when squared, makes this gravitational attraction so small as to be almost without meaning. Even the other eight planets of the solar system exert very small gravitational pulls on the Earth. However, the sun, although quite distant, has a mass so great, and the moon, although quite small, is so close, that these two heavenly objects account for nearly all of the exterior gravitational influence on the Earth. In the same manner, nearly all of the external energy received by the Earth is generated by the sun or reflected by the moon. Thus, it is apparent that these two objects deserve our close attention.

Surface Gravity of the Earth

The Newtonian principle of gravitation also governs the surface gravity of the Earth. However, at the Earth's surface, it is the Earth itself that offers the only appreciable mass. See Figure 1-6. The mass of an object (whatever it may be) being attracted toward the center of the Earth is so small by comparison that its gravitational attraction to its own center of mass (m_2) can be ignored. Thus, the *surface gravity of the Earth* (g), that is, the force attracting surface bodies toward the center of the Earth, may be calculated from the equation of gravitational attraction given below (where m_1 is the mass and d is the radius of the surface body):

$$g = k\frac{m_1}{d^2}$$
$$= 32.15 \text{ ft/s}^2 \text{ or } 9.8 \text{ m/s}^2$$

Notice that gravity is expressed as a *force of acceleration* of about 32 feet (9.8 m) per second per second. This means that, ignoring the frictional resistance offered by the air itself, an item dropped from 1000 feet (305 m) would be falling at the rate of 32 feet (9.8 m) per second at the end of the first second, at 64 feet (19.6 m) per second at the end of the next second, and so on, adding another 32 feet (9.8 m) per second of speed for each additional second of fall.

Every living organism on the Earth's surface, plant or animal, has adapted its development to the force of gravity. The size, shape, and strength of the stems and trunks of plants as well as the outer or inner skeletons of animals would have to be different on a planet with a greater or lesser surface gravity. The physical processes that change the surface of the Earth itself, such as running water or glaciers, are also significantly dependent on gravity. And all the structures built by mankind, such as buildings, bridges, and aircraft, have to be constructed to accommodate the force of gravity.

Size and Shape of the Earth

Is the Earth large or small? The answer to this question depends on one's frame of reference. In the setting of the universe, the Earth is almost infinitely small. The radius of our planet is only about 4000 miles (6400 km). In comparison, the moon is 230,000 miles (370,000 km) from the Earth, the sun is 93,000,000 miles (150,000,000 km) away, and the nearest star is 25,000,000,000,000 miles (40,000,000,000,000 km) distant.

From a human viewpoint, however, the Earth is a very impressive mass. Its highest point is almost 30,000 feet (9000 m) above sea level, and the deepest spot in the oceans is about 36,000 feet (11,000 m) below sea level. See Figure 1-7. The vertical distance between these two extremes is less than 13 miles (21 km), which is minuscule when contrasted with the 8000-mile (12,875-km) diameter of the Earth. To illustrate this relationship, if we had a globe with the diameter of a 40-story building, the difference in elevation between the highest point of the

Figure 1-6. The strength of surface gravity depends largely on the mass of the body beneath the surface. By way of illustration, a person who can high jump 6 feet on Earth would be able to leap considerably higher on a smaller heavenly body but not nearly so high on a larger one.

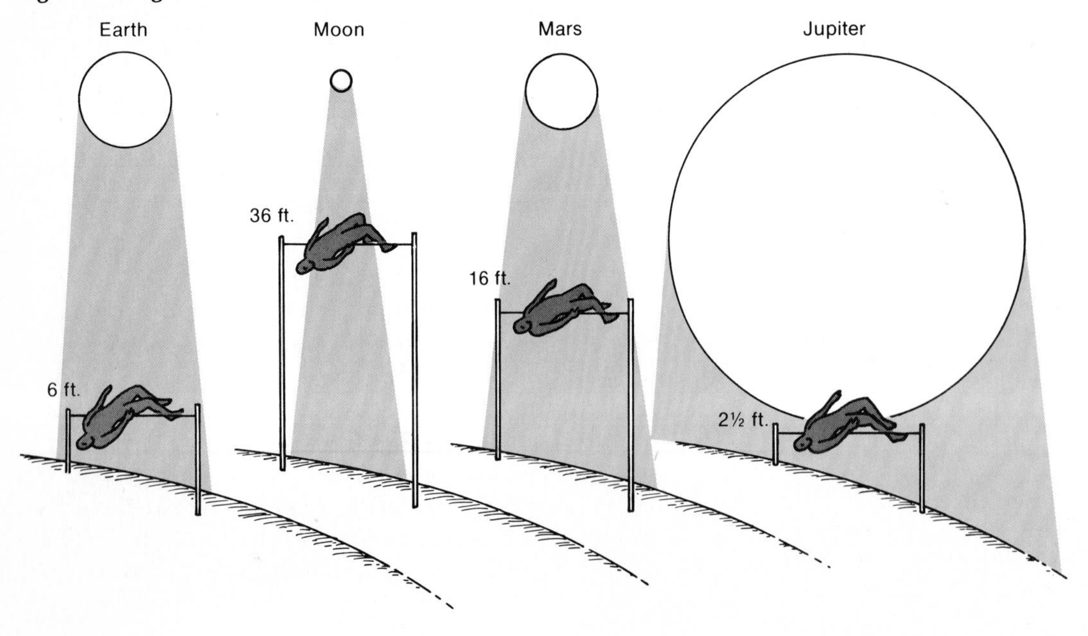

Figure 1-7. The terrain of the Earth's surface is very impressive when one is close to it, but in comparison with the vast size of our planet, the surface irregularities seem very minor. This is a satellite view southeasterly across the world's highest mountains, the Himalayas. [Courtesy NASA.]

continents and the lowest point in the ocean basins would be represented by the width of a single brick on the top of the building. See Figure 1-8.

From time to time, various people have maintained that the Earth is flat or even disk-shaped. (There is even a Flat Earth Society that still exists in the United States today). However, as early as the sixth century B.C., some Greek scholars correctly believed the Earth to have a spherical shape. About 200 years before Christ, the director of the Greek library at Alexandria, Eratosthenes, calculated the circumference of the Earth trigonometrically. He determined the angle of the noon sun rays at Alexandria and at Syene (a place in Upper Egypt thought to be located on the Tropic of Cancer), measured the distance between the two localities, and from these angular and linear distances was able to estimate an Earth circumference of almost 26,700 miles (43,000 km), which is reasonably close to the actual figure of 24,900 miles (40,000 km). Historians believe that Eratosthenes made errors in his calculations, errors that partially cancelled one another and made his result more accurate than it deserved to be. Be that as it may, several other Greek scholars made independent calculations of the Earth's circumference during this period, and all of their results were close to reality, which indicates the antiquity of accurate knowledge of the size and shape of the Earth.

Even the most casual observer, however, can see that the surface of the Earth is not smooth, which means that our planet does not have a perfectly spherical shape. Moreover, there are occasional reminders in the media that the Earth is a "spheroid," or "pear-shaped," or some other deviation from a perfect sphere. Although perfect sphericity is lacking, the Earth is much more nearly spherical than many other objects that we do not challenge as perfect spheres. Any rotating body has a tendency to bulge around its equator and flatten at the polar ends of its rotational axis. This tendency of matter to move away from the center of a rotating object is a response to *centrifugal force.* Although the rocks of the Earth may seem quite rigid and immovable to us, they are sufficiently pliable to allow the Earth to develop a bulge in its midriff.

Figure 1-8. The Earth's maximum relief is about 65,000 feet (more than 20 km), from the top of Mt. Everest to the bottom of the Mariana Trench. In comparison with the diameter of the Earth, however, this would be equivalent in vertical dimension to a single brick on top of a 40-story building.

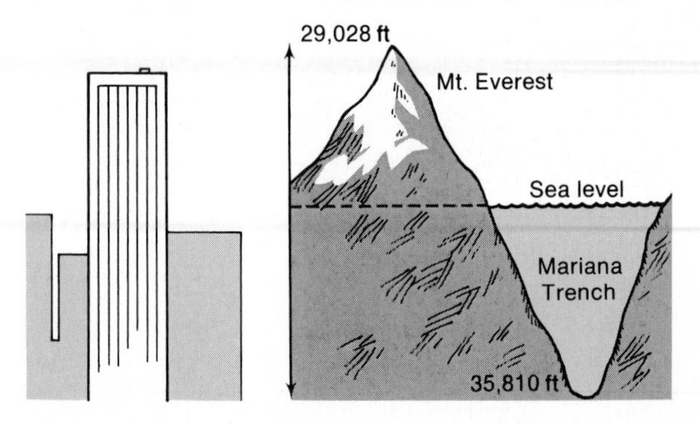

Focus on Proofs of the Earth's Spherical Shape

Various kinds of evidence can be used to prove the sphericity of the Earth. Some of the more prominent proofs include the following:

1. *Disappearance of ships into the ocean*
   As a ship sails away toward the horizon, it appears to sink into the ocean. Slowly the level of the ocean seems to rise to cover first the hull, then the decks, the mast, and the sails, until the vessel is swallowed up on the horizon. Since the ship is not actually sinking beneath the water, the ocean surface must be curving away from the viewer. Nevertheless, a large number of carefully recorded observations, noting the proportion of the ship covered at various distances and directions from many different ports, would be needed to prove that the shape of the curving surface on all oceans is a part of a single sphere.
2. *Circular shadow of the Earth during a lunar eclipse*
   It is a geometric axiom that a sphere is the only three-dimensional object that will always cast a circular shadow. Since the Earth approaches the moon from many different angles in a series of lunar eclipses, the fact that the shadow of the Earth is always a circle proves that the Earth is a sphere.
3. *Varying position of the North Star on the Earth's surface*
   At the North Pole, the North Star (Polaris) is almost directly overhead. As a night passes, the other stars appear to circle the celestial dome of the sky while the North Star apparently remains motionless overhead. At New York City (latitude 40° north), Polaris appears 40° above the north horizon with the other stars circling it. At the equator, Polaris is on the north horizon. A similar set of observations may be made on a minor star directly over the South Pole. We can conclude from the many observations of these relationships made by countless mariners in all the oceans over the centuries that only on a sphere is such a set of results possible.
4. *Photographs from space*
   Perhaps the most dramatic and convincing proofs that the Earth is a sphere are the sets of photographs of our planet taken by astronauts in space. The successive views of the rotating Earth dramatically establish beyond any doubt that we live on a rotating sphere.

A ship appears to sink into the sea as it passes beyond the horizon line, indicating the spherical shape of the Earth. [After A. N. Strahler and Alan H. Strahler, *Elements of Physical Geography*, 2nd ed. (New York: John Wiley & Sons, Inc., 1979), p. 10. © 1979 by John Wiley & Sons. Used by permission of John Wiley & Sons.]

An *Apollo 11* photo leaves no doubt that the Earth is spherical in shape. [Courtesy NASA.]

The slightly flattened polar diameter of the Earth is 7900 miles (12,714 km), whereas the slightly bulging equatorial diameter is 7927 miles (12,757 km), a difference of only about 0.3 percent. Thus our planet is properly described as an *oblate spheroid* rather than a true sphere. See Figure 1-9.

Moreover, the shape of the Earth has obvious deviations from true sphericity that are the result of topo-

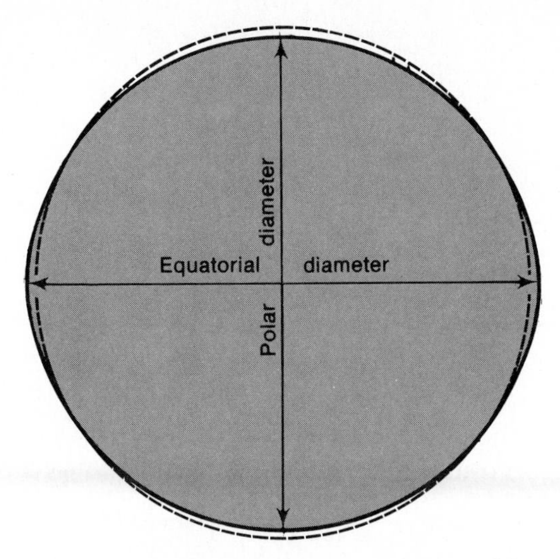

Figure 1-9. The Earth (surface represented by a solid line) is an oblate spheroid. The dashed line represents the shape of a perfect sphere. The variation from sphere to spheroid is so small that it has been exaggerated in this diagram in order to make it visible.

graphic irregularities on its surface. This surface varies in elevation from the highest mountain peak, Mt. Everest at 29,000 feet (8850 m) above sea level, to the presumed deepest ocean trench, the Mariana Trench of the Pacific Ocean, at 35,800 feet (10,915 m) below sea level, a total difference in elevation of 65,000 feet (19,800 m). Although prominent on a human scale of perception, this difference is minor on a planetary scale. If the Earth were the size of a basketball, Mt. Everest would be an imperceptible pimple no greater than 0.00069 inch (about seven ten-thousandths of an inch) or 0.018 mm in height. Similarly, the Mariana Trench would be a tiny crease only 0.0009 inch (0.023 mm), or about one-thousandth of an inch, deep. This represents a depression of a lesser depth than the thickness of a sheet of notebook paper.

Our perception of the relative size of these topographic irregularities has been further confused by three-dimensional wall maps and globes that emphasize such landforms. To portray any noticeable appearance of topographic variation, the vertical distances on such maps usually are exaggerated 8 to 20 times their actual proportional dimensions.

The relative variation from true sphericity, then, is exceedingly minute. For most purposes, the Earth may be properly considered as a sphere. Indeed, it is a more nearly perfect sphere than most of the "spheres" with which we come into frequent contact, such as basketballs, bowling balls, baseballs, or ping-pong balls.

The Geographic Grid

Any understanding of the distribution of phenomena over the Earth's surface requires some system of accurate location. Such a system should be capable of pinpointing with mathematical precision the position of any spot on the surface. The simplest technique for achieving this precision is to design two unvarying sets of lines that intersect at right angles, thus permitting the location of any point on the surface to be described (mathematically) by the appropriate intersection. Such a network of intersecting lines is referred to as a *grid system*. For purposes of location on the Earth's surface, a geographic grid, consisting of east-west lines and north-south lines, has been devised.

If the Earth were an absolutely perfect sphere, the problem of describing precise surface locations would be much more formidable than it is. With no deviation from sphericity, no natural points of reference would exist from which measurements and surveys could be started. Our rotating Earth, however, with its slightly bulging "middle" and slightly flattened "ends," provides the natural reference points for a systematic locational system.

The Earth rotates continually about an *axis*, which is a diameter line that connects the points of maximum flattening on the Earth's surface. These points are called the *North Pole* and the *South Pole*. If we visualize an imaginary plane that passes through the Earth halfway between the poles and perpendicular to the axis of rotation, we will then have another valuable reference feature: *the plane of the equator*. Where this plane intersects the Earth's surface is the imaginary midline of the Earth, called simply the *equator*. We can now use the North Pole, South Pole, rotational axis, and equatorial plane as natural reference points for measuring and describing locations on the Earth's surface. Without these, a set of reference points would have to be chosen arbitrarily to establish an accurate locational system. Imagine the difficulty of trying to describe the location of a particular point on a ping-pong ball. If we stick a needle through its diameter and draw a circle around the ball midway between the ends of the needle, we will have produced "poles" and an "equator," and thus provided arbitrary reference points for a locational system.

Any plane that is passed through the center of a sphere bisects that sphere (divides it into two equal halves) and creates a *great circle* where it intersects the surface of the sphere. Planes passing through any other part of the sphere, but not through its center, produce *small circles* where they intersect the surface. See Figure 1-10. Great circles have two principal properties of special interest:

1. A great circle is the largest circle that can be drawn on a sphere; it represents the circumference of that sphere and divides its surface into two equal halves, or *hemispheres*. For example, the sun illuminates one-half of the Earth at any given moment. The edge of the sunlit hemisphere, called the *circle of illumination*, is a great circle that divides the Earth between a light half and a dark half.

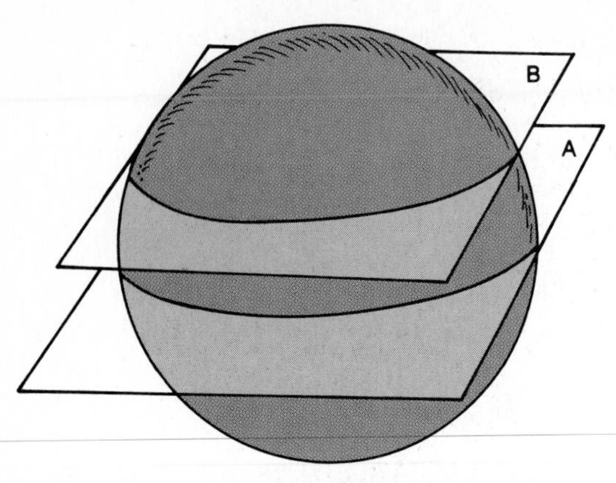

Figure 1-10. Comparison of great and small circles. A great circle represents the intersection of the Earth's surface with a plane (A) that passes through the exact center of the Earth. A small circle results from the intersection of the Earth's surface with any plane (B) that does not pass through the center of the Earth.

2. Only one great circle can be constructed to include any two given points (not at the same latitude or diametrically opposite) on the Earth's surface. The segment, or *arc*, of the great circle connecting those two points is always the shortest route between the points. Such routes are known as *great circle routes*. They will appear as curved lines on most maps.

Latitude Latitude is distance measured north and south of the equator. As shown in Figure 1-11, we can project a line from a given point on the Earth's surface to the center of the Earth. The angle between this line and the equatorial plane is the measure of the latitude of the point. Like any other angular measurement, latitude is expressed in degrees, minutes, and seconds. There are 360 degrees (°) in a circle; 60 minutes (′) in a degree; and 60 seconds (″) in a minute. We must keep these expressions in mind when using angular data because our general use of the decimal system is likely to leave us with the unfounded impression that all measurements are based on 10.

Figure 1-11. The latitude of any place on the Earth's surface is determined by measuring the angle between a radius from that place to the center of the Earth and the equatorial plane.

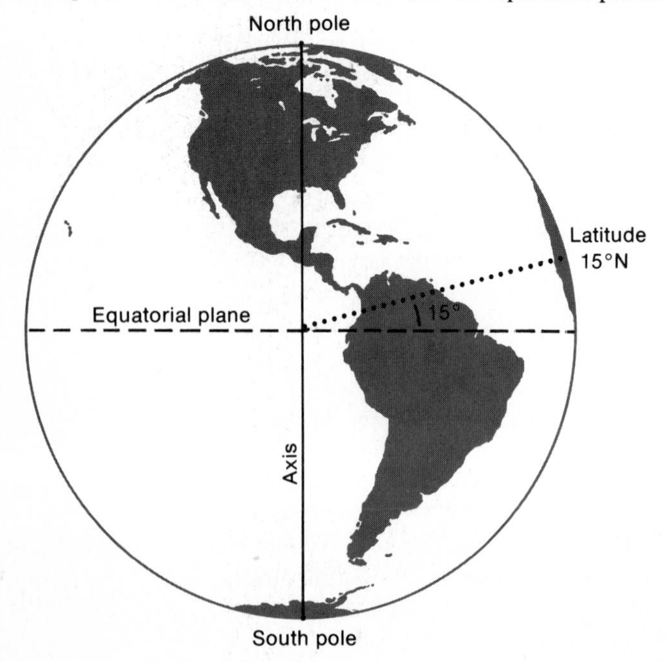

Latitude varies from 0° at the equator to 90° at the North Pole or South Pole. When a direct line to the center of the Earth is projected from a point lying on the surface between the equator and the North Pole, the resulting angular distance between that line and the equatorial plane is *north* latitude. The angle for such a line from the Naval Observatory at Washington, D.C., for example, is 38 degrees, 55 minutes, and 14 seconds; thus, the latitude of the Naval Observatory is expressed: 38°55′14″N. The latitude for any point in the Northern Hemisphere will lie between 0° and 90°N; conversely, any point within the Southern Hemisphere will have a latitude between 0° and 90°S.

A line connecting all points of the same latitude is called a *parallel* because it is parallel to all other lines of latitude. The equator is the parallel of 0° latitude, and it, alone of all parallels, constitutes a great circle. All other parallels are small circles, and all are aligned in true east-west directions on the Earth's surface. Because parallels are imaginary lines, there can be an infinite number of them. There can be a parallel for each degree of latitude, or for every minute, or for every second, or for any fraction of a second. However many latitude lines that one visualizes, their common property is that they are all parallel to each other.

Although it is possible to construct, or visualize, an unlimited number of parallels, seven of them are of particular significance in a general study of the Earth; we will emphasize their importance in this chapter. See Figure 1-12. The seven are:

1. Equator—0°
2. North Pole—90°N
3. South Pole—90°S

(2 and 3: These two are actually points, but can be thought of as parallels.)

4. Tropic of Cancer—23 ½°N
5. Tropic of Capricorn—23 ½°S
6. Arctic Circle—66 ½°N
7. Antarctic Circle—66 ½°S

A degree of latitude has a length of about 69 miles (see Table 1-1). The actual distance varies slightly with latitude, due to the polar flattening of the Earth.

Longitude Latitude comprises one-half of the Earth's grid system, the north-south component. The other half is represented by *longitude*, which is distance measured

The United States is currently in the process of establishing the mandatory use of the metric system for the following categories of measurement:

Category	Metric Base Unit
1. Length (including area and volume)	Meter
2. Mass (weight)	Kilogram
3. Temperature	Celsius
4. Electric current	Ampere
5. Luminous intensity	Candela
6. Amount of substance	Mole

Since the metric system was legalized for both interstate and intrastate trade in 1866, the current "metrication" of the United States represents the phase-out and eventual prohibition of the traditional (or English) system as one of two alternative legal measuring systems for commerce. You have undoubtedly noticed that this book gives measurements in both currently legal systems.

The metric system was originally designed by French scientists in the midst of the French Revolution (1790s), responding to a directive by their government to "deduce an invariable standard for all the measures and all the weights" that would be both simple and scientific. The designers chose a standard unit of length from which could be derived measures for volume and weight, thereby relating these basic units of the system to one another and to nature. Moreover, the larger and smaller components of each unit could be obtained simply by moving the decimal point (i.e., multiplying or dividing by multiples of 10). The basic unit chosen was the *metre* (which we spell *meter*), a word derived from the Greek *metron*, meaning "a measure." A meter was defined as one ten-millionth of the distance from the North Pole to the equator along the central meridian of France. The basic unit of weight was the *gram*, the mass of one cubic centimeter of water at its temperature of maximum density. *Liter*, or the capacity of a cubic decimeter, was the basic unit of volume.

France officially adopted the metric system in 1840, and other nations successively accepted it until by 1900, most of Europe and Latin America had "gone metric." During the twentieth century, international conventions revised many of the details, but not the basic concepts, of the metric system, and the name of the modernized system was changed to International System of Units, abbreviated SI. In 1975, the United States became the last significant country in the world to begin adoption of the system.

Although the impetus for this changeover is from businesses that seek metric standardization in order to increase their economies of scale when competing in foreign markets where metric units are standard, the considerable costs of this transition are often justified in terms of the greater logic, consistency, and uniformity of having all of our measurements decimal, that is, base-10. However, even those countries that mandated a change to the metric system more than a century ago still retain many nondecimal measurement units.

The Gregorian calendar, which we use, and the measurement of time, now fairly standard throughout the world, represent quite a potpourri of numerical equivalencies. For instance, there are 60 seconds in a minute, 60 minutes in an hour, 24 hours in a day, and 7 days in a week. A year has 365.242199 days, or 52.177455 weeks, or 12 months, with each month having, variously, 28, 29, 30, or 31 days. Similarly, a circle has 360 degrees, each degree has 60 minutes, and each minute has 60 seconds. Why are these measurements not also being revised to a base-10 (decimal) standard? Although the changeover of our timekeeping, calendars, and all angular measurements would be costly and psychologically unsettling, it would not be more expensive or troublesome than the changes already projected.

The remaining nondecimal units of measurement are commonly expressed in multiples of 60 or 12. What is the origin of this tendency toward 12's and 60's? As early as 4500 B.C., traders carried Egyptian ideas of numeration systems, including base-10, to Babylonia. However, the Babylonians reorganized the Egyptian system so that written numbers would have a positional, or place, value. As far as we know, the Babylonians were the first people to develop this system, which is basic to all modern numeration. Also, since the Babylonians concentrated on studying astronomy, they were justifiably impressed by the increased number of even divisors in a system based on 12. For example, 10 has three divisors: 2, 5, and 10; whereas 12 has five divisors: 2, 3, 4, 6, and 12. Moreover, although 100 (and any other base-10 number ending in two zeros) is divisible by 2, 4, 5, 10, 25, 50, and 100, the number 144 (the "100" of a base-12 system) is evenly divisible by twice as many divisors: 2, 3, 4, 6, 8, 9, 12, 16, 18, 24, 36, 48, 72, and 144. For this reason, the Babylonians first chose 12 as the base for their number system. Later, they combined the advantages of both the base-10 and the base-12 systems by changing the base of their system to 60 (12 × 5).

Because the Greeks began their study of geometry with knowledge from the Babylonians, they adopted the Babylonian base-60 for measuring angles. Through time, most of the world's cultures have developed or returned to a numeration system based on 10, probably because we have 10 fingers on our hands and we first learn to count by using these fingers. However, most of the world's cultures have retained the dividing properties of 12 in measuring time and angles, as well as in an array of everyday commercial measurements such as:

Dozen	12
Gross	144 (12 × 12)
Great gross	1728 (144 × 12)
Quire	24 sheets (paper)
Ream	20 quire = 480 sheets (now often 500)
Karat	$\frac{1}{24}$ pure gold (pure gold = 24 karats; one-half gold = 12 karats)

A table of equivalents between the English and metric systems can be found in Appendix A.

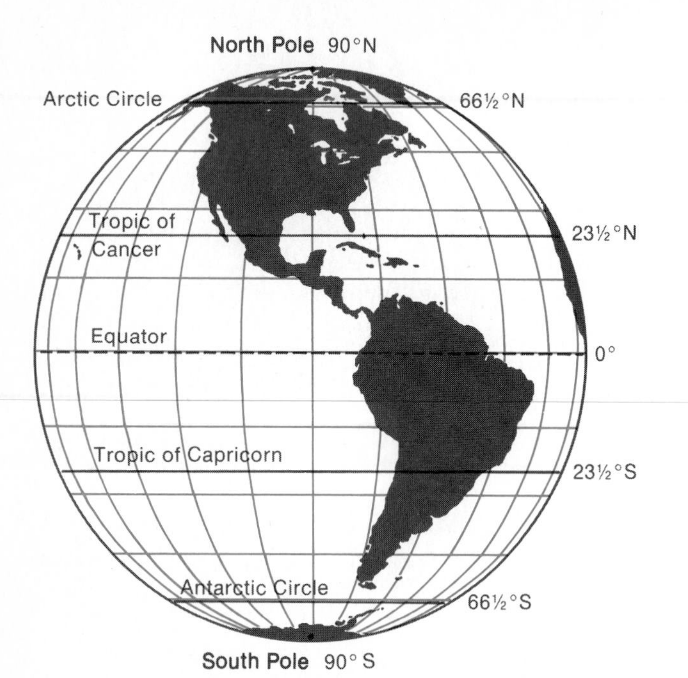

Figure 1-12. The significant parallels of latitude.

east and west on the Earth's surface. Again, this is angular distance, so longitude is also measured in degrees, minutes, and seconds.

Longitude is represented by imaginary lines extending from pole to pole, and crossing all parallels at right angles. These lines, called *meridians*, are not parallel to one another except where they cross the equator. They are farthest apart at the equator, becoming increasingly close together northward and southward, and finally converge completely at the poles. A meridian, then, is a great semicircle that extends from one pole to the other, crossing all parallels perpendicularly, and being aligned in a true north-south direction.

The distance between meridians varies significantly but predictably. At the equator, the length of one degree is about the same as that of one degree of latitude (see Table 1-1). However, since meridians converge toward the poles, the length of a degree of longitude decreases poleward, diminishing to zero at the poles where all meridians meet at a point.

The equator is a "natural" line to serve as a baseline from which to measure latitude, but no such natural reference line exists for longitudinal measurement. Consequently, for most of recorded history there was no accepted longitudinal baseline; each country would arbitrarily select its own "prime meridian" as the reference line for east-west measurement. Thus the French measured from the meridian of Paris, the Italians from the meridian of Rome, the Mexicans from the meridian of Mexico City, etc. Not until about a century ago was standardization finally achieved. In 1884, an international conference of geographers and engineers, meeting in Washington, D.C., chose the meridian passing through the Royal Observatory at Greenwich (England), just east of London, as the *prime meridian* for all longitudinal measurement. The prolific and prominent work of British map-makers, as well as the political power and influence of Great Britain at that time, led to this decision.

TABLE 1-1 Lengths of Degrees of Latitude and Longitude

Latitude (degrees)	Length of 1° of: Latitude, as measured along a meridian — Miles	Kilometers	Longitude, as measured along a parallel — Miles	Kilometers
0	68.703	110.567	69.172	111.321
10	68.726	110.605	68.129	109.641
20	68.789	110.705	65.026	104.649
30	68.883	110.857	59.956	96.488
40	68.998	111.042	53.063	85.396
50	69.121	111.239	44.552	71.698
60	69.235	111.423	34.674	55.802
70	69.328	111.572	23.729	38.188
80	69.387	111.668	12.051	19.394
90	69.407	111.699	0	0

Thus, an imaginary north-south plane passing through Greenwich and through the Earth's axis represents the plane of the prime meridian. The angle between this plane and a plane passed through any other point and the axis of the Earth is a measure of longitude. For example, the angle of a plane passing through Washington, D.C., is 77 degrees, 3 minutes, and 57 seconds. Since the angle is formed *west* of the plane of the prime meridian, the longitude of Washington is written: 77°03′57″ W.

Longitude is measured both east and west of the prime meridian to a maximum of 180° in each direction. Exactly halfway around the globe from the prime meridian, in the middle of the Pacific Ocean, is the 180th meridian, which is 180° of longitude removed from the prime meridian. All places on Earth, then, have a location that is either east longitude or west longitude, except for points exactly on the prime meridian (described simply as 0° longitude) or exactly on the 180th meridian (described as 180° longitude).

The network of intersecting parallels and meridians creates a *geographic grid* over the entire surface of the Earth. See Figure 1-13. The location of any place on the Earth's surface can be described with great precision by reference to detailed latitude and longitude data. For example, at the 1964 World's Fair in New York City, a time capsule (a container filled with records and memorabilia of contemporary life) was buried. For reference purposes, the U.S. Coast and Geodetic Survey determined that the capsule was located at 40°28′34.089″ north latitude and 73°43′16.412″ west longitude. At some time in the future, if a hole were to be dug at the spot indicated by those coordinates, it would be within 6 inches (15 cm) of the capsule.

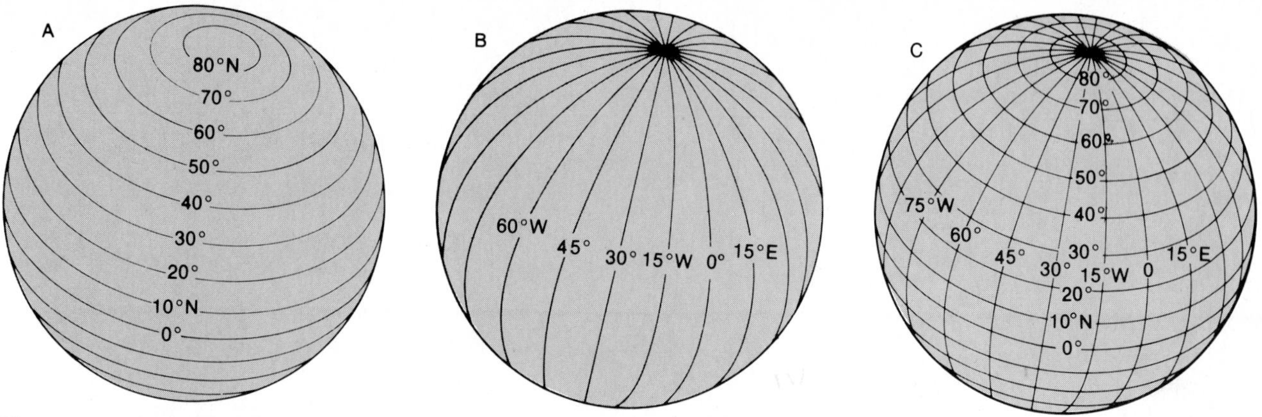

Figure 1-13. Development of the geographic grid: (A) parallels of latitude; (B) meridians of longitude; (C) the completed grid system.

Telling Time

In prehistoric times, the rising and setting of the sun were probably the principal means of telling time. As civilizations developed, however, more precise timekeeping was required. Early agricultural civilizations in Egypt, Mesopotamia, India, China, and even England, as well as in Aztec and Mayan civilizations in the New World, were significantly concerned with observing the sun and the stars in order to tell time and to keep accurate calendars.

Local solar noon can be determined by watching for the moment when objects cast their shortest shadows. The Romans used sundials to tell time and gave great importance to the noon position, which they called the *meridian*—the sun's highest (*meri*) point of the day (*diem*). Our use of A.M. (*ante meridian*: before noon) and P.M. (*post meridian*: after noon) was derived from the Roman world. When nearly all transportation was by foot, horse, or sailing vessel, it was difficult to compare time at different localities. In those days, each community set its own time by correcting its clocks to high noon at the moment of the shortest shadow. A central public building, such as a cathedral in Europe, a town hall in New England, a county courthouse in the Midwest, or a mission chapel in Spanish California, usually had a large clock or loud bells to toll the hour. Periodically, this time was checked against the shortest shadow.

Standard Time As the telegraph and railroad began to speed words and passengers between cities, the use of local solar time created increasing problems. A cross-country rail traveler in the United States a century ago might have experienced as many as 24 different times between the Atlantic and Pacific coasts. Indeed, in the Pittsburgh rail terminal alone, six different times were in use for a while. Eventually, the railroads stimulated the development of a standardized time system. At the previously mentioned 1884 International Meridian Conference in Washington, the major nations agreed to divide the world into 24 standard time zones, each extending over 15° of longitude. The local solar time of the Greenwich (prime) meridian was chosen as the standard for the entire system. The prime meridian became the center of a time zone that extends $7\frac{1}{2}$° of longitude both to the west and to the east of the prime meridian. Similarly, the meridians that are multiples of 15 (15, 30, 45, 60, 75, 90, 105, 120, 135, 150, 165, and 180), both east and west of the prime meridian, were set as the *central meridians* for the 23 other time zones, each of which is 15° of longitude in extent. See Figure 1-14.

In international waters, these time zones are shown exactly 7°30′ to the east and 7°30′ to the west of the central meridians. Over land areas, however, the actual eastern and western boundaries of time zones vary to coincide with appropriate political and economic constraints. Most of the nations of the world are sufficiently small east-west so as to lie totally within a single time zone. However, large countries may encompass several zones: the U.S.S.R. occupies 11 time zones; the United States occupies 7; Canada, 5; and Australia, 3. Although China extends clearly across four 15° zones, the entire nation, at least officially, observes the time of the 120th east meridian, which is closest to Beijing. National time varies by 30 or 45 minutes from the standard zone pattern in about a dozen countries, to suit their own convenience, and one country (Saudi Arabia) has no standard time in the accepted sense of the term.

From the map of time zones of the United States (Figure 1-15), we can recognize a great deal of manipulation of the time zone boundaries for economic and political convenience. For example, the Central Standard Zone, centered on 90° W, extends all the way to 105° W (which is the central meridian of the Mountain Standard Zone) in Texas in order to keep most of that state within the same zone. By contrast, El Paso, Texas, is officially within the Mountain Standard Time Zone, revealing its role as a major market center for southern New Mexico, which observes Mountain Standard Time. In the same vein, northwestern Indiana is in the Central Standard Time Zone with Chicago.

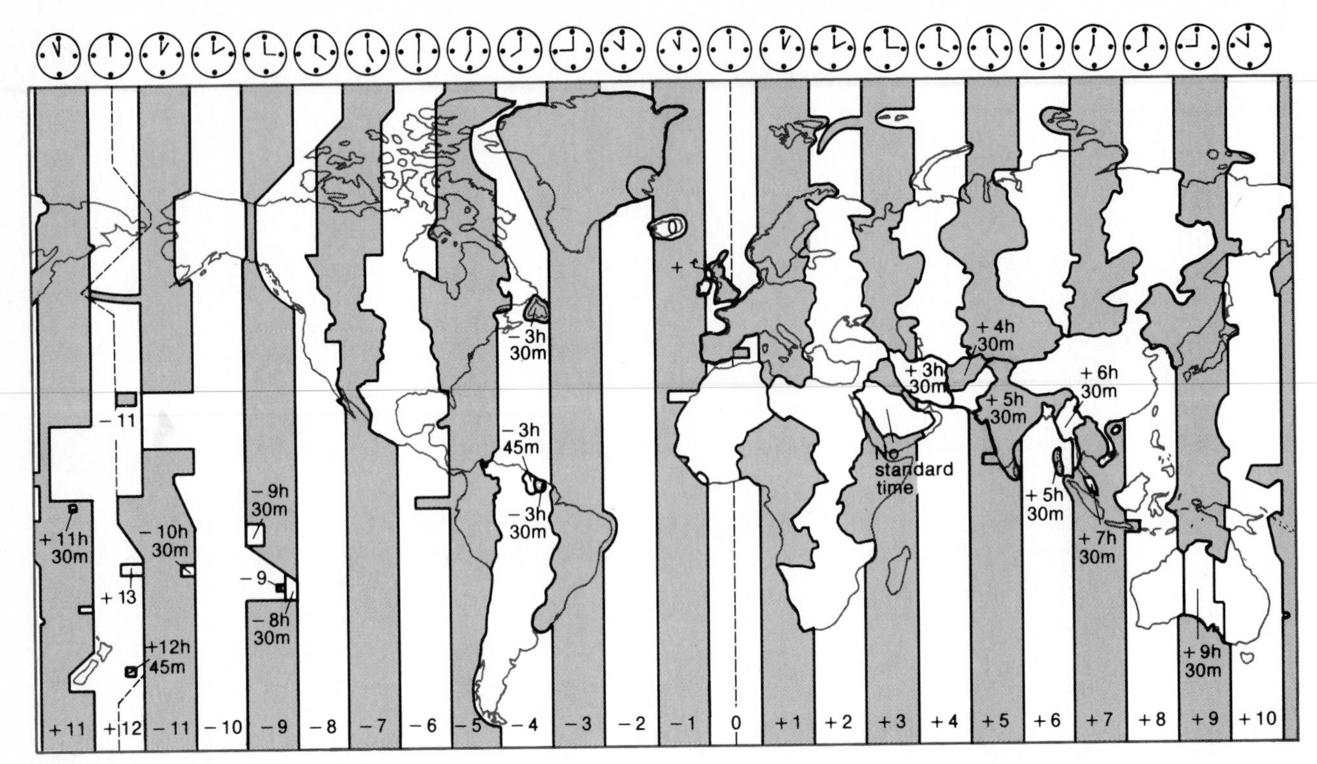

Figure 1-14. Standard time zones of the world.

The International Date Line One advantage of establishing the Greenwich meridian as the prime meridian is that its opposite arc is in the Pacific Ocean. The 180th meridian, transiting the sparsely populated mid-Pacific, was chosen as the meridian at which new days would begin and old days would exit from the surface of the Earth. The international date line deviates from the 180th meridian in the Bering Sea to include all of the Aleutian Islands of Alaska within the same day and again in the South Pacific to keep islands of the same group (Fiji, Tonga) within the same day. See Figure 1-16.

The new day first appears on Earth at midnight at the international date line. For the next 24 hours, the new day advances westward around the world, finally covering the entire surface for an hour at the end of this period. For the next 24 hours, this day leaves the Earth, one hour at a time, making its final exit 48 hours after its first appearance. Except at Greenwich noon, two days exist on Earth: the more recent one (e.g., January 2nd) extending from the international date line westward to the current position of midnight, and the older one (e.g., January 1st) extending the rest of the way to the date line. Thus, when you cross the international date line going from west to east, the day becomes "earlier" (e.g., from January 2nd to January 1st); whereas when you move across the line from east to west, the day becomes "later" (e.g., from January 1st to January 2nd).

It should be noted that the international date line is in the middle of a time zone. Consequently, there is no time zone (i.e., hourly) change at that point; there is only a change on the calendar, not on the clock.

Daylight-Saving Time To conserve energy during World War I, Great Britain ordered all clocks set forward by one hour. This practice allowed the citizenry to "save" one hour of daylight by extending the daylight period into the usual evening hours, thus reducing the consumption of electricity for lighting. The United States began a similar policy in 1918, but many localities declined to observe "summer time" until the Uniform Time Act made the practice mandatory in all states that had not deliberately exempted themselves. Arizona, Hawaii, and part of Indiana have exempted themselves from observance of daylight-saving time under this act.

The Soviet Union has adopted permanent daylight-saving time. Figure 1-14 shows that the time zones of the U.S.S.R. are an hour later than the time indicated by their central meridian. In recent years, most of the nations of western Europe, Canada, Australia, and New Zealand have also adopted daylight-saving time. In the Northern Hemisphere, many of them, like the United States, begin daylight time early Sunday morning of the last weekend in April and resume standard time early Sunday morning of the last weekend of October. In the tropics, the lengths of day and night change little seasonally, and there is not much twilight. Consequently, daylight-saving time would offer little or no "savings" for tropical areas.

The 24-Hour Clock The U.S. military forces and most European nations prefer to use the 24-hour clock. Most international transportation and communication schedules are also expressed in this manner. In the 24-hour

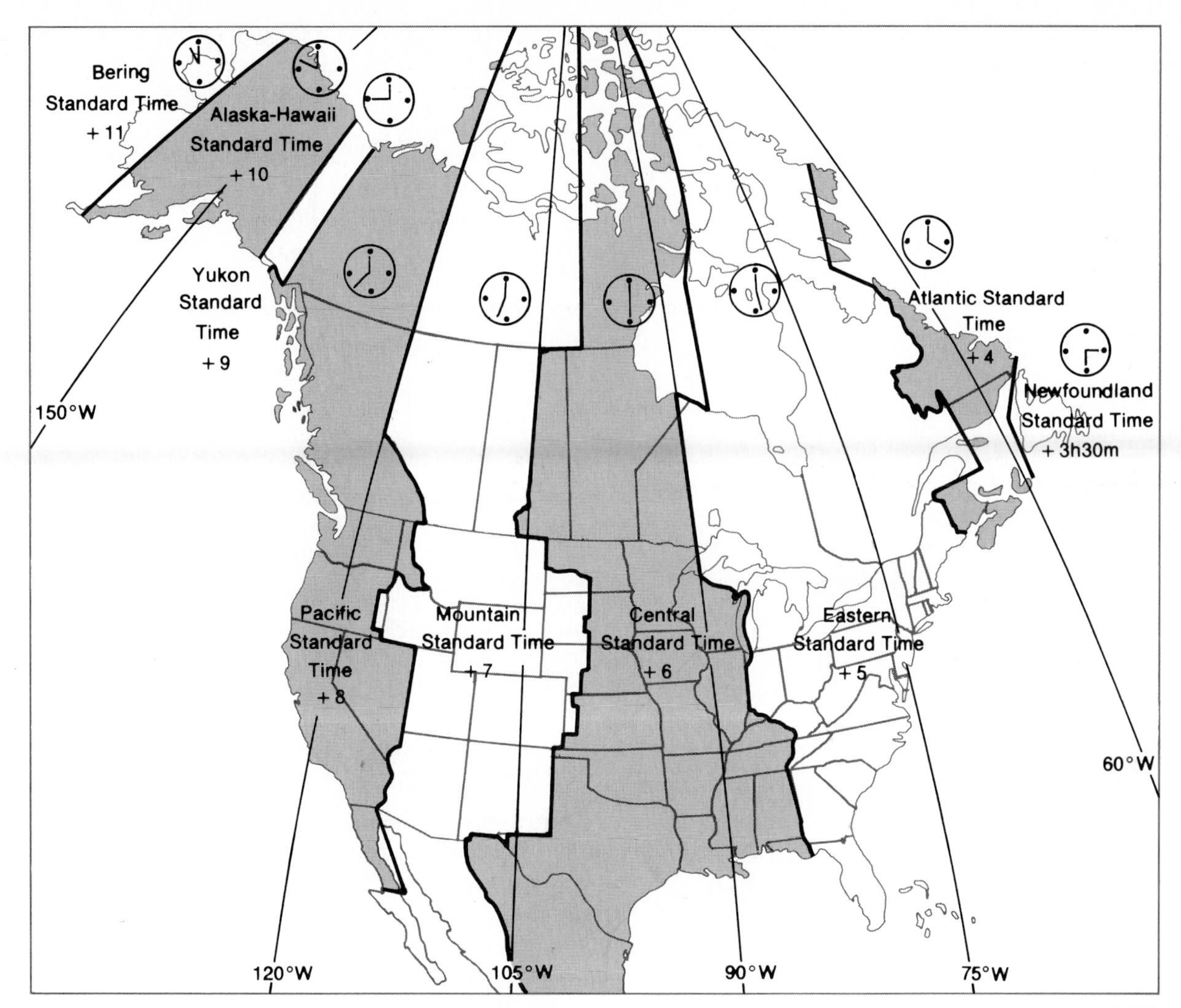

Figure 1-15. Standard time zones of Anglo-America.

Figure 1-16. The international date line generally follows the 180th meridian, but it zigzags around various island groups to maintain their continuity.

clock, the times prior to noon are identical to conventional usage. After noon, however, the 24-hour clock continues the count from the morning hours so that 1:00 P.M. becomes 13 hours, 2:00 P.M. is 14 hours, and so on. To convert from traditional time to the 24-hour clock, you should add 12 to traditional P.M. times. To convert from the 24-hour clock to traditional time, merely subtract 12 from any hour greater than twelve and designate the remainder as "P.M." Thus 22:45 would become 10:45 P.M. The U.S. military forces have the custom of referring to 23:00 as "twenty-three hundred hours" or merely "twenty-three hundred." One should be cautious in mimicking this usage, however, since we are not really dealing with "hundreds" in the 24-hour clock. As there are 60 minutes in an hour, 1 minute after 22:59 is 23:00 and does not involve a "hundred" of any type. The proper expression for 21:37 is "twenty-one hours and 37 minutes"; for 23:00, it would be "twenty-three hours."

Although Greenwich mean time (GMT) is now referred to as *coordinated universal time* (CUT), the prime meridian is still the reference for standard time. Since it is always the same number of minutes after the hour in all time zones, except those situations noted above, we usually need to know only the correct time at Greenwich and the number of hours that our local time zone is later or earlier than the Greenwich meridian in order to know exact local time. Figure 1-14 shows the number of hours "fast" (later) or "slow" (earlier) for coordinated universal time in each time zone of the world.

Earth and Sun Relations

Life on Earth is dependent on solar energy, and the Earth's functional relationship to the sun is of vital importance. This relationship is not static because of the perpetual motions of the Earth, which continually change the geometric perspective between the two bodies.

Movements of the Earth

The Earth makes three basic types of movements, but only two of them—rotation and revolution—change its position with respect to the sun and are thus of interest to geographers.

Rotation on Its Axis The Earth rotates toward the east on its axis. See Figure 1-17. A complete rotation requires one *solar day* of 24 hours. As viewed with respect to the stars, a full rotation takes only 23 hours, 56 minutes, and 4.099 seconds. This shorter period is called the *sidereal day.*

The apparent "motion" of the sun, the moon, and the stars is, of course, just the opposite of the true direction of earthly rotation. They appear to rise in the east and set in the west, which is an illusion created by the steady easterly spin of the Earth.

Figure 1-17. The Earth rotates on its axis toward the east.

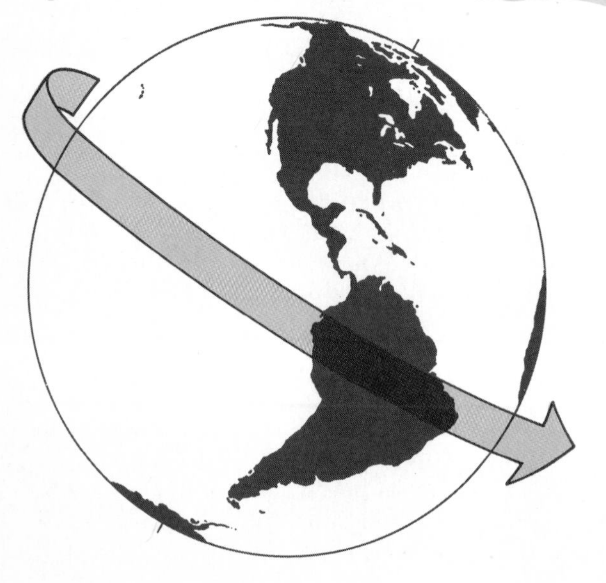

Rotation causes all parts of the Earth surface, except the poles, to move in a circle around the Earth's axis. Table 1-2 shows the speed of this motion at various latitudes. Although the velocity of rotation varies from place to place, it is constant at any given place. This is the reason that we experience no sense of motion. Often one can get the same impression in a modern jetliner, where a smooth flight at cruising speed is much like sitting in a comfortable living room. Only during take-off and landing is the sense of motion quite apparent. Similarly, the motion and speed of earthly rotation would become apparent to us only if that rotation rate suddenly increased or decreased—a very unlikely event.

Rotation has several striking effects on the physical characteristics of the Earth's surface. Most important are the following:

1. The constancy of the Earth's rotation in the same direction causes apparent deflections in the flow path of both air and water. The deflection is invariably toward the right in the Northern Hemisphere and toward the left in the Southern Hemisphere. This phenomenon is called the *Coriolis effect* and will be discussed in detail in Chapter 4.
2. The rotation of the Earth brings varying parts of the surface through the increasing and then decreasing gravitational pull of the moon and the sun. Although the land areas of the Earth are too rigid to be noticeably moved by these oscillating gravitational attractions, oceanic waters move onshore and then recede in a rhythmic pattern as a result of the interplay of earthly rotation with these gravitational forces. The rise and fall of water level constitutes the *tides,* which will be discussed further in Chapter 9.
3. Undoubtedly the most important effect of earthly rotation is the diurnal (daily) alternation of light and darkness, as portions of the Earth's surface are turned toward, and then away from, the sun. This variation in exposure to sunlight greatly influences local temperature, humidity, and even wind movements. Except

TABLE 1-2 Speed of Rotation of The Earth's Surface at Selected Latitudes

Latitude	Miles per hour	Kilometers per hour
0	1037.6	1669.9
10	1021.9	1642.0
20	975.4	1569.7
30	899.3	1447.3
40	795.9	1280.9
50	668.3	1075.5
60	520.1	837.0
70	355.9	572.8
80	180.8	291.0
90	0	0

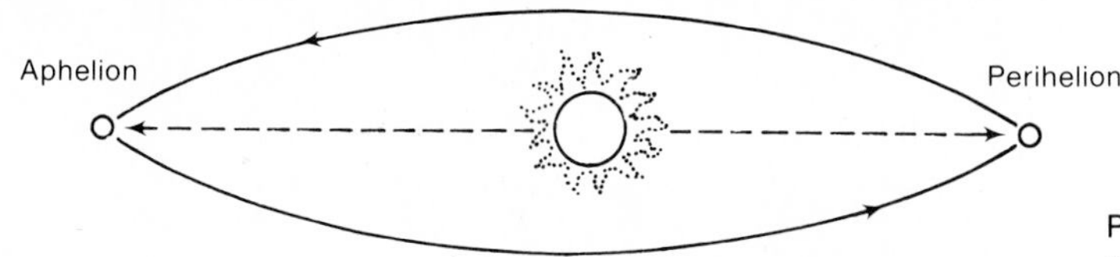

Figure 1-18. An exaggerated view of the Earth's elliptical orbit around the sun. The perihelion (closest) position occurs on January 3rd; the aphelion (furthest) position is on July 4th. [After Robert E. Gabler et al., *Essentials of Physical Geography,* 2nd ed. (Saunders College Publishing, 1982), p. 12. Copyright © 1982 by CBS College Publishing. Reprinted by permission of CBS College Publishing.]

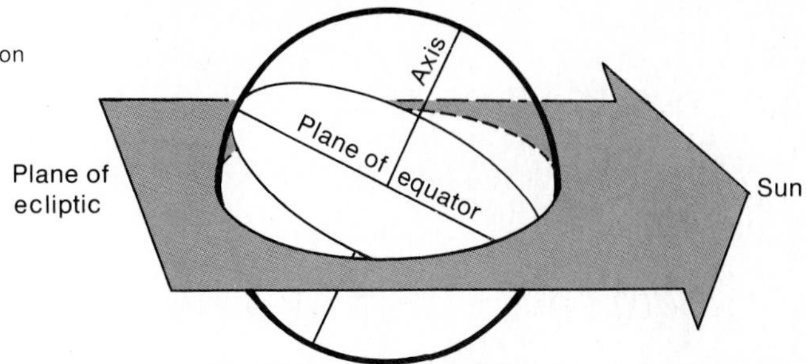

Figure 1-19. The plane of the ecliptic is an imaginary surface that is visualized as passing through the sun and through the Earth at all positions in the Earth's orbit around the sun. It does *not* coincide with the plane of the equator.

for the organisms of caves or the ocean deeps, all forms of life have adapted to this sequential pattern of light and darkness. Even human beings fare poorly when their circadian (daily) rhythms are misaligned as the result of high-speed air travel that significantly interrupts the normal sequence of daylight and darkness. We are left with a sense of fatigue and psychological distress known as *jet lag*, which can include unpleasant changes in our usual patterns of appetite and sleep.

Revolution Around the Sun Another significant Earth motion is *revolution*. The Earth revolves around the sun once every 365 days, 5 hours, 48 minutes, and 46 seconds; or 365.242199 days. This is known officially as the *tropical year*, and for practical purposes is usually simplified to $365\frac{1}{4}$ days. Astronomers define the year in other ways as well, but the duration is very close to that of the tropical year and need not concern us here.

The path followed by the Earth in its journey around the sun is not a true circle but an ellipse. See Figure 1-18. The average distance between the centers of the Earth and the sun, the astronomical unit (AU), is 92,955,806 miles (149,597,892 km). The actual distance varies from 91,445,000 miles (147,166,480 km) at the *perihelion* position on January 3rd to 94,555,000 miles (152,171,500 km) at the *aphelion* position on July 4th. The Earth is actually 3.3 percent closer to the sun during the Northern Hemisphere winter than in the summer, an indication that the varying distance between the Earth and the sun is not an important determinant of seasonal temperature fluctuations.

The Annual March of the Seasons

If one were to imagine a plane that passes through the sun and through the Earth at every position in its orbit around the sun, this would be the *plane of the ecliptic* (Figure 1-19). Our sense of orderly geometric relationships tends to make us think that the plane of the equator should coincide with the plane of the ecliptic. However, this incorrectly assumes that the axis of earthly rotation would be perpendicular to the plane of the ecliptic. Such a positional relationship between the Earth and the sun would create a very different world from the one in which we live. In this case, the rays of energy from the sun would always strike the Earth, at noon, at the same angle at any one place. Daylight and darkness, caused by the rotation of the Earth on its axis, would be the only temporal variations in the amount of energy reaching the Earth. The result would be a world without seasons.

The plane of the equator, however, is *not* coincidental with the plane of the ecliptic, nor is the axis of the Earth perpendicular to it. Rather, the axis is tilted at about $66\frac{1}{2}°$ above the plane of the ecliptic, or about $23\frac{1}{2}°$ away from the perpendicular (Figure 1-20). This is referred to as the *inclination* of the axis. Moreover, the axis always points toward Polaris, the North Star, no matter where

Figure 1-20. The Earth's axis is inclined at an angle of 66½° from the plane of the ecliptic, but this inclination is normally described as being 23½° from the vertical (a line perpendicular to the plane of the ecliptic).

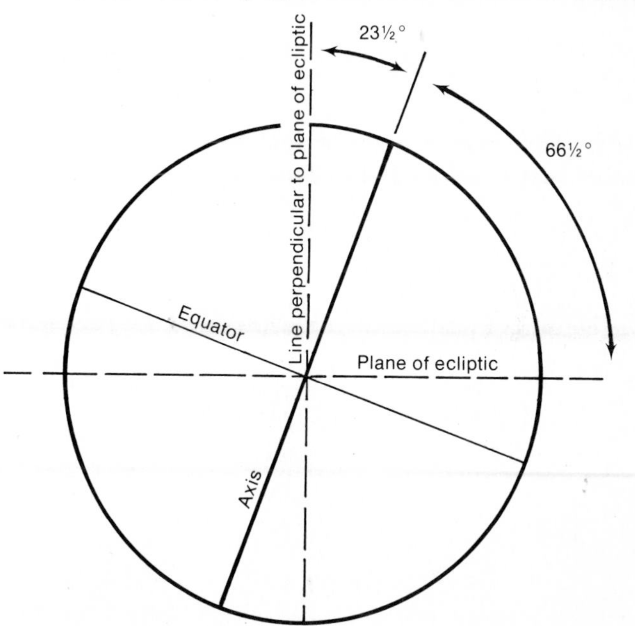

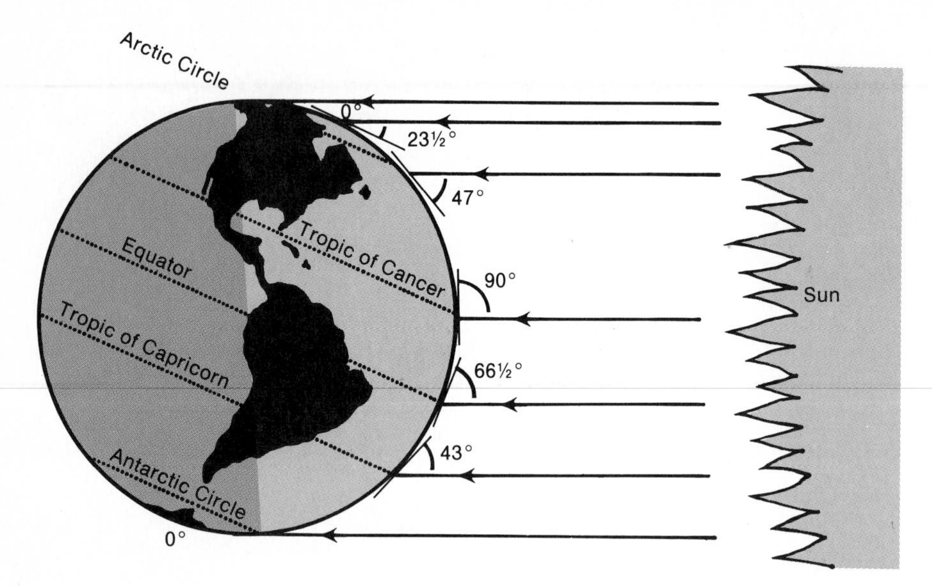

Figure 1-21. An example of differing solar angles at different latitudes. This diagram represents the Northern Hemisphere summer solstice. Noon sunlight angles are indicated for various latitudes.

the Earth is in its orbit around the sun. In other words, at any time during the year the Earth's axis is parallel to its orientation at all other times. This is called the *polarity*, or *parallelism*, of the axis.

The combined effect of rotation, revolution, inclination, and polarity is such that the angle at which the sun's rays strike the Earth changes throughout the year. This is a critical determinant of the amount of solar energy delivered from the sun to any spot on Earth. The basic generalization is that the more direct the angle at which the sunlight strikes the Earth, the more effective is the resultant heating. In Figure 1-21, the area between the equally spaced parallel lines represents equal amounts of energy coming from the sun. Where the sun's rays strike the Earth perpendicularly, solar energy is concentrated over the smallest possible surface area. By contrast, where the sun's rays strike the Earth obliquely (at an angle nearer to the horizon), the same amount of energy is spread over a much larger surface area. Thus the amount of energy reaching a particular surface area is significantly smaller. The angle at which the sun's rays strike the Earth is fundamental in determining the amount of *in*coming *so*lar radi*ation*—*insolation*—reaching any given point on Earth.

Solstices At all times, the sun illuminates one-half of the Earth. See Figure 1-22. On June 22nd, the noon rays from the sun are perpendicular at the latitude line lying 23½° north of the equator. This parallel, the *Tropic of Cancer*, marks the northernmost location reached by the vertical rays of the sun in the annual cycle of the Earth's revolution. This occurs once a year on about June 22nd, although the exact date varies slightly from year to year.

At this time, the outer limit of the lighted hemisphere, the *circle of illumination*, reaches to within 23½° of each pole. See Figure 1-23. As the Earth rotates on its axis, all points lying north of 66½° N (within 23½° of the North Pole) remain continuously within the circle of illumination, thus experiencing 24 continuous hours of daylight. By contrast, all points within 23½° of the South Pole, that is, south of 66½° S, are outside the circle of illumination and thus have 24 continuous hours of darkness. These parallels, which represent the points of tangency of the circle of illumination on the solstice dates, are the *polar circles*. The northern polar circle at 66½° N is the *Arctic Circle*; the southern polar circle at 66½° S is the *Antarctic Circle*.

On or about December 22nd (slightly variable from year to year), the perpendicular rays of the sun strike the parallel of 23½° S, the *Tropic of Capricorn*. Once again, the circle of illumination is tangent to the polar circles, but this time the illuminated hemispheres are reversed. Areas north of the Arctic Circle now lie outside the circle of illumination and thus are in darkness, whereas areas within the Antarctic Circle are continuously within the circle of illumination and thus in daylight for 24 hours.

Although the angle of the incoming rays of the sun shifts 47 degrees from June 22nd to December 22nd, the relationships between the Earth and the sun are very similar on those days, which are called the *solstices*. In the Northern Hemisphere, December 22nd is called the *winter solstice*, and June 22nd is referred to as the *summer solstice*. (In the Southern Hemisphere, the seasonal designations are reversed.) The following conditions are true on both solstices:

1. The vertical rays of the sun strike 23½° poleward from the equator:
   (a) 23½° N on June 22nd
   (b) 23½° S on December 22nd
2. The circle of illumination is tangent to the polar circles, which enclose an area of continuous light or continuous darkness:

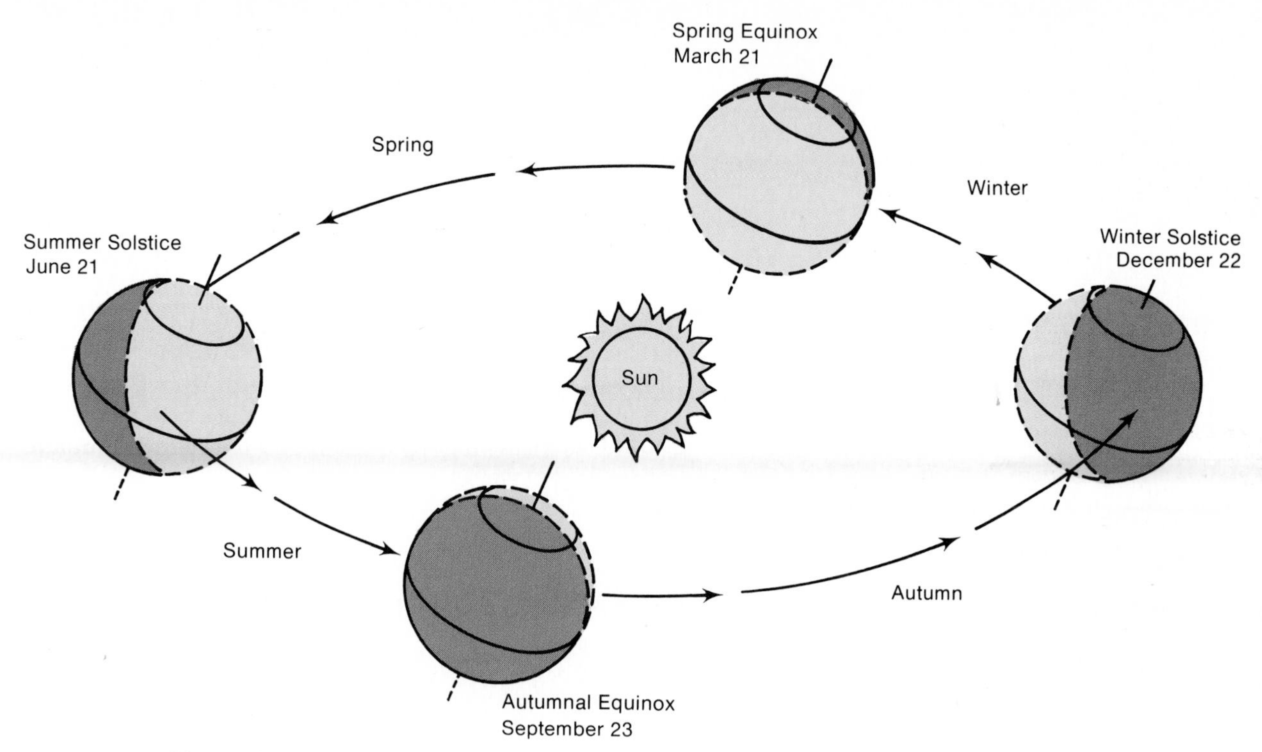

Figure 1-22. The Earth's elliptical orbit around the sun, with the four seasonal positions illustrated. Descriptive terms are from a Northern Hemisphere perspective.

Figure 1-23. At the time of the solstices, the sun's noon rays strike directly at 23½° latitude. The June (Northern Hemisphere summer) solstice is a time when sunlight is concentrated in the Northern Hemisphere; the December (Northern Hemisphere winter) solstice is a time of Southern Hemisphere concentration. The arrows on the accompanying drawings give an indication of the sun's angle at various latitudes.

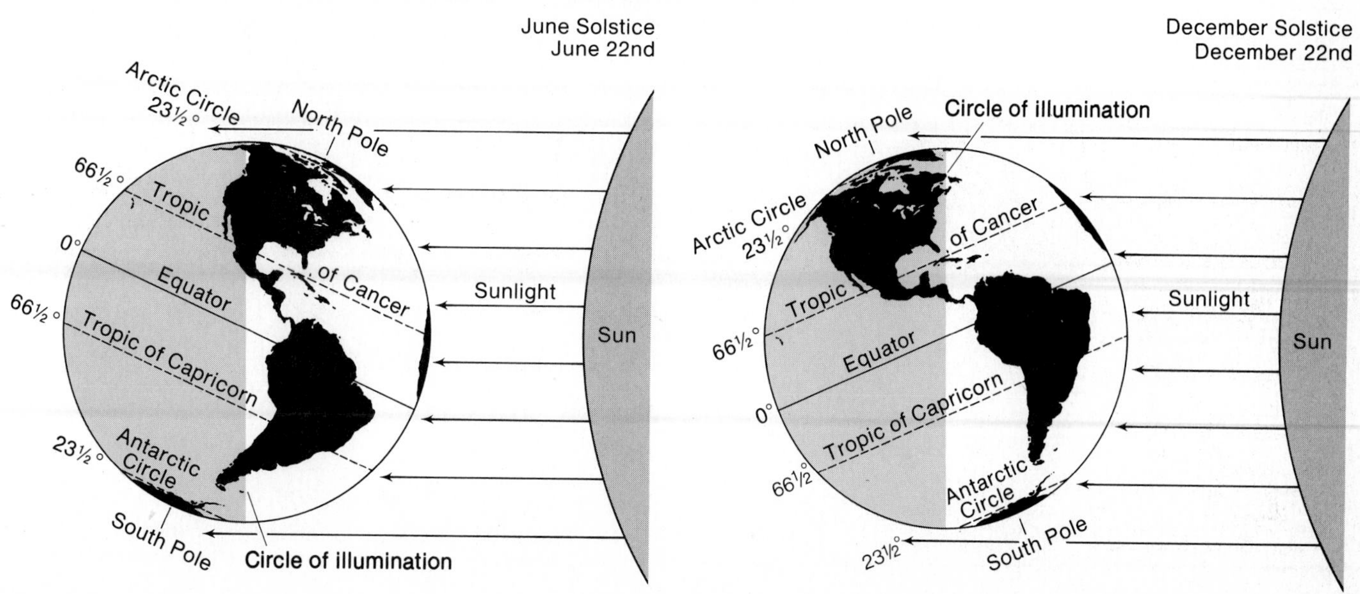

	June 22	*December 22*
Arctic Circle	Continuous light	Continuous darkness
Antarctic Circle	Continuous darkness	Continuous light

3. These two dates, June 22nd and December 22nd, represent the most poleward extent of the direct (perpendicular) rays of the sun.

Equinoxes Figure 1-24 provides a diagram of the relationship between the Earth and the sun and the positions of the circle of illumination on approximately March 21st and September 23rd (dates slightly variable from year to year), which are the *equinoxes*. March 21st is the *vernal* (spring) *equinox* in the Northern Hemisphere, and September 23rd is the *autumnal equinox*; the seasonal terms are reversed in the Southern Hemisphere. The positional relationships of the Earth and the sun are virtually identical on these two dates, and the following characteristics prevail:

1. The perpendicular rays of the sun strike the equator.
2. The circle of illumination just touches both poles.
3. The periods of daylight and darkness are each 12 hours long all over the Earth, a situation that occurs only on these two dates. (The word *equinox* derives from the Latin *equus*, equal, and *nox*, nights.)
4. The equinoxes represent the midpoints in the shifting of direct days of the sun between the Tropic of Cancer and the Tropic of Capricorn.

Figure 1-24. At the time of both equinoxes, the direct rays of the noon sun strike the equator. On those two days only, the length of daylight and darkness is equal all over the Earth. The arrows in this diagram indicate the solar angle at various latitudes.

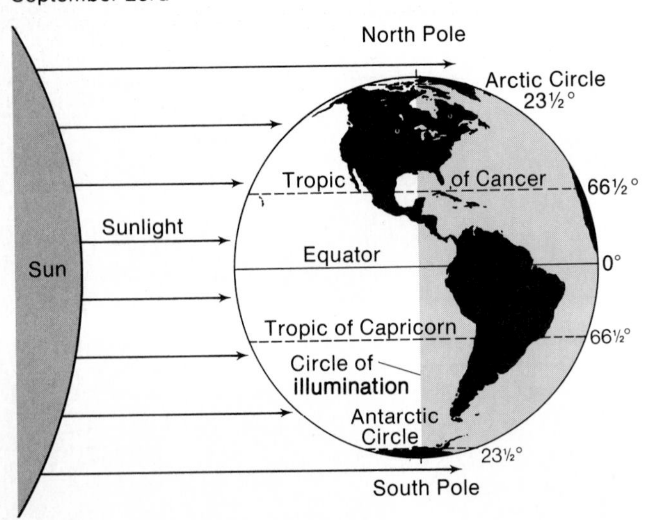

Changes in Daylight and Darkness From the March equinox until the June solstice, the period of daylight gradually increases everywhere north of the equator, reaching its maximum on June 22nd, the longest day of the year in the Northern Hemisphere. On this date, north of the equator the period of daylight is progressively longer with increased latitude until it reaches a maximum of 24 hours, or continuous daylight, everywhere north of the Arctic Circle. Conversely, the period of daylight diminishes for all points south of the equator during this period, reaching a minimum on June 22nd, which is the longest night of the year south of the equator. In the Southern Hemisphere, each succeeding parallel has a shorter period of daylight, reaching a June 22nd minimum of no daylight everywhere south of the Antarctic Circle. From June 22nd till September 23rd, these changes are reversed, with the days getting shorter in the Northern Hemisphere and longer in the Southern Hemisphere until September 23rd, when day and night are again equal over all the Earth.

After the September equinox, the days of the Northern Hemisphere continue to become shorter and those of the Southern Hemisphere continue to lengthen. By the time of the December solstice, the days are shortened progressively northward, reaching a minimum of no daylight north of the Arctic Circle. During this same interval, the length of the period of daylight gets longer south of the equator, reaching a maximum of 24 hours of continuous daylight south of the Antarctic Circle. Again, this pattern is reversed following the December solstice, and periods of daylight and darkness become equalized over the entire world by the time of the March equinox.

The duration of daylight and darkness has an important influence on the amount of insolation that reaches any point on the Earth's surface. Both the length of the period of daylight and the angle at which the sun's rays strike the Earth are principal determinants of the amount of insolation received at any particular latitude.

Most of this first chapter has been concerned with the celestial setting of the Earth in the universe and in our solar system, with emphasis on the significance of this setting on the physical geography of our planet. In succeeding chapters we will focus attention on the Earth itself, considering each of the major components of its environment.

Review Questions

1. Identify and briefly define the four environmental "spheres."
2. Explain the concept of a *light-year*.
3. In what ways do the inner and outer planets of our solar system differ?

4. Compare the gravitational influence of the sun, moon, and Earth on objects at the Earth's surface.
5. Is the Earth large or small? Explain.
6. How do we know that the Earth is spherical in shape?
7. The sphericity of the Earth is not perfect. Is this fact important in our study of physical geography?
8. What are the major differences between *parallels* and *meridians*?
9. Explain why the *plane of the ecliptic* does not coincide with the *equatorial plane*.
10. What would be the effect on the annual march of the seasons if the Earth's axis did not maintain *parallelism* during earthly revolution?

Some Useful References

ABELL, G. O., *Exploration of the Universe*, 3rd ed. New York: Holt, Rinehart and Winston, 1975.

CARTER, D. B., T. H. SCHMUDDE, AND D. M. SHARPE, *The Interface as a Working Environment: A Purpose for Physical Geography*. Commission on College Geography, Technical Paper No. 7. Washington, D.C.: Association of American Geographers, 1972.

CLOUD, P., *Cosmos, Earth, and Man*. New Haven, Conn.: Yale University Press, 1978.

KING, I. R., *The Universe Unfolding*. San Francisco: W. H. Freeman & Company, Publishers, 1976.

Scientific American Editors, *The Solar System: A Scientific American Book*. San Francisco: W. H. Freeman & Company, Publishers, 1975.

2 Portrayal of the Earth

The surface of the Earth is the focus of the geographer's interest. Its enormity and complexity create a milieu that would be difficult to comprehend and analyze without tools and equipment to aid in systematizing and organizing the varied data. Many kinds of tools are used in geographical studies, but probably the most important and most universal are maps. The mapping of the phenomena under study is normally essential for an understanding of spatial distributions and relationships.

Maps are very useful to many people other than geographers. In the most rudimentary of societies, of course, maps may be irrelevant. A band of primitive people may function satisfactorily with their spatial knowledge stored entirely as mental maps, retained solely in their minds and perhaps never even sketched in the sand. At any level beyond the most primitive, however, the depiction of spatial relationships in some visible form becomes at least a convenience, and often a necessity.

Our concern here is essentially with the usefulness of maps to geographers, i.e., to students of geography. In some cases, geographers deal with maps as an end in themselves, but more often than not, maps serve geographers as a means to some end. This book is a case in point. It contains numerous maps of various kinds, each inserted in the book to further the understanding of some fact, concept, or relationship.

It is important at the outset to realize that maps have their faults. Indeed, we can begin with the negative axiom that no map is perfectly accurate. Most people understand that just because something is written in a book, it is not necessarily true—the printed word is not infallible simply because it exists in black and white. These same people, however, may accept the information portrayed on a map as being the gospel truth. Most people have insufficient experience with maps to transfer their skepticism about the written word to an understanding of the same problem with maps. Nevertheless, the inaccuracy of maps is ubiquitous, simply because a map is an attempt to portray an impossible geometrical relationship—for example, a curved surface on a flat piece of paper. We will explore this problem further in a subsequent section.

Our purpose in this chapter is threefold:

1. To develop an understanding of the basic characteristics of maps.
2. To unfold an appreciation of the capabilities and limitations of maps.
3. To learn something of the various ways that the Earth's surface can be portrayed—through map projections, globes, photographs, and remotely sensed imagery.

The Nature of Maps

A map is a depiction of an area in graphic form. It is a two-dimensional representation of the spatial distribution of selected phenomena. In essence, a map is a scaled drawing of a portion of the Earth's surface; it is a representation of an area at a reduced scale in which only selected data are shown. A map serves as a surrogate (a substitute) for a part of the Earth's surface that we wish to portray or to study. It is always smaller, usually extremely smaller, than the portion of the surface that is represented.

Maps display horizontal dimensions. Their basic attribute is the capability of showing distance, direction, size, and shape in their horizontal spatial relationships. In addition to these fundamental pieces of graphic data, most maps are designed to show other kinds of information as well. Maps nearly always have a special purpose, and that purpose is usually to show the distribution of one or more phenomena. Thus, a map may be designed to show street patterns, or the distribution of Texans, or the ratio of sunshine to clouds, or the number of earthworms per square meter of soil, or any of an infinite number of other facts or combinations of facts. It is easy to see that maps are indispensable tools for geographers because they provide graphic displays of *what* is *where*, and they are often helpful in providing clues as to *why* such a distribution occurs.

The Matter of Scale

A map is always smaller than the portion of the Earth's surface it represents, so any understanding of area relationships (distance, size) depends on the proper use of scale. The *scale* of a map gives the relationship between length measured on the map itself and the corresponding distance on the ground. Knowing the scale of a map makes it possible to measure distance, determine area, and make comparisons of size.

Scale can never be represented with perfect accuracy on a map because of the impossibility of rendering the curve of a sphere on the flatness of a sheet of paper without distortion. Therefore, the scale of a map cannot be constant (i.e., the same) all over the map. If the map is of a small area, however, the scale will be so nearly perfect that it can be accepted uncritically throughout the map. On the other hand, if the map represents a large portion, or all, of the Earth's surface, there will be enormous scale variations because of the significant distortions involved.

Maps are made in an infinite variety of sizes and styles and are designed to serve a limitless diversity of purposes. Some are general-reference maps, and some are thematic maps, which show or emphasize the location or distribution of particular phenomena. All maps, however should have a few basic components to facilitate their use. Omission of any of these essential components decreases the clarity of the map and makes it more difficult to interpret. The essential components are as follows:

Title The title is in essence a brief summary of the map. It should identify the area covered and provide some indication of the map content, as "Road Map of Kenya," or "River Discharge in Northern Europe," or "Seattle: Shopping Centers and Transit Lines."

Date Most maps depict conditions or patterns that are temporary, or even momentary. To be meaningful, the reader should be informed when the data were gathered, as this will give an indication of how much out-of-date the map is. Some maps list both the date to which the information pertains and the date of publication of the map.

Legend Most maps make use of symbols, colors, shadings, or other devices to represent features or the amount, degree, or proportion of some quantity. Some symbols are self-explanatory, but it is usually necessary to include a legend box in a corner of the map to explain the symbolization.

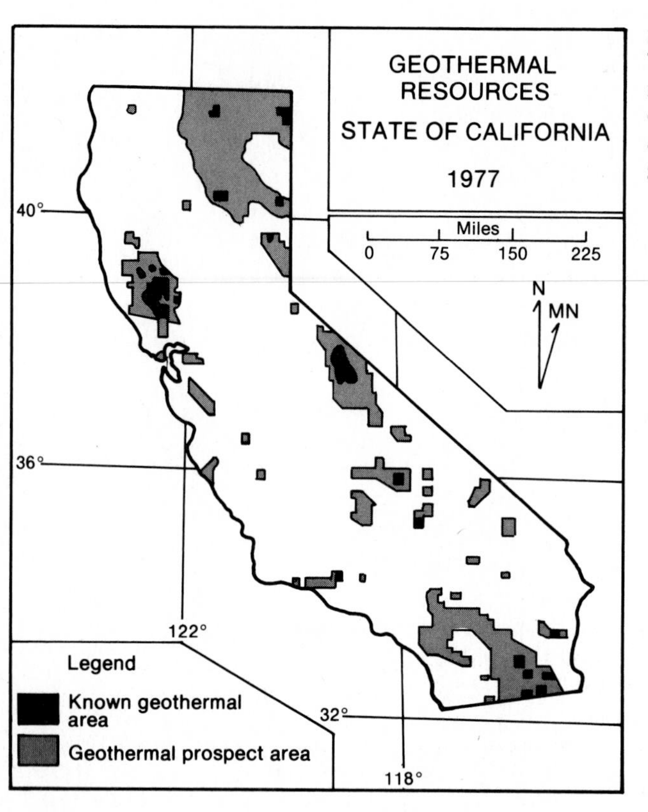

Some map essentials. This example of a thematic map contains title, date, legend, scale, north arrow, and geographic grid.

Scale Any map that serves as more than a pictogram must be drawn to scale, at least approximately. Scale is discussed in some detail on pages 23–25.

Direction Direction is normally shown on a map by means of the geographic grid. Meridians are supposed to extend north-south, and parallels are east-west

Graphic and Fractional Scales

There are several ways to express or portray scale on a map, but only two are widely used: a graphic scale and a fractional scale. See Figure 2-1.

A *graphic scale* makes use of a line marked off in graduated distances. The advantage of a graphic scale is the simplicity of determining approximate distances on the ground by measuring them directly on the map. Moreover, a graphic scale remains correct even if the map is reproduced in a larger or smaller size because the graduated line will be changed in size precisely as the map is changed.

Figure 2-1. Examples of fractional and graphic scales.

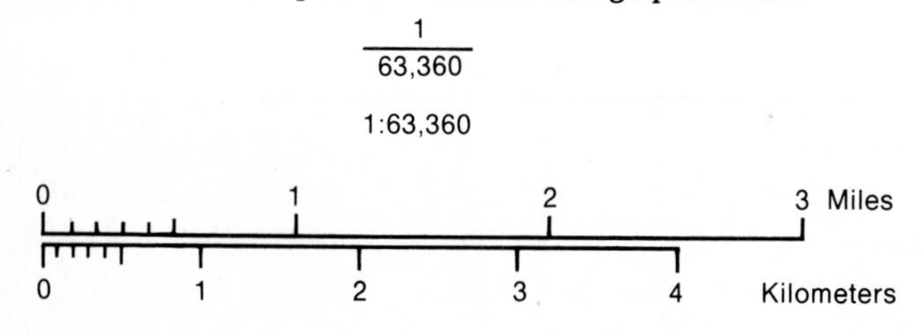

A *fractional scale* compares map distance with ground distance by proportional numbers and is expressed as a fraction or ratio. It is often referred to as a *representative fraction*. For example, a common fractional scale is 1/63,360, which is usually expressed as a ratio, 1:63,360. This means that 1 unit of distance on the map represents 63,360 of the same units on the ground. Thus 1 inch on the map represents 63,360 inches on the ground, or 1 centimeter on the map represents 63,360 centimeters on the ground, etc. With a fractional scale, the same units of measurement must be used in both numerator and denominator, or on both sides of the ratio, unless the required mathematical transition is performed. Accordingly, a scale of 1:63,360 could also be expressed as 1 in.:1 mile, because a mile contains 63,360 inches. In this case, the understanding is that 1 inch on the map represents 1 mile on the ground.

lines. In addition, a map often is marked with a straight arrow pointing northward, which is called a *north arrow*. A north arrow is aligned with the meridians and thus points toward the North Pole, which is more accurately described as the *north geographic pole*.

An element of some confusion concerning map direction is caused by the fact that *geographic* (*true*) *north* usually does not coincide with *magnetic north*, which is the direction that a magnetic compass needle points. The magnetic field of the Earth is not uniform, and the north and south magnetic poles are actually hundreds of miles away from the north and south geographic poles. In addition, the magnetic poles are in very slow but continuous motion, so their positions shift with time. The needle of a compass points toward magnetic north rather than true north. At any point on the Earth, the angle between a magnetic north line and a true north line is known as the *magnetic declination*, and is expressed as degrees east or west of the meridian of the point in question. Magnetic declination can be readily mapped, and is normally shown by lines connecting points of equal declination; these *isogonic lines* are identified by the number of degrees of declination east or west. There is a single around-the-world isogonic line on which there is no declination because of the exact alignment of true north and magnetic north; this is called the *agonic line*.

Magnetic declination varies widely from place to place, but it also varies slightly through time. Consequently, isogonic maps become inaccurate and must be updated periodically. Thus it is important to know the date of an isogonic map.

Location The standard system for locating places on a map is by use of the geographic grid showing latitude and longitude. In addition, a number of other coordinate (or grid) systems have been devised for specifying locations because the latitude/longitude system, with its angular subdivisions, is cumbersome to use. These alternative systems are devised like the x and y coordinates of a graph. Some maps display more than one coordinate system.

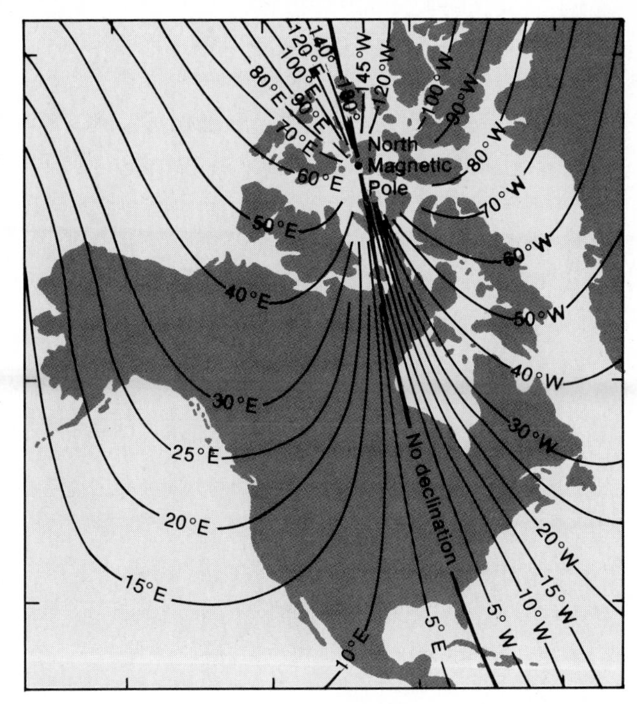

Isogonic map of North America. Magnetic declination, as of 1970, is shown by isogonic lines.

Large and Small Scales

The adjectives *large* and *small* are often used to describe a map scale. This usage represents an important concept, but the words are used in a comparative, rather than an absolute, sense. Scales are large or small in comparison with other scales.

A *large-scale* map is one that has a relatively large representative fraction, which means that its denominator is small. Thus, 1/10,000 is a larger fraction than, say, 1/1,000,000, so a scale of 1:10,000 is large in comparison with one of 1:1,000,000, and a map at a scale of 1:10,000 would be called a large-scale map. A large-scale map portrays only a small portion of the Earth's surface, but in considerable detail. See Figure 2-2. For example, if the page you are reading now were covered with a map at a scale of 1:10,000, it would be able to show just a small part of a single county, but it would be capable of doing so in great detail.

A *small-scale* map, on the other hand, involves a small representative fraction, i.e., one with a large denominator. So a map scale of 1:10,000,000 would be classed as a small-scale map. If this page were covered with a map at a scale of 1:10,000,000, it would be able to portray about one-third of the United States. Hence, a small-scale map can show a large portion of the Earth's surface, but in limited detail.

The Role of Globes

For a proper portrayal of the sphericity of the Earth, there is no substitute for a model globe. See Figure 2-3. If care is used in its manufacture, a globe can be an accurate representation of the shape of our planet. The only thing changed in the transition from the immensity of the Earth to the manageable proportions of a model globe is the size. A globe is capable of maintaining the correct geomet-

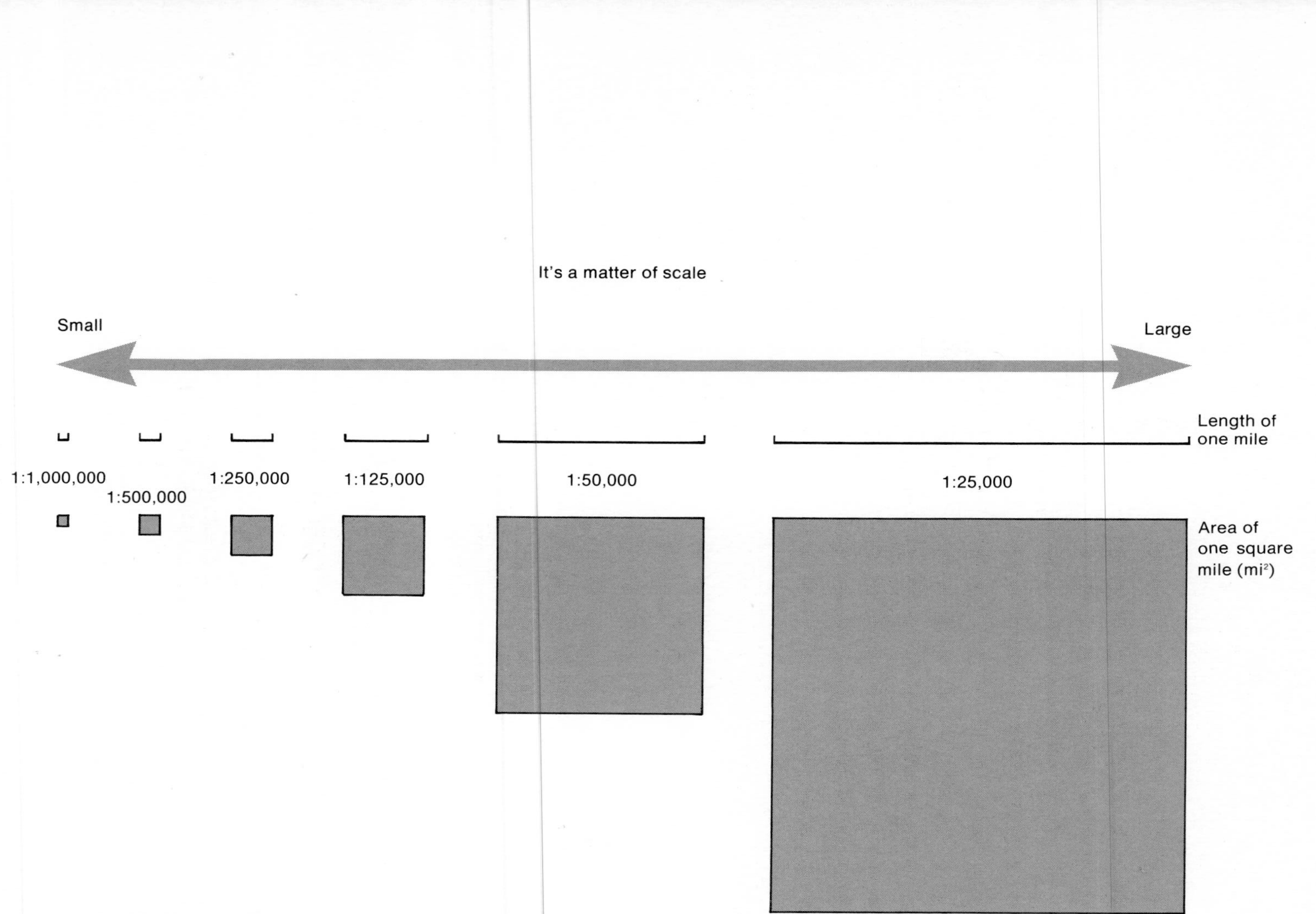

Figure 2-2. Comparisons of distance and area with different map scales. A small-scale map can portray a large part of the Earth's surface, whereas a large-scale map can show only a small part of the surface.

Figure 2-3. A model globe provides a splendid broad representation of the Earth at a very small scale, but no details can be portrayed. [The Rand McNally Earth and Moon Photo Library.]

Figure 2-4. The Earth is indeed a globe. This deep space view from *Apollo 11* shows mostly water and clouds. We are looking at the Pacific Ocean, with Australia in lower left and North America in upper right. [Courtesy NASA.]

rical relationships of meridian to parallel, of equator to pole, of continents to oceans. It can show comparative distances, comparative sizes, and accurate directions. It can represent, essentially without distortion, the spatial relationships of the features of the Earth's surface.

A globe is not without disadvantages, however. For example, only half of it can be viewed at one time. See Figure 2-4. Moreover, almost any globe must be constructed at a very small scale, which means that it is incapable of portraying much detail. The principal problem with a globe, however, is that it is too cumbersome for almost any use other than classroom study or quiet contemplation. Maps are much more portable and versatile than globes, even though they may lack the accuracy. Accordingly, there are literally billions of maps in use over the world—for an extraordinary diversity of purposes. But globes are extremely limited both in numbers and variety.

Map Projections

The challenge to the cartographer (map-maker), then, is to combine the geometrical exactness of the globe with the convenience of the flat map. This has been attempted for many centuries, and further refinements continue to be made. The fundamental problem remains: How to transfer data from a spherical surface to a flat piece of paper with a minimum of distortion?

A map *projection* is a systematic representation of all or part of the three-dimensional Earth's surface onto a two-dimensional flat surface. A piece of paper cannot be closely fitted to a sphere without wrinkling or tearing, so data from a globe (parallels, meridians, continental boundaries, etc.) cannot be directly transferred to a map without distortion. There are many ways that a cartogra-

Although some map projections were devised centuries ago, there has been a continuing development and refining of projections right up to the present day. Thus it is remarkable that the most famous of all projections was "invented" more than four hundred years ago and is still in common usage today without significant modification. This is the Mercator projection, originated in 1569 by a Flemish geographer and cartographer.

Gerhardus Mercator (who was born Gerhard Kremer, but is better known by the latinized version of his name) produced some of the best maps and globes of his time; however, his place in history is based largely on the fact that he developed a special-purpose projection that became inordinately popular for general-purpose use. The Mercator projection was essentially a navigational chart of the world, designed to facilitate oceanic navigation; his instructions accompanying the map stated clearly that its proper use would guide the mariner by simple compass direction to his destination, but that it would not necessarily be the quickest way to get there.

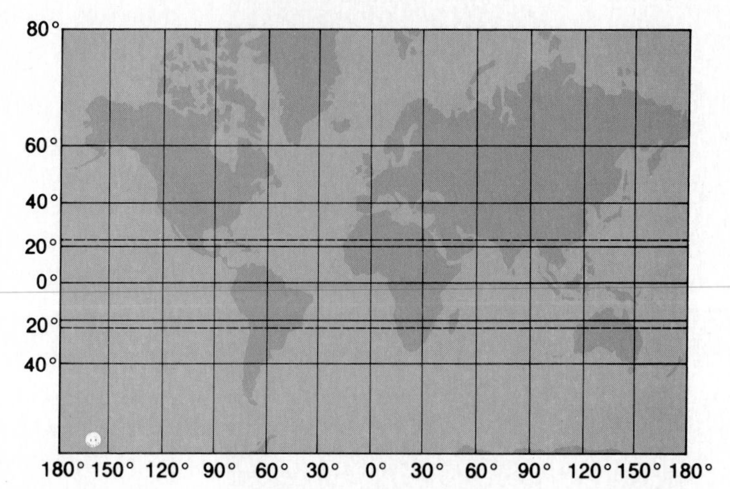

The Mercator projection, in all its simplicity and exaggeration.

The prime advantage of a Mercator map is that it shows *loxodromes* as straight lines. A loxodrome, also called a *rhumb line*, is a true compass heading, a line of constant compass direction. A navigator, whether on a ship or a plane, can plot the shortest distance between origin and destination (a great circle) on some projection in which great circles are shown as straight lines, and then transfer that route to a Mercator projection by marking spots on the meridians that the great circle crosses. These spots can then be connected by straight-line loxodromes, which approximate arcs of the great circle but consist of constant compass headings. This allows the navigator to chart an approximately shortest-distance route between origin and destination by making periodic changes in the compass course of the airplane or ship as it generally follows a great circle.

The Mercator map is a cylindrical projection mathematically adjusted to attain conformality. Parallels and meridians form a perfectly rectangular grid on the map, with the equator as the circle of tangency. It is accurate at the equator and relatively undistorted in the low latitudes, but distortion of size increases rapidly in the middle and high latitudes. In the natural concept of this projection, it can be considered that a light in the center of the globe projects the network of the geographic grid onto a tangent cylinder. This causes the meridians to appear as parallel lines rather than

pher can manipulate the data in an attempt to mitigate the problem. For example, the grid system can be arranged on a plane surface so that one or more of the globe's geometric properties are retained, or so that the principal areas of distortion fall in the "less important" areas of the map, or so that the map is interrupted by blank spaces in oceanic regions in order to minimize distortion on the continents.

The basic principle of projecting a map is direct and simple. Imagine a transparent globe on which are drawn meridians, parallels, and continental boundaries, and which has a light bulb in its center. A piece of paper in the form of a geometric figure such as a cylinder or a cone is placed over the globe. See Figure 2-5. When the bulb is lighted, the lines drawn on the globe will be "projected" outward onto the paper cylinder or cone. These lines can then be sketched on the paper, the paper can be laid out flat, and a map projection has been produced.

In actuality, very few map projections have ever been constructed by the direct projection of data from a globe to a piece of paper. Nearly all projections have been derived by mathematical computation. Their common feature is that each projection shows the correct location of latitude and longitude on the Earth's surface. In other words, each projection consists of an orderly rearrangement of the geographic grid (meridians and parallels) transposed from the globe to the map. The arrangement of the grid, however, varies from projection to projection. Indeed, the difference among projections is the difference among the grid layouts. There is no possible way to avoid distortion completely, so no map projection is perfect. Each of the many hundreds of different projections has been designed as a compromise from reality

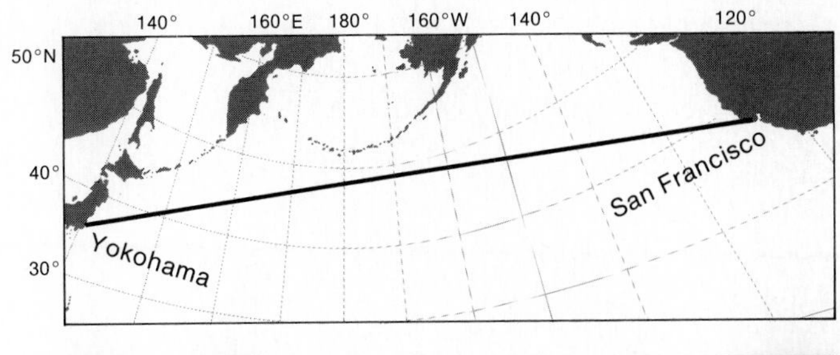

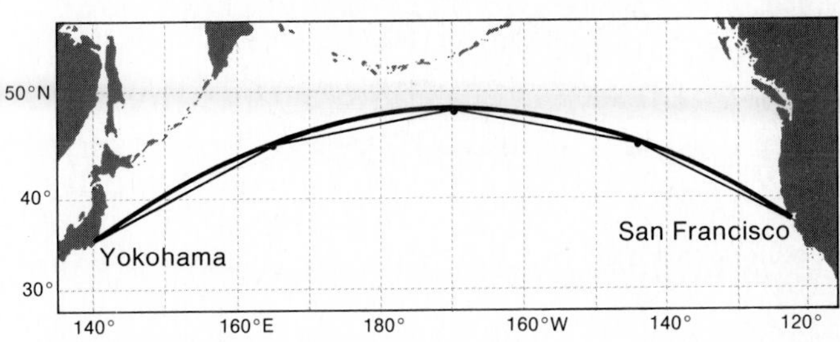

The prime virtue of the Mercator projection is for straight-line navigation. The shortest distance between San Francisco and Yokohama can be plotted on any of several projections in which great circles are shown as straight lines (top map). This route can then be transferred to a Mercator projection (bottom map) with mathematical precision. Straight-line loxodromes can then be substituted for the curved great circle, allowing the navigator to maintain constant compass headings from point to point, approximating the curve of the great circle.

converging at the poles, resulting in extreme east-west distortion in the higher latitudes. In order to maintain conformality, Mercator compensated for the east-west stretching by spacing the parallels of latitude increasingly further apart so that north-south stretching occurs at the same rate. This procedure allowed shapes to be approximated with reasonable accuracy, but at great expense to proper areal relationships. Area is distorted by 4 times at the 60th parallel of latitude and by 36 times at the 80th parallel. If the North Pole were shown on a Mercator projection, it would be as extensive as the equator; rather than a single point, it would become a line 25,000 miles (40,250 km) long.

It is clear then that the Mercator projection is excellent for straight-line navigation, but it has serious imperfections for most other uses. However, despite the obvious flaws associated with areal distortion in the high latitudes, Mercator projections have been widely used in American classrooms and atlases. Indeed, several generations of American students have passed through school with their principal view of the world provided by a Mercator map. This has created many misconceptions, not the least of which is confusion about the relative sizes of high-latitude landmasses. For example, on a Mercator projection, the island of Greenland appears to be as large or larger than four of the recognized continents (Africa, Australia, Europe, South America). In actuality, however, Africa is 14 times larger than Greenland, South America is 9 times larger, even Australia is $3\frac{1}{2}$ times larger; it just doesn't appear that way on a Mercator map.

The Mercator projection, then, is a map that was devised several centuries ago for a specific purpose, and it still serves that purpose well. Its fame, however, is significantly due to its misuse.

to achieve some purpose. Each projection, then, has some advantage over the others, but it also has its own particular limitations.

The Major Dilemma: Equivalence vs. Conformality

The cartographer wants the map to portray accuracy of distance and direction, in order to provide correct size and shape. However, such perfection is impossible, so a compromise must be struck. Which to emphasize: size or shape? Which to sacrifice: shape or size? This is the central problem in constructing or choosing a map projection. The projection properties that are involved are called *equivalence* and *conformality*.

Equivalence In an equivalent projection, the size ratio between any area on the map and the corresponding area on the ground will be the same all over the map. To illustrate: Suppose you have a world map before you, and place four coins of the same denomination at different places (perhaps one on Brazil, one on Australia, one on Siberia, and one on South Africa). Calculate the extent of the area covered by each of the coins. If it is the same (say, 5000 square miles or 13,000 km²) in each case, there is a good chance that the map is equivalent—that there are equal areal relationships all over the map. Equivalent maps are very desirable because areas are shown in correct proportion to one another, and misleading impressions of size are avoided. The world maps in this book are mostly drawn on equivalent projections because they are so useful in portraying distributions.

Equivalent maps are by no means perfect, however. Equivalence is difficult to achieve on small-scale maps

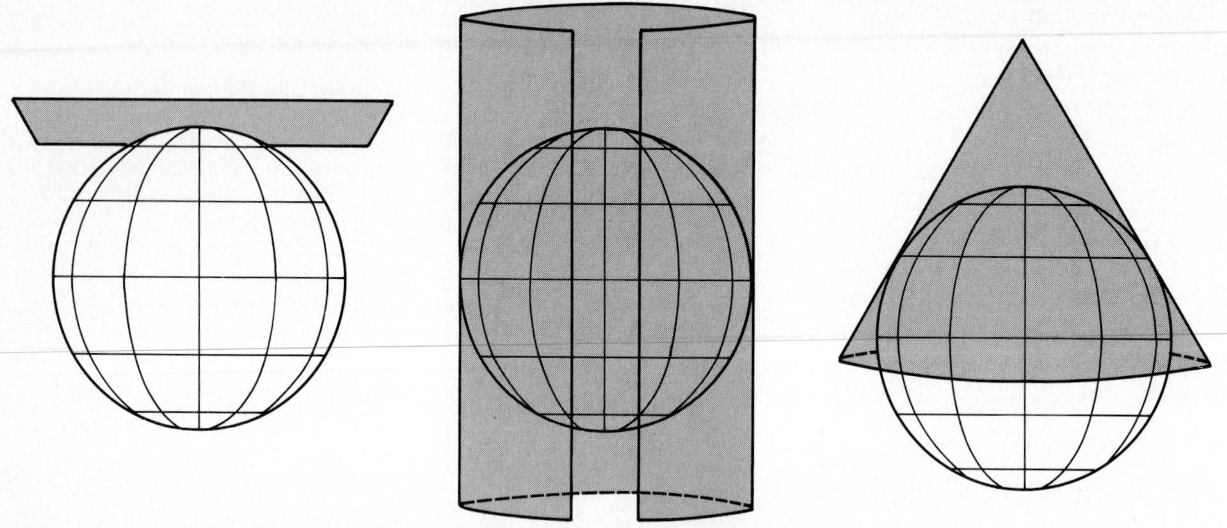

Figure 2-5. Illustrating the theory of map projection. The three common geometric figures used in projections are a plane, a cylinder, and a cone.

because shapes must be sacrificed to maintain proper areal relationships. Most equivalent world maps display terribly disfigured shapes around the edge of the map: For example, Greenland is usually squatty, New Zealand is long and stretched, and Alaska is both squeezed and attenuated.

Equivalence, then, is a very desirable property that can be achieved, but at a sacrifice.

Conformality A conformal projection is one in which shapes are truly represented. This means that the proper angular relationships are maintained so that the shape of something on the map is the same as its actual shape on the Earth. It is impossible to depict true shapes for large areas, such as a continent or sizable country, but they can be approximated; and small areas can be shown in true shape on a conformal map. All conformal projections have meridians and parallels crossing each other at right angles, just as they do on a globe.

The outstanding problem with conformal maps is that the size of an area often must be considerably distorted in order to depict the proper shape. Thus the scale

Figure 2-6. Portraying the whole world on a flat map requires considerable distortion. A *conformal* projection (a) displays correct shapes, but enormously exaggerates sizes in high latitude areas. An *equivalent* projection (b) can portray accurate sizes, but shapes are extremely distorted around the edges of the map. [Adapted from Henry M. Kendall et al., *Introduction to Physical Geography,* 2nd ed. (New York: Harcourt Brace Jovanovich, Inc., 1974), p. 359. Copyright © 1974 by Harcourt Brace Jovanovich, Inc. Reproduced by permission of the publisher.]

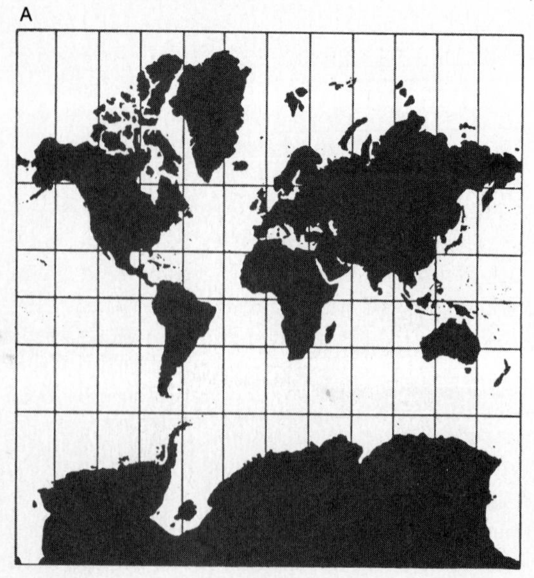

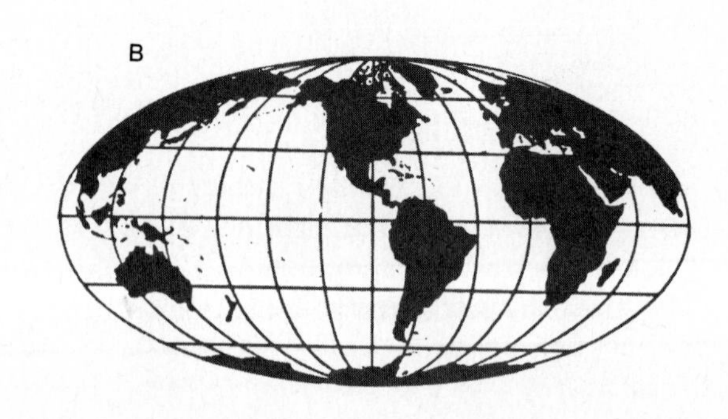

necessarily changes from one region to another. For example, a conformal map of the world normally greatly enlarges sizes in the higher latitudes.

Except for maps of very small areas (very large-scale maps), conformality and equivalence are mutually exclusive properties, i.e., they cannot both be maintained on the same projection. See Figure 2-6. It can be stated that some projections are conformal, some are equivalent, none are both conformal and equivalent, and most maps are neither conformal nor equivalent but reach a compromise between the two.

Families of Projections

More than a thousand different map projections have been devised for one purpose or another. Most of them can be grouped into just a few families, based on their derivation. Projections in the same family generally have similar properties and related distortion characteristics. See Figure 2-7 and Table 2-1.

TABLE 2-1 The Families of Projections, with Examples of Each

A. *Cylindrical*:
- Mercator's
- Central Perspective Cylindrical
- Gall's Stereographic
- Gnomonic Cylindrical
- Lambert's Cylindrical Equal Area
- Equirectangular

B. *Elliptical*:
- Sinusoidal
- Mollweide's
- Goode's Homolosine
- Denoyer Semielliptical
- Aitoff's

C. *Azimuthal*:
- Orthographic
- Stereographic
- Gnomonic
- Lambert's Equal Area
- Azimuthal Equidistant

D. *Conic*:
- Simple Conic
- Lambert's Conformal Conic
- Polyconic
- Alber's Conic Equal Area
- Bonne

Figure 2-7. Some sample projections: (1) Lambert's Cylindrical Equal Area projection is an example of the cylindrical family. (2) Mollweide's Projection is in the elliptical group. (3) The Stereographic Projection is a member of the azimuthal family. (4) Alber's Conic Equal Area is a conic projection.

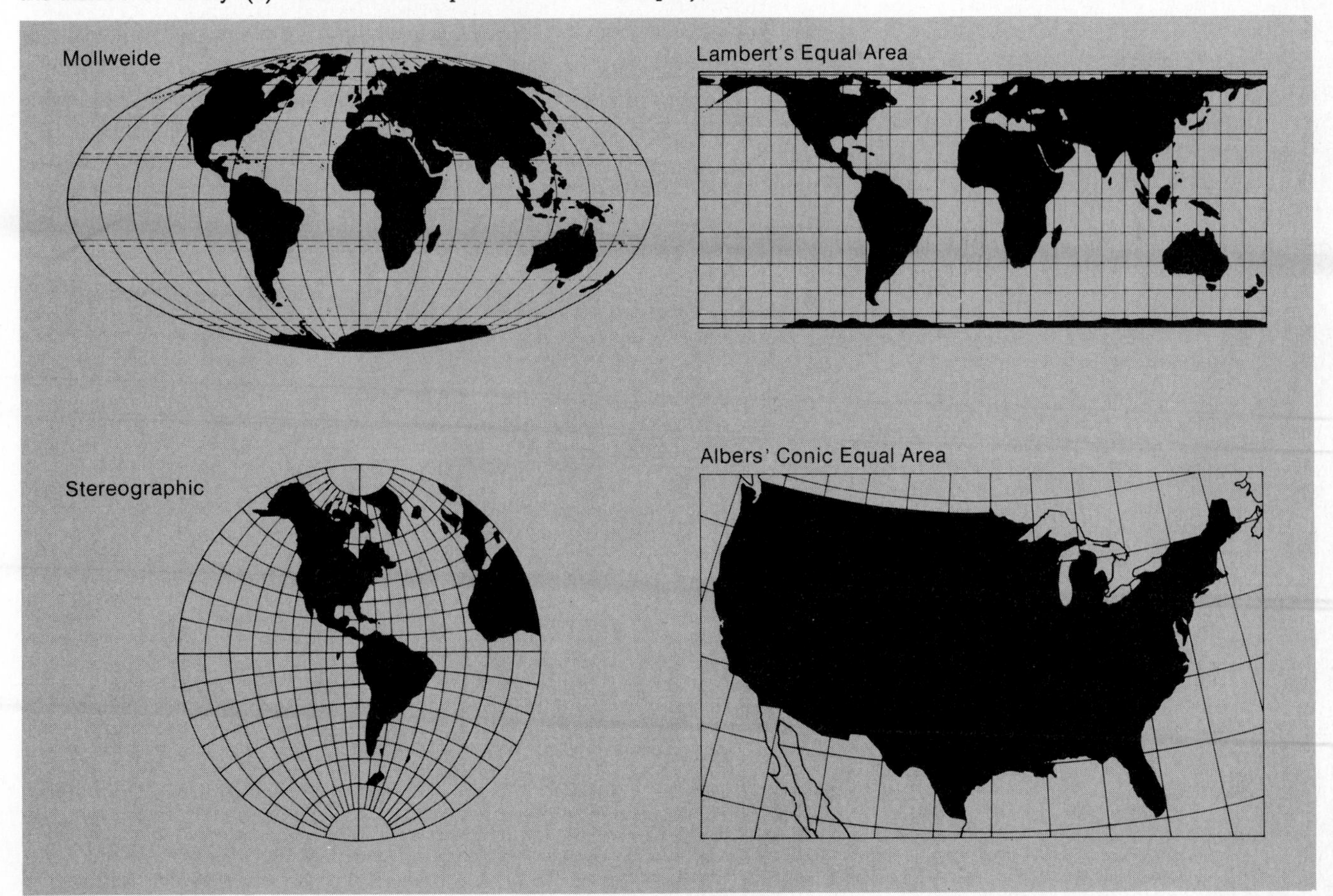

Cylindrical Projections A cylindrical-type projection is derived from the concept of projection onto a paper cylinder that is tangential to, or intersecting with, a globe. Most cylindrical projections are designed so that the cylinder is tangent to the globe (i.e., just touches the globe) at the equator. This produces a right-angled grid network (which means that meridians and parallels meet at right angles) on a rectangular map. There is no distortion at the circle of tangency, where the globe makes contact with the cylinder, but distortion increases progressively away from this circle. Thus in most cylindrical projections, there is little distortion in low latitudes but enormous distortion in the polar regions. (This is clearly exemplified by the Mercator projection.) Cylindrical projections are generally used for maps of the whole world.

Elliptical or Oval Projections The elliptical or oval type contains a number of roughly football-shaped projections on which the entire world is displayed. In most elliptical projections, a central parallel (usually the equator) and a central meridian (generally the prime meridian) cross at right angles in the middle of the map, which is a point of no distortion. Distortion normally increases progressively toward the outer margins of the map. Parallels are usually arranged parallel to one another, but meridians (apart from the central meridian) are shown as curved lines.

Azimuthal Projections An azimuthal projection (also called *plane* or *zenithal*) is derived by the perspective extension of the geographic grid from a globe to a plane that is tangent to the globe at some point. The point of tangency can be any spot on the globe, but it usually is either the North or South Pole or some point on the equator. There is no distortion immediately around the point of tangency, but distortion increases progressively away from this point. No more than half the Earth (a hemisphere) can be displayed on an azimuthal projection.

Conic Projections In conic projections, it is conceived that one or more cones is set tangent to, or intersecting, a portion of the globe, and the geographic grid is projected onto the cone(s). Normally the apex of the cone is considered to be above a pole, which means that the circle of tangency coincides with a parallel, which becomes the standard parallel of the projection. Distortion increases progressively away from the standard parallel and is least in the vicinity of the extent of that parallel. Consequently, these projections are best suited for regions of east-west orientation in the middle latitudes, which makes them particularly useful for maps of the United States, Europe, or China. It is impractical to use conic projections for more than one-fourth of the Earth's surface (a semi-hemisphere), and they are particularly well adapted for mapping relatively small areas, such as a state or county.

Other Projections There are dozens of other projections that do not fit into any of the four families discussed above. Each has specific uses and specific limitations, but due to space limitation, will not be discussed here.

Manipulation of Projections

As most projections are based on a systematic layout of the geographic grid through mathematical derivation, it is easy to visualize that there is almost unlimited scope for manipulation of the geometry to produce refinements that might be useful for one purpose or another. Most such refinements involve the rotating of the orientation of a projection, so that its long dimension extends, for example, north-south rather than east-west.

Figure 2-8. An interrupted projection of the world. The purpose of interruption is to portray certain areas (usually continents) more accurately, at the expense of portions of the map (usually oceans) that are not important to the map's theme.

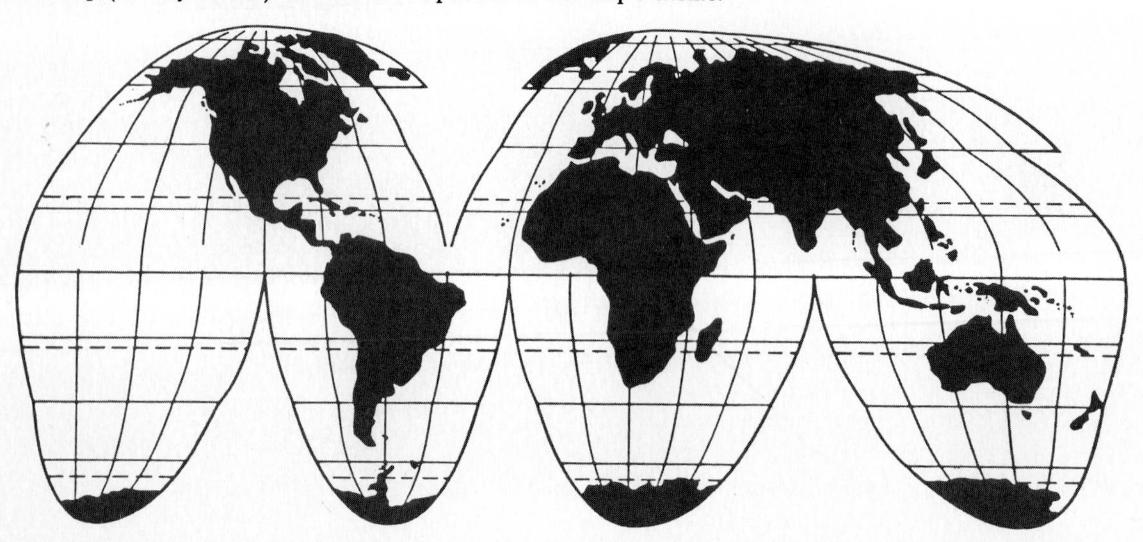

The most striking manipulation, and one that serves very effectively on maps showing world distribution of phenomena, is the *interruption* of a projection. See Figure 2-8. For global distribution maps, the continents are often much more important than the oceans, yet the oceans would occupy most of the map space in a normal projection. Hence, the projection can be interrupted in the three principal oceanic areas (Pacific, Atlantic, Indian) and can be based on standard meridians that are central to each of the major landmasses. This means that no land area is far from a central meridian, which greatly decreases the distortion of shape or size for the landmasses. The interruption of the projection in the oceans literally creates a void in the map, one that is simply filled with information that is not part of the map. The result is that some of the oceanic portions of the map are torn apart and otherwise distorted, but the major landmasses can be shown with relatively little distortion, and the overall accuracy of the distribution pattern is enhanced.

Remote Sensing

Throughout most of history, maps have been the only tools available to humanity that could depict anything more than a tiny portion of the Earth's surface with any degree of accuracy. More recently, however, sophisticated technology has been developed that permits precision recording instruments to operate from high-altitude vantage points, providing a remarkable new set of tools for the study of the Earth.

Remote sensing, broadly considered, is the measurement or acquisition of information by a recording device that is not in physical contact with the object under study. Included under this heading is Air Photo Interpretation (API), which was not only the major but almost the only form of remote sensing used for geographical purposes for the first eighty years during which the technology existed. An air or aerial photograph is one taken from an elevated platform such as a balloon, airplane, rocket, or satellite. See Figure 2-9. The earliest air photos were taken from balloons in France in 1858 and in the United States in 1860. A platform (the balloon) was available before a practical method of photography was developed in the 1830s and 40s, but it was another two decades before the first aerial photos were actually made.

Aerial Photographs

A major problem with photographs taken from balloons—the lack of control of the platform—was overcome in the early twentieth century by the development of heavier-than-air craft (airplanes). By the time of the First World War (1914–1918), systematic aerial photographic coverage was possible. Specific flight lines could be planned, and photographs taken at designated intervals. In terms of their geometry, there are two major types of aerial photos: oblique and vertical. See Figure 2-10. Oblique photos can be further subdivided into those in which the horizon is visible (high obliques) and those in which it is not (low obliques). The great advantage of oblique photographs of both kinds is that features are seen from a more or less familiar point of view. The disadvantage is that, because of perspective, measurement is more difficult on oblique than on vertical photos—those taken with the optical axis of the camera approximately perpendicular to the surface of the Earth.

The possibility of precise measurement on vertical aerial photos was realized almost as soon as the first such images were available. Subsequently, a new discipline was developed—*photogrammetry*, the science of obtaining reliable measurements from photographs and, by extension, mapping from aerial photos. The workhorse of photogrammetry and of much interpretation up to the present time has been the black and white vertical aerial photograph taken automatically at regular intervals—called "air" or "camera stations." Figure 2-11(A) is a diagram of two photographs taken sequentially at camera station 1 and camera station 2, so that the resulting vertical images of the Earth overlap by 60 percent in the direction of flight [Figure 2-11(B)]. The fundamental formula for obtaining the scale (S) of a vertical aerial photograph is

$$S = \frac{f}{H}$$

where f is the focal length of the camera lens and H is the height of the camera above the ground at the moment the photo was taken [Figure 2-11(A)]. The shaded portion of Figure 2-11(B) is the area of stereoscopic overlap on two adjacent vertical aerial photos.

Although the stereoscopic effect is possible with ground photos and oblique aerial photos, it is best obtained by vertical aerials, and is of the greatest utility for the photogrammetrist and air photo interpreter. With the aid of a stereoscope (a binocular optical instrument) and two overlapping photographs, an observer can view an object simultaneously from two different perspectives to obtain the mental impression of a three-dimensional model. Vertical distance (height) can be measured from the model and contours can be plotted, so that together with two-dimensional measurements obtained with the aid of an interpreter's ruler when the scale is known, the length, breadth, and height of features can readily be ascertained. Extremely complicated apparatus has been developed for photogrammetry, but good measurements can be made with quite simple and inexpensive equipment.

Color, Color Infrared, and Ultraviolet

Experiments were undertaken with conventional color photography in the 1920s, and a viable system of color photogrammetry was developed in the 1940s. In contrast

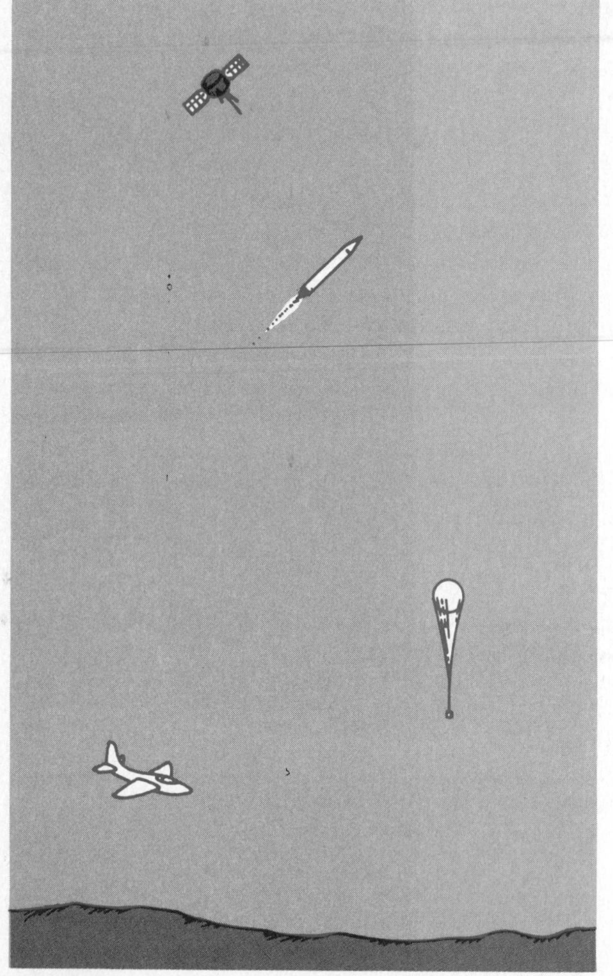

Figure 2-9. Various kinds of flying or floating objects can serve as "platforms" from which images of the Earth's surface can be recorded.

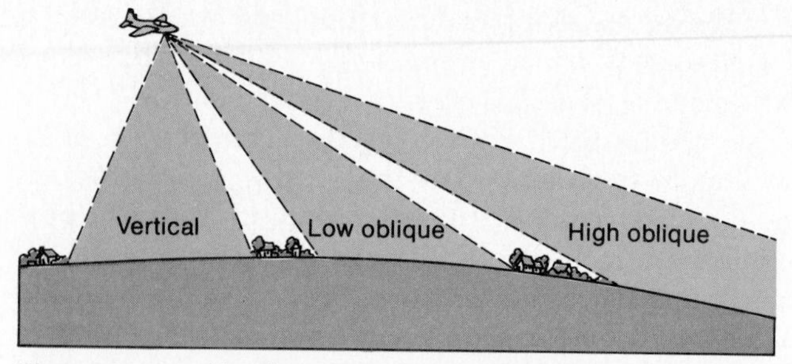

Figure 2-10. Air photos can be classified on the basis of their angular relationship to the Earth's surface—vertical, low oblique, or high oblique.

to black and white images, where only a few tones can be recognized in different photo environments, a large number of colors (hues) can be discriminated on color photographs. In spite of this advantage, color film initially met with considerable resistance due to its much greater cost and the innate conservatism of interpreters. However, under the pressure of World War II, color film was used increasingly for aerial photography and *c*olor *i*nfra*r*ed (CIR) film was introduced. Although it was first known as camouflage detection film because of its ability to discriminate living vegetation from withering vegetation used to hide or disguise objects, CIR has now become one of the most versatile tools available to the interpreter.

Since CIR senses, in part, beyond the visible spectrum, it is desirable at this point to review the concept of the electromagnetic spectrum, which is basic to an understanding of all types of imagery. Figure 2-12 is a simplified diagram of a portion of this spectrum. The visible part is a very narrow band separating shorter wave radiation on the blue side and longer wave radiation on the red side. Like the human eye, panchromatic black and white and conventional color films sense only in the visible portion, from about 0.36 to 0.72 micrometers (millionths of a meter). However, special scanners have been developed to sense in shorter wavelengths, allowing production of photolike images that give the viewer a picture of what the eye would see if it could sense in the ultraviolet (UV) range. Such UV images have been useful in pollution studies, but along with much shorter wave lengths such as X-rays and gamma rays, UV has proved to be more useful in medicine and other studies than in the interpretation of the Earth. On the other hand,

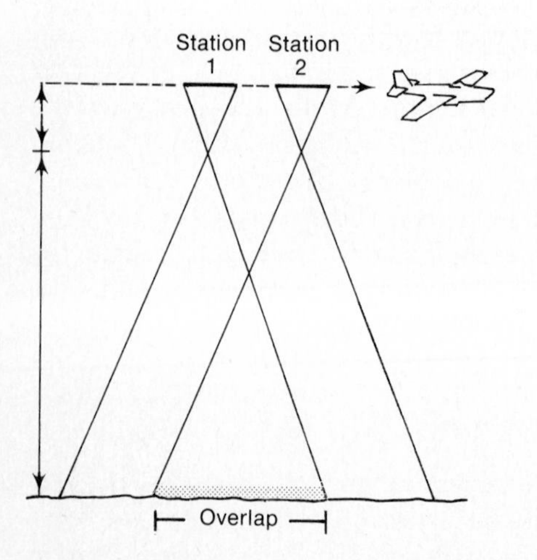

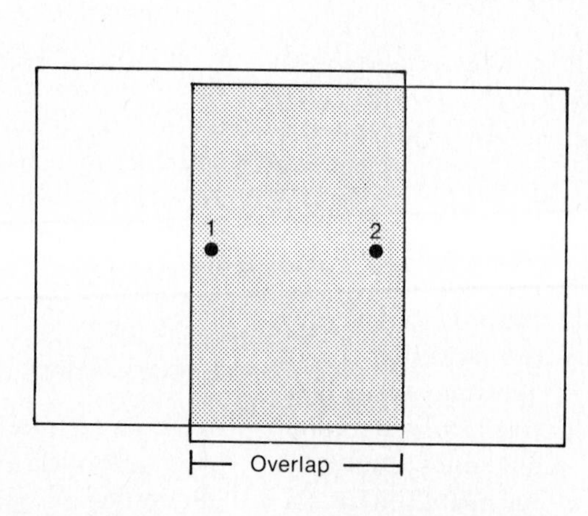

Figure 2-11. Overlapping aerial photographs: (a) horizontal sketch of the taking of two consecutive photographs in a flight line; (b) vertical images of the same two photographs. The overlap is designed to be about 60 percent.

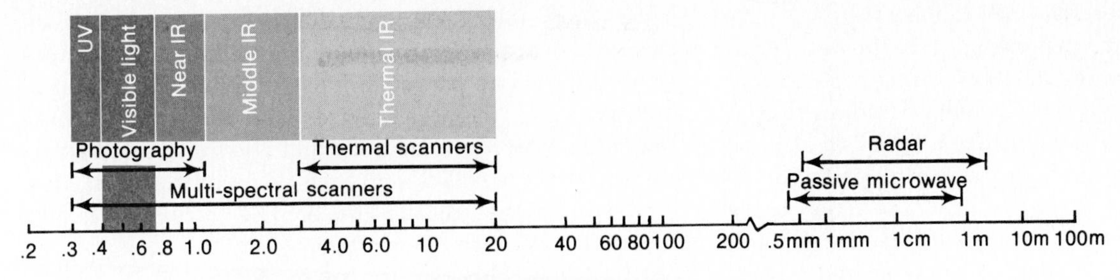

Figure 2-12. The electromagnetic spectrum, in part. Conventional photography can use only a small portion of the total spectrum. Various specialized remote sensing scanners are capable of utilizing a much greater expanse of the infrared portion of the spectrum.

the longer wavelengths, beyond the red side of the visible spectrum, have been of great value in expanding our ability to interpret features of the Earth.

As shown in Figure 2-12, the *infrared* (IR) is divided into near, middle, and far regions relative to the visible. Color infrared (CIR) films have been produced that sense in the near IR up to 1.0 micrometer. With such CIR film, blue is filtered out and photographic infrared added. This produces false-color imagery in which the color of objects is unfamiliar (for example, living vegetation appears red), but which is nevertheless extremely valuable. See Figure 2-13. Thus, on panchromatic black and white or on conventional color film, it might be difficult to distinguish between Astroturf and grass, but on CIR imagery, the Astroturf would appear blue, and the grass pink to red (depending on the condition of the grass). See Figure 2-13. One of the major uses of CIR imagery is to evaluate the health of crops and trees. See Figure 2-14.

Much of the usable portion of the near infrared cannot be detected by photographic emulsions. As a re-

Figure 2-13. A false-color CIR air photo of a portion of the U.C.L.A. campus in Los Angeles. The football practice field near the center of the photo is mostly vegetated with natural grass, which shows red in this imagery. Near the upper right corner of the football field is a small rectangle of synthetic turf, which takes on a dark green color. [Photo courtesy of Norman Thrower/NASA.]

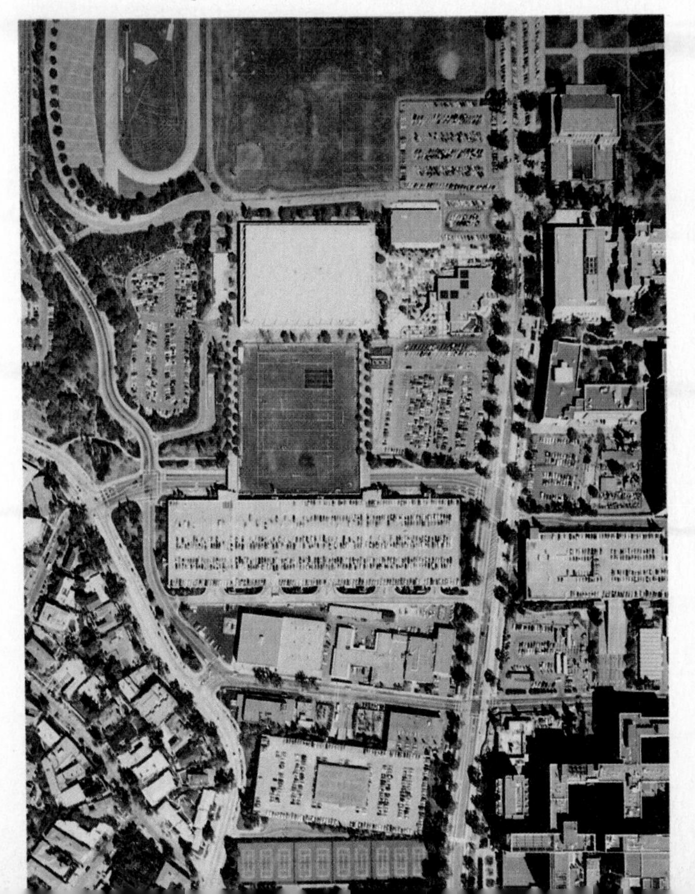

Figure 2-14. A color infrared photo from *Apollo 9,* orbiting 110 miles above the Earth's surface. The tiny red squares are irrigated farm fields. The big farming area at lower left is California's Imperial Valley. Near the bottom of the photo, the color of the Imperial Valley changes abruptly as the international border is reached. On the Mexican side of the boundary, irrigation is much less intensive and the crop plants show up less prominently. The smaller irrigated valley in the upper center of the photo is the Palo Verde Valley around Blythe, California. [Courtesy NASA.]

sult, various optical-mechanical scanner systems have been developed that can sense further into the near infrared and at the same time simulate CIR photographic imagery. The most widely used such system is Landsat, a series of unmanned satellites that orbit the Earth at an altitude of 570 miles (915 km) and can image all parts of the Earth except the polar regions every nine days (using any two satellites together) in four distinct wavelength bands. See Figure 2-15. Except for a low spatial resolution—minimum size of object sensed—of 260 feet (79 m), the multiband Landsat system with its continuous observation capability approaches the ideal in the production of imagery. As the technology improves further, better systems will be developed for both unmanned and manned satellite platforms.

Thermal Infrared

None of the middle or far infrared, i.e., the *t*hermal *i*nfra*r*ed (TIR), can be sensed with film, and as a result, special supercooled scanners are needed. Thermal scanning senses the radiant temperature of objects and may be carried out either day or night. Photolike images can be produced that are particularly useful for showing diurnal temperature differences between land and water (including clouds), between bedrock and alluvium, for thermal water pollution studies, and even for detecting forest fires.

By far the greatest use of TIR scanning systems to date has been on meteorological satellites such as Tiros and Nimbus. Although the spatial resolution is only on the order of a few kilometers, it is more than sufficient to provide details for far more accurate and complete weather forecasting than has ever before been possible.

Microwave

Other, even longer, wavelength sensing systems are used in earth sciences, including microwave radiometry, which senses in the 100 micrometer to 1 meter range. Although such systems have low spatial resolution, they are particularly useful for showing subsurface characteristics such as moisture and radio temperature. Like all the systems

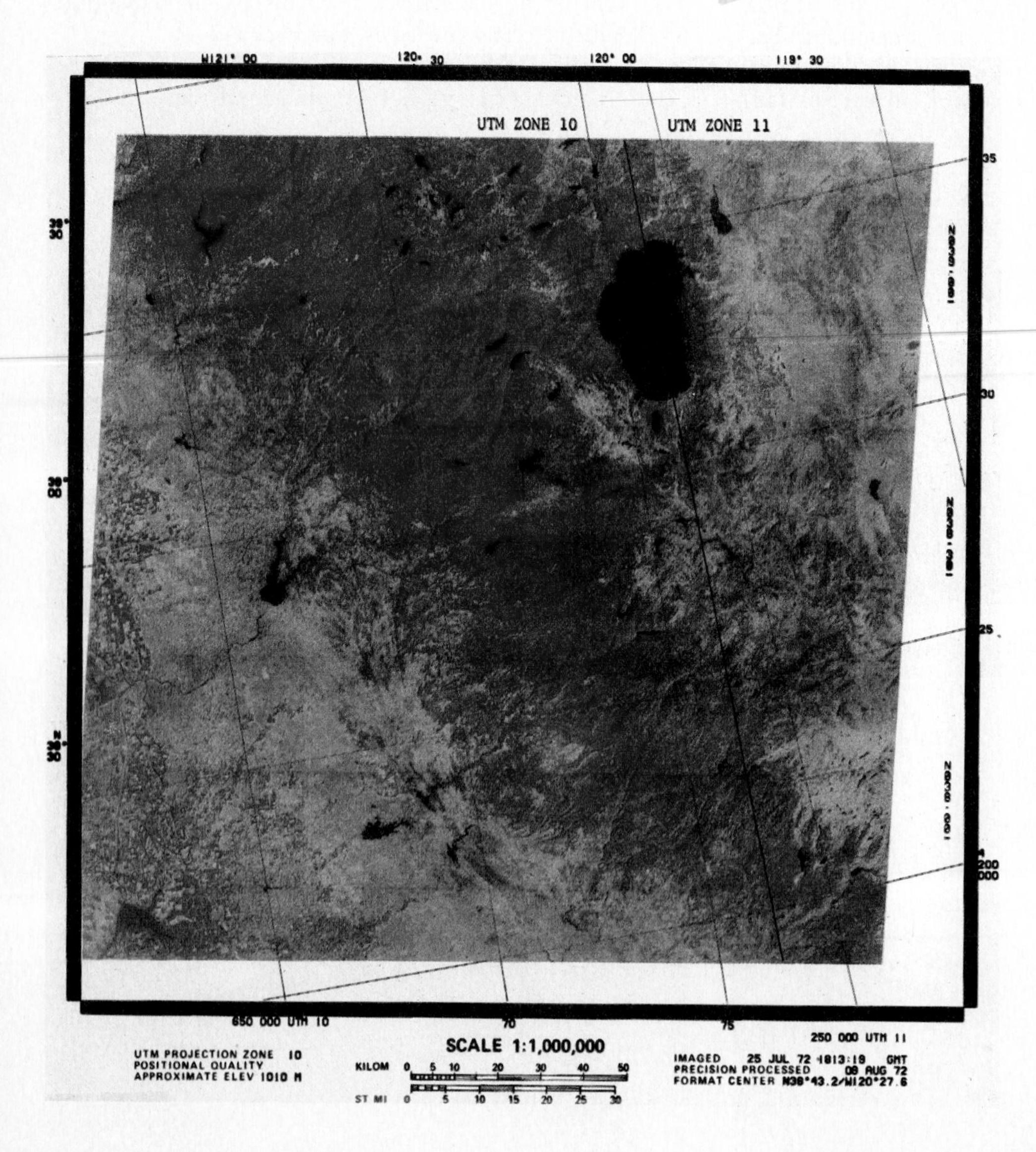

Figure 2-15. A representative Landsat image of Lake Tahoe and vicinity in California and Nevada. This actually represents three different images that have been merged into a false-color composite. The nonsquare, parallelogram shape is a trademark of Landsat images. [Courtesy of EROS.]

Figure 2-16. A radar image of the Los Angeles lowland acquired by NASA's Seasat satellite in 1978. Radar images differ in various technical details from photographic imagery, but the result often appears to be much the same. [Courtesy of Ron Wasowski/NASA.]

mentioned so far, microwave is a passive system; i.e., it senses natural radiation emitted or reflected from a target. This stands in contrast to an active system, i.e., one having its own source of electromagnetic radiation.

Radar and Sonar

The most important active sensing system used in the earth sciences is radar, the acronym for *ra*dio *d*etection *a*nd *r*anging. Radar senses in wavelengths longer than 1 millimeter, using the principle that the time it takes for an emitted signal to reach and return from a target can be converted into distance information. Initially, radar images were only viewed on a screen, but they are now available in photolike form. See Figure 2-16. In common with some other sensors, radar is capable of operating by day or night, but it is unique in its ability to penetrate atmospheric moisture. Thus, some wet tropical areas that could never be sensed by other systems now have been imaged by radar. Radar imagery is particularly useful for terrain analysis.

Other remote sensing systems such as sonar (*so*und *na*vigation *r*anging) permit underwater imaging, so that at last the true surface form of the lithosphere can be appreciated.

Acquisition

A vast array of imagery of various types and of different areas is available to the interpreter, ranging from panchromatic photography to the output of exotic sensing systems now only in the experimental stage of development. Many foreign governments place restrictions on the use of coverage of their territory, but the United States has made imagery easily accessible to its own citizens and to others. For example, all images from the U.S. civilian space program are unclassified and are obtainable at a modest cost. These include: Gemini conventional color photos from the mid-1960s; Apollo conventional color and CIR from the late 1960s; Skylab conventional color and CIR from 1973–1974; and scanner imagery from *Landsat* (ERTS) *I* launched in 1972, *II* launched in 1975, and *III* launched in 1978 (to replace *I*).

Interpretation

In order to interpret imagery of all kinds, workers in the earth sciences and other fields have developed various tools and procedures. The stereoscope, more useful for low-altitude vertical aerial photos than for satellite imagery, has already been mentioned. The *transferscope*, which allows the operator to bring a map and photo, or photo and other image, to a common scale for comparison, is another piece of equipment useful to the geographer. The grey scale, a monochrome strip of shades ranging from white to black, and the solid color have both proved useful. With the help of these tools, the tone or hue of photo elements can be better understood. Tone and/or hue, texture, pattern, shape, shadow, size, and association are all characteristics of imagery that should be taken into account, along with scale. Most serious interpretation is done by utilizing several or all of these characteristics. With the help of various tools and procedures combined with the ability to enhance images by optical means and the use of computers, interpretation has greatly advanced in the past three decades.

Role of the Geographer

In using this photographic equipment, the geographer should never lose sight of the major objective of better understanding the Earth. The new imagery available to us is an adjunct to field study, geographical description, and maps, but not a substitute for any of these. To the extent that it can help us better appreciate the complexity of the Earth, remote sensing imagery will be increasingly employed.

Certain types of imagery are useful for particular purposes; no one sensing system has universal applicability for all problems. Accordingly, each interpreter must select and obtain the best type of imagery for his/her special needs. For providing an overview of the lithosphere, high-altitude space imagery has been of particular value, and important discoveries have been made through its use. But such imagery might have limited value in detailed terrain studies, where large-scale color or black and white photos would be more appropriate. For the hydrosphere, images of different scales have proved useful. Satellite images of an entire hemisphere can tell us much about the water content in clouds, air masses, glaciers, and snowfields at a given time, but detailed conventional color photos might be better for discriminating a complicated shoreline because of scale and the penetrating ability of the film. On the other hand, the biosphere and especially its vegetation often are best appreciated on CIR imagery of several scales—overall vegetation patterns on satellite images and detailed imagery down to 1:5,000 for crop and forest inventory studies. Features of human creation are generally not evident on very high-altitude imagery, but they become increasingly clear as one approaches the Earth. Thus, survey patterns, transportation lines, rural settlements, and cities are best interpreted on imagery of intermediate or large scale and of various types, including TIR.

Mapping, then, has been revolutionized through the use of aerial imagery. A new product in which map accuracy standards are attained while retaining the richness of detail of the aerial imagery is the *orthophotomap*. See Figure 2-17. This is accomplished through computerized rectification of the imagery. Thus, maps and aerial images of various kinds (or combinations of these) give us an increasingly graphic and accurate portrayal of the Earth.

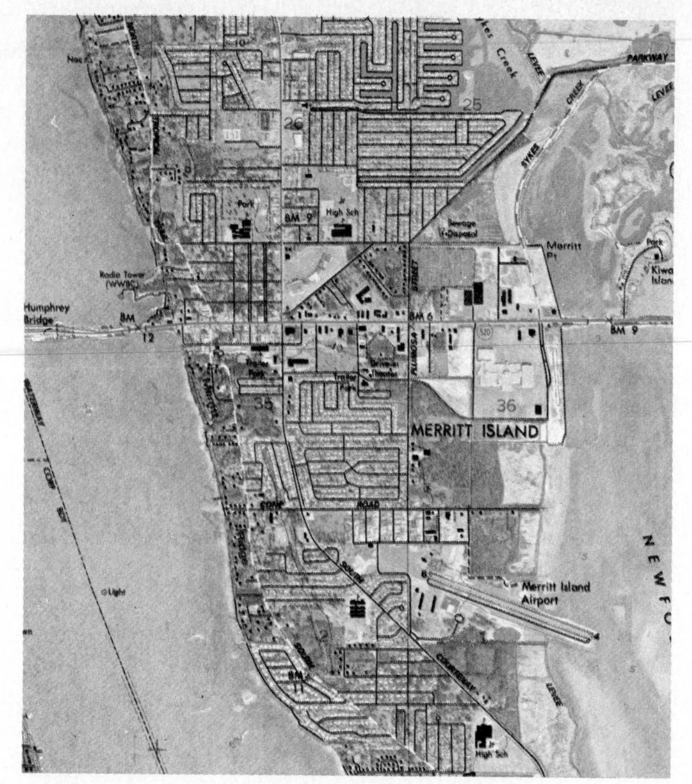

Figure 2-17. An example of an orthophotomap. This is part of the Cocoa quadrangle in Florida.

Review Questions

1. Explain the implications of the statement, "No map is totally accurate."
2. Explain the difference between large-scale and small-scale maps.
3. Which is more useful, a *graphic scale* or a *fractional scale*? Why?
4. A globe can portray the Earth's surface more accurately than a map, but globes are rarely used in comparison with maps. Why?
5. Explain the concept of *equivalence*.
6. Explain the concept of *conformality*.
7. A cylindrical map of the world always has considerable distortion, yet cylindrical projections are widely used. Why?
8. Compare the advantages and disadvantages of vertical and oblique aerial photographs.
9. What are the advantages of color infrared imagery?
10. What are the advantages of thermal infrared imagery?

Some Useful References

COLWELL, R. N., ed., *Manual of Photo Interpretation*. Washington, D.C.: American Society of Photogrammetry, 1960.

MUEHRCKE, P. C., *Map Use: Reading, Analysis, and Interpretation*. Madison, Wisconsin: JP Publications, 1978.

REEVES, R. G., ed., *Manual of Remote Sensing*, 2 vols. Falls Church, Virginia: American Society of Photogrammetry, 1975.

ROBINSON, A. H., R. D. SALE, and J. L. MORRISON, *Elements of Cartography*, 4th ed. New York: John Wiley & Sons, Inc., 1978.

RUDD, R. D., *Remote Sensing: A Better View*. North Scituate, Mass.: Duxbury Press, 1974.

SABINS, F. J., JR., *Remote Sensing: Principles and Interpretation*. San Francisco: W. H. Freeman & Company, Publishers, 1978.

THROWER, N. J. W., *Maps and Man: An Examination of Cartography in Relation to Culture and Civilization*. Englewood Cliffs, N.J.: Prentice-Hall, Inc., 1973.

3 Introduction to the Atmosphere

The Earth is different from other known planets in a variety of ways. One of the most notable differences is the presence around our planet of a substantial atmosphere with components and characteristics that are distinctive from those of other planetary atmospheres. It is our atmosphere that makes life possible on Earth. It supplies most of the oxygen that animals must have to survive, as well as the carbon dioxide needed by plants. It helps maintain a water supply, which is essential to all living things. It serves as an insulating blanket to ameliorate temperature extremes and thus provide a livable environment over most of the Earth. It also shields the Earth from much of the sun's ultraviolet radiation, which otherwise would be fatal to most earthly life forms. See Figure 3-1.

Air, generally used as a synonym for *atmosphere*, is not a specific gas but rather a mixture of gases that have various physical and chemical properties. Included in this gaseous mixture are minor but varying quantities of solid and liquid particles that can be thought of as impurities. The individual particles are mostly so minute as to be submicroscopic in size, and these are held in suspension in the air.

The gaseous atmosphere is invisible; its various components are generally colorless, odorless, and tasteless. The impurities, on the other hand, often can be smelled and may become visible due to the growth or coalescing of tiny particles until they can reflect or scatter sunlight. Clouds, by far the most conspicuous visible features of the atmosphere, represent a concentration of nongaseous impurities.

The atmosphere completely surrounds the solid and liquid Earth. It can be thought of as a gaseous envelope with the Earth tucked inside, or as a vast ocean of air with the Earth at its bottom. It is held to the Earth by gravitational attraction, and therefore accompanies our planet in all its celestial motions, such as rotation and revolution. The attachment of Earth and atmosphere, however, is a loose one, and the latter has actions of its own, doing things that the Earth does not (and cannot) do.

Although the atmosphere extends outward at least 6000 miles (10,000 km), most of its mass is concentrated at very low altitudes. Essentially, the atmosphere is composed of molecules of gas separated by space. These molecules are packed much closer together at lower levels, and their density diminishes rapidly upward. More than half of the mass of the atmosphere is below the elevation of North America's highest peak. Nearly all the atmosphere (more than 98 percent) lies within 16 miles (26 km) of sea level. We must realize, then, that the mass of the atmosphere is shallow, and that its horizontal dimensions are probably more important than the vertical ones.

In addition to its upward reach above the Earth's surface, the atmosphere also extends slightly downward. Air expands to fill empty spaces, so it penetrates into caves and crevices in the rocks and soil. Moreover, it is dissolved in the waters of the Earth and in the bloodstreams of organisms.

The atmosphere interacts significantly with other components of the earthly environment, and it is instru-

Figure 3-1. The most obviously visible portion of the atmosphere is represented by clouds. These cumulus clouds tower above the Acropolis in Athens, Greece. [Courtesy of NOAA.]

mental in providing a hospitable milieu for life. Whereas we often speak of human beings as creatures of the Earth, it is perhaps more accurate to consider them as creatures of the atmosphere. As surely as a crab crawling on the sea bottom is a resident of the ocean, so a person living at the bottom of the ocean of air is a denizen of the atmosphere.

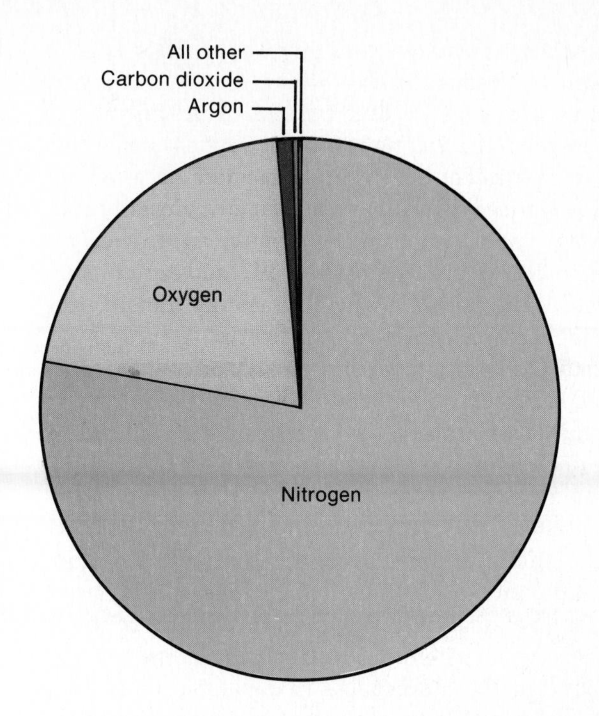

Figure 3-2. Proportional volume of the gaseous components of the atmosphere. Nitrogen and oxygen are the dominant elements.

Constituents of the Atmosphere

The atmosphere is composed of a mixture of discrete gases and an immense number of tiny suspended particles in solid or liquid form. The chemical composition of pure, dry air at lower elevations is simple, uniform, and basically unvarying through time. Certain minor gases and nongaseous particles, however, vary markedly from place to place and from time to time, as does the amount of moisture in the air.

The Gases

Most of the volume of the atmosphere is provided by two chemical elements: *nitrogen* and *oxygen*. Nitrogen comprises more than 78 percent of the total, and oxygen makes up nearly 21 percent. See Table 3-1 and Figure 3-2. Nitrogen is added to the air by the decay and burning of organic matter, by volcanic eruptions, and by the chemical breakdown of certain rocks. It is removed from the atmosphere by certain biological processes and by being washed away in rain or snow. Overall, the addition and removal of nitrogen gas are balanced, and the quantity remains constant. Oxygen is produced by vegetation and is removed by a variety of organic and inorganic processes; its total quantity also apparently remains stable. The remaining 1 percent of the atmosphere's volume consists mostly of the inert gaseous element *argon*. These three principal atmospheric elements—nitrogen, oxygen, argon—are of minimal importance in their effect on weather and climate, and therefore need no further consideration here.

Several other gases occur in sparse, although highly variable, quantities in the atmosphere, but their influence on weather and climate is prominent. *Water vapor* is the gaseous phase of moisture in the air and represents the humidity of the atmosphere. It is largely absent from the upper atmosphere, but near the surface it is often present in notable proportions. It is most common in those

TABLE 3-1 Principal Gases of the Earth's Atmosphere

Component	Percent of volume of dry air	Concentration in parts per million of air
Uniform gases:		
Nitrogen (N_2)	78.084	
Oxygen (O_2)	20.948	
Argon (A)	0.934	
Neon (Ne)	0.00182	18.2
Helium (He)	0.00052	5.2
Methane (CH_4)	0.00015	1.5
Krypton (Kr)	0.00011	1.1
Hydrogen (H_2)	0.00005	0.5
Important variable gases:		
Water vapor (H_2O)	0–4	
Carbon dioxide (CO_2)	0.0353	353
Carbon monoxide (CO)		<100
Ozone (O_3)		<2
Sulfur dioxide (SO_2)		<1
Nitrogen dioxide (NO_2)		<0.2

portions of the atmosphere overlying warm, moist surface areas, such as tropical oceans, where it may amount to as much as 4 percent of total volume. Over deserts and in polar regions, the amount of water vapor is but a tiny fraction of 1 percent. For the atmosphere as a whole, the total component of water vapor remains virtually constant. Water vapor is inordinately significant to weather and climate in that it is the source of all clouds and precipitation and is intimately involved in energy transfer (the storage, movement, and release of heat).

Carbon dioxide is important to life processes because of its role in photosynthesis (see Chapter 10), but it also has a significant influence on climate. This is primarily due to its potent ability to absorb infrared radiant energy, which maintains the warmth of the lower atmosphere. It is distributed fairly uniformly in the lower layers, but its accumulation has been increasing for the last century or so, and the rate of accumulation has been accelerating, presumably because of the increased burning of fossil fuels. The long-range effect of increasing amounts of carbon dioxide in the atmosphere is debatable, but many scientists believe that it will warm up the lower atmosphere sufficiently to produce major global climatic changes. The proportion of carbon dioxide in the atmosphere has been increasing at a rate of about 0.0007 percent (7 parts per million) per year, and at present is about 353 ppm.

Another minor but vital gas in the atmosphere is *ozone*. Ozone is mostly concentrated a few miles above the Earth, particularly at the level between 10 and 25 miles (16 and 40 km) high. Ozone is an excellent absorber of ultraviolet solar radiation; it filters out enough of these burning rays to protect earthly life from potentially deadly effects. In recent years there has been considerable concern that emissions from high-flying supersonic jet engines and the addition of certain chemicals into the air from aerosol spray cans might dissipate the "ozone layer" sufficiently to allow injurious amounts of ultraviolet rays to reach the Earth's surface. The validity of these worries is still unclear, but several countries, including the United States, have enacted laws to reduce greatly the use of aerosol sprays, just in case.

Several other minor gases exist in the atmosphere in minuscule quantities. Some of them—particularly carbon monoxide, sulfur dioxide, nitrogen oxides, and various hydrocarbons—are increasingly introduced into the atmosphere by emission from machines that serve humanity, such as furnaces, factories, and automobiles. All these gaseous pollutants are hazardous to life and at least potentially influential on the climate.

The Particles

In addition to the various gases that comprise the bulk of the atmosphere, there are vast quantities of solid and liquid particles. The larger ones are composed mostly of water and ice, which form clouds, rain, snow, sleet, and hail; these will be discussed in Chapter 6. There are also dust particles large enough to be visible, which are sometimes kept aloft in the turbulent atmosphere in sufficient quantity to cloud the sky, but they are too heavy to remain long in the air. Smaller particles, invisible to the naked eye, may remain suspended in the atmosphere for months or even years.

These microscopic particles have innumerable sources, some natural and some the result of human activities. See Figure 3-3. Volcanic eruptions, windblown soil and pollen grains, meteor debris, smoke from wildfires, and salt spray from breaking waves are examples of natural sources. Human sources mostly consist of emissions

Figure 3-3. Major sources of particulate matter in the atmosphere include volcanic eruptions, wind-lifted dust, fires, ocean spray, smokestacks, and automobile exhausts.

from machines and engines, as well as smoke and soot from fires of human origin.

These tiny particles are most numerous near their places of origin—above cities, seacoasts, and active volcanoes. They may, however, be carried great distances, both horizontally and vertically, by the restless atmosphere. They affect weather and climate in three major ways:

1. Many are water-absorbent, and they collect water vapor as they float around. This is a critical element in the formation of clouds, as we will see in Chapter 6.
2. Some absorb or reflect sunlight, thus decreasing the amount of solar energy that reaches the Earth's surface. See Figure 3-4.
3. Many have an optical effect on low-angle sunlight, producing colorful sunrises and sunsets.

Vertical Structure of the Atmosphere

In our efforts to comprehend the physical geography of the Earth's surface, the atmosphere is an important component for consideration. The next five chapters will deal with atmospheric processes and behavior and their influence on climatic patterns. Our attention in these chapters, however, will be devoted primarily to the lower portion of the atmosphere, which is the zone in which most weather phenomena occur. The upper layers of the atmosphere are involved less significantly and more indirectly in affecting the environment of the Earth's surface. Nevertheless, it is useful to have some understanding of the total atmosphere; thus we will note some of the general characteristics of the atmosphere in its total vertical extent, emphasizing vertical patterns of pressure, temperature, and composition.

Pressure

Atmospheric pressure can be thought of, for simplicity's sake, as the weight of the overlying air. (In Chapter 5, the concept of pressure will be explored in much greater detail.) The greater the amount of overlying air, the greater the "weight" that is exerted on the air below. Air is highly compressible, so the lower layers of the atmosphere become compressed by the air above, which increases both the pressure and density of the lower lay-

Figure 3-4. Dust particles sometimes cloud the sky for a short time over a limited part of the Earth's surface. This is a springtime scene at El Paso International Airport, Texas. [USDA-Soil Conservation Service photo.]

TABLE 3-2 Proportion of Average Sea-level Pressure at Various Altitudes

Altitude		Percentage of average sea-level pressure
Miles	Kilometers	
0	0	100
3.5	5.5	50
10	16	10
20	32	1
30	48	0.1
50	80	0.001
60	96	0.00001

ers. Conversely, air in upper layers is subjected to less compression, and therefore has lower pressure and density. Pressure is normally highest where the Earth's surface is lowest—at sea level and below. With increasing altitude, pressure decreases rapidly.

The change of pressure with altitude, however, is not constant. As a generalization, pressure decreases upward at a decreasing rate. In the lower atmosphere, the rate of pressure change is rapid, but the rate diminishes significantly at higher levels. Average pressure variation with altitude is illustrated numerically and graphically in Table 3-2 and Figure 3-5. One-half of the atmosphere lies below $3\frac{1}{2}$ miles (5.6 km), and 90 percent of the atmosphere is concentrated in the first 10 miles (16 km) above sea level. Pressure becomes so slight in the upper layers that above about 50 miles (80 km), there is not enough to register on an ordinary barometer, the normal instrument for pressure measurement. Above this level, atmospheric molecules are so scarce that air pressure is less than that which would occur in the most perfect man-made vacuum at sea level. Yet, traces of atmosphere extend for literally thousands of miles higher. The "top" of the atmosphere is an abstruse concept rather than a reality, with no true boundary between atmosphere and outer space.

The vast majority of the gases that comprise the atmosphere, then, are concentrated very near the Earth's surface, and the atmosphere merges gradually with the emptiness of space.

Temperature

Most of us have had some personal experience with temperature variations associated with altitudinal change. As we go up on a mountain or in an airplane, we can readily sense a decrease in temperature. Until about a century ago, it was generally assumed that temperature decreased throughout the atmosphere, from bottom to top, just as we know to be true in the lower atmosphere. Such is not the case, however. The vertical pattern of temperature is complex, consisting of a series of layers in which temperature alternately decreases and increases with height. See Figure 3-6.

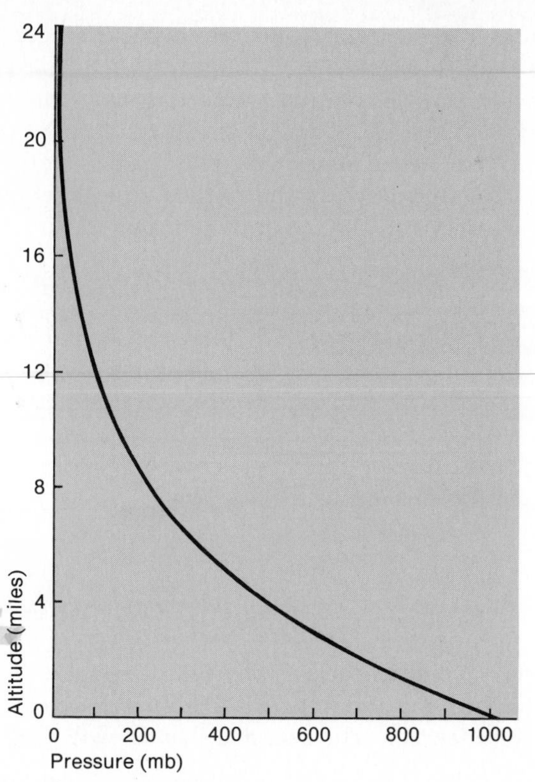

Figure 3-5. Average altitudinal variation in pressure in the atmosphere.

The lowest layer, in which temperature decreases generally with height, is called the *troposphere*. The name is derived from the Greek word *tropos* ("turn"), and implies an "overturning" of air in this zone due to the vertical mixing and turbulence that are characteristic. The depth of the troposphere is variable, both in time and place. It is deepest over tropical regions and shallowest over the poles, but its height also varies with the seasons (higher in summer than in winter) and with the passage of warm and cold air masses. On the average, the top of the troposphere is about 11 miles (18 km) above the equator and about 5 miles (8 km) over the poles. Within the troposphere there is a general, but not constant or uniform, decrease of temperature with increasing height. The top of the troposphere, called the *tropopause*, marks the level at which temperature ceases to decrease with height. The tropopause usually occurs as a transition zone rather than as an abrupt boundary surface.

Above the tropopause is the *stratosphere*, a name derived from the Latin *stratum* ("a cover"), and it implies a layered or stratified condition without vertical mixing. Within the stratosphere the temperature remains relatively constant for some distance above the tropopause and then begins to increase. Maximum temperature is reached at an elevation of about 30 miles (48 km), which is considered to be the top of this thermal layer and is called the *stratopause*.

Upward of the stratopause is the *mesosphere*, from

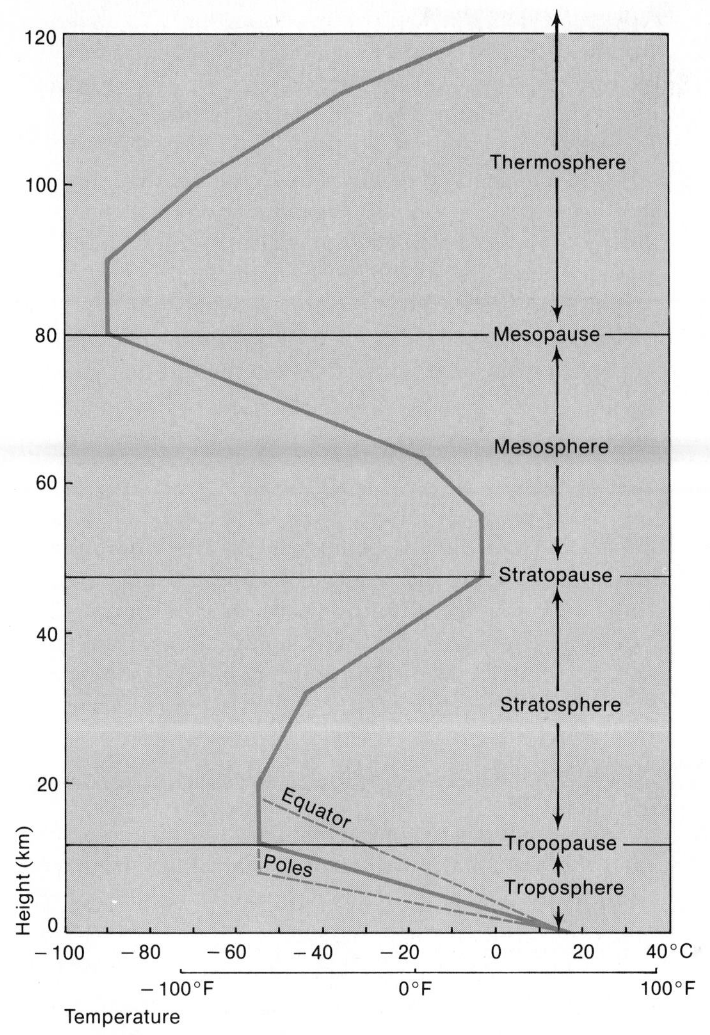

Figure 3-6. The generalized thermal structure of the atmosphere.

the Greek *meso* ("middle"), in which temperature again decreases with height. The top of the mesosphere, the *mesopause*, marks the level of minimum temperature, at an elevation of about 50 miles (80 km) above sea level.

Above the mesopause, is the *thermosphere* (from the Greek *therm*, or "heat"). Within this zone the temperature remains relatively uniform for several miles and then increases continually with height; at about 125 miles (200 km) above the Earth's surface, it is considerably warmer than any temperature in the troposphere.

The thermosphere merges gradually into the *exosphere*, which in turn blends into interplanetary space.

Although there are many interesting physical relationships in the stratosphere, mesosphere, and thermosphere, our attention will be directed almost entirely to the troposphere. In this lowest thermal layer are found all precipitation and almost all clouds. Storms and essentially all other phenomena that we call "weather" occur here. Only occasionally will we consider atmospheric conditions above the troposphere.

Composition

There is vertical variation in the distribution of a variety of other atmospheric features, and other kinds of layers have been defined in addition to the thermal layers discussed above. For our purposes, however, the most significant vertical pattern not already described involves the distribution of atmospheric components.

The principal gases of the atmosphere have a remarkably uniform vertical distribution throughout the lowest 50 miles (80 km) or so of the atmosphere. This zone of homogenous composition is referred to as the *homosphere*. The sparser atmosphere above does not display this uniformity; rather, the gases tend to be layered in accordance with their molecular weights—nitrogen below, with oxygen, helium, and hydrogen successively above. This higher zone is called the *heterosphere*. See Figure 3-7.

Water vapor, a quantitatively minor but very important constituent of the atmosphere, also varies in its vertical distribution. Most water vapor is found near the Earth's surface, with general diminishment upward. Above 10 miles (16 km) above sea level there is rarely enough moisture to provide the raw material to make even a wisp of a cloud.

Two other vertical patterns are worthy of mention here:

1. The ozone layer (mentioned on page 42) is a zone of relatively rich concentrations of ozone, and is some-

Figure 3-7. Various vertical zones are generally recognized in the atmosphere.

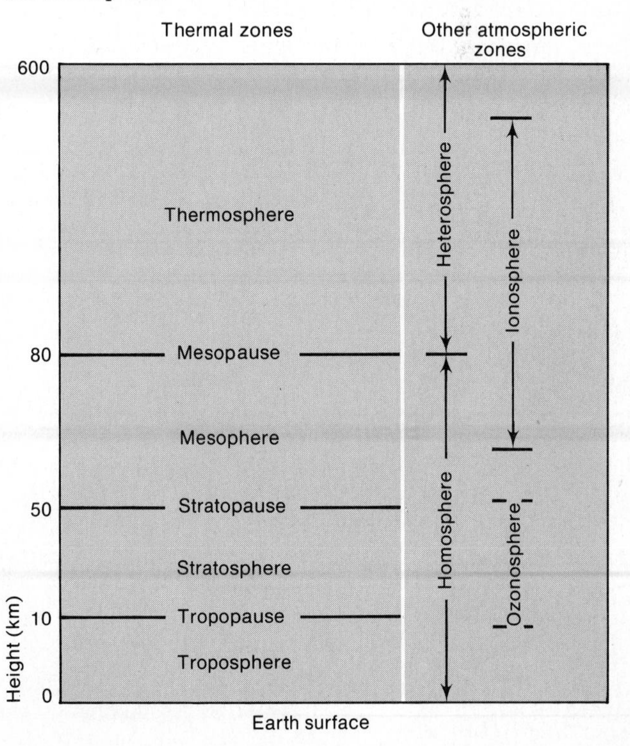

times distinguished by the name *ozonosphere*. The shielding effect of ozone has already been discussed.

2. The *ionosphere* is a deep layer of electrically charged molecules and atoms in the upper mesosphere and lower thermosphere, between about 40 and 250 miles (60 and 400 km). The ionosphere is significant because it aids long-distance communication by reflecting radio waves back to Earth. It is also known for its auroral displays (northern lights, etc.).

Weather and Climate

Our vast and invisible atmospheric envelope, energized by solar radiation, stimulated by earthly motions, and affected by contact with the Earth's surface, reacts by producing an infinite variety of conditions and phenomena known collectively as *weather*. The term *weather* is a temporary concept that refers to short-run atmospheric conditions for a given time and a specific area. It is the sum total of temperature, humidity, cloudiness, precipitation, pressure, winds, storms, and other atmospheric variables for a short period of time. Thus we speak of the weather of the moment or the week or the season, or perhaps even of the year or the decade. See Figure 3-8.

Weather is in an almost constant state of change, sometimes in seemingly erratic fashion. Yet in the long-run view, it is possible to generalize the variations into a composite pattern, which is termed *climate*. Climate is the aggregate of day-to-day weather conditions over a long period of time. It encompasses not only the average characteristics but also the variations and extremes. To describe the climate of an area requires weather information over an extended period, normally several decades at least.

The concepts of weather and climate, then, are related but not synonymous. The distinction between them is the difference between immediate specifics and protracted generalities. As the country philosopher said, "Climate is what you expect; weather is what you get." Or, stated more sarcastically, "It is the climate that attracts people, and the weather that makes them leave."

The Study of Climate

Climate is probably the component of the environment that interests mankind most. It has direct and obvious influences on agriculture, transportation, and human life in general. Moreover, climate is significantly interrelated with the other facets of the environment. Indeed, general climatic characteristics are enormously influential on the gross development of all major aspects of the physical landscape—soils, vegetation, animal life, hydrography, and topography.

In our efforts to comprehend the significance of the atmosphere as a major portion of the earthly environment, the ultimate goal is to achieve an understanding of the distribution and characteristics of climatic types over the Earth. To accomplish this it is necessary to consider in

Figure 3-8. A photo of weather, in this case, a 1982 snowstorm in New York City. [Courtesy Marc P. Anderson.]

some detail many of the processes that function in the atmosphere, including their interrelationships and results. Our concern is primarily with long-run atmospheric conditions, but we must have some appreciation for the dynamics involved in the momentary state of the atmosphere.

The Elements of Weather and Climate

The atmosphere is a complex medium, and its mechanisms and processes are sometimes very involved. Its nature, however, is generally expressed in terms of only a few variables, which are measurable. The data thus recorded provide the raw materials for understanding both temporary (weather) and long-term (climate) atmospheric conditions.

These variables can be thought of as the *elements* of weather and climate. The most important are (a) temperature, (b) moisture, (c) pressure, and (d) wind. These are the basic ingredients of weather and climate, and measurement of their variations in time and space makes it possible to decipher at least partly the complexities of weather dynamics and climatic patterns.

Climatic Controls

Variations in the climatic elements are frequent, if not continual, over the Earth. Such variations are caused by, or at least strongly influenced by, certain semipermanent attributes or features of our planet, which are often referred to as *climatic controls*. In other words, the condition, quantity, or extent of the climatic elements is a reflection of complex atmospheric interactions that are more or less determined by a relatively few climatic controls. The principal controls are briefly described below.

Latitude We noted in Chapter 1 that the continually changing positional relationship between the sun and the Earth brings continually changing amounts of sunlight, and therefore radiant energy, to different parts of the Earth's surface on the basis of latitudinal location. Thus the basic distribution of heat and cold over the Earth is first and foremost a function of latitude. This pattern, in turn, has considerable influence on patterns of other climatic elements in addition to temperature.

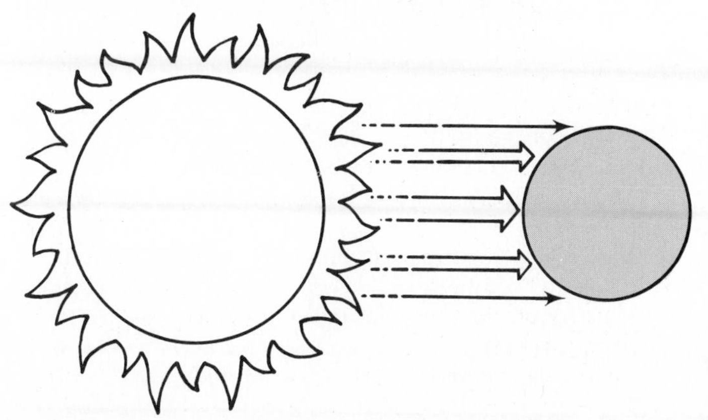

Distribution of Land and Water Probably the most fundamental distinction that can be made concerning the geography of climate is that between continental climates and maritime climates. Oceans heat and cool more slowly and to a lesser degree than do landmasses (as we will see in the next chapter), which means that maritime areas experience milder temperatures in both summer and winter than continental areas. Also, oceans are a much more prolific source of moisture for the atmosphere; thus maritime climates are normally more humid than continental areas. The uneven distribution of continents and oceans over the world, then, is a prominent control of gross climatic characteristics.

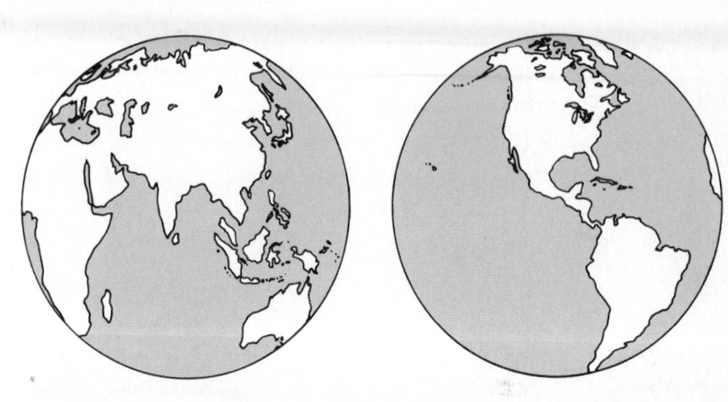

General Circulation of the Atmosphere The atmosphere is in constant motion, with flows that range from transitory local breezes to vast regional wind regimes. At the planetary scale, a semipermanent pattern of major wind and pressure systems dominates the troposphere and greatly influences most aspects of weather and climate. As a simple example, most surface winds in the tropics come from the east, whereas the middle latitudes are characterized by flows that are mostly from the west. We will learn more about atmospheric circulation in Chapter 5.

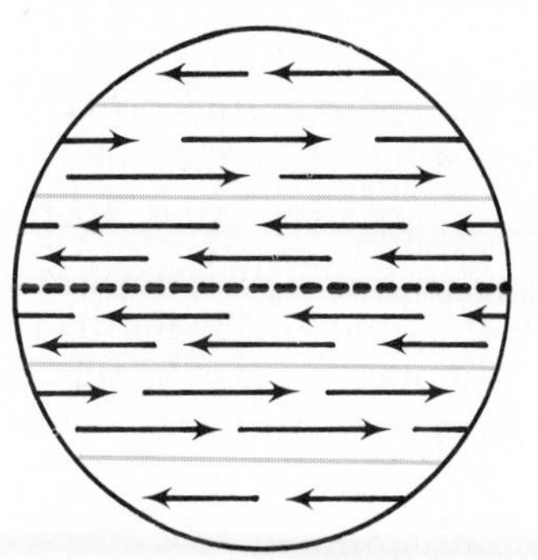

General Circulation of the Oceans Somewhat analogous to atmospheric movements are the motions of the oceans. Like the atmosphere, the oceans have many minor motions but also a broad pattern of generalized circulation of currents. These currents assist in heat transfer by moving warm water poleward and cool water equatorward. Although the influence of currents on climate is much less than that of atmospheric circulation, it is not inconsequential. For example, warm currents are found off the east coasts of continents and cool currents occur off west coasts, a distinction that has a profound effect on coastal climates. Oceanic circulation will be discussed in Chapter 4.

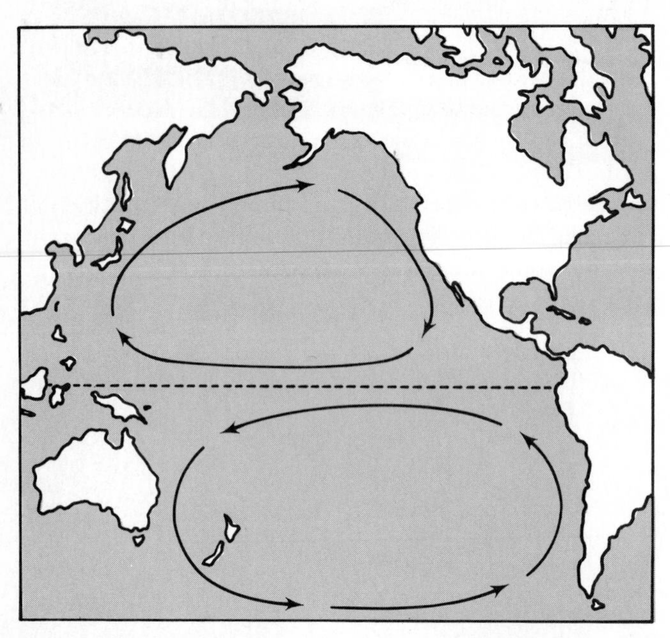

Altitude We have already noted that temperature, pressure, and moisture generally decrease upward within the troposphere. These simple relationships have significant ramifications for many climatic characteristics, particularly in mountainous regions.

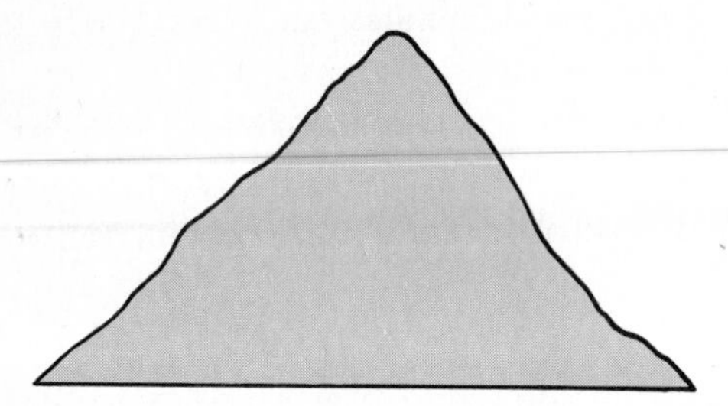

Topographic Barriers In addition to the altitude aspect of mountains and hills, they sometimes have prominent effects on climate by blocking or diverting wind flow. The exposed ("windward") side of a mountain range, for example, is likely to have a vastly different climate than that of the sheltered ("leeward") side, largely because of the more abundant precipitation of the former.

Storms Various kinds of storms occur over the world; some have very widespread distribution, whereas others are quite localized. All of them create specialized weather circumstances, and some are prominent and frequent enough to affect the climatic pattern. The principal storms will be discussed in Chapter 7.

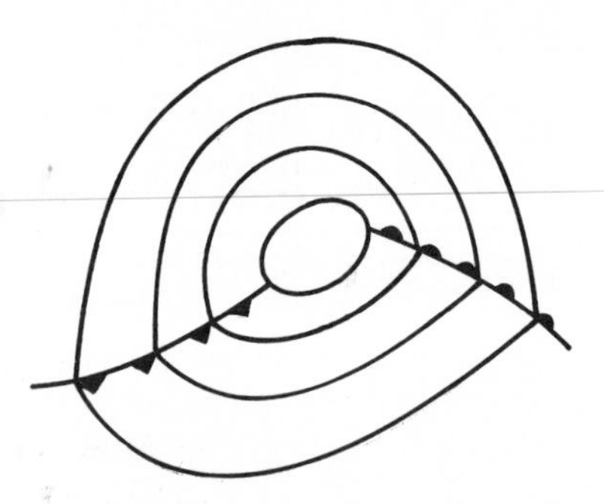

Earthly Rotation The continual rotation of the Earth from west to east produces an apparent deflection in the direction of movement of winds and of ocean currents. This is referred to as the *Coriolis effect* and will be discussed in Chapter 4.

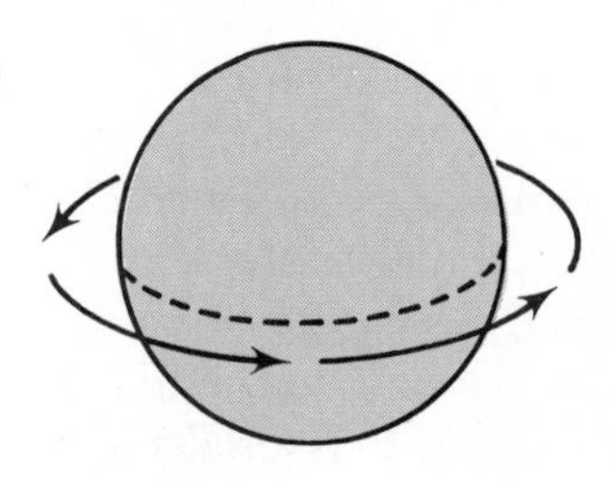

In our study of the atmosphere, the climatic elements will provide the basic organizing framework. We will first consider each of the elements—temperature, pressure, wind, moisture—separately, but it is important to keep in mind that they are all closely interrelated. Following individual treatment of the elements (in Chapters 4, 5, and 6) will be an integrative chapter devoted to air masses, fronts, and storms, in which the interactions of climatic elements, climatic controls, and atmospheric processes will be emphasized. This will set the stage for a consideration of the climatic pattern of the world.

Review Questions

1. Describe the composition of the atmosphere.
2. Why is the question, "How deep is the atmosphere?" difficult to answer?
3. What is *ozone* and why is it important to life on Earth?
4. In our study of physical geography, why do we concentrate primarily on the *troposphere* rather than on other zones of the atmosphere?
5. What is the difference between *weather* and *climate*?

Some Useful References

BYERS, R. H., *General Meteorology*, 4th ed. New York: McGraw-Hill Book Company, 1974.

GEDZELMAN, S. D., *The Science and Wonders of the Atmosphere*. New York: John Wiley & Sons, Inc., 1980.

RIEHL, H., *Introduction to the Atmosphere*, 3rd ed. New York: McGraw-Hill Book Company, 1978.

TREWARTHA, G. T., AND L. H. HORN, *An Introduction to Climate*, 5th ed. New York: McGraw-Hill Book Company, 1980.

4 Insolation and Temperature

Temperature is probably the weather element with which we are most familiar. Our skin is notably sensitive to heat and cold, and even small changes in air temperature are likely to be recognized promptly. The temperature at any time and/or place in the atmosphere is the result of the interaction of a variety of complex factors. In this chapter, our attention will be focussed on the energetics of the atmosphere—the important processes involved in bringing heat to the atmosphere, in determining the extent of heating (and cooling) that takes place, and in transferring heat from one place to another. This will lead us to a general understanding of the distribution of temperature over the Earth.

Solar Energy

The sun is the only important source of energy for the Earth and its atmosphere. Millions of other stars radiate energy, but they are too far away to affect the Earth. Energy is also released within the Earth, primarily from the decay of radioactive minerals, and this heat flows upward to the crust; some also escapes to the atmosphere through volcanoes and geysers, but its quantity is insignificant. Thus the sun supplies essentially all the energy that supports life on Earth, and it energizes most of the atmospheric processes.

The sun is a star of average size and average temperature, but its relative proximity to the Earth gives it a far greater influence on our planet than all other celestial bodies combined. The sun functions as an enormous nuclear reactor, producing energy by fusion, which burns only a very small portion of the sun's mass but provides an immense and continuous flow of radiant energy that is dispersed in all directions.

A vast and relatively constant amount of radiant energy flows outward from the sun, mostly in the form of *electromagnetic waves*. These are waves that can transport energy without requiring a medium (the presence of matter) to pass through. They traverse the great voids of space in unchanging form. The beams travel outward from the sun in straight lines at the speed of light—186,000 miles (300,000 km) per second.

Figure 4-1. The sun emits radiant energy in all directions. Only a tiny fraction of the total output is intercepted by the Earth.

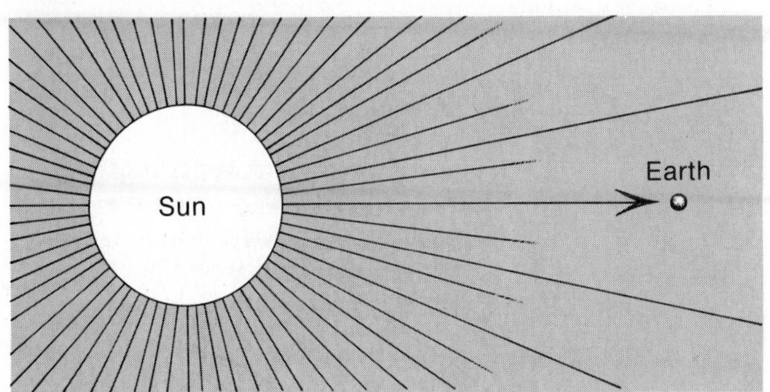

Only a tiny fraction of the sun's radiant output is intercepted by the Earth. See Figure 4-1. The waves travel through space without loss of energy, but since they are diverging from a spherical body, their intensity per unit area continually diminishes with increased distance from the sun. As a result of this divergence and the distance separating the Earth from the sun, less than one two-billionth of total solar output reaches the outer limit of the Earth's atmosphere, having traveled 93,000,000 miles (150,000,000 km) in just over 8 minutes.

Insolation

Solar energy that reaches the Earth and/or its atmosphere is called *insolation*. Although it consists of only a miniscule portion of total solar output, in absolute terms the amount is enormous. The solar energy received by the Earth in one second is approximately equivalent to all the electric energy generated on the Earth in one week.

Electromagnetic waves are classified on the basis of wavelength, which can be thought of as the distance from the crest of one wave to the crest of the next. Wavelengths vary enormously; some standard radio waves have lengths of several hundred kilometers, whereas some gamma rays are less than a picometer (a millionth of a millionth of a meter) in length. For our purposes, the most important distinction is between shortwave and long-wave radiation, the dividing line between the two being a wavelength of about 4 micrometers. As a basic generalization, hot bodies radiate mostly shortwaves, and cool bodies radiate mostly long waves. The sun is the ultimate hot body, so most of its radiation is in the shortwave spectrum, and solar radiation is often referred to simply as *shortwave radiation*. Thus insolation reaching the Earth and its atmosphere is mostly shortwave.

The total amount of insolation received at the top of the atmosphere is believed to be constant when averaged over a year, although it may vary slightly with fluctuations in the sun's temperature. The carefully calculated amount of incoming insolation—referred to as the *solar constant*—is slightly less than 2 calories per square centimeter per minute. (A *calorie* is the amount of heat required to raise the temperature of 1 gram of water by 1 degree Celsius.) This is more properly referred to as 2 langleys per minute, because the accepted measure of radiation intensity is the *langley*, which is 1 calorie per square centimeter.

The receipt of insolation waves by the atmosphere is just the beginning of a complex series of events within

All radiant energy exists in the form of electromagnetic waves, which can be classified and displayed in a table based on their wavelengths; the table is known as the *electromagnetic spectrum*. Within the spectrum there are five broad groupings of wavelengths, only three of which are of much significance to physical geography. The five groupings are defined here:

1. *Very long waves* are those with a wavelength exceeding about 1 millimeter. They include (in order of decreasing wavelength) electric waves, long radio waves, short radio waves, radar waves, and microwaves.
2. Between about 1.0 millimeter and 0.7 micrometer are *infrared waves*, which are too long to be seen by the human eye. They are generally emitted by hot objects, and are sometimes referred to as "heat rays." They have a variety of uses that are dependent in part on their ability to pass through materials that block ordinary light rays but still produce heat inside the blocking material, as with infrared heat lamps. Earth radiation is entirely infrared, but only a small fraction of solar radiation is in the infrared spectrum.
3. *Visible light* is concentrated entirely in the narrow band between about 0.4 and 0.7 micrometers. Only about 3 percent of all electromagnetic waves are in the visible light spectrum. Visible light colors range from red (longest) through orange, yellow, green, and blue, to violet (shortest).
4. Between wavelengths of about 0.4 and 0.01 micrometers are the *ultraviolet waves*. The sun is a prominent natural source of ultraviolet rays, and solar insolation reaching the top of our atmosphere contains a considerable amount. However, much of it is absorbed in the atmosphere, and the shortest ultraviolet rays do not reach the Earth's surface, where they could cause considerable damage to most living organisms.
5. *Very short waves* are those with wavelengths less than about 0.01 micrometers. They include primarily X-rays, gamma rays, and cosmic rays.

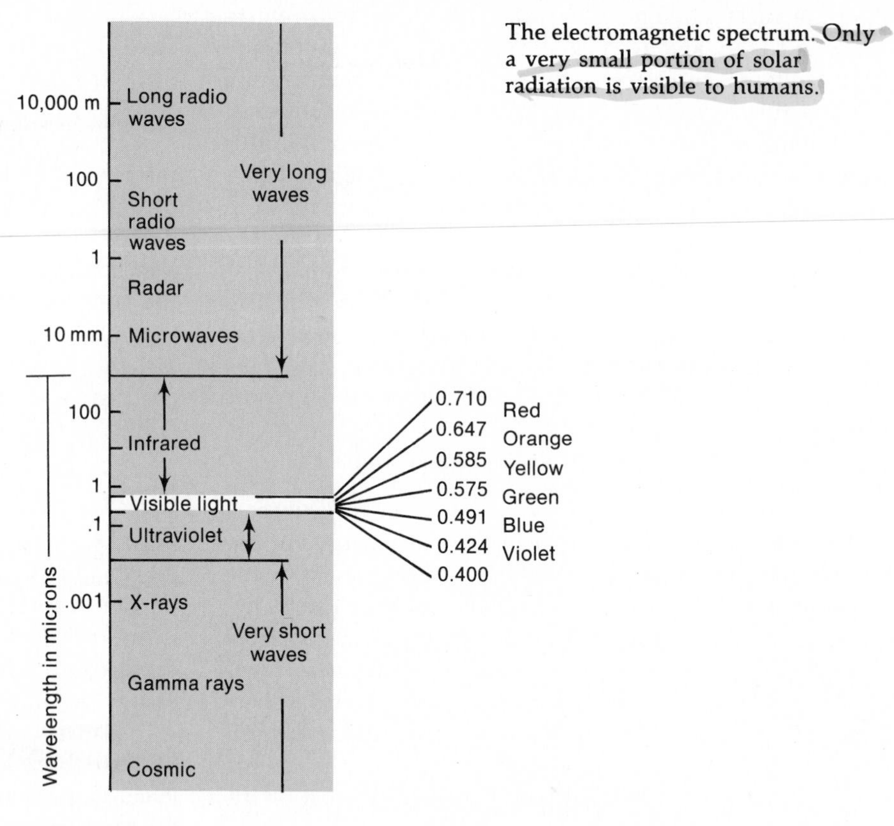

The electromagnetic spectrum. Only a very small portion of solar radiation is visible to humans.

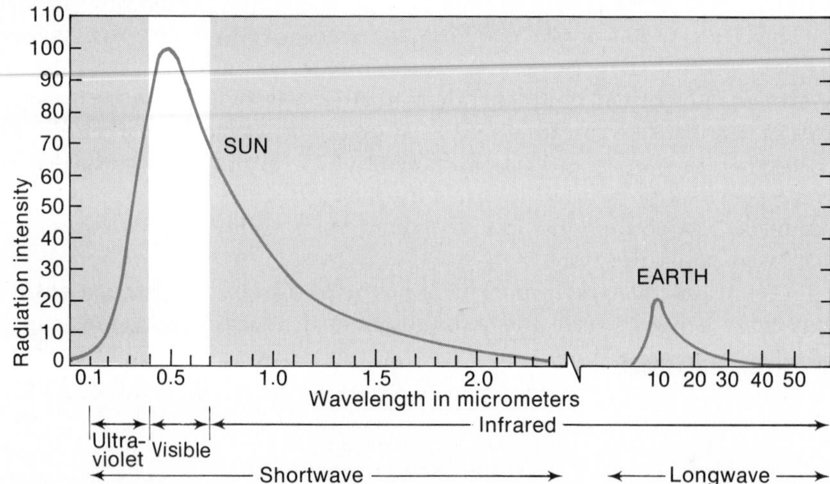

Comparsion of solar and terrestrial radiation intensity, by wavelength.

Our interest is largely in incoming solar radiation and outgoing terrestrial radiation. Solar radiation is concentrated in the visible light spectrum, but it is also important in the shorter infrared and the longer ultraviolet wavelengths. Terrestrial radiation is entirely in the infrared spectrum. A wavelength of about 4 micrometers is considered to separate long waves from shortwaves; thus all terrestrial radiation is long-wave, and almost all solar radiation is shortwave.

the atmosphere and at the Earth's surface. Some of the insolation is rejected by the atmosphere; the rest is transformed. Some passes through the atmosphere to the Earth's surface; some does not. The mixed reception of the insolation waves and the energy cascade that results will be discussed after a brief digression to define our terms.

Basic Processes in Heating and Cooling the Atmosphere

Prior to consideration of the actual sequence of events that transpires when the stream of radiant energy comes to the Earth from the sun, it is useful to note the various physical processes that are involved. These are the prominent processes that affect the reception of the electromagnetic waves in the earthly environment and govern the transfer of heat from one place to another in the atmosphere.

Radiation

The process by which energy is emitted from a body is called *radiation*. It generally involves the flow of radiant energy from the body in some sort of waves. All bodies radiate, but hotter bodies are more potent radiators than cooler ones. The hotter the object, the more intense is its radiation and the shorter its wavelength. Thus a hot body produces more radiation and a higher proportion of shortwave radiation than a cold body.

Temperature, however, is not the only control of radiation effectiveness. Objects at the same temperature may vary considerably in their radiating capability. The nature and the substance of the surface of the objects are important determining factors. A body that emits the maximum amount of radiation possible, at every wavelength, for its temperature, is called a *blackbody*.

Our interest in radiation mostly concerns the sun and the Earth. Both the sun and the Earth function essentially as blackbodies, that is, as perfect radiators. They radiate with almost 100 percent efficiency for their respective temperatures. The sun, however, is exceedingly hotter than Earth, so the sun emits 2 billion times more energy than the Earth. This enormous difference in temperature also means that solar radiation is mostly in the shortwave spectrum, whereas terrestrial (earthly) radiation consists of long waves. Maximum terrestrial radiation waves are approximately 20 times longer than maximum solar radiation waves.

Absorption

In the context of the energetics of the atmosphere, *absorption* is the ability of an object to assimilate energy from waves that strike it. When an insolation wave strikes an object and is absorbed, the temperature of the absorbing surface is thereby increased.

Different objects and surfaces are quite variable in the absorptive capabilities, with the variations depending in part on temperature and wavelength. The basic generalization is that a good radiator is also a good absorber, whereas a poor radiator is a weak absorber. If an object effectively radiates at a particular wavelength, it also is an efficient absorber at that wavelength. Both the sun and the Earth, then, are efficient absorbers as well as radiators.

At a larger scale, it can be seen that there is a multitude of different objects and materials on the Earth's surface, and their absorptive capabilities are quite varied. Mineral materials (rock, soil) are generally excellent absorbers; snow and ice are poor absorbers; water surfaces are variable in their absorbing efficiency. One important distinction concerns color. Dark-colored surfaces are much more efficient absorbers in the visible portion of the spectrum than are light-colored surfaces.

Reflection

For our purposes, *reflection* can be considered as the ability of an object to repel waves without altering either the object or the waves. Thus in some cases, an insolation wave will strike a surface in the atmosphere or on the Earth and be bounced away, unchanged, in the general direction from which it came.

In this context, reflection is a complementary process to absorption. If the wave is reflected, it cannot be absorbed. Hence, an object that is a good absorber is likely to be a poor reflector, and vice versa. A simple example of this principle is the existence of a snowy surface on a warm, sunny spring day. Although the air temperature may be well above freezing, the snow does not melt rapidly because its white surface reflects away a large share of the solar energy that strikes it.

Scattering

Tiny particles and gas molecules in the air sometimes deflect light waves and redirect them, in a process known as *scattering*. This involves a change in direction, but not in wavelength, of the light beam. Some of the waves are backscattered into space and thus are lost to the Earth, but most of them continue through the atmosphere in altered directions.

The amount of scattering that takes place depends on the wavelength of the beam as well as on the size, shape, and composition of the molecule or particle. Shorter waves are more readily scattered than longer ones. This means that the violets and blues in the visible part of the spectrum are more likely to be redirected than are the oranges and reds. Thus, on a clear day, the sky appears very blue because so much (often more than 50 percent) of the blue radiation is scattered, whereas most of the reds and oranges pass through without scattering. The usual blueness of the Earth's sky contrasts markedly with the blackness of the sky above celestial bodies (such as our moon) that have no atmosphere and therefore no scattering.

The yellowish or reddish tint of the sun near the horizon (i.e., early or late in the day) is a further manifestation of scattering. In such a position, sunlight must pass through a great deal of atmosphere, and most of the blue radiation is scattered out, leaving only the longer waves (reds and oranges) to be seen.

Larger particles, such as those associated with haze or smog, scatter light more equally in all wavelengths. The result is that no sky color predominates, and the sky appears white. One implication is that bluer skies indicate fewer large particles in the air; so the bluer the sky, the cleaner the air.

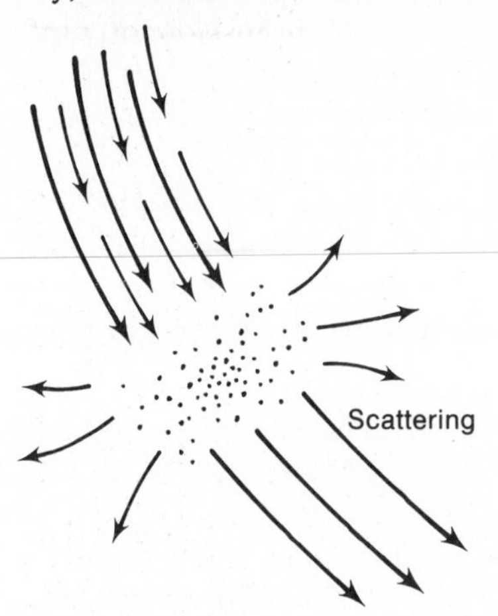

Transmission

Transmission is the ability of a medium to allow rays to pass through it. There is obviously considerable variability among different mediums in their capacity to transmit rays. Earth materials, for example, are very poor transmitters of insolation; sunlight is absorbed right at the surface of rock or soil and does not penetrate at all. Water, on the other hand, permits considerable transmission of sunlight. Even in very murky water, light will penetrate some distance below the surface, and in clear water, sunlight may illuminate to considerable depths.

In some cases, transmission depends on the wavelength of the rays. For example, glass has high transmissivity for shortwaves but not for long waves. Thus a closed automobile, if left parked in the sunlight, will experience a conspicuous heat buildup inside because shortwave insolation is transmitted through the window glass, but the long waves that are reradiated from the interior of the car cannot escape in similar fashion. The result is that the inside of the vehicle gets hotter and hotter as long as the sunlight reaches it.

The relevance of this situation for our study of climate is that the atmosphere allows for considerable transmission of shortwave solar radiation, but it is not nearly as effective a transmitter of long-wave terrestrial radiation. This characteristic is emphasized when the air is laced with such "impurities" as clouds, water vapor, and dust. In simplest terms, this means that solar energy readily penetrates to the Earth's surface, but that reradiated terrestrial energy is mostly "trapped" in the lower troposphere, and much of it is again reradiated back toward the ground. This keeps the Earth's surface and lower troposphere at a higher average temperature than would be the case if there were no atmosphere.

The circumstances just described are referred to as the *greenhouse effect* because it was long thought that greenhouses maintained heat in the same manner—the glass roof of the greenhouse transmitting shortwave solar energy in but inhibiting the passage of long-wave radiation out. More recently it has been shown that this is not the full story; for example, greenhouses with windows of rock salt, which permit equal transmission of both long waves and shortwaves, rather than glass, experienced a heat buildup approximately as great as that of ordinary glass greenhouses. Greenhouses with regular glass windows maintain high temperatures largely because the warm air in the building is trapped and does not dissipate through mixing with the cooler air outside. Thus the term *greenhouse effect* is based on a misconception, and the trapping of heat in the lower troposphere because of differential transmissivity for shortwaves and long waves should probably be called something else; *atmospheric effect* has been suggested, but *greenhouse effect* continues to be the customary term. Whatever the name, however, the principle is exceedingly important in understanding the heating of the atmosphere, and it will be considered more fully later in this chapter.

Conduction

The movement of energy from one molecule to another without changes in their relative positions is called *conduction*. It enables the transfer of heat among different parts of a stationary body, or between different objects that are in contact. Conduction comes about through molecular collision. A hot molecule becomes increasingly agitated and collides against another, transferring energy to it. In this manner, the heat is passed on from one place to another. The principle is that when two molecules of unequal temperature are in contact with one another, heat passes from the warmer to the cooler until they both attain the same temperature or until the source of heat is shut off.

The ability of different substances to conduct heat is quite variable. For example, most metals are excellent conductors, as can be demonstrated by pouring hot coffee into a metal cup and then touching your lips to the edge of the cup. The heat of the coffee is quickly conducted throughout the metal of the cup and will burn the lips of the incautious drinker.

Of major interest is the fact that the Earth's land surface warms up rapidly during the day because it is a good absorber, and some of that warmth is transferred by conduction. Earth materials are not good conductors, and the heat of the warm ground surface is conducted downward only slowly. In contrast, heat is transferred more readily to the very lowest portion of the atmosphere by conduction from the top of the ground. Air, however, is a poor conductor, so only the layer in immediate contact with the ground is heated very much. Physical movement of the air is required to spread the heat around.

Moist air is a slightly more efficient conductor than dry air. If you are outdoors on a winter day, you will stay warmer if there is little moisture in the air. Damp air will conduct your body heat away more rapidly, whereas dry air will allow it to be more effective in keeping you warm.

Advection

Heat is transferred laterally in the atmosphere by horizontal wind movements in a process called *advection*. Warmth or coolness are thus shifted from one place to another as masses of air are moved horizontally.

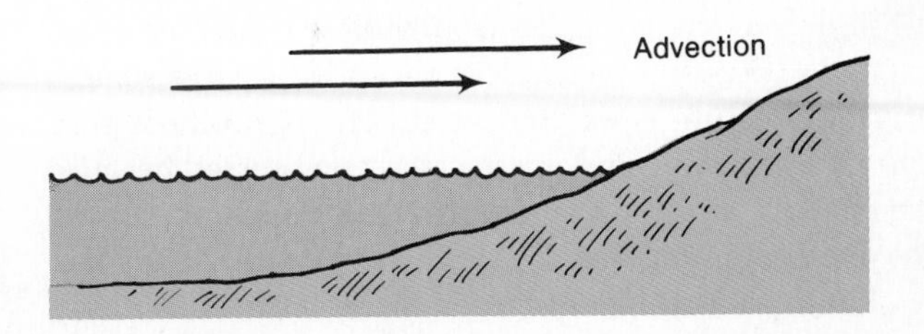

Convection

Convection is a process whereby heat is transferred from one point to another by bodily movement of a liquid or a gas. It involves the actual displacement of molecules. Some horizontal motion is included, but the principal action is vertical; warm air rises and cold air sinks.

If a pan of water is heated from below, the heated portion expands, rising directly upward in the direction of lower pressure. This decreases its density and pressure, and surrounding air moves in to equalize the situation. Cooler water from above descends to replace that which has moved in, and a cellular circulation is established—up, out, down, and in.

A similar convective pattern frequently develops in the atmosphere. Unequal heating (for a variety of reasons) may cause surface air to become warmer at one place than the surrounding air. This causes it to expand in volume, which means that it will move upward, in the direction of lowest pressure. The cooler surrounding air then moves in toward the heat source where pressure is now lower, and air from above sinks down to replace that which has moved in, thus establishing a complete convective system. The prominent elements of the system are an updraft of warm air and a downdraft of cool air. Convection is common over much of the world during the summer and throughout the year in the tropics.

Convection is not the only cause of vertical air movements; they can also be initiated by mechanical turbulence and other forms of irregular air motions. Whatever the cause, however, the updrafts and downdrafts that result produce vertical mixing that spreads the heat from the lowest portion of the troposphere to considerable heights.

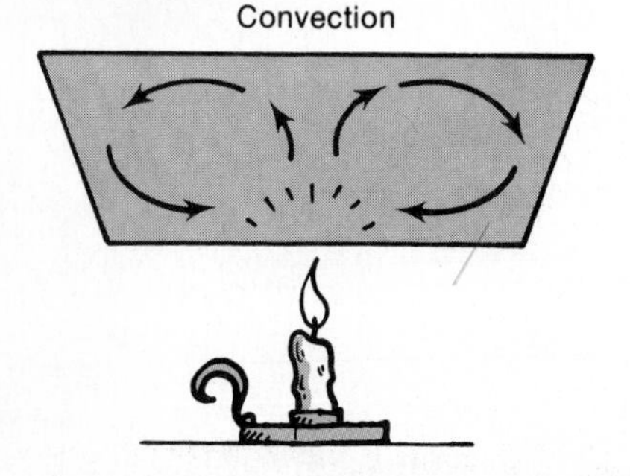

The Adiabatic Process

Whenever air ascends or descends, it experiences a temperature change. This invariable result of vertical movement is due to the variation in pressure. When air rises, it expands because there is less air above it, and so less overlying pressure is exerted on it. The opposite is true of the reverse; when air descends, it is compressed because there is more air above it and it therefore comes under more overlying pressure.

The *expansion* that occurs in rising air is inevitably a cooling process, even though no heat is taken away. Spreading the molecules over a greater volume of space requires energy, which slows them down somewhat and decreases their frequency of collision. The result is a drop in temperature. This is *adiabatic cooling*—cooling by expansion in rising air.

Conversely, when air descends, it must become warmer. The descent causes *compression* as the air comes under increasing pressure. The molecules draw closer together and collide more frequently. The result is a rise in temperature, even though no heat is added from external sources. This is *adiabatic warming*—warming by compression in descending air.

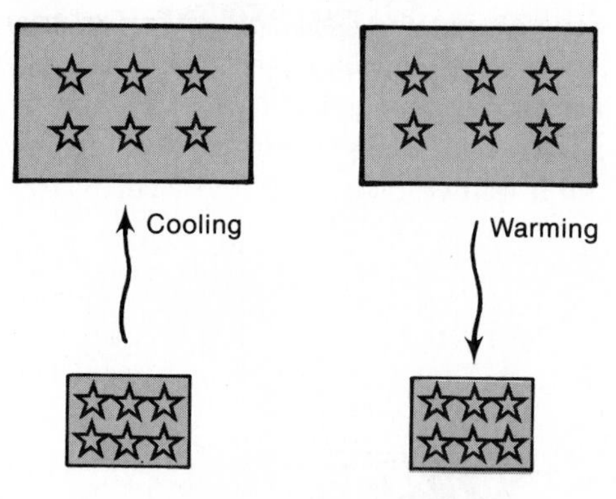

Latent Heat

We will see in Chapter 6 that moisture in the atmosphere is frequently changed in state, involving liquid (water), solid (ice), and gas (water vapor). The changes involve either the storage or release of energy, depending on the process. Most prominent are *evaporation*, in which liquid water is converted to gaseous water vapor, and *condensation*, in which the reverse (water vapor to water) takes place. In the former process, energy is stored as *latent heat*, and in the latter, the latent heat is released.

The Heating of the Atmosphere

With these processes—radiation, absorption, reflection, scattering, transmission, conduction, convection, advection, expansion, contraction, evaporation, condensation—in mind, we can now turn our attention to the specifics of atmospheric heating. What happens to solar radiation when it enters the Earth's atmosphere? How is it received and distributed? What are the dynamics of converting electromagnetic waves to atmospheric heat?

The basic generalization is that most insolation received by the atmosphere does not contribute directly to its heating. A significant portion is reflected back into space, and the majority of the remainder passes on through to the Earth's surface. From the Earth, energy is sent back into the atmosphere in the form of latent heat and long-wave terrestrial radiation, and it is from this that the atmosphere receives most of its heat. For the most part, then, the atmosphere is heated directly from the Earth and only indirectly from the sun, although the sun is the original source of the energy.

The situation is modeled in Figure 4-2. Of the total amount of insolation intercepted by the atmosphere, one-third is redirected back into space without having any effect on the Earth's heat balance. Approximately 25 percent of total insolation is reflected or scattered into space by the atmosphere (largely by clouds), and another 8 percent is transmitted through the atmosphere to the Earth's surface where it is reflected directly back into space. This fraction (33 percent) of total radiation that is bounced back, unchanged, into space is called the *albedo*.

About 19 percent of total insolation is absorbed directly by the atmosphere. Perhaps one-sixth of this 19 percent is absorbed by ozone and oxygen in the stratosphere; most of the remainder is absorbed by water vapor, clouds, and dust particles in the troposphere. This means that slightly less than half of the total insolation passes through the atmosphere—mostly by transmission, some by scattering—and is absorbed by the Earth's surface.

More than three-fourths of all sunshine falls on a water surface when it reaches the Earth. Much of this energy is utilized in evaporating water, which is stored as latent heat in the resulting water vapor. This energy is subsequently released when condensation takes place, and it is a major source of heat energy for the atmosphere.

Some of the solar energy that is absorbed at the Earth's surface is reradiated, largely in the long-wave spectrum. Only a small fraction of this terrestrial radiation—about 5 percent—is transmitted through the atmosphere and lost to space. The rest is absorbed by the atmosphere, particularly by water vapor, dust, carbon dioxide, and clouds. This, in turn, is mostly reradiated, with

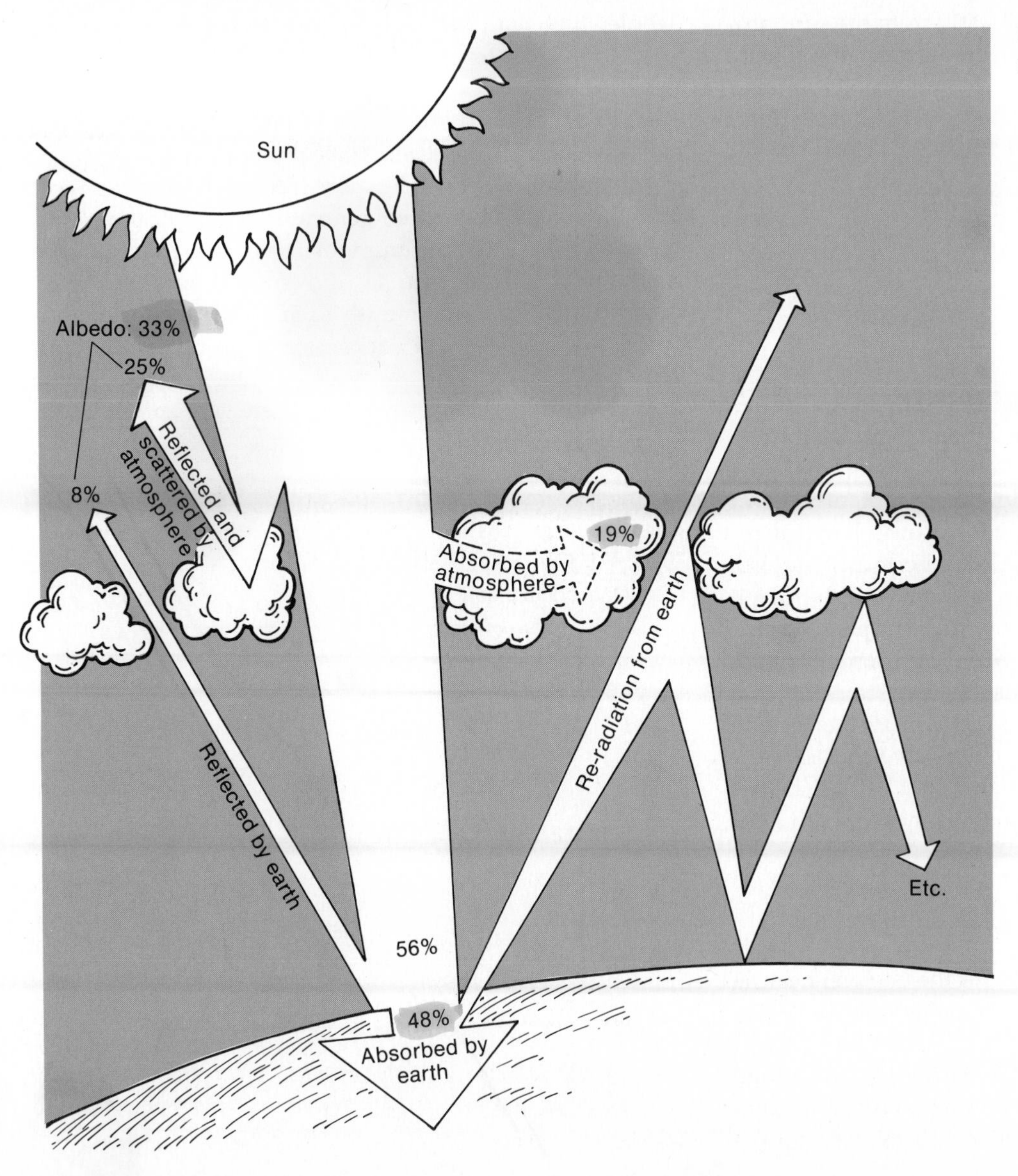

Figure 4-2. The Earth's solar radiation budget.

the bulk of it returning to Earth. This sequence of reradiation between Earth and atmosphere continues indefinitely, as the "greenhouse effect."

A small proportion—less than 10 percent—of the energy absorbed at the Earth's surface is then conducted back into the atmosphere, where most of it is dispersed by convection and turbulence.

In the long run, there is an apparently perfect balance in this complex energy budget. As much energy leaves the atmosphere for space as enters it from the sun. And as much energy reaches the Earth's surface from the atmosphere as leaves it for the atmosphere.

This complicated sequence of atmospheric heating has many ramifications. One of the most striking is that the atmosphere is mostly heated directly from below rather than from above, i.e., from the Earth rather than from the sun. The result is a troposphere in which cold air overlies warm air. This unstable situation (to be explored further in Chapter 5) creates an environment of almost constant convective activity and vertical mixing. If the atmosphere were heated directly from the sun, producing warm air at the top and cold air near the Earth's surface, the situation would be stable, essentially without vertical air movements. The result would be a troposphere that was largely motionless, apart from the effects of the Earth's rotation.

Spatial and Seasonal Variations in Heating

It has already been noted that world weather and climate differences are fundamentally based on the unequal heating of the Earth and its overlying atmosphere. There are significant regional variations in the receipt of heat at different latitudes and in different seasons. Tropical regions achieve an energy surplus throughout the year; the midlatitudes experience a surplus in summer and a deficit in winter; and polar regions attain a small surplus in summer and an enormous deficit in winter. It is only because there are secondary mechanisms for heat transfer that the low latitudes do not become increasingly hotter and the high latitudes increasingly colder. Nevertheless, heat is particularly concentrated, on an annual basis, in the tropics and subtropics, and coldness is most pronounced in the high latitudes.

Latitudinal Differences

There are only a few basic reasons for the unequal heating of different latitudinal zones, as discussed below.

Angle of Incidence As the Earth's surface is curved and there is a continual shifting of the positional relationship between the Earth and the sun, there is also a continuing change in the angle at which the sun's rays strike the Earth's surface, referred to as the *angle of incidence*. This changing angle is the primary determinant of the intensity of solar radiation received at any spot on the Earth. If the sunbeam strikes at a high angle, the energy is concentrated in a small area; if the angle of incidence is low, the energy is diffused over a larger portion of the surface. The more nearly perpendicular the ray is, the more effective is the heating.

Figure 4-3 portrays the noon sun striking the Earth on the date of one of the equinoxes. Since the sun is so far from the Earth, the rays can be considered to be parallel to one another. At the equator, the beams are vertical; i.e., the angle between ray and surface is 90°. At the poles, on the other hand, the angle is 0°. Between these extremes, the rays reach the surface at an intermediate angle, which can be calculated by subtracting the latitude from 90°. The black area on the globe's surface indicates the comparative spread of rays at the angular extremities. It is clear that, considering the year as a whole, high latitude regions receive much less intense insolation than tropical areas.

Length of Day The duration of sunlight is another important factor in explaining latitudinal inequalities of heating. Longer days allow more insolation to be received, and thus more heat to be generated. In tropical regions, this factor is relatively unimportant because the number of hours between sunrise and sunset does not vary significantly from one month to another; at the equator, of course, daylight and darkness are essentially equal in length (12 hours each) every day of the year. In middle and high latitudes, however, there are pronounced

Figure 4-3. Comparative angles of the sun's rays. The more direct the angle, the more concentrated is the energy and the more effective is the heating.

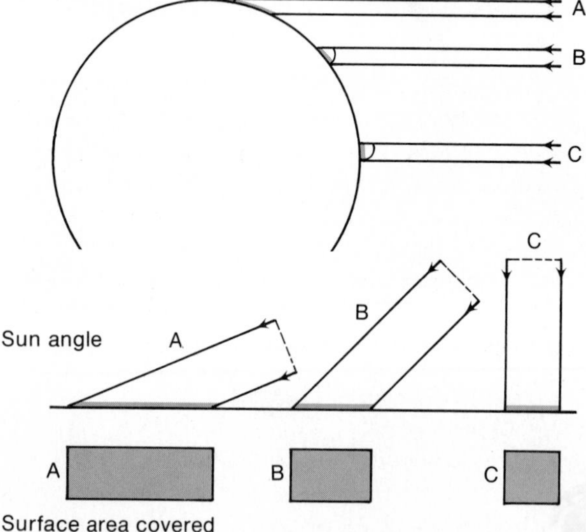

latitudes than in the low latitudes; thus there are smaller losses of energy in the tropical atmosphere than in the polar atmosphere.

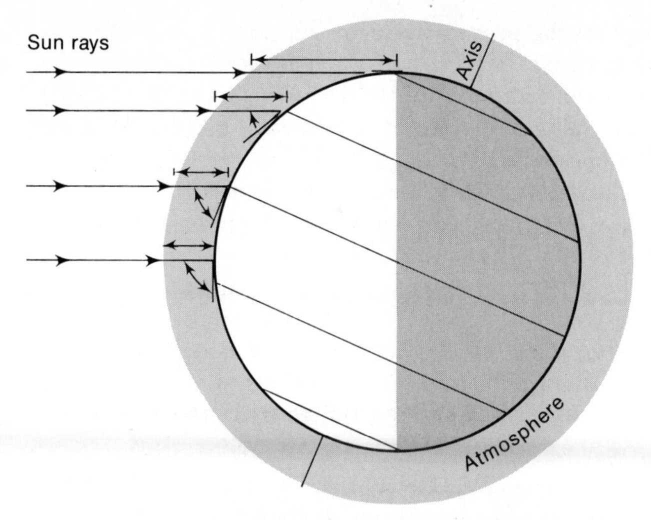

Figure 4-4. Atmospheric obstruction of sunlight. Low-angle rays must pass through more atmosphere than high-angle rays; thus the low-angle rays are subject to more depletion through reflection, scattering, and absorption.

seasonal variations in the duration of sunlight. The conspicuous buildup of heat in summer in these regions is largely a consequence of the long hours of daylight, and the winter cold is an obvious manifestation of limited insolation being received because of the brevity of the daylight period.

Atmospheric Obstruction We have already noted that the atmosphere has a debilitating effect on insolation passing through it. Clouds, particles, and molecules in the atmosphere intercept some of the solar energy and either prevent its passage by absorption and reflection or diffuse it through scattering. The result is to reduce the intensity of the energy received at the Earth's surface. On the average, sunlight received at the Earth's surface is only about half as strong as it is outside the Earth's atmosphere.

This weakening effect varies from time to time and from place to place, depending on two factors: the amount of atmosphere and its transparency. See Figure 4-4.

1. The length of passage of a solar beam through the atmosphere is determined by the angle of incidence. A high-angle beam will traverse a shorter course through the atmosphere than a low-angle one. A tangent ray (one striking the Earth at the lowest possible angle) must pass through nearly 20 times as much atmosphere as a direct ray (one striking the Earth at a 90° angle).
2. The transparency of the atmosphere is highly variable, determined largely by the amount of cloud cover.

The total effect of atmospheric obstruction on the distribution of solar energy at the Earth's surface is to reinforce the pattern established by the varying angle of incidence. Solar beams are more depleted in the high latitudes than in the low latitudes; thus there are smaller losses of energy in the tropical atmosphere than in the polar atmosphere.

Latitudinal Radiation Balance We have seen that variations in the angle of incidence, length of day, and amount of atmospheric obstruction cause the unequal heating of different latitudinal zones. See Figure 4-5. As the direct rays of the sun shift northward and southward across the equator during the course of the year, the belt of maximum solar energy swings back and forth through the tropics. Thus in the low latitudes there is an energy surplus, with more incoming than outgoing radiation. Conversely, in the high latitudes there is an energy deficit, with more radiant loss than gain. The latitudinal radiation balance has a profile that shows positive values equatorward of about 34° latitude and negative values poleward of that parallel. The Earth as a whole receives a net gain of radiant energy between 34° north latitude and 34° south latitude, and registers a net loss of heat poleward of those latitudes.

We know, however, that there is a balance between incoming and outgoing radiation for the Earth/atmosphere complex as a whole; i.e., the net radiation balance for the Earth is zero. The mechanisms for exchanging heat between the surplus and deficit regions involve the general circulation patterns of the atmosphere and oceans, which will be discussed subsequently.

Land and Water Contrasts

The atmosphere is heated, for the most part, directly from the Earth rather than from the sun; thus the heating of the Earth's surface is a primary control of the heating of the air above. In order to comprehend variations in

Figure 4-5. Generalized latitudinal variation of the solar energy budget. Low-latitude regions have a surplus; high-latitude regions experience a deficit. [After A. N. Strahler and A. H. Strahler, *Elements of Physical Geography,* 2nd ed. (New York: John Wiley & Sons, Inc., 1979), p. 60. ©1979 by John Wiley & Sons. Used by permission of John Wiley & Sons.]

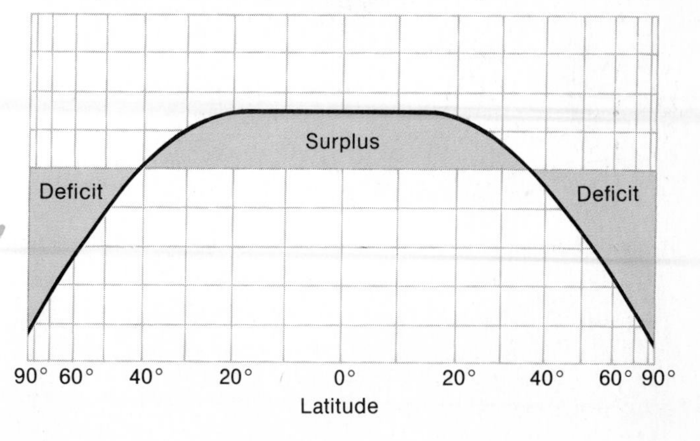

air temperatures, it is useful to understand how different kinds of surface on the Earth react to solar energy; i.e., what are the heating properties of different surfaces. There is considerable variation in the absorbing and reflecting capabilities of the almost limitless kinds of surfaces that are found on Earth—soil, water, grass, trees, cement, rooftops, etc. Their varying receptivity to incoming solar radiation in turn causes differences in the temperature of the overlying air.

By far the most significant contrasts from a general climatic standpoint are those between land and water surfaces. Oceans respond very differently from continents to the arrival of solar radiation. The generalization is that *land heats and cools faster and to a greater degree than water*. See Figure 4-6.

Heating A land surface will heat more rapidly and will reach a higher temperature than a comparable water surface even though both are subject to the same intensity of insolation. There are several significant reasons for this reaction:

1. Water has a higher *specific heat* than land. Specific heat is a physical property of matter that is defined as the amount of energy required to raise the temperature of a unit mass of a substance by 1°C. The specific heat of water is about five times as great as that of land, which means that water can absorb much more solar energy without showing it (by a rise in temperature) than land can.
2. Sun rays *penetrate* water more deeply than they do land; i.e., water is a better transmitter than land. Even in very murky water, sunlight will penetrate for several inches or feet, whereas in clear water, penetration may extend for several tens of yards or meters. Not so on land, however. The land is opaque, and all heat is absorbed right at the surface rather than being diffused through a broad layer.
3. Water is highly *mobile*, so that turbulent mixing and ocean currents disperse the heat both broadly and deeply. Land, of course, is inert, so dispersal of heat can only take place by conduction, and land is a very poor conductor.
4. The unlimited availability of moisture on a water surface means that *evaporation* is much more prevalent than it is on a land surface. We will see in Chapter 6 that the evaporation process requires heat that is temporarily stored (latent heat). It draws this from the water body from which evaporation takes place, and from the immediate surroundings, causing a drop in temperature. Thus the cooling effect of evaporation slows down any heat buildup on a water surface.

For these reasons, then, most land areas experience a relatively rapid buildup of heat during the summer, and to a lesser degree on sunny winter days, than do comparable water areas at similar latitudes. In essence, a thin layer of land is heated to relatively high temperatures, whereas a thick layer of water is heated more slowly to moderate temperatures.

Cooling In winter, a land surface cools more rapidly and to a lower temperature than a water surface. During winter the shallow heated layer of land radiates its heat away quickly. Water, on the other hand, loses its heat more gradually because the heat has been stored deeply and is brought only slowly to the surface for radiation. As the surface of the water cools, the chilled surface water sinks, to be replaced by warmer upwellings from below. The entire water body must be cooled before the surface temperatures decrease significantly. In winter, then, land surfaces cool rapidly to low temperatures, whereas water surfaces undergo less of a temperature decline and at a slower rate.

Implications The broader significance of these contrasts between land and water is that both the hottest and coldest areas of the Earth are found in the interiors of continents, distant from the influence of oceans. In the study of the atmosphere, probably no single geographic relationship is more important than the distinction between *continental* and *maritime* climates. A continental climate experiences greater seasonal extremes of temperature—hotter in summer, colder in winter—than a maritime climate.

A simplistic expression of these differences is shown in Figure 4-7, which portrays average monthly temperatures for two U.S. cities. San Diego, California, and Dallas, Texas, are at approximately the same latitude and experience almost identical lengths of day and angles of incidence. Although their annual average temperatures are almost the same, the monthly averages are significantly at variance. Dallas, in the interior of the continent, experiences notably warmer summers and cooler winters than

Figure 4-6. Some contrasting characteristics of the heating of land and water.

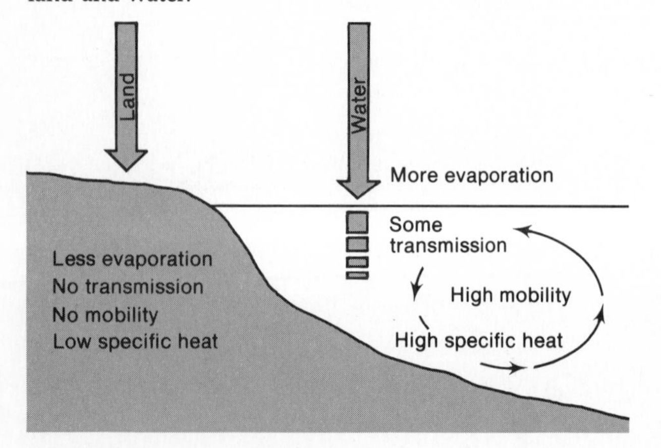

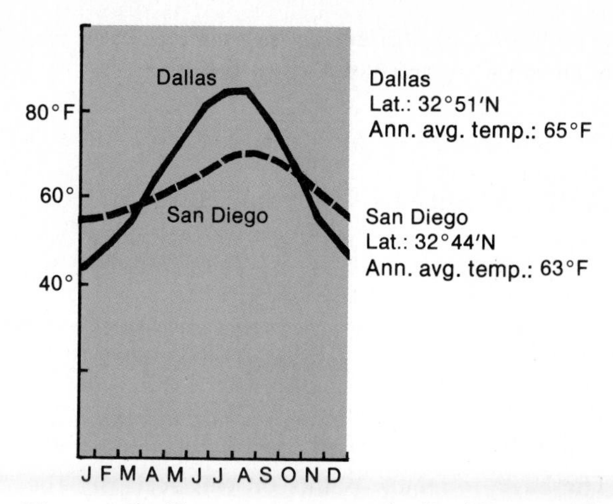

Figure 4-7. Temperature curves for San Diego and Dallas. San Diego, situated on the coast, experiences milder temperatures in both summer and winter than inland Dallas.

San Diego, which enjoys the moderating influence of an adjacent ocean.

The oceans, in a sense, act as great reservoirs of heat. In summer they absorb heat and store it. In winter they give off heat and warm up the air. Thus they function as a sort of global thermostat, moderating temperature extremes. Oceanic temperatures do not vary a great deal from summer to winter, in contrast to the notable changes of continental temperatures.

The ameliorating influence of the oceans can also be demonstrated, on a totally different scale, by comparing latitudinal temperature variations in the Northern and Southern Hemispheres. The former is often thought of as a "land hemisphere," because 39 percent of its area is land surface; the latter is a "water hemisphere," with only 19 percent of its area as land. Table 4-1 shows the average annual temperature range (difference in average temperature of the coldest and warmest months) for comparable parallels in each hemisphere. It is obvious that the "water hemisphere" experiences milder temperatures, and the "land hemisphere" has greater extremes.

TABLE 4-1 Average Annual Temperature Range by Latitude (in degrees Celsius)

Latitude	Northern hemisphere	Southern hemisphere
0	0	0
15	3	4
30	13	7
45	23	6
60	30	11
75	32	26
90	40	31

From Frederick K. Lutgens and Edward J. Tarbuck, *The Atmosphere: An Introduction to Meteorology*. (Englewood Cliffs, N.J.: Prentice-Hall, Inc., 1979), p. 52. Used by permission of Prentice-Hall.

Temperature: Expression of Heat

Thus far in this book we have referred to *energy* and *heat* more or less interchangeably, and have mentioned *temperature* several times without definition. It is important to clarify these terms and the concepts they represent.

Energy is essentially the capacity to do work, and can take various forms. *Heat* is one form of energy; it is associated with the random motion of molecules, and makes things hotter by the collision of the moving molecules. *Temperature* is simply the degree of hotness or coldness of a substance; it is an expression of how hot or how cold the substance is.

Most people have a ready sensitivity to temperature. The nerves in our skin respond to variations in heat and coldness quite readily and make us conscious of temperature changes. In this book, the adjectives *hot*, *cold*, *warm*, and *cool* will be used frequently, and like other descriptive words that will appear (*high*, *low*, *wet*, *dry*, *fast*, *slow*, etc.), they will often be used in a comparative sense; something is *hot* or *cold* in comparison with something else.

As far as temperature is concerned, however, the sensitivity of our skin is not a reliable indicator. Our nerves may be "misled" by other factors, such as moisture or air movement, factors that are quite separate from temperature but affect our perception of temperature. For example, on a cool day when a strong wind is blowing, we may perceive that the weather is colder than the temperature would indicate, simply because our body heat is dissipated so rapidly by the air movement. The term *sensible temperature* is applied to this phenomenon. It is the "sensation" of temperature that we feel in response to the total condition of the air around us; it may or may not be representative of the actual air temperature. We will refer further to this concept in subsequent chapters.

Instruments

A *thermometer* is an instrument designed for the measurement of temperature. The design principle is that most substances expand when heated and contract when cooled. If this change in volume can be systematically calibrated, it can provide a precise measurement of temperature changes.

Most thermometers are of the *liquid-in-glass* variety, in which a liquid is sealed in a glass tube that has an enlarged bulb at the bottom to store the liquid. The liquid, normally mercury or alcohol, expands much more than the glass when heated, and contracts much more than the glass when cooled. Thus it rises or falls in the tube in response to temperature changes. The length of the column of liquid in the tube thereby indicates the temperature of the surrounding air. The thermometer can be calibrated permanently at a few fixed temperatures

Figure 4-8. A typical installation of maximum and minimum thermometers in a weather shelter. [Courtesy NOAA.]

(such as freezing point and boiling point), and accurately interpolated for other values.

A special type of thermometer, called the *minimum thermometer*, is used to record the lowest temperature experienced in a given period. See Figure 4-8. In this instrument a small glass filament or wire, shaped like a tiny dumbbell and called an *index*, is placed in the column of alcohol (which is used instead of mercury because it is transparent and allows the index to be seen). The thermometer is kept horizontal, and when the temperature decreases, the alcohol contracts and drags the index down by surface tension. When the temperature increases, the alcohol expands and rises in the tube, but the index is left behind as clear evidence of the lowest temperature reached during the period, such as the overnight minimum. After this value is read by the observer, the thermometer can be reset by inverting it so that the index drops to the end of the column again.

In contrast, the *maximum thermometer* has no index but instead a narrow *constriction* in the glass tube just above the bulb. It also is kept horizontal. As the temperature rises, the mercury expands and is forced through the constriction one tiny drop at a time. When the temperature subsequently falls, however, nothing forces the mercury back through the constriction, so the extent of the column indicates the maximum temperature attained during the period. The maximum thermometer can be reset by shaking or whirling it to force the mercury back into the bulb. This is the same principle that is used in a *clinical thermometer*, which is inserted orally or rectally to determine body temperature.

Other, more specialized thermometers used by meteorologists depend on differential expansion of bimetal strips or the electrical resistance of metallic wires.

Scales

In the United States, three scales of temperature are in concurrent use. Each of them permits a precise measurement, but the existence of three different scales creates an unfortunate degree of confusion.

Fahrenheit The scale that is most widely understood by the general public is called Fahrenheit. Public weather reports from the National Weather Service and the news media usually state temperature data in degrees Fahrenheit. The basic reference points on this scale are the sea-level freezing and boiling points of pure water, which are 32° and 212°, respectively. The United States is the only major country in the world still using the Fahrenheit scale.

Celsius In other countries the Celsius scale is used either exclusively or predominantly. It is an accepted component of the SI (the revised metric system) because it is a decimal scale with 100 units (degrees) between the freezing and boiling points of water. It has long been used for scientific work, including upper air temperature measurements, in the United States, and is now slowly being established to supersede the Fahrenheit scale in public consciousness.

To convert from Fahrenheit to Celsius, the following formula is used:

$$F = \tfrac{9}{5}C + 32°$$

To convert from Celsius to Fahrenheit, the formula is:

$$C = \tfrac{5}{9}(F - 32°)$$

Kelvin For many scientific purposes the kelvin scale has long been used because it has no negative values but still maintains a 100° range between the freezing and boiling points of water. It is not normally used by climatologists and meteorologists, however, and we will ignore it in this book except to demonstrate its comparability to the Fahrenheit and Celsius scales.

Kelvin and Celsius units are exact equivalents; however they are numbered differently. Thus, very simple formulae can be applied to convert from one to the other:

$$C = K - 273 \qquad \text{and} \qquad K = C + 273$$

Data

At thousands of locations (called *stations*) throughout the world, temperature data are recorded on a regular basis. These data provide important raw material for contemporary weather reports and forecasts, as well as for long-run climatic analyses.

At some stations, the temperature is recorded every hour by a human observer, but more often a *thermograph*

Focus on the Historical Development of Temperature Scales

Most of us take thermometers and temperature readings for granted. Recognition of air and body temperatures is so much a part of our daily life as to be commonplace and unremarkable. Nevertheless, the development of instruments to measure temperature and the value scales for such measurement was a lengthy process involving contributions and modifications from many scientists.

There are allegations that the liquid-in-glass thermometer was invented by Galileo Galilei (1564–1642), the renowned Italian physicist and astronomer, but there is no clear evidence of this fact. Instead, the first thermometer was probably developed about 1654 by Ferdinand II, Grand Duke of Tuscany in northern Italy. Modifications were devised by other scientists in succeeding decades, and more than a dozen different temperature scales were in use by the early 1700s, as each instrument maker used his own scale without standardization.

In 1714, Daniel Gabriel Fahrenheit (1686–1736), a German physicist who had moved to Holland to establish an instrument business, designed the first mercury-in-glass thermometer and devised the temperature scale that bears his name. His original scale had two fixed points: 0° was the lowest temperature that he could achieve in an ice/water/salt solution; and he selected normal body temperature as 96° (changed from an initial value of 90°). From this scale he determined that the freezing point of water (called the *ice point*) was 32°, and subsequent workers established the boiling point of water (the *steam point*) at 212°. As his original 0° point was difficult to reproduce and his body temperature point was found to be inaccurate (it should have been 98.6° rather than 96°), the Fahrenheit scale is now calibrated in terms of the ice and steam points.

The Fahrenheit scale was accepted as the standard of temperature measurement in several countries, most notably Great Britain, where it was incorporated into the English system of weights and measures and was spread throughout the world in British colonies and dominions.

Anders Celsius (1701–1744), a Swedish astronomer, marked a thermometer in 1741 with 0° for boiling and 100° for freezing. His friend and countryman Carl von Linne, or Linneaus (1707–1748), the biologist who conceived the Linnean classification of plants and animals, reversed Celsius' scale (after Celsius' death), so that it ran from 0° at freezing to 100° at boiling. At almost the same time, Jean Pierre Christin (1683–1755) in France ordered from instrument-maker Pierre Casati thermometers with an identical 100-unit scale from 0° at freezing to 100° at boiling. After the French Revolution in 1789, this "centigrade" (which means hundred-unit) scale was adopted as part of the metric system in France, from where it has spread, with metrication in general, throughout the world.

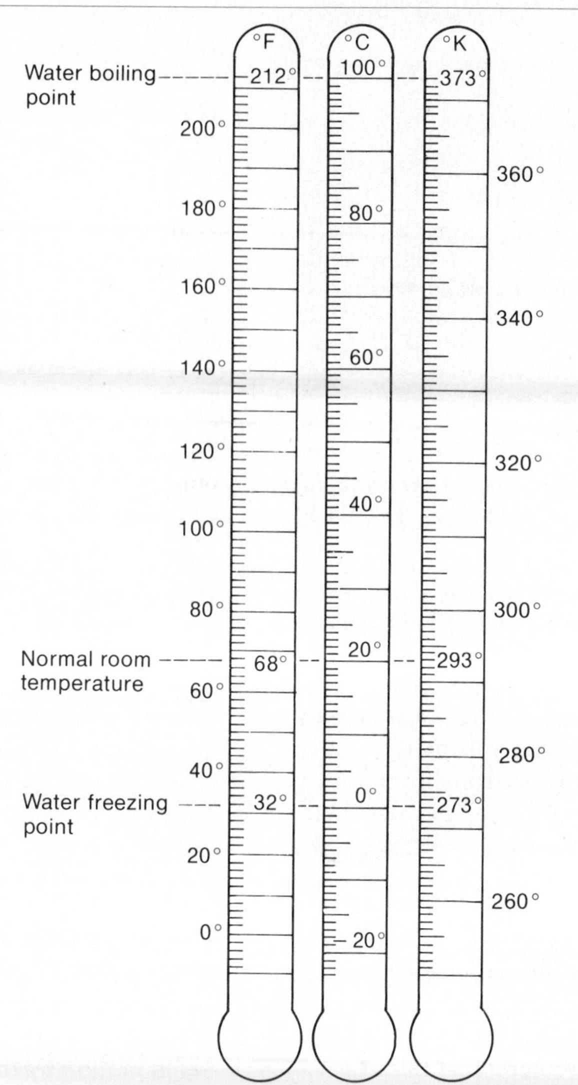

Comparison of Fahrenheit, Celsius, and Kelvin temperature scales.

The kelvin scale was first proposed in 1854 by the English physicists William Thomson, titled as Lord Kelvin (1824–1907), and James Prescott Joule (1818–1889). The kelvin scale runs upward from *absolute zero* (0°), which is the hypothetical temperature at which all atomic and molecular motions cease. Degree separations were exactly the same on the kelvin and centigrade scales, with 100° separating the ice point and steam point for both. On the kelvin scale, however, freezing occurs at 273° (0° on the centigrade scale) and boiling takes place at 373° (100° centigrade).

In 1948, by international agreement, the centigrade thermometer scale of the metric system was renamed Celsius, in honor of the apparent first user of a 0–100 scale. This attained conformity with other scales, all of which were named for people. Thus, degrees on the Celsius scale are designated as "C," which would be apropos for centigrade, Christin, and Casati, as well as Celsius. The unit of temperature, however, is properly called the *kelvin*, not "Celsius degree." In the SI system, a degree is strictly a unit of angular measure.

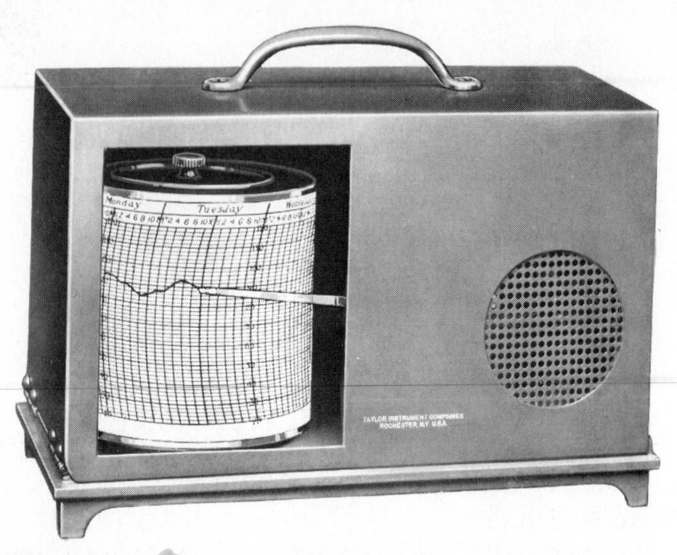
Figure 4-9. A thermograph records temperature continuously by means of an inked trace on a rotating drum. [Courtesy Sybron/Taylor Instruments.]

(a continuously recording thermometer) is in operation. See Figure 4-9. At most stations only the highest and lowest temperatures are recorded for each 24-hour period, generally by means of maximum-minimum thermometers. The *daily mean temperature* is then calculated by averaging the 24 hourly readings, or is estimated from the average of the maximum and minimum. From these data can also be calculated the *daily* or *diurnal temperature range*, which is the number of degrees difference between the highest and lowest recorded values of the 24-hour period.

The *monthly mean temperature* for each station is determined by averaging the daily means for each day in the calendar month. The *annual mean temperature* represents the average of the 12 monthly means. The difference between the means of the warmest and coldest months is the *annual temperature range*.

Mechanisms of Heat Transfer

We have seen that there is a significant inequality in the receipt of heat at different locations over the Earth's surface, and that this unequal distribution is largely a function of latitude. If there were not important mechanisms for secondary transfer of heat in a general poleward direction in both hemispheres, the tropics would become progressively warmer and the high latitudes continually colder. Such temperature trends do not occur because there is a persistent shifting of warmth toward the high latitudes and of coldness toward the low latitudes. This is accomplished by movements of air and water, i.e., by circulation patterns in the atmosphere and in the oceans. The broad-scale, or planetary, circulation of these two mediums moderates the buildup of heat in equatorial regions and cold in polar regions, and thereby makes both of those latitudinal zones more habitable. Both the atmosphere and the oceans act as enormous thermal engines, with their latitudinal imbalance of heat serving to drive the currents of air and water, which, in turn, transfer heat and somewhat modify the imbalance.

Atmospheric Circulation

Of the two mechanisms of global heat transfer, by far the more important is the general circulation of the atmosphere. Air movements take place in almost infinite diversity, but there is a broad planetary circulation pattern that serves as a general framework for moving warm air poleward and cool air equatorward. Some 75 to 80 percent of all horizontal heat transfer is accomplished by atmospheric circulation.

Our discussion of atmospheric circulation will be withheld until Chapter 5, following consideration of some fundamentals concerning pressure and wind.

Oceanic Circulation

There is a close relationship between the general circulation patterns of the atmosphere and oceans. Air blowing over the surface of the water is the principal drive and guide for the flow of major ocean currents, and the heat energy stored in the oceans has important effects on atmospheric circulation.

Various kinds of oceanic water movements are categorized as *currents*. They may be horizontal, vertical, or oblique in their flow. The deeper flows may be activated by density contrasts associated with variations in temperature and salinity, but surface currents are nearly all set in motion by the frictional drag of wind moving horizontally over the ocean surface. For our purposes in understanding heat transfer, we are concerned primarily with the broad-scale surface currents that make up the general circulation of the oceans.

These major currents respond to changes in wind direction, but they are so broad and ponderous that the response time normally amounts to many months. In essence, ocean currents reflect average wind conditions over a period of several years, with the result that the major components of oceanic circulation are closely related to major components of atmospheric circulation.

Ocean Basins All the oceans of the world are interconnected, but due to the location of landmasses and the pattern of atmospheric circulation, it is convenient to visualize five relatively separate ocean basins—North Pacific, South Pacific, North Atlantic, South Atlantic, and South Indian. Within each of these basins there is a similar pattern of surface current flow, based on a general similarity of prevailing wind patterns.

The Basic Pattern Despite variations based on the size and shape of the various ocean basins, and on the season of the year, a single simple pattern of surface currents is characteristic of all of the basins. See Figure 4-10. It consists of an enormous closed elliptical loop, elongated east-west and centered approximately at 30° of latitude. These loops, called *gyres*, have a clockwise direction of flow in the Northern Hemisphere and a counterclockwise movement in the Southern Hemisphere. The positions of the gyres are roughly accordant with the locations of semipermanent subtropical high-pressure cells in the atmosphere above (see Chapter 5).

On the equatorward side of each of these subtropical gyres is the *Equatorial Current*, which moves steadily toward the west. Near the western margin of each ocean basin, the general current curves poleward, which produces a north-flowing current in the Northern Hemisphere and a south-flowing current in the Southern Hemisphere. As these currents approach the poleward margins of the ocean basins, they curve easterly, and as they reach the eastern edges of the basins, they curve back toward the equator, producing a closed loop in each basin.

The movement of these currents, although energized by the wind, is impelled by the deflective force of the Earth's rotation, which is called the *Coriolis effect*. This apparent force (discussed in the accompanying focus box) dictates that the ocean currents will be deflected to the right in the Northern Hemisphere and to the left in the Southern Hemisphere. A glance at the basic pattern shows that the current movement around the gyres responds precisely to the Coriolis effect.

One further component of the basic pattern is the *Equatorial Countercurrent*, which is an east-moving flow approximately along the equator in each of the oceans. Each Equatorial Current has an average position 5° to 10° north or south of the equator, and "feeds" the Equatorial Countercurrent near its western margin in each basin. Water from the Equatorial Countercurrent, in turn, drifts poleward to feed the Equatorial Current near the eastern end of its path.

Northern and Southern Variations In the two Northern Hemisphere basins—North Pacific and North Atlantic—the bordering continents lie so close together at the poleward margin of the basins that the bulk of the current flow is prevented from entering the Arctic Ocean. The North Pacific has only very limited flow northward between Asia and North America, whereas in the North Atlantic, a larger proportion of the flow escapes northward between Greenland and Europe.

In the Southern Hemisphere, on the other hand, the continents are far apart. Thus the poleward links (westerly flow) of the gyres in the South Pacific, South Atlantic, and South Indian oceans are connected as a continuous flow in the uninterrupted belt of ocean that extends around the world in the vicinity of 60° south latitude. This circumpolar flow is called the *West Wind Drift*.

Current Temperatures Of utmost importance to our understanding of latitudinal heat transfer are the temperatures of the various currents. Each of the major currents can be characterized as "warm" or "cool," but

Figure 4-10. The major surface ocean currents. (1) North Pacific Drift, (2) Alaska Current, (3) California Current, (4) Equatorial Current, (5) West Wind Drift, (6) Humboldt Current, (7) Gulf Stream, (8) Labrador Current, (9) North Atlantic Drift, (10) Canaries Current, (11) Brazil Current, (12) Benguela Current, (13) Agulhas Current, (14) West Australian Current, (15) East Australian Current, (16) Kuroshio (Japan Current), and (17) Oyashio (Kamchatka Current).

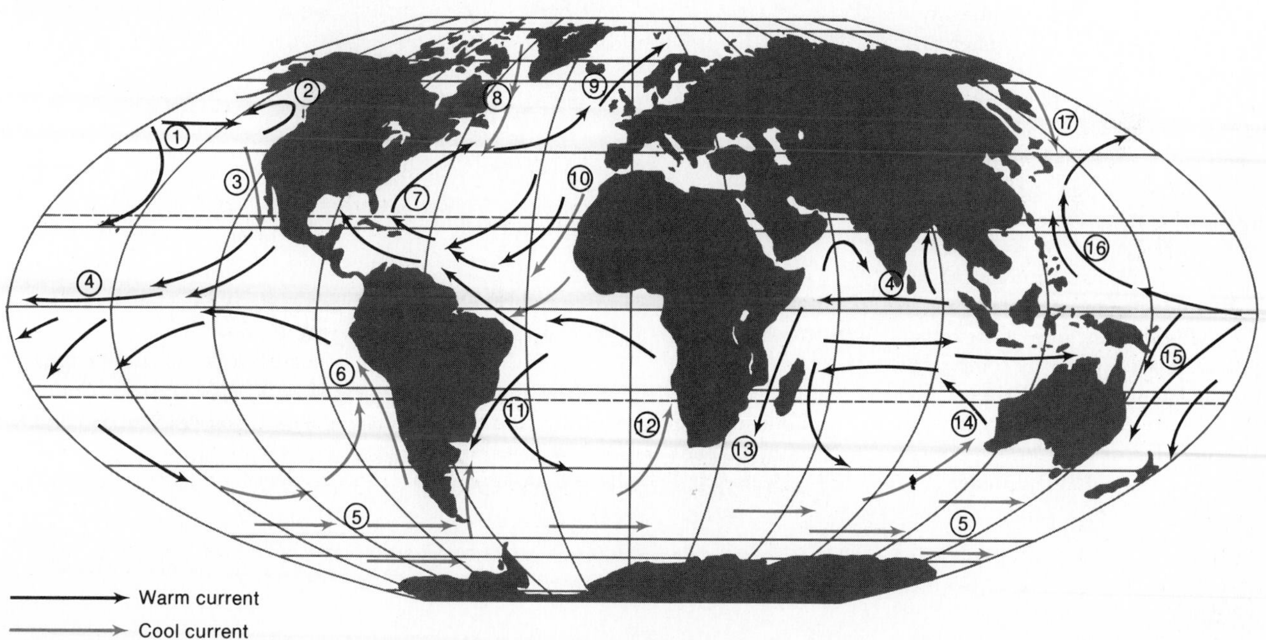

Everyone is familiar with the unremitting force of gravity. Its powerful pull toward the center of the Earth exerts a prominent vertical influence on any motions that take place on the Earth's surface or in the atmosphere. Much less well known, however, because of its inconspicuous nature, is a pervasive horizontal influence on earthly motions. All things that move over the surface of the Earth or in the Earth's atmosphere drift sidewise from their direct path of movement as a result of the Earth's rotation.

This sidewise deflection makes it appear as if a force were acting on the moving object to divert it from its normal path. Such is not the case, however; no "force" is involved, and the deflection is apparent only to an observer on Earth. From an earthly standpoint, the object is deflected sidewise, but an observer in space would see that the Earth simply rotated beneath the moving object; there is no deviation from a direct line. The deflection, then, pertains only in reference to a coordinate system (such as the latitude/longitude grid) that is fixed with respect to the Earth, and rotates with it.

George Hadley described this apparent deflection in the 1730s, but it was not explained carefully and quantitatively until Gaspard G. de Coriolis (1792–1843), a French mathematician, did so a century later. The phenomenon is called the *Coriolis effect* in his honor.

The Coriolis principle applies to any free-moving object—ball, bullet, airplane, automobile, even a person walking. For these and other short-range movements, however, the deflection is so minor and so counterbalanced by other "forces" (such as friction, air resistance, and initial impetus) as to be insignificant. Long-range movements, on the other hand, can be significantly influenced by the Coriolis effect. The accurate firing of artillery shells or launching of rockets and spacecraft require careful compensation for the Coriolis effect if the projectiles are to reach their targets.

Recognition of the Coriolis principle is simple, although a full understanding of it is much more complicated. The four basic points to remember are:

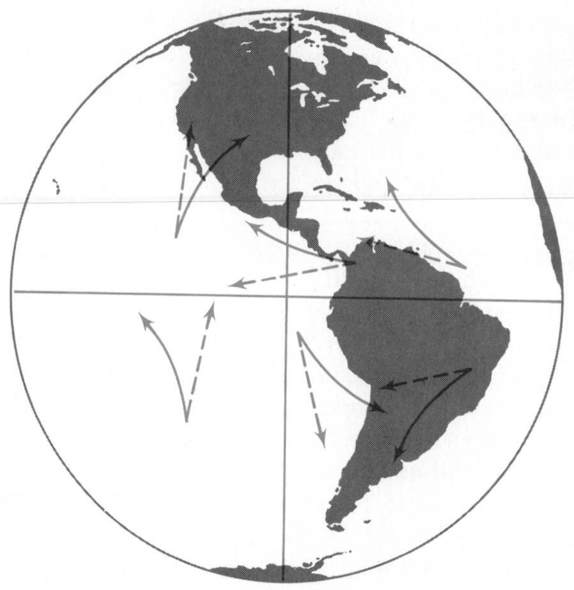

The apparent deflection of the Coriolis effect is to the right in the northern Hemisphere and to the left in the Southern Hemisphere. The dashed lines represent the gradient force; the solid lines represent actual movement.

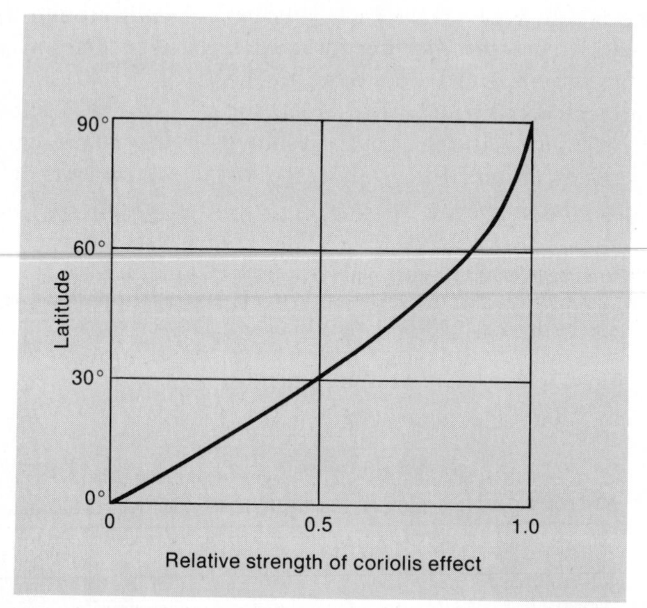

The Coriolis effect is much stronger in high latitudes than in low latitudes.

1. Regardless of the initial direction of motion, any free-moving object in the Northern Hemisphere will appear to be deflected to the right, and any free-moving object in the Southern Hemisphere to the left.
2. The apparent deflection is strongest at the poles and decreases progressively toward the equator, where it is zero (no deflection).
3. The Coriolis effect is also proportional

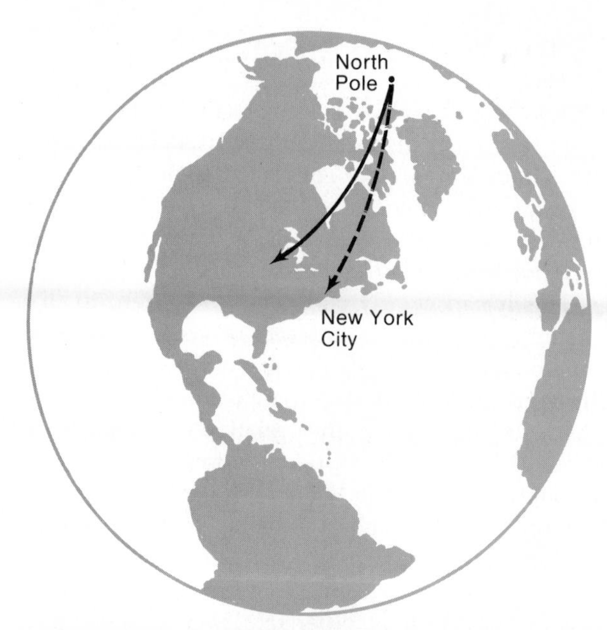

A rocket fired from the North Pole toward New York City would land well to the west unless earthly rotation were included in the ballistic computation.

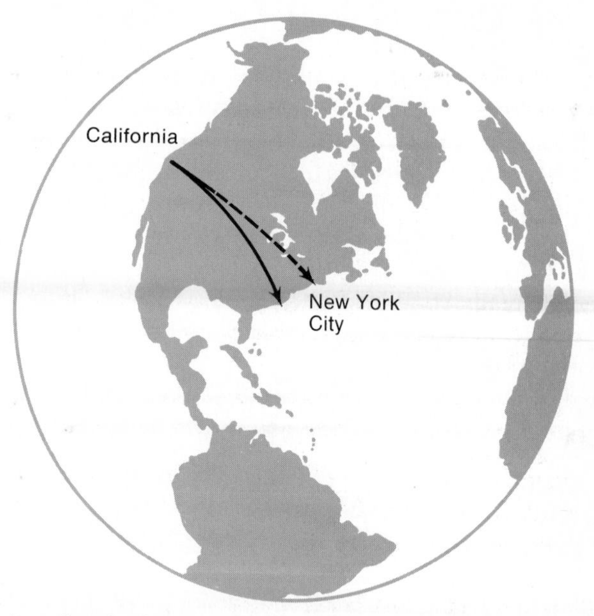

A rocket fired toward New York City from the same latitude in California would appear to curve to the right because both firing and target points would rotate easterly while the rocket was in the air.

to the speed of the object, so a fast-moving object will seem to be deflected more than a slower one.

4. The Coriolis effect influences direction of movement only; it has no influence on speed.

The nature of the apparent deflection due to the Coriolois effect can be demonstrated by imagining a rocket fired at New York City from the North Pole. During the few minutes that the rocket is in the air, earthly rotation will have moved the target some miles to the east. If the Coriolis effect was not included in the ballistic computation, the rocket would land some miles to the west of New York City. The flight path, then, would appear to have been deflected to the right.

This rightly deflection (in the Northern Hemisphere) applies no matter which direction the rocket moves. For example, if a rocket were aimed at New York City from a location on the same parallel of latitude—say, northern California—there would also be a drift to the right. During the time the rocket is in motion, both the California firing point and the New York target will have rotated easterly. Thus the landing point will be to the southwest of New York, and the trajectory of the rocket will appear to earthly observers to have followed a rightward curve.

The major importance of the Coriolis effect in our study of climate involves its influence on winds and ocean currents:

1. All winds are affected by the Coriolis principle. This is a matter of great significance, as we will see in Chapter 5.
2. The more ponderous movements of ocean currents are also deflected by the Coriolis effect. This is an important component of the dynamics of the general circulation of the oceans, with the Northern Hemisphere currents trending to the right and the Southern Hemisphere currents sidling to the left. The Coriolis drift is also a causative factor in the upwelling of cold water that takes place in subtropical latitudes where cool currents veer away from continental coastlines. The surface water that moves away from the shore is replaced by cold water rising from below.

One phenomenon that the Coriolis effect does not appear to influence is the circulation pattern of water that drains out of a washbowl. There is an "old wives' tale" that claims that Northern Hemisphere washbowls drain in a clockwise pattern and Southern Hemisphere washbowls empty in a counterclockwise whirl. The time involved is so short and the velocity of the water so slow, however, that Coriolis control cannot be postulated for these movements. The characteristics of the plumbing system, the shape of the washbowl, and pure chance are more likely to determine the flow patterns.

the adjective is always used in a relative sense—meaning that the current is warmer or cooler than the surrounding water at that latitude. The generalized temperature characteristics are as follows:

1. Low-latitude currents (equatorial current, equatorial countercurrent) are warm.
2. Poleward-moving currents on the western sides of ocean basins carry warm water toward higher latitudes.
3. Northern components of the Northern Hemisphere gyres (Kuroshiro/North Pacific Current in the Pacific and Gulf Stream/North Atlantic Drift in the Atlantic) carry warm water toward the north and east.
4. Southern components of the Southern Hemisphere gyres (generally combined into the West Wind Drift) are strongly influenced by Antarctic waters and are essentially cool.
5. Equatorward-moving currents on the eastern sides of ocean basins carry cool water toward the equator.

In summary, then, the significant aspect of heat transfer accomplished by the general circulation of the oceans is a poleward flow of warm tropical water along the western edge of each ocean basin and an equatorward movement of cool high-latitude water along the eastern margin of each basin.

Rounding Out the Pattern Two other relatively minor aspects of oceanic circulation are influential in heat transfer:

1. The northwestern portions of Northern Hemisphere ocean basins receive an influx of cool water from the Arctic Ocean. A prominent cool current from the vicinity of Greenland comes southward along the Canadian coast, and a smaller flow of cold water issues from the Bering Sea southward along the coast of Siberia to Japan.
2. Wherever an equatorward-flowing cool current pulls away from a subtropical west coast, a pronounced and persistent upwelling of cold water occurs. See Figure 4-11. This is most striking off South America but is also notable off the coast of North America, northwestern Africa, and southwestern Africa. It is much less developed off the coast of western Australia.

Figure 4-11. Schematic diagram of a coastal upwelling. The most persistent upwellings occur where cool currents, flowing generally toward the equator, pull away from coastlines. Colder water then rises from the ocean depths.

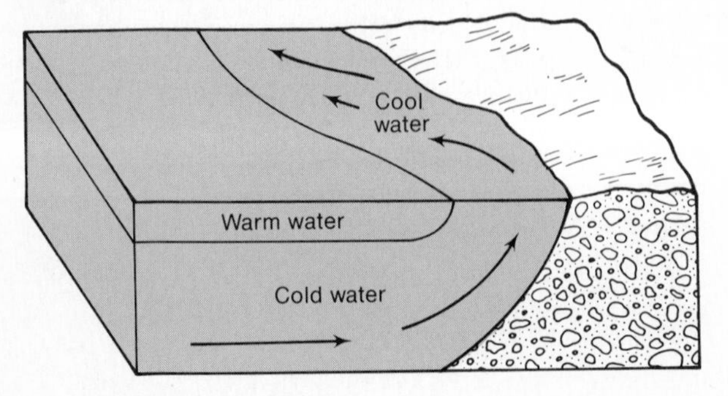

Vertical Temperature Patterns

As we study the geography of climate—or anything else, for that matter—most of our attention is directed to the horizontal dimension; we are concerned with the spatial distribution of phenomena over the surface of the Earth. From time to time, however, we must focus on vertical patterns or processes because of their influence on surface features. This is particularly pertinent with regard to climatic phenomena because of the ease of vertical motions and the frequency of vertical activity in the atmosphere.

The vertical dimension of temperature within the troposphere is significant and relatively predictable. Throughout the troposphere, under normal conditions, there is a general decrease in temperature with increasing altitude. However, this is not a universal truth; there are many exceptions. Indeed, the rate of vertical temperature decline can vary due to season, time of day, amount of cloud cover, and a host of other factors. In some cases, there is even an opposite trend, with the temperature increasing upward for a limited distance. As a generalization, however, as one goes higher in the troposphere, it gets colder.

Lapse Rate

The rate of temperature decrease with height is variable, particularly in the lowest few hundred feet of the troposphere, but the normal expectable rate of decrease is about 3.6° F per 1000 feet (6.5° C per km). This is the *average lapse rate* or *normal vertical temperature gradient*. Thus, if a thermometer measures a temperature 1000 feet above a previous measurement, the reading will be, on the average, 3.6° F cooler; if the second measurement is 1000 feet below the first, it would register about 3.6° F warmer.

This refers to the sampling of temperatures at different elevations, with no vertical air movement involved. When considering a lapse-rate temperature change, only the thermometer is moved; the air is at rest. If the air is moving vertically, expansion or contraction will cause an adiabatic temperature change, which is totally different from a lapse-rate temperature change. This concept will be explored more fully in Chapter 6.

Temperature Inversion

The most prominent exception to a normal lapse-rate condition is a *temperature inversion*, which is a situation in which temperature increases upward; thus the normal condition is inverted. See Figure 4-12. Inversions are relatively common in the troposphere but are usually of brief duration

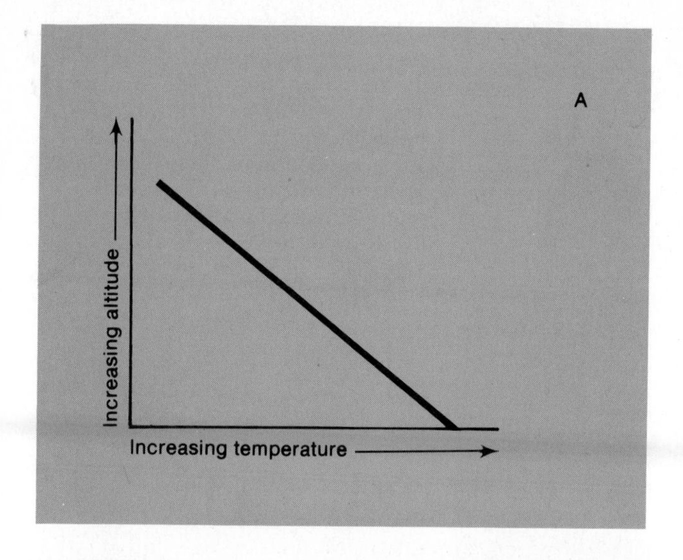

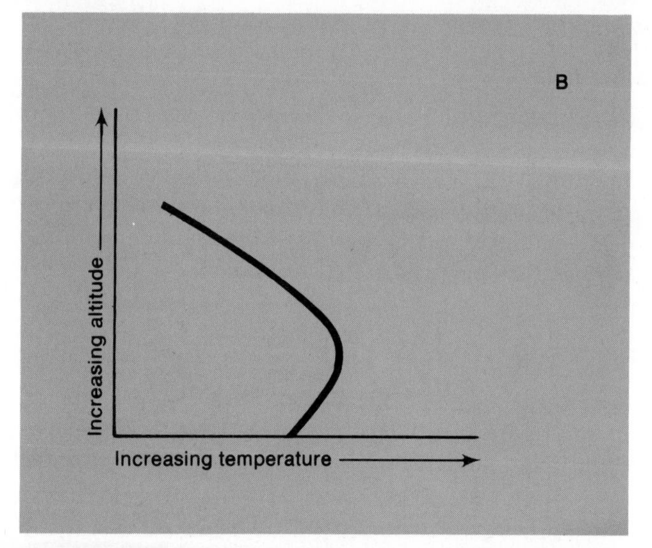

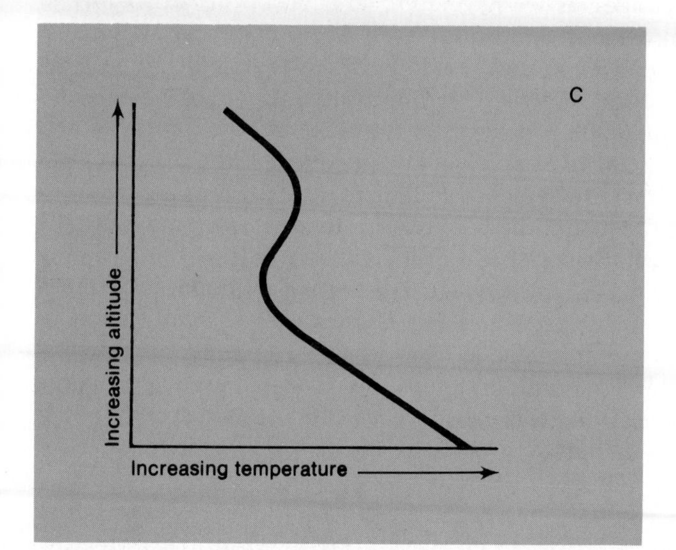

Figure 4-12. A comparison of normal and inversion lapse rates. (A) Tropospheric temperature normally decreases with height. (B) With a surface inversion there is a low-level temperature increase with height. (C) With an upper-air inversion there is an upward temperature increase at some level above the surface.

and restricted depth. They can occur near the Earth's surface or at higher levels (Figure 4-13).

Surface Inversions The most readily recognizable inversions are those found at ground level. These are usually classed as *radiational inversions* because they result from rapid radiational cooling. They occur typically on a long cold winter night when a land surface (which is a very efficient radiator) rapidly emits long-wave radiation into a clear, calm sky. The ground is soon colder than the air above it, which is further cooled by conduction. In a relatively short time, the lowest few hundred feet of the troposphere become colder than the air above, and a temperature inversion is in effect. Radiational inversions are primarily winter phenomena because there is only a short daylight period for incoming solar heating and a long night for radiational cooling. They are therefore much more prevalent in high latitudes than elsewhere.

An inverted surface temperature gradient can also be the result of an *advectional inversion*, in which there is a horizontal inflow of cold air into an area. This is usually produced either by cold air "draining" downslope into a valley or by cool maritime air blowing into a coastal locale. Advectional inversions are usually short-lived (typically overnight) and shallow, and are more common in winter than in summer.

Upper-Air Inversions Temperature inversions sometimes occur well above the surface, nearly always as the result of air sinking from above. These *subsidence inversions* are usually associated with high-pressure cells, which are particularly characteristic of subtropical latitudes throughout the year and of Northern Hemisphere conti-

Figure 4-13. Temperature inversions can occur at any height in the atmosphere.

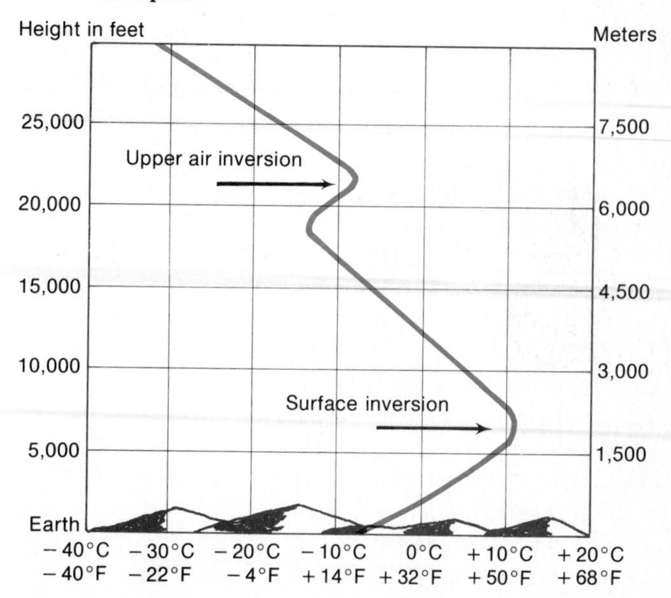

Figure 4-14. A temperature inversion usually functions also as a stability lid, which inhibits air circulation. Thus air pollutants cannot be dispersed normally. The result often is an accumulation of smog. This 1960 scene in Los Angeles has been repeated many hundreds of times. [Courtesy U.S. Public Health Service.]

nents in winter. A subsidence inversion can be fairly deep (sometimes several thousand feet), and its base is usually a few thousand feet above the ground, as low-level turbulence prevents it from sinking lower.

Significance In addition to the obvious reversal of the normal vertical temperature pattern, inversions have other important influences on weather and climate. As we will see in Chapter 6, an inversion inhibits vertical air movements and greatly diminishes the possibility of precipitation. Inversions also contribute significantly to increased air pollution because they create stagnant air conditions that greatly limit the natural dispersal of urban/industrial pollutants. See Figure 4-14.

Global Temperature Patterns

Thus far in this chapter we have been examining a variety of facts, concepts, and processes associated with insolation and temperature. The ultimate goal of this and the two succeeding chapters is to gain an understanding of the world distribution pattern of the climatic elements—temperature, pressure, wind, and moisture. With the preceding pages as background, we will now turn our attention to the worldwide distribution of temperature.

The basic maps of global temperature patterns display the seasonal extremes rather than the annual average. January and July are the months of lowest and highest temperatures for most places on the Earth, so maps portraying the average temperatures of these two months provide a simple but meaningful expression of thermal conditions in winter and summer. Figures 4-15 and 4-16 represent these two basic maps. Temperature distribution is shown by means of *isotherms*, which are lines joining points of equal temperature. Although the maps are on a very small scale, they permit a broad understanding of temperature patterns for the world.

Prominent Controls of Temperature

Figures 4-15 and 4-16 show that the gross patterns of temperature are controlled largely by four factors—altitude, latitude, land/water contrasts, and ocean currents.

Altitude Temperature responds sharply to altitudinal changes, as we have seen, so it would be misleading to plot *actual* temperatures on these maps, as stations at higher elevations would invariably be colder than low-altitude stations. The complexity introduced by hills and mountains would make the maps more complicated and difficult to comprehend. Consequently, the data for most maps displaying world temperature patterns are modified by reducing the temperature to what it would be if the station were at sea level. This can be done most simply by using the average lapse rate. It produces artificial temperature values but eliminates the complication of terrain differences. These maps are useful in showing world patterns, but they are not satisfactory for indicating actual temperatures for specific locations that are not close to sea level.

Latitude Clearly the most conspicuous feature of these maps is the general east-west trend of the isotherms, roughly following the parallels of latitude. If the Earth had a uniform surface and did not rotate, the isotherms probably would coincide exactly with the parallels, showing a regular decrease of temperature, progressively,

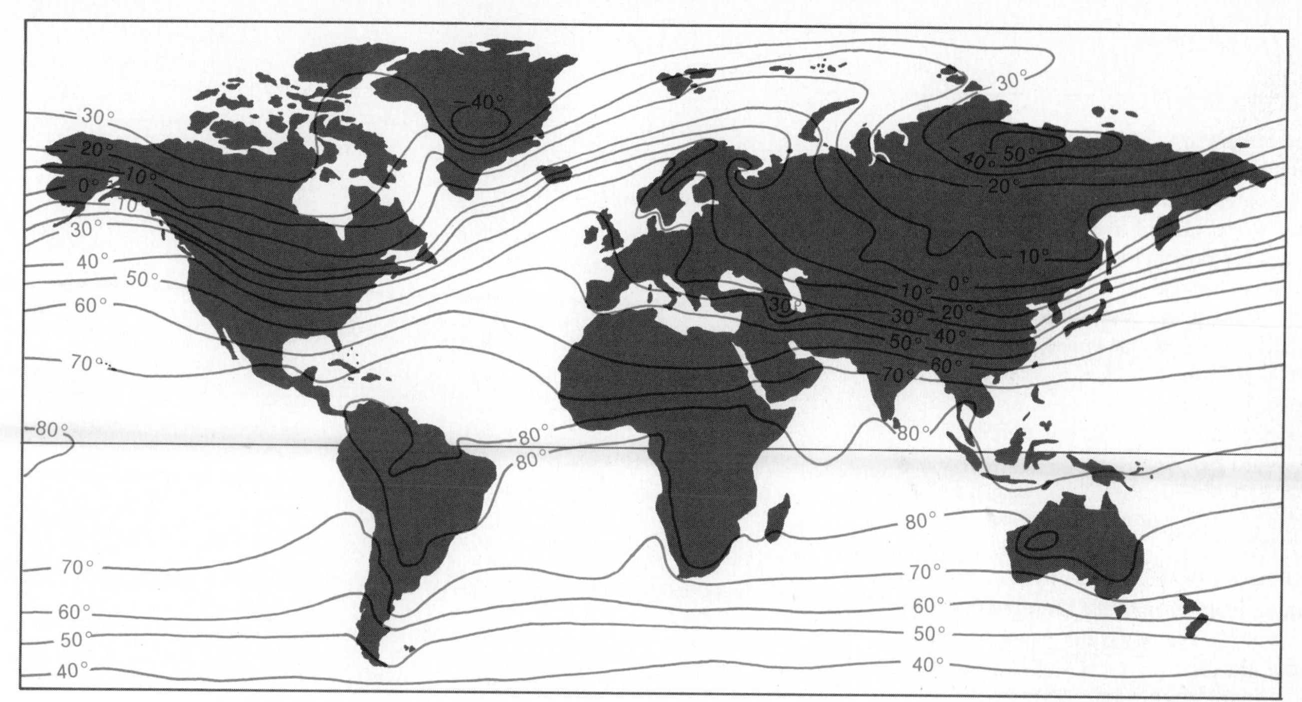

Figure 4-15. Average January sea-level temperatures.

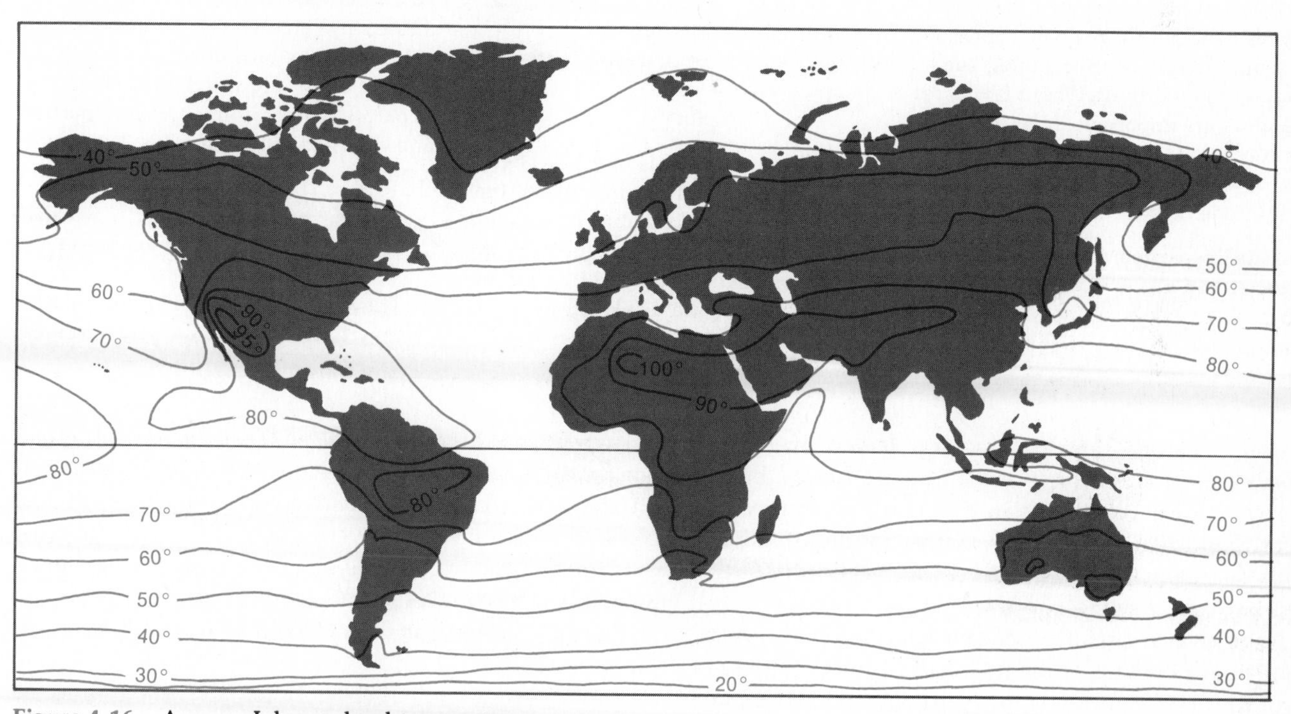

Figure 4-16. Average July sea-level temperatures.

poleward from the equator. However, the Earth does rotate, and it has ocean waters that circulate and land that varies in elevation. Consequently, there is no precise temperature correlation with latitude. Nevertheless, the fundamental cause of temperature variation the world over is the effectiveness of solar insolation, which is governed primarily by latitude.

Land/Water Contrasts The differential heating and cooling characteristics of land and water are also reflected conspicuously on the maps. Summer temperatures are higher over the continents than over the oceans, as shown by the poleward curvature of the isotherms over continents in the respective hemispheres (July in the Northern Hemisphere; January in the Southern).

Geographers are avid users of maps of many kinds, and they employ a variety of cartographic devices to display data on the maps. One of the most widespread devices for portraying the spatial distribution of *quantities* of some phenomena is the *isoline* (from the Greek *isos*, "equal"), which is also called by a variety of related terms (*isarithm, isogram, isopleth, isometric line*), all of which can be considered as synonymous for our purposes. The word *isoline* (and its synonyms) is a generic term that refers to a line that joins points of equal value of something. More than one hundred different kinds of isolines have been identified by name, ranging from *isoamplitude* (amplitude of variation) to *isovapor* (water vapor content of the air).

Some isolines represent tangible surfaces, such as contour lines, which are indicative of elevations on the land. Most, however, signify such intangible features as temperature or precipitation, and some express relative values, such as ratios or proportions. Only a few types of isolines are important in an introductory consideration of physical geography. Each of the following appears in this book in one or more maps:

Contour line—a line joining points of equal elevation.

Isobar—a line joining points of equal atmospheric pressure.

Isogonic line—a line joining points of equal magnetic declination.

Isohyet—a line joining points of equal quantities of precipitation.

Isotherm—a line joining points of equal temperature.

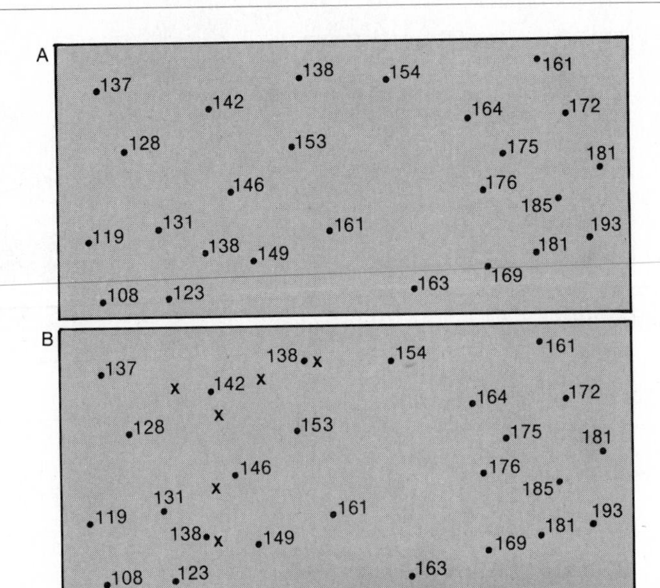

To construct an isoline it is always necessary to estimate or interpolate for values that are not available. As a simple example, the above figure illustrates the basic steps in construction of an isoline map. In part (A) are displayed temperatures recorded at a variety of stations. In part (B), small crosses are placed at the estimated points where a value of 70° would be found; these points are located by interpolating among the values that are recorded. In part (C), the isolines of 70° are sketched in smoothly. In part (D), other isolines are interpolated and sketched at 10° intervals. In part (E), shading is added to clarify the pattern.

The basic characteristics of isolines are:

1. They are always closed lines; that is, they have no ends. On any specific

Conversely, winter temperatures are lower over the continents than over the oceans; the isotherms bend equatorward over continents in this season (January in the Northern Hemisphere; July in the Southern). Thus at both seasons, isotherms make greater north-south shifts over land than over water.

Another manifestation of the land/water contrast is the regularity of the isothermal pattern in the midlatitudes of the Southern Hemisphere, in contrast to the situation in the Northern Hemisphere. There is very little land in these Southern Hemisphere latitudes, so that contrasting surface characteristics are absent.

Ocean Currents Some of the most obvious bends in the isotherms occur in near-coastal areas of the oceans where prominent warm or cool currents reinforce the isothermal curves caused by land/water contrasts. Cool currents compel isotherms to be deflected equatorward, whereas warm currents cause a shift toward higher latitudes. Cool currents produce the greatest isothermal bends in the warm season: Note the January situation off the west coast of South America and the southwest coast of Africa or the July conditions off the west coast of North America. Warm currents have their most prominent effects in the cool season: Witness the isothermal pattern in the North Atlantic Ocean in January.

Seasonal Patterns

Apart from the general east-west trend of the isotherms, probably the most conspicuous feature of the two seasonal temperature maps (Figures 4-15 and 4-16) is the latitudinal shift of the isotherms from one map to the other. The isotherms follow the changing balance of insolation during the course of the year, moving northward from January to July and returning southward from July

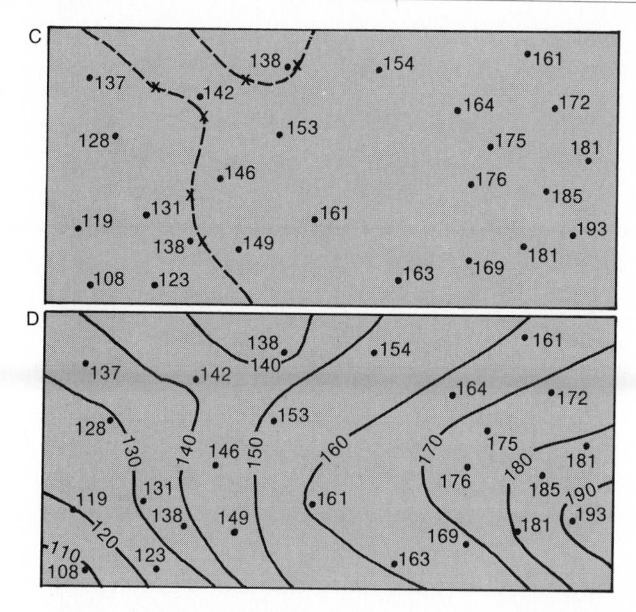

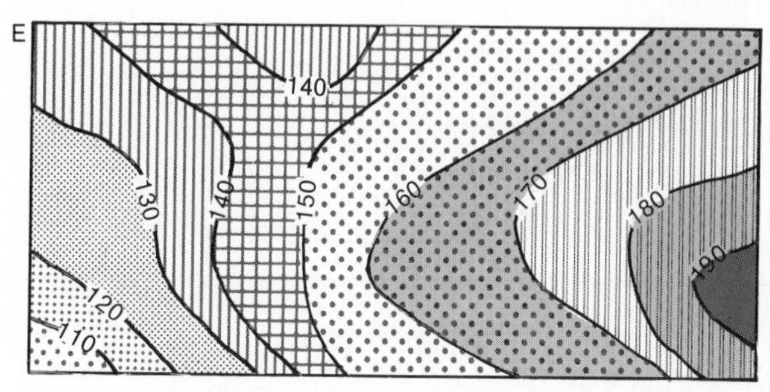

An example of the technique of drawing isolines. (A) The value of each point is determined and recorded. (B) The approximate location of one or more critical isolines (in this case the value of 140) is interpolated and marked by x's. (C) The critical isolines are drawn. (D) The other isolines are interpolated and drawn. (E) Shading is added for clarity.

map, of course, an isoline is likely to extend beyond the edge, so that closure may not appear on that map.

2. They represent gradations in quantity, so they can never touch or cross one another except under rare and unusual circumstances.
3. The numerical difference between one isoline and the next is called the *interval*. Although intervals can be varied according to the wishes of the map-maker, it is normally more useful to maintain a constant isoline interval all over the map.
4. Where isolines are close together, a steep gradient (i.e., a rapid change, horizontally) is indicated, in comparison with an area where isolines are further apart.

Edmund Halley (1656–1742), an English astronomer and cartographer, was not the first person to use isolines, but he published a map in 1700 that apparently was the first time that isolines appeared in print. This map showed isogonic lines in the Atlantic Ocean. Isoline maps are now commonplace and are considered to be very useful to geographers, even though an isoline is an artificial construct; that is, it does not occur in nature. An isoline map, however, can reveal spatial relationships that might otherwise go undetected. Patterns that are too large or too abstract or too detailed for ordinary comprehension are often significantly clarified by the use of isolines.

to January. This isothermal shift, however, is much more pronounced in high latitudes than in low, and over the continents in comparison with the oceans. Thus tropical areas, particularly tropical oceans, show relatively small displacement of the isotherms from January to July, whereas over middle- and high-latitude landmasses, the position of an isotherm may migrate northward or southward more than a thousand miles (1,600 km). See Figure 4-17.

Isotherms are also more tightly packed in the "winter" hemispheres—Northern Hemisphere in January and Southern Hemisphere in July. This indicates that the temperature gradient (rate of temperature change with horizontal distance) is steeper in winter than in summer, which, in turn, reflects the greater contrast in radiation balance in winter. The temperature gradient is also steeper over continents than over oceans.

The coldest places on Earth are seen to be over landmasses in the higher latitudes. During July, the polar region of Antarctica shows clearly as the dominant area of coldness. In January, on the other hand, the coldest temperatures occur many hundreds of miles south of the North Pole, in subarctic portions of Siberia, Canada, and Greenland. The principle of greater cooling of land than water is clearly demonstrated.

The highest temperatures also are found over the continents. The locations of the warmest areas, however, are not equatorial. Rather they are in subtropical latitudes where descending air maintains clear skies most of the time, allowing for almost uninterrupted influx of solar radiation. Frequent cloudiness precludes such a condition in the equatorial zone. Thus the highest July temperatures occur in northern Africa and in the southwestern portions of Asia and North America, whereas the principal areas of January heat are in subtropical parts of Australia, southern Africa, and South America.

The ice-covered portions of the Earth—Antarctica and Greenland—remain quite cold throughout the year.

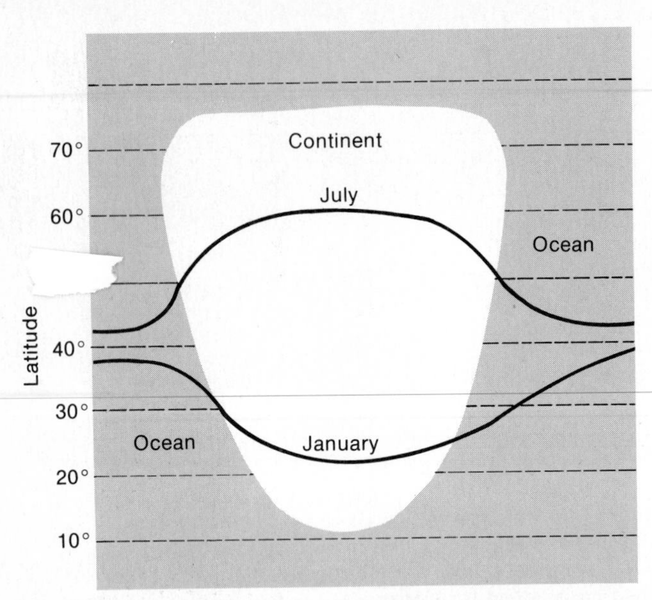

Figure 4-17. Idealized seasonal migration of the 60°F isotherm over a hypothetical Northern Hemisphere continent. The latitudinal shift is greatest over the interior of the continent and least over the adjacent ocean.

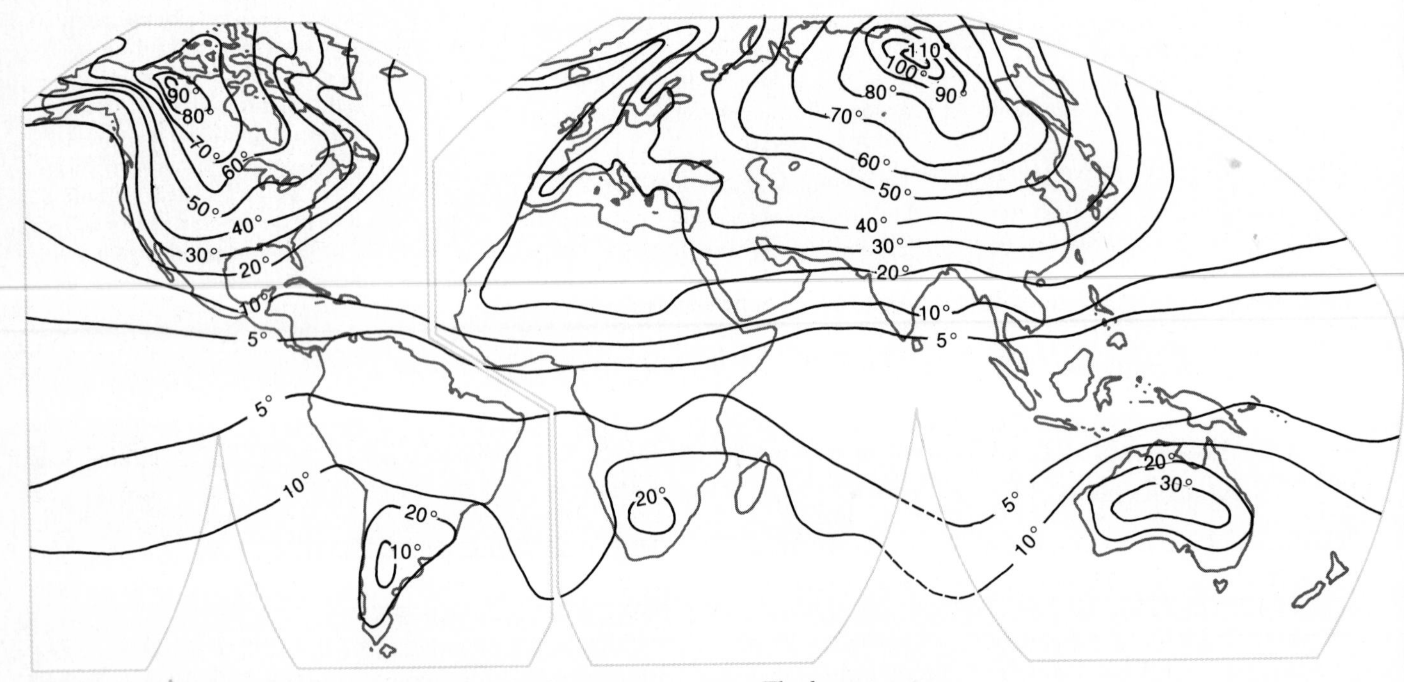

Figure 4-18. The world pattern of average annual temperature range. The largest ranges occur in the interior of high-latitude landmasses.

Annual Temperature Range

Another map that is very useful in helping to understand the global pattern of air temperature is one that portrays the *average annual range of temperatures*; see Figure 4-18. Annual temperature range is the difference in degrees between the average temperatures of the warmest and coldest months (normally July and January). The data are portrayed on the map by isolines that resemble isotherms.

Enormous seasonal variations in temperature occur in the interiors of high-latitude continents, and continental areas in general experience much greater ranges than do equivalent oceanic latitudes. At the other extreme, the average temperature fluctuates only slightly from season to season in the tropics, particularly over tropical oceans.

Temperature is a basic element of weather and climate, the existence of which depends largely on an external source of energy (solar radiation). It is a critical component of the earthly environment and has a fundamental relationship with the climatic elements—pressure and wind—that are discussed in the following chapter.

Review Questions

1. Why is the sun so dominant a source of energy for the Earth?
2. Distinguish between long-wave and shortwave radiation.
3. Explain the concept of the *solar constant*.
4. Explain the difference between *conduction* and *transmission*; between *absorption* and *reflection*.
5. "The atmosphere is mostly heated directly from the Earth rather than the sun." Comment on the validity of this statement.
6. Why do high latitudes receive less insolation than low latitudes?
7. Explain why land heats faster than water.
8. Describe the pattern of flow of major surface ocean currents in a typical ocean.
9. Is the temperature lapse rate of the troposphere steady and constant? Explain.
10. Why are the hottest parts of the Earth *not* in equatorial regions?

Some Useful References

BUDYKO, M. I., *The Heat Balance of the Earth's Surface*. Washington, D.C.: U.S. Department of Commerce, Weather Bureau, 1958.

BOODY, R. M., AND J. C. G. WALKER, *Atmospheres*. Englewood Cliffs, N.J.: Prentice-Hall, Inc., 1972.

LUTGENS, F. K., AND E. J. TARBUCK, *The Atmosphere: An Introduction to Meteorology*. Englewood Cliffs, N.J.: Prentice-Hall, Inc., 1979.

ROBINSON, N., ed., *Solar Radiation*. Amsterdam, N.Y.: American Elsevier Publishing Co., 1966.

5 Atmospheric Pressure and Wind

For most people, atmospheric pressure is the most difficult of the four basic climatic elements to comprehend. The other three—temperature, wind, and moisture—are more readily understood because our bodies are much more sensitive to them. We can "feel" heat, air movement, and humidity, and we are quick to recognize variations in their characteristics. Pressure, on the other hand, is a phenomenon of which we are usually unaware; its variations are considerably less noticeable to our senses. Pressure usually impinges on our sensitivity only when we experience rapid vertical movement, as in an elevator or an airplane. This creates an unpleasant sensation because of the difference in pressure inside and outside of our ears, which is sometimes relieved only by "popping" them.

Despite its inconspicuousness, however, pressure is a very important feature of the atmosphere. It is tied closely to the other weather elements, acting on them and responding to them. Pressure has an intimate relationship with wind; spatial variations in pressure are significantly responsible for air movements. Hence, pressure and wind are often discussed together, as will be done in this chapter.

The Nature of Air Pressure

Because the atmosphere is essentially a mixture of gases, it is composed of gas molecules and behaves like a gas. Gas molecules, unlike those of a solid or a liquid, are not strongly bound to one another. They are in continual motion, frequently colliding with one another and with adjacent surfaces. When a collision occurs, a push is exerted and the molecules rebound like elastic balls. The force of millions and millions of these pushes is called *pressure*. See Figure 5-1.

Relationship of Pressure to Density and Temperature

Atmospheric pressure is closely related to density and temperature. Variations in any one of the three can cause variations in the other two.

Density The atmosphere exerts pressure on every solid or liquid surface that it touches. The pressure is omnidirectional, which means that it is exerted usually in all directions—up, down, sideways, or obliquely. In other words, it is not simply a "weight" from above, but it is just as forceful on the surface regardless of the position of the surface.

The atmosphere is held to the Earth by the force of gravity, which prevents the gaseous molecules from escaping into space. At lower levels in the atmosphere, the molecules are packed more densely together because of greater gravitational pull. See Figure 5-1. Hence there are more molecular collisions and therefore higher pressure at lower levels. At higher elevations, the air is less dense, and there is a corresponding decrease in pressure. At any level in the atmosphere, then, the pressure is equivalent to the "weight" of the air directly above; thus the lower the elevation, the greater the pressure.

At sea level, under average conditions, pressure is calculated to amount to slightly less than 15 pounds per square inch (slightly more than 1 kg per cm^2). This means that every square inch of any exposed surface—animal, vegetable, or mineral—at sea level is subjected to that much pressure. Humans are not sensitive to this ever-present burden of pressure because the fluids and cells of our bodies contain air at the same pressure; in humans, there is an exact balance between the outward pressure and the inward pressure from the atmosphere.

Temperature If air is heated, as we noted in the preceding chapter, the molecules become more agitated and their speed is increased. This produces a greater force

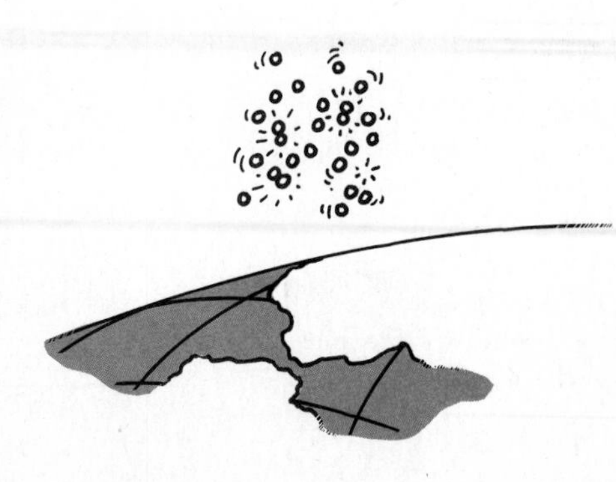

Figure 5-1. In the upper atmosphere, gaseous molecules are far apart and collide infrequently, which produces relatively low pressure. In lower layers, the molecules are closer together, and there are many more collisions, which produces higher pressure.

to their collisions and results in higher pressure. If other conditions remain the same, an increase in temperature will produce an increase in pressure, and a decrease in temperature causes a decrease in pressure.

Complications From the above, we can see that the pressure of a gas is proportional to its density and temperature. This relationship can be explained by several equations that are sometimes collectively referred to as the *gas law*. The cause-and-effect association, however, is complex. For example, when air is heated, it tends to expand, which decreases its density as well as increasing the speed of molecular movement. Thus the increase in temperature may actually be accompanied by a decrease in pressure because of the decrease in density.

The point of all this is that atmospheric pressure is affected by both density and temperature. The relationship among the three variables is intricate. It is important for us to be alert to these linkages, but it is difficult to predict how a change in one will influence the others in a specific instance.

Millibars and Isobars

Air pressure is measured in diverse ways and is expressed in a variety of terms. For our purposes, however, it is logical to concentrate on the system that is standard for weather maps used in the United States and most other countries. An instrument that measures atmospheric pressure is called a *barometer*, and since 1940, the unit of measure of all U.S. weather maps has been the *millibar*.

The millibar is an "absolute" measure of pressure, which means that it has the same dimensions as the quantity it is used to express. It is derived from the physical unit of force called the *dyne*, which is defined as the force needed to accelerate one gram of mass one centimeter per second. Average sea-level pressure is reckoned to be 1013.25 millibars or 101.325 kilopascals.

Weather stations normally record pressure, either continuously or periodically, in millibars. When this information is plotted on weather maps, it is then possible to draw isolines, called *isobars*, of equal pressure. See Figure 5-2. The pattern of the isobars reveals the horizontal

Focus on the Measurement of Pressure

Instruments All instruments that measure atmospheric pressure are *barometers* of some sort. The liquid barometer was first developed in 1643 by Evangelista Torricelli (1608–1647), an Italian physicist and student of Galileo, but the name *barometer* (meaning "pressure measure") was not coined until 1666 by Robert Boyle (1627–1691), an Irish physicist. The instrument is simple, consisting of a tube filled with liquid and inverted so that the mouth is submerged in a container holding more of the same liquid. The level of liquid in the tube then falls until its downward pressure balances the atmospheric pressure acting on the open surface of liquid in the container around the tube. This means that the weight of liquid in the tube equals the weight of a similar-diameter column of air extending from the ground to the top of the atmosphere. When air pressure increases, the height of the liquid in the tube rises; when pressure diminishes, the liquid level declines. A variety of different fluids, including water, can be used in liquid barometers, but by far the most common is mercury.

The other widely used instrument for measuring pressure is the *aneroid* (dry) *barometer*, which is less cumbersome and more portable than the liquid barometer but has the disadvantage of "drifting" from its setting and thus requires frequent recalibration. The aneroid consists of an airtight metal box from which part of the air has been evacuated. The flexible sides of the box respond to variations in air pressure by bulging out when pressure decreases and contracting inward when pressure rises. These changes are registered by a pointer that moves across a scale on the surface of a dial.

A barometer can be connected to a rotating drum to make a *barograph*, which functions similarly to a thermograph in producing a continuous recording of air pressure.

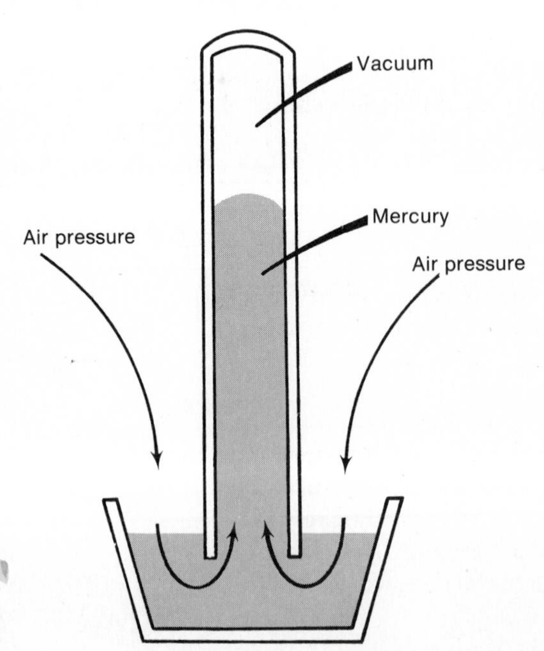

In a mercury barometer, the height of the column of mercury is sustained by the force of the outside atmospheric pressure on the mercury in the open container.

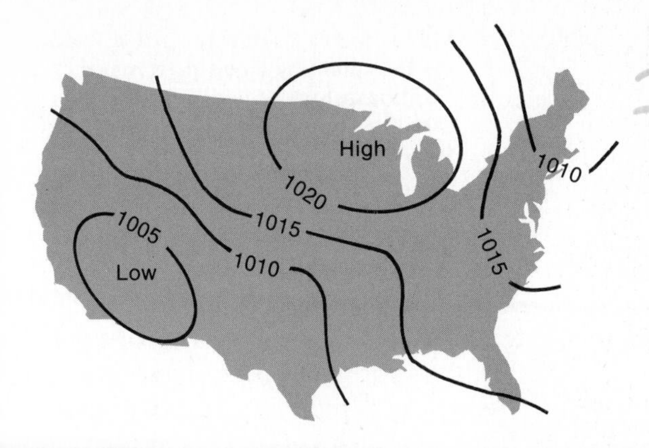

Figure 5-2. Isobars are lines connecting points of equal atmospheric pressure. When they have been sketched on a weather map, it is easy to determine the location of high- and low-pressure centers. This simplified weather map shows pressure in millibars.

distribution of pressure in the region under consideration. Prominent on such maps are roughly circular or oval areas that are characterized by "high" and "low" pressure. These highs and lows represent relative conditions—pressure that is higher or lower than the surrounding areas.

On most maps of air pressure, actual pressure readings are revised to represent a common elevation, usually sea level. This is done because pressure decreases rapidly with altitude, and significant variations in simultaneous pressure readings are likely at different weather stations simply because of differences in elevation. See Figure 5-3. This precaution allows pressure readings from widely separated points to be rendered comparable and not simply the result of varying elevations of the stations.

An important adaptation of the aneroid has been its use as an *altimeter*, an instrument used in aircraft to indicate altitude and by surveyors to determine heights. Because air pressure decreases with height, it is possible to mark an altimeter in feet or meters and then read off the altitude directly from the pressure variation. Altimeters must be recalibrated frequently to compensate for changes in pressure.

Scales Until early in this century, nearly all recordings of atmospheric pressure data were in a scale based on the height of a column of mercury in a liquid barometer. Thus the readings were in terms of *inches of mercury* (millimeters of mercury in most countries), with average sea-level pressure reckoned as 29.92 inches of mercury. This scale of measurement is still sometimes used, despite the incongruity of stating a force in terms of a linear measurement.

The common scale of pressure measurement today is based on an expression of force per unit area, specifically, dynes per square centimeter. (A *dyne* is the force required to accelerate one gram of mass one centimeter per second.) This scale is defined in bars and expressed in millibars. A *bar* represents one million dynes per square centimeter, which is approximately equal to average sea-level pressure, so it is more convenient to record in smaller units. A *millibar* (mb) is one-thousandth part of a bar, or 1,000 dynes per square centimeter. As noted elsewhere, average atmospheric pressure at sea level is 1013.25 mb. See the accompanying table.

Atmospheric scientists are now shifting their emphasis, somewhat slowly, from millibars to the corresponding SI unit: the *pascal*, named for Blaise Pascal (1623–1662), French mathematician and physicist who first demonstrated that atmospheric pressure decreases with height. A *pascal* (abbreviated Pa) is the pressure of one newton per square meter. (A *newton* is the force that must be exerted on a mass of 1 kg in order to accelerate it at a rate of one meter per second per second.) A *kilopascal* (kPa) is equivalent to 10 millibars.

As the metric system is phased into American life, we can anticipate an eventual change from the millibar scale to the pascal scale for expressing measurements of atmospheric pressure. For contemporary general usage, however, the millibar scale is still standard.

The U.S. Standard Atmosphere (normal variation of pressure with altitude)

Height in feet	Height in kilometers	Inches of mercury	Millibars (mb)	Kilopascals (kPa)
100,000	49	0.32	1	0.1
90,000	44	0.50	2	0.2
80,000	39	0.83	3	0.3
70,000	34	1.39	5	0.5
60,000	29	2.17	11	1.1
50,000	24	3.44	25	2.5
40,000	20	5.54	55	5.5
30,000	15	8.88	121	12.1
20,000	10	13.75	265	26.5
10,000	5	20.58	540	54.0
Sea level	0	29.92	1013	101.3

Figure 5-3. A meteorologist at work. He is examining cloud patterns of a Caribbean hurricane as recorded by satellite photographs. [Courtesy NOAA.]

As with other types of isolines, the relative closeness of the isobars indicates the horizontal rate of pressure change, or *gradient*. The gradient can be thought of as representing the "steepness" of the pressure slope, which, in turn, has a direct influence on the speed of the wind.

The Nature of Wind

The atmosphere is virtually always in motion. Air is free to move in any direction, its specific movements being shaped by a variety of factors. Some airflow is lackadaisical and brief; in other cases, it is strong and persistent. Atmospheric motions are often three-dimensional, involving both horizontal and vertical displacement. Small-scale vertical motions, however, are normally referred to as *updrafts* and *downdrafts*; large-scale vertical motions are *ascent* and *subsidence*; the term *wind* is applied only to horizontal movements. Although both vertical and horizontal motions are important in the atmosphere, much more air is involved in the latter than in the former. Wind, then, is horizontal air movement; it has been characterized as "air in a hurry."

Direction of Movement

The ultimate driving force of wind is solar radiation, because the unequal heating of the Earth and the atmosphere produces horizontal differences in air pressure that set winds in motion. Winds represent nature's attempt to balance out the uneven distribution of air pressure over the Earth. The generalization is that air sets out to flow from areas of high pressure to areas of low pressure. Indeed, if the Earth did not rotate and if there were no such thing as friction, that is precisely what would happen—a direct movement from high pressure to low. However, rotation and friction both exist, and they impinge on wind flow significantly. The direction of wind movement is determined principally by the interaction of three factors: pressure gradient, the Coriolis effect, and friction. See Figure 5-4.

Pressure Gradient Pressure gradient is the basic activating force for wind. If there is higher pressure on one side of a parcel of air than on the other, the parcel will be "pushed" away from the higher toward the lower. If one were to visualize a high-pressure area as a pressure "hill" and a low-pressure area as a pressure "valley," it is not difficult to imagine air flowing down this pressure "slope" in the same manner that water would flow down a topographic slope from hill to valley. The pressure gradient force acts at right angles to the isobars in the direction of the lower pressure. If there were no other factors to consider, that is the way the air would move; but such a flow rarely occurs in the atmosphere, so there are clearly other important influences.

The Coriolis Effect We saw in Chapter 4 that any free-moving object is deflected to the right in the Northern Hemisphere and to the left in the Southern Hemisphere as a result of the Earth's rotation. This Coriolis effect has an important influence on the direction of wind flow. The practical result is that the wind would be shifted 90° from its pressure gradient path and would flow parallel to the isobars, *providing* no other factor impinges.

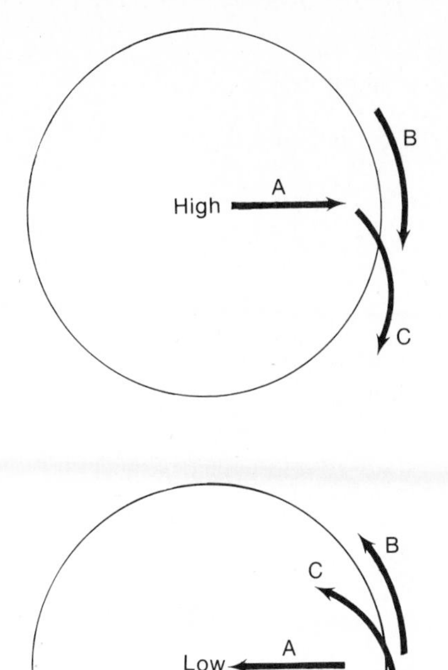

Figure 5-4. The direction of wind flow is determined by the balance among three "forces": (A) The pressure gradient dictates that movement will be perpendicular to the isobars. (B) The Coriolis effect causes movement to parallel the isobars. (C) Friction determines that an intermediate course will be followed, moving across the isobars at an acute angle.

There is an eternal battle, then, between the pressure gradient force and the Coriolis effect. The Coriolis effect keeps the wind from flowing down the pressure gradient (toward lower pressure), and the pressure gradient prevents the Coriolis effect from turning the wind back up the pressure slope (toward higher pressure). Where these two factors are in balance, the wind moves parallel to the isobars and is called a *geostrophic wind*. In actuality, most winds are almost geostrophic; i.e., they flow nearly parallel to the isobars. Only near the ground is there another significant factor to further complicate the situation.

Friction In the lower portion of the troposphere, a third "force" influences wind direction—friction. The frictional drag of the Earth's surface acts both to slow down wind movement and to modify its direction of flow. Instead of blowing directly across the isobars (in response to the pressure gradient force) or parallel to the isobars (due to the Coriolis effect), the wind takes an intermediate course between the two and crosses the isobars at a small angle. As a general rule, the frictional influence is greatest near the Earth's surface and diminishes progressively upward. Thus the angle of wind flow across the isobars is greatest (closest to 90°) at low levels and becomes smaller at increasing elevations. The effect of friction ("the friction layer") extends to only about 5,000 feet (1,500 m) above the ground. Higher than that, most winds follow a geostrophic course, i.e., parallel to the isobars.

Cyclones and Anticyclones

Distinct and predictable wind-flow patterns develop around all high- and low-pressure centers. These patterns are determined by the combined influence of the factors discussed above: pressure gradient, Coriolis effect, and friction in the lower levels of the troposphere; and pressure gradient and Coriolis effect at elevations above the friction layer. Eight possible combinations result—four associated with high-pressure cells and four with low-pressure centers. See Figure 5-5.

Figure 5-5. The eight basic patterns of air circulation around pressure cells.

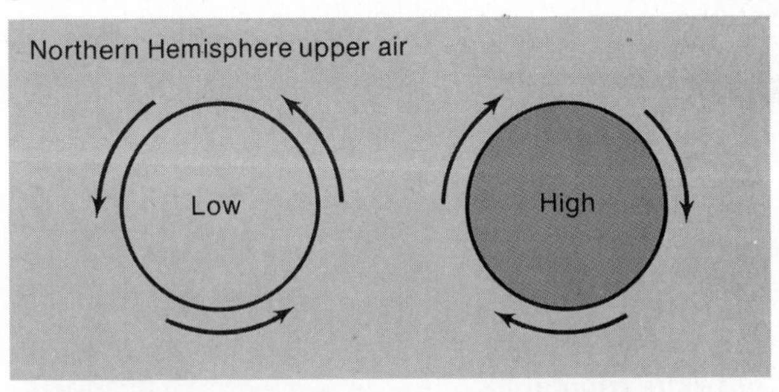

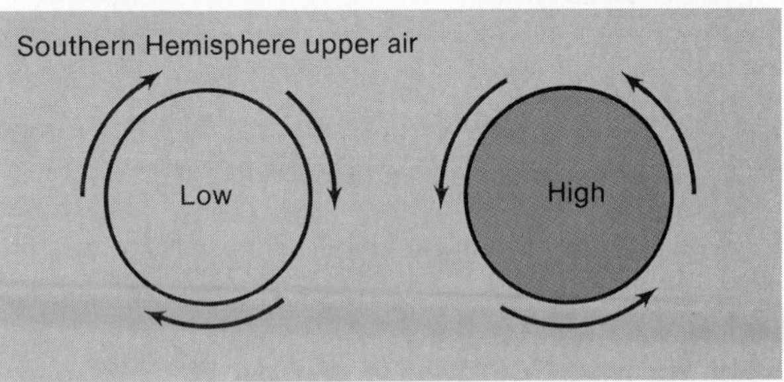

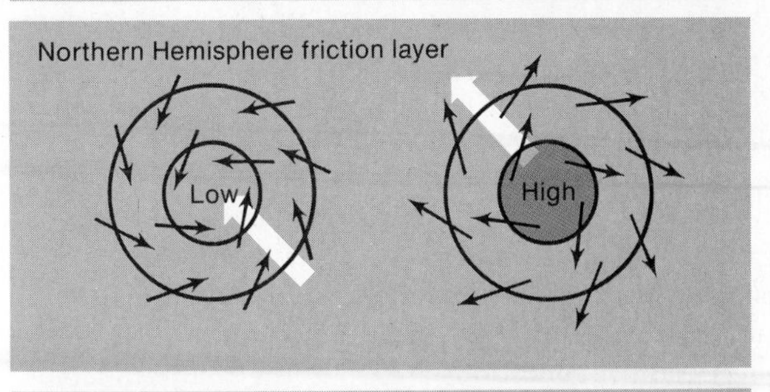

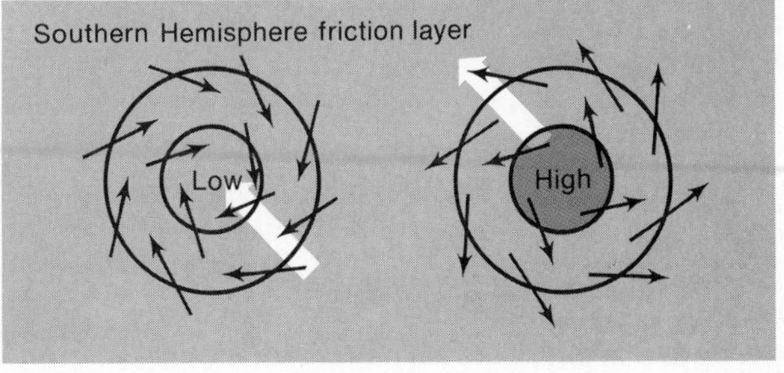

High-Pressure Circulation Patterns A high-pressure center is known as an *anticyclone*, and the flow of air associated with it is *anticyclonic*. The four patterns of anticyclonic circulation are listed here and are shown in Figure 5-5.

1. In the lower troposphere of the Northern Hemisphere, there is a divergent clockwise flow, with the air spiraling out away from the center of the anticyclone.
2. Above the friction layer in the Northern Hemisphere, the winds move parallel to the isobars in clockwise fashion.
3. In the lower troposphere of the Southern Hemisphere, the pattern is a mirror image of example 1 above. The air diverges in a counterclockwise pattern.
4. Above the friction layer in the Southern Hemisphere, there is a counterclockwise movement parallel to the isobars.

Low-Pressure Circulation Patterns Low-pressure centers are called *cyclones*, and the associated wind movement is said to be *cyclonic*. As with anticyclones, Northern Hemisphere cyclonic circulations are mirror images of their Southern Hemisphere counterparts. Again see Figure 5-5.

5. In the lower troposphere of the Northern Hemisphere, a converging counterclockwise flow exists.
6. Above the friction layer in the Northern Hemisphere, air movement parallels the isobars in a counterclockwise direction.
7. In the lower troposphere of the Southern Hemisphere, the winds move inward in a clockwise spiral.
8. In the upper air of the Southern Hemisphere, there is clockwise flow paralleling the isobars.

Diagrams of cyclonic patterns in the lower troposphere—examples 5 and 7 above—may at first glance appear to be misleading because the arrows seem to deny the Coriolis effect. In the Northern Hemisphere case (example 5), the arrows bend to the left, whereas we know that the Coriolis deflection is to the right. The point to remember is that the arrows portray the general flow pattern, but the Coriolis deflection pertains to individual parcels of air. Each individual parcel is deflected to the right of a direct path across the isobars toward the center of the cyclone, but the resultant flow is an inspiral to the left. Similar conditions prevail in the Southern Hemisphere case (example 7), where a leftward Coriolis deflection has a rightward appearance.

A prominent vertical component of air movement also is associated with pressure centers. See Figure 5-6. Air descends in anticyclones and rises in cyclones. Such motions are particularly notable in the lower layers of the troposphere. The anticyclonic pattern can be visualized as upper air sinking down into the center of the high and then diverging near the surface. Opposite conditions prevail in a low-pressure cell, with the air converging horizontally into the cyclone and then rising.

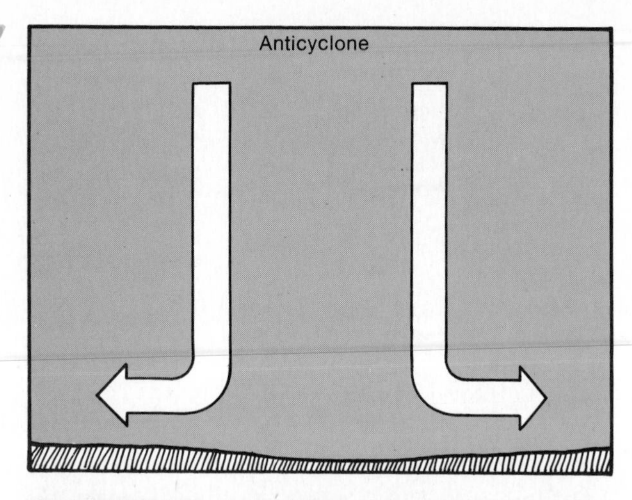

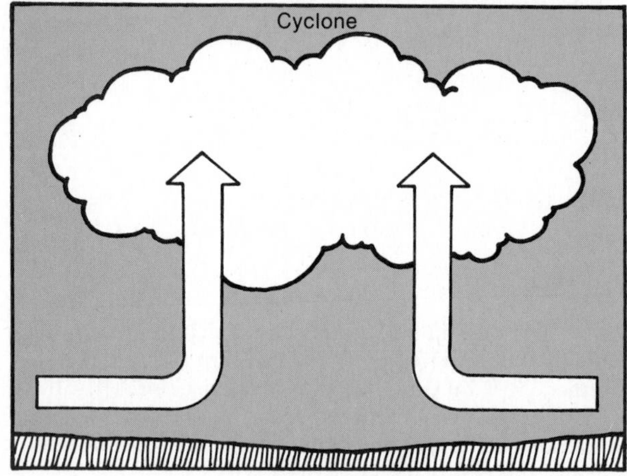

Figure 5-6. In an anticyclone (high-pressure cell), air subsides and diverges; in a cyclone (low-pressure cell), air converges and rises.

Wind Velocity

Thus far, we have been considering direction of wind movement and paying little attention to speed. Although some complications are introduced by centrifugal force

Figure 5-7. The velocity of wind flow is determined by pressure gradient, which is shown on maps by the spacing of isobars. Where isobars are close together, wind speed is high; where isobars are far apart, wind velocity is low.

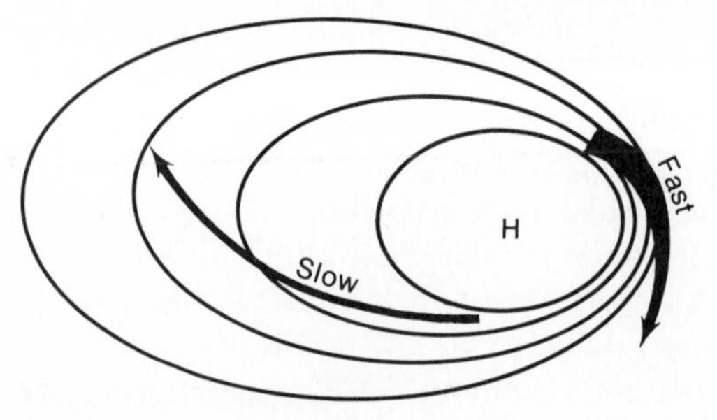

and other factors, it is accurate to say that the velocity of wind flow is determined primarily by the pressure gradient. If the gradient is steep, the air moves swiftly; if the gradient is gentle, movement is slow. This relationship can be portrayed in a simple diagram (Figure 5-7), which represents an asymmetrical high-pressure cell. The closeness of the isobars indicates the steepness of the pressure gradient. Thus we can visualize high-velocity

Focus on Wind Measurement and Terminology

For most purposes, only two dimensions of wind flow are measured and recorded: direction and speed.

Direction is easily determined by a *wind vane*, which is probably the most common of all weather instruments. Although some are much more elaborate, a wind vane normally consists of a structure shaped like an arrow, with a point at one end and a flattened vertical panel at the other. Air movement acts on the panel to align the arrow so that it always points in the direction from which the wind blows. The vane can be attached to a dial, the needle of which gives an indication of wind direction in terms of points of the compass, or azimuthal degrees. The vane can also be connected to a revolving drum or chart, which makes a continuous recording of wind direction.

Speed is normally measured by an *anemometer*, which consists of cups mounted on (usually three) arms extending from a central pivot point. The cups revolve freely in the wind, the velocity of which can be read from a dial much like an automobile speedometer; normally in miles or kilometers per hour, or knots (a knot is a unit of speed, equivalent to one nautical mile—6076 feet—per hour). Like the wind vane, the anemometer can be connected to a revolving drum or chart for continuous recording.

Descriptive terminology of winds is standardized and simple. A wind is named by the direction from which it comes. Thus a wind that blows from north to south is called a *north wind*. Other common terms used to describe wind-related conditions include the following:

—A *windward* location faces the direction from which the wind comes, whereas a *leeward* position faces toward the direction in which the wind is blowing.

—An *onshore* wind is one that blows from water to land; an *offshore* wind blows from land to water.

The enclosed figure is an attempt to clarify these terms. It is a sketch map of Lake Michigan between the states of Wisconsin on the west and Michigan on the east. In this example, a west wind is blowing, creating an offshore wind along the Wisconsin coast, which is the leeward side of Wisconsin; whereas the adjacent waters represent the windward side of Lake Michigan. On the eastern side of the lake an onshore wind is blowing from the leeward side of Lake Michigan onto the windward coast of the state of Michigan.

One other frequently used wind adjective is *prevailing*. Winds blowing from one direction significantly more frequently than from others are called *prevailing winds*.

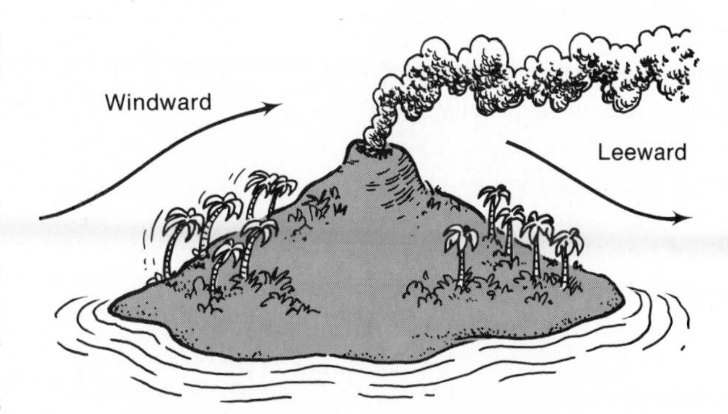

The windward side of anything faces the direction from which the wind comes; the leeward side faces in the direction the wind is going.

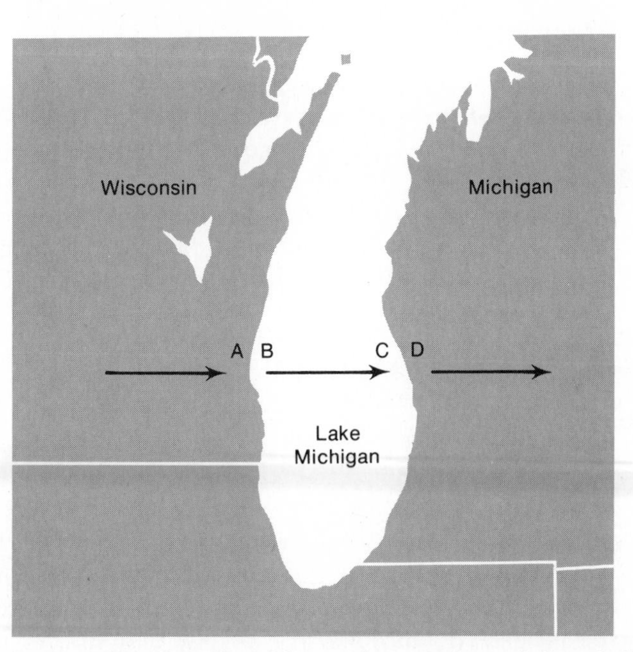

An example of windward/leeward terminology. Arrows indicate direction of wind flow. Winds are "offshore" as regards Wisconsin; "onshore" as regards Michigan. (A) is on the leeward side of Wisconsin; (B) is on the windward side of Lake Michigan; (C) is on the leeward side of Lake Michigan; (D) is on the windward side of Michigan.

winds on the right side of the cell, and gentle zephyrs on the left side.

Vertical Variations in Pressure and Wind

Although most of our attention in this chapter is directed toward an understanding of the spatial distribution of pressure and wind in the horizontal dimension, it is worthwhile to note major features of the vertical patterns as well.

Atmospheric pressure, with only minor localized exceptions, decreases rapidly with height. See Table 5-1. The pressure change is most rapid at lower elevations, the rate diminishing significantly above about 10,000 feet (3 km). It should be noted that prominent surface pressure centers (anticyclones and cyclones) often lean with height, which is to say that they are not absolutely vertical in orientation.

The velocity of wind is quite variable from place to place and from time to time, and usually increases with height. Winds tend to move faster above the friction layer; as a generalization, winds aloft are speedier than winds near the surface. As we will see in subsequent sections, the very strongest tropospheric winds usually are found at intermediate levels in so-called jet streams, or under more localized conditions in violent storms near the Earth's surface.

The global pattern of surface wind flow is very complicated and will be examined in some detail later in this chapter. Winds in the upper portion of the troposphere, however, have a much simpler generalized arrangement. See Figure 5-8. Over the tropical portions of the Earth, between the equator and 20° to 25° of latitude, winds aloft are generally from the east. Everywhere else in the upper troposphere, the winds are generally westerly, except over polar areas where easterly winds again occur.

TABLE 5-1 The Vertical Variation of Pressure with Height (based on the U.S. Standard Atmosphere)

Height (km)	Pressure (mb)	Height (km)	Pressure (mb)
70.0	0.06	8.0	356.5
60.0	0.23	7.0	411.1
50.0	0.78	6.0	472.2
40.0	2.87	5.0	540.4
35.0	5.75	4.0	616.6
30.0	11.97	3.5	657.6
25.0	25.49	3.0	701.2
20.0	55.29	2.5	746.9
18.0	75.65	2.0	795.0
16.0	103.5	1.5	845.6
14.0	141.7	1.0	898.8
12.0	194.0	.5	954.6
10.0	265.0	0	1013.2
9.0	308.0		

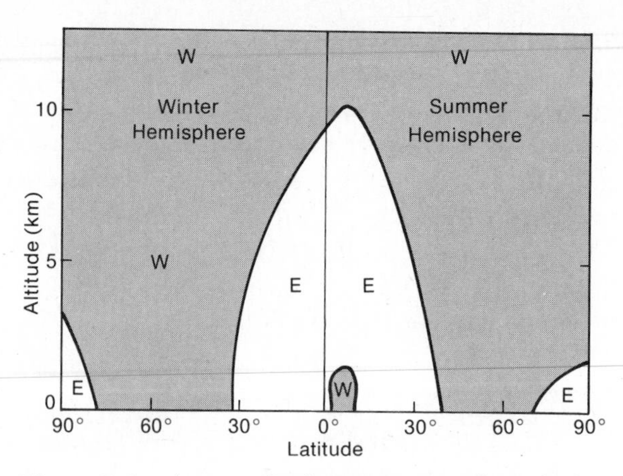

Figure 5-8. A generalized cross section through the troposphere, showing the horizontal component of air movement: "E"—easterly winds; "W"—westerly winds.

The General Circulation of the Atmosphere

The Earth's atmosphere is an extraordinarily dynamic medium. It is constantly in motion, responding to the various "forces" that we have described previously, as well as to a variety of more localized conditions. Some atmospheric motions are broad scale and sweeping; others are minute and momentary. Most important to an understanding of geography is the general pattern of circulation, which involves major semipermanent conditions of both wind and pressure. This circulation is the principal mechanism for both longitudinal and latitudinal heat transfer, and is exceeded only by the global pattern of insolation receipt as a determinant of world climates. See Figure 5-9.

There are seven principal surface components to the general pattern of atmospheric circulation:

Subtropical high pressure
Trade winds
Intertropical convergence zone
The westerlies
Polar high pressure
Polar easterlies
Subpolar low pressure

Subtropical High Pressure

Tropospheric circulation can be thought of as a closed system, with neither a beginning nor an end, so its description can be initiated almost anywhere within the system. It seems logical, however, to begin a discussion

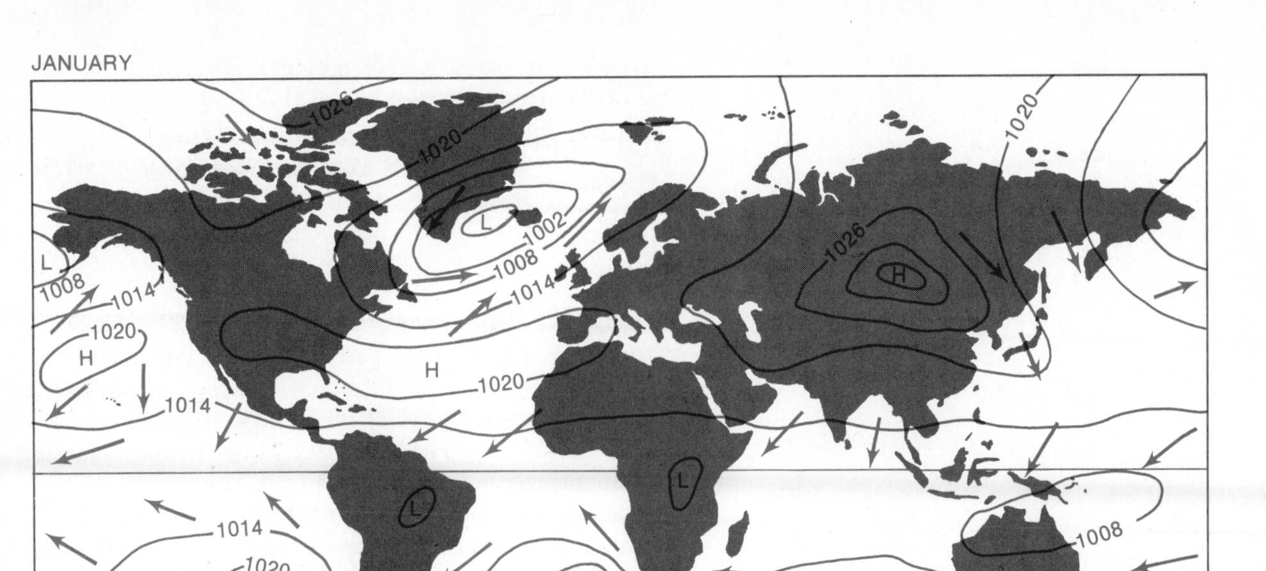

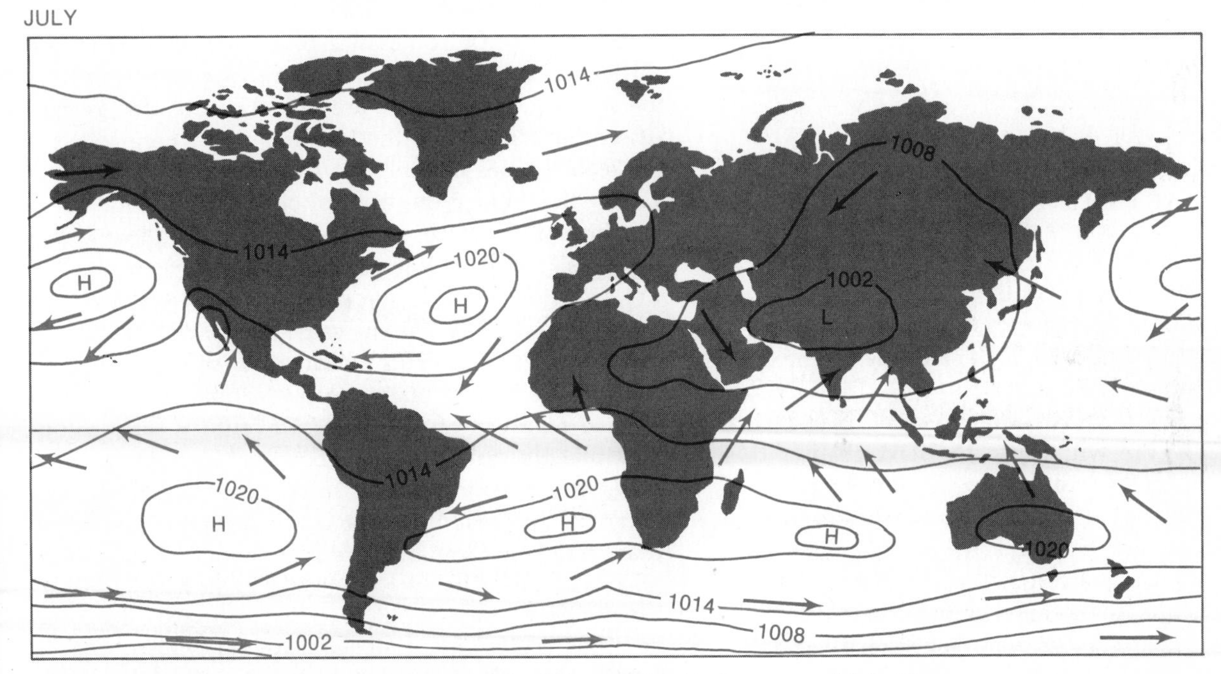

Figure 5-9. Average atmospheric pressure and wind conditions in January and July. Pressure is reduced to sea level and shown in millibars. Arrows indicate generalized surface wind movements.

of atmospheric circulation similarly to that of oceanic circulation, in the subtropical latitudes of the five major ocean basins.

Each of the five basins (North Pacific, South Pacific, North Atlantic, South Atlantic, South Indian) has a large semipermanent high-pressure cell centered at about 30° of latitude. These gigantic anticyclones, with an average diameter of perhaps 2000 miles or 3200 km, usually are elongated east-west and tend to be centered in the eastern portions of their respective basins. Their latitudinal positions vary from time to time, shifting a few degrees poleward in summer, equatorward in winter.

The *subtropical highs* (*STHs*) are so persistent that each has been given a proper name (the Azores STH in the North Atlantic, for example). From a global standpoint, the STHs represent intensified cells of high pressure in two general ridges of high pressure that extend around the world in these latitudes, one in each hemisphere. The high-pressure ridges are significantly broken up over the continents, especially in summer, but the STHs normally

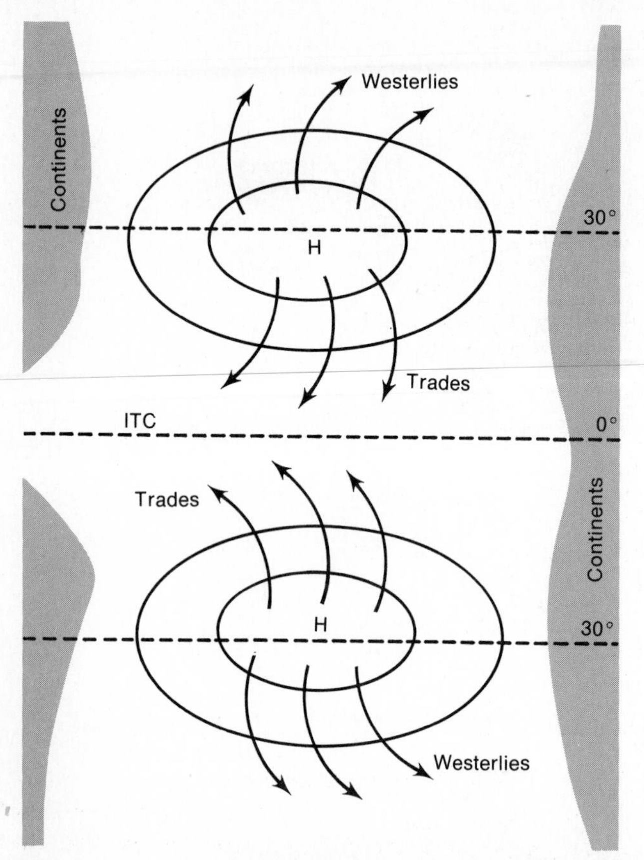

Figure 5-10. Subtropical high-pressure cells dominate the five major oceans in subtropical latitudes. They are centered at 25° to 30° and, in a sense, serve as the source for surface trade winds and westerlies.

persist over the ocean basins throughout the year. In association with these high-pressure conditions, there is a general subsidence of air from higher altitudes in the form of a broad-scale, gentle downdraft.

Within the STHs, the weather is nearly always clear, warm, and calm. We will see in the next chapter that subsiding air is totally inimical to the development of clouds or the production of rain. Instead, these areas are characterized by warm, tropical sunshine and an absence of wind. These regions are sometimes called the "horse latitudes," presumably because sixteenth- and seventeenth-century sailing ships were sometimes becalmed there and their cargos of horses were thrown overboard to conserve drinking water.

The air circulation pattern around the STHs is typically anticyclonic: divergent clockwise in the Northern Hemisphere and divergent counterclockwise in the Southern Hemisphere. In essence the STHs can be thought of as gigantic wind-wheels whirling in the lower troposphere, fed with air sinking down from above, and spinning off winds horizontally in all directions. The winds, however, are not dispersed uniformly around the STHs; they are concentrated on the northern and southern sides. Wind flow is less pronounced on the eastern and western sides of the STHs because of the previously mentioned high-pressure ridge in those latitudes.

Although the global flow of air is essentially a closed circulation from a viewpoint at the Earth's surface, the STHs can be thought of as the source regions for two of the world's three major wind systems: the trade winds and the westerlies.

Trade Winds

Issuing from the equatorward sides of the STHs and diverging toward the west and toward the equator is the major wind system of the tropics—the *trade winds*. These winds cover most of the Earth between about 25° north latitude and 25° south latitude. They are particularly prominent over oceans but tend to be significantly interrupted and modified over landmasses. Because of the vastness of the Earth in tropical latitudes, and the fact that most of this expanse is oceanic, the trade winds dominate more of the globe than any other wind system.

The trade winds are predominantly easterly; that is, they generally flow toward the west. There is also a latitudinal component to their movement, however. In the Northern Hemisphere, they usually come from the northeast (and are sometimes called the *northeast trades*); and south of the equator, they are from the southeast (the *southeast trades*). There are exceptions to this general pattern, especially over the Indian Ocean where westerly winds sometimes prevail, but for the most part there is easterly flow above the tropical oceans.

Indeed, the trade winds are by far the most "reliable" of all winds. They are extremely persistent as to both direction and velocity. They blow most of the time in the same direction at the same speed, day and night, summer and winter. This steadiness is reflected in their name: "trade winds" really means "winds of commerce." Colonial mariners of the sixteenth century early recognized that the quickest and most reliable route for their sailing vessels from Europe to America lay in the belt of northeasterly winds of the southern part of the North Atlantic Ocean. Similarly, the trade winds were used by galleons in the Pacific Ocean, and the name became generally applied to these tropical easterly winds.

The trades originate as warming, drying winds, which are capable of holding an enormous amount of moisture. As they blow across the tropical oceans, they evaporate vast quantities of moisture, and therefore have a tremendous potential for storminess and precipitation. They do not release the moisture, however, unless they are forced to do so by being uplifted by a topographic barrier or some sort of pressure disturbance. See Figure 5-11. Low-lying islands within the trade wind zone often are truly "desert" islands because the moisture-laden winds pass over them without dropping any rain. If there is even a slight topographic irregularity, however, the air that is forced to rise may release abundant precipitation. Some of the wettest places in the world are windward slopes in the trade winds, such as in Hawaii.

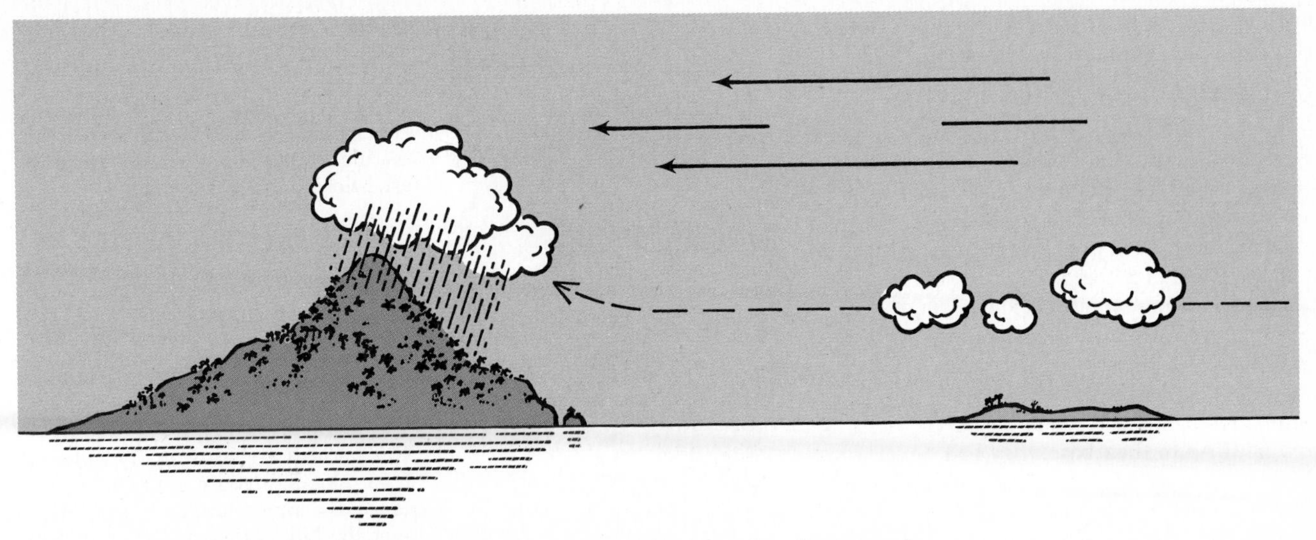

Figure 5-11. Trade winds usually are heavily laden with moisture, but they do not produce clouds and rain unless they are caused to rise. Thus they may blow across a low-lying island with little or no visible effect. A "high" island, on the other hand, causes ascent of the air that may produce heavy rainfall.

Intertropical Convergence

The region between the two sets of trade winds is a zone of convergence and weak horizontal airflow. The northeast trades and the southeast trades come together in the general vicinity of the equator, although the latitudinal position shifts northward and southward following the sun. This zone of meeting of air from the two hemispheres is usually called the *Intertropical Convergence Zone* (ITC), but it is also referred to by such names as equatorial front, intertropical front, and *doldrums* (this last name due to the fact that sailing ships were often becalmed in these latitudes).

From the standpoint of wind flow, the ITC is characterized by weak and erratic air movement. See Figure 5-12. This is a globe-girdling zone of low pressure, associated with instability and rising air. The warm, humid, ascending air produces towering clouds, clammy humidity, and frequent rainfall, often heavy.

The Westerlies

The fourth component of the general atmospheric circulation is represented by the arrows that issue from the poleward sides of the STHs in Figure 5-10. This is the great wind system of the midlatitudes, commonly called the *westerlies*. They flow basically from west to east around the world in the latitudinal zone between about 30° and 60° both north and south of the equator. Because the globe is smaller at these latitudes than in the tropics, the westerlies are less extensive than the trades; nevertheless they cover much of the Earth.

Near the surface, the westerlies are much less constant and persistent than the trades, which is to say that surface winds often do not actually flow from the west but may come from any point of the compass. The winds aloft, however, are very prominently from the west. Moreover, there is a remarkable core of high-velocity winds, called a *jet stream*, that usually occupies a position 30,000 to 40,000 feet (9 to 12 km) high in the westerlies. The belt of the westerlies can be thought of as a meandering "river" of air moving generally from west to east around the world in the midlatitudes, with the jet stream as its fast-moving core, and a slower westerly flow around it. Near the margins of the westerly "river," particularly its bottom (i.e., near the Earth's surface), there are inter-

Figure 5-12. The intertropical convergence zone is characterized by instability, vertical air movement, towering cumuliform clouds, and considerable rainfall.

During the last months of World War II, the first land-based American bombing raid over Japan encountered an atmospheric phenomenon never before met by man. As the west-bound bomber squadron leveled out over Tokyo targets at an air speed of 250 mph (400 km/hr), it experienced headwinds of equivalent velocity, which meant that its ground speed had been reduced to zero. This made the bombers extremely vulnerable to antiaircraft fire, and when the survivors returned to their base, they had some choice words for the chief meteorologist because he had not warned them to expect prodigious upper air winds. Such an event was beyond anticipation, however. That bombing mission was history-making not only as an air strike but also as the first human experience with a jet stream.

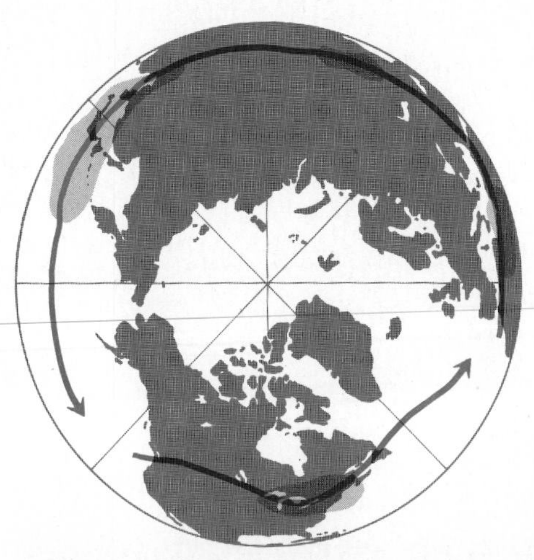

The average position of the Northern Hemisphere westerly jet is shown by the black line. The usual areas of highest velocity winds are shaded.

The westerly jet occasionally generates linear clouds that are conspicuous evidence of its presence, as in this case over the Nile Valley and the Red Sea. [*Gemini XII* photo, courtesy NASA.]

A *jet stream* is a high-level, high-velocity flow of air. It is officially defined as a "strong narrow current, concentrated along a quasi-horizontal axis in the upper troposphere or in the stratosphere, characterized by strong vertical and lateral wind shears and featuring one or more velocity maxima." A jet stream is a discontinuous flow; that is, it does not extend completely around the world. Typically, its length is measured in thousands of miles (or kilometers), its width is measured in hundreds of miles (or kilometers), and its depth is only a few miles (or kilometers). Its speed is variable; the maximum recorded is about 300 mph (500 km/hr), and the defined minimum is 50 knots (57.5 miles or 92.5 km) per hour.

As mentioned above, the first jet stream was discovered at an elevation of about 30,000 feet (9000 m) in the midlatitudes of the Northern Hemisphere. However, the strong possibility of high velocity westerlies at high levels had been predicted since the early years of this century, and details of a westerly jet had been modelled theoretically (although not recognized) by Scandinavian meteorologists in the 1930s.

It is now known that there are several jet streams in the atmosphere, all apparently located in the upper troposphere or lower stratosphere. By far the most important are the two so-called polar front jets, which function as the core of the westerly flow. In the Northern Hemisphere, this westerly jet is most persistent over eastern Asia, North America, and Europe. Its Southern Hemisphere counterpart is less well understood primarily because it occurs mostly over the ocean where upper-air weather records are sparse.

Two subtropical jets are found at about 30° of latitude (roughly above the center of the STHs) and at somewhat higher elevations than the midlatitude jets. They consist of high-velocity westerly winds, which are more strongly developed in winter than summer. Their influence on world weather and climate is not at all clear.

There is spotty evidence of other jet streams in the atmosphere, at higher latitudes and at higher altitudes. The picture is as yet murky, however; more observations are needed for clarification.

What is clear is that the jet streams of the middle latitudes are extremely influential on the weather and climate of that part of the world, although all the ramifications are not understood. Other jets may or may not be significant to an understanding of climatic patterns. Time will tell.

ruptions and modifications of the westerly flow, which can be likened to eddies and countercurrents in a river of water. These interruptions are caused by surface friction, topographic barriers, and especially by migratory pressure systems, which produce airflow that is not westerly.

This jet stream core of the westerlies shifts its position with some frequency, which has considerable influence on the trajectory of general airflow in the midlatitudes. Although the basic direction of movement is west to east, frequently undulations develop that produce a meandering path that wanders widely to north and south. The undulations are very large and are generally referred to as *long waves* or *Rossby waves* (after the Chicago meteorologist C. G. Rossby, who first explained their nature). See Figure 5-13. At any given time there are usually from three to six Rossby waves in the westerlies of each hemisphere. These waves can be thought of as separating cold polar air from warmer tropical air. When the jet stream path is more directly west-east, there is a "zonal" nature to the weather, with cold air poleward of warm air. However, when the jet stream begins to oscillate and the Rossby waves develop significant amplitude (which means a prominent north-south component of movement, or "meridional" flow), cold air is brought equatorward and warm air moves poleward, bringing frequent and severe weather changes to the midlatitudes.

All things considered, no other portion of the Earth experiences such short-run variability of weather as the midlatitudes. These variations are not caused by the westerlies themselves but by the Rossby waves and by the migratory pressure systems and storms that are associated with westerly flow, and which will be discussed in a subsequent chapter.

Figure 5-13. Rossby waves are major undulations in the general flow (particularly the upper-air flow) of the westerlies. When there are few waves and their amplitude (north-south component of movement) is small, cold air usually remains poleward of warm air, as shown in diagram #1. When the waves have greater amplitude (as in diagram #3), cold air pushes equatorward, and warm air moves in a poleward direction.

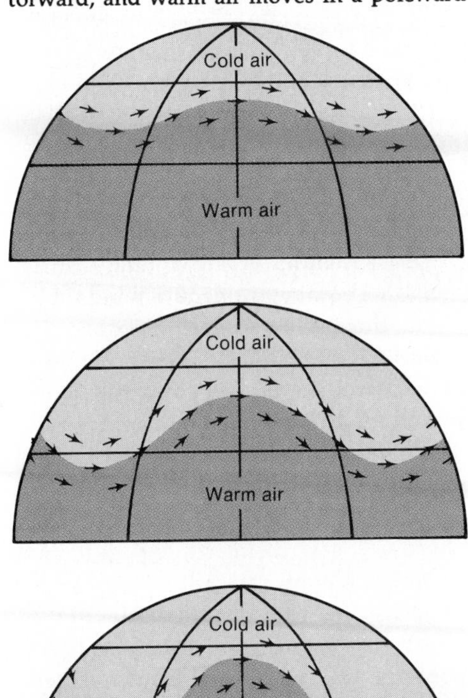

Polar High Pressure

Situated over both polar regions are high-pressure cells, called *polar highs*. The Antarctic high is strong, persistent, and almost a permanent feature above the Antarctic continent. The Arctic high is much less pronounced and is more transitory in nature, particularly in winter. It tends to form over northern continental areas rather than over the Arctic Ocean. Air movement in association with these cells is typically anticyclonic. Air from above sinks down into the high and diverges horizontally near the surface, clockwise in the Northern Hemisphere and counterclockwise in the Southern Hemisphere.

Polar Easterlies

The third broad-scale global wind system occupies most of the area between the polar highs and about 60° of latitude. The winds move generally from east to west, and are called the *polar easterlies*. They are typically cold and dry but quite variable.

Subpolar Low-Pressure Zone

The final surface component of the general pattern of atmospheric circulation is a zone of low pressure at about 50° to 60° of latitude in both Northern and Southern hemispheres. It is commonly called the *subpolar low*, and often contains the *polar front*. It is a meeting ground and zone of conflict between the cold winds of the polar easterlies and the warmer westerlies. The subpolar low of the Southern Hemisphere is virtually continuous over the uniform ocean surface of the cold seas surrounding Antarctica. In the Northern Hemisphere, however, the low-pressure zone is discontinuous, being interrupted by the large continental landmasses. It is much more prominent in winter than in summer and is best developed over the northernmost reaches of the Pacific and Atlantic oceans.

This polar front area is a convergent battleground between easterly and westerly wind systems. It is characterized by rising air, widespread cloudiness, precipitation, and generally unsettled or stormy weather conditions. Many of the migratory storms that travel with the westerlies have their origin in the conflict zone of the polar front.

Modifications of the General Circulation

The seven components described above comprise the generalized pattern of global wind and pressure systems. See Figure 5-14. There are, of course, many variations to the pattern. For a proper comprehension of actual conditions, it is necessary to discuss three important modifications of the generalized scheme.

Seasonal Variations in Location

The major systems previously described experience latitudinal shifts in their locations with the changing seasons. When sunlight is concentrated in the Northern Hemisphere (Northern Hemisphere summer), all the components are displaced northward; at the opposite season (Southern Hemisphere summer), everything is shifted southward. The displacement is greatest in the low latitudes and least in the polar regions. The intertropical convergence zone, for example, can be found as much as 20° north of the equator in July and 10° south of the equator in January. The polar highs, on the other hand, experience little or no latitudinal displacement from season to season.

Monsoons

By far the most significant disturbance of the pattern of general circulation is the development of monsoons in certain parts of the world, particularly southern and eastern Asia. The word *monsoon* is derived from the Arabic (*mausim*: "season") and has come to mean a seasonal reversal of winds, a general onshore movement in summer and a general offshore flow in winter. Associated with the wind pattern is a very distinctive seasonal precipitation regime—heavy summer rains derived from the moist maritime air of the onshore flow and a pronounced winter dry season when continental air moving seaward dominates the circulation. Where monsoonal systems prevail, the summer rains are absolutely vital to agricultural success and therefore to human survival.

It would be convenient to explain the development of monsoonal circulation on the basis of the unequal heating of continents and oceans. A strong thermal (i.e., heat-produced) low pressure generated over a continental landmass in summer would attract oceanic air onshore; similarly, a prominent thermal anticyclone in winter over a continent would produce an offshore circulation. Despite the attractiveness of this reasoning, it is insufficient to explain what actually occurs. Most of the world's monsoons occur in tropical or subtropical areas where the winter is not severe enough to generate a thermal anticyclone, and there are also other problems with the simplistic explanation suggested above.

Monsoon winds essentially represent unusually large latitudinal migrations of normal trade winds and westerly flow. The explanation for such extensive migrations, however, is not clear, and we are left with the realization that the actual origin of monsoons is still not understood, although there is increasing evidence that it is associated with upper-air phenomena, particularly jet stream behavior.

The characteristics of monsoons, on the other hand, are well known, and it is possible to describe the monsoonal patterns with some precision. There are two major monsoonal systems (in South Asia and East Asia), two minor monsoons (in Australia and West Africa), and several other regions where monsoonal tendencies develop (especially in Central America and southeastern United States).

Figure 5-14. The general circulation of the atmosphere, disregarding the effect of major landmasses.

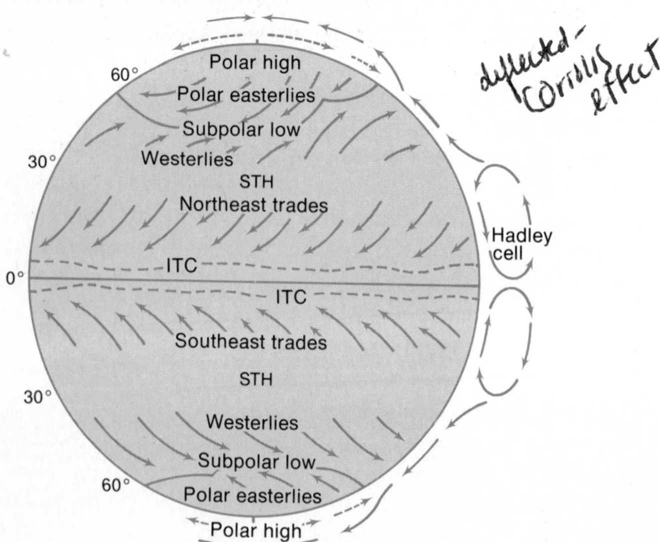

South Asian Monsoon The most notable environmental event each year in India is the annual "burst" of the summer monsoon. Prominent onshore winds spiral into South Asia from the Indian Ocean, bringing life-giving rains to the parched subcontinent. See Figure 5-15. The winds approach southern Asia from the southwest and can be thought of as Southern Hemisphere trade winds that have been deflected to the right by the Coriolis effect after crossing the equator. They tend to converge in counterclockwise fashion toward a low-pressure area situated over northern Pakistan, bringing high humidity and relatively abundant showery rainfall.

In winter, the Indian subcontinent is dominated by outblowing dry air diverging generally from the northeast. This flow is not very different from normal northeast trades except for its low humidity content.

East Asian Monsoon Winter is the more prominent season in the East Asian monsoonal system, which primarily affects China, Korea, and Japan. See Figure

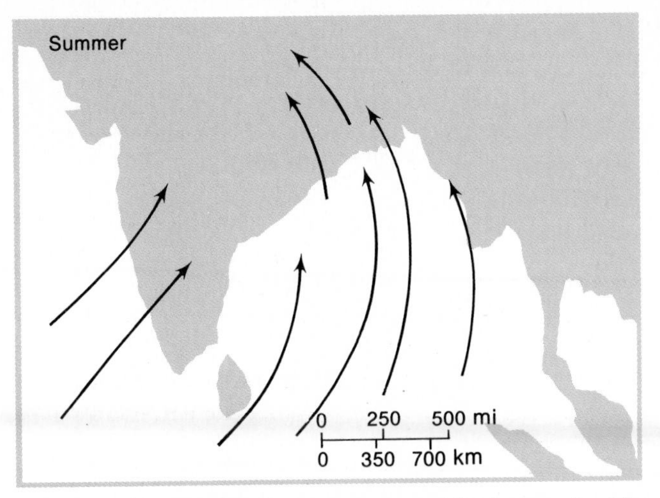

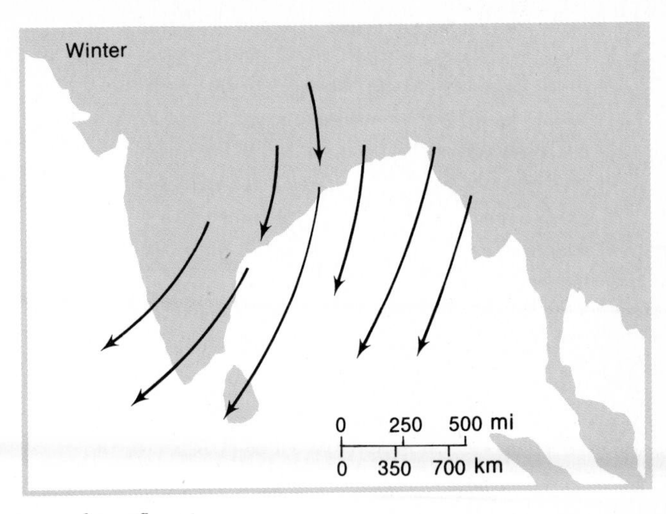

Figure 5-15. The South Asian monsoon is characterized by a strong onshore flow in summer and a somewhat less pronounced offshore flow in winter.

5-16. A strong outflow of dry continental air, largely from the northwest, is associated with anticyclonic circulation around a massive thermal high-pressure cell over western China. The onshore flow of maritime air in summer is not as notable as that in the South Asian monsoon, but it does bring southerly and southeasterly winds, as well as considerable moisture, to the region.

Although the two Asian monsoonal circulations overlap in the southeastern portion of the continent (Vietnam, Thailand, etc.), the systems are essentially separate. Apparently the massive Himalaya Mountains serve as an effective barrier between the two.

Australian Monsoon The northern quarter of the Australian continent experiences a distinct monsoonal circulation. See Figure 5-17. Moist, northerly, onshore air moves southward during the height of the Australian summer (December through March) and is replaced by dry, southerly, offshore flow during most of the rest of the year. The islands north and northwest of Australia (New Guinea, Indonesia) are also affected by this monsoonal pattern. It is likely that there is some interconnection between the Australian and East Asian monsoonal systems, although clear-cut evidence is yet lacking.

West African Monsoon The so-called Guinea Coast (the south-facing portion of western Africa) is dominated by a monsoonal circulation within about 400 miles (650 km) of the coastline. Moist oceanic air flows onshore from the south and southwest during summer, and dry northerly continental flow prevails in winter. See Figure 5-18.

Vertical Components of the General Circulation

Although the troposphere is a zone of considerable turbulence and vertical mixing, most of the broad-scale air movements that comprise the general circulation are horizontal. Important vertical components of the pattern are associated with the semipermanent belts and centers of high and low pressure, with air sinking into the highs and rising out of the lows.

Apparently only the tropical regions have a complete vertical circulation cell; such cells have been postu-

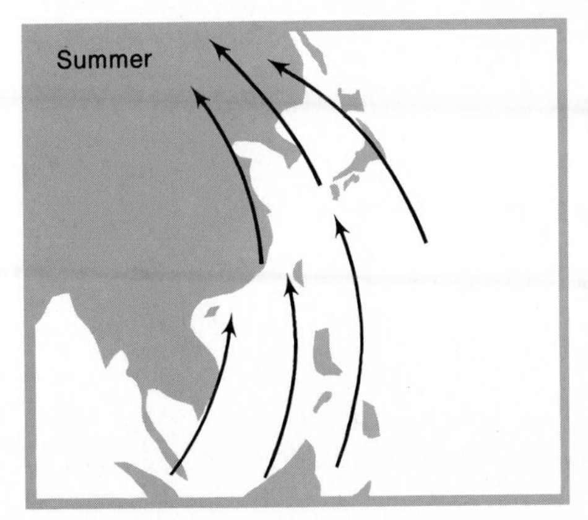

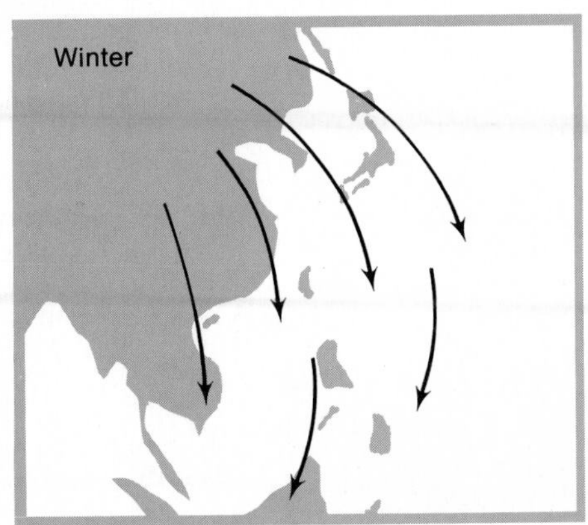

Figure 5-16. In East Asia, the outblowing winter monsoon is stronger than the onshore flow of the summer monsoon.

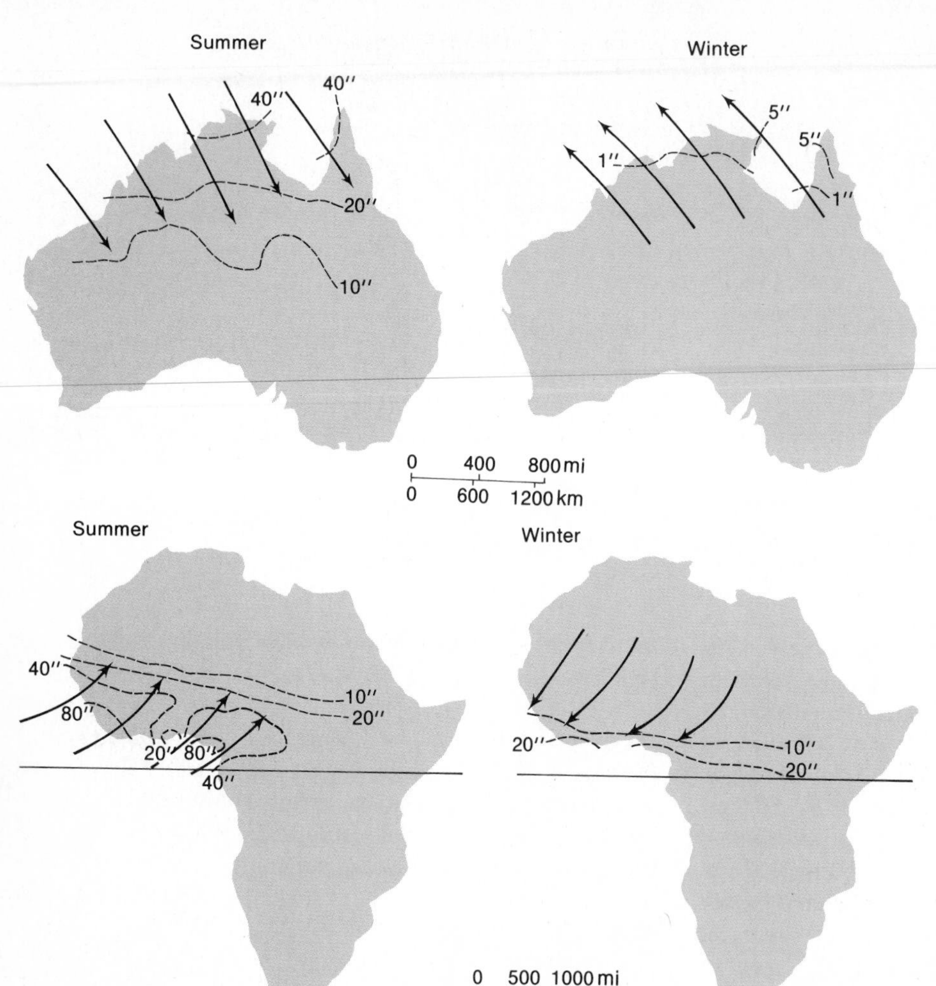

Figure 5-17. The monsoonal pattern in Australia. Northwesterly summer winds bring the wet season to northern Australia; dry southeasterly flow dominates the rest of the year. Three-month seasonal isohyets are shown.

Figure 5-18. The West African monsoon. Summer flow is from the southwest; winter winds are from the northeast. Three-monthly isohyets are shown.

lated for the middle and high latitudes, but observations of actual air movements indicate that the cells either do not exist or are weakly and sporadically developed. The low latitude cells—one north and one south of the equator—can be thought of as gigantic convection systems. The trade winds move toward the west and toward the equator, converging at the ITC zone. There much of the warm air rises to great heights, as in a convective updraft. In the upper troposphere, a poleward flow begins, moving both north and south. In the vicinity of 25° to 30° of latitude, much of the air subsides into the STHs and the general ridge of high pressure that circles the globe there. These two prominent tropical circulations are called *Hadley cells*, after George Hadley (1685–1768), English meteorologist who first conceived the idea of enormous convective circulation cells in 1735. See Figure 5-19.

Localized Wind Systems

All winds originate from the same basic sequence of events: Unequal heating of different parts of the Earth's surface brings about temperature differences, which generate pressure gradients. The immediately preceding sections dealt only with the broad-scale wind systems that comprise the global circulation and are major influences on the climatic pattern of the world. There are many kinds of lesser winds, however, that are of considerable significance to weather and climate at a more localized scale. Such winds are the result of local pressure gradients that develop in response to topographic configurations in the immediate area, sometimes in conjunction with broad-scale circulation conditions.

Figure 5-19. Clear-cut cells of vertical circulation occur within tropical latitudes; they are called *Hadley cells.* The equatorial air rises to 40,000 or 50,000 feet (12,000 to 15,000 m) before spreading poleward.

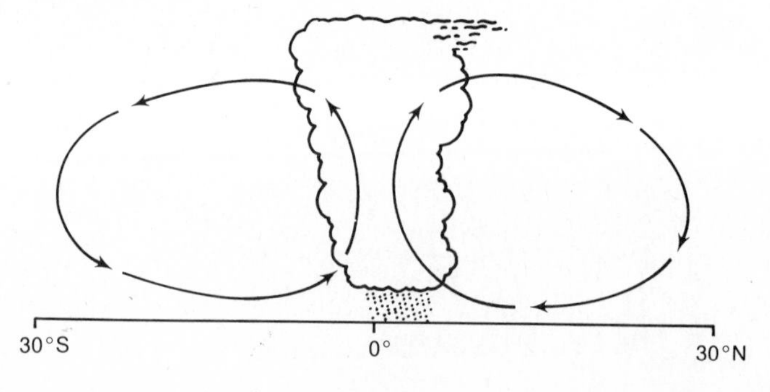

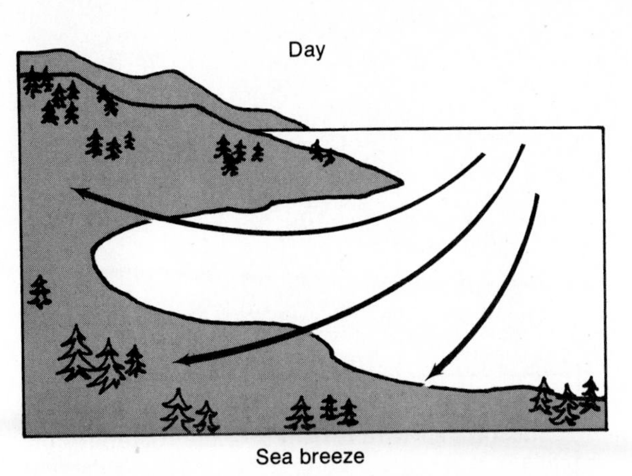

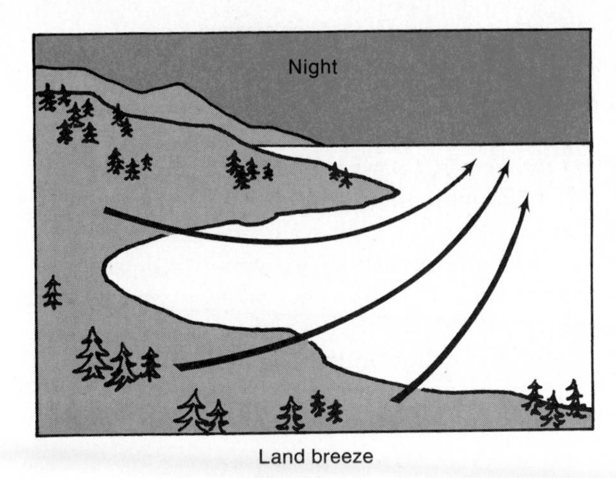

Figure 5-20. In a typical sea/land breeze cycle, diurnal heating over the land produces relatively low pressure, which attracts an onshore flow of air from the sea; whereas nocturnal cooling over the land yields relatively high pressure, which creates an offshore flow of air.

Sea and Land Breezes

A very common local wind system along tropical coastlines and to a lesser extent during the summer in midlatitude coastal areas is the cycle of *sea breezes* during the day and *land breezes* at night. See Figure 5-20. This is essentially a convectional circulation that is based on the differential heating of land and water surfaces. The land warms up rapidly during the day, heating the air above by conduction and reradiation. This causes the air to expand and rise, creating low pressure that attracts surface breezes from over the adjacent water body. The onshore flow is relatively cool and moist, and has the effect of holding down daytime temperatures in the coastal zone, as well as providing moisture for possible afternoon showers. Sea breezes are sometimes strong, but they rarely are influential for more than 10 to 20 miles (16 to 32 km) inland. They often provide welcome relief to the immediate seashore or lakeshore zone from summer or tropical heat.

The reverse flow at night is normally considerably weaker than the daytime wind. The land and the air above it cool more quickly than the adjacent water body, producing relatively higher pressure over land. Thus air flows offshore in a *land breeze*.

Valley and Mountain Breezes

Another notable diurnal (daily) cycle of airflow is characteristic of many hill and mountain areas. During the day, greater heating of air near the slopes (particularly slopes that receive sunlight at a more direct angle of incidence) than at comparable elevations away from the slopes (as over the adjacent valley floor) produces active convectional ascent of air up the topographic slope in a relatively warm wind called a *valley breeze*. See Figure 5-21. The rising air often causes clouds to form around the peaks, and afternoon showers may be common in the high coun-

Figure 5-21. Daytime heating along mountain slopes engenders convective activity and an upslope flow of valley breezes. The slopes experience radiational cooling at night, which chills the adjacent air and causes it to move downslope as mountain breezes.

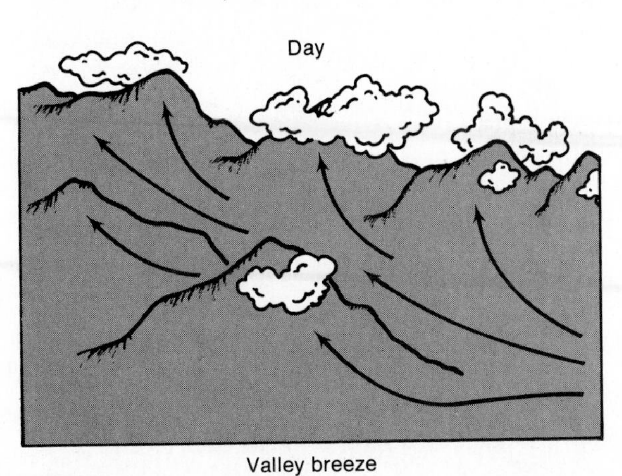

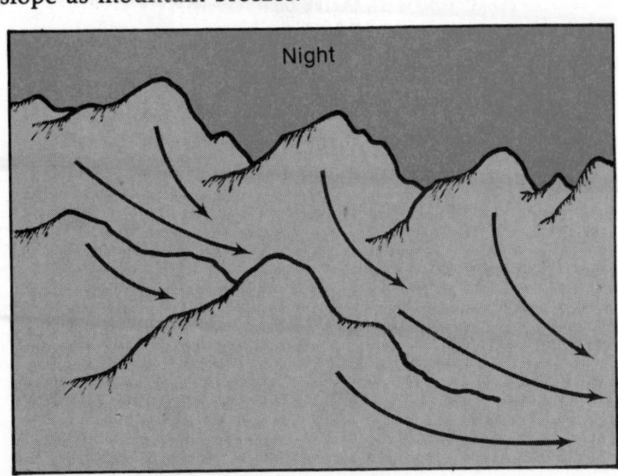

try as a result. After dark, the pattern is reversed. The mountain slopes lose heat rapidly through radiation, which chills the adjacent air, causing it to slip downslope as a *mountain breeze*.

Valley breezes are particularly prominent in summer when solar heating is most intense. Mountain breezes, on the other hand, are often weakly developed in summer and are likely to be more prominent in winter. Indeed, a frequent winter phenomenon in areas of even gentle slope is *air drainage*, which is simply the nighttime sliding of cold air downslope to collect in the lowest spots; this is a modified form of mountain breeze.

Katabatic Winds

Related to simple air drainage is the more general and more powerful spilling of air downslope in the form of *katabatic winds*. These winds originate in cold upland areas and cascade toward lower elevations under the influence of gravity; they are sometimes referred to as *gravity-flow winds*. The air is dense and cold, and although warmed adiabatically as it descends, it is usually colder than the air it displaces in its downslope flow.

Katabatic winds are particularly common in Greenland and Antarctica, especially where they come whipping off the edge of the high, cold ice sheets. Sometimes a cold katabatic wind will become channeled through a narrow valley where it may develop high velocity and considerable destructive power. An infamous example of this phenomenon is the *mistral*, which sometimes surges down France's Rhone Valley from the Alps to the Mediterranean Sea. Similar winds are called *bora* in Yugoslavia and *taku* in southeastern Alaska.

Foehn/Chinook Winds

Another variety of localized downslope wind is sufficiently widespread and distinctive to warrant discussion. It does not have a generic name and is simply referred to by the local name in its area of occurrence: *foehn* (pronounced as in "fern" but with a silent "r") in the Alps and *chinook* in the Rocky Mountains area. See Figure 5-22. These winds develop only when certain pressure conditions prevail, normally a significant low-pressure trough on the eastern side of the mountains that pulls air over the range and down the leeward slope. Under these conditions, the air is dry and relatively warm, having not only lost its moisture through precipitation on the windward side of the range but also having retained a relatively high temperature by the release of latent heat in the condensation process. As the wind blows down the leeward slope, it is further warmed adiabatically, so it arrives at the base of the range as a warming, drying wind. It can produce a remarkable rise of temperature leeward of the mountains in just a few minutes. It is known along the Rocky Mountains front as a "snow-eater" because it not only melts the snow rapidly but quickly dries the resulting mud.

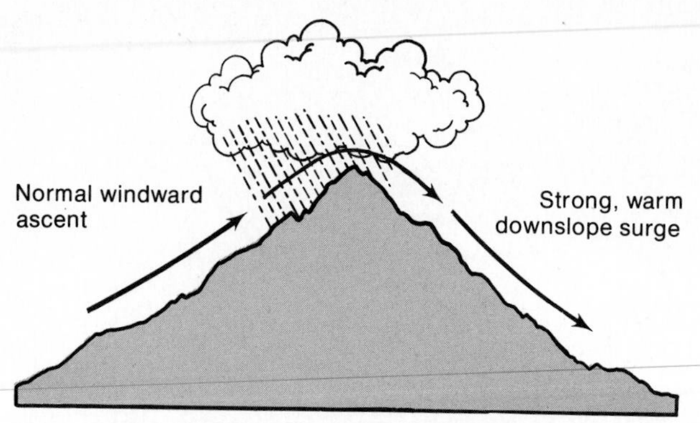

Figure 5-22. A chinook wind is a rapid downslope movement of warming air associated with certain pressure conditions on the leeward side of a mountain range.

Wind, activated by pressure, is a very active component of the atmosphere. Even more dynamic, however, is moisture, which evaporates and condenses, freezes and thaws, rises and falls, and produces the most tangible aspects of the atmosphere—clouds, rain, snow, sleet, and hail. The complexities of atmospheric moisture will be introduced in the next chapter.

Review Questions

1. Explain the relationship of atmospheric pressure to density and temperature.
2. What is the relationship of *millibars* to *isobars*?
3. Pressure gradient is a pervasive force in the atmosphere. Why do winds *not* simply flow down the pressure gradient?
4. What is the relationship of the *Coriolis effect* to *geostrophic winds*?
5. Describe the four patterns of anticyclonic flow that occur in the atmosphere.
6. Does atmospheric pressure always decrease with increasing height? Explain.
7. Discuss the location and persistence of subtropical high-pressure cells.
8. What are the principal differences between *trade winds* and *westerlies*?
9. What are the principal factors that interfere with the "normal" pattern of general atmospheric circulation?
10. What are the principal characteristics of monsoonal circulations?

Some Useful References

BATTAN, L. J., *Weather*. Englewood Cliffs, N.J.: Prentice-Hall, Inc., 1974.

HARE, F. K., *The Restless Atmosphere*, rev. ed. New York: Harper & Row Publishers, 1961.

HIDY, G. M., *The Winds*. New York: Van Nostrand Reinhold Company, 1967.

RAMAGE, C. S., *Monsoon Meteorology*. New York: Academic Press, Inc., 1971.

REITER, E. R., *Jet Streams*. New York: Doubleday & Co., Inc., 1967.

6 Atmospheric Moisture

The fourth, and final, basic element of weather and climate is moisture. It might seem that this would be a familiar feature since our skin is sensitive to water, and we have immediate sensual recognition of being wet. In actuality, however, most atmospheric moisture does not occur as liquid water but rather as water vapor, which is much less conspicuous and much less familiar.

Water is a unique substance in the earthly environment. One of the most distinctive attributes is that moisture occurs in the atmosphere in all three states in which matter can occur—as a solid, liquid, or gas:

1. The solid state of moisture is represented by solid forms of precipitation—snow, hail, sleet—as well as by tiny ice crystals that combine to form some of the high clouds.
2. The liquid portion is commonplace in our experience; it consists of rain and of the tiny droplets of water that aggregate to form most clouds.
3. The gaseous aspect of moisture is water vapor, which is both the least obvious and the most important of the three states insofar as the dynamics of the atmosphere is concerned.

Water Vapor and the Hydrologic Cycle

Water vapor is a colorless, odorless, tasteless, invisible gas that mixes freely with the other gases of the atmosphere. In terms of quantity, it is a minor constituent of the atmosphere, and its occurrence is quite variable from place to place and from time to time. Essentially, it is restricted to the lower part of the troposphere. More than half of all water vapor is found within a mile (1.6 km) of the Earth's surface, and only a tiny fraction exists above about 4 miles (6.4 km). Even at the lowest levels of the atmosphere, however, the distribution of water vapor is very uneven; it is virtually absent in some places and is sufficiently concentrated in others to comprise as much as 4 percent of the total volume.

This erratic distribution is a reflection of the ease with which moisture can change from one state to another at the pressures and temperatures that are characteristic of the lower troposphere. Moisture can leave the Earth's surface as a gas and return to it as a liquid or a solid. Indeed, there is a continuous interchange of moisture between the Earth and the atmosphere. See Figure 6-1. This unending circulation of our planet's water supply is referred to as the *hydrologic cycle*, and it will be discussed in more detail in Chapter 9.

The essential feature of the cycle is that liquid water (primarily from the oceans) evaporates into the gaseous air above and subsequently is converted again to a liquid

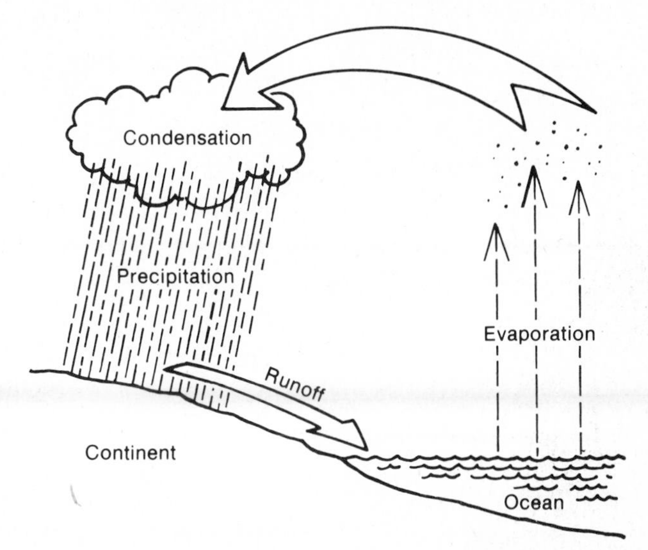

Figure 6-1. The hydrologic cycle is a continuous interchange of moisture between the atmosphere and the Earth.

(or solid) state and returns to the Earth as rain or some other form of falling moisture. The movement of moisture through the cycle is intricately related to many atmospheric phenomena and is an important determinant of climatic characteristics through its role in rainfall distribution and temperature modification.

Any understanding of the role of moisture in the atmosphere is predicated upon a consideration of three processes, processes that are integral to the hydrologic cycle—evaporation, condensation, and precipitation. The first two processes involve not only a change of state for moisture but also a change of availability for heat. Evaporation requires and stores heat, and condensation releases it.

Evaporation

The conversion of moisture from liquid to gas—from water to water vapor—is called *evaporation*. This process involves molecular escape; molecules of water become detached from the liquid surface and escape into the overlying or surrounding air. See Figure 6-2. Water molecules are continually in motion and frequently collide with one another. We noted in a previous chapter that the activity of molecules is speeded up by heat. Higher temperatures cause molecules to move faster and collide more forcefully. The impact of such collisions near the water surface may provide sufficient energy to allow the molecules to break free from the water and enter the air.

The energy absorbed by the escaping molecules is stored, normally to be released subsequently when the vapor changes back to a liquid. The stored energy is called

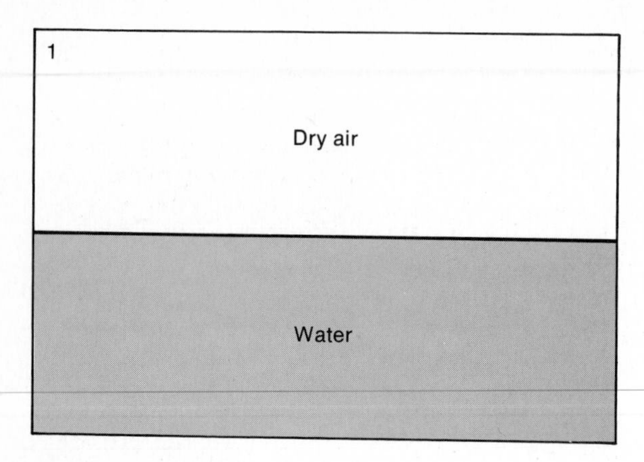

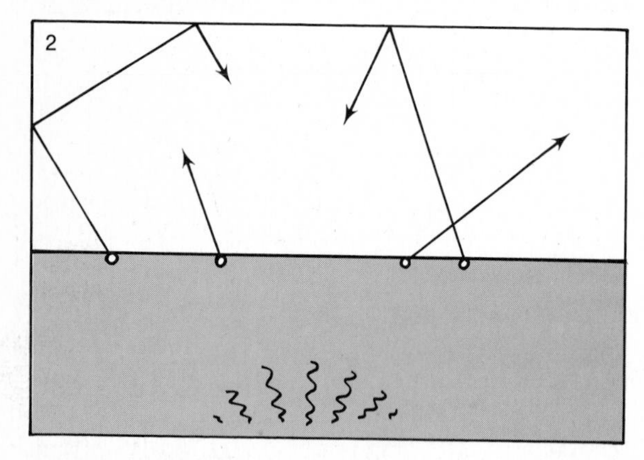

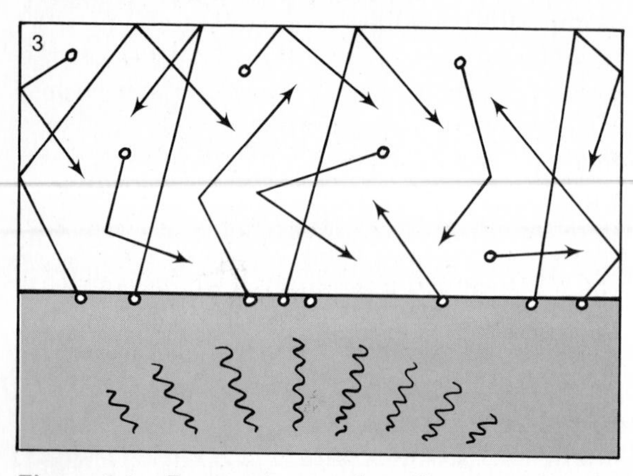

Figure 6-2. Evaporation involves the escape of water molecules from a liquid surface into the air. This can be illustrated schematically by the heating of a container of dry air overlying water (1). As the temperature rises (2), molecules become more agitated, and some break free from the water surface to enter the air. Continued heating increases molecular activity (3) so much that the air may become saturated with water vapor.

latent heat of vaporization. The heat thus removed with the escaping molecules is subtracted from the remaining water, reducing its temperature in an activity popularly referred to as *evaporative cooling*. The effect of evaporative cooling is experienced when a swimmer leaves a swimming pool on a warm day. The dripping wet body immediately loses moisture through evaporation to the surrounding air, and the skin feels the consequent drop in temperature.

The amount and rate of evaporation from a water surface depends essentially on two factors: the temperature and the degree of saturation of the air.

Temperature

In warm water the molecules are more agitated than in cool water; thus there is more evaporation. Warm air also promotes evaporation because warm air can "hold" more water vapor than cool air. Just as high water temperature produces more agitation among water molecules, so high air temperature encourages increased activity among molecules of water vapor. Thus they bounce around to a greater extent and excite the entry of more molecules from any adjacent water surface.

In technical terms this involves an increase in vapor pressure. As with any gas, water vapor exerts pressure, which is called *vapor pressure*. There is a direct relationship between increasing temperature and increasing vapor pressure. The higher the temperature, the higher the vapor pressure can be, thus allowing for more water vapor molecules in the air.

Degree of Saturation

If the air overlying a water surface is close to saturation with water vapor, very little further evaporation can take place, regardless of the temperature. Under these conditions, water molecules enter the air at about the same rate that air molecules enter the water. If the air remains calm and the temperature doesn't change much, net evaporation will be at a standstill. If the air is in motion, however, through windiness and/or turbulence, the molecules of moisture are dispersed more widely. This reduces the degree of saturation of the layer of air adjacent to the water surface and permits an increased rate of evaporation.

To summarize, the rate of evaporation from a water surface is determined by the temperature of the water, the temperature of the air, and the degree of windiness. Higher temperatures and greater windiness engender more evaporation.

Evapotranspiration

Although most evaporation into the air comes from water surfaces, a relatively small amount of evaporation also originates from the land. A minor proportion of this consists of direct evaporation of water from the soil and other inanimate substances, but the bulk of it involves transfer of moisture from plant leaves to the atmosphere, which is called *transpiration*. Because so many different kinds of surfaces and other variables impinge on the situation, the transfer of moisture from land to air is often referred to collectively as *evapotranspiration*.

Humidity Measuring humidity is both more complicated and less accurate than measuring the other three weather elements that we have discussed so far. No simple devices have yet been invented to make precise measurements of humidity. Moreover, the direct quantitative expressions of water vapor in the air—absolute humidity and specific humidity—are the most difficult to measure. Normally these two measures are computed mathematically from their relationship with temperature and relative humidity, which can be measured directly. Tables and graphs are available to permit computation of absolute humidity or specific humidity from temperature and relative humidity data.

A *hygrometer* is an instrument for measuring humidity. One type of hygrometer is a *psychrometer*, which consists of two thermometers mounted side by side. One of these is an ordinary thermometer (called a *dry bulb* in this case), which simply measures air temperature. The other thermometer (called a *wet bulb*) has its bulb encased in a covering of muslin or gauze, which is saturated with distilled water prior to use. The two thermometers are then thoroughly ventilated either by being whirled around (this instrument has a handle around which the thermometers can be whirled and is referred to as a *sling psychrometer*) or by fanning a current of air past them. This ventilation encourages evaporation of water from the covering of the wet bulb at a rate that is directly related to the humidity of the surrounding air. Evaporation is a cooling process, and the temperature of the wet bulb drops. In dry air there is more evaporation, and therefore more cooling, than in moist air. The difference between the resulting wet-bulb and dry-bulb temperatures is an expression of the relative saturation of the surrounding air. A large difference indicates low relative humidity; a small difference means that the air is near saturation. If the air is completely saturated, no net evaporation will take place; thus the two thermometers would have identical readings.

Relative humidity can also be measured by other kinds of hygrometers. The *hair hygrometer* operates on the principle that human hair expands when the humidity is high and shrinks when it is low. A group of hairs can be attached to a dial so that the needle indicates the relative humidity directly. Such hygrometers are not very accurate, even with frequent calibration. The more complex *electric hygrometer* is based on differences in electrical resistivity of certain moisture-absorbing chemicals; it is often used in upper-air instrument packages.

Hygrometers can be attached to rotating drums so that a continuous recording of relative humidity is obtained; such a device is called a *hygrograph*. In many weather-recording installations, a hygrograph and a thermograph are combined in a single instrument (the *hygrothermograph*) so that continuous and comparable recordings of relative humidity and temperature appear together on the rotating drum.

Precipitation Precipitation is a tangible feature that can be measured simply and directly. The principal difficulty with measuring precipitation is that atmospheric motion and surface impediments interfere with the vertical fall of rain or snow. Wind and turbulence and the blocking effect of buildings and trees inhibit the collection of a representative quantity of precipitation, so instruments must be positioned with care.

Almost any open container can be used as a *rain gauge*. The depth of rain that collects in the container can be measured directly to determine the depth of rainfall for any given period. Sophisticated modifications of the rain gauge make possible the automatic recording of the amount of rainfall per unit of time; there are many types of these recording rain gauges.

Depth of snowfall is even simpler to measure, except for the problems introduced by the wind blowing and drifting the snow. Depth can be determined easily by poking a calibrated rod vertically down to the bottom of the accumulated snow; the real problem is to pick a representative site for inserting the rod.

A more important snow measure is its actual moisture content. A sample vertical column of snow can be dug out and melted to obtain the liquid equivalent. A value of 1 unit of water content for 10 units of snow depth is a generally accepted average, but actual conditions may fluctuate widely from this average. Fluffy dry snow may require 30 units to produce a single unit of water, whereas the ratio with clinging wet snow may be as little as 4:1.

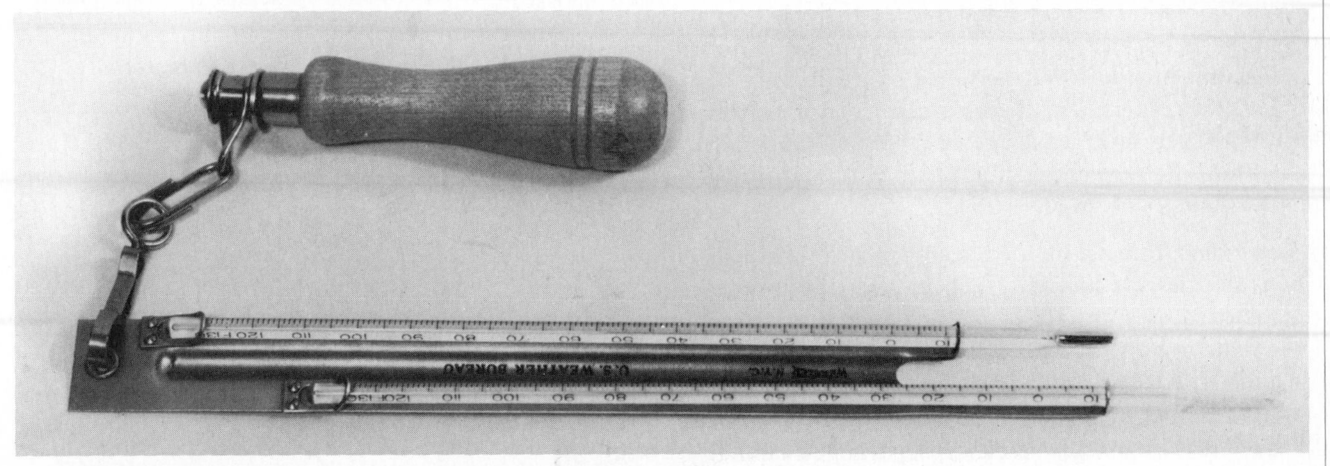

A sling psychrometer. [Courtesy NOAA.]

Measures of Humidity

The amount of water vapor in the air is referred to as *humidity*. It can be measured and expressed in a number of ways, each of which is useful for certain purposes.

Absolute Humidity

A direct measure of the water vapor content of air is *absolute humidity*. This is expressed as the weight of water vapor in a given volume of air, normally in terms of grams of vapor per cubic meter of air (1 gram is approximately 0.035 ounces, and 1 cubic meter is about 35 cubic feet). This is a clear-cut quantitative measurement of the actual amount of water vapor in the air at a given time and place. As with any comparative measure, its value fluctuates when the comparative parameter varies. Thus if the volume of air changes, particularly when warming or cooling causes expansion or contraction, the value of the absolute humidity also changes, even though there has been no change in the vapor content. See Figure 6-3.

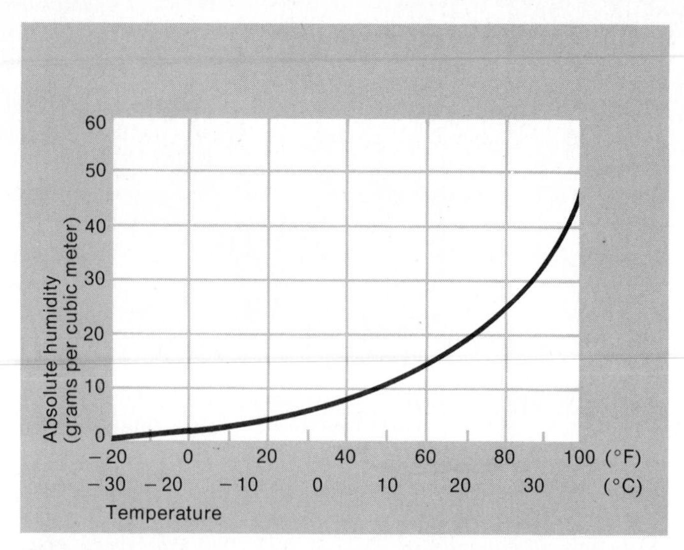

Figure 6-3. Maximum absolute humidity is governed by temperature. This curve portrays the relationship. Cold air can hold very little water vapor, but as the temperature increases, the air's capacity for holding moisture rises rapidly.

Specific Humidity

A related direct measure of water vapor content is *specific humidity*. This is the mass (weight) of water vapor in a given mass of air, and is usually expressed in terms of grams of vapor per kilogram of air. Specific humidity is difficult to measure accurately, but it has the advantage of changing only as the quantity of water vapor varies; it is not affected by variations in temperature or pressure. Specific humidity is particularly useful in studying the characteristics and movements of air masses (see Chapter 7).

Relative Humidity

The most familiar of humidity measures is *relative humidity*, which is an expression of the amount of water vapor in the air in comparison with the total amount that could be there if the air were saturated. It is not a direct measure of quantity; rather, it is a ratio that is expressed as a percentage. In essence, it is the percentage of saturation.

Relative humidity is determined by the balance between vapor content and the air's capacity for holding moisture, which is in turn primarily dependent on its temperature. Relative humidity varies if either part of the ratio changes, as follows:

1. *Vapor content*. If moisture is added to the air by evaporation, the relative humidity will increase. If moisture is removed from the air, by condensation or dispersal, the relative humidity will decline.
2. *Capacity for holding moisture*. If the air's capacity for water vapor is diminished, normally by cooling, the relative humidity will rise. If the capacity is increased, usually by heating, the relative humidity will decline.

These varying relationships are exemplified in Figure 6-4, which demonstrates the fluctuations between relative humidity and temperature during a typical day. In the early morning hours, the temperature is low and the relative humidity is high because the air's capacity for holding moisture is limited. As the day warms up, the relative humidity declines because capacity is increased. With the approach of evening, temperature decreases, capacity diminishes, and relative humidity therefore rises.

The concept of relative humidity is simple, but, as with most measures of atmospheric phenomena, it must be used with an awareness of its limitations. Completely saturated air should have a relative humidity of 100 percent; any value less than 100 percent is a proportional indication of how near the air is to being saturated. In most parts of the world, temperature fluctuates significantly on a daily basis, as well as periodically and seasonally. Relative humidity, then, will vary similarly. Air that is near saturation (and thus considered to be damp) at a cool temperature might be far from saturation (and thus considered to be dry) at a warmer temperature.

Related Humidity Concepts

Two other concepts related to relative humidity are useful in a study of physical geography.

Dew Point As we have seen, when air is cooled, its percentage of saturation increases. If the cooling is sufficient to reach complete saturation (100 percent relative humidity), condensation normally begins. The critical temperature at which saturation is reached is called the *dew point*. The dew point varies, of course, with the humidity content of the air.

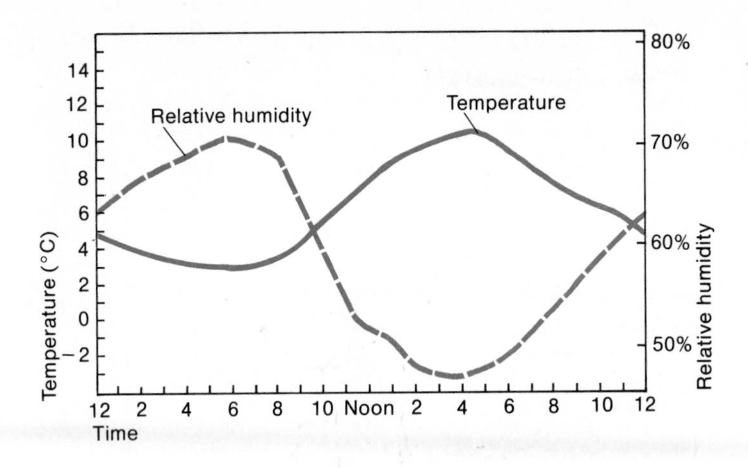

Figure 6-4. Typically there is an inverse relationship between temperature and relative humidity on any given day. As the temperature rises, the relative humidity decreases. Thus relative humidity tends to be lowest in midafternoon and highest just before dawn.

Sensible Temperature The term *sensible temperature* is a concept of the relative temperature that is sensed by a person's body. It involves not only the actual temperature but also other atmospheric conditions, particularly relative humidity and wind, that influence our perception of heat and cold. On a warm, humid day, for example, the air seems hotter than the thermometer indicates, and the sensible temperature is said to be "high." This is because the air is relatively near saturation, and perspiration on the human skin does not evaporate readily. Thus there is little evaporative cooling, and the air seems warmer than it actually is. On a warm, dry day, however, evaporative cooling is effective, and the air seems to be cooler than it actually is; in this case the sensible temperature is called "low." See Figure 6-5.

On a cold day, if the relative humidity is high, the coldness seems more piercing because body heat is conducted away more rapidly in damp air; the sensible temperature is described as "low." On a cold, dry day, however, body heat will be more effective because it is not conducted away as fast; the temperature will appear to be warmer than it actually is, and the sensible temperature is "high."

The amount of wind movement also affects sensible temperature, primarily by its influence on evaporation and body heat.

Figure 6-5. The general relationships among temperature, relative humidity, and sensible temperature. Plus signs mean "high"; minus signs signify "low."

Temperature	Relative humidity	Sensible temperature
+	+	+
+	−	−
−	+	−
−	−	+

Condensation

Condensation is the opposite of evaporation. It is the process whereby water vapor in the atmosphere is converted to water. It is a change in state from gas to liquid.

The Process

For condensation to take place, the air must be saturated. In theory this can come about through the addition of water vapor to the air, but in practice it is usually the result of the air being cooled to a temperature below the dew point. Saturation alone, however, is not enough. There also must be a surface on which the condensation can take place. If no surface is available, no condensation will occur. In such a situation, the air would become *supersaturated* if cooling were to continue.

Normally, plenty of surfaces are available for condensation. At ground level this is obviously no problem. In the air above the ground there is also usually an abundance of "surfaces," as represented by tiny particles of dust, smoke, salt, pollen, and other compounds. Most such particles are submicroscopic and totally invisible. These minute particles are most common over cities, seacoasts, and volcanoes, which are the source of much particulate matter, but they are generally widespread throughout the troposphere. They are referred to as *condensation nuclei*, and they serve as collection centers for water molecules. From the human viewpoint, these particles are incredibly small. See Figure 6-6. Most have diameters less than 0.1 micrometer. But they occur in countless billions in the atmosphere.

As soon as the temperature cools to the dew point, water vapor molecules begin to condense around these nuclei. The droplets grow rapidly by accretion of increasing amounts of moisture, and as they become larger, they may bump into one another and coalesce. Continued growth can make them large enough to be visible, forming haze or cloud particles. The diminutive size of these particles can perhaps best be indicated by the realization that a single raindrop may contain a million or more condensation nuclei with their associated moisture.

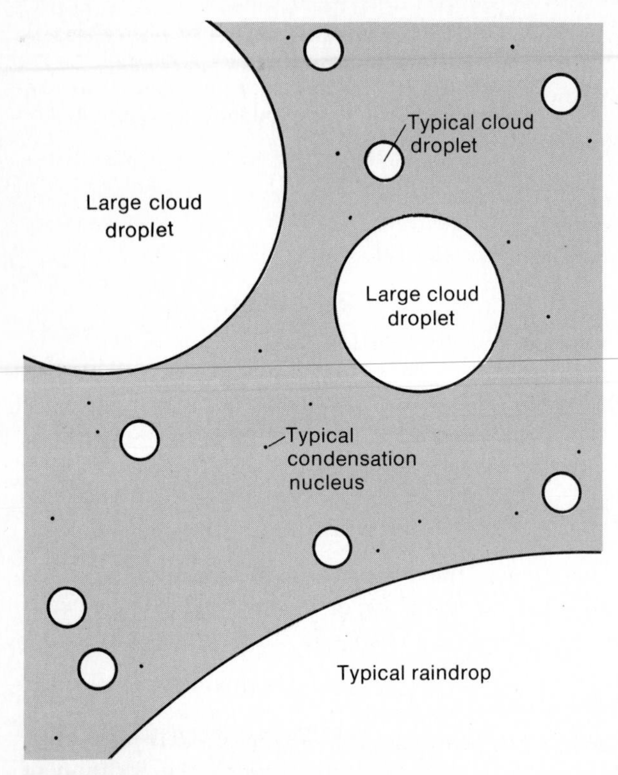

Figure 6-6. Comparative sizes of condensation and precipitation particles.

Clouds often are composed of water droplets even when their temperatures are below freezing. Although water in large quantity freezes at 32°F (0°C), if it is dispersed as fine droplets it can remain in liquid form at temperatures as much as 70°F (40°C) below the freezing point. Water that persists in liquid form at temperatures below freezing is said to be *supercooled*. Supercooled droplets are important to the condensation process because they promote the growth of ice particles in cold clouds by freezing around them or by evaporating into vapor from which water molecules are readily added to the ice crystals.

Adiabatic Activities

One of the most significant facts in the study of physical geography is that the only way in which large masses of air can be cooled to the dew point is by expansion in rising air. The only prominent mechanism for the development of clouds and the production of rain is through such adiabatic cooling. As we noted in Chapter 4, when air rises it reaches a lower pressure, so it expands and cools. If it rises sufficiently, it cools to the dew point. This permits condensation to begin and clouds to form.

As a parcel of air rises, it cools at the relatively steady rate of $5\frac{1}{2}$°F per 1000 feet (10°C per km). This is known as the *dry adiabatic lapse rate* or simply as the *adiabat*. As the rising air cools, its capacity for holding moisture decreases and its relative humidity increases. If it rises high enough and cools sufficiently, it will reach 100 percent relative humidity at the dew point temperature. The altitude at which this occurs is known as the *lifting condensation level* (LCL). At this point in the uplift, condensation begins. Under normal circumstances the LCL is clearly visible as the base of the clouds that form.

As soon as condensation begins, latent heat is released (this heat was stored originally as the latent heat of vaporization). If the air continues to rise, cooling will continue, but release of the latent heat will slacken the rate of cooling. This diminished rate of cooling is at the *wet adiabatic lapse rate*, or *pseudoadiabat*, which depends on temperature but averages about 2°F per 1000 feet (6.5°C per km).

Adiabatic warming occurs when air is caused to descend. Any descent immediately produces warming through compression as the air comes under higher pressure. This increases the capacity of the air for holding moisture and produces an unsaturated condition. Therefore, any descending air will warm at the dry adiabatic lapse rate.

In any consideration of adiabatic temperature changes, we should remember that we are dealing with air that is rising or descending. The adiabatic lapse rates are not to be confused with the *average lapse rate*, which pertains to air that is not in motion. As an analogy, an adiabatic change can be likened to a rising balloon, in which a parcel of air is in vertical motion. A thermometer within the balloon would record temperature change within the rising air (adiabatic gradient), whereas a thermometer attached outside the balloon would record temperature change in the surrounding, nonmoving air (lapse-rate gradient). See Figure 6-7.

Figure 6-7. Hypothetical relationships of vertical cooling rates. Unsaturated rising air cools at the dry adiabatic lapse rate. Saturated rising air cools at the wet adiabatic lapse rate. Nonrising air experiences vertical cooling at various rates, generalized as the average lapse rate.

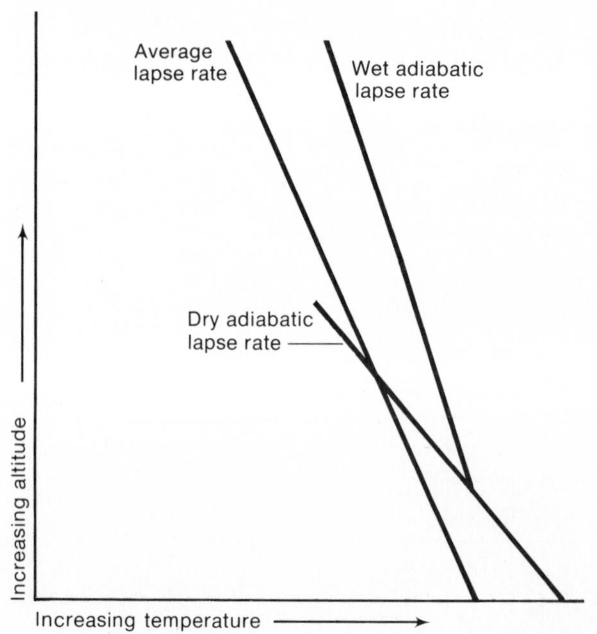

Clouds

Most atmospheric activities are invisible to the human eye. The majority of the processes and phenomena thus far discussed have no visible manifestation at all. Clouds, on the other hand, are not only prominent but are sometimes spectacular features in the sky. Clouds serve as the visible expression of condensation and often provide perceptible evidence of other things that are happening in the atmosphere.

The basic importance of clouds is that they serve as the source of precipitation. Not all clouds precipitate, but all precipitation comes from clouds. Clouds are also significant because of their influence on radiant energy. Both insolation from above and terrestrial radiation from below are received by clouds, which results variously in absorption, reflection, scattering, and reradiation of the energy. The function of clouds in the global energy budget is important.

Formation A cloud consists of a visible collection of minute droplets of water or tiny crystals of ice. With a few exceptions, cloud formation results from adiabatic cooling in rising air. When the air is cooled to saturation, condensation, as previously described, is initiated. As more and more water vapor condenses, the accumulation of water or ice particles becomes sufficiently large and dense to become visible, and a cloud begins to appear. The base of a cloud is often sharp and clear-cut; it marks the lowest level at which condensation began. The cloud then grows upward and outward from this base.

Classification Clouds are good indicators of the state of the atmosphere, providing at a glance some understanding of the present weather and often serving as harbingers of what is to come. Although they occur in an almost infinite variety of shapes and sizes, certain patterns of form recur commonly. Moreover, the various cloud forms are normally found at certain generalized elevations. On the basis of these two factors—appearance and altitude—clouds can be classified into types, which is a useful tool in interpreting the atmosphere.

The international classification of clouds recognizes three prominent cloud forms, and ten basic cloud types. See Figure 6-8. The types overlap, and the cloud forms

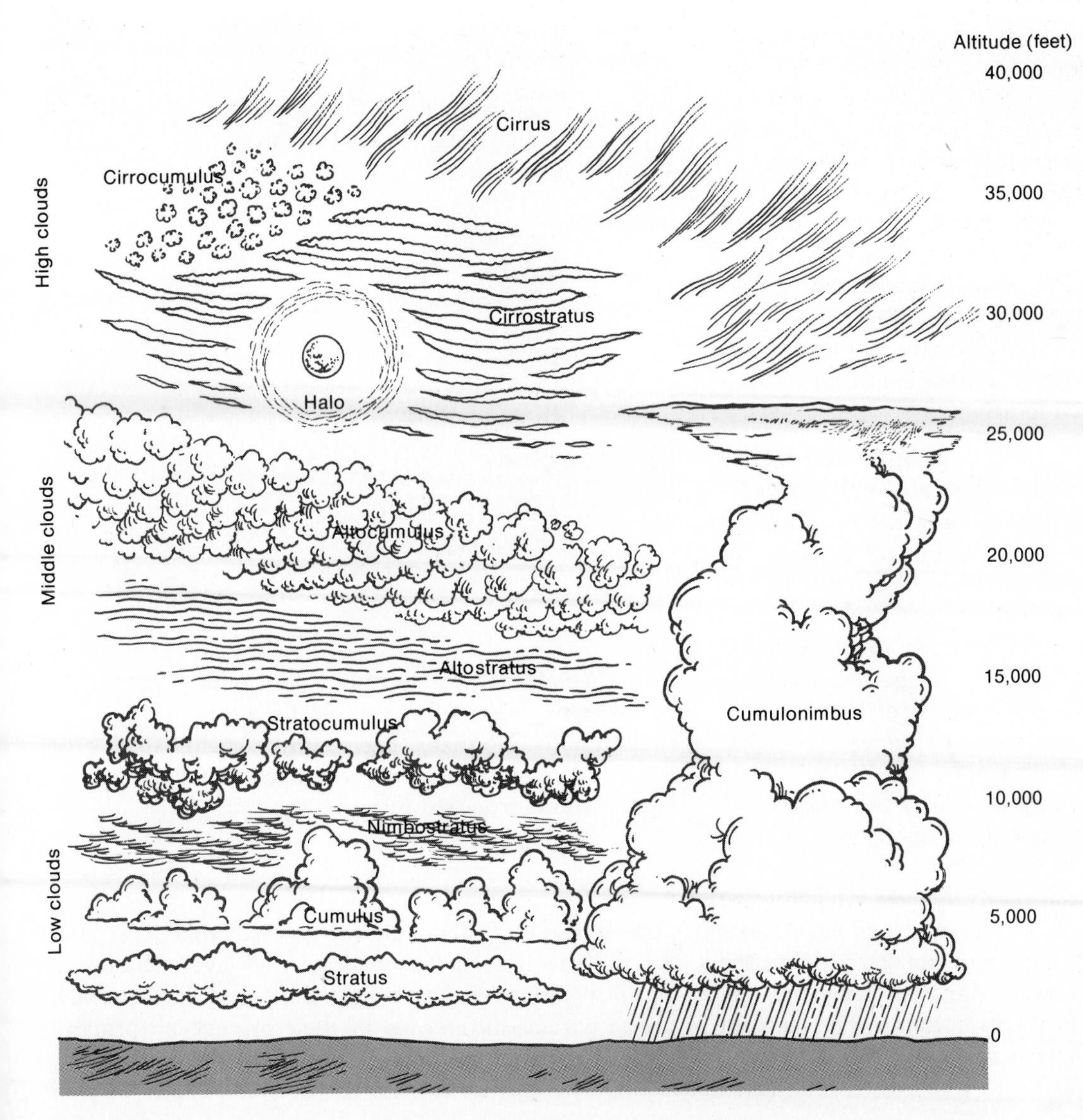

Figure 6-8. Idealized diagram of typical shapes and altitudes of the principal cloud types.

Figure 6-9. Multiple cumulus clouds in the foreground with a cumulonimbus developing an anvil top in the background. [TLM photo.]

are in an almost constant state of change, so that one type may evolve into another.

Cirriform clouds (Latin *cirrus*, "a lock of hair") are thin and wispy, composed of ice crystals rather than water particles; they are found at high elevations. *Stratiform clouds* (Latin *stratus*, "spread out") appear as grayish sheets or layers that cover most or all of the sky, rarely being broken up into individual cloud units. *Cumuliform clouds* (Latin *cumulus*, "mass" or "pile") are massive and rounded, usually with a flat base and limited horizontal extent, but often billowing upward to great heights. See Figure 6-9.

Four families of clouds are identified on the basis of their general altitudinal location:

1. *High clouds* are generally found above 20,000 feet (6 km). Because of the small amount of water vapor and low temperatures at such altitudes, these clouds are thin and white and are composed of ice crystals. Included in this family are *cirrus*, *cirrocumulus*, and *cirrostratus*. See Figure 6-10.
2. *Middle clouds* normally occur between about 6500 and 20,000 feet (2 and 6 km). They may be either layered or puffy in form and are composed of liquid water. Included types are *altocumulus* and *altostratus*.
3. *Low clouds* usually are below 6500 feet (2 km). They sometimes occur as individual clouds but more often appear as a general overcast. Low cloud types include *stratus*, *stratocumulus*, and *nimbostratus*.
4. A fourth family, *clouds of vertical development*, grow upward from low bases to heights of as much as 50,000 feet (15 km) or even more. Their horizontal spread, on the other hand, is usually very restricted. They indicate very active vertical movements in the air. The relevant types are *cumulus* and *cumulonimbus*.

Figure 6-10. Typical cirrus clouds are high, wispy, and composed of ice crystals. [Courtesy NOAA]

Precipitation comes only from clouds that have the root *nimb* in their name, specifically nimbostratus or cumulonimbus. Normally these types develop from other types; i.e., cumulonimbus develops from cumulus and nimbostratus from stratus.

Fog

Fog is simply a cloud on the ground, that is, a cloud whose base is at or very near ground level. There is no physical difference between a cloud and fog, but there are important differences in their process of formation and in their location. Most clouds develop due to adiabatic cooling in rising air, but only rarely is uplift involved in fog formation. From a global standpoint, fogs represent a minor form of condensation. Their importance to mankind, however, is inordinate because they can hinder visibility enough to make surface transportation (land, sea, or at airports) hazardous or even impossible.

Most fogs are formed by cooling, but some result from the addition of enough water vapor to the air to bring about saturation. Four general types of fogs can be identified (see Figure 6-11):

1. A *radiation fog* results from radiational cooling, usually at night. It is characteristic of the middle and high latitudes on clear, cold winter nights. Outgoing terrestrial radiation during the night makes the ground colder and colder, and this coldness is transferred by conduction to the air immediately above. If the air is cooled to the dew point, condensation will produce a fog. Radiation fog is most likely to develop in valleys, where downslope air drainage concentrates colder and denser air. When the sun begins to shine, radiation fogs usually burn off rapidly.
2. An *advection fog* can develop when warm, moist air moves horizontally over a cold surface, such as snow-covered ground or a cool ocean current. The air is chilled from below by conduction and radiation. If cooled enough to reach the dew point, fog will occur. Advection fogs are more common than any other type and are particularly associated with coastal areas. They are also usually more widespread and longer lasting than other fogs.
3. An *upslope fog* is sometimes created by adiabatic cooling when humid air is caused to ascend a topographic slope. Such a situation usually produces clouds, but over sloping ground, with gentle movement and stable air, extensive fog may occur instead.
4. *Evaporation fogs* result from the addition of water vapor to cold air that is already near saturation. There are several ways that this can occur in nature, but most common is the evaporation of water from a relatively warm lake or ocean surface into colder overlying air in a northern environment. The rising water vapor sometimes recondenses immediately and rises with the air that is being warmed from below in what is sometimes referred to as a "steam fog" or "Arctic smoke." See Figure 6-12.

Dew

Dew is another nocturnal feature that originates from terrestrial radiation. Nighttime radiation cools objects (grass, pavement, automobiles, or whatever) at the Earth's surface, and the adjacent air is in turn cooled by conduction. If the air is cooled enough to reach saturation, tiny beads of water will collect on the cold surface of the object, as dew. If the temperature is below freezing, ice crystals (*white frost*) rather than water droplets are formed.

Figure 6-11. The most common types of fog.

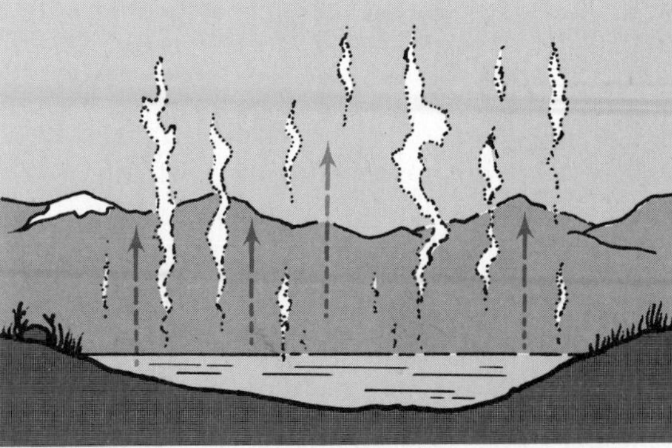

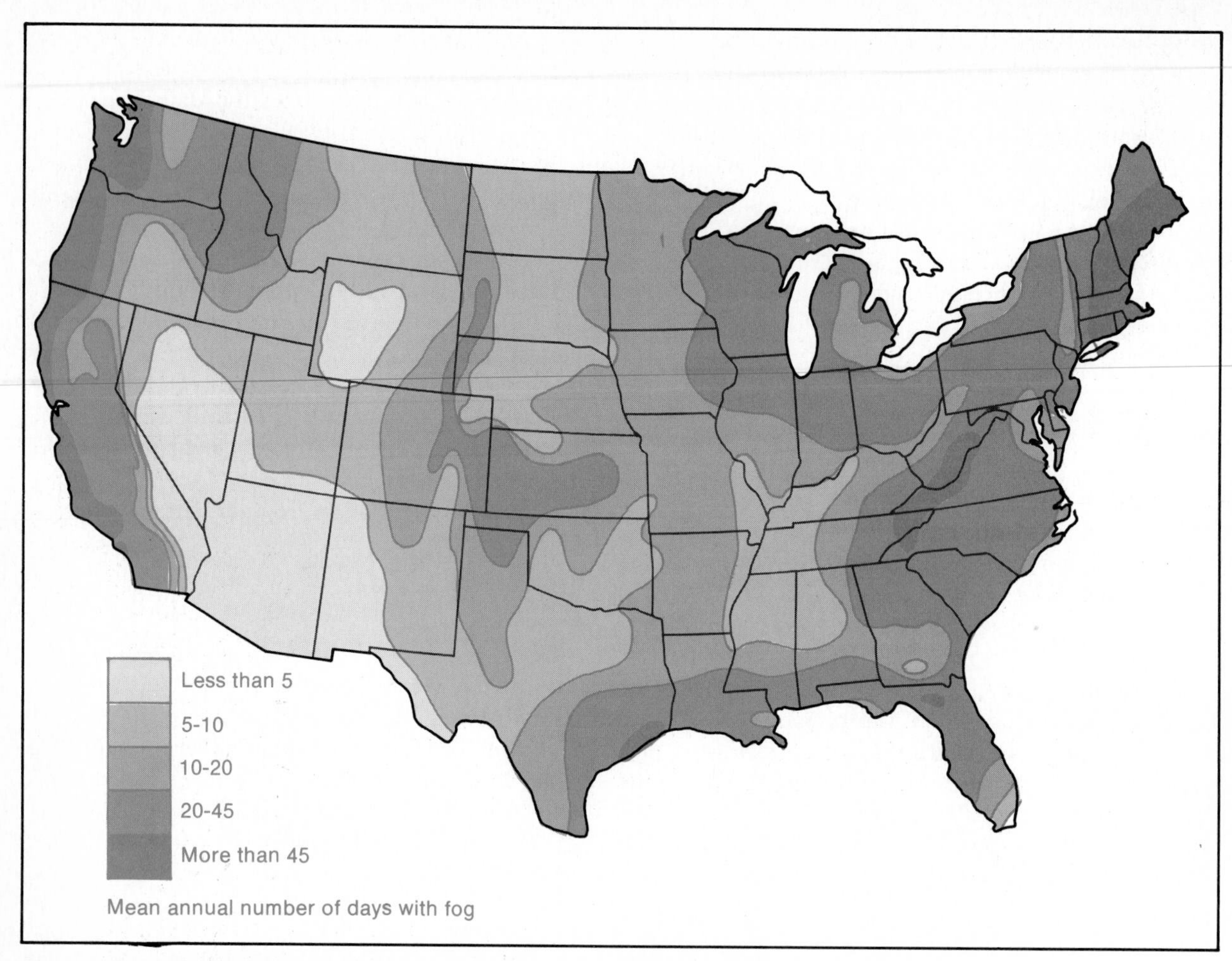

Figure 6-12. Distribution of fog in the conterminous states. The most frequent occurrences of fog are in the Pacific coast states.

The Buoyancy of Air

We have seen that most condensation and almost all precipitation are the result of air rising. It is also clear by now that under certain circumstances air rises more freely and more extensively than at other times. Conditions that promote or hinder upward movements in the troposphere are obviously of great importance to weather and climate. The concept of the buoyancy of air is one of the most significant and complex in physical geography.

The Concept of Stability

As with other gases, air tends to seek its own level. This means that a parcel of air will move vertically until it reaches a level at which the surrounding air is of equal density. If a parcel of air is warmer, and thus lighter, than the surrounding air, it will tend to rise. If the parcel is cooler than the surrounding air, it would tend to sink or at least to resist uplift.

Air can be forced to rise by various environmental conditions. See Figure 6-13. If it resists vertical movement, it is said to be *stable*. If stable air is forced to rise, it will do so only as long as the force is applied. If the impelling force is removed, stable air will sink back to

Figure 6-13. The stability of rising air is determined by its relationship to the lapse rate of the surrounding air. If the lapse rate of the surrounding air is less than the adiabat (the rate of cooling of the rising air), conditions are stable and the air will rise only if forced to do so. If the lapse rate of the surrounding air is greater than the adiabat, conditions are unstable and the rising will be buoyant without an uplifting force.

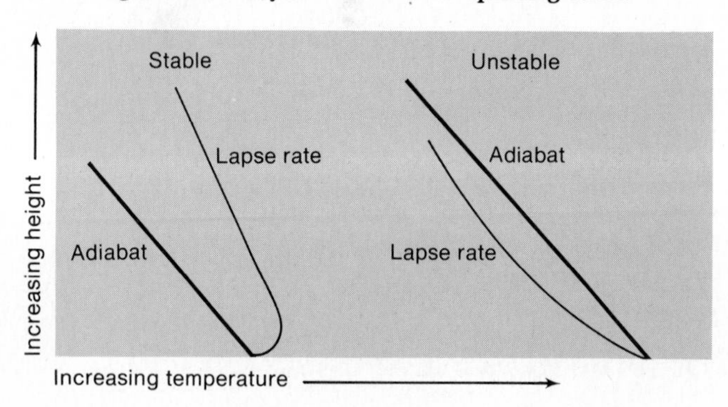

its former position. In other words, stable air is nonbuoyant. Air is said to be *unstable* if it rises of its own accord, even without an impelling force, or after the impelling force has ceased to function. Unstable air is intrinsically buoyant.

In the atmosphere, stability is promoted when cold air is beneath warm air, a condition most prominently displayed when a temperature inversion is in effect. With colder, denser air below warmer, lighter air, upward movement is unlikely. The air resists vertical displacement. A cold winter night is a typical stable air situation, although stability can also occur in the daytime. Stable air obviously provides little opportunity for adiabatic cooling unless there is some sort of forced uplift. Stable air is normally not associated with cloud formation and precipitation.

When air is heated enough so that it is warmer than the surrounding air, it will become unstable. This is a typical condition on a warm summer afternoon. Unstable air is prone to uplift. It will rise until it reaches an altitude where the surrounding air has the same temperature. See Figure 6-14. While ascending, it will be cooled adiabatically. In this situation, clouds are likely to form.

Figure 6-14. As stable air blows over a mountain, it rises only as much as it is forced up by the mountain slope; on the leeward side it descends downslope. When unstable air is forced up a mountain slope, it is likely to continue rising of its own accord until it reaches surrounding air of similar temperature and density; if it rises to the condensation level, clouds will form.

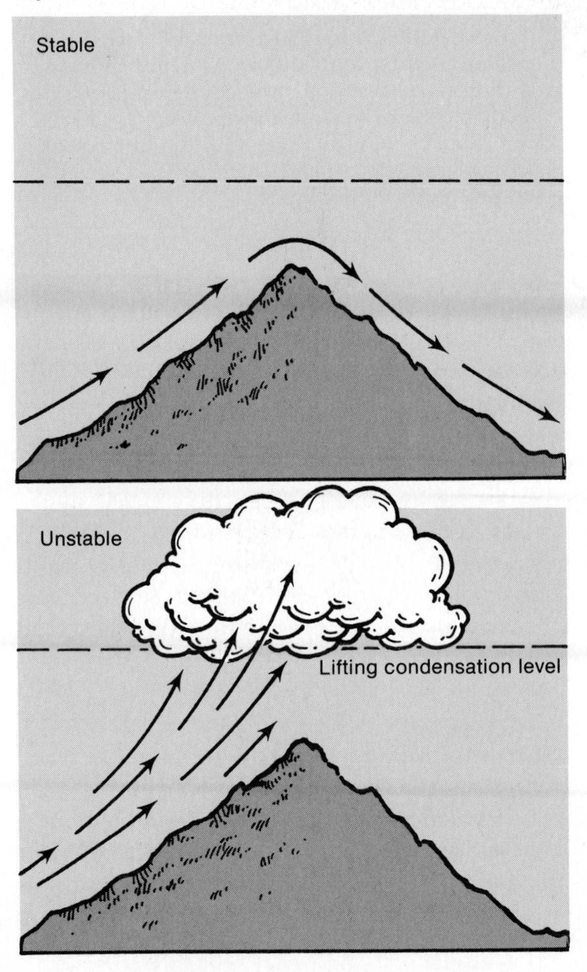

Determination of Stability

Recognition of the relative buoyancy of air is important to an understanding of the dynamics of the atmosphere. An accurate determination of stability depends on temperature measurements, but a rough indication often can be obtained simply by observing the state of the sky (primarily the cloud forms).

Determination by Measurement The temperature of the rising air is compared with the temperature of the surrounding nonrising air. The rising air cools (at least initially) at the dry adiabatic lapse rate of about $5\frac{1}{2}$°F per 1000 feet (1°C per 100 m). The lapse rate (vertical temperature change) of the surrounding air may vary on the basis of many conditions, so it could be either more or less than the dry adiabatic lapse rate.

If the lapse rate of the surrounding air is less than that of the rising air, the latter will be cooler than the former, and the air will be stable. Under such conditions, the air will rise only as far as it is forced to do so. If the lifting force is removed, the air will cease to rise and may actually sink.

Instability results if the lapse rate is greater than the dry adiabatic lapse rate. In this case, the rising air is warmer than the air around it, so it will rise even without an impelling force until it reaches an elevation where the surrounding air is of similar temperature.

The situation becomes more complicated if the rising air is cooled to its dew point. This would cause condensation to begin, and the rising air would then begin to cool at a slower rate (the wet adiabatic lapse rate) because latent heat of condensation would be released. This would increase the tendency toward instability and would reinforce the rising trend.

Visual Determination The cloud pattern is often indicative of stability conditions. Unstable air is associated with distinct updrafts, and this is likely to produce clouds of prominent vertical development. Thus the presence of cumulus clouds suggests instability, and a towering cumulonimbus cloud is an indicator of pronounced instability.

Stable air will not rise unless forced; so it will not, of itself, form any clouds at all. However, stable air is often forced to rise; in which case, horizontally developed clouds, most notably stratiform types, may be formed.

Regardless of stability conditions, no clouds will be formed unless the air is cooled to the dew point. Thus the mere absence of clouds is not certain evidence of stability; it is only an indication.

Relationships

The general relationships of air buoyancy to atmospheric moisture can be summarized as follows:

1. *Stability*⟶no clouds or stratiform clouds⟶no precipitation or drizzle. In stable air, either no clouds will be formed or stratiform clouds will develop. This will result in either no precipitation or drizzly precipitation (small raindrops or snowflakes falling gently).
2. *Instability*⟶no clouds or cumuliform clouds⟶no precipitation or showers. In unstable air, either no clouds will be formed (if the air is not cooled to the dew point), or vertically developed clouds will be produced. This will result in either no precipitation or showers (large raindrops or snowflakes or hailstones falling rapidly).

Precipitation

Condensation readily forms clouds, and all precipitation originates in clouds. However, most clouds do not yield precipitation. Moreover, exhaustive experiments have demonstrated that simple condensation, under normal circumstances, is insufficient in itself to produce raindrops. The tiny droplets that comprise clouds cannot fall to the ground as rain because they are so small that they are very buoyant, and the normal turbulence of the atmosphere keeps them aloft. Even in still air, their fall would be so slow that it would take many days for them to reach the ground from even a low cloud. Besides that, most droplets would evaporate in the drier air below the cloud before they made a good start downward.

Despite these difficulties, rain and other forms of precipitation are commonplace in the troposphere. What is it, then, that produces precipitation in its various forms?

The Processes

An average-sized raindrop contains several million times as much water as an average-sized cloud particle. Consequently, great multitudes of cloud particles must join together in order to form a drop large enough to overcome both turbulence and evaporation, and thus be able to fall to Earth.

It is still not well understood why most clouds do not produce precipitation. The mechanisms that cause raindrops to grow are obviously missing from most clouds. In order for precipitation to take place, water droplets must coalesce into a relatively large mass, which can then respond to the pull of gravity and descend from the cloud. Two separate mechanisms are believed to be principally responsible for producing precipitation particles: (1) ice crystal formation, and (2) collision and coalescence of water droplets.

Ice Crystal Formation See Figure 6-15. Many clouds or portions of clouds extend high enough to have temperatures well below the freezing point. In this situation, ice crystals and supercooled water droplets often coexist in the cloud. These two are in direct competition for the available water vapor that is not yet condensed. There is lower vapor pressure around the ice crystals, so they attract most of the vapor, and the water droplets, in turn, tend to evaporate to replenish the diminishing supply of vapor. So the ice crystals grow at the expense of the water droplets until they are large enough to fall. As the crystals descend through the lower, warmer portions of the cloud, they pick up more moisture and become still larger. They may then precipitate from the cloud as snowflakes, or they may be melted and fall as raindrops.

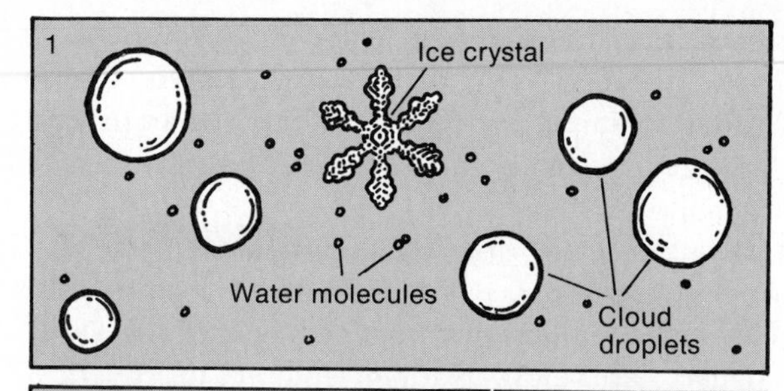

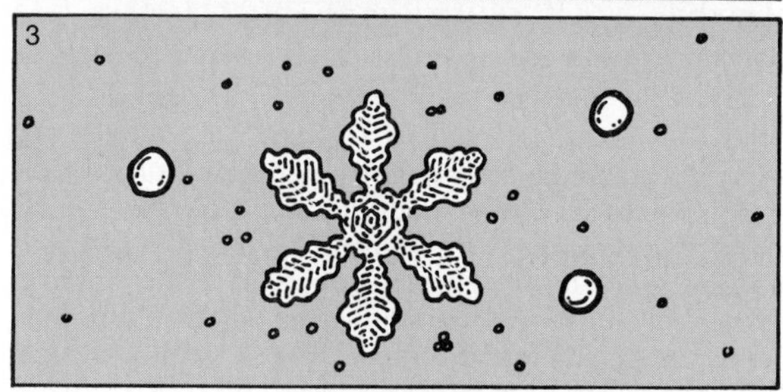

Figure 6-15. Precipitation through ice crystal formation in clouds. Ice crystals grow by attracting water vapor to them, causing cloud droplets to evaporate, thereby replenishing the water vapor supply. The process of growing ice crystals and shrinking cloud droplets may continue until the ice crystals are large and heavy enough to fall. Particle sizes are greatly exaggerated in this and the next figure.

Collision/Coalescence See Figure 6-16. In many cases, particularly in the tropics, clouds have temperatures too high for the formation of ice crystals. In such clouds, rain is produced by the collision and merging of water droplets that fall at different velocities. The larger droplets fall faster, overtaking and often coalescing with smaller ones, which are swept along in the descent. Not all colli-

Focus On Cloud Seeding

The ancient dictum that "everybody talks about the weather but nobody does anything about it," is, like many tidbits of conventional wisdom, erroneous. From time immemorial, people have attempted to change the weather by various means, ranging from the simple to the bizarre. Most such efforts have been aimed at bringing rain. At various times through the centuries, people have sought to "make" rain fall by praying, dancing, sacrificing maidens, ringing bells, making loud noises, flailing at the sky with sticks, and other activities.

Scientific weather modification is essentially a product of the middle of the twentieth century. The first breakthrough came in 1946 when Vincent J. Schaefer of General Electric Laboratories developed the principle of cloud seeding. He demonstrated that the introduction of dry ice into a supercooled cloud encouraged the growth of ice crystals. In such clouds, ice particles will grow at the expense of water droplets, and if they become large enough, they will fall as precipitation. Soon Bernard Vonnegut discovered that silver iodide has a similar crystal structure to that of ice, so it could be used in cloud seeding to provide an increased number of freezing nuclei. Moreover, silver iodide was easier to disperse because it could be supplied from burners on the ground, whereas dry ice had to be scattered from an aircraft or shot into the air in a projectile.

Cloud seeding has become increasingly sophisticated, but it is still successful in only limited circumstances. In the first place, clouds must be present; seeding cannot produce them. Moreover, only certain clouds are susceptible to seeding; at least the upper portion of the cloud must be supercooled. Attempts to seed "warm" clouds with such things as salt and water have been problematical. At the present state of the art, the seeding of susceptible clouds with freezing or condensation nuclei (still mostly silver iodide or dry ice) is successful in producing precipitation only about 10 percent of the time. Even in these successful instances, the amount of precipitation is variable, and it often falls at some location other than where the seeding actually took place.

However, cloud seeding techniques have been adapted for purposes other than rainmaking. They have been used most successfully in *fog dispersal*. Introducing ice nuclei into a supercooled fog can induce the formation of ice crystals that fall to the ground, producing a "hole" in the fog. This is particularly valuable at airports, where it can provide sufficient visibility for aircraft operations in an otherwise foggy area. Most fogs, however, are not supercooled and are therefore not susceptible to seeding.

The same principle has been applied in attempting cloud dispersal, hail suppression, hurricane modification, and even lightning suppression, although thus far with very limited success. As more is learned about cloud dynamics and as techniques are improved, however, the success ratio is likely to grow.

The potential ramifications of cloud seeding are enormous and complex. As mankind gains greater control of the volatile atmosphere, how should this power be exercised? Does cloud seeding actually produce rain for an area, or does it "hijack" rain that would eventually fall somewhere else? The ethical and legal implications are far from clear.

A fog with a hole in it. An Air Force experiment with dry ice pellets dispersed fog from this three-mile-wide "hole" in one hour. [Courtesy U.S. Air Force.]

sions result in coalescence, however. The droplets often fail to merge when they collide. Apparently coalescence is assured only if atmospheric electricity is favorable; i.e., if a positively-charged droplet collides with a negatively-charged one. This sequence of events favors the continued growth of the larger particles.

Forms of Precipitation

Several forms of precipitation can result from the processes just described. The form that results depends largely on the temperature of the air and on its degree of turbulence.

Rain By far the most common and widespread form of precipitation is *rain*, which consists of drops of liquid water. Most rain is the result of condensation and precipitation in ascending air that has temperatures above freezing, but some results from the thawing of ice crystals as they descend through warmer air.

Snow The general name given to solid precipitation in the form of ice crystals, small pellets, or flakes is *snow*. It is formed by direct conversion of water vapor into ice, without an intermediate water stage. However, the vapor may have evaporated from supercooled liquid cloud droplets.

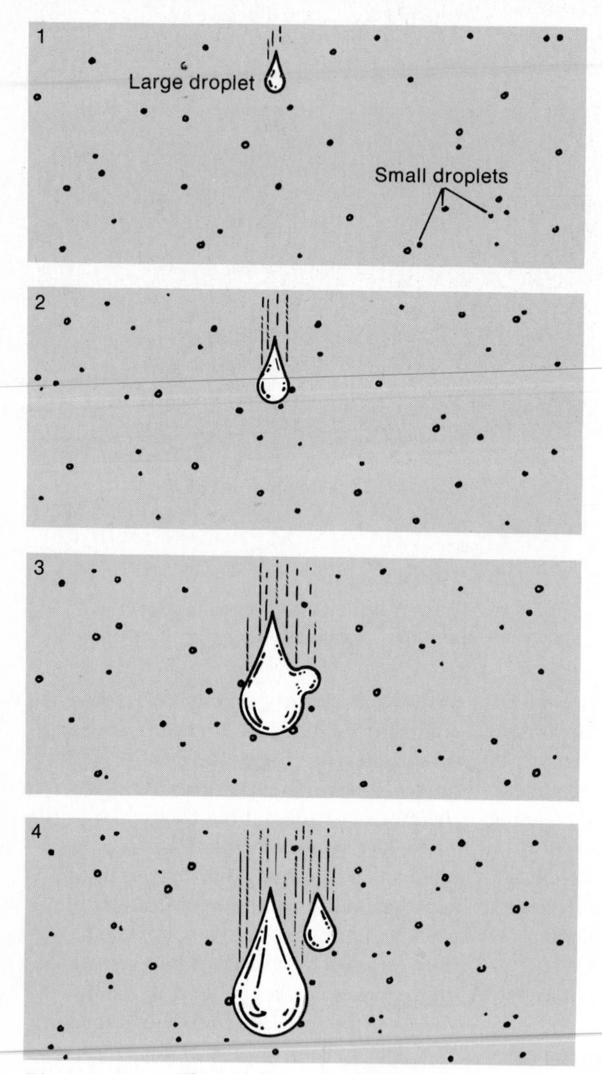

Figure 6-16. The process of collision/coalescence. Large droplets fall more rapidly than small ones, coalescing with some and sweeping others along in their descending path. As droplets become larger during descent, they sometimes break apart.

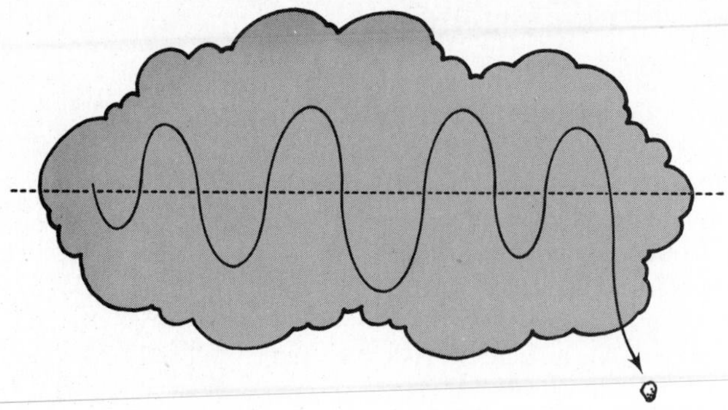

Figure 6-17. The formation of hailstones is fairly complex but essentially involves repetitive vertical movement of ice particles, engendered by active turbulence and vertical air currents. The more often the hailstone rises and falls within the cloud, the larger it is likely to grow. The dashed line represents the freezing level. [From Robert Gabler, et al., *Introduction to Physical Geography* (New York: Rinehart Press, 1975), p. 191. Copyright © 1975 by Rinehart Press. Reprinted by permission of Holt, Rinehart and Winston, CBS College Publishing.]

Sleet In the United States, *sleet* refers to small raindrops that freeze during descent. In other countries, the term is often applied to a mixture of rain and snow.

Hail The precipitation form with the most complex origin is *hail*, which consists of rounded or irregular pellets or lumps of ice. Large hailstones are usually composed of roughly concentric layers of clear and cloudy ice. The opaque portions contain numerous tiny air bubbles among small crystals of ice, whereas the clearer parts are made up of large ice crystals. Hail is produced in cumulonimbus clouds as a result of active turbulence and vertical air currents. See Figure 6-17. A powerful current of rising air carries small ice particles upward, where they grow by collecting moisture from supercooled cloud droplets. When the particles become too large to be supported in the air, they fall, accruing more moisture on the way down. If they encounter another strong updraft, they may be carried skyward again, only to fall another time. This sequence may be repeated several times.

The hailstone normally continues to grow whether it is rising or falling, providing it passes through portions of the cloud that contain supercooled droplets. If there is considerable moisture available, the icy hailstone is surrounded by a wet layer that freezes relatively slowly, producing large ice crystals and forcing the air out of the water, which forms clear ice. If there is a more limited supply of supercooled droplets, the water may freeze almost instantly around the hailstone. This produces small crystals with tiny air bubbles trapped among them, forming opaque ice.

The eventual size of the hailstone depends on the amount of supercooled water in the cloud and the length of the trajectory of the ice pellet through the cloud (up, down, and sideways). The largest known hailstone, which fell in Kansas in 1970, weighed 766 grams (1.67 lb).

Types of Atmospheric Lifting and Precipitation

The role of rising air and adiabatic cooling has been stressed in this chapter. Only through these events can any significant amount of precipitation originate. It remains for us to consider the *causes of rising air*. What are the natural forces that can engender the ascent of extensive masses of air sufficiently to reach the condensation level? There are four principal types of atmospheric lifting, as described below. See also Figure 6-18. One type is spontaneous, and the other three require forceful lifting. However, two or more types may be interrelated. They operate in conjunction more often than not, which means that more than one type of lifting may be responsible for the precipitation event.

Convective Lifting and Precipitation Due to unequal heating of different surface areas, a parcel of air near the ground may be warmed by conduction more than

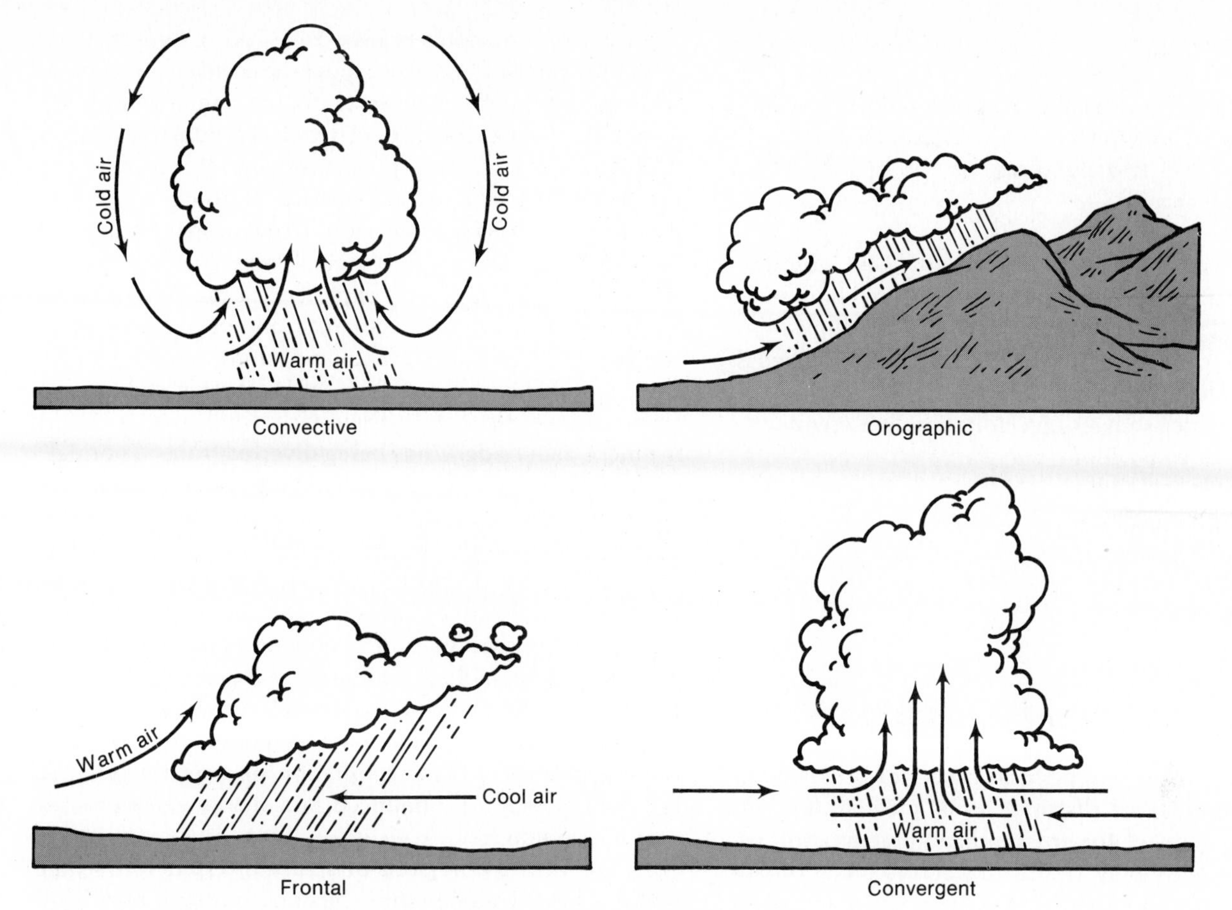

Figure 6-18. The four basic types of atmospheric lifting and precipitation.

the air around it. The heated air expands and rises vertically toward lower pressure. This spontaneous uplift is particularly notable if unstable air is involved, which is often the case on a warm summer day. Adiabatic cooling takes place, and the air is cooled to the dew point. Condensation then begins, and a cumulous cloud will form. With the proper humidity, temperature, and stability conditions, the cloud is likely to grow into a towering cumulonimbus thunderhead, with a downpour of showery rain and/or hail, accompanied sometimes by lightning and thunder.

An individual convective cell is likely to cover only a small horizontal area, although sometimes multiple cells are in juxtaposition with one another. Convective precipitation typically is showery, with large raindrops falling fast and furiously but for only a short duration.

Various other kinds of forced uplift, as well as differential heating, can trigger a convective response if the air tends toward instability. Thus convective uplift often accompanies other kinds of uplift. Convective precipitation is particularly associated with the warm parts of the world or the warm season of the year.

Orographic Lifting and Precipitation Topographic barriers that block the path of horizontal air movements are likely to cause large masses of air to travel upslope. This kind of forced ascent can produce orographic precipitation if the ascending air is cooled to the dew point. If significant instability has been triggered by the upslope motion, the air will keep rising when it reaches the top of the slope, and the precipitation will continue. More commonly, however, the air will descend the leeward side of the barrier. As soon as it begins to move downslope, adiabatic cooling is replaced by adiabatic warming, and condensation/precipitation will cease. Thus the windward slope of the barrier is the wet side, the leeward slope is the dry side, and the term *rain shadow* is applied to both the leeward slope and the area beyond as far as the drying influence extends.

Orographic precipitation can occur at any latitude, any season, any time of day. The only requisite conditions are a topographic barrier of some sort (hill, mountain, escarpment, even a general land slope) and moist air to move over it. Orographic precipitation is more likely to be general and prolonged than showery and brief.

Frontal Lifting and Precipitation When unlike air masses meet, they do not mix. Rather, a zone of discontinuity called a *front* is established between them, and the warmer air inevitably rises over the cooler air. As the warmer air is forced to rise, it may be cooled to the dew point with resulting clouds and precipitation.

We will discuss frontal precipitation in greater detail in the next chapter. It tends to be widespread and protracted, but frequently it is also associated with convective showers.

Frontal activity is most characteristic of the midlatitudes, so frontal precipitation is particularly notable in those regions. It is less significant in the high latitudes and rare in the tropics.

Convergent Lifting and Precipitation Less common than the other three types, but nevertheless significant in some situations, is convergent lifting and precipitation. Whenever air converges, it results in a general uplift because of the crowding. This enhances instability, and it is likely to produce showery precipitation. It is frequently associated with convection.

Convergent precipitation is particularly characteristic of the low latitudes. It is common, for example, in the Intertropical Convergence zone (ITC) and is notable in such tropical disturbances as hurricanes and easterly waves.

Global Distribution of Precipitation

The most important geographical aspect of atmospheric moisture is the spatial distribution of precipitation. As with other features of the environment, precipitation occurs very unevenly over the Earth's surface. The broad-scale zonal pattern is based on latitude, but many other factors are involved, and the overall pattern is one of considerable complexity.

This concluding section of the chapter focuses on a series of maps that illustrate several major components of worldwide and United States precipitation distribution. The cartographic device used on these maps is the *isohyet*, a line joining points of equal quantities of precipitation.

Average Annual Precipitation

The amount of precipitation that occurs on any part of the Earth's surface is fundamentally determined by the nature of the air mass involved and the degree to which that air is uplifted. The characteristics—particularly the humidity, temperature, and stability—of the air mass are mostly dependent on where the air mass originated (e.g., over land or water, or in high or low latitudes) and the trajectory that it has followed in its movements. Whether or not lifting takes place and the amount of that lifting are determined largely by zonal pressure patterns, the effect of topographic barriers, and the occurrence of storms and other atmospheric disturbances.

The most conspicuous feature of the worldwide precipitation pattern is that the tropical latitudes contain most of the wettest areas. The trade winds are capable of carrying enormous amounts of moisture, and where they are caused to rise, very heavy rainfall is usually produced. Equatorial regions, particularly, reflect these conditions, as warm, moist, unstable air is uplifted in the intertropical convergence zone. Also, where trade winds are forced to rise by topographic obstacles, considerable precipitation may result. As the trades are easterly winds, it is the east coasts of tropical landmasses where this orographic effect is most pronounced, for example, the east coast of Central America, northeastern South America, and Madagascar. Where the normal trade wind pattern is modified by monsoonal developments, the onshore trade wind flow may experience a complete reversal of direction. Thus the wet areas on the west coast of southeast Asia (Burma, Thailand, and vicinity), India, and the so-called Guinea Coast of West Africa are caused by the onshore flow of southwesterly winds that are nothing more than trade winds being diverted from a "normal" pattern by the South Asian and West African monsoons.

The only other regions of conspiciously high precipitation shown on the world map are narrow zones along the west coasts of North and South America between about 40° and 60° of latitude. See Figure 6-19. These areas reflect a combination of frequent onshore westerly airflow, considerable storminess, and mountain barriers athwart the direction of the prevailing westerly winds. The presence of these north-south mountain ranges near the coast restricts the precipitation to a relatively small area and creates a pronounced rain shadow effect to the east of the ranges.

The principal regions of sparse precipitation on the world map are found in three types of locations, as defined below:

1. Dry lands are most prominent on the western sides of continents in subtropical latitudes (centered at 25° or 30°). See Figure 6-20. High-pressure conditions dominate at these latitudes, particularly on the western sides of continents, which are closer to the normal positions of the subtropical high-pressure cells (STHs). High pressure means subsiding air, which is inimical to condensation and precipitation. In north Africa and Australia, these dry zones are much more extensive, primarily because of the blocking effect of landmasses or highlands to the east, which inhibits any moisture being brought in from that direction.
2. Dry regions in the midlatitudes are most extensive in central and southwestern Asia, but they also occur in western North America and southeastern South America. In each case, the dryness is due to lack of access for moist air masses. In the Asian situation, this is essentially a function of distance from any ocean where onshore airflow might occur. In North and South America, there are rain shadow situations in regions of predominantly westerly airflow.
3. In the high latitudes there is not much precipitation anywhere. Water surfaces are scarce and cold, so little opportunity exists for evaporation of moisture into the air. Accordingly, polar air masses are dry and precipitation is slight. These regions can be referred to accurately as "cold deserts."

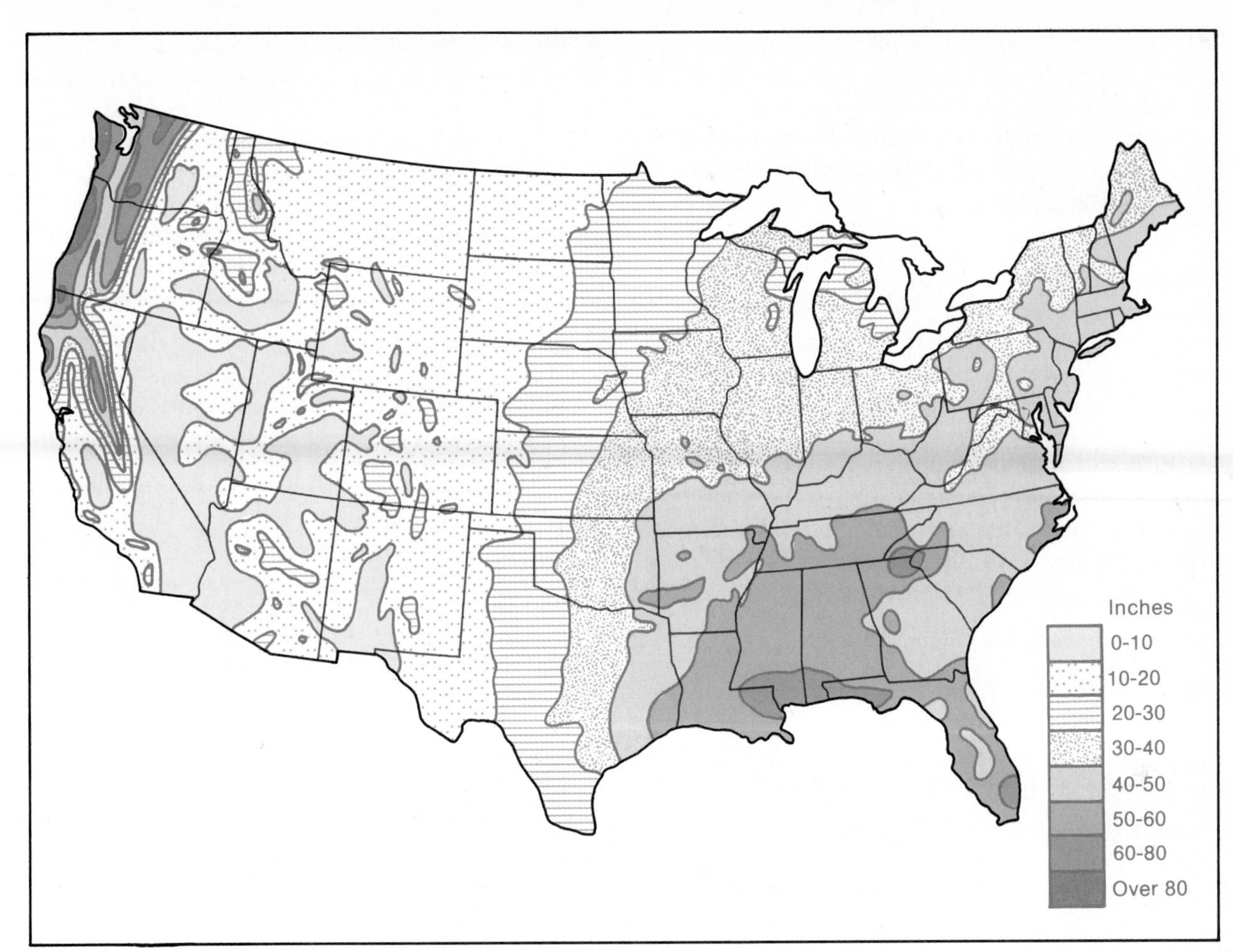

Figure 6-19. Average annual precipitation in the conterminous states. The western half of the country receives sparse precipitation, except along the north Pacific coast and in some mountain areas. The eastern half is generally more humid, particularly in the southeast.

Figure 6-20. A subtropical desert landscape in central Arizona, with the Superstition Mountains in the background. [USDA-Soil Conservation Service photo by B. Brixner.]

One further generalization on precipitation distribution is the contrast between continental margins and interiors. Coastal regions usually receive more precipitation than the interior due to their relative closeness to sources of moisture. This is a significant difference between "continental" and "maritime" climates.

Seasonal Precipitation Regimes

A geographic understanding of climate requires knowledge of seasonal as well as annual precipitation patterns. Most parts of the world experience considerable variation from summer to winter in the amount of precipitation received. This variation is most pronounced over continental interiors, where strong summer heating at the surface induces greater instability and the potential for greater convective activity. Thus, most of the year's precipitation occurs during summer months, and winter is generally a time of anticyclonic conditions with diverging airflow. Coastal areas, however, often have a more balanced seasonal precipitation regime, which is again a reflection of their nearness to moisture sources.

The maps of average January and July precipitation provide a reliable indication of winter and summer rain/snow conditions over the world. See Figure 6-21. Prominent generalizations that can be derived from the maps include the following:

1. The seasonal shifting of major pressure and wind systems, following the sun (northward in July and southward in January) is mirrored to a lesser extent in the displacement of wet and dry zones. This is seen most clearly in tropical regions, where the heavy rainfall belt of the ITC clearly migrates north and south in different seasons.

Figure 6-21. Average January and July precipitation.

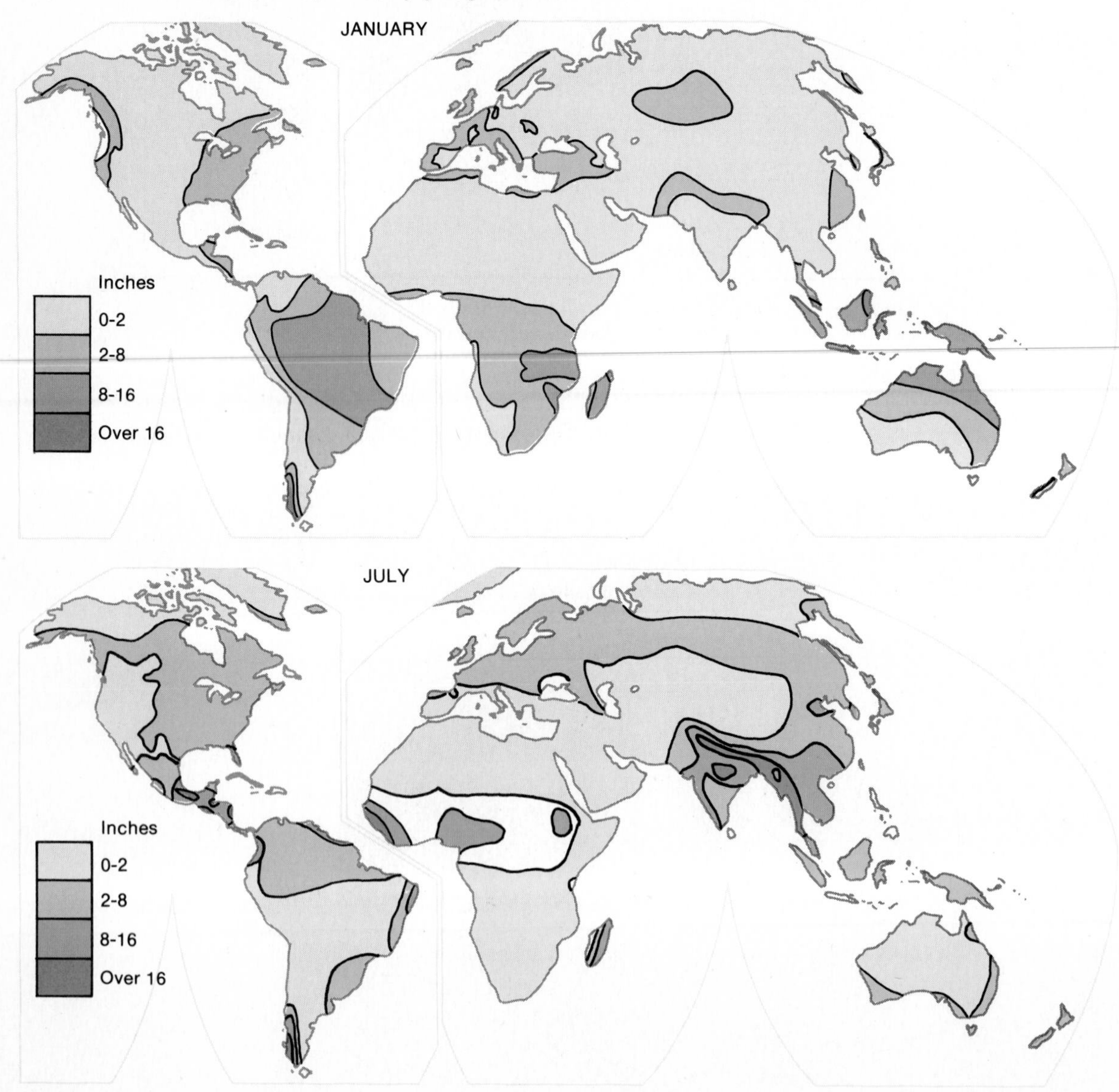

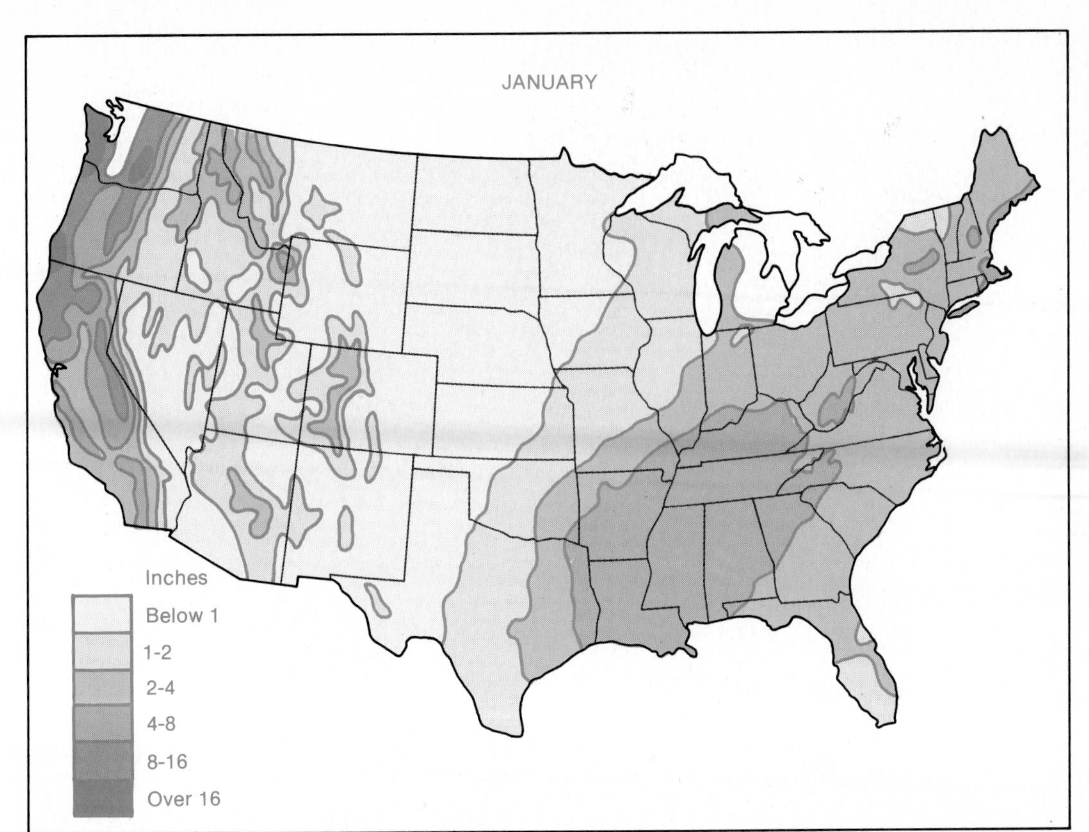

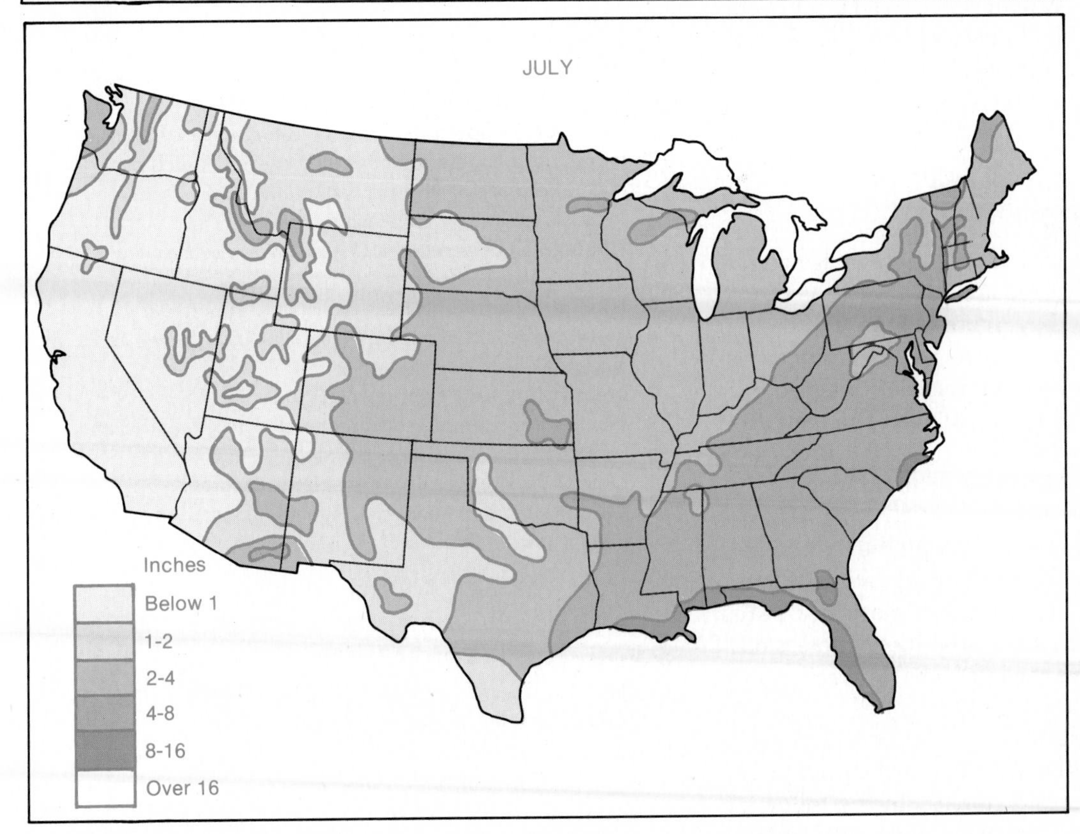

Figure 6-22. Average monthly precipitation in the conterminous United States for January and July. Winter precipitation occurs mostly in the west; summer rainfall is more general.

2. Summer is the time of maximum precipitation over almost all the world. Northern Hemisphere regions experience heaviest rainfall in July, and Southern Hemisphere locations receive most precipitation in January. See Figure 6-22. The only important exceptions to this generalization occur in relatively narrow zones along

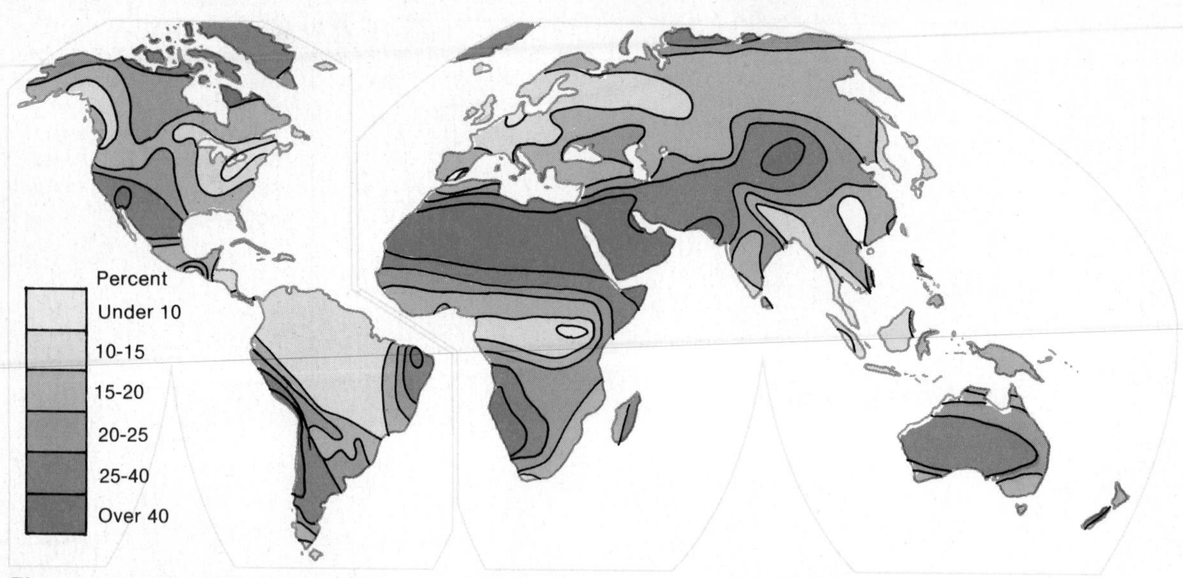

Figure 6-23. Precipitation variability. This map shows the percentage departure from average precipitation over the world. It is readily apparent that dry regions experience greater variability than humid areas.

west coasts between about 35° and 60° of latitude, as exemplified in North and South America, New Zealand, and southernmost Australia.

3. The most conspicuous variation in seasonal precipitation is found, predictably, in monsoon regions (principally southern and eastern Asia, northern Australia, and West Africa), where summer tends to be very wet, and winter is very dry.

Precipitation Variability

The maps thus far considered all portray average conditions. The data on which they are based were gathered over decades, and thus they represent an abstraction rather than a reality. In any given year, or any given season, the actual amount of precipitation may or may not be similar to the long-term average. This variation from average is another important facet of climate and is often critical for the organisms (including people) inhabiting a region.

Comparison of the maps of precipitation variability (Figure 6-23) and average annual precipitation (Plate I in back of book) reveals that regions of normally heavy precipitation experience the least variability, whereas normally dry regions also have the least reliable precipitation on a year-to-year basis.

This completes our individual consideration of the basic elements of weather and climate. In the following chapter we will attempt to pull these facts and concepts together into a more unified understanding of atmospheric dynamics, as represented by the complex behavior of various atmospheric disturbances, which can be referred to collectively as *storms*.

Review Questions

1. How does evaporation involve the storage of heat?
2. What are the factors that determine how much evaporation will take place at any particular place and time?
3. What is meant by *evapotranspiration*?
4. Distinguish between *absolute humidity*, *specific humidity*, and *relative humidity*.
5. How is it possible for air to become supersaturated?
6. What is the difference between *adiabatic lapse rate* and *average lapse rate*?
7. Identify the four families of clouds.
8. How is atmospheric stability related to adiabatic temperature changes?
9. How is hail related to atmospheric instability?
10. Is it possible to have a *rain shadow* in association with frontal precipitation? Explain.

Some Useful References

Barry, R. G. and R. J. Chorley, *Atmosphere, Weather and Climate*. New York: Holt, Rinehart and Winston, 1970.

Battan, L. J., *Fundamentals of Meteorology*. 2nd ed. Englewood Cliffs, N.J.: Prentice-Hall, Inc., 1984.

Neiburger, M., J. G. Edinger, and W. D. Bonner, *Understanding Our Atmospheric Environment*. San Francisco: W. H. Freeman & Company, Publishers, 1973.

7 Transient Atmospheric Flows and Disturbances

We have explored (in Chapter 5) the principal components of atmospheric circulation—the broad-scale wind and pressure systems of the troposphere. Over most of the Earth, however, particularly in the midlatitudes, day-to-day weather conditions are closely identified with phenomena that are more limited in both magnitude and permanence. These phenomena involve the more or less cohesive flow of a body of air that moves as a unit (called an *air mass*) as well as a variety of atmospheric disturbances (which are usually referred to as *storms*). Both the flows and the disturbances are secondary features within the general circulation of the atmosphere. They move in and with the general circulation as migratory entities that persist for a relatively short time before dissipating. Air masses and storms, then, are transient and temporary, but in some parts of the world, they are so frequent in occurrence and so dominating in influence that their interactions are major determinants of weather conditions and, to a lesser extent, of climatic characteristics.

Air Masses

Although the troposphere is a continuous body of mixed gases that surrounds the entire planet, it is by no means a uniform blanket of air. Instead, it is composed of many large, variable parcels of air, which are often distinct from one another in their characteristics. Such parcels are referred to as *air masses*.

Characteristics

To be distinguishable as an air mass, an air parcel must meet three requirements:

1. It must be large. Characteristically an immense body of air is involved. A typical air mass is more than 1,000 miles (1,600 km) in diameter and can be thought of as subcontinental in size. It is several miles or kilometers deep, and except under unusual circumstances, it is immediately adjacent to the Earth's surface.
2. It must have relatively uniform properties in the horizontal dimension. This means that at any given altitude within the mass, its physical characteristics—primarily temperature, humidity, and stability—will be relatively homogeneous.
3. It must appear and travel as a recognizable entity. Thus it must be distinct from the surrounding or neighboring air, and when it moves, it must retain its basic original characteristics and not be torn apart by differences in airflow.

Origin

An air mass develops its characteristics by remaining over a relatively uniform land or sea surface for enough time to acquire the temperature/humidity/stability characteristics of the underlying surface. For an air mass to form, air must stagnate over a large area that has a uniform surface with distinctive properties. The stagnation only needs to last for a few days if the underlying surface has prominent temperature and moisture characteristics.

The formation of air masses is normally associated with the concept of *source regions*. The idea is that certain parts of the Earth's surface are particularly well suited to generate air masses. Such regions must be extensive, physically uniform, and associated with air that is relatively stationary or anticyclonic (cyclonic conditions involve convergence, which is not likely to produce an air mass). Ideal source regions are represented by ocean surfaces and by extensive flattish land areas that have a uniform surface of snow, forest, or desert.

Figure 7-1 portrays the principal recognized source regions for air masses that affect North America. Warm air masses can form over the waters of the southern North Atlantic, the Gulf of Mexico/Caribbean Sea, and the southern North Pacific, and in summer over the deserts of southwestern United States/northwestern Mexico. Cold air masses are developed over the northern portions of the Atlantic and Pacific oceans and over the snow-covered lands of north-central Canada.

It may well be that the concept of source regions for air mass genesis is of more theoretical than actual value. A broader view would show that air masses can originate almost anywhere in the low or high latitudes, but almost never in the midlatitudes. See Figure 7-2. Thus,

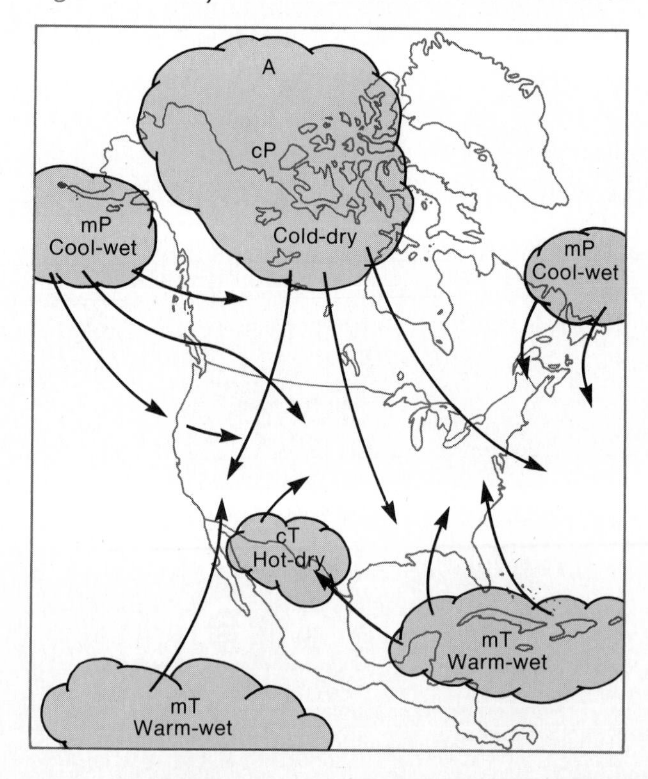

Figure 7-1. Major air masses that affect North America.

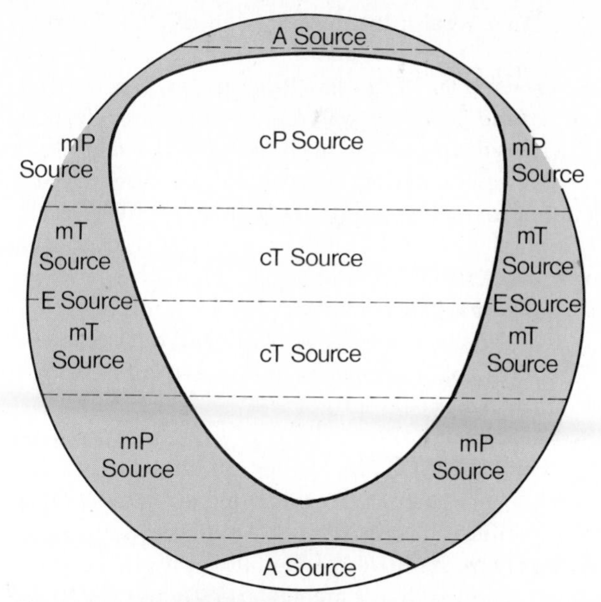

Figure 7-2. Schematic world map of air mass source regions. The tropics and subtropics are important source regions, as are the high latitudes. Air masses do not originate, however, in the middle latitudes.

air masses have their origin either in tropical/subtropical or in polar/subpolar regions, from which locations they are often injected into the midlatitudes.

Classification

Air masses are broadly identified and classified on the basis of their source regions, from which their original characteristics are derived. The latitudinal position of the source region indicates the temperature of the air mass, and the nature of the underlying surface strongly influences the humidity content of the air mass. Thus a low-latitude air mass is warm; a high-latitude one is cold. If the air mass develops over a continental surface, it is likely to be dry; if it originates over an ocean, it will be moist.

A two-letter code is used to identify air masses. Although some authorities recognize other categories, the basic classification is sixfold: arctic (A), continental polar (cP), maritime polar (mP), continental tropical (cT), maritime tropical (mT), and equatorial (E). Table 7-1 illustrates the classification with an indication of source region properties.

Movement and Modification

Some air masses remain in their source regions for long periods, even indefinitely. In such cases, the weather persists with little variation. Our interest, however, is focussed primarily on air masses that leave their source regions and move into other regions, particularly into the midlatitudes.

When an air mass departs from the area from which it acquired its original characteristics, its structure begins to change, due in part to thermal modification (heating or cooling from below), in part to dynamic modification (uplift, subsidence, convergence, turbulence), and perhaps also in part due to addition or subtraction of moisture. Thus the air mass is modified when it leaves its source region.

More importantly, the air mass modifies the weather of the region into which it moves. It takes source region characteristics into other regions. A classic example of this modification is displayed in Figure 7-3, which diagrams a situation that may occur one or more times every winter. In this case, a midwinter outburst of cP air from northern Canada sweeps down across the central part of North America. With a source region temperature of −50° F (−46° C) around Great Bear Lake, the air mass has warmed to −30° F (−34° C) by the time it reaches

TABLE 7-1 Simplified Classification of Air Masses

Type	Code	Source regions	Source region properties
Arctic (Antarctic)	A	Antarctica; Arctic Ocean and its fringes; Greenland	Very cold, very dry, very stable
Continental polar	cP	High-latitude plains of Eurasia and North America	Cold, dry, very stable
Maritime polar	mP	Oceans in the vicinity of 50° to 60° latitude	Cool, moist, relatively unstable
Continental tropical	cT	Low-latitude deserts, especially the Sahara and Australian deserts	Hot, very dry, unstable
Maritime tropical	mT	Tropical and subtropical oceans	Warm, moist, variable stability
Equatorial	E	Oceans near the equator	Warm, very moist, unstable

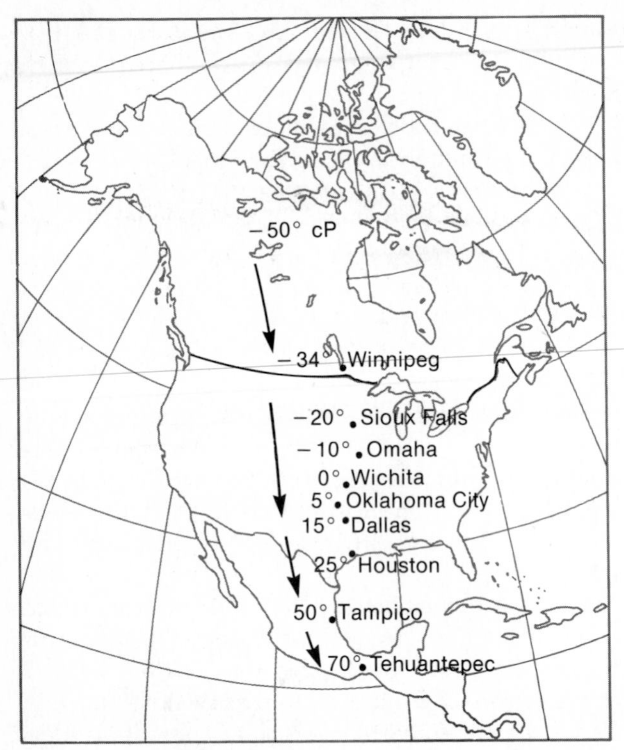

Figure 7-3. An example of temperatures resulting from a strong southerly outburst of cP air from Canada in midwinter.

Winnipeg, and it continues to warm as it moves southward:

−20° F (−29° C) at Sioux Falls, South Dakota
−10° F (−23° C) at Omaha, Nebraska
0° F (−17° C) at Wichita, Kansas
5° F (−15° C) at Oklahoma City, Oklahoma
15° F (−10° C) at Dallas, Texas
25° F (−4° C) at Houston, Texas
50° F (10° C) at Tampico, Mexico
70° F (21° C) at Tehuantepec, Mexico

Throughout its southward course, the air mass is becoming warmer, but it also is bringing some of the coldest weather that each of these places will receive all winter. Thus the air mass is being modified, but it is also modifying the weather in all regions that it traverses. Moreover, temperature is only one of the characteristics involved; there are modifications of humidity and stability.

North American Air Masses

The North American continent is a prominent area of air mass interaction. The lack of east-west trending mountains permits extensive meridional movement, so polar air can sweep southward and tropical air can flow northward unhindered by terrain, particularly over the eastern two-thirds of the continent. Major ranges in the West, however, impede the movement of the Pacific air masses, causing significant modification of their characteristics.

Continental polar (cP) air develops in central and northern Canada, and *arctic* (A) air originates farther north. These two are similar in characteristics except that the latter is even colder and drier than the former. They are dominant features in winter with their cold, dry stable nature. Sporadically cP air (and rarely A air) surges southward or southeastward, and occasionally southwestward, bringing cold waves wherever it goes. In summer, cP air is much less well developed and prominent, but occasionally it provides cooling relief to portions of eastern and central North America.

Maritime polar (mP) air that affects this continent normally originates as cold, dry cP air that has moved off the land and stagnated or has moved slowly over the North Pacific or North Atlantic, acquiring its mP characteristics only in the lower portion of the air mass. Consequently, the lower layers are cool, moist, and relatively unstable, whereas conditions aloft may be cold, dry, and stable. Pacific mP air normally brings widespread cloudiness and heavy precipitation to the mountainous coastal regions, but it is often severely modified in crossing the western ranges; and by the time it reaches the interior of the continent, it has moderate temperatures and clear skies. In summer, the ocean is colder than the land, so Pacific mP air produces fog and low stratus clouds along the coast but takes no distinctive weather conditions to the interior.

Air masses that develop over the North Atlantic are also cool, moist, and unstable. Except for occasional incursions into the mid-Atlantic coastal region, however, Atlantic mP air does not affect North America because the prevailing circulation is westerly.

Maritime tropical (mT) air from the Atlantic/Caribbean/Gulf is characteristically warm, moist, and unstable. It strongly influences weather and climate east of the Rockies in the United States and southern Canada, serving as the principal source of moisture for most precipitation in this broad region. It is more prevalent and extensive in summer than in winter, bringing periods of uncomfortable humid heat.

Pacific mT air originates over water in areas of anticyclonic subsidence, so it is cooler, drier, and more stable than Atlantic mT air. Its influence is felt only in southwestern United States and northwestern Mexico, where in winter it produces some coastal fog and occasional moderate orographic rainfall if forced to ascend mountain slopes. It is also the source of some summer rains in the southwestern interior.

Continental tropical (cT) air is relatively unimportant in North America because its source region is not extensive and consists of varied terrain. There is no winter air mass genesis in this northern Mexico/southwestern United States source region; but in summer, hot, very dry, unstable cT air develops. It surges into the southern

Great Plains area on occasion, bringing heat waves and drought conditions.

Equatorial (E) air affects North America only in association with hurricanes. It is similar to mT air except that it provides an even more copious source of rain because of its high humidity and instability.

Fronts

When unlike air masses meet, they do not mix readily; instead, a boundary zone develops between them. This boundary is a *front*. A front should not be thought of as a simple two-dimensional boundary surface, nor as an extensive zone of gradual transition. Rather, it consists of a relatively narrow zone of discontinuity within which the rate of change of air mass properties is rapid. A typical front is several miles or even tens of miles in width.

The basic significance of a front is that it separates air masses of differing characteristics; it functions as a barrier between the two masses, preventing their mingling except in the narrow transition zone. The most conspicuous difference between air masses is usually in terms of temperature, so a frontal surface usually separates warm air from cool air. At any given level in the air there may be relatively uniform warm temperatures on one side of the front and relatively uniform cool temperatures on the other, with a fairly steep and abrupt temperature gradient through the frontal zone (Figure 7-4).

Air masses may differ in other factors than temperature, of course. They may vary significantly in density, humidity, stability, and other characteristics. Frequently all these factors show a steep gradient of change through the frontal zone.

Although fronts may be several miles in width, they are quite narrow in comparison with the vast extent of the adjacent air masses. Their position on weather maps is usually shown by a line that marks the intersection of the front with the Earth's surface. On a typical weather map's scale, it is not seriously misleading to represent the surface position of a frontal zone by a broad line.

An important attribute of frontal surfaces is that they lean with height. Indeed, they slope at such a low angle that they are much closer to horizontal than vertical in their orientation. The normal slope of a front varies between about 1:50 and 1:300, with an average of about 1:150; this means that 150 miles away from the surface position of the front, the frontal surface will be at a height of only one mile above the ground. Because of this very low angle of slope (less than 1°), the steepness shown in most vertical diagrams of frontal surfaces is greatly exaggerated. A front always slopes so that the warmer air overlies the cooler air.

Figure 7-4. Vertical cross section through a hypothetical front, with the cooler air on the left and the warmer air on the right. The solid lines represent isotherms.

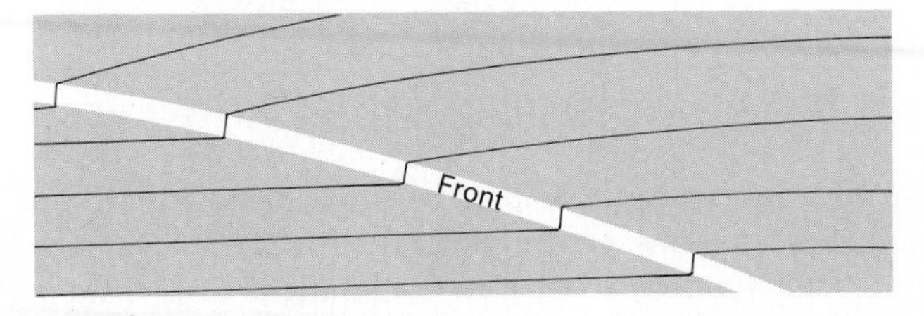

In some cases, a front remains stationary, without moving over the Earth's surface, for a few hours or even for a few days. More commonly, however, the front is in more or less constant motion, shifting the position of the boundary between the air masses but maintaining its function as a barrier between them. Usually one air mass is actively displacing the other; thus the front is advancing in the direction dictated by the movement of the more active air mass.

The frontal concept was developed by Norwegian meteorologists at the time of World War I, and the term *front* was coined because they considered that the clash between unlike air masses was analogous to a confrontation between opposing armies along a battle front. As the more "aggressive" air mass advances at the expense of the other, there is some mixing of the two within the frontal zone, but for the most part, the air masses retain their separate identities as one is displaced by the other. Regardless of which air mass is advancing, it is always the warmer that rises over the cooler. The warmer, lighter air is inevitably forced aloft, and the cooler, denser mass functions as a wedge upon which the lifting occurs. An "aggressor" warm air mass actively overrides the cooler air, and an advancing cool air mass actively underrides the warmer air. Fronts are usually identified on the basis of these situations.

Warm Fronts

The leading edge of an advancing warm air mass represents a *warm front*. The slope of a warm front is extremely gentle, averaging about 1:200. The warm air ascends over the retreating cool air, its temperature decreasing adiabatically as it rises. The usual result is clouds and precipitation. The frontal uplift is very gradual, so that cloud formation normally is slow, expansive, and not very turbulent. High-flying cirrus clouds may serve as advance harbingers of the approaching front many hours before the surface front arrives. As the surface front comes closer, the clouds become lower, thicker, and more extensive. Precipitation usually occurs in a broad band; it is likely to be protracted and gentle, without much convective activity. As illustrated in Figure 7-5, most precipitation falls in advance of the surface position of the front. If the rising air is inherently unstable, however, precipitation can be showery and even violent.

Figure 7-5. Along a warm front, warm air rises above cooler air. Often widespread cloudiness and precipitation develop along and in advance of the surface position of the front. Higher and less dense clouds often appear dozens or hundreds of miles ahead of the surface position of the front.

Figure 7-7. A cold front is usually steeper than a warm front. The warm air is forced upward by the advancing cold air behind the front. This often creates cloudiness and relatively heavy precipitation along and immediately behind the surface position of the front.

The surface position of a warm front is portrayed on a weather map either by a red line (if color is used) or (more typically) by a solid black line along which black semicircles are located at regular intervals with the semicircles extending in the direction of the cool air. See Figure 7-6.

Cold Fronts

The leading edge of a cool air mass actively displacing warm air is a *cold front*. Frictional contact with the Earth's surface retards the advance of the lower portion of the front, so that it tends to become steeper as it progresses and usually develops a protruding "nose" a few hundred yards (meters) above the ground. The average cold front has a slope that is twice as steep as the average warm front. See Figure 7-7. Moreover, cold fronts normally move faster than warm fronts. This combination of steeper slope and faster advance makes for more rapid lifting of the warm air, which enhances instability and produces more blustery and violent weather along the cold front. Vertically developed clouds are common, with considerable turbulence and showery precipitation. Both clouds and precipitation tend to be concentrated along and immediately behind the surface position of the front. Precipitation is usually of higher intensity but shorter duration than that associated with a warm front.

On a weather map, the surface position of a cold front is shown either by a blue line or by a solid black line studded at intervals with solid triangles that extend in the direction of the warm air.

Figure 7-6. Weather map symbols for fronts.

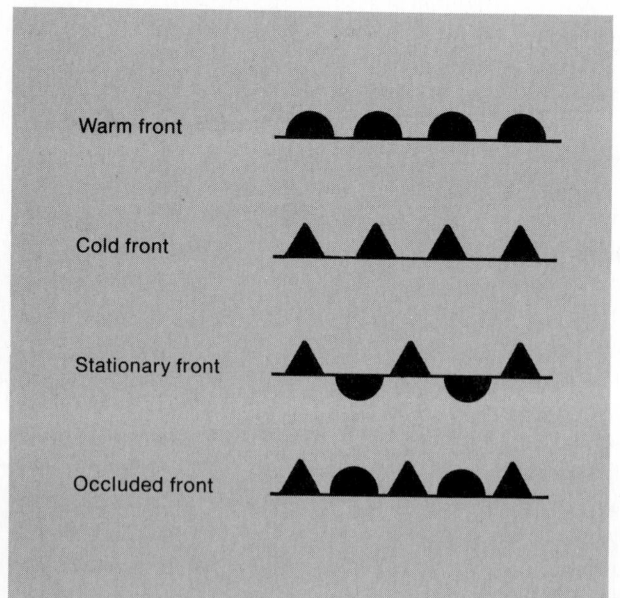

Stationary Fronts

When neither air mass displaces the other, their common "boundary" is a *stationary front*. Weather is not readily predictable along such a front, but often gently rising warm air produces limited precipitation of the warm-front type. Stationary fronts are portrayed on a weather map by a combination of warm and cold front symbols: either an alternating red and blue line, or a black line with alternating triangles on one side and semicircles on the other.

Occluded Fronts

A more complex type of front is formed when a cold front overtakes a warm front, forming an *occluded front*. The development of occluded fronts will be discussed later in this chapter. An occluded front is portrayed on a weather map either by an alternating red and blue line or by a black line with alternating semicircles and triangles on the same side of the line.

Atmospheric Disturbances

The broad-scale dynamics of the troposphere have been discussed in terms of the general circulation. We now turn our attention to the various kinds of disturbances that occur within the general circulation. These are migratory in nature and more limited in magnitude but are of great significance to both weather and climate. Most of these disturbances involve unsettled and even violent atmospheric conditions, and can be referred to as *storms*. Some, however, produce calm, clear, quiet weather that is quite the opposite of stormy. Their common characteristics are:

1. They are smaller than the components of the general circulation, although they are extremely variable in size.
2. They are migratory and transient.
3. They have a relatively brief duration, persisting for only a few minutes, a few hours, or a few days.
4. They produce characteristic and relatively predictable weather conditions that are readily recognizable in our daily lives.

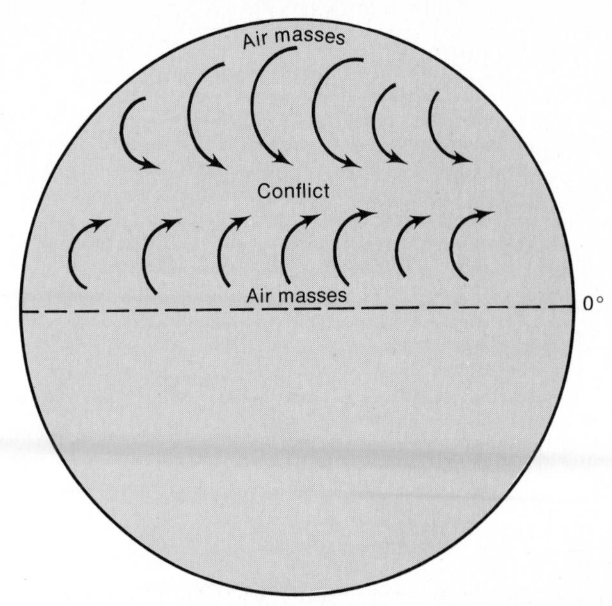

Figure 7-8. The midlatitudes are zones of air mass conflict.

Major Midlatitude Disturbances

The middle latitudes represent the principal battleground of tropospheric phenomena, where the air masses from polar and tropical regions meet and come in conflict, where most frontal conditions occur, and where weather is most dynamic and changeable from season to season and from day to day. See Figure 7-8. A considerable variety of atmospheric disturbances is associated with the midlatitudes, but two of these are much more important than the others because of their size and prevalence. They are the *extratropical cyclones* and *anticyclones*.

Extratropical Cyclones

Probably most significant of all atmospheric disturbances are the *extratropical cyclones*. Everywhere poleward of 30° latitude, they dominate weather maps, are basically responsible for most day-to-day weather changes, and bring precipitation to much of the populated portions of the planet. They consist of large migratory low-pressure systems that occur within the middle latitudes and portions of the high latitudes. In Europe, they are usually referred to as *depressions*; in the United States, they are sometimes called *lows* or *wave cyclones*.

These disturbances are associated primarily with air mass convergence and conflict in regions between about 35° and 70° of latitude. Thus they are found almost entirely within the zone of westerly winds. Their general path of movement is toward the east, which explains why weather forecasting in the midlatitudes is essentially a west-facing vocation; that is, the principal weather systems move from west to east with the general flow of the westerlies.

Extratropical cyclones can properly be considered as storms, for they bring changeable, unsettled weather that normally includes widespread, abundant, and often intensive precipitation. Their characteristics are well known, but their variations are almost infinite. Each one differs from the others in greater or lesser detail, so that any description can only be generalized and cannot be expected to fit a specific cyclone. The discussions that follow, then, pertain to "typical" or idealized conditions. Moreover, these conditions are presented as Northern Hemisphere phenomena. For Southern Hemisphere application, the patterns of isobars, fronts, and wind flow should be visualized as mirror images of the Northern Hemisphere patterns.

Characteristics of a Well-Developed Extratropical Cyclone A typical mature extratropical cyclone is very extensive; it has a diameter of 1000 miles (1600 km) or so. It is essentially a vast cell of low pressure—characteristically, the surface pressure in the center of the low is between 990 and 1000 millibars, but it can be somewhat higher or significantly lower under extreme conditions—with a normal cyclonic circulation pattern—converging counterclockwise in the Northern Hemisphere. The system (as shown by the closed isobars on a weather map) usually tends toward an oval shape with the long axis trending northeast-southwest. Often a clear-cut pressure trough or "valley" extends southwesterly from the center of the low.

Figure 7-9 depicts a mature extratropical cyclone in the middle of the United States, with its center approximately over St. Louis. The cyclonic wind-flow pattern attracts cool air from the Great Lakes area and southern Canada into the northern and western portions of the system, and warm air from the Gulf Coast into the southern and eastern portions. The irregular convergence of these unlike air masses characteristically creates two

Figure 7-9. A typical mature extratropical cyclone in the Northern Hemisphere has a cold front trailing to the southwest and a warm front extending toward the east.

fronts: a cold front extending southwesterly from the center of the low in the pressure trough previously mentioned, and a warm front extending generally toward the east from the center of the low in another pressure trough.

These two fronts serve to divide the cyclone into two "sectors": a cool sector north and west of the center of the low, and a warm sector to the south and east. At the surface, as shown in Figure 7-10, the cool sector is the larger of the two, but aloft the warm sector is more extensive. This is because both frontal surfaces lean with height over the cool air. Thus the cold front slopes upward toward the northwest, and the warm front slopes upward toward the north.

Along both frontal surfaces, warm air rises. It is uplifted by the advancing cold air along the cold front and actively overrides the cold air as it ascends the warm front. The typical result is two zones of cloudiness and precipitation that overlap around the center of the storm and extend outward in general association with the fronts. Along and immediately behind the surface position of the cold front, a band of cumuliform clouds usually yields showery precipitation. The air rising more gently along the warm frontal surface will produce a more extensive expanse of horizontally developed clouds, perhaps with widespread, protracted, low-intensity precipitation. In both cases, most of the precipitation originates in the warm air rising above the fronts (i.e., in warm sector air) and falls down through the frontal surface to reach the ground in the cool sector of the storm.

This does not mean that all the cool sector has unsettled weather and that the warm sector experiences clear conditions throughout. Although most frontal precipitation falls within the cool sector, the general area to the north, northwest, and west of the center of the low is frequently cloudless as soon as the cold front conditions have moved on. Thus much of the cool sector is typified

Figure 7-10. Cross section through an idealized mature extratropical cyclone. The arrows indicate the direction of frontal movements.

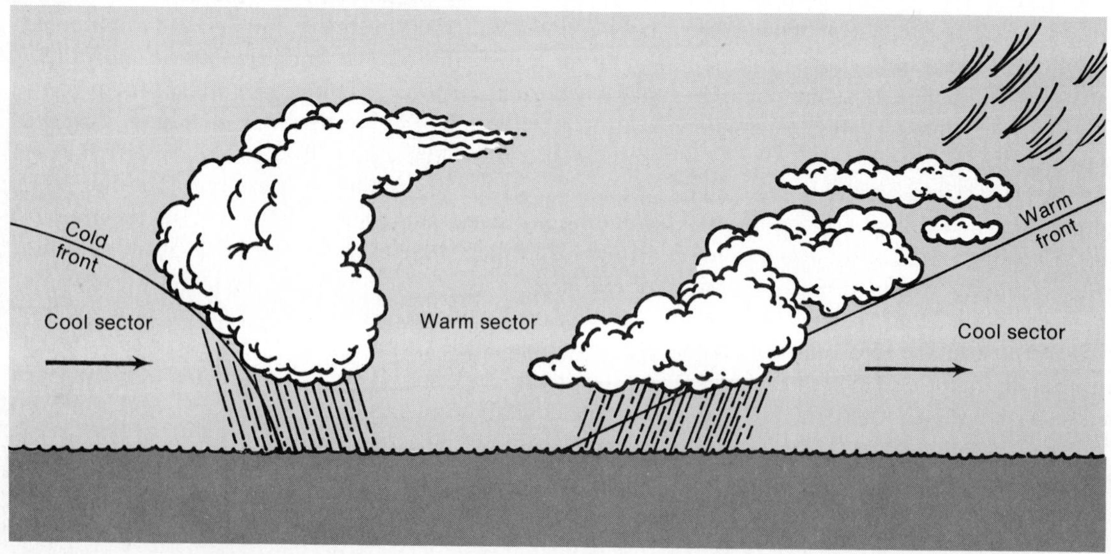

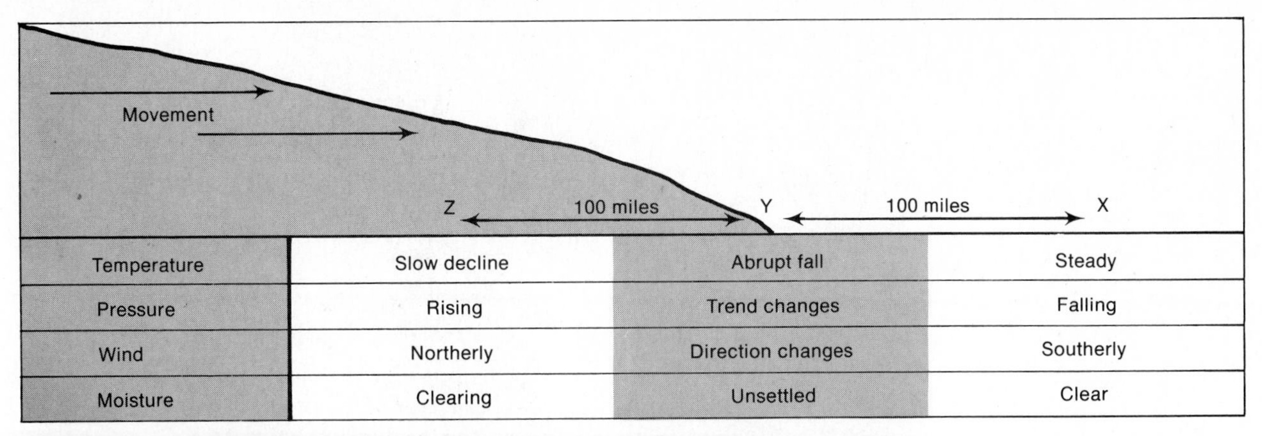

	Z	Y	X
Temperature	Slow decline	Abrupt fall	Steady
Pressure	Rising	Trend changes	Falling
Wind	Northerly	Direction changes	Southerly
Moisture	Clearing	Unsettled	Clear

Figure 7-11 Schematic diagram of weather changes associated with the passage of a cold front. As the position of the front moves from left to right across the diagram, the weather at "X" becomes the weather at "Y," which becomes the weather at "Z." All four weather elements experience a significant change as the front passes.

by clear, cold, stable air. In contrast, the air of the warm sector is often moist and tending toward instability, so thermal convection may produce sporadic thunderstorms. Also sometimes one or more squall lines may develop in advance of the cold front within the warm sector. This is a line of intense thunderstorms generated by the cold front that parallels it but precedes its advance by several miles or tens of miles.

When a front passes over an area, very abrupt weather changes normally occur. For example, with passage of a cold front (see Figure 7-11), we note the following:

1. The temperature decreases sharply.
2. Winds shift from southerly ahead of the front to northerly following it.
3. The front is in a pressure trough, so pressure falls as the front approaches and rises after it passes.
4. Generally clear skies are replaced by cloudiness and precipitation.

Similar changes, although usually of lesser magnitude, occur with the passage of a warm front.

Movements Extratropical cyclones are essentially transient features that are on the move throughout their existence. This transiency is complicated by the fact that four different kinds of movement are involved (see Figure 7-12); these are related to one another but can be analyzed as discrete entities:

1. The system has a cyclonic wind circulation, with air generally converging from all sides in a counterclockwise pattern in the Northern Hemisphere.
2. The entire storm system moves as a major disturbance in the westerlies, traversing the middle latitudes generally from west to east. The rate of movement averages 20 to 30 miles (32 to 48 km) per hour, which means that the storm can cross the United States in three to four days. It moves faster in winter than in summer.
3. The cold front normally advances faster than the storm is moving. Thus it swings from its pivot in the center of the low, increasingly moving into and displacing the warm sector of the storm.
4. The warm front usually advances slower than the storm itself. This causes it to lag behind and has the effect of swinging the warm front backwards in the direction of the advancing cold front.

Life Cycle Extratropical cyclones are ephemeral phenomena. Some exist for only a few hours, but most survive for several days or up to a maximum of about two weeks. A typical system develops from origin to maturity

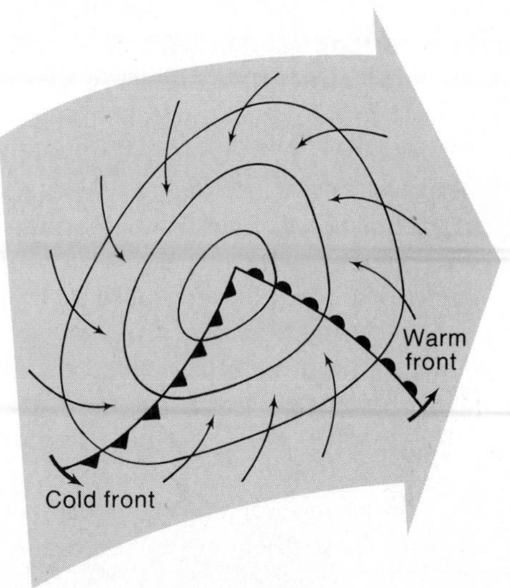

Figure 7-12. Four different varieties of motion occur in a typical extratropical cyclone. These are: (1) generally convergent airflow, (2) cold front advance, (3) warm front advance, and (4) movement of the entire system in the general flow of the westerlies.

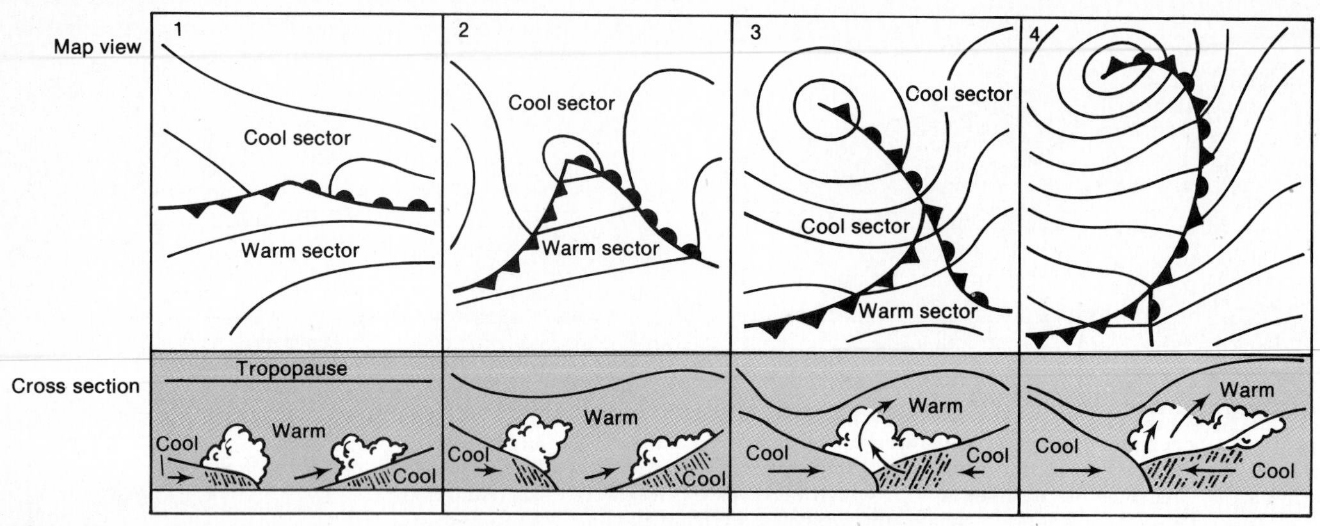

Figure 7-13. Stages in the life cycle of an extratropical cyclone: (1) early development, (2) maturity, (3) partial occlusion, (4) full occlusion.

in four to seven days, and from maturity to dissipation in about the same length of time. See Figure 7-13.

Cyclogenesis (the origin of cyclones) is complex and not completely understood. Most extratropical cyclones begin as waves along the contact line between unlike air masses in the subpolar low-pressure zone, a line that is sometimes called the *polar front*. The opposing airflows tend to initiate a counterclockwise rotation, producing a small perturbation that may develop into a full-fledged cyclone. The critical conditions for cyclogenesis apparently are found in the upper troposphere; in other words, pressure and wind characteristics aloft determine whether or not a surface wave will develop.

Whatever the mechanics of origin, once the wave begins to develop, it produces flows of warm air poleward and of cold air equatorward. As the contrasting air masses come into conflict, fronts become clearly established, and the storm grows in intensity and size. The mature stage of the storm has been described previously.

Ultimately, the storm dissipates because the cold front overtakes the warm front. As the two fronts come closer and closer together (Figure 7-13), the warm sector is increasingly displaced at the surface, forcing more and more warm air aloft. When the cold front actually catches up with the warm front, warm air is no longer in contact with the Earth's surface, and an occluded front is formed. See Figure 7-14. This *occlusion* process usually results in a short period of intensified activity (precipitation and wind), until eventually all the warm sector is forced aloft, and the surface low-pressure center is surrounded on all sides by cool or cold air. This weakens the pressure gradient and shuts off the storm's energy, so that it dies out.

It is not unusual, however, for another storm to develop from the remnant of the previous one. As occlusion progresses, a new wave sometimes forms along the trailing end of the cold front, off to the southwest of the original storm center. This process may be repeated so that a third or fourth cyclone could eventuate from a single frontal system. See Figure 7-15.

Occurrence and Distribution Extratropical cyclones are continually present in some numbers. At any given time, from 6 to 15 may exist in the Northern Hemisphere midlatitudes, and an equal number in the Southern Hemisphere. They occur at scattered but irregular intervals throughout the zone of the westerlies.

In each hemisphere, these migratory disturbances are more numerous, better developed, and faster moving in winter than in summer. Also, they follow much more equatorward tracks in the cold part of the year. In the Southern Hemisphere, the Antarctic continent provides a prominent year-round source of cold air, so that vigorous cyclones are almost as numerous in summer as in winter; but they are further poleward, over the Southern Ocean, with little effect on land areas.

The specific route of travel of any individual cyclone is likely to be undulating and erratic, although it moves generally from west to east. Certain general tracks, how-

Figure 7-14. Cross section of an occluded front.

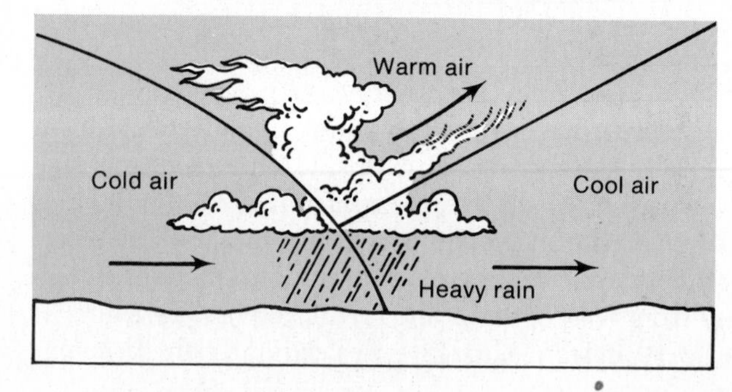

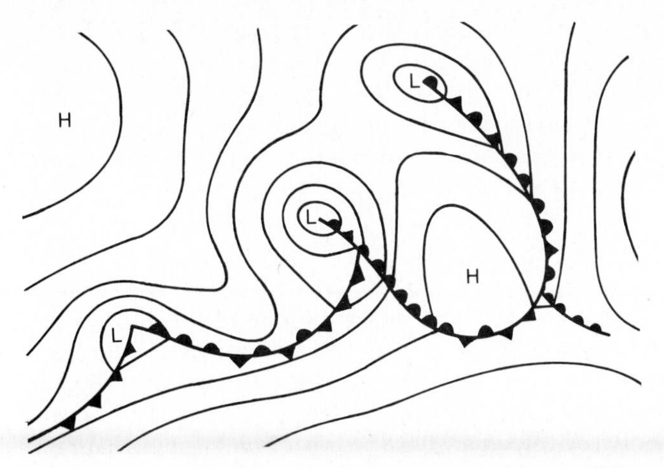

Figure 7-15. Schematic diagram of a series of wave cyclones. [After Morris Neiburger, James G. Edinger, and William D. Bonner, *Understanding Our Atmospheric Environment,* 2nd ed. (San Francisco: W. H. Freeman & Company, Publishers, 1982), p. 187. Copyright © 1982 by W. H. Freeman and Company. Used by permission of the publisher. All rights reserved.]

ever, are followed more commonly than others. In the United States, for example, most cyclones enter the continent across the Pacific Northwest, and almost all depart across New England. See Figure 7-16. Across the central part of the country, their paths are much more dispersed, but the northerly tracks are the most frequented. In addition, extratropical cyclones often are generated on the leeward (eastern) side of the Rocky Mountains, especially in Colorado and Alberta, and along the central Atlantic (Virginia/Carolina) coast.

Extratropical Anticyclones

Another major disturbance in the general flow of the westerlies is the *extratropical anticyclone*, frequently referred to simply as a *high*. This is an extensive, migratory high-pressure cell of the midlatitudes. See Figure 7-17. Typically it is somewhat larger than an extratropical cyclone and moves with the westerlies generally west to east.

As with any high-pressure center, an extratropical anticyclone has air subsiding into it from above and diverging at the surface, clockwise in the Northern Hemisphere. No air mass conflict or convergence is involved,

Figure 7-16. Typical paths of extratropical cyclones across the United States.

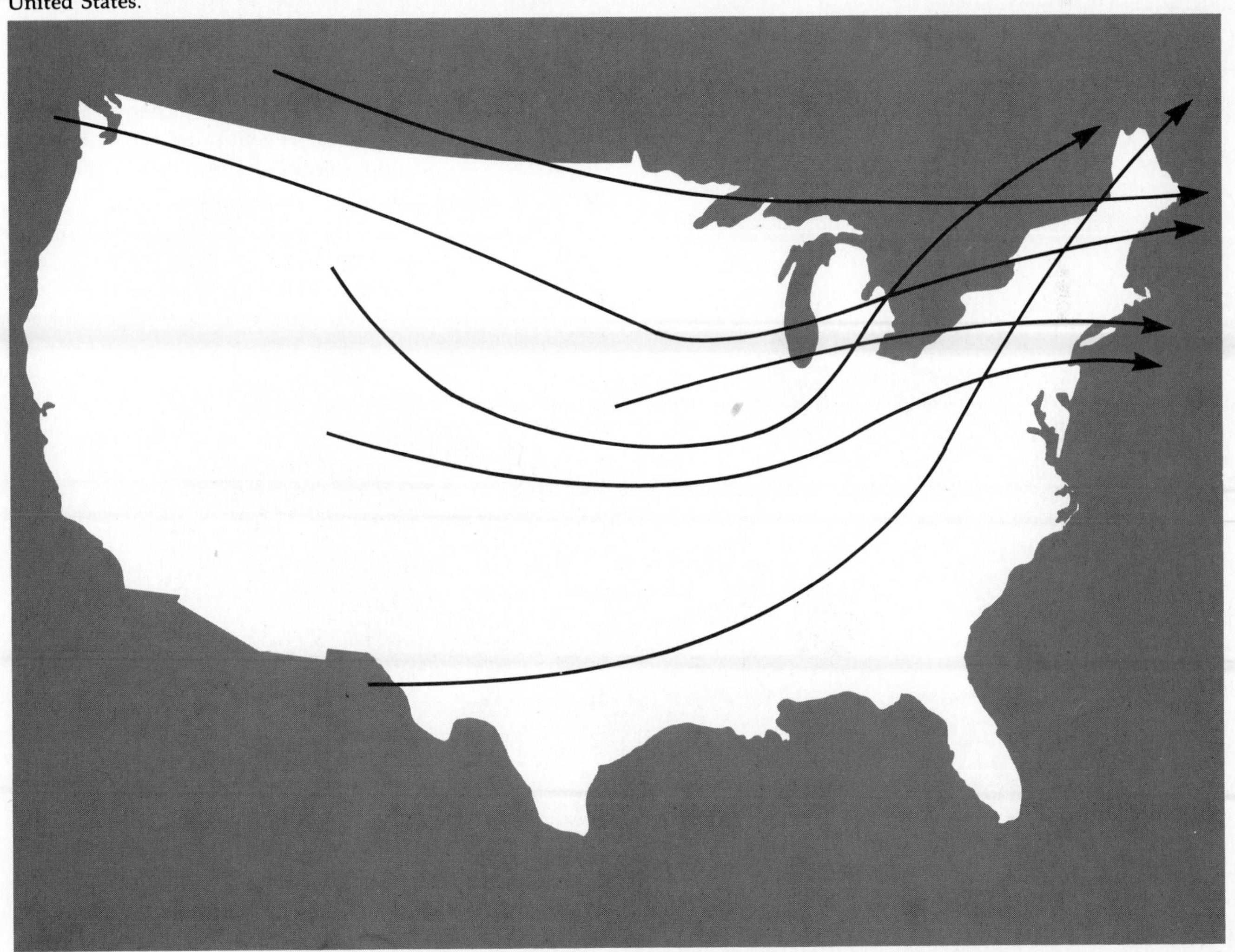

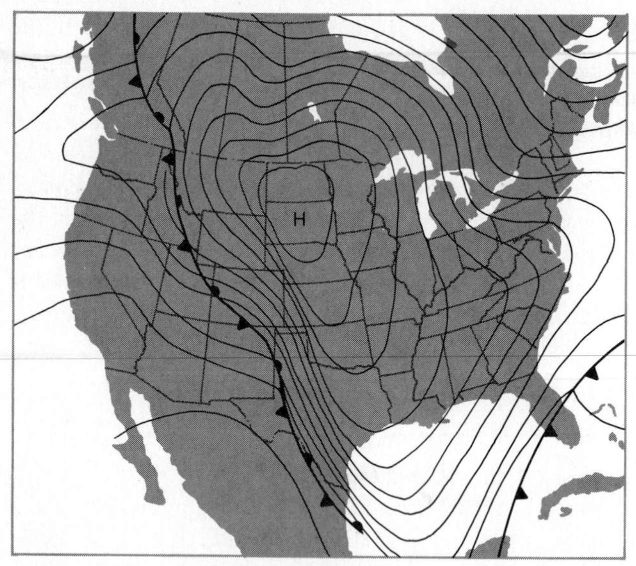

Figure 7-17. A typical well-developed extratropical anticyclone, centered over the Dakotas.

so anticyclones contain no fronts. The weather is clear and dry with little or no opportunity for cloud formation. Wind movement is very limited near the center of an anticyclone, but it increases progressively outward. Particularly along the eastern margin (the leading edge) of the system, there may be strong winds. In winter, anticyclones are characterized by very cold temperatures.

Anticyclones move toward the east at about the same rate as, or a little slower than, extratropical cyclones. Unlike cyclones, however, they are occasionally prone to stagnate and remain over the same region for several days. This brings clear, stable, dry weather to the affected region, which enhances the likelihood of air pollutants becoming concentrated. Such stagnation also may block the eastward movement of cyclonic storms, causing protracted precipitation in some other region while the anticyclonic region remains dry.

Relationships of Cyclones and Anticyclones

Extratropical cyclones and anticyclones function as migratory perturbations in the westerlies, alternating with one another in irregular sequence around the world in the midlatitudes. See Figure 7-18. Each can occur independently of the other, but there is often a functional relationship between them. This can be seen most clearly, perhaps, when an anticyclone closely follows a cyclone, as diagrammed in Figure 7-19. The winds diverging from the eastern margin of the high fit precisely into the flow of air converging into the western side of the low. It is easy to visualize the anticyclone as a polar air mass with its leading edge represented by the cold front of the cyclone.

The development of cyclones and the paths followed by both cyclones and anticyclones are significantly determined by conditions aloft in the westerlies. The undulations of the westerly jet and the position of the Rossby waves are closely associated with the movements of these migratory pressure systems. Weather forecasting in the midlatitudes largely depends on predicting the characteristics and movements of these lows and highs. As our knowledge of upper tropospheric conditions expands, we will increasingly be better equipped to under-

Figure 7-18. At any given time, the midlatitudes are dotted with extratropical cyclones and anticyclones. This map depicts a hypothetical situation in January.

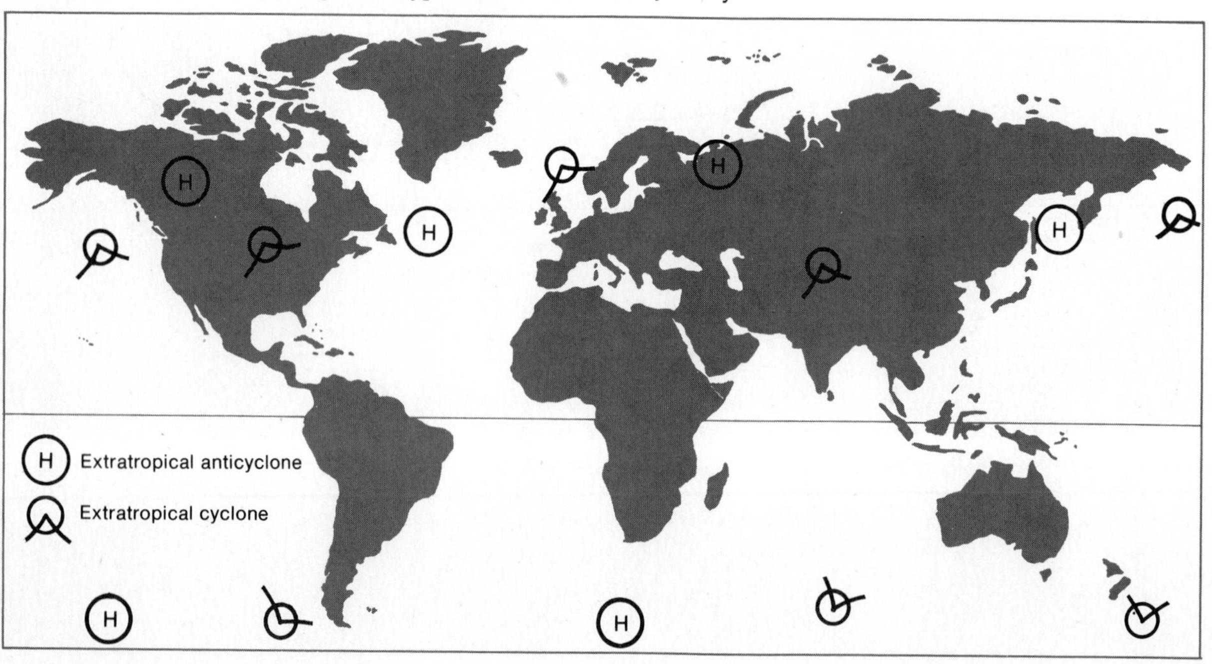

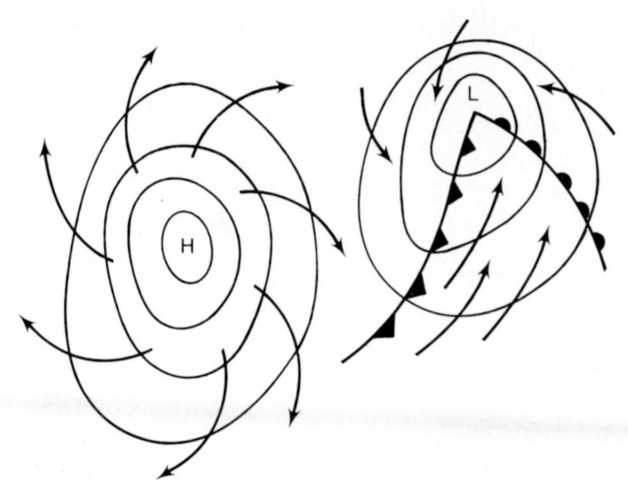

Figure 7-19. Extratropical cyclones and anticyclones often occur in juxtaposition to one another in the middle latitudes.

stand the intricacies of extratropical cyclones and anticyclones.

These lows and highs are by far the dominant features of winter weather in the middle latitudes; their significance in summer is somewhat less. Beyond that, however, they are also vital as an influence on broader climatic characteristics, due to their size, frequency, distribution, and weather patterns.

Hurricanes: Major Tropical Disturbances

Whereas the midlatitudes constitute a zone of dynamic, changeable weather, the low latitudes are characterized by monotony—repetitive weather conditions day after day, week after week, month after month. Almost the only breaks in this monotonous regime are provided by transient atmospheric disturbances, of which by far the most significant affecting the tropics and subtropics are the *tropical cyclones*. These storms are known by various

Figure 7-20. A hurricane as it appears on a simplified weather map.

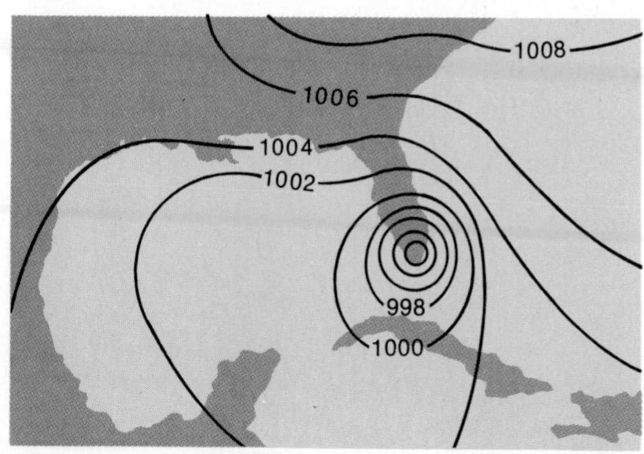

Figure 7-21. The remarkable cloud spiral of a well-developed hurricane over the Gulf of Mexico. This is hurricane Gladys in 1968. [*Apollo 7* photo, courtesy NASA.]

names in different portions of the low latitudes: those affecting North and Central America are called *hurricanes*; they are known as *typhoons* in the western North Pacific region, except in the Philippines where they are *baguios*; and they are simply called *cyclones* in the Indian Ocean and Australia. Whatever the name, these are intense, revolving, rain-drenched, migratory, destructive storms that occur erratically in certain regions of the tropics and subtropics. See Figures 7-20 and 7-21.

Hurricanes (as they will be referred to in this chapter to avoid confusion) consist of prominent low-pressure centers that are essentially circular in shape and have a steep pressure gradient outward from the center. A mature hurricane will have a 50–100-millibar difference in pressure between its center and its periphery. As a result, strong winds spiral inward in cyclonic fashion. See Figure 7-22. Winds must reach a speed of 74 miles (119 km) per hour for the storm to be considered officially as a hurricane, but winds in a well-developed hurricane often double that velocity and occasionally triple it. See Figure 7-23. The speedy flow of the winds causes them to converge at a low angle across the isobars, with convergent uplift increasing inward.

These lows are considerably smaller than extratropical cyclones, typically having a diameter of between 100 and 600 miles (160 and 1,000 km). A remarkable feature of hurricanes is the presence of a nonstormy *eye* in the center of the storm. The winds do not converge to a central point but reach their highest speed at the *eye wall*, which is the edge of the eye. The eye itself has a diameter of from 10 to 25 miles (16 to 40 km) and is a singular

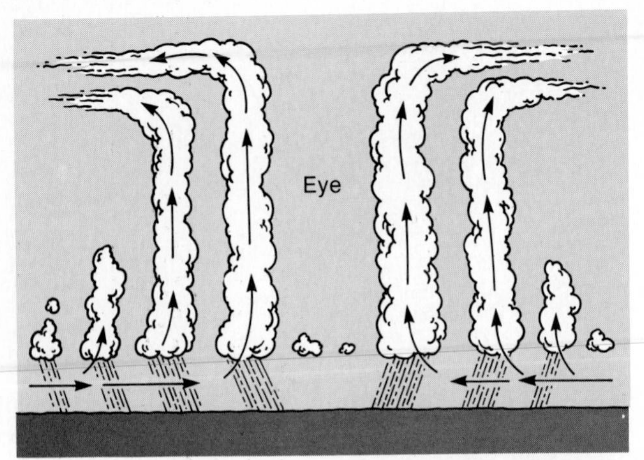

Figure 7-22. Idealized cross section through a well-developed hurricane. Air spirals into the storm horizontally and rises rapidly to produce towering cumulus and cumulonimbus clouds that yield torrential rainfall.

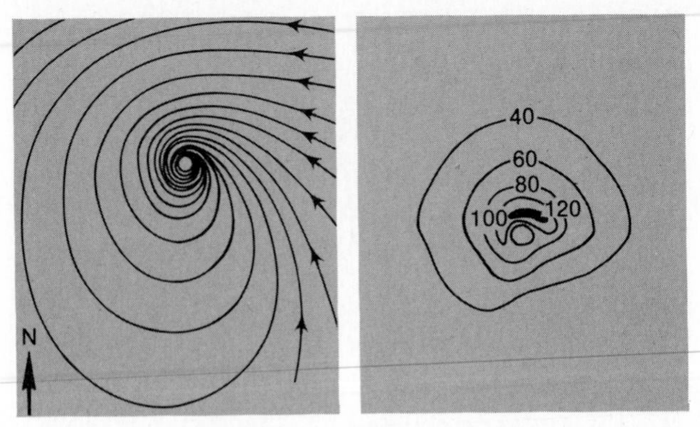

Figure 7-23. Hypothetical models of surface wind-flow pattern and wind velocity in a Northern Hemisphere hurricane. Wind velocity is shown in miles per hour.

area of calmness in the maelstrom that whirls around it. Updrafts are common throughout the hurricane, becoming most prominent around the eye wall. Near the top of the storm, outward flow dominates except in the eye, where a downdraft inhibits cloud formation.

The weather pattern within a hurricane is relatively symmetrical around the eye. Bands of dense cumulus and cumulonimbus clouds spiral in from the edge of the storm to the eye wall, producing heavy rains that generally increase in intensity inward. The clouds of the eye wall tower to heights that may exceed 10 miles (16 km). Within the eye, however, there is no rain or low cloud, and scattered high clouds may part to let in intermittent sunlight.

The origin of hurricanes is not yet understood in detail. They form only over warm oceans in the tropics, but at least 5° north or south of the equator. Hurricanes require an enormous supply of energy, which is provided by the latent heat released during condensation of water vapor evaporated from tropical ocean surfaces. The exact mechanism of formation is not clear, but they develop from smaller tropical disturbances (especially easterly waves, discussed subsequently), which provide low-level convergence and lifting. Less than 10 percent of these minor disturbances grow into hurricanes.

Tropical cyclones occur in a half-dozen low-latitude regions. See Figure 7-24. They are most common in the North Pacific basin, originating largely in two areas: east of the Philippines and west of southern Mexico. The third most notable region of tropical cyclone development is in the west-central portion of the North Atlantic basin,

Figure 7-24. Principal tropical cyclone tracks.

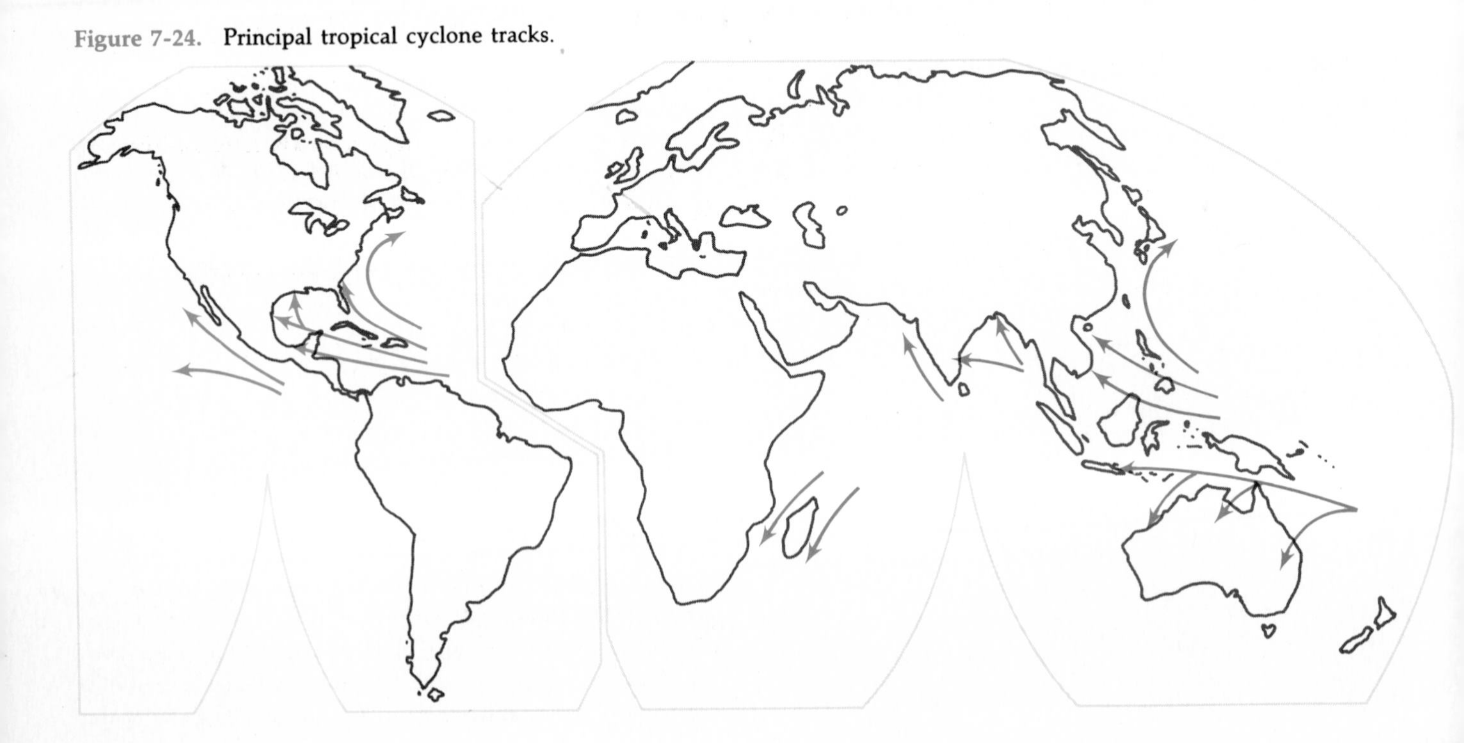

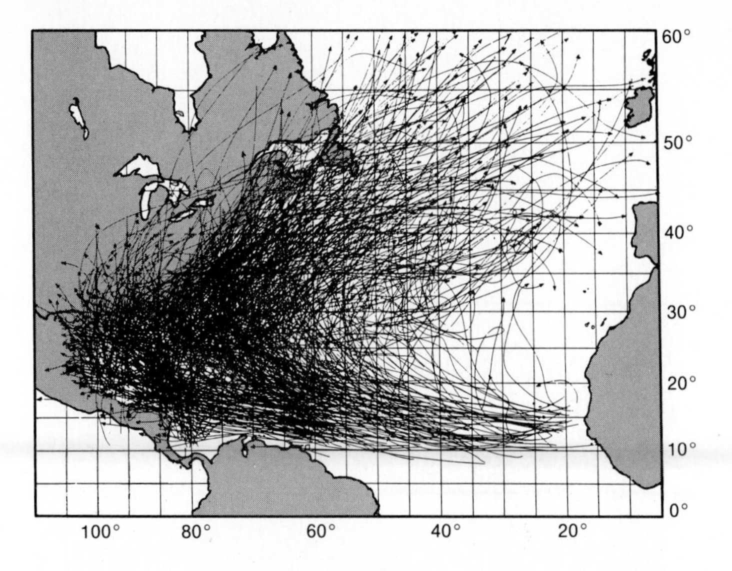

Figure 7-25. The tracks of 761 known North Atlantic hurricanes from 1886 through 1977.

extending into the Caribbean Sea and Gulf of Mexico. See Figure 7-25. They also are found in the western portions of the South Pacific and South Indian Ocean basins as well as in the North Indian Ocean both east and west of the Indian peninsula. They are totally absent from the South Atlantic Ocean.

Once formed, hurricanes follow irregular tracks within the general flow of the trade winds. A specific path is very difficult to predict in advance, but the general pattern of movement is highly predictable. Roughly one-third of all hurricanes travel directly from east to west without much latitudinal change. The rest, however, begin on an east-west path and then curve prominently poleward, where they either dissipate over the adjacent continent or become enmeshed in the general flow of the midlatitude westerlies. The speed of hurricane movement is quite variable, but it averages from 10 to 20 miles (16 to 32 km) per hour.

Whichever path they follow, they do not last long. The average hurricane exists for only about a week, with three weeks as maximum duration. See Figure 7-26. As soon as a hurricane leaves the ocean and moves over land, it begins to die, for its energy source (warm moist air) is cut off. If it stays over the ocean but moves into the midlatitudes, it dies more slowly as it penetrates a cooler environment. It is not unusual for a tropical hurricane that moves into the midlatitudes to diminish in intensity but grow in magnitude until it develops into an extratropical cyclone that travels with the westerlies.

There is a marked seasonality to hurricane formation. See Figure 7-27. They are largely restricted to late summer and fall, presumably because this is the time that ocean temperatures are highest and the intertropical convergence zone is shifted furthest poleward. In the Caribbean/Atlantic region, 80 percent of all hurricanes occur during August, September, and October—averaging four per year. On the other hand, from December through April, there has been only one hurricane every thirty years, on the average, and most of these have been in December.

Hurricanes are best known for their destructive capabilities. Some of the destruction comes from high winds and torrential rain, but the overwhelming cause of damage and loss of life is from the high seas whipped up beneath the storm. The low pressure in the center of the storm allows the ocean surface to form a bulge as much as 3 feet (1 m) high. To this is added a *storm surge* of wind-driven water as much as 25 feet ($7\frac{1}{2}$ m) above normal

Figure 7-26. Duration of North Atlantic/Caribbean hurricanes, based on observations from 1886 through 1977. Most persist for from 3 to 12 days.

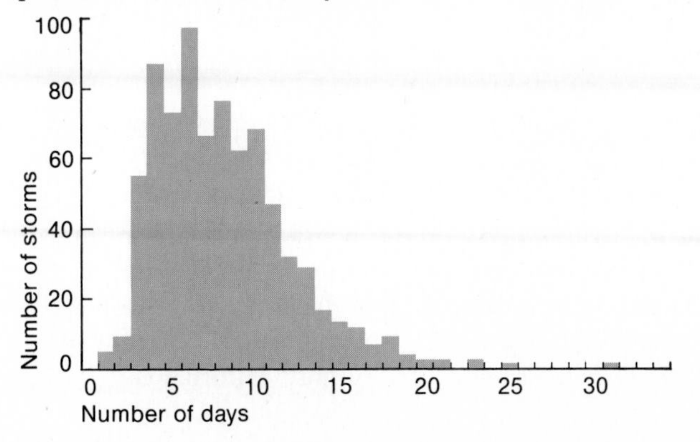

Figure 7-27. Seasonality of North Atlantic/Caribbean hurricanes observed from 1886 through 1977. Fall is hurricane season, with a prominent maximum in September.

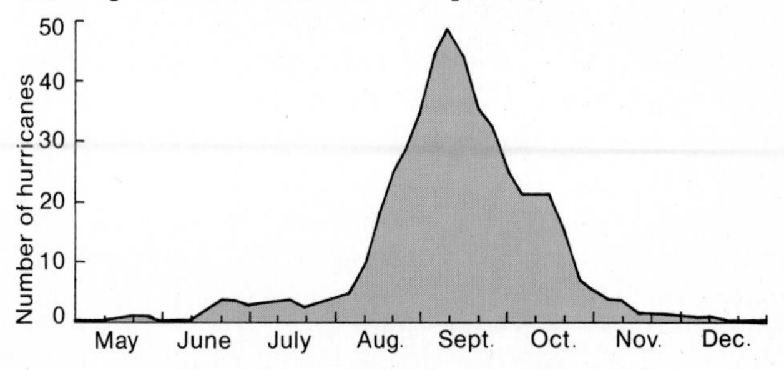

In any given year, about 600 minor disturbances with the potential to grow into hurricanes originate over the tropical oceans. Some 50 of these develop into full-fledged tropical cyclones. The North Atlantic/Caribbean region over the last hundred years has had an average of half a dozen hurricanes annually. During the twentieth century, only three of these storms have achieved a classification of 5 on the Saffir-Simpson Hurricane Scale, which categorizes on the basis of intensity and potential destructiveness. Five is the highest classification on the scale, which is described as "catastrophic." The first storm of the 1980 hurricane season, named Allen, received a classification of 5.

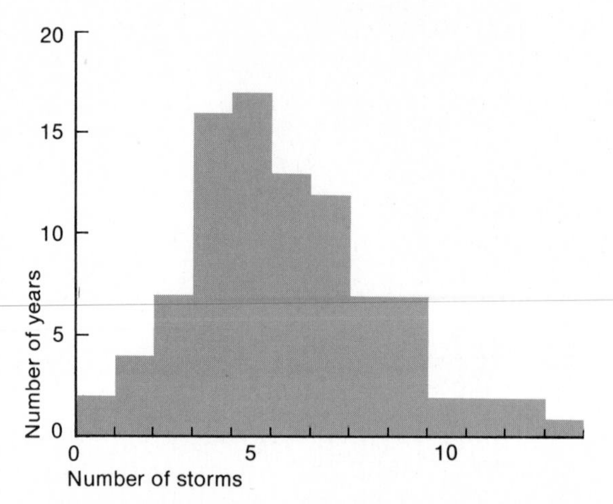

Annual frequency of hurricanes in the North Atlantic/Caribbean. In most years, from four to seven hurricanes occur.

Allen is generally considered to be the mightest storm ever recorded in the Caribbean region. Wind speed in intensive storms cannot be measured accurately because the winds tend to destroy anemometers. However, surface winds of 185 miles (297 km) per hour were recorded for Allen on one island, and a hurricane-hunter aircraft that penetrated the storm measured gusts of 215 mph (345 km per hour) as well as the second lowest central pressure ever measured in a hurricane. By any standard, Allen had the potential to leave enormous devastation in its wake.

Allen made its appearance on August 3rd at 12° north latitude, about halfway between Africa and Trinidad, when a minor pressure perturbation blossomed into an intensified low and its wind speeds first exceeded 74 miles (119 km) per hour. It continued to gain force as its path bore down on the heavily populated island of Barbados. It swung slightly north of Barbados, and then threaded its way directly west between St. Lucia and St. Vincent, causing considerable damage on all three islands but hitting none of them directly. In its path across the Caribbean, it veered west and missed Haiti, then north and mostly avoided Jamaica, then west again and missed western Cuba. This put it on a direct line to crash into the crowded tourist resorts of Mexico's Yucatan Peninsula. But again it dodged northward, avoiding Yucatan. At about the 90th meridian, it abruptly swerved further north, this new trajectory sending it toward Galveston and Houston.

By Saturday morning, August 9th, Allen's massive cloud formations almost filled the entire Gulf, from Louisiana to Yucatan to the east coast of Mexico. Then suddenly it veered erratically away from

tide level when the hurricane pounds into a shoreline. Thus a low-lying coastal area can be severely inundated, and most deaths actually result from drowning. The greatest hurricane disaster in American history occurred in 1900 when Galveston Island (Texas) was overwhelmed by a 20-foot (6-m) storm surge that killed 6000 people, nearly one-sixth of Galveston's population. The most devastating natural disasters in all the world have been the result of tropical cyclones roaring out of the Bay of Bengal to inundate the flat delta of the Ganges and Brahmaputra rivers in India and Bangladesh; on each of three separate occasions within the last three centuries (1737, 1876, and 1970), approximately 300,000 people were drowned.

Destruction and tragedy are not the only legacies of hurricanes, however. Such regions as northwestern Mexico, northern Australia, and southeastern Asia rely on tropical storms for much of their water supply. Hurricane-induced rainfall is often a critical source of moisture for agriculture. Although crops within the immediate path of the storm may be devastated or damaged by winds and flooding, a much more extensive area may be nurtured by the life-giving rains.

Minor Atmospheric Disturbances

Several kinds of lesser atmospheric disturbances are common in various parts of the world. Some are locally of great significance, and some are destructive in nature. All, however, occur at a much more localized scale than those discussed previously. Three of the more prominent types will be discussed below as "minor" disturbances.

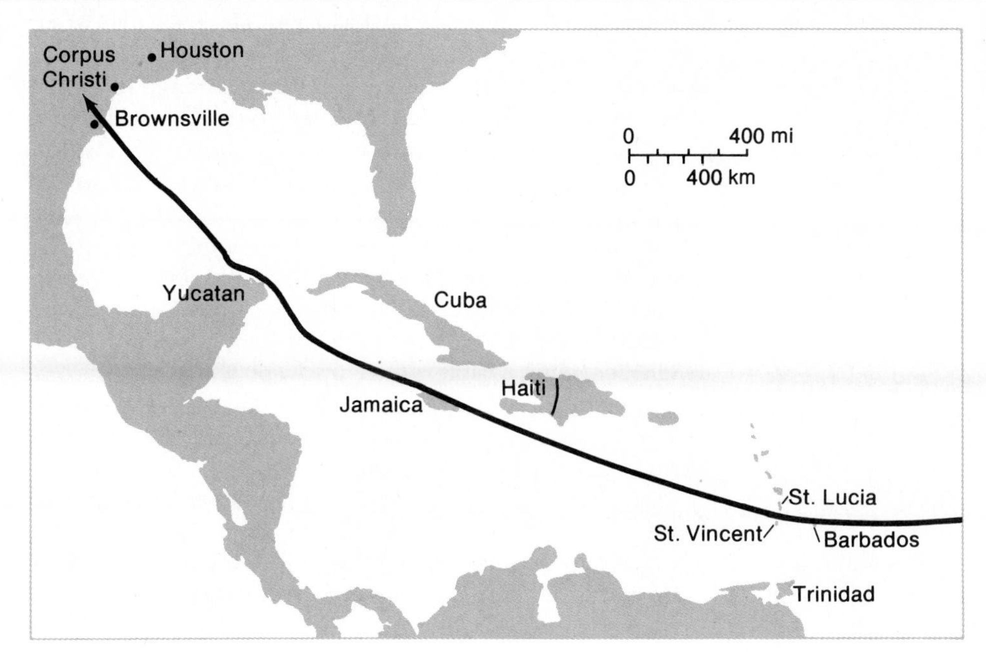

The path of hurricane Allen, early August 1980.

the track toward Houston and aimed at Corpus Christi. Violent storm warnings were issued for all of south Texas. Finally, in the predawn hours of the following morning, it left the Gulf and crossed the shoreline on the least populated portion of the Texas coast—between Corpus Christi and Brownsville. If an ideal location had been chosen to minimize damage to humanity, that would have been the place!

Once over land, the hurricane died rapidly. It caused flooding in southern Texas and brought rain as far north as Kansas, but by Tuesday, it was no longer discernible as a distinct entity on the weather map.

Allen was widely heralded as the "storm of the century." It had the potential for taking thousands of lives and causing billions of dollars worth of property damage. In actuality, it caused about 275 deaths in the Caribbean islands and 4 on the Texas coast. This was not an inconsequential result, but it was significantly better than could have been postulated in even a "best case" scenario for a catastrophic hurricane that crossed the entire Caribbean and then smashed into the southern coast of the United States. Somehow it dodged island after island. Every time it was headed for massive destructive impact, it changed course and went elsewhere. It is likely that Allen will be remembered as the most benevolent major hurricane in history.

Easterly Waves

More common but much less significant to low-latitude weather patterns than tropical cyclones are various weak disturbances that produce unsettled weather and interrupt the long stretches of monotonous conditions that are typical. Best known of these is the *easterly wave*, which is a long but weak migratory low-pressure trough that may occur almost anywhere between about 5° and 30° of latitude. See Figure 7-28. Easterly waves are usually several hundred miles long and nearly always oriented north-south. They drift slowly westward in the flow of the trade winds, bringing with them a characteristic suite of weather conditions. Ahead (to the west) of the wave is fair weather with airflow that tends to be divergent. Behind the wave (to the east), convergent conditions prevail, with moist air being uplifted to yield convective thunderstorms and sometimes widespread cloudiness. There is little or no temperature change with the passage of an easterly wave. Easterly waves occasionally intensify into hurricanes.

Tornadoes

Although very small in size and localized in effect, the *tornado* is the most destructive of all atmospheric disturbances. It is the most intense vortex in nature, a deep cyclonic low-pressure cell surrounded by a whirling cylinder of wind spinning so violently that the centrifugal force creates a partial vacuum within the funnel. These are tiny storms, generally less than $\frac{1}{4}$ mile (800 m) in diameter, but they have the most extreme pressure gradients known, amounting to as much as 100-millibars difference from the center of the tornado to the air imme-

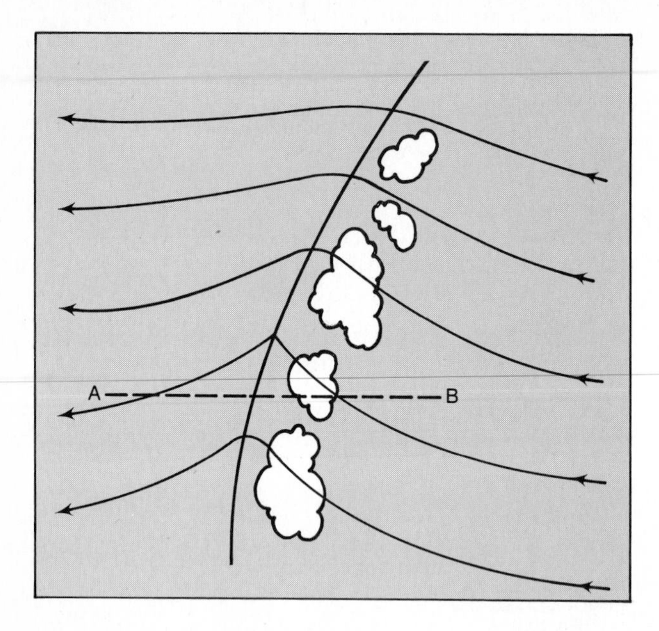

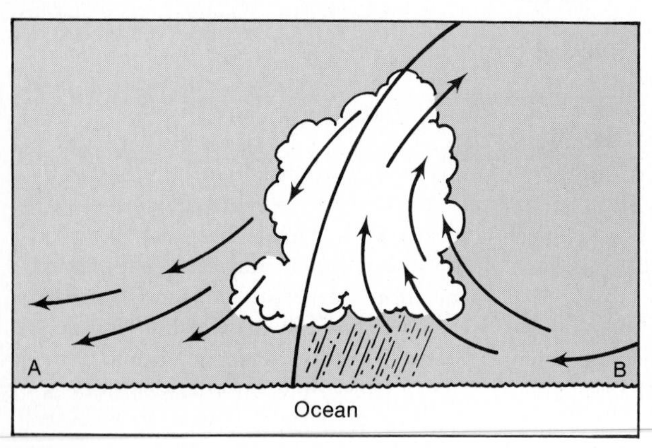

Figure 7-28. Diagrammatic map view and cross section of an easterly wave. The arrows indicate general direction of airflow.

diately outside the funnel. This produces winds of extraordinary velocity. Actual measurements are unknown because any tornado that has come close enough to an anemometer to be recorded has blown the instrument to bits. Maximum wind speed estimates range from 200 to 500 miles (320 to 800 km) per hour. Air sucked into the vortex also rises at an inordinate rate. Such a storm clearly has the capability of transporting a little girl to the land of Oz! See Figure 7-29.

Tornadoes are very erratic phenomena, occurring at random, moving quixotically, jumping from place to place, and persisting briefly. They usually originate a few hundred feet or yards above the ground, the rotating vortex becoming visible when upswept water vapor condenses into a funnel-shaped cloud dangling downward. The tornado generally advances at a speed of 15 to 30 miles (25 to 50 km) per hour along an irregular track that generally extends from southwest to northeast in the United States. Sometimes the funnel sweeps along the ground, devastating everything in its path, but its trajectory is usually twisting and dodging and includes frequent intervals in which the funnel lifts completely off the ground only to settle earthward at a nearby location. A tornado lasts from a few minutes to an hour or so, with maximum longevity recorded as about 8 hours.

The dark, twisting funnel of a tornado displays not only cloud but also dust and debris sucked into the storm in its devastating path along the ground. Damage is caused largely by the strong winds and swirling updraft, but structures are often destroyed by actually exploding due to the abrupt pressure contrast when the center of the storm (very low pressure) passes over a closed building (relatively high pressure).

As with most storms, the exact mechanism of tornado formation is not well understood. It usually develops

Figure 7-29. Sequential movement of a tornado funnel over Nebraska. [U.S. Weather Bureau photo by Homer from the American Red Cross.]

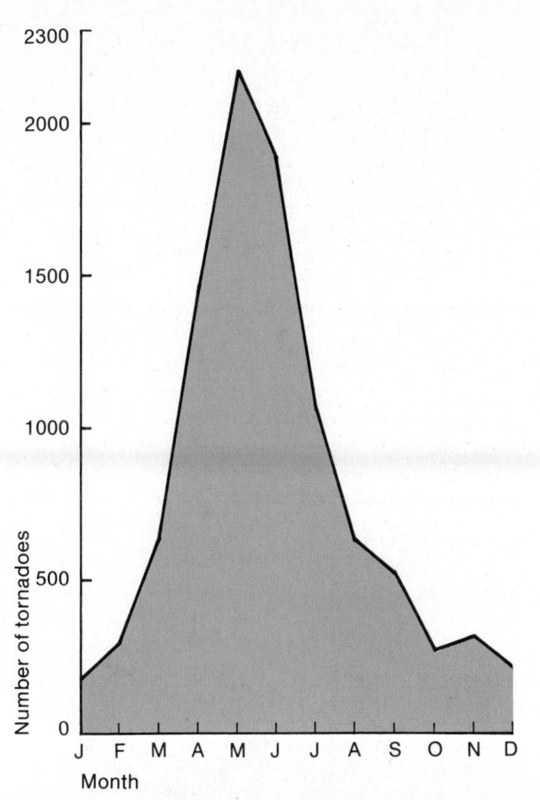

Figure 7-30. Seasonality of tornadoes in the United States, 1953–1967. Late spring and early summer comprise the prime time.

in warm, moist, unstable air associated with an extratropical cyclone. It is most often spawned along a squall line that precedes a rapidly advancing cold front, or along the cold front itself; thus temperature contrasts and instability apparently are important to genesis. Spring and early summer are favorable for tornado development because of considerable air mass contrast in the midlatitudes at that time; a tornado can form in any month, however. See Figure 7-30. Most occur in midafternoon, at the time of maximum heating.

Tornadoes are known in various portions of the middle latitudes and subtropics, but more than 90 percent of all such storms are reported from the United States. See Figure 7-31. Such concentration in a single country presumably reflects optimum environmental conditions, the relatively flat terrain of central and southeastern United States providing an unhindered zone of interaction between prolific source regions for Canadian cP and Gulf mT air masses. Between 800 and 1200 tornadoes are recorded annually in the United States, but the actual total may be considerably higher than that since many small tornadoes that occur briefly in uninhabited areas are not reported. See Figure 7-32. Most U.S. tornadoes are found in the interior and Gulf states, the greatest frequency occurring in parts of Oklahoma and Kansas.

True tornadoes apparently are restricted to land areas. Similar appearing phenomena over the ocean ("waterspouts") have a lesser pressure gradient, gentler winds, and reduced destructive capability.

Thunderstorms

A *thunderstorm* is a relatively violent convective storm accompanied by thunder and lightning. It is usually small and localized in extent and short-lived in duration. It is always associated with vertical air motion and instability, a condition that produces a towering cumulonimbus cloud and (nearly always) showery precipitation.

Figure 7-31. Anticipated distribution of tornadoes over the world in any given half-decade.

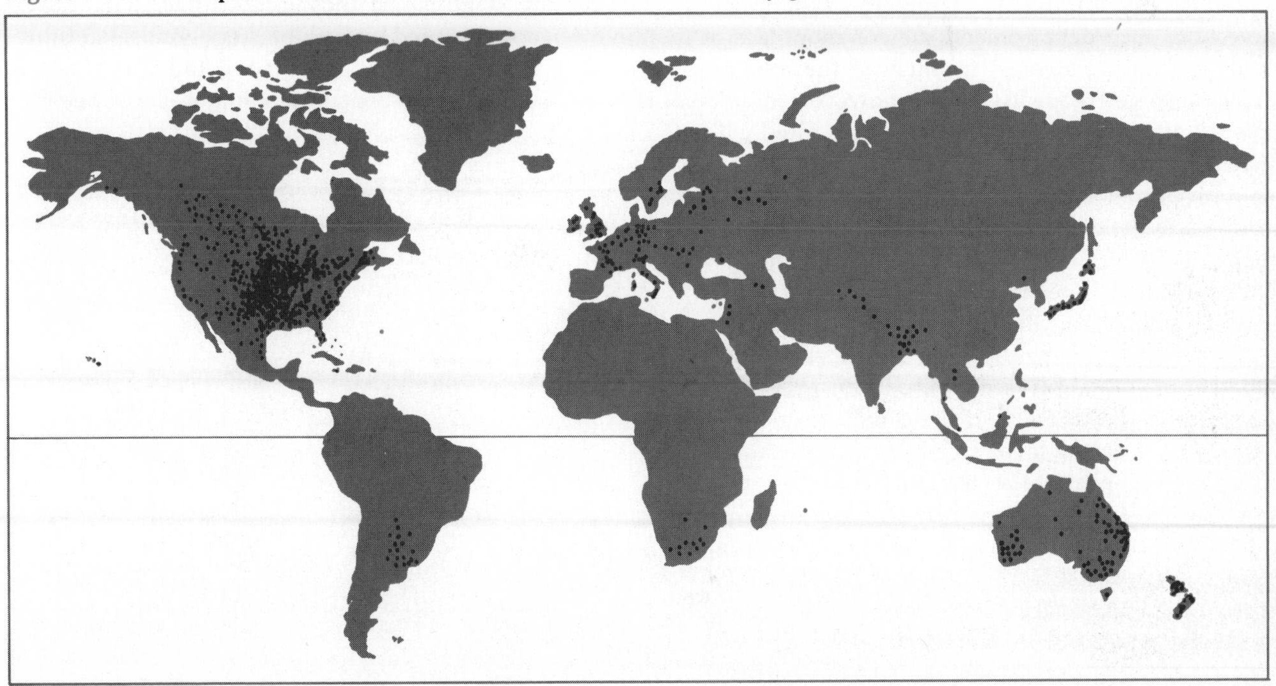

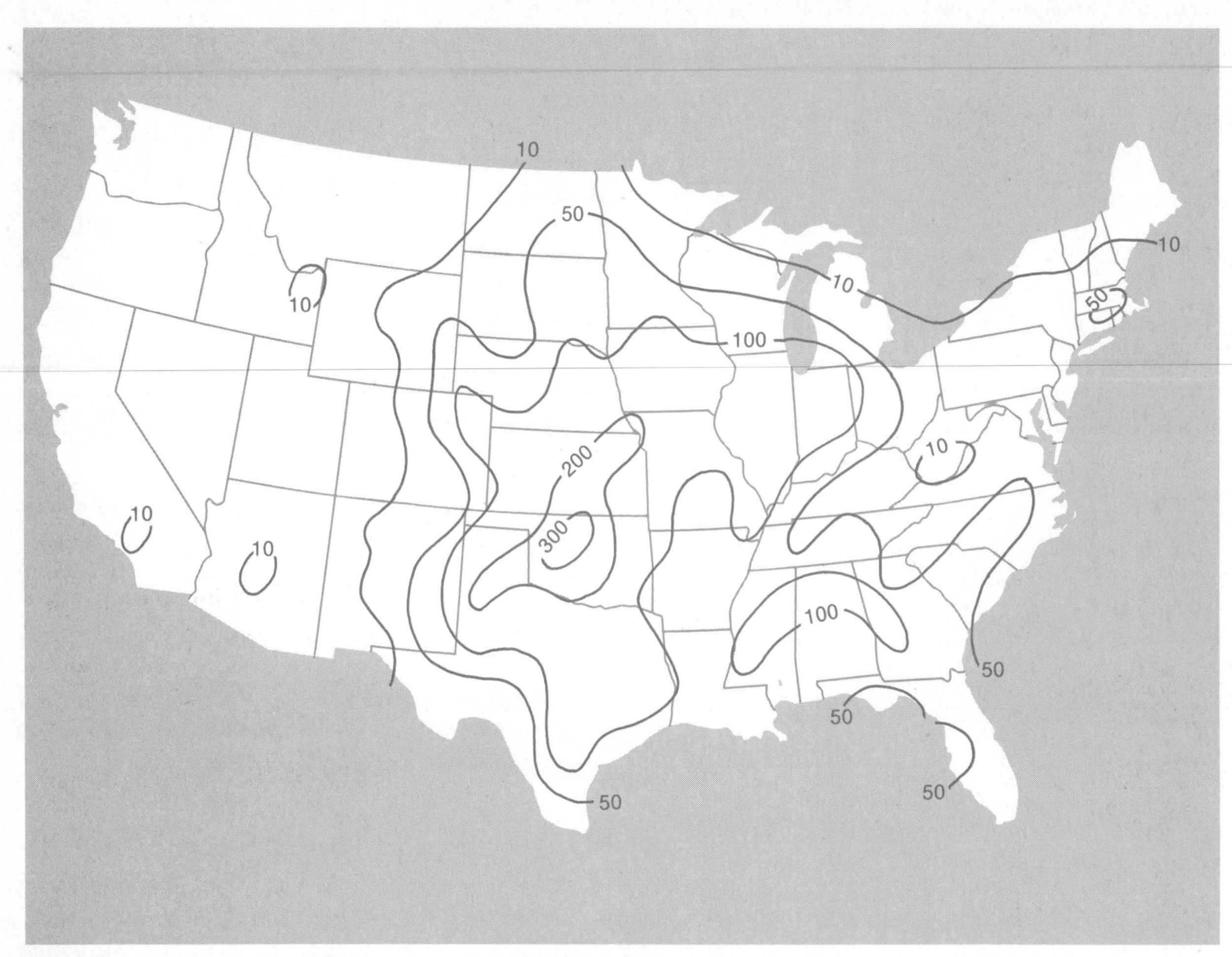

Figure 7-32. Distribution of tornadoes in the conterminous United States. Isolines refer to average number of observed tornadoes on an annual basis.

Thunderstorms sometimes occur as individual clouds, produced by nothing more complicated than thermal convection; such developments are commonplace in the tropics and during summer in much of the midlatitudes. They also are frequently found, however, in conjunction with other kinds of storms or are associated with other mechanisms that can trigger unstable uplift. Thus thunderstorms often accompany hurricanes, tornadoes, fronts (especially cold fronts) in extratropical cyclones, and orographic lifting that forces unstable buoyancy.

The uplift, by whatever mechanism, of warm moist air must release enough latent heat of condensation to sustain the continued rise of the air. In the early stage of thunderstorm formation (see Figure 7-33), called the *cumulus stage*, updrafts prevail and the cloud continues to grow. Above the freezing level, supercooled water droplets and ice crystals begin to accumulate, and when they become too heavy to be supported by the updrafts, they begin to fall. These falling particles drag air with them, initiating a downdraft. When the downdraft with its accompanying precipitation leaves the bottom of the cloud, the thunderstorm enters the *mature stage*, in which updrafts and downdrafts coexist as the cloud continues to enlarge. The mature stage is the most active time, with heavy rain often accompanied by hail, blustery winds, lightning, thunder, and the growth of an anvil top composed of ice crystals on the massive cumulonimbus cloud. See Fig-

Figure 7-33. Sequential development of a thunderstorm cell.

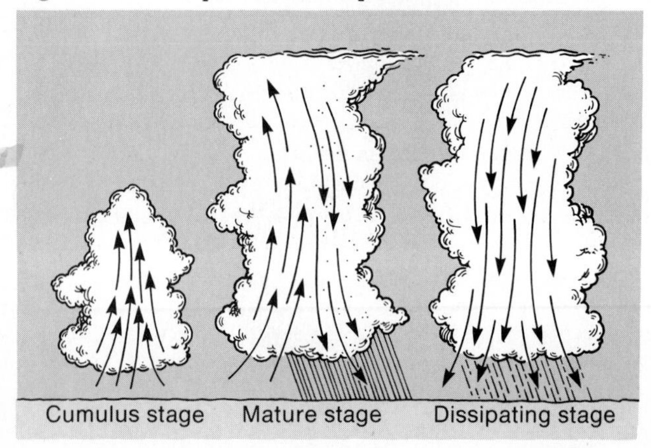

At any given moment, some 2000 thunderstorms exist over the Earth. These storms produce about 6000 flashes of lightning every minute, or more than $8\frac{1}{2}$ million lightning bolts daily. A lightning flash heats the air along its path to as much as 18,000° F (10,000° C), and can develop 100,000 times the amperage used in household electricity.

With such frequency and such power, lightning clearly poses a significant potential danger for humanity. In the United States, more than 125 deaths are blamed on lightning annually, which is more than that caused by all other weather events combined. The most dangerous places to be during a lightning storm are under a tree, in a boat, on a tractor, and playing golf. However, two out of three people struck by lightning in this country are not killed. It is a massive but brief shock, and quick first aid (mouth-to-mouth resuscitation and heart massage) can save most victims.

The sequence of events that leads to lightning discharge is known, but the mechanism of electrification is not. Development of a large cumulonimbus cloud causes a separation of electrical charges. Positively charged droplets or crystals rise to the icy upper layers of the cloud, while negatively charged particles concentrate near the cloud base. The growing negative charge in the lower part of the cloud attracts a growing positive charge on the Earth's surface immediately below. The contrast between the two (cloud base and ground surface) builds to tens of millions of volts before the insulating barrier of air that separates the charges is overcome.

Finally, a finger of negative current flicks down from the cloud and meets a positive charge darting upward from the ground. This makes an electrical connection of ionized air from cloud to ground, and a surge of electrical power explodes downward as the first lightning flash. Other flashes may follow in relatively quick succession, until all or most of the negative charges have been drained out of the cloud base.

In addition to such ground-to-cloud discharges, less spectacular but more frequent lightning is exchanged between adjacent clouds or between the upper and lower portions of the same cloud.

Typical arrangement of electrical charges in a thunderstorm cloud. Positively charged particles are mostly high in the cloud, whereas negatively charged particles tend to be concentrated near the base.

Lightning and thunder occur almost instantaneously, but the lightning is perceived first because its image travels much faster (the speed of light) than the noise of the thunder (the speed of sound).

To describe the separation of electrical charges in a cloud is one thing; to explain it is another. The basic cause of thundercloud electrification is still not understood. The most popular theory involves updrafts carrying positively charged particles to the top of the cloud while falling ice pellets gather negative charges and transport them downward. Another promising line of inquiry points to a triggering action by cosmic rays from deep space. There is obviously still much to learn.

The abrupt heating occasioned by a lightning bolt produces instantaneous expansion of the air, which creates a shock wave that becomes a sound wave that we hear as *thunder*. The lightning and thunder occur simultaneously, but we perceive them at different times. Lightning is seen at essentially the instant it occurs because its image travels at the speed of light. Thunder, however, travels at the much slower speed of sound. Thus it is possible to estimate the distance of a lightning bolt by timing the interval between sight and sound. A 5-second interval indicates that the lightning flash is about a mile away; 3 seconds means about a kilometer distant. Rumbling thunder is indicative of a long lightning trace some distance away, with one portion being nearer than another to the hearer. If no thunder can be heard, the lightning is far away—probably more than a dozen miles (19 km).

Figure 7-34. An enormous thunderstorm cell over the Amazon lowland of Brazil, as seen from directly above by Apollo 9. The dark space in the center that appears to be a hole is actually a shadow from the cumulonimbus tower just to the east. [Courtesy NASA.]

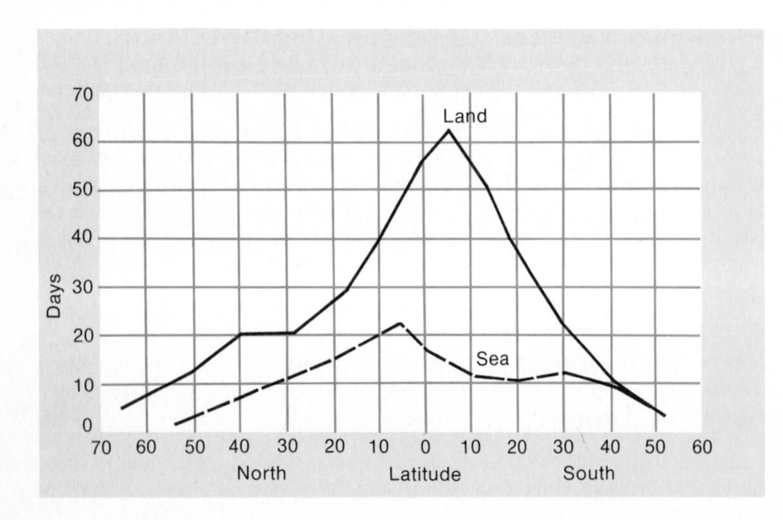

Figure 7-35. Average number of days with thunderstorms, as generalized by latitude. Most thunderstorms obviously are in the tropics. Land areas experience many more thunderstorms than ocean areas because they warm up much more in summer.

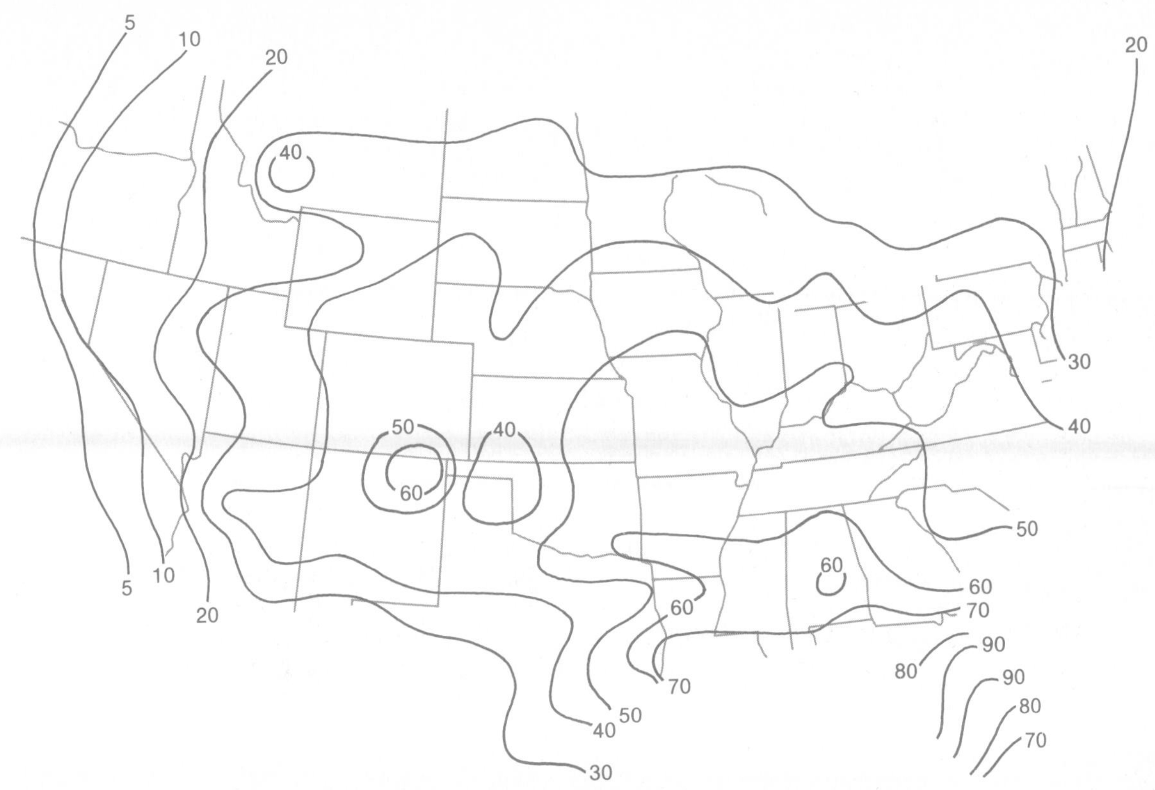

Figure 7-36. Distribution of thunderstorms in the conterminous United States. The Gulf coast is the area of principal activity. The isolines refer to the average number of thunderstorms on an annual basis.

ure 7-34. Eventually, downdrafts dominate the cloud and the *dissipating stage* is reached, with light rain ending and turbulence ceasing.

Thunderstorms are most common where there are high temperatures, high humidity, and high instability. This combination is typical of the intertropical convergence zone, so it is no surprise that the greatest number of thunderstorms occurs in the equatorial region. There is a general decrease away from the equator, and thunderstorms are virtually unknown poleward of 60° of latitude. There is much greater frequency of thunderstorms over land than water because summer temperatures are higher over land, and most thunderstorms occur in the summer. See Figure 7-35. In the United States, thunderstorms are most numerous in the Southeast, especially along the Gulf coast, which has the highest frequency of thunder outside the tropics. See Figure 7-36. The scarcity of thunderstorms on the Pacific coast is related to the prevalence of stable, anticyclonic air in summer.

Review Questions

1. What regions of the Earth are least likely to produce air masses? Why?
2. Why are mP air masses from the Atlantic Ocean less important to the United States than mP air masses from the Pacific Ocean?
3. Explain the differences between a *warm front* and a *cold front*.
4. Explain the various motions associated with an extratropical cyclone.
5. Briefly explain what happens to *all four* weather elements when a cold front passes.
6. What parts of the world are most affected by extratropical cyclones?
7. Why are there no fronts in an extratropical anticyclone?
8. Why are there no fronts in a tropical cyclone?
9. Hurricanes usually are destructive storms. Do they have any "beneficial" effects? Explain.
10. Which type of storm is most common over the world? Why?

Some Useful References

ANTHES, R. A., et al., *The Atmosphere*, 2nd ed. Wembley, England: Merrill Publications, 1978.

COURT, A., *Tropical Cyclone Effects on California*, NOAA Technical Memorandum NWS WR-159. U.S. Department of Commerce, 1980.

EAGLEMAN, J. R., *Meteorology: The Atmosphere in Action*. New York: D. Van Nostrand Company, 1980.

HARMAN, J. R., *Tropospheric Waves, Jet Streams, and United States Weather Patterns*, Commission on College Geography Resource Paper No. 11. Washington, D.C.: Association of American Geographers, 1971.

SIMPSON, R. H., AND H. RIEHL, *The Hurricane and Its Impact*. Baton Rouge: Louisiana State University Press, 1981.

8 Climatic Regions

The ultimate goal of the geographical study of climate, as of the geographical study of anything, is to understand its spatial characteristics—its distribution over the Earth. This is exceedingly difficult to do because so many variables are involved. It is relatively easy to describe and even to map the distribution of such uncomplicated phenomena as giraffes, rose bushes, Tasmanians, voting patterns, religious preferences, crop yields, and a host of other concrete entities or simple relationships. Climate, however, encompasses the integration of a number of different elements that are, for the most part, continuously and independently variable. Temperature, for instance, is one of the simplest of the elements of climate, yet it fluctuates in time and place so frequently that, for example, in the *National Atlas of the United States* (U.S. Department of the Interior, Geological Survey, Washington, D.C., 1970), 56 different maps are used to portray various aspects of temperature.

Although temperatures can be measured precisely, some form of abstraction is necessary for tabulation and analysis. Most widely used is the *arithmetic mean* (or average), but for some purposes the *most probable* (mode) temperature, or some expression of *variation* (e.g., standard deviation), is more useful. Indeed, in some situations, extremes are more important than averages. The point at issue is that climate involves almost continuous variation, not only of temperature but also of a host of other factors. Thus a satisfactory synthesis of all these data involves a great deal of generalization and subjectivity.

Climatic Classification

To cope with the great diversity of information encompassed by the concept of *climate*, the most meaningful aspects must be selected, and some systematic way must be found to classify the data. Classifiers generally have chosen temperature and precipitation as the most significant and understandable features, as well as the most available, to serve as the framework of their classifications.

Purpose

Classification of climates is an intellectual device based on human perception. It is not a natural environmental occurrence; it is a product of the human mind. It is useful only insofar as it fills a need. In other words, its value depends on its intended use. However, a classification that is useful for one purpose is not necessarily relevant for another.

Our need in this book for a climatic classification is clear-cut and straightforward: We need a classification that will be useful as a learning tool. We need a device to simplify, organize, and generalize a vast array of climatic data into a comprehensible system that will help us understand the distribution and characteristics of climates over the Earth. We cannot comprehend all the details, so we need a "shorthand" system that will reduce the complexities and enhance our comprehension.

Suppose a beginning geography student in Georgia is asked to describe the climate of southeastern China. The student is likely to be bewildered by such an assignment and feel incapable of a satisfactory performance without considerable research. A world map that displays a reputable climatic classification, however, can show the student that southeastern China has a climate of the same type as that in southeastern United States. Thus the student's familiarity with the home climate in Georgia provides a base for understanding the general characteristics of the climate of southeastern China.

One function of a climatic classification is to facilitate the understanding of world climates by *analog* (recognition of similarities between familiar and unfamiliar features). At a broader level, such a classification will be seen to have a relatively orderly arrangement over the Earth. At certain latitudes and at similar positions on continents, climatic patterns repeat. This is because similar atmospheric characteristics are likely to be found where there are similar environmental relationships. Thus the general distribution of climatic types on our planet is readily comprehensible and even predictable.

Many Classifications

The earliest known climatic classification originated with the ancient Greeks, in the first or second century B.C. Although the "known world" was very small at that time, some Greek scholars were aware of the shape and approximate size of the Earth. They knew that at the southern limit of their "world," up the Nile River and along the southern coast of the Mediterranean, the climate was much hotter and drier than on the islands and northern coast of that sea. At the other end, up the Danube River and on the north coast of the Black Sea, things were much colder, especially in winter.

So the Greeks spoke of three climatic regions: the Temperate Zone of the midlatitudes, in which they lived (Athens is at 38 north latitude), the Torrid Zone of the tropics to the south, and the Frigid Zone of the "far" north. See Figure 8-1. These are three of the five basic climatic zones recognized today. However, the Greeks did not know that near the equator, beyond their hot and dry Torrid Zone, lay the warm and rainy region of the Intertropical Convergence Zone (see Chapter 5), or that beyond their Frigid Zone, which we now call "cool temperate," were the snowy climates of the Arctic. Instead, because they knew the Earth was a sphere, they

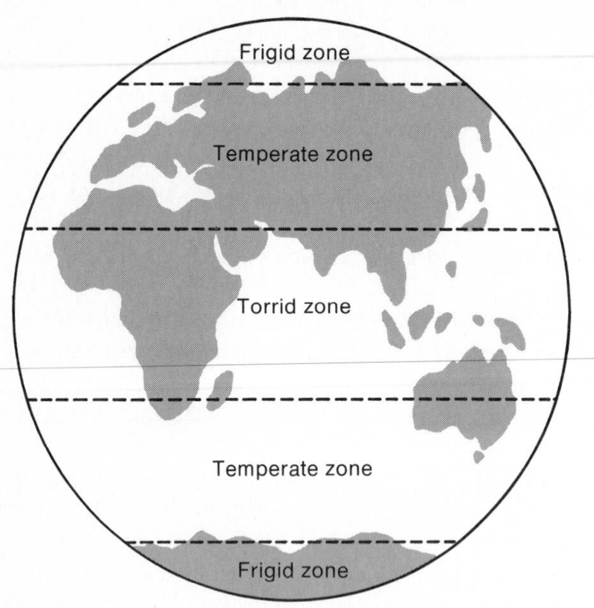

Figure 8-1. Climatic regions as recognized by ancient Greek scholars.

suggested that the Southern Hemisphere would have similar Temperate and Frigid Zones, making five climatic zones in all.

For many centuries, these notions were handed down from scholar to scholar and from scribe to scribe. Gradually these five climatic zones were confused with, and eventually their climates ascribed to, the five astronomical zones of the Earth, bounded by the Tropics of Cancer and Capricorn and the Arctic and Antarctic Circles. This revision put the equatorial rainy zone in with the hot arid region in the Torrid Zone, extended the Temperate Zone to include much of what the Greeks had called Frigid, and moved the Frigid Zone poleward to the polar circles. This simplistic but unrealistic classification persisted for more than a thousand years and was discarded finally only in the present century.

Today we recognize five basic climates in the world: equatorial warm-wet, tropical hot-dry, subtropical warm temperate, midlatitude cool temperate, and high-latitude cold. See Figure 8-2. The first two of these zones are differentiated by rainfall amount and frequency. The two temperate zones differ primarily in whether summer or winter is the dominant season: In Atlanta, houses are built to be cool in summer, whereas in Minneapolis, keeping warm in winter is more important. The cold zone has hardly any summer—not enough to grow important crops.

Many climatologists have tried to refine the distinction between these five basic climatic zones and to subdivide the zones in various ways. Most efforts have relied primarily on natural vegetation as an indicator of climate. Classifications with the greatest pedagogic (learning) value have three important attributes:

1. They are relatively simple to comprehend and to use.
2. They show some sort of orderly pattern over the Earth.
3. They give some indication of genesis (reasons underlying development).

The Köppen System

The Köppen system meets these criteria reasonably well and is by far the most widely used climatic classification. Wladimir Köppen (1846–1940) was a Russian-born German climatologist who was also an amateur botanist. The first version of his climatic classification appeared in 1918, and he continued to modify and refine it for the rest of his life, the last version being published in 1936. The Köppen system uses an objective, numerical basis of classification, but the boundaries are located with regard to vegetation patterns. He did not attempt to identify "natu-

Figure 8-2. Weather variety reflects climatic differences. Summer monsoonal rains in Benares, India, contrast with winter snowfall in New York City. [Paolo Koch from Rapho/Photo Researchers, Inc.; Marc P. Anderson.]

Focus on the Thornthwaite Classification

Although many geographers and climatologists have criticized aspects of the Köppen classification, few have produced any significant alternatives. The most important alternative system developed in the United States was first proposed by climatologist C. Warren Thornthwaite in 1931, and was totally revised in 1948; the revised system should be associated with Thornthwaite's name.

The Thornthwaite system defines five climatic types and a host of subtypes primarily on the basis of their water budgets. This involves the balance between precipitation effectiveness, seasonal distribution of precipitation, and thermal efficiency. A critical element of the Thornthwaite scheme is the role of evapotranspiration and potential evapotranspiration. Evapotranspiration represents the combined water loss through evaporation from the soil surface and transpiration from plants. Potential evapotranspiration refers to the maximum amount of moisture that would be lost from soil and vegetation if the water were available. This is estimated from temperature and length of day. Thornthwaite considered vegetation as a part of the climatic process, more or less as an antithesis to clouds in precipitation, in that plants constitute an important medium of evaporation and transpiration.

The Thornthwaite classification is purely empirical (*rational* was the term he used), and it was developed independently of any nonclimatic variable. Thus it is considered a descriptive classification, with no relationship to genesis. It is a refined analytical tool, but it has important limitations in pedagogic usefulness.

The principal problem is that it is too complex to serve as a teaching device. Its permutations are too numerous. For example, California displays seven types in the Köppen system, but about 40 types in the Thornthwaite system. Moreover, the complications of the formulae suggest a degree of accuracy that is greater than the relatively crude nature of the data from which they are calculated.

No world maps of the Thornthwaite climatic classification have been published; the complexities are too great. It is nevertheless an important alternative to Köppen and has been received with great approval by agronomists, biologists, and others particularly concerned with soil moisture, growing conditions of plants, and related matters.

ral" climatic regions (if there are such things), but he considered that the close association between natural vegetation and climate permitted determination of climatic boundaries from the distribution of major plant associations. Thus the boundaries in the Köppen system represent climatic expressions of floristic limits.

The entire Köppen system uses as a data base only the mean annual and monthly values of temperature and precipitation, combined and compared in a variety of ways. Consequently, the necessary statistics are commonly tabulated and easily acquired. Data for any location ("station") on Earth can be used to determine the precise classification of that place, and the areal extent of any recognized climatic type can be determined and mapped. This means that the classification system is functional at both specific and general levels.

Köppen defined four of his five major climatic groups by temperature characteristics, the fifth on the basis of moisture. Each group was then subdivided according to various temperature and precipitation relations. The system's distinctive feature is the use of a symbolic nomenclature to designate the various climatic types. This nomenclature consists of a combination of letters, with each letter having a precisely defined meaning.

The Köppen system is comprehensive and relatively flexible. Although its types and boundaries are empirically determined and were not chosen with genesis in mind, they coincide surprisingly well with broad features of the general circulation; thus some genetic relationships are apparent. The Köppen system has been widely used, although often modified, and its terminology has been even more widely adopted.

There are some deficiencies in the Köppen system, as there are in any other climatic classification that has been devised. Köppen himself was unsatisfied with his last version and did not consider it as being a finished product. Thus many geographers and climatologists have used the Köppen system as a springboard to devise systems of their own or to provide modifications of the "final" Köppen classification.

Modified Köppen

The system of climatic classification used in this book is modified Köppen. It follows the basic design of the Köppen system, but with a variety of minor modifications for simplication. Some of these modifications follow the lead of Glen Trewartha, geographer/climatologist at the University of Wisconsin. No attempt will be made to distinguish between pure Köppen and modified Köppen in this system; our purpose here is to comprehend the general pattern of world climate, not to learn a specific system or to nit-pick about boundaries.

Within this system are the usual five major climatic groups, classified into various types and subtypes, and a sixth for "highland" climates (see Table 8-1). Some of the subtypes are considered together in the table and the subsequent discussion because they fit together logically and/or because they occupy only small portions of the continents. In all, 15 significant climatic types are recognized.

Climatic Diagrams

Probably the most useful tool in a general study of world climatic classification is a simple graphic representation

TABLE 8-1 Climatic Classification (modified Köppen)

Letters* 1st	2nd	3rd	Derivation	Description	Definition	Types
A			Alphabetical	Low-latitude humid climates	Average temperature of each month above 64° F (18° C)	Tropical wet (Af) Tropical monsoonal (Am) Tropical savanna (Aw)
	f		German: *feucht* ("moist")	No dry season	Average rainfall of each month at least 2½ in. (6 cm)	
	m		Monsoon	Monsoonal; short dry season compensated by heavy rains in other months	1 to 3 months with average rainfall less than 2½ in. (6 cm)	
	w		Winter dry	Dry season in "winter" (low sun season)	3 to 6 months with average rainfall less than 2½ in. (6 cm)	
B			Alphabetical	Dry climates; evaporation exceeds precipitation		Subtropical desert (BWh) Subtropical steppe (BSh) Midlatitude desert (BWk) Midlatitude steppe (BSk)
	W		German: *Wuste* ("desert")	Arid climates; "true deserts"	Average annual precipitation less than approx. 15 in. (38 cm) in low latitudes; 10 in. (25 cm) in midlatitudes	
	S		Steppe, or semiarid	Semiarid climates; steppe	Average annual precipitation between about 15 in. (38 cm) and 30 in. (76 cm) in low latitudes; between about 10 in. (25 cm) and 25 in. (64 cm) in midlatitudes; *without* pronounced seasonal concentration	
		h	German: *heiss* ("hot")	Low-latitude dry climate	Average annual temperature more than 64° F (18° C)	
		k	German: *kalt* ("cold")	Midlatitude dry climate	Average annual temperature less than 64°F (18°C)	
C			Alphabetical	Mild midlatitude climates	Average temperature of coldest month less than 65° F (18° C); at least 8 months average temperature more than 50° F (10° C)	Mediterranean (Csa, Csb) Humid subtropical (Cfa, Cwa) Marine west coast (Cfb, Cfc)
	s		Summer dry	Dry summer	Driest summer month has less than ⅓ the average precipitation of wettest winter month	
	w		Winter dry	Dry winter	Driest winter month has less than $\frac{1}{10}$ the average precipitation of wettest summer month	
	f		German: *feucht* ("moist")	No dry season	Does not fit either s or w above	
		a	Alphabetical	Hot summers	Average temperature of warmest month more than 72° F (22° C)	
		b	Alphabetical	Warm summers	Average temperature of warmest month below 72° F (22° C); at least 4 months with average temperature above 50° F (10° C)	
		c	Alphabetical	Cool summers	Average temperature of warmest month below 72° F (22° C); less than 4 months with average temperature above 50° F (10° C)	
D			Alphabetical	Humid midlatitude climates with severe winters	4 to 8 months with average temperatures more than 50° F (10° C)	Humid continental (Dfa, Dfb, Dwa, Dwb)

TABLE 8-1 Continued

Letters*						
1st	2nd	3rd	Derivation	Description	Definition	Types
2nd and 3rd letters same as in C climates						Subarctic (Dfc, Dfd, Dwc, Dwd)
		d	Alphabetical	Very cold winters	Average temperature of coldest month less than −36° F (−38° C)	
E			Alphabetical	Polar climates; no true summer	No month with average temperature more than 50° F (10° C)	Tundra(ET) Ice cap (EF)
	T		Tundra	Tundra climates	At least one month with average temperature more than 32° F (0° C) but less than 50° F (10° C)	
	F		Frost	Ice cap climates	No month with average temperature more than 32° F (0° C)	
H			Highland	Highland climates	Significant climatic changes within short horizontal distances due to altitudinal variations	Highland (H)

* The code letters in the first column designate the various climatic types according to a modified Köppen system as described in the text.

of monthly temperature and precipitation for a specific weather station. See Figure 8-3. The customary graph has 12 columns, one for each month, with a temperature scale on the left side and a precipitation scale on the right. Average monthly temperatures are connected by a curved line near the top, and average monthly precipitation is represented by bars extending upward from the bottom. Miscellaneous information is recorded above the graph itself.

The value of a climatic diagram is twofold: (1) it displays precise details of important aspects of the climate of a specific place, and (2) it can be used to determine the climatic classification of that station.

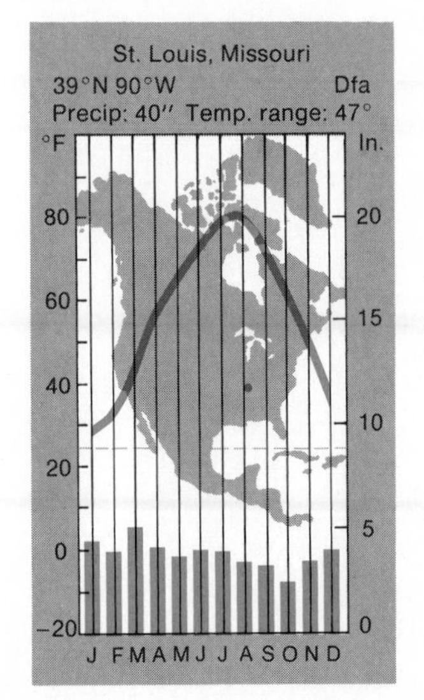

Figure 8-3. A specimen climatic diagram.

World Distribution of Major Climatic Types

Most of the remainder of this chapter will be devoted to a discussion of the major climatic types. Their distribution pattern shows a certain degree of regularity and predictability, which is very helpful in comprehending the broad climatic characteristics of the Earth. Our attention will be focused primarily on three questions:

1. Where are the various climatic types?
2. What are the characteristics of each type?
3. Why are these characteristics found in these locations?

Tropical Humid Climates (Group A)

The tropical humid climates occupy almost all the land area of the Earth within 15° to 20° of the equator, in both the Northern and Southern hemispheres. See Figure 8-4. This globe-girdling belt of A climates is interrupted slightly here and there by mountains or small zones of aridity, but it dominates the equatorial regions and extends poleward to beyond the 25° parallel in some windward coastal lowlands.

As a group, the A climates are noted not so much for warmth as for lack of coldness. These are the only truly winterless climates of the world. They are characterized by moderately high temperatures throughout the year, as is to be expected from their near-equatorial location. The sun is high in the sky every day of the year, and even the shortest days are not appreciably shorter

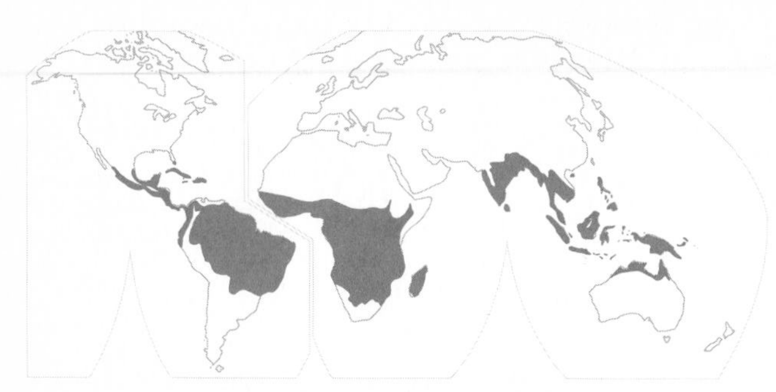

Figure 8-4. World distribution of A climates.

than the longest ones. These are climates of perpetual warmth, although they do not experience the world's highest temperatures. The fundamental character of the A climates, then, is molded by their latitudinal location.

Of almost equal prominence as a typifying characteristic of the tropical humid climates is the prevalence of moisture. Although not universally rainy, much of the A climate area is among the wettest in the world. Warm, moist, unstable air masses frequent the oceans of these latitudes, and the Intertropical Convergence Zone inhabits the region for much of the year. Moreover, onshore winds and thermal convection are commonplace phenomena. Thus the A climate zone has not only abundant sources of moisture but also mechanisms for uplift. High humidity and considerable rainfall are expectable results.

The tropical humid climates are classified into three types on the basis of the quantity and regime of annual rainfall. The tropical wet type (Af in the modified Köppen system) experiences relatively abundant rainfall (several inches) in every month of the year. See Figure 8-5. The tropical monsoonal type (Am) has a short dry season but a very rainy wet season. The tropical savanna type (Aw) is characterized by a longer dry season and a prominent but not extraordinary wet season.

Tropical Wet Climate (Af) The Af climate type is essentially equatorial in location. It characteristically occurs in an east-west sprawl astride the equator, extending 5° to 10° poleward on either side. In some east coast situations, it may extend as much as 20° or even 25° away from the equator. The largest areas of Af climate occur in the upper Amazon basin of South America, the northern Congo (Zaire) basin of Africa, and the islands of the East Indies.

The single most descriptive word that can be applied to the tropical wet climate is *monotonous*. In terms of weather conditions, every day is like every other day, every week like every other week, every month like every other month. It is a seasonless climate, with endless repetition on a daily basis.

Warmth prevails. Every month has an average temperature close to 80° F (27° C), usually with no perceptible seasonal variation. The terms *summer* and *winter* are meaningless. The annual temperature range (fluctuation between the average temperatures of the coolest and warmest months) is minuscule, typically only 2° or 3°, and only rarely over 7° or 8°. This represents by far the smallest annual temperature range in any climatic type.

Daily temperature variations are somewhat greater, although still not impressive. This is one of the few climates in which the average daily temperature range exceeds the average annual temperature range. See Figure 8-6. On a typical afternoon, the temperature will rise to the high 80s° F (low 30s° C), dropping to the middle or low 70s° F (low 20s° C) in the coolest period just before dawn. The temperature rarely extends much into the 90s° F (mid-30s° C), even on the hottest days, and equally unusual is much cooling off at night.

Regardless of the thermometer reading, however, the weather feels warm in this climate because high humidity makes for high sensible temperatures, except perhaps where a coastal sea breeze flows. This is a "hot-

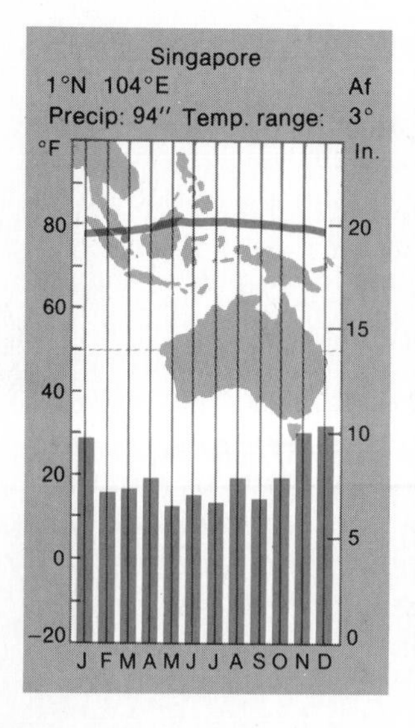

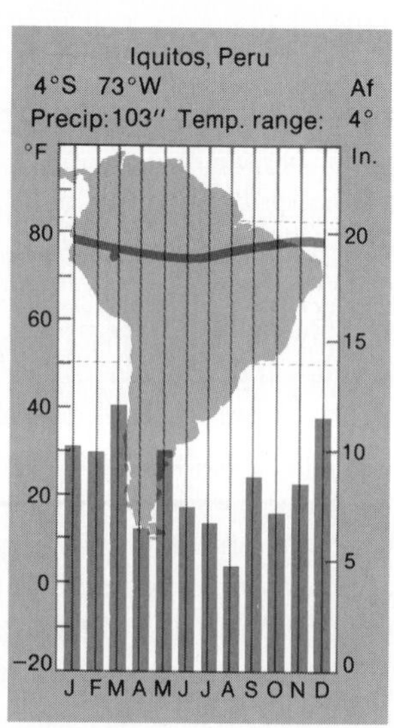

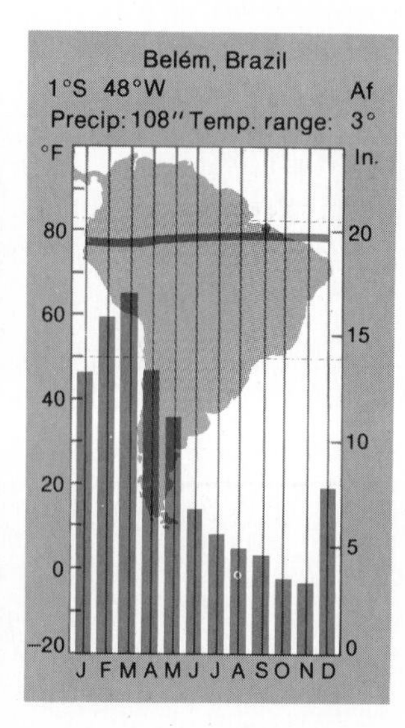

Figure 8-5. Climatic diagrams for typical tropical wet stations.

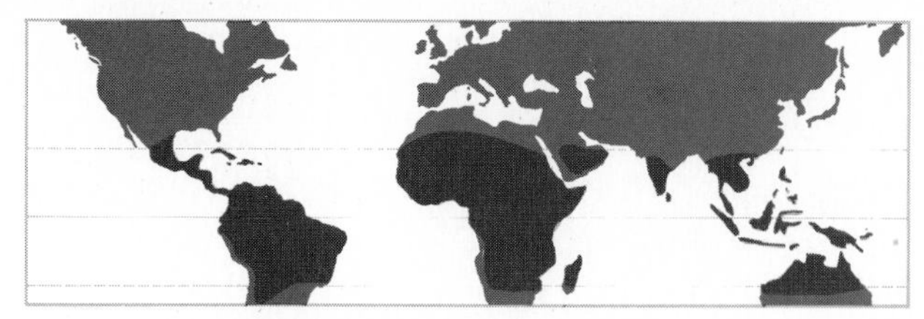

Figure 8-6. The shaded areas are the land portions of the world where the average daily temperature range exceeds the average annual temperature range. This characteristic is typical of A climates but is rare in other climatic types.

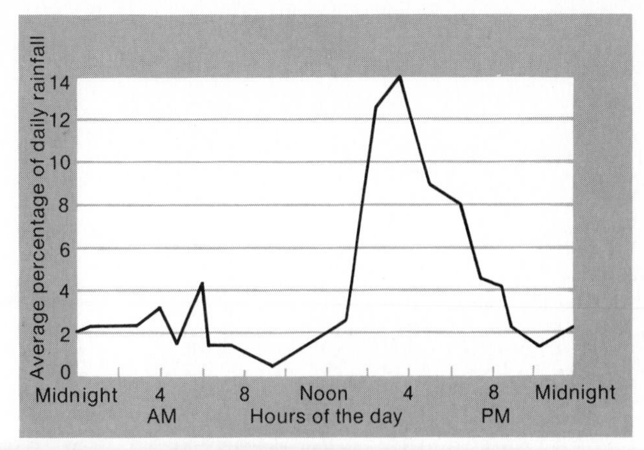

Figure 8-7. A typical temporal regime of rainfall in a tropical wet climate. This example, from Malaysia, shows a heavy concentration in midafternoon with a small secondary peak at or shortly before dawn.

house" situation—a combination of relatively high temperature and high humidity, producing enervating conditions.

Moisture is abundant almost all the time. See Figure 8-7. Both absolute and relative humidity are notable, and rain can be expected just about every day—sometimes twice or three times a day. Rainfall is usually of the unstable, showery variety, often coming from thunderstorms that yield heavy rain for a short time. A typical morning dawns bright and clear. Cumulus clouds build up in the forenoon and develop into cumulonimbus thunderheads, producing a furious convective rainstorm in early afternoon. Then the clouds usually disperse, so that by late afternoon there is a partly cloudy sky and a glorious sunset. The clouds often recur at night to create a nocturnal thunderstorm, followed by dispersal once again. The next day dawns bright and clear, and the sequence begins to repeat.

Each month receives several inches of rain, and the annual total normally is between 60 and 100 inches (150 and 250 cm), although in some locations it is considerably greater. Yearly rainfall in the Af climate exceeds that of all but two other climatic types.

One other general climatic characteristic is lack of wind movement. The Af climate is poorly ventilated, except along coast lines where sea breezes may be frequent or even persistent.

The reasons that these climatic conditions occur where they do are relatively straightforward. The principal climatic control is latitude. High sun conditions throughout the year make for relatively uniform receipt of insolation, so there is little opportunity for seasonal temperature variation. This extensive heating produces considerable thermal convection, which accounts for a portion of the raininess. More importantly, trade wind convergence in the intertropical convergence zone leads to widespread uplift of warm, humid, unstable air. Also, persistent onshore wind movements along trade wind (i.e., east-facing) coasts provide a consistent source of moisture and add still another mechanism for precipitation—orographic ascent. Indeed, the maximum poleward extent of Af climate is found along such trade wind coasts, as in Central America and Madagascar.

Tropical Monsoonal Climate (Am) The tropical monsoonal climate is more specialized in origin and more limited in occurrence than the tropical wet climate. It is most extensive on the windward (west-facing) coasts of southeastern Asia (primarily in India, Burma, and Thailand), but it also occurs in more restricted coastal regions of western Africa, northeastern South America, the Philippines, and some islands of the East Indies.

The distinctiveness of this climatic type is shown primarily in its rainfall regime. See Figure 8-8. During the high-sun period, an enormous amount of rain is released in association with the summer monsoon. It is not unusual to experience more than 20 or 30 inches (50 or 75 cm) of rain in each of two or three months. The annual total for a typical Am station is between 100 and 200 inches (250 and 500 cm). An extreme example is Cherrapunji (in the Khasi hills of India), which has an annual average of 425 inches (1065 cm); Cherrapunji has been inundated with 84 inches (210 cm) in three days, with 366 inches (930 cm) in one month, and with a memorable 1,042 inches (2647 cm) in its record year.

World Record—Greatest Average Annual Precipitation

472" (11,989 mm)—Mt. Waialeale, Hawaii—Af

World Record—Greatest One-Year Precipitation

1042" (26,467 mm)—Cherrapunji, India—1860/1861—Am

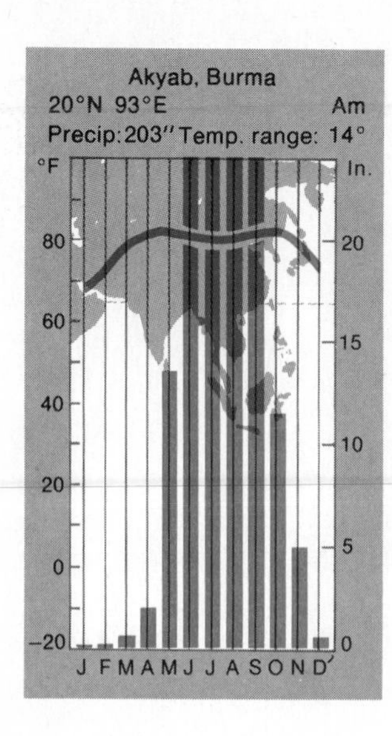

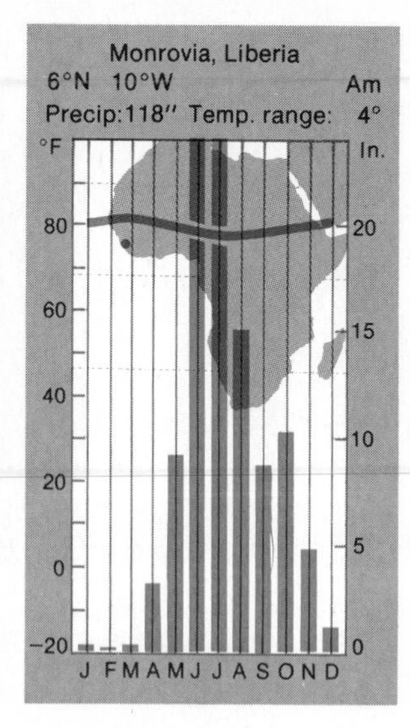

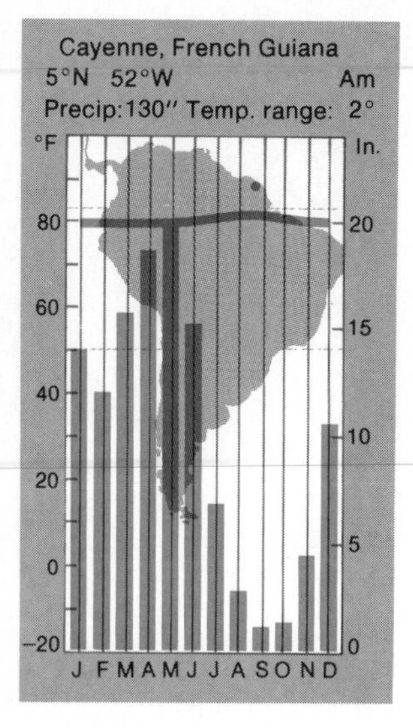

Figure 8-8. Climatic diagrams for representative tropical monsoonal stations.

Focus on Environmental Relationships

One of the most important recurring themes in physical geography is the intertwining relationship of the various components of the environment. Time after time we will note situations in which one aspect of the environment affects another—sometimes conspicuously, sometimes subtly. Although such effects often can be seen on a micro-scale, in this book your attention is directed mostly toward broad-scale patterns.

Climate is, in itself, a very significant environmental component, but it takes on added importance because of its notable effects on other components. For our purposes, this dual role is especially conspicuous in terms of gross distribution patterns. It is premature at this point to carry out a systematic analysis of these myriad relationships because thus far we have explored the fundamentals of only one "sphere"—the atmosphere. Consideration of the fundamentals of hydrosphere, biosphere, and lithosphere will take place in subsequent chapters. It seems useful, however, to interject in this focus box an example of some of the environmental associations that interest the physical geographer.

Our attention will be directed to a single climatic type—the tropical wet—as an exemplar of these relationships. The specific regions of tropical wet climate—equatorial zones and windward coasts in tropical latitudes—are characterized by fairly specific environmental conditions. It is tempting to state that the Af climate "creates" these conditions, but the cause-and-effect relationship is not that simple. It is proper, however, to note the correlation between a particular climatic type and relatively predictable patterns of other environmental conditions. If climate is not *the* cause, it is at least a prominent influence on such development.

Flora Regions with an Af climate normally are covered with an association of natural vegetation that is unexcelled in luxuriance and variety as a response to the high levels of temperature and moisture. For the most part, this takes the form of a tropical rainforest or *selva*, which is a broadleaf evergreen forest with numerous tree species. Many of the trees are very tall, and their intertwining tops form a relatively continuous canopy that virtually prohibits sunlight from shining on the forest floor. Often shorter trees form a second, and even a third, partial canopy at lower elevations. Most of the trees are smooth-barked and have no low limbs, although there is a profusion of vines and lianas that entangle the trunks and dangle from higher limbs. The dimly lit forest floor is relatively clear of growth because lack of sunlight inhibits survival of bushes and shrubs.

Where much sunlight reaches the ground, as along the edge of clearings or the banks of streams, a maze of undergrowth can prosper. This sometimes impenetrable tangle of bushes, shrubs, vines, and small trees is called a *jungle*.

Fauna This is the realm of flyers, crawlers, creepers, and climbers. Larger species, particularly hoofed animals, are not common. Birds and monkeys inhabit the forest canopy, often in great quantity and diversity. Snakes and lizards are common both on the forest floor and in the trees. Rodents are sometimes numerous at ground level, but the sparser population of larger mammals typically is secretive in behavior and nocturnal in habit. Aquatic life, particularly fish and amphibians, is usually abundant. Invertebrates, especially insects and arthropods, are characteristically superabundant.

Soil The copious, warm, year-round rains provide an almost continuous infiltration of water downward, with the result that soils are usually deep but highly leached and infertile. Leaves, twigs, flowers, and branches frequently fall from the trees to the ground, where they are rapidly decomposed by the abundant earthworms, ants, bacteria, and microfauna of the soil. The accumulated litter is continually being incorporated into the soil, where some of the nutrients are taken up by plants and the remainder is carried away by the infiltrating water. Laterization is the principal soil-forming process; it produces a thin layer of fertile topsoil that is rapidly used by plants, and a deep subsoil that is largely an infertile mixture of such insoluble constituents as iron, aluminum, and magnesium compounds. These minerals typically impart a reddish color to the soil. River floodplains tend to develop soils of higher fertility because of flood-time deposition of silt.

Hydrography The abundance of runoff water on the surface feeds well-established drainage systems. There is usually

During the low-sun season, the Am climate regions are dominated by offshore winds. The winter monsoon produces little precipitation; from one to four months may record less than 2 inches (5 cm) per month, and one or two months may be absolutely rainless.

A lesser distinction of the Am climate relates to its annual temperature curve. Although the annual temperature range may be only slightly greater than in a tropical wet climate, the highest temperatures normally occur in late spring, prior to the onset of the summer monsoon. The heavy cloud cover of the wet monsoon period shields out some of the insolation, resulting in slightly lower temperatures in summer than spring.

Apart from these monsoonal modifications, the climatic characteristics of Af and Am locations are similar.

Tropical Savanna Climate (Aw) The most extensive of the A climates, the tropical savanna type, generally fringes the Af and Am areas to the north and south. It is also widespread in Africa to the east of the tropical wet climate region and extends poleward to about 20° of latitude, north and south. It occurs broadly in South America, central Africa, and southern Asia, and is found significantly in northern Australia and middle America.

The distinctiveness of this climatic type is in its clear-cut seasonal alternation of wet and dry periods; it is often referred to as "tropical wet-and-dry climate." See Figure 8-9. This characteristic is explained by the intermediate position of Aw climates between unstable, converging air on the equatorial side and stable, anticyclonic conditions on the poleward side. As the global wind and pressure systems shift latitudinally with the sun, the Aw regions are overlain with extreme contrasts from "winter" to "summer." During the low-sun season ("winter"), all systems shift toward the opposite hemisphere, so that savanna regions are dominated by subtrop-

A tropical rain-forest scene in Colombia. [Courtesy John Littlewood, United Nations.]

Monkeys are typical rain-forest mammals. [Courtesy Florida Department of Commerce, Division of Tourism.]

A typical rain-forest stream carries lots of water and lots of sediment. This is the Atrato River in Colombia. [Courtesy John Littlewood, United Nations.]

a dense network of streams, most of which carry both a great deal of water and a heavy load of sediment. Lakes are not common because there is enough erosion to drain them naturally. Where the land is very flat, swamps sometimes develop through inadequate drainage and the rapid growth of vegetation.

The environmental relations summarized above represent some of the more conspicuous features associated with a tropical wet climate. They may not apply to any specific location, but they serve as generalizations for such regions. Similar generalizations can be made, with varying validity, for each of the other climatic types discussed in this chapter.

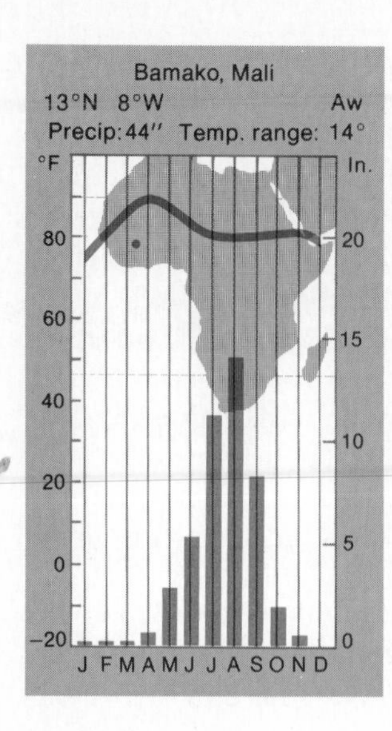

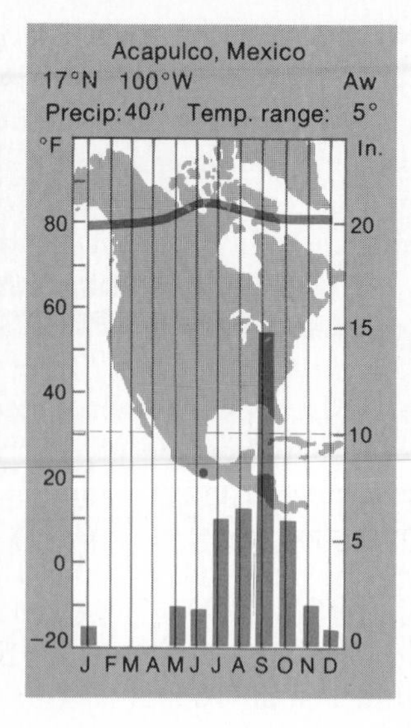

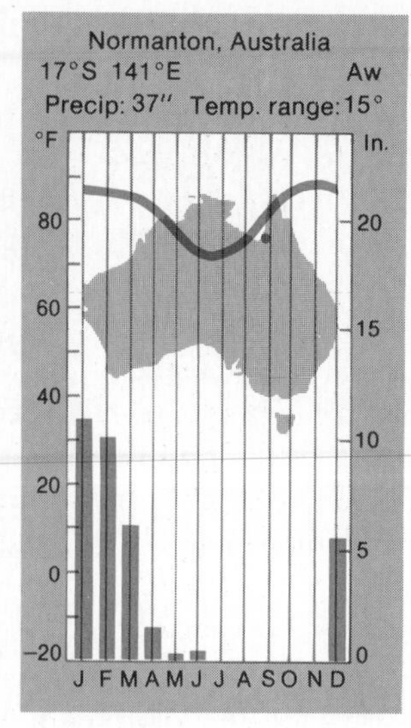

Figure 8-9. Climatic diagrams for representative tropical savanna stations.

ical high-pressure conditions of subsiding and diverging air, which produce clear skies. In "summer," the systems shift in the opposite direction, bringing the ITC zone and its wet tropical weather patterns into the region. See Figure 8-10. The poleward limits of Aw climate are approximately equivalent to the poleward maximum migration of the ITC zone.

Annual rainfall totals in the tropical savanna climate are generally less than in the other two A climatic types; typical averages are between 35 and 70 inches (90 and 180 cm). The high-sun season is conspicuously wet, with one to four months receiving as much as 10 inches (25 cm) each. The low-sun season, on the other hand, is distinctly a time of drought, sometimes with three or four months that are absolutely rainless. See Figure 8-11.

Average annual temperatures are about the same as in the wet tropics but with a somewhat greater month-to-month variation. "Winter" is a little cooler, and "summer" is a little hotter, giving an annual temperature range generally between 15° and 30°. As in tropical monsoon climates, the hottest time of the year is likely to be in late "spring," just before the onset of the summer rains.

From a practical standpoint, the tropical savanna climate can be thought of as having three seasons rather than two. The wet season is much like the wet season anywhere in the tropics, with high sensible temperatures, muggy air, and frequent convective showers. The early part of the dry season is a period of clearing skies and slight cooling. The later part of the dry season is a time of fire; the dessicated grass and shrubs experience wildfire almost every year, unless it is suppressed artificially.

Dry Climates (Group B)

The dry climates are even more extensive than the tropical humid climates. See Figure 8-12. They cover about 30 percent of the land area of the Earth, which is more than any other climatic group. Although at first glance their distribution pattern may appear to be erratic and complex, it actually has a considerable degree of predictability. The largest expanses of dry areas are associated with subtropical high-pressure zones, particularly in the western and central portions of continents. In the midlatitudes, the B climates are found in extreme continental locations,

Figure 8-10. The average poleward locations of the ITC zone are shown here. Its migration follows the sun northward and southward.

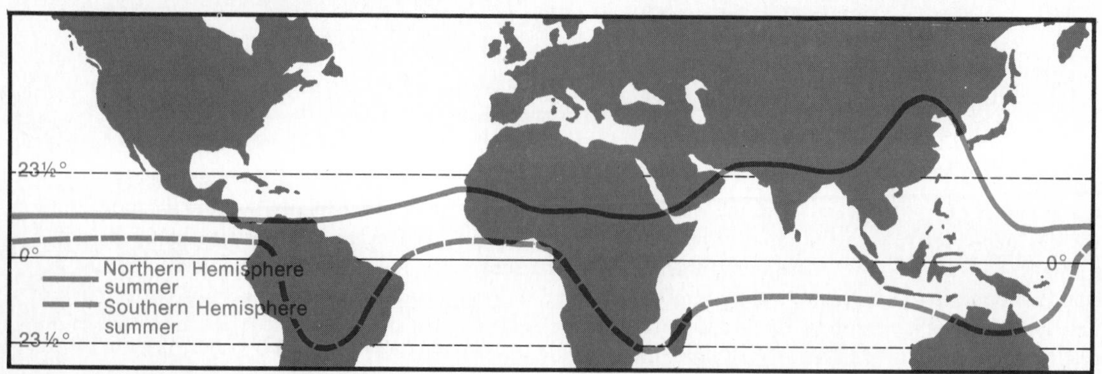

Figure 8-11. Giraffes in a tall-grass savannah in Kenya. Savannahs often are dotted with shrubs or occasional low trees, but in many localities there is no vegetation other than grass as far as the eye can see. [TLM photo.]

which means areas that are remote from sources of moisture, either because of distance or topographic barriers. In Asia and North America, the subtropical and midlatitude dry lands merge to form continuous regions of moisture deficiency.

The concept of a "dry climate" is not simple because it involves the balance between incoming (precipitation) and outgoing (evapotranspiration) moisture. Climatic dryness depends not only on rainfall but also on temperature. The basic generalization is that higher temperature engenders greater potential evapotranspiration, so hot regions can receive more precipitation than cool ones and still be "drier."

Köppen established precise quantitative boundaries to separate B climates from others and also to subdivide the two principal types of B climates (BW and BS). The appropriate formulae are displayed in graphic form in

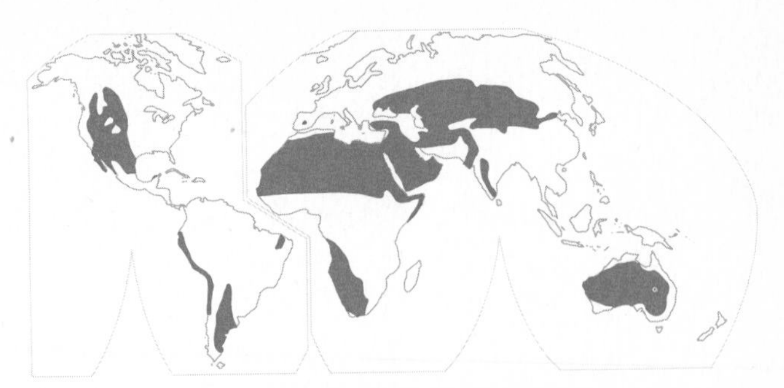

Figure 8-12. World distribution of B climates.

Figure 8-13. For our purposes, however, such precision seems unnecessary.

From the conceptual standpoint, deserts are arid, and steppes are semiarid. Normally the deserts of the world are large core areas of aridity, surrounded by a transitional fringe of slightly less dry steppe. This arrangement is particularly marked in the subtropical dry lands, but is less apparent in the midlatitudes where the steppe may be more expansive than the desert.

The B climates can be classified conveniently into four types: subtropical desert (BWh), subtropical steppe (BSh), midlatitude desert (BWk), and midlatitude steppe (BSk). Our discussion of climatic characteristics will focus on the deserts as they represent the epitome of dry conditions—the arid extreme. Most of what is stated about deserts will apply also to steppes, but in modified intensity.

Subtropical Desert Climate (BWh) The subtropical low-latitude deserts coincide in position with the subtropical zones of high pressure in both the Northern and Southern hemispheres. See Figure 8-14. Anticyclonic conditions are strongest in the eastern portions of the subtropical highs (STHs), so arid conditions reach to the western coasts of all continents in these latitudes. Subsi-

Figure 8-13. Dry climate boundaries according to the Köppen formulae. [After A. N. Strahler and A. H. Strahler, *Elements of Physical Geography,* 2nd ed. (New York: John Wiley & Sons, 1979), p. 527. © 1979 by John Wiley & Sons. Used by permission of John Wiley & Sons.

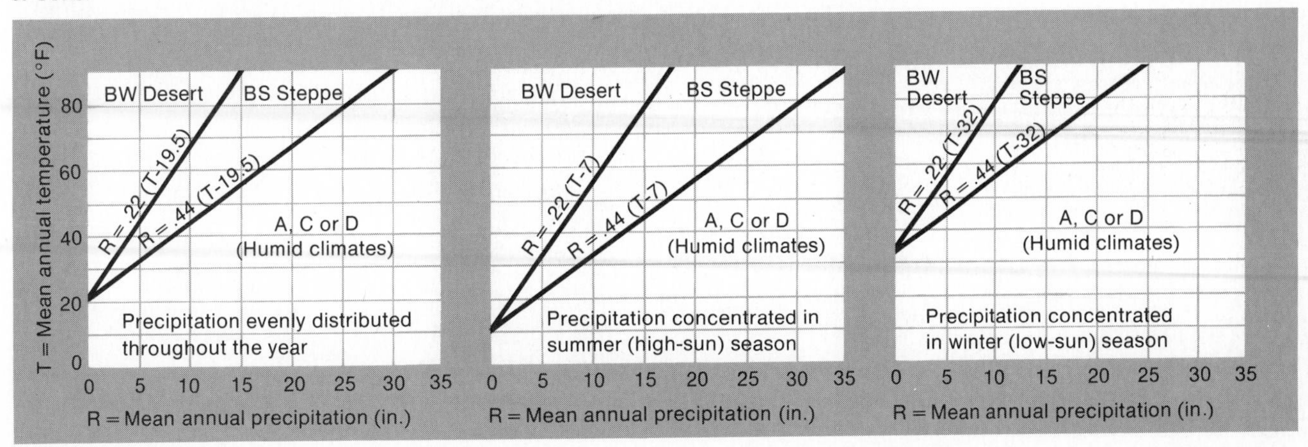

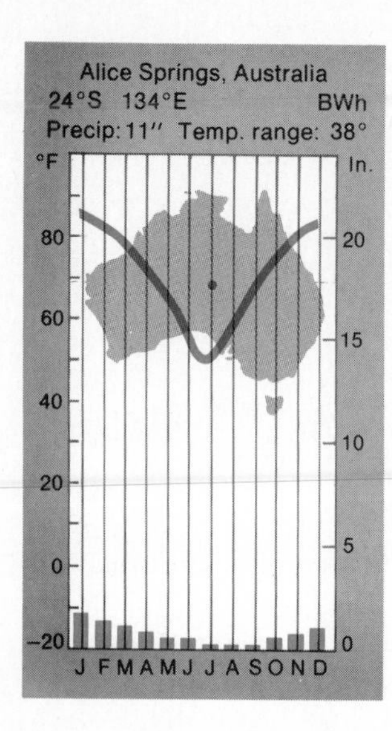

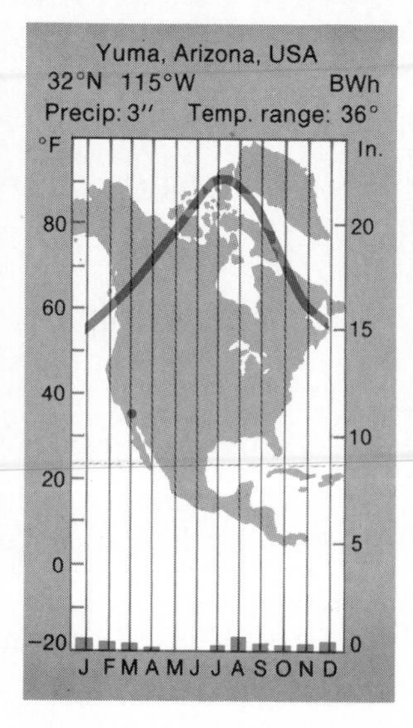

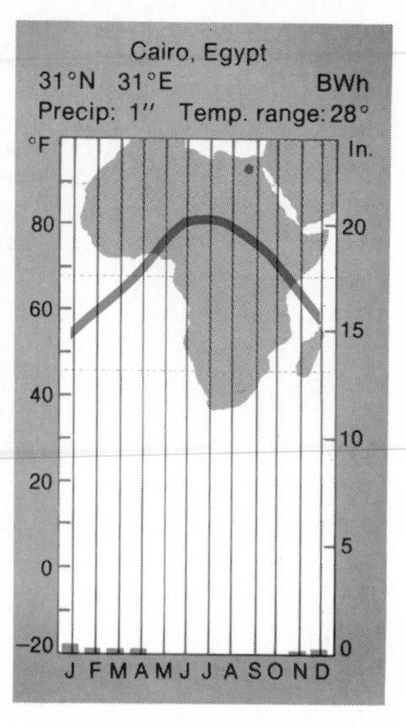

Figure 8-14. Climatic diagrams for representative subtropical desert stations.

dence is weaker on the western sides of the STHs, with the result that the arid zone does not extend to the east coast of any continent in these latitudes, except in North Africa, which is sheltered by the Arabian peninsula from oceanic influence in that direction.

The enormous expanse of BWh climate in North Africa and southwestern Asia represents more desert area than is found in the rest of the world combined. Such an extensive development is explained by the year-round presence of anticyclonic conditions and the remoteness from any upwind source of moisture. The adjacency of the Asian landmass has the climatic effect of making Africa a continent without an east coast north of 10° north latitude. The desert climate is also very expansive in Australia (50 percent of the continental area) because the mountains that parallel the east coast of that continent are just high enough to prevent Pacific Ocean winds from penetrating; most of Australia is in the rain shadow of those eastern highlands.

Subtropical deserts have a much more restricted longitudinal extent in southern Africa, South America, and North America, but they are elongated latitudinally along the coast because of the presence of cold offshore waters. See Figure 8-15. The greatest attenuation occurs along the western side of South America, where the Atacama Desert is not only the "longest" but also the driest of the dry lands. The Atacama is sandwiched in a double rain shadow position: Moist winds from the east are denied by the presence of one of the world's great mountain ranges (the Andes), and air from the nearby Pacific Ocean is thoroughly chilled and stabilized by passage over the world's most prominent cold current (the Humboldt).

The distinctive climatic characteristic of deserts is lack of moisture. Three adjectives, however, are particularly applicable to precipitation conditions in subtropical deserts: *scarce, unreliable, intense.*

1. *Scarce.* These deserts are the most nearly rainless regions on Earth. Some are claimed to have experienced

Figure 8-15. Although some desert landscapes contain considerable vegetation, they are usually open, sparse, and xerophytic in nature. This is a southern Arizona scene. [Courtesy Arizona-Sonora Desert Museum.]

three or four years without a single drop of moisture falling from the sky. Most BWh regions, however, are not totally without precipitation. Annual totals of between 2 and 8 inches (5 and 20 cm) are characteristic, and some places receive as much as 15 inches (38 cm).

2. *Unreliable*. An important climatic axiom is that the less the mean annual precipitation, the greater its variability. The very concept of an "average" yearly rainfall in a BWh location is misleading because of the fluctuations that can take place from one year to the next. Yuma, Arizona, for example, has a long-term average rainfall of 2.7 inches ($6\frac{1}{2}$ cm), but within the last decade it has received as little as 0.5 inches (1.3 cm) and as much as 7 inches (18 cm) within a given year. The percentage deviation from the norm tends to be very high in the BWh climate.
3. *Intense*. Most precipitation in these regions falls in vigorous convective showers, which are localized and of short duration. Thus the rare rains may bring brief floods to regions that have been bereft of surface moisture for months.

Temperatures in the BWh climate also have certain distinctive characteristics. The combination of low-latitude location and lack of cloudiness permits a great deal of insolation to reach the surface, and nocturnal terrestrial radiation is likewise appreciable. Summers are interminably long and blisteringly hot, with monthly averages in the middle to high 90s°F (high 30s°C). Midwinter months have average temperatures in the 60s°F (high teens °C), which gives moderate annual temperature ranges of 15° to 25°. Daily temperature ranges, on the other hand, are sometimes astounding. Summer days are so hot that the nights do not have time to cool off significantly, but during the transition seasons of spring and fall, a 50° fluctuation between the heat of the afternoon and the cool of the following dawn is not unusual.

Subtropical deserts experience much windiness during daylight, but nights are usually much quieter. This horizontal air movement apparently is related to rapid daytime heating and strong convective activity, which accelerates surface currents. The persistent winds are largely undeterred by soil and vegetation, with the result that a great deal of dust and sand is frequently carried along.

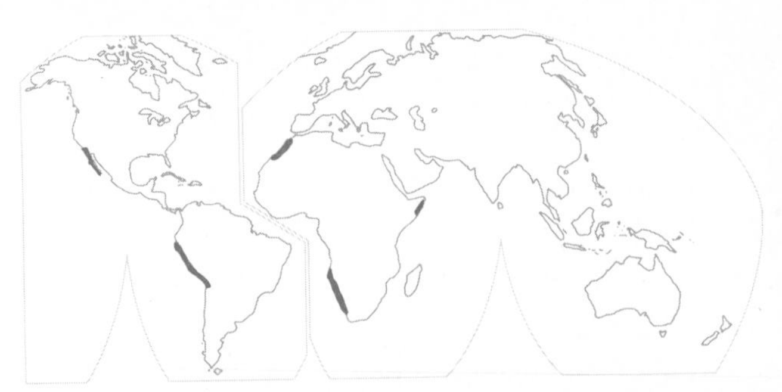

Figure 8-16 Cool and foggy west coast deserts are found along coasts paralleled by cool ocean currents and cold upwellings. They are mostly in subtropical west coast locations, with two exceptions: they are absent from the west coast of Australia, and they occur on the east coast of the "horn" of Africa (Somalia).

Specialized and unusual conditions prevail along west coasts in subtropical deserts. See Figure 8-16. The cold waters offshore (currents and upwellings) chill any air that moves across them in typical advective fashion. This produces high relative humidity as well as frequent fog and low stratus cloud. Precipitation almost never results from this advective cooling, and the influence normally extends only a few miles inland. The immediate coastal zone, however, is characterized by such abnormal desert conditions as relatively low summer temperatures (typical hot-month averages in the low 70s °F or low 20s °C), continually high relative humidity, and greatly reduced annual and daily temperature ranges. See Figure 8-17.

Subtropical Steppe Climate (BSh) The BSh climates characteristically surround the BWh climates (except on the western side), separating them from the more humid climates beyond. Temperature and precipitation conditions are not significantly unlike the deserts previously

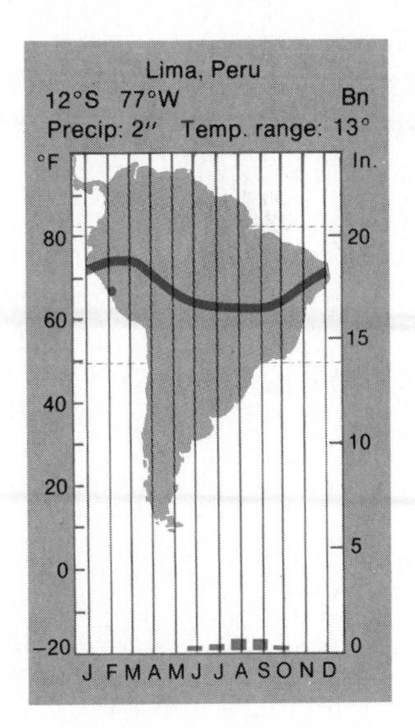

Figure 8-17 Climatic diagram for a cool west coast desert station.

World Record—Highest Absolute Temperature

136° F (58° C)—El Azizia, Libya—1922—BWh

World Record—Greatest Temperature Range in One Day

100° F (56° C)—In Salah, Algeria—1927—BWh

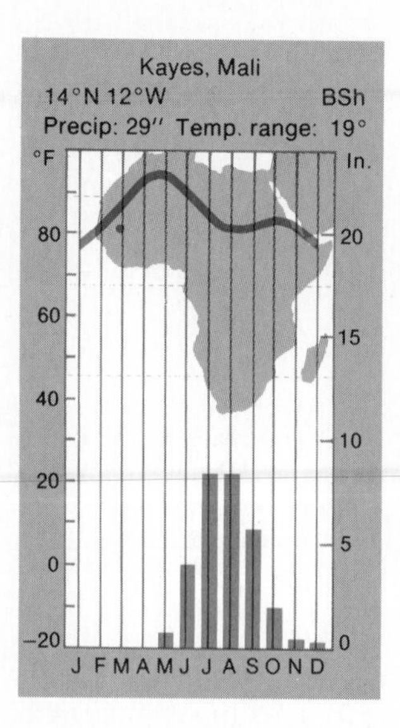

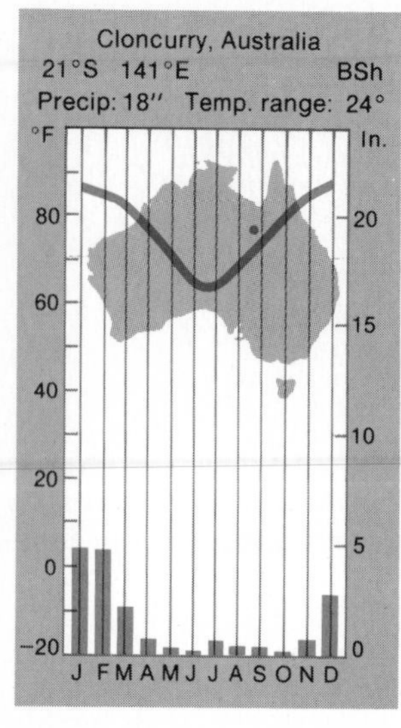

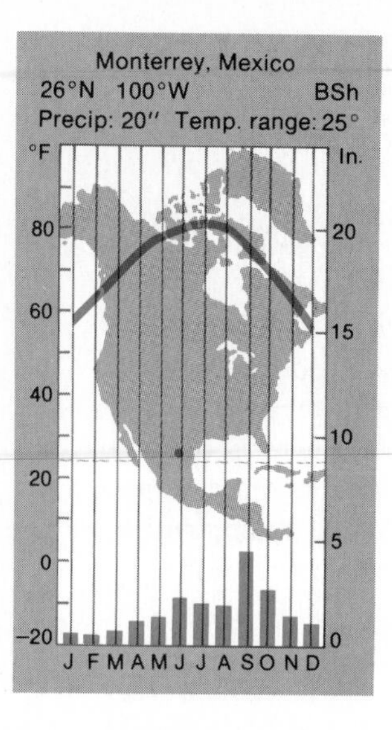

Figure 8-18. Climatic diagrams for representative subtropical steppe stations.

described except that the extremes are more muted in the steppes. See Figure 8-18. Thus rainfall is somewhat greater and more reliable, and temperatures are slightly moderated. Moreover, the meager precipitation tends to have a seasonal concentration. On the equatorward side of the desert, rain occurs in the high-sun season; on the poleward side, it is concentrated in the low-sun season.

Midlatitude Desert Climate (BWk) The BWk climates do not owe their existence to persistent anticyclonic subsidence, as do their BWh counterparts. Instead, the midlatitude deserts occur primarily in the deep interiors of continental landmasses, where they are either far removed geographically or blocked functionally from oceanic influence. See Figure 8-19. The largest expanse of midlatitude dry climates, in central Asia, is not only distant from any ocean but is also protected on the south by massive mountain ranges from any contact with the Indian summer monsoon. In North America, high mountains closely parallel the west coast, and moist maritime air from the Gulf of Mexico affects the eastern half of the continent, so the dry climates are displaced well to the west. The only other significant BWk area occurs in southern South America, where the desert actually reaches the east coast of Patagonia (southern Argentina). This anomalous situation has developed because the continent is so narrow in these latitudes that all of it lies in the rain shadow of the Andes to the west.

Precipitation conditions in midlatitude deserts are much like those of subtropical deserts—meager and erratic. Differences lie in two aspects: seasonality and intensity. Most BWk regions receive the bulk of their precipitation in summer, when warming and instability are common. Winter is usually dominated by low temperatures and anticyclonic conditions, which are inimical to precipitation. Although most BWk precipitation is of the unstable, showery variety, there are also some periods of general overcast and protracted drizzle.

The principal climatic differences between midlatitude and subtropical deserts are in temperature, especially

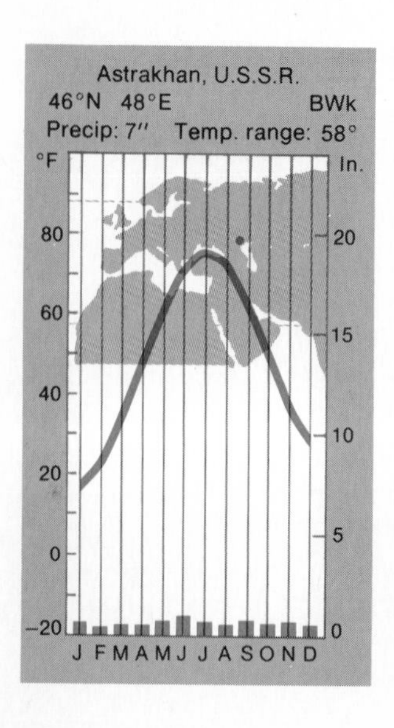

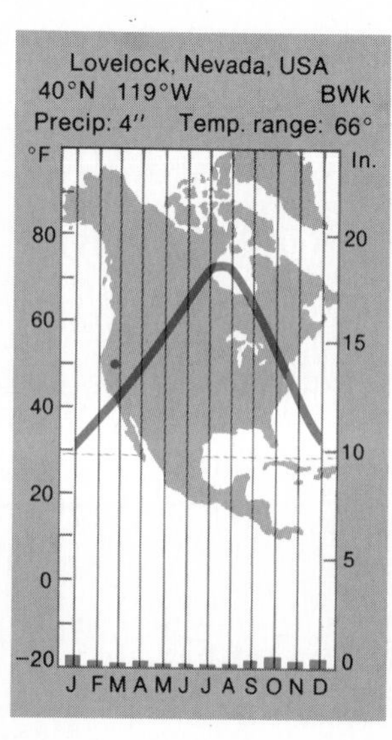

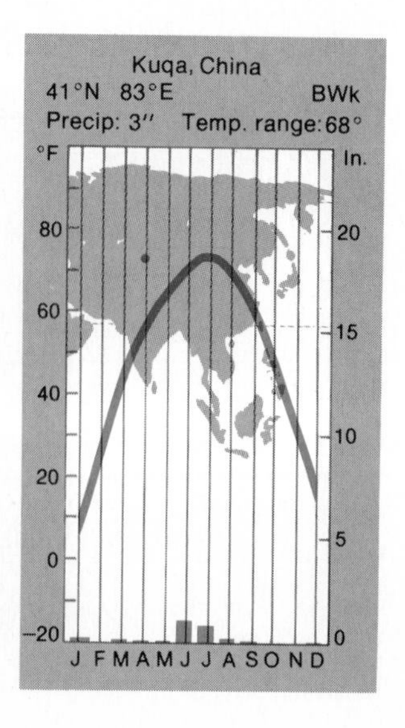

Figure 8-19. Climatic diagrams for representative midlatitude desert stations.

Figure 8-20. Midlatitude deserts are characterized by cold winters. Here a fresh snowfall covers such typical desert plants as creosote bush and yucca in northern Arizona. [TLM photo.]

winter temperature. See Figure 8-20. BWk regions have severely cold winters. The average cold-month temperature is normally below freezing, and some BWk stations have six months with below-freezing averages. This produces much lower average annual temperatures than in BWh regions and greatly increases the annual temperature range; variations exceeding 50° between January and July average temperatures are not uncommon.

Midlatitude Steppe Climate (BSk) As in the subtropics, midlatitude steppes generally occupy transitional positions between deserts and humid climates. See Figure 8-21. Typically they have somewhat more precipitation and slightly subdued temperature extremes.

In western North America, the steppe climate is much more extensive in area than in the desert, so that the desert-core-and-steppe-fringe model is somewhat misleading. Semiarid conditions are broadly prevalent; only in the interior southwest of the United States is the climate sufficiently arid to be classified as desert.

Mild Midlatitude Climates (Group C)

We have noted previously that the middle latitudes have the greatest weather variability on a short-run (day-to-day or week-to-week) basis. Seasonal contrasts are also marked in these latitudes. They lack the constant heat of the tropics and the almost-continual cold of the polar regions. This is a zone of air mass contrast, with frequent alternating incursions of tropical and polar air producing more convergence than is found anywhere else except in the immediate equatorial zone. This air mass conflict creates a kaleidoscope of atmospheric disturbances and weather changes. The seasonal rhythm of temperature is usually more prominent than that for precipitation. Whereas in the tropics the seasons are called "wet" and "dry," in the midlatitudes they are clearly "summer" and "winter."

The mild midlatitude climates (Group C) occupy the equatorward margin of the middle latitudes, occasionally extending into the subtropics and being elongated

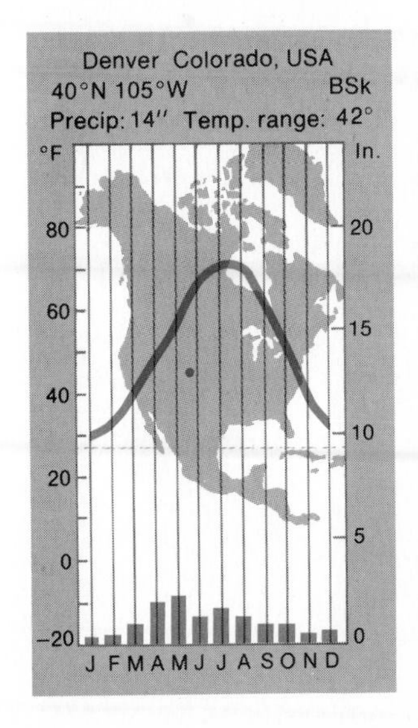

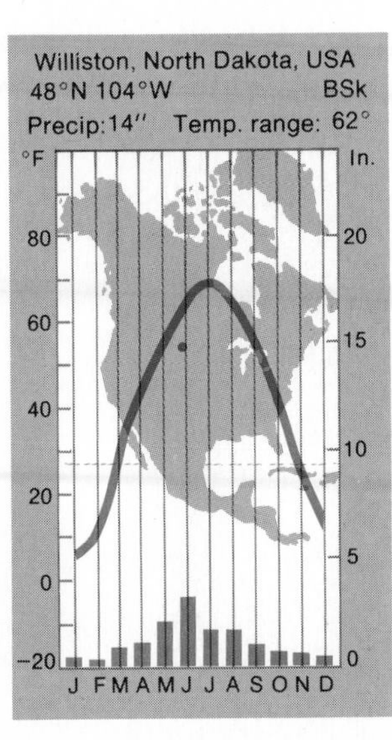

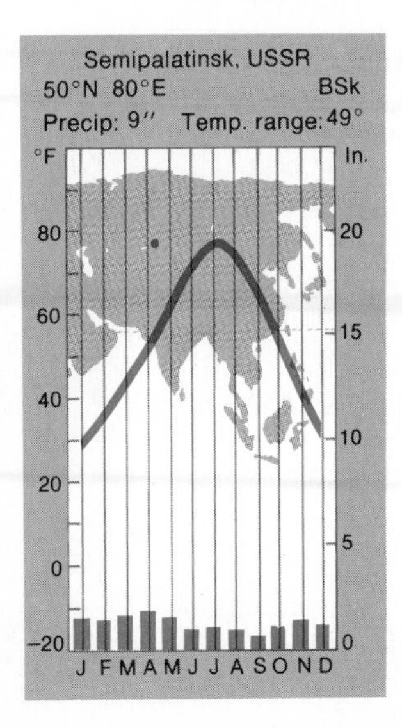

Figure 8-21. Climatic diagrams for representative midlatitude steppe stations.

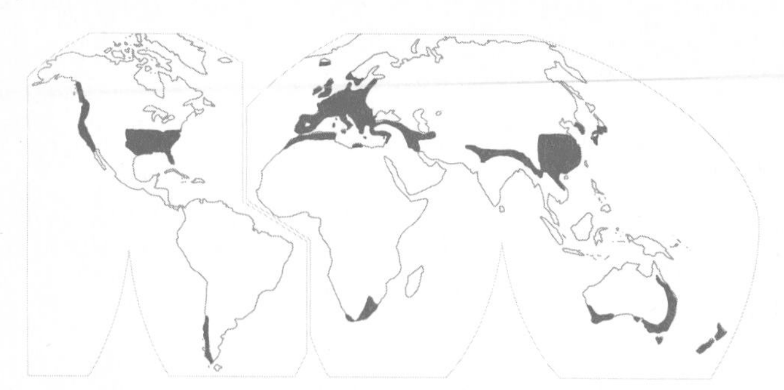

Figure 8-22. World distribution of C climates.

poleward in some west coastal situations. See Figure 8-22. They constitute a transition between the warmer tropical climates on one side and the severe midlatitude climates on the other.

Summers in the C climates are long and usually hot; winters are short and relatively mild. These regions, in contrast to the A climates, experience occasional winter frosts and therefore do not have a year-round growing season. Precipitation conditions are highly variable in the C climates, both as to total amount and seasonal distribution. Year-round moisture deficiency is not characteristic, but there are sometimes pronounced seasonal deficiencies.

The C climates are subdivided into three types, primarily on the basis of precipitation seasonality, and secondarily on the basis of summer temperatures.

1. *Mediterranean climates* (Csa, Csb) have a pronounced winter-rain/summer-drought regime.
2. *Humid subtropical climates* (Cfa, Cwa) have hot summers and either no dry season or a winter dry season.
3. *Marine west coast climates* (Cfb, Cfc) experience mild summers along with either no dry season or a limited winter dry season.

Mediterranean Climate (Csa, Csb) The Cs climates are sometimes referred to as *dry subtropical*, but the more widely used designation is *mediterranean*. (The proper terminology is without capitalization because it is a generic term for a type of climate; Mediterranean, capitalized, refers to a specific region around the Mediterranean Sea.) This climatic type is found on the western side of continents, centered at about 35° of latitude, north and south. See Figure 8-23. With one exception, all mediterranean climatic regions are small in extent, being restricted mostly to coastal areas by interior mountains or limited landmasses at this latitude. These small regions are in central and southern California in North America, central Chile in South America, the southern tip of Africa, and the two southwestern "corners" of Australia. The only extensive area of mediterranean climate is around the borderlands of the Mediterranean Sea in southern Europe, north-coastal Africa, and southwestern Asia.

This climatic type has three distinctive characteristics:

1. The modest annual precipitation is concentrated entirely in winter, summers being virtually rainless.
2. Winter temperatures are unusually mild for the midlatitudes, and summers vary from hot to warm.
3. Clear skies and abundant sunshine are typical, especially in summer. Average annual precipitation is slight, ranging from about 15 inches (38 cm) on the equatorward margin to about 25 inches (64 cm) at the poleward extreme. The midwinter months normally experience from 3 to 5 inches (8 to 13 cm) each, and two or three midsummer months are absolutely dry. No other climatic type has such a concentration of precipitation in winter. See Figure 8-24.

Most of the mediterranean climate is classified as Csa, which means that summers are hot, with midsummer monthly averages of between 75° and 85° F (24° to 29° C), and frequent maxima above 100° F (38° C). Average cold-month temperatures are about 50° F (10° C), with occasional minima below freezing.

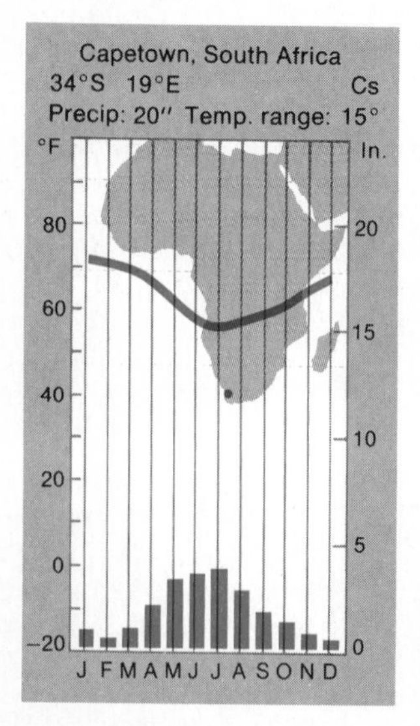

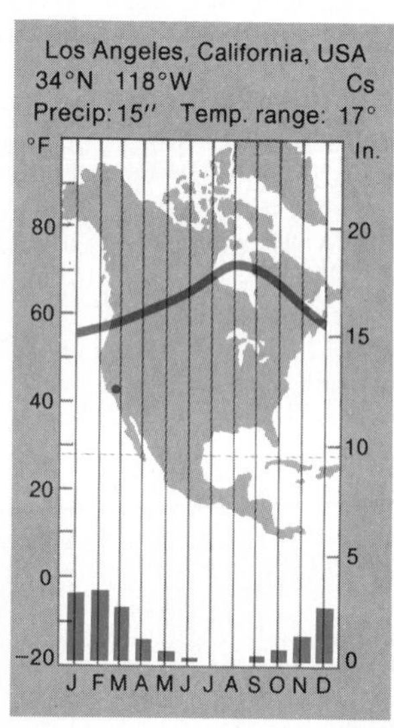

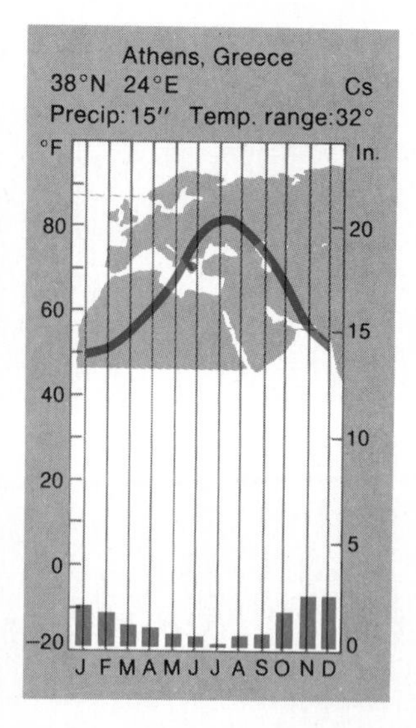

Figure 8-23. Climatic diagrams for representative mediterranean climate stations.

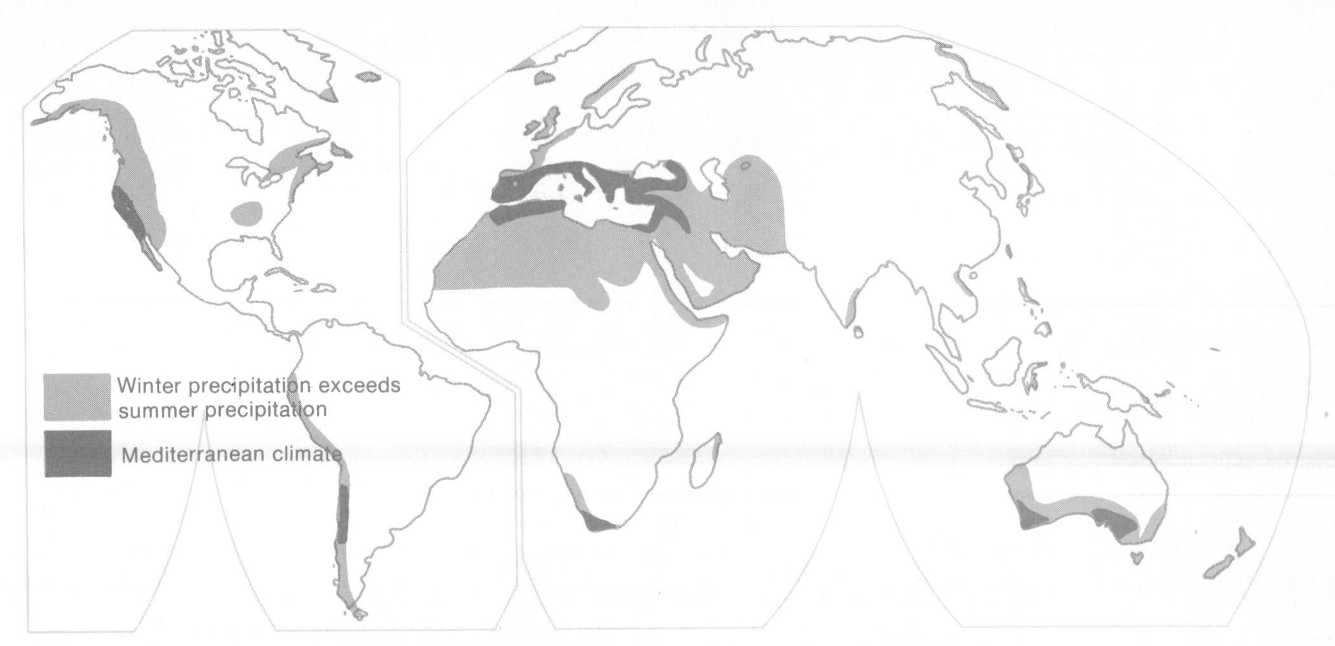

Figure 8-24. Lands that receive winter precipitation maxima are mostly associated with mediterranean and marine west coast climates and regions adjacent to them. Most parts of the world experience their wet season in summer.

The immediate mediterranean coastal areas have much milder summers than inland areas due to sea breezes; these climates are classified as Csb. See Figure 8-25. In such areas, the average hot-month temperature is between 60° and 70° F (16° and 21° C). Winters are slightly milder in these coastal areas, with cold-month averages of about 55° F (13° C). Csb areas also have higher humidity, frequent nocturnal fog, and occasional development of low stratus overcast.

The genesis of mediterranean climates is clear-cut. These regions are dominated in summer by dry, stable, subsiding air from the eastern portions of STHs. In winter, the wind and pressure belts shift equatorward, and mediterranean regions come under the influence of the westerlies with their migratory extratropical cyclones and associated fronts. Almost all precipitation comes from these cyclonic storms, except for occasional tropical influences in California.

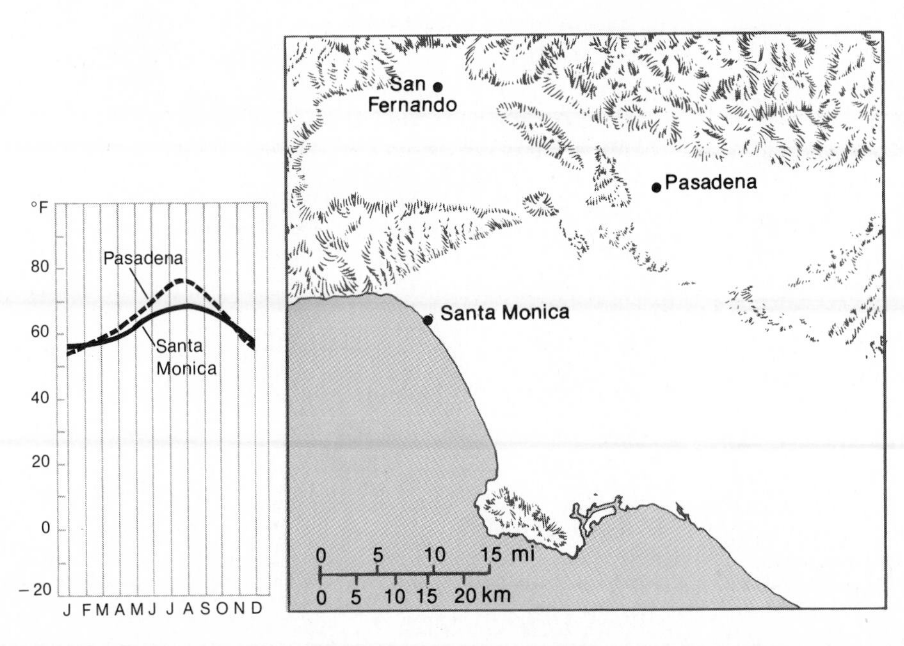

Figure 8-25. There are often significant temperature differences between coastal and inland mediterranean areas. This climatic diagram shows the annual temperature curve for Santa Monica (a coastal station; Csb climate) and the annual temperature curves for Pasadena and for San Fernando (both are inland stations and have exactly the same average temperatures for each month; Csa climate). The physiographic map shows the topographic and coastal relationships of the three stations.

Focus on Climatic Distribution in Africa: A Practically Perfect Pattern

Africa is a fascinating continent for geographic study, for many reasons. It has unusual geologic features, a hydrographic pattern unlike that of any other continent, a remarkable fauna, a long and dramatic human history, an extraordinary diversity of cultures, an extremely volatile political situation, and various other unusual attributes that attract attention. For certain aspects of physical geography, however, Africa is particularly interesting because of its normality. It is the only continent that is approximately bisected by the equator and thereby has the same latitudinal extent in both the Northern and Southern Hemispheres. This produces a regularity of pattern for some environmental features that is almost like a textbook model.

Most notable in this regard is the arrangement of climatic types, which can be displayed across the map of Africa as a model almost without blemish. For the geographer seeking predictable patterns, it is a joy to behold.

In the center of the continent is a hot, wet core region (Af), and at the proper subtropical latitudes are two hot, dry core regions (BWh). Between these precipitation extremes are the two expectable transition climates (Aw and BSh), and beyond the deserts on the poleward side, the transition (BSh) again occurs, leading into the moister climates of the middle latitudes (C). It is a practically perfect pattern through the A climates and the subtropical B climates into the fringe of the C climates on the poleward extremities of the continent.

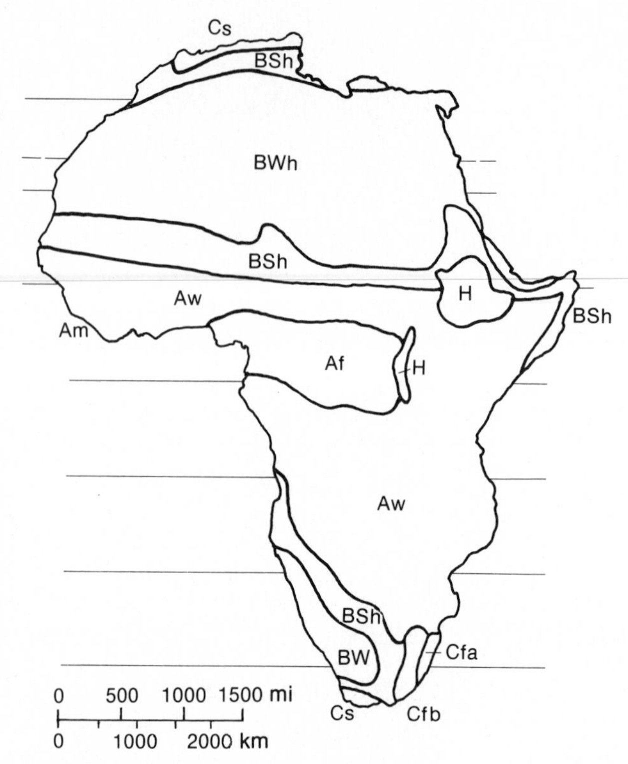

The climatic regions of Africa.

The inevitable anomalies, for the most part, can be explained fairly easily:

1. Most striking is the vast extent of dry climate in North Africa. The "normal" expectation is for desert to extend from the west coast for a considerable distance inland, but in this case it carries all the way to the eastern border of the continent. This results from the presence of the Asian landmass immediately to the east, which excludes maritime air and preserves the characteristic of "continentality" clear across North Africa.
2. The arrangement of climatic types in the equatorial portion of East Africa is quite different from what might be

Humid Subtropical Climate (Cfa, Cwa)

mately the same latitude on the other (eastern) side of continents is found the humid subtropical climate. This type is more extensive in area, both latitudinally and longitudinally. In some places it reaches equatorward to almost 20° of latitude, and extends poleward to about 40°. Its east-west extent is greatest in North America, Asia, and South America.

The humid subtropical climates differ from mediterranean climates in several important respects, as shown in the Table 8-2. Summer temperatures are generally warm to hot, with the highest monthly averages between 75° and 80° F (24° and 27° C). This is not dissimilar to the situation in mediterranean climates, but the humid subtropical regions are characterized by much higher hu-

TABLE 8-2 Principal Differences between Humid Subtropical Climates and Mediterranean Climates

Characteristic	Humid Subtropical	Mediterranean
Precipitation seasonality	Summer maximum	Winter maximum
Dry season	None or only short winter dry season	Distinct summer dry season
Total precipitation	Relatively abundant	Sparse
Winter temperatures	Generally mild; occasional cold spells	Generally mild
Annual temperature range	15° to 40°	10° to 35°

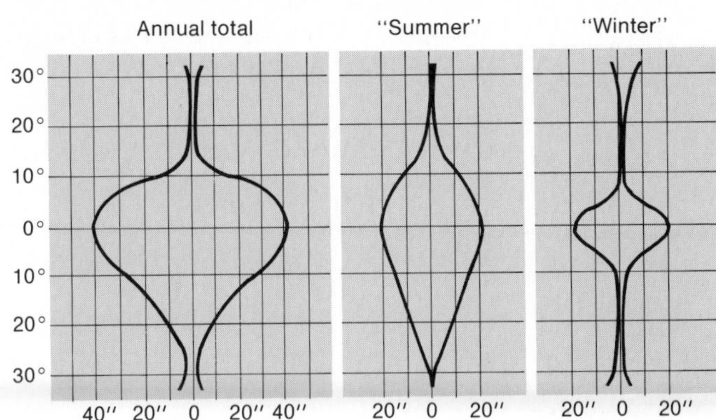

Schematic diagram of rainfall variation along the central meridian (20° east longitude) of Africa. The left portion represents the annual total; the central portion portrays the high-sun season in both hemispheres; the right portion is representative of the low-sun season in both hemispheres.

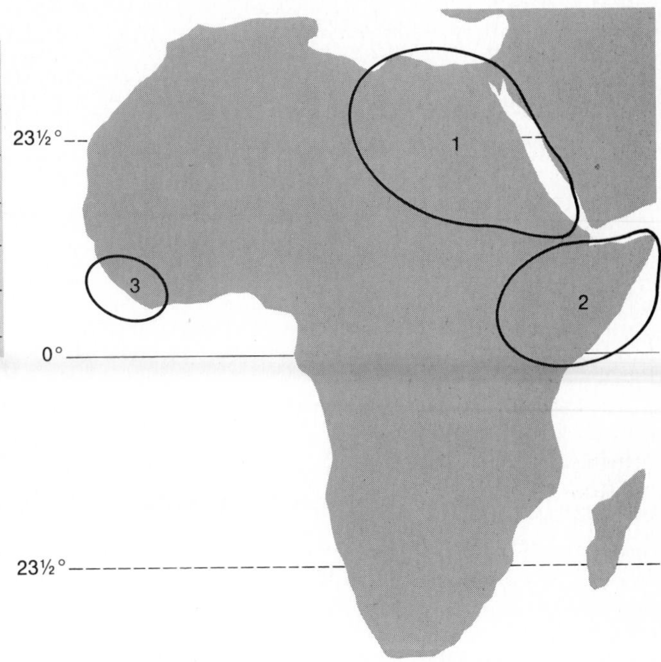

The pattern of climates in Africa is mostly regular and "normal." The three principal exceptions to that normality are shown on this map: (1) a greatly expanded area of B climate in the northeast; (2) an unusual and complex climatic pattern in East Africa; and (3) monsoonal development on the Guinea coast of West Africa.

expected. The tropical wet (Af) climate is absent from this region, being replaced by Aw, BS, and BW. This is due in part to the higher elevations of most of East Africa, which produces lower temperatures. More important is the pattern of airflow along the coast north of about 5° south latitude. Influenced by the South Asian monsoon, winds blow parallel to this coast (northeasterly in winter and southwesterly in summer) throughout the year, which means that maritime air rarely is carried inland, and dryness prevails.

3. A tropical monsoonal climate (Am) is developed along the western Guinea Coast of West Africa. Although the full story of monsoon origin is not clear, there is a definite seasonal reversal of wind flow in this region, with maritime southwesterlies bringing a heavy summer rainy season and dry northeasterlies dominating for the remainder of the year.

An explanation of the general distribution of climatic types over the world varies from the simple to the complex. If one is trying to comprehend the pattern, Africa is probably the best place to begin.

midity in summer, and so sensible temperatures are higher. The days tend to be hot and sultry, and often night brings little relief. See Figure 8-26. Winter temperatures are mild on the average, but winter is punctuated sporadically by cold waves that bring severe weather for a few days at a time. Absolute minimum temperatures can be 10° to 20° lower in the humid subtropics than in mediterranean regions, which means that killing frosts occur much more frequently, especially in North America and Asia. The importance of this factor can be shown by agricultural adjustments. For example, the northernmost limit of commercial citrus production in eastern United States (Florida) is at about 29° north latitude, but in western United States (California), it is at 38° north latitude, 625 miles (1,000 km) farther north (Figure 8-27).

Figure 8-26. Cotton is a major commercial crop in regions of humid subtropical climate. This is an October harvest scene in Arkansas. [USDA-SCS photo by Morgan Prince.]

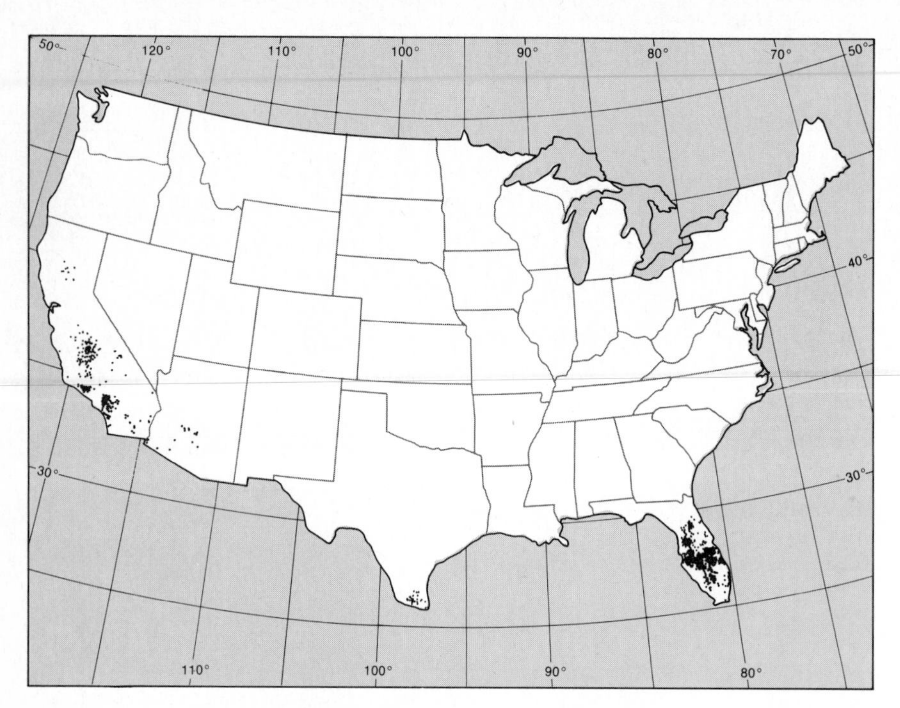

Figure 8-27. Distribution of citrus production in the conterminous United States. Each dot represents 1,000 acres; 1978 data. Winter cold spells prohibit commercial citrus growing north of central Florida in eastern United States; the lack of severe freezes in much of California allow citrus to be grown much further north. [After U.S. Census of Agriculture, 1978.]

The mild-summer characteristics of mediterranean Csb coasts have no counterpart in the humid subtropical regions because the offshore waters are warm rather than cool. Mediterranean climates are all adjacent to cool currents, whereas warm currents wash the humid subtropical coasts.

Annual precipitation is generally abundant and in some places copious, with a general decrease from east to west, i.e., interiorward. See Figure 8-28. Averages mostly are between 40 and 60 inches (100 and 150 cm), although some locations record as little as 30 inches (75 cm), and some receive up to 100 inches (250 cm). Summer is usually the time of precipitation maximum, associated with onshore flows of maritime air and frequent convection. Winter is a time of diminished precipitation, but it is not really a dry season except in China where monsoonal conditions dominate. The rain and occasional snow of winter are the result of extratropical cyclones passing through the region. In the North American and Asian coastal areas, a late summer/autumn bulge in the precipitation curve is due to rainfall from tropical cyclones.

Marine West Coast Climate (Cfb, Cfc) Immediately poleward of the mediterranean climate is the marine west coast type. See Figure 8-29. As its name implies, it is situated on the western side of continents, which is a windward location in these latitudes, between about 40° and 65°. Only in the Southern Hemisphere where landmasses are small in these regions (as New Zealand and southernmost South America) does this oceanic climate extend across into east coast locales. The most extensive area of marine west coast climate is in western and central

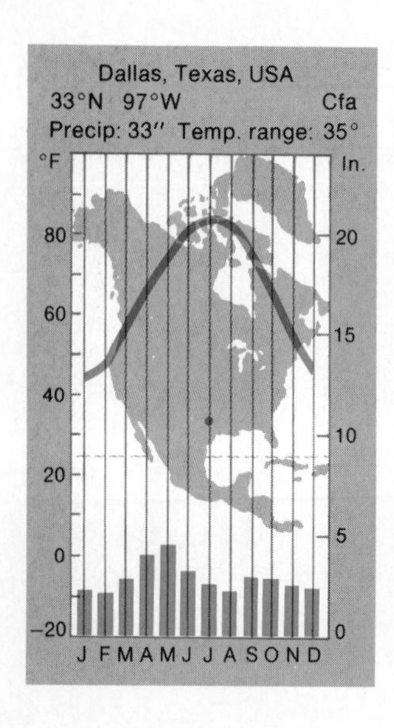

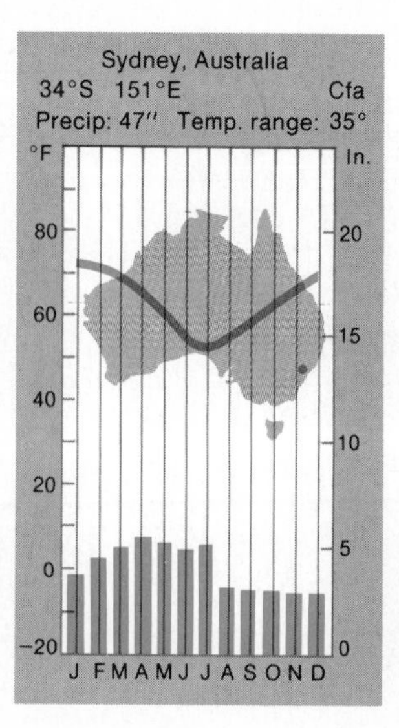

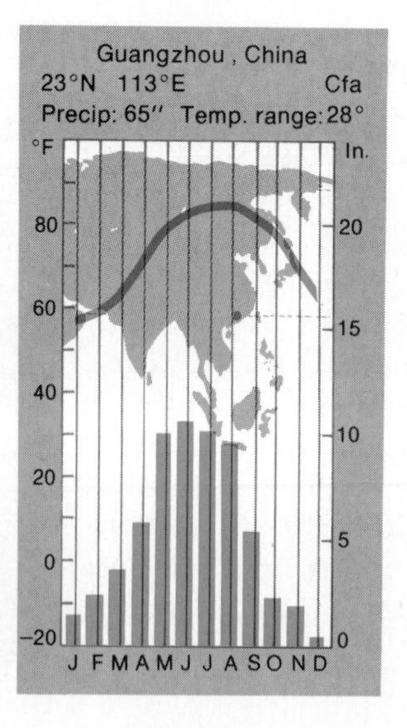

Figure 8-28. Climatic diagrams for representative humid subtropical stations.

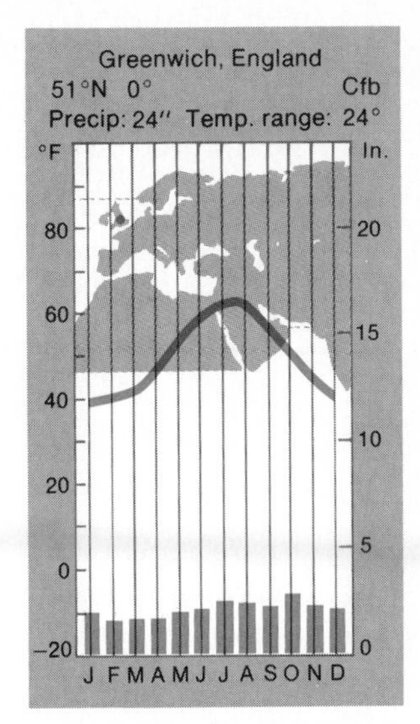

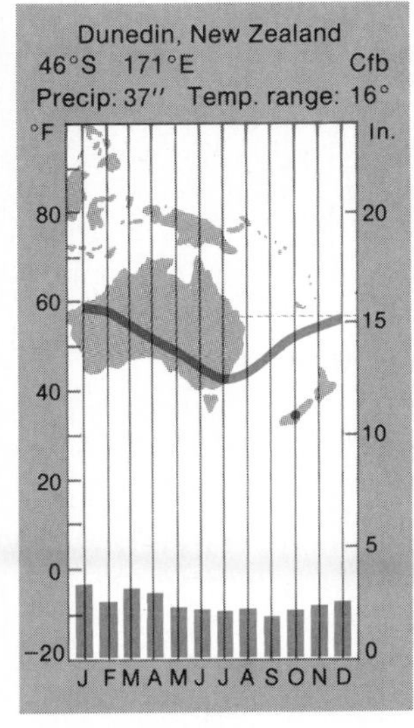

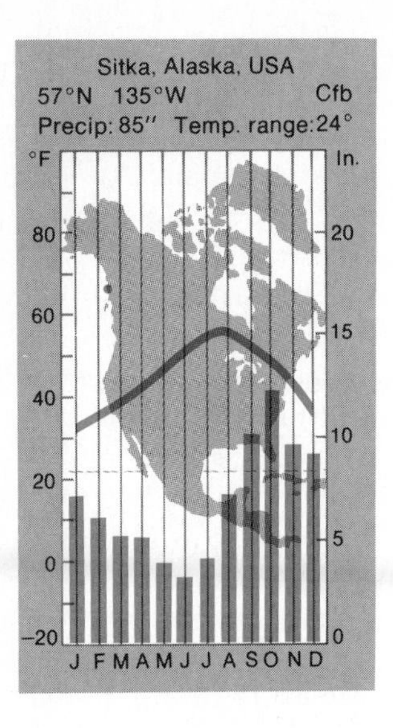

Figure 8-29. Climatic diagrams for representative marine west coast stations.

Europe, where the maritime influence can be carried some distance inland without hindrance from topographic barriers. The North American region is much more restricted interiorward by the presence of formidable mountain ranges athwart the direction of onshore flow.

These regions lie under the influence of the westerlies throughout the year, and the persistent onshore air movement ensures that the maritime influence permeates throughout. This creates an extraordinarily temperate climate considering the latitude; it is the classic example of a "maritime" climate on land. Its distinctive characteristics are lack of extreme temperatures, consistently high humidity, much cloudiness, and a high proportion of days with some precipitation.

The oceanic influence ameliorates temperatures most of the time; this moderation is particularly noticeable when considering the extremes—daily and seasonal maxima and minima. Isotherms tend to parallel the coastline rather than following their "normal" east-west paths. See Figure 8-30. Indeed, temperatures on a west coast in these latitudes decrease poleward less than half as fast as on an east coast. Average hot-month temperatures are generally between 60° and 70°F (16° and 21°C), with cold months averaging between 35° and 45°F (2° and 7°C). There are occasionally very hot days (Seattle has recorded a temperature of 98°F or 37°C, and Paris has reached 100°F or 38°C), but prolonged heat waves are unknown. Similarly, very cold days occur upon occasion, but low temperatures rarely persist. Frosts are relatively infrequent, except in interior European areas; London, for example, at 52° north latitude, experiences freezing temperatures on less than half the nights in January. There is also an abnormally long growing season for the latitude; Seattle, for instance, has a month longer growing season than Atlanta, which is 14° of latitude further equatorward.

This is one of the wettest regions of the middle latitudes, although the total amount of precipitation is

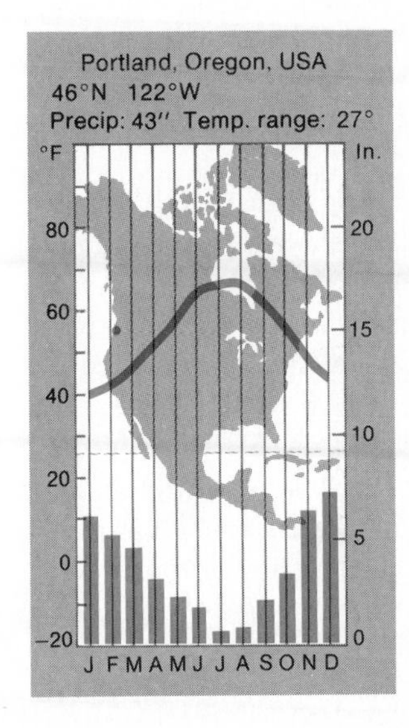

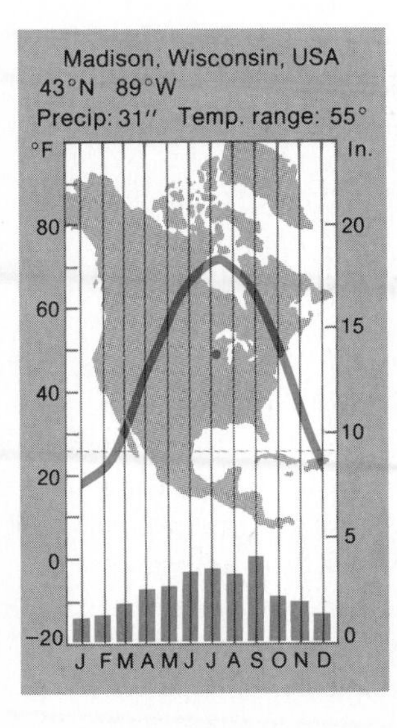

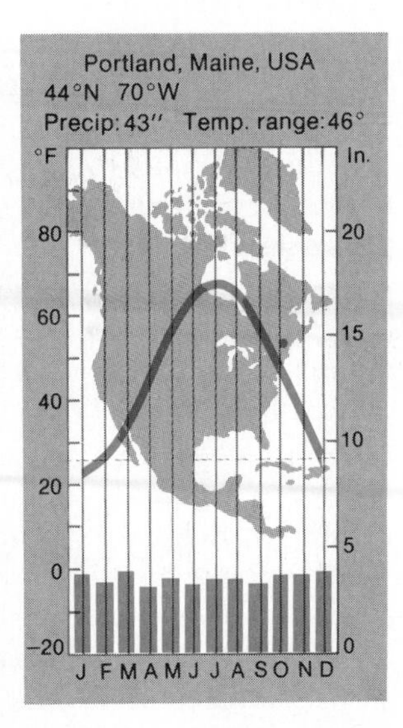

Figure 8-30. Climatic comparisons for three stations at the same latitude in northern United States. Temperatures are mildest on the west coast (Portland, Oregon) and least mild in the interior (Madison, Wisconsin). Note also the seasonal precipitation variations: Portland, Oregon, has a winter maximum; Madison has a summer maximum; Portland, Maine, experiences a very even year-round regime.

not remarkable except in upland areas. Average precipitation is usually described as *adequate,* but the definition of that term is not precise. Some localities receive as little as 20 inches (50 cm), but a range of between 30 and 50 inches (75 and 125 cm) is more typical. Much higher totals are recorded on exposed slopes, reaching 100 to 150 inches (250 to 375 cm) in some places. Snow is uncommon in the lowlands, but higher, west-facing slopes receive some of the heaviest snowfalls in the world.

Perhaps more important to an understanding of the character of the climate than total precipitation is its frequency. Rainfall probability and reliability are high, but intensity is rather low. Frontal precipitation of the drizzly type is characteristic. Humidity is high, with much cloudiness. Seattle, for example, receives only 43 percent of total possible sunshine each year, in contrast to 70 percent in Los Angeles. This is caused by a great many overcast days and days with measurable precipitation. London has experienced as many as 72 consecutive days with rain. (Is it any wonder that the umbrella is the civic symbol?) Indeed, some places on the west coast of New Zealand's South Island have recorded 325 rainy days in a single year.

In summary, winter tends to be stormy, overcast, and dreary, but not particularly cold, while summer is likely to be pleasant, variable, and stimulating.

Severe Midlatitude Climates (Group D)

The severe midlatitude climates occur only in the Northern Hemisphere, because the Southern Hemisphere has no landmasses at the appropriate latitudes, 40° to 70° This climatic group extends broadly across North America (encompassing northeastern United States and much of Canada and Alaska) and Eurasia (from eastern Europe through most of the Soviet Union to the Pacific Ocean). See Figure 8-31.

"Continentality" is a keynote in the D climates. Landmasses are broader at these latitudes than anywhere else in the world. Even though these climates extend to the east coasts of the two continents, they experience little maritime influence because the general atmospheric circulation is westerly.

The most conspicuous result of continental dominance is the broad fluctuation of temperatures during the year. These climates have four clearly recognizable seasons: a long cold winter, a relatively short summer that varies from warm to hot, and stimulating transition periods in spring and fall. Annual temperature ranges are very large, particularly at more northerly locations where the winters are most severe. Precipitation is moderate, although throughout the D climates it exceeds the potential evapotranspiration. Summer is the time of precipitation maximum, but winter is by no means completely dry, and snow cover lasts for many weeks or months.

The severe midlatitude climates are subdivided into two types on the basis of temperature. The humid continental variety (Dfa, Dfb, Dwa, Dwb) has relatively long warm summers. The subarctic type (Dfc, Dfd, Dwc, Dwd) is characterized by short summers and very cold winters.

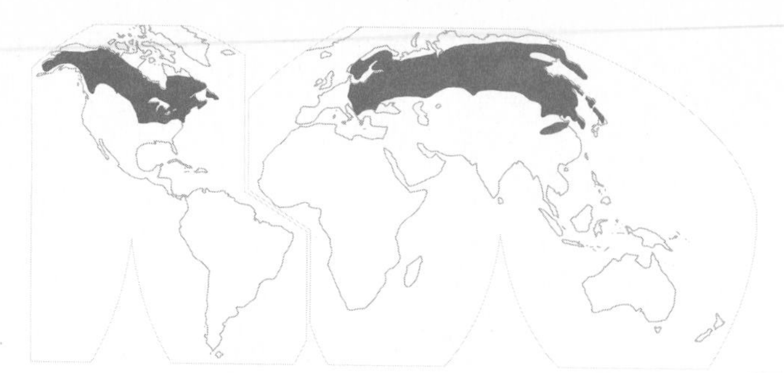

Figure 8-31. World distribution of D climates.

Humid Continental Climate (Dfa, Dfb, Dwa, Dwb)

The three sizable regions of humid continental climate are on the eastern sides of continents, extending interiorward for a few hundred miles. Their latitudinal range in North America and Asia is between 35° and 55°. The European region spreads from 35° to 60° in central Europe and tapers easterly through Russia and into Siberia. See Figure 8-32.

This climatic type is dominated by the westerly wind belt throughout the year, which means that there are frequent weather changes associated with the passage of migratory pressure systems, especially in winter. Variability, then, is a prominent characteristic, both seasonally and on a day-to-day basis. Because of the variability, the humid continental climate is often referred to as "stimulating."

Hot-month temperatures generally average in the mid-70s°F (mid-20s°C), so summers are similar in warmth to those of the humid subtropical climate to the south, although summer is shorter. The average cold-month temperature is usually between 10° and 25°F (−12° and −4°C), with from one to five months averaging below freezing. Winter temperatures decrease rapidly northward in this climatic type, and the growing season diminishes from about 200 days on the southern margin to about 100 days on the northern edge. See Figure 8-33.

This is considered to be a humid climate, but precipitation is not copious. Annual totals average between 20 and 50 inches (50 and 125 cm), with highest values on the coast and a general decrease interiorward. There is also a decreasing tendency from south to north. See Figure 8-34. Both of these horizontal trends reflect increasing distance from warm moist air masses. Summer is distinctly the wetter time of the year, but winter is not totally dry, and in coastal areas the seasonal variation is muted. Summer rain is mostly convective or monsoonal in origin. Winter precipitation is associated with extratropical cyclones, and much of it falls as snow. During a typical winter, snow covers the ground for only two

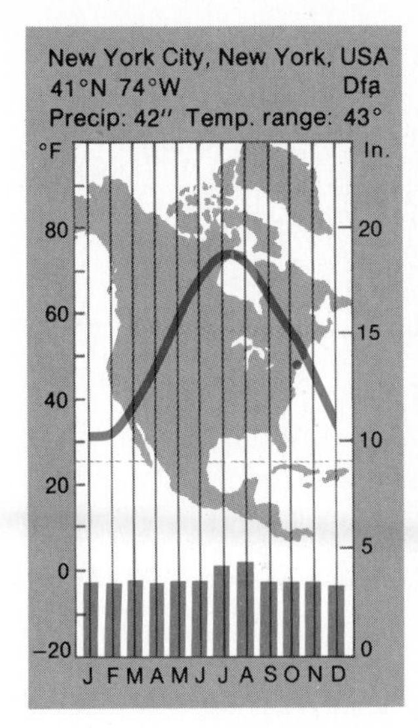

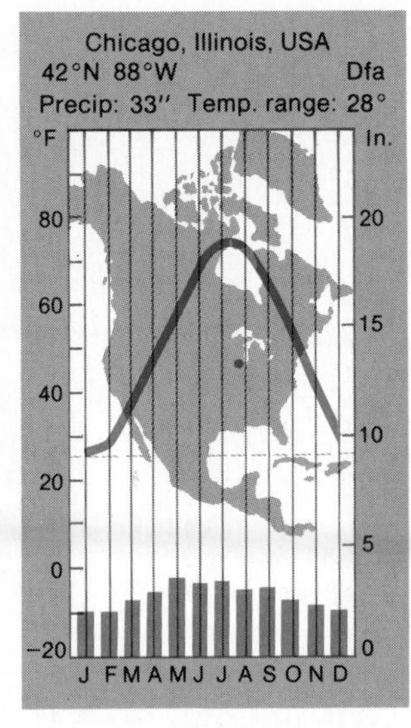

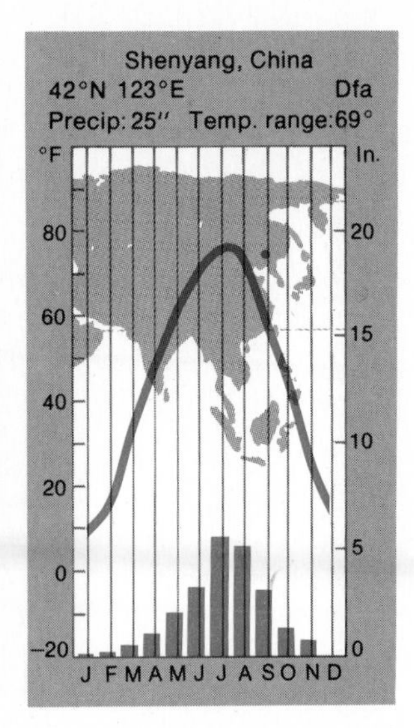

Figure 8-32 Climatic diagrams for representative humid continental stations.

or three weeks in the southern part of these regions, but as many as five months can have continual snow cover in the northern portions. See Figure 8-35.

Day-to-day variability and dramatic changes are prominent features of the weather pattern. These are regions of cold waves, heat waves, blizzards, thunderstorms, tornadoes, and other dynamic atmospheric phenomena.

Subarctic Climate (Dfc, Dfd, Dwc, Dwd) The subarctic climate occupies the higher middle latitudes, generally between 50° and 70°. See Figure 8-36. It occurs as two vast, uninterrupted expanses across the broad northern landmasses: from western Alaska across Canada to the east coast in Newfoundland, and across Eurasia from Scandinavia to easternmost Siberia. The name *boreal* is sometimes applied to this climatic type in Canada; in Eurasia it is often called *taiga* after the Russian name for the forest in that region.

The key word in the subarctic climate is *winter*, which is long, dark, and bitterly cold. In most places, ice begins to form on the lakes in September or October and doesn't thaw until May or later. For six or seven months, the average temperature is below freezing, and the coldest months have averages below −36° F (−38° C).

Figure 8-33. North-south temperature variation in the midlatitudes is much sharper in winter than in summer. These maps of eastern United States show a very steep north-south January temperature gradient but only limited north-south temperature contrasts in July.

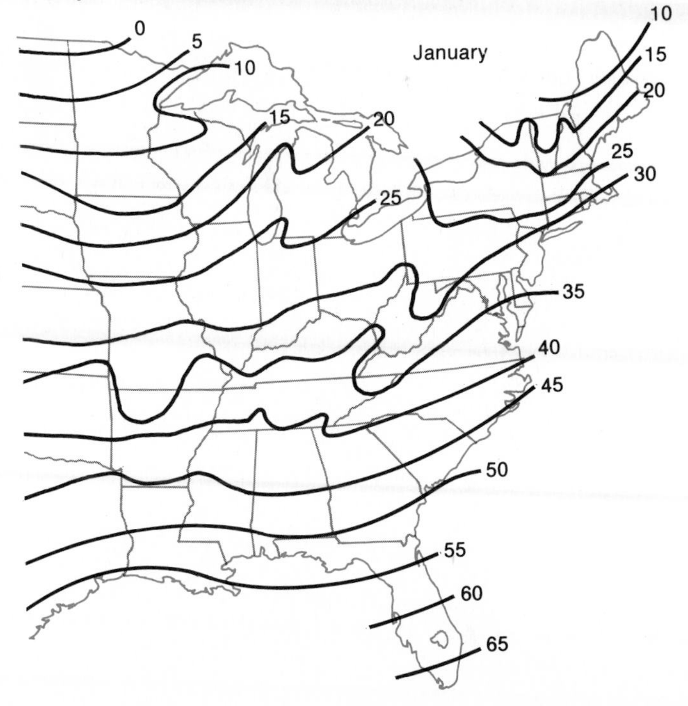

Figure 8-34. Annual precipitation in eastern North America. The isohyet values are in inches of moisture. Precipitation generally decreases inland and northward.

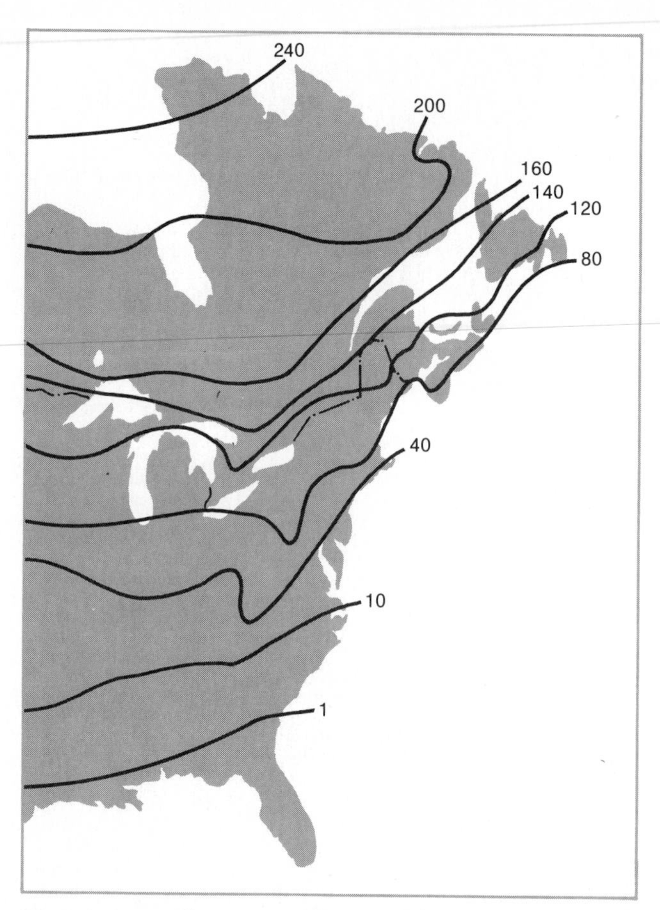

Figure 8-35. The average duration of snow cover in eastern North America. The numbers on the isolines represent the average annual number of days with a snow cover of 1 inch or more.

The world's coldest temperatures, apart from the Antarctic and Greenland ice caps, are found in the subarctic climate; the records are −90° F (−68° C) in Siberia and −82° F (−62° C) in Alaska.

Summer, on the other hand, warms up remarkably despite its short duration. Although the intensity of the sunlight is low (because of the low angle of incidence), summer days are very long and nights are too short to permit much radiational cooling. Average hot-month temperatures are typically in the high 50s or low 60s° F (mid-teens or low 20s° C), but occasional frosts may occur in any month. Annual temperature ranges in this climate are the largest in the world. Variations from average hot-month to average cool-month temperatures frequently exceed 80° F (45° C) and in some places are more than 100° F. The *absolute* annual temperature variation (fluctuation from the very coldest to the very hottest ever recorded) sometimes reaches unbelievable magnitude; the world record is 188° F (minus 90° to plus 98°) in Verkhoyansk, Siberia.

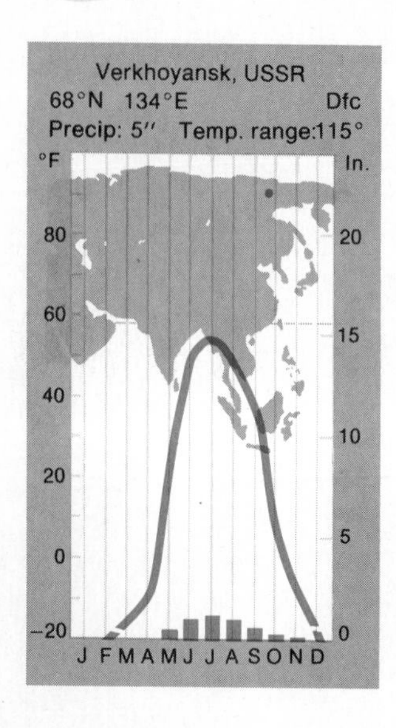

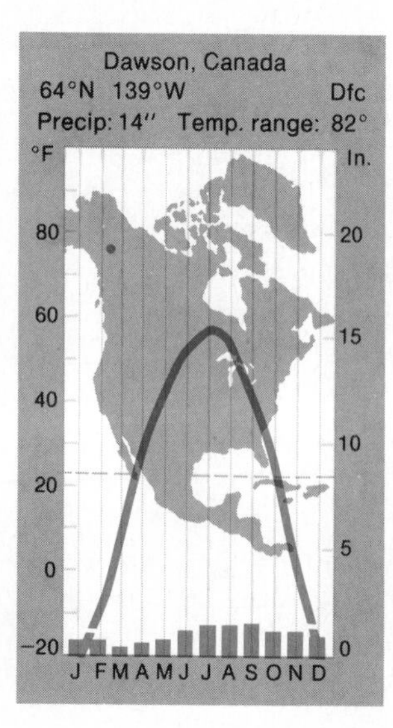

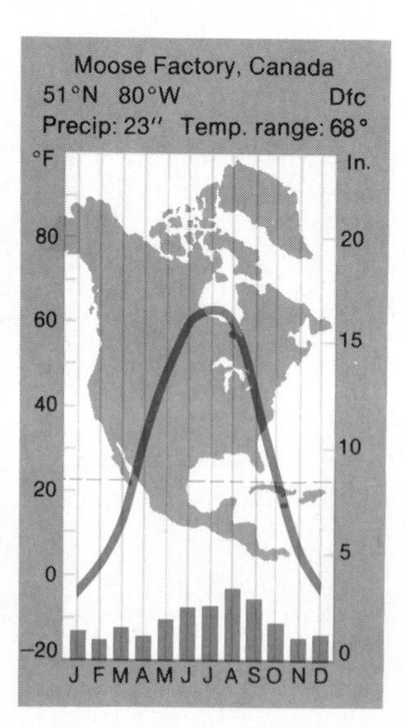

Figure 8-36. Climatic diagrams for representative subarctic stations.

World Record—Greatest Average Annual Temperature Range

112° F (62° C)—Yakutsk, Siberia—Dfd

Spring and fall, then, are brief transition seasons that slip by rapidly, usually in April/May and September/October. Summer is short, and winter is dominant.

Precipitation is usually meager. Annual totals range from only 5 inches (13 cm) to about 20 inches (50 cm), with somewhat more in immediate coastal areas. The low temperatures allow for little humidity in the air, and anticyclonic conditions predominate. Despite these sparse totals, the evaporation rate is low and the soil is frozen for much of the year, so moisture is adequate to support a forest. Summer is the "wet" season, and most precipitation comes from scattered convective showers. Winter experiences only light snowfalls (except near the coasts), which may accumulate to depths of 2 or 3 feet (60 to 90 cm). The snow that falls in October is likely to be still on the ground in May because little melts over the winter. Thus a continuous snow cover exists for many months despite the sparseness of actual snowfall.

Polar Climates (Group E)

The polar climates are the most remote from heat. See Figure 8-37. Furthest removed from the equator, they receive inadequate insolation for any significant warming. By definition, no month has an average temperature of more than 50° F (10° C). If the wet tropics represent conditions of monotonous heat, the polar climates are known for their enduring cold. They have the coldest summers and the lowest annual and absolute temperatures of the world. They are also extraordinarily dry, but evaporation is so minuscule that the group as a whole is classified as humid.

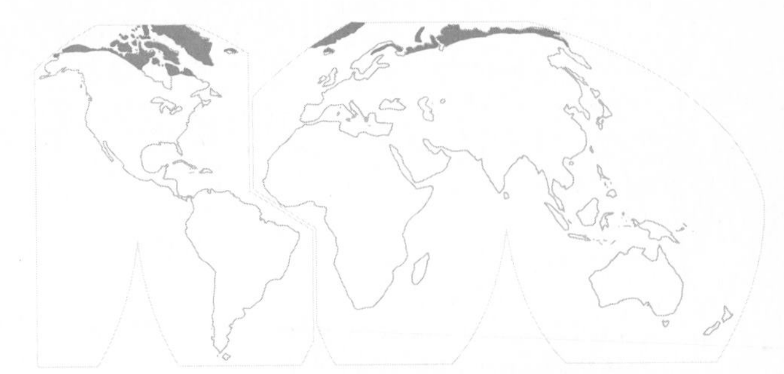

Figure 8-37. World distribution of E climates.

The two subcategories of the polar climates are distinguished by summer temperature. The tundra climate (ET) has at least one month with an average temperature exceeding the freezing point. The ice cap climate (EF) does not.

Tundra Climate (ET) The generally accepted equatorward edge of the polar climates as a group, and therefore of the tundra climate specifically, is the 50° F (10° C) isotherm for the average temperature of the warmest month. See Figure 8-38. This same isotherm corresponds approximately with the poleward limit of trees, so that the boundary between D and E climates is the "tree line," or the equatorward boundary of the tundra climate.

At the other—poleward—margin, the ET climate is bounded by the isotherm of 32° F (0° C) for the warmest month, which approximately coincides with the extreme limit for growth of any plant cover. More than for any other climatic type, the delimitation of the tundra climate demonstrates Köppen's contention that climate is best delimited in terms of plant communities.

The tundra climate is mostly a Northern Hemisphere feature. It occupies the fringes of the Arctic Ocean—along

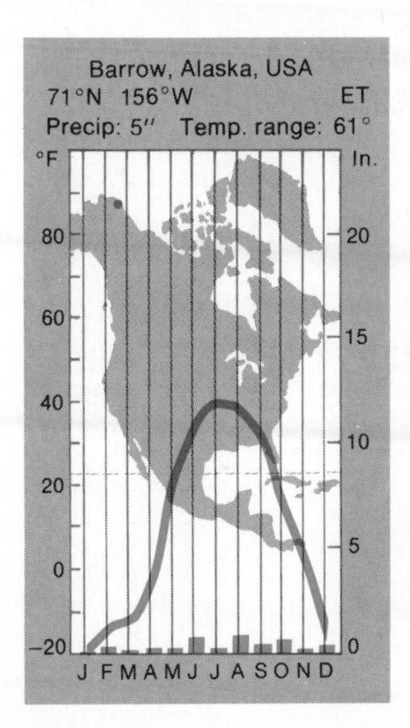

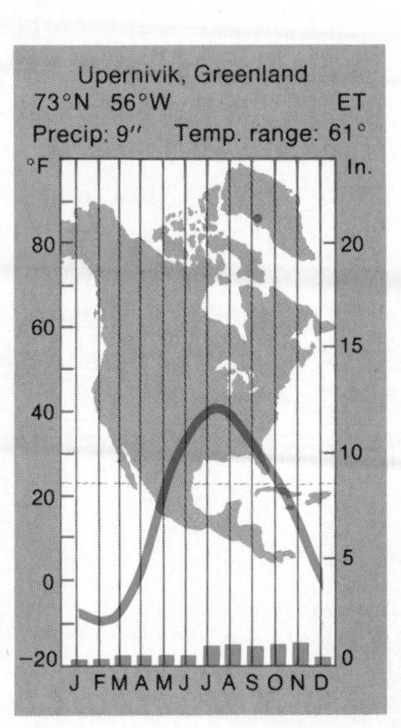

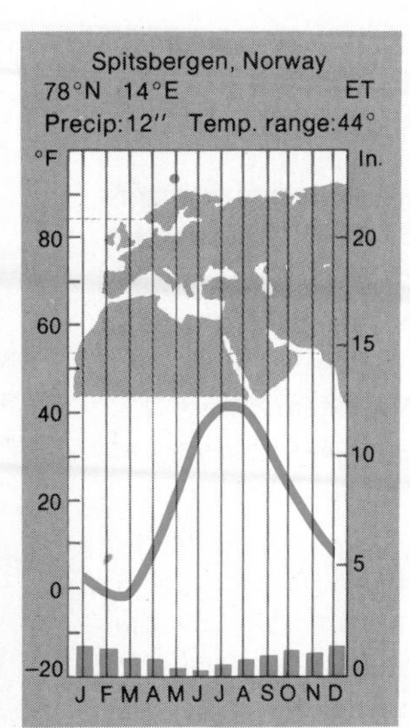

Figure 8-38. Climatic diagrams for representative tundra stations.

the north coast of Eurasia and North America, the islands of the Canadian Arctic Archipelago, and the coastal margins of Greenland and northern Iceland. The Southern Hemisphere has little land in these latitudes; ET climate is restricted to the northernmost peninsula of the Antarctic continent, the southernmost tip of South America, and a few small subantarctic islands.

Long, cold, dark winters and brief, cool summers characterize the tundra. Only one to four months experience average temperatures above freezing, and the average "hot"-month temperature is in the 40s °F (between 4° and 10° C). Freezing temperatures can occur at any time, and frosts are likely every night except in midsummer. Although winter is bitterly cold, it is not as severe as in the subarctic climate further south because it is less continental in position. Indeed, coastal stations in the tundra often have cold-month average temperatures of only about 0° F (−18° C), whereas an inland ET location is more likely to average −25° or −30° F (−32° or −35° C). Annual temperature ranges are fairly large, commonly between 40° and 60°. Daily temperature ranges, on the other hand, are small because the sun is above the horizon for most of the time in summer and is below the horizon for most of the time in winter; thus nocturnal cooling is limited in summer, and daytime warming is almost nonexistent in winter.

Moisture availability is very restricted in these regions despite the proximity of an ocean. The air is simply too cold to hold much moisture, so the absolute humidity is almost always very low. Moreover, anticyclonic conditions are common, with little uplift of air to encourage condensation. Annual total precipitation is generally less than 10 inches (25 cm) but is somewhat greater in eastern arctic Canada. Generally, more precipitation falls in the warm season than in winter, although the total amount in any month is small, and the month-to-month variation is minor. Winter snow is often dry and granular; it appears to be more than it actually is because there is no melting and because winds swirl it horizontally even when no snow is falling. Radiation fogs are fairly common throughout this climatic type, and sea fogs are sometimes prevalent for days along the coast.

Ice Cap Climate (EF) The most severe of the Earth's climates is restricted to Greenland (all but the coastal fringe) and most of Antarctica, the combined extent of these two regions amounting to more than 9 percent of the world's land area. See Figure 8-39. It is a climate of perpetual frost where vegetation cannot grow, and the landscape consists of a permanent cover of ice and snow.

The extraordinary severity of the temperatures is emphasized by the fact that both Antarctica and Greenland are ice plateaus, so that relatively high altitude is added to high latitude as a thermal factor. All months have average temperatures below freezing, and in the most extreme locations, the average temperature of the

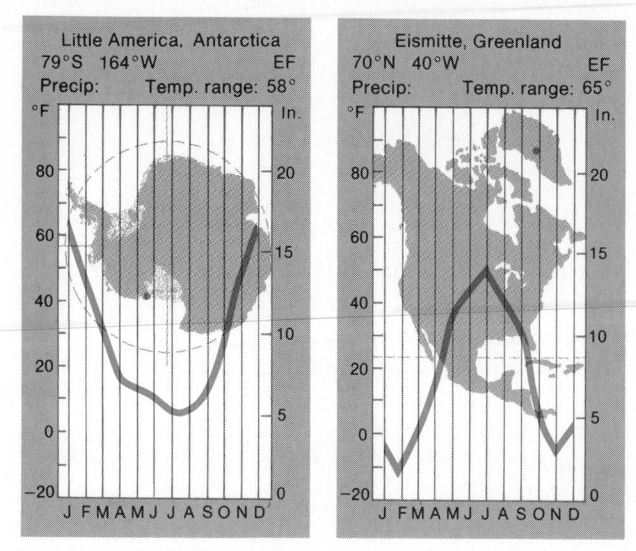

Figure 8-39. Climatic diagrams for representative ice cap stations.

warmest month is below zero F (−18° C). Cold-month temperatures average between −30° and −60° F (−34° and −51° C), and extremes well below −100° F (−73° C) have been recorded at interior antarctic weather stations.

The air is chilled so intensely from the underlying ice that strong surface temperature inversions prevail most of the time. Heavy, cold air often flows downslope as a vigorous katabatic wind. A characteristic feature of the ice cap climate, particularly in Antarctica, is the frequency of strong winds and blowing snow.

Actual precipitation is very limited. These regions are polar deserts; most places receive less than 5 inches (13 cm) of moisture annually. The air is too dry and too stable, with too little likelihood of uplift, to permit much precipitation. Evaporation, of course, is minimal, so an actual moisture surplus may be added to the ice.

Highland Climate (H)

Highland climate is not defined in the same sense as the other climatic types that have been discussed. Climatic conditions in mountainous areas have almost infinite variations from place to place, and many of the differences extend over very limited horizontal distances. Köppen did not recognize highland climate as a separate type, although most of his modifiers have added such a category. Highland climates are delimited on the world map (Figure 8-40) in this book to identify relatively high uplands (mountains and plateaus) with great complexity of local climatic variation within small areas.

World Record—Lowest Absolute Temperature

−127° F (−88° C)—Vostok, Antarctica—
1960—EF

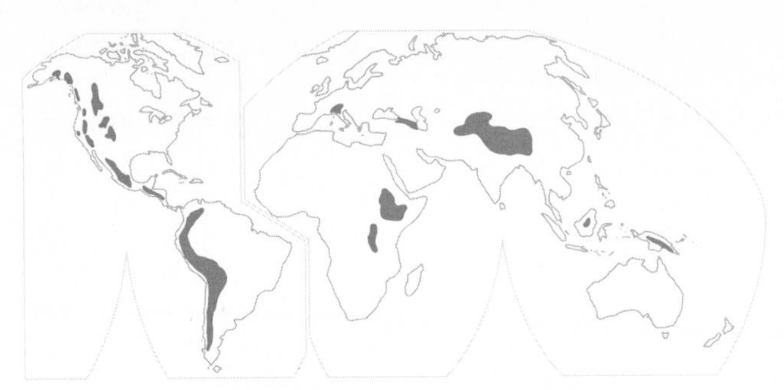

Figure 8-40 World distribution of H climates.

Climates of highlands usually are closely related to those of the adjacent lowland regions, particularly with regard to seasonality of precipitation. Some aspects of highland climate, however, differ significantly from those of the surrounding lowlands. Latitude becomes less important as a climatic control than altitude and exposure. The critical climatic controls on a mountain slope are usually relative elevation and angle of exposure to sun and wind.

Altitude variations influence all four elements of weather and climate. A vertical temperature gradient of about $3\frac{1}{2}$° F (2.3° C) per 1000 feet (300 m) generally prevails. Atmospheric pressure also decreases rapidly with increased elevation. Air movement is less predictable in highland areas, but it tends to be brisk and abrupt, with many local wind systems. Precipitation is characteristically heavier in highlands than in surrounding lowlands, so that the mountains usually stand out as moist "islands" on a rainfall map. See Figure 8-41.

In a real sense, altitude compensates for latitude in highland areas, so that a pattern of vertical zonation is usually present. The steep vertical gradients of climatic change are expressed as horizontal bands along the slopes, whereby an increase of a few hundred feet in elevation may be equivalent to a journey of several hundred miles poleward insofar as temperature and related environmental characteristics are concerned. Vertical zonation is particularly prominent in tropical highlands.

The *exposure* of a slope or peak or valley also has a profound influence on its climatic characteristics. A windward aspect experiences ascending air with a likelihood of relatively heavy precipitation, whereas a leeward location is sheltered from moisture or has predominantly downslope wind movement with limited opportunity for precipitation. The angle of exposure to sunlight is also of utmost climatic import, especially outside the tropics. Slopes that face equatorward receive very direct sunlight, which makes them warm and dry (through more rapid evapotranspiration); adjacent slopes facing poleward may be much cooler and more moist simply because of a smaller angle of solar incidence and more shading. Similarly, west-facing slopes receive direct sunlight in the hot afternoon, but east-facing slopes are sunlit during the cooler morning hours.

Changeability is perhaps the most conspicuous single characteristic of highland climate. The thin, dry air permits rapid influx of insolation by day and rapid loss of radiant energy at night, so that daily temperature ranges are abnormally large. There is frequent and rapid oscillation between freeze and thaw conditions. Daytime upslope winds and convection cause rapid cloud development and abrupt storminess. Travelers in highland areas are well advised to be prepared for sudden changes from

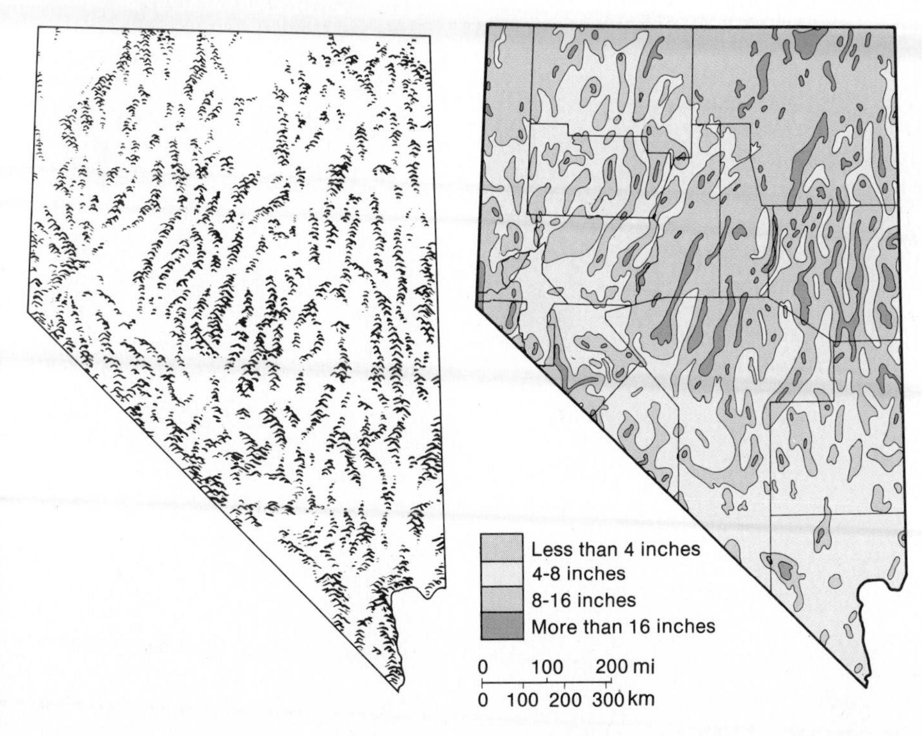

Figure 8-41. Comparison of annual precipitation and topography in the state of Nevada. Where there are mountains, there is more precipitation; where there are basins, there is less precipitation. The relationship is simple and straightforward, and is typical of highland environments.

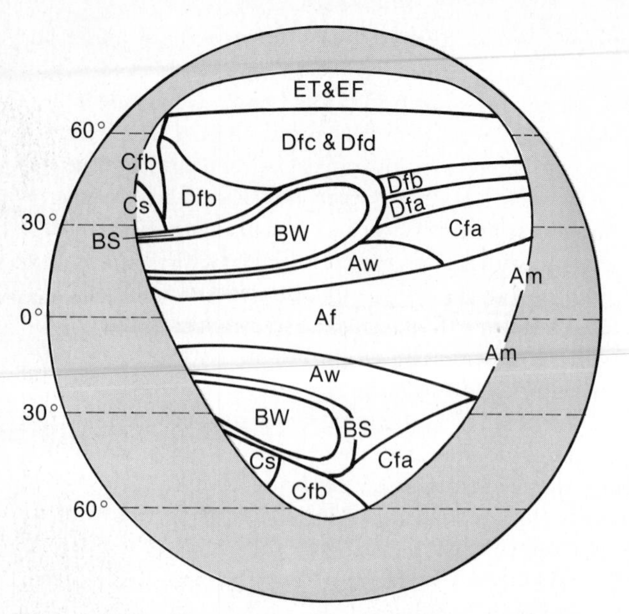

Figure 8-42. The presumed arrangement of Köppen climatic types on a hypothetical continent.

hot to cold, from wet to dry, from clear to cloudy, from quiet to windy, and vice versa.

The Global Pattern Idealized

In the preceding pages, the basic characteristics and distribution of 15 major climatic types have been presented. From these data it is possible to summarize the global climatic pattern by constructing a model of the distribution of climatic types on a hypothetical continent. See Figure 8-42.

The model portrays the generalized distribution of five of the six groups and 14 of the 15 types; highland climate is not included because its location is determined solely by topography, which is not a predictable environmental component. Four of the climatic groups (A, C, D, and E) are defined by temperature, which means that their boundaries are strongly latitudinal because they are determined by insolation. The fifth climatic group (B) is defined by moisture conditions, and its distribution cuts across those of the thermally defined groups.

The "real world" holds many refinements and modifications to this global pattern, but the schematic model shows the general alignments with considerable validity.

This ends our specific discussion of the atmosphere, which is probably the most important of the earthly "spheres" in the study of physical geography because of the pervasive influence of climate on the other components of the environment. In succeeding sections of this book we will consider, in succession, the hydrosphere, biosphere, and lithosphere.

Review Questions

1. In our study of physical geography, why are we interested in climatic classification?
2. Are the five climatic zones of the ancient Greeks of any relevance today? Explain.
3. Explain the basic concept of the Köppen system.
4. What is shown on a climatic diagram?
5. Explain the differences between the three types of tropical humid (A) climates.
6. Distinguish between desert and steppe climates.
7. What are the specialized climatic characteristics of west coast subtropical deserts? What causes these anomalies?
8. Why are mediterranean climates less extensive than humid subtropical climates?
9. In what climatic types is winter the season of precipitation maximum?
10. Why is precipitation so sparse in E climates?

Some Useful References

BARRY, R. G., *Mountain Weather and Climate*. London: Methuen Publications, 1981.

CRITCHFIELD, H. J., *General Climatology*, 3rd ed. Englewood Cliffs, N.J.: Prentice-Hall, Inc., 1974.

KENDREW, W. C., *Climates of the Continents*, 4th ed., London: Oxford University Press, 1953.

ORVIG, S., ed., *Climates of the Polar Regions*. Amsterdam, N.Y.: American Elsevier Publishing Co., 1970.

RILEY, P., AND L. SPOLTON, *World Weather and Climate*. Cambridge, England: Cambridge University Press, 1974.

RUMNEY, G. R., *Climatology and the World's Climates*. New York: Macmillan Publishing Co., Inc., 1968.

9 The Hydrosphere

The hydrosphere is the least well defined of the four "spheres" that make up the environment of the Earth. The concept of hydrosphere encompasses all the moisture of our planet, in all its forms. Thus it includes the surface waters in oceans, lakes, rivers, and swamps; underground water wherever it may be; frozen water in the form of ice, snow, and high cloud crystals; water vapor in the atmosphere; and the moisture that is temporarily stored in the tissues and organs of all plants and animals.

Thus it is apparent that the hydrosphere impinges on and overlaps significantly with the other three spheres. Water, ice, and even water vapor occur in the soil and some of the crustal rocks of the lithosphere. Water vapor and cloud particles composed of water and ice are important constituents of the lower portion of the atmosphere. And water is a critical component of every living organism, thus occurring throughout the biosphere. It is through the medium of moisture, then, that the interrelationships of the four "spheres" are most conspicuous and pervasive.

The Nature of Water: Commonplace but Unique

Water is the most common substance on the face of the Earth. Surface water occupies more than 70 percent of the area of our planet, and it occurs to greater or lesser degree in the ground, air, and organisms everywhere.

Apparently the amount of moisture in existence is finite and remains constant through time—past, present, and future. There is now as much water as there ever was, or ever will be. It changes from one form to another, and moves from one place to another, but it is neither created nor destroyed. Theoretically it is possible that some of the water of your morning shower is the same that was used to wash Jesus' feet two thousand years ago, or was drunk by a brontosaurus 50 million years in the past.

Water is a remarkable substance and absolutely unique in several important characteristics. For example, it is the only known substance that is found naturally in all three states in which matter can occur: liquid, solid, gaseous. The great majority of the Earth's moisture is in the liquid form of water. Water can be changed to the gaseous form (water vapor) by *evaporation* or to the solid form (ice) by *freezing*. Water vapor can be converted to water by *condensation* or directly to ice by *sublimation*. Ice is convertible to water by *melting* or to water vapor by *sublimation*. Moisture in plants can be passed through leaves into the air as vapor by a process called *transpiration*.

Life is impossible without water; every living thing depends on water for continuance of its life processes. Watery solutions in organisms dissolve or disperse nutrients for nourishment. Chemical reactions that can take place only in a solution create energy from the nutrients. Most waste products are carried away in solutions. Indeed, the total mass of every living thing is more than half water, the proportion ranging from about 60 percent for some animals to more than 95 percent for some plants.

Water is a substance that has many unusual properties. One of the most striking is its liquidity at ordinary temperatures. It remains in a liquid state at temperatures found at most places on the Earth's surface. No other common substance is liquid at ordinary Earth temperatures. The liquidity of water is thus a normal state and greatly enhances its versatility as an active agent in the atmosphere, lithosphere, and biosphere.

Another environmentally important characteristic of water is its great heat capacity. When something is warmed, it absorbs energy and its temperature rises as a result. When water is warmed, however, it can absorb an enormous amount of energy without showing it by a rise in temperature. Water's heat capacity is exceeded by no other common substance except ammonia. The practical result, as we have seen in Chapter 4, is that bodies of water are very slow to warm up during the daytime or in summer, and conversely, are very slow to cool off during the night or in winter. Thus water bodies have a moderating effect on surrounding temperatures by serving as reservoirs of warmth during winter and having a cooling influence in summer.

Most substances contract as they grow colder, but when water freezes into ice, it expands. This quality makes ice less dense than water, and it allows ice to float on and near the surface of water, as icebergs and ice floes do. If ice were denser than water, it would sink to the bottom of lakes and oceans where melting would be virtually impossible, and eventually many water bodies would become ice-choked.

Water normally responds to the pull of gravity and moves downward, but it is also capable of moving upward against the attraction of gravity under certain circumstances. This is because water has extremely high surface tension, which means that it can stick to itself and pull itself together. The water molecules stick closely together, and they wet the surfaces with which they come in contact. The high surface tension and wetting ability combine to allow water to climb upward. This climbing capability is most notable in situations where water is confined in small pore spaces or narrow tubes. In such restricted confinement, water can sometimes climb upward for many inches or even feet, in an action called *capillarity*. Capillarity enables water to circulate upward through rock and soil, or through roots and stems of plants.

Of all its interesting and unusual attributes, however, perhaps the most significant capability of water as an active agent in the landscape is its ability to dissolve other substances. Water can dissolve almost any substance, and it is sometimes referred to as the "universal solvent." It functions in effect as a mild acid, dissolving some things quickly and in large quantities; others slowly and in minute proportions. As a result, water in nature is nearly always impure; i.e., it contains various other chemicals in addition to its hydrogen and oxygen atoms. As it moves around through the atmosphere, over the surface of the Earth, within soil and rocks, and through plants and animals, it carries with it a remarkable diversity of dissolved minerals and nutrients as well as tiny solid particles in suspension.

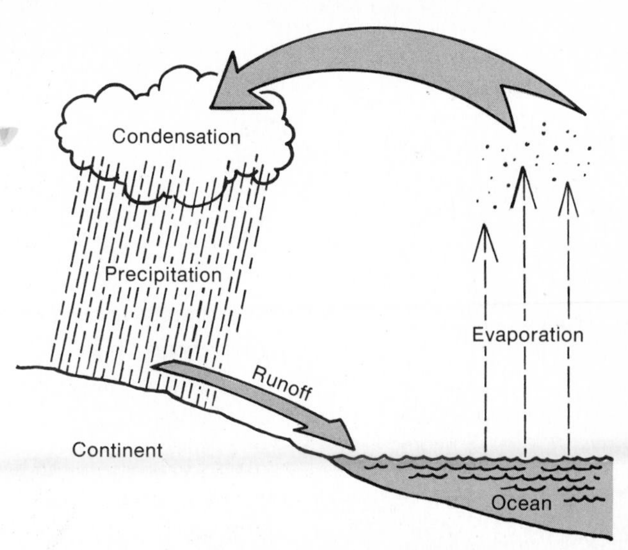

Figure 9-1. In simplest form, the hydrologic cycle involves the movement of moisture from surface to air and back again.

The Hydrologic Cycle

This unique substance—water—essential to life and finite in amount, is distributed very unevenly on, in, and above the Earth. The great bulk of all moisture, more than 99 percent of it, is in storage—in oceans and large lakes, locked up as glacial ice, or held in rocks below the surface. The proportional amount of moisture in these various storage reservoirs is relatively constant and does not vary greatly over thousands of years. Only during an "ice age" is there a notable change in these components: During periods of glaciation, the oceans become smaller as the ice sheets grow, and atmospheric water vapor diminishes; then during deglaciation, the ice diminishes, and the oceans expand.

The remaining small fraction—less than 1 percent—of the Earth's total moisture is involved in an almost continuous sequence of movement and change, the effects of which are absolutely critical to life on our planet. The constancy of the storage components is maintained by the effectiveness of the system of circulation. This is the *hydrologic cycle*, which can be viewed as a series of storage areas interconnected by various transfer processes, in which there is a ceaseless interchange of moisture in terms both of its geographical location and its physical state. In essence, the movement is a circular one, from sea to air to land to sea, but with many permutations.

A simplistic model of the cycle is shown in Figure 9-1. Moisture evaporates as water vapor from the Earth's surface into the atmosphere, where it condenses and precipitates as water or ice back to the surface, from which it runs off into storage areas for evaporation into the atmosphere once again. In detail, of course, the hydrologic cycle is much more complex. Figure 9-2 presents a schematic representation of what actually occurs. As it is a closed, circular system, we can begin the discussion at any point. It is perhaps clearest to start with the movement of moisture from the Earth's surface into the atmosphere.

Surface to Air

Most of the moisture that enters the atmosphere from the Earth's surface does so through *evaporation*. The oceans, of course, are the principal sources of water for evaporation. They occupy 71 percent of the Earth's surface, have unlimited moisture available for ready evaporation, and are extensive in low latitudes where considerable heat and wind movement facilitate the conversion of water to water vapor. As a result, an estimated 84 percent of all evaporated moisture is derived from ocean surfaces. The 16 percent of evaporated moisture that comes from land surfaces actually includes the twin processes of evaporation and transpiration. Evaporation can take place from any moist surface, though not as freely as from a water surface. *Transpiration* involves the transfer of water vapor to the air from plants after it has been brought up through the plants' internal circulation system from the roots.

When the water vapor enters the atmosphere, it remains there a relatively short time—usually only a few hours or days. During that interval, however, its position may be shifted considerably. It may be moved vertically through *convection*, or it may be transported horizontally through *advection* by wind currents.

Air to Surface

Sooner or later—usually sooner—water vapor in the atmosphere is condensed into water or sublimated into ice to form cloud particles. Under the proper circumstances (see Chapter 6), the clouds may drop precipitation in the form of rain, snow, sleet, or hail. Over a lengthy period of time, such as several years, total worldwide precipitation will be almost exactly equal to total worldwide evaporation/transpiration.

Although these totals balance in time, they do not

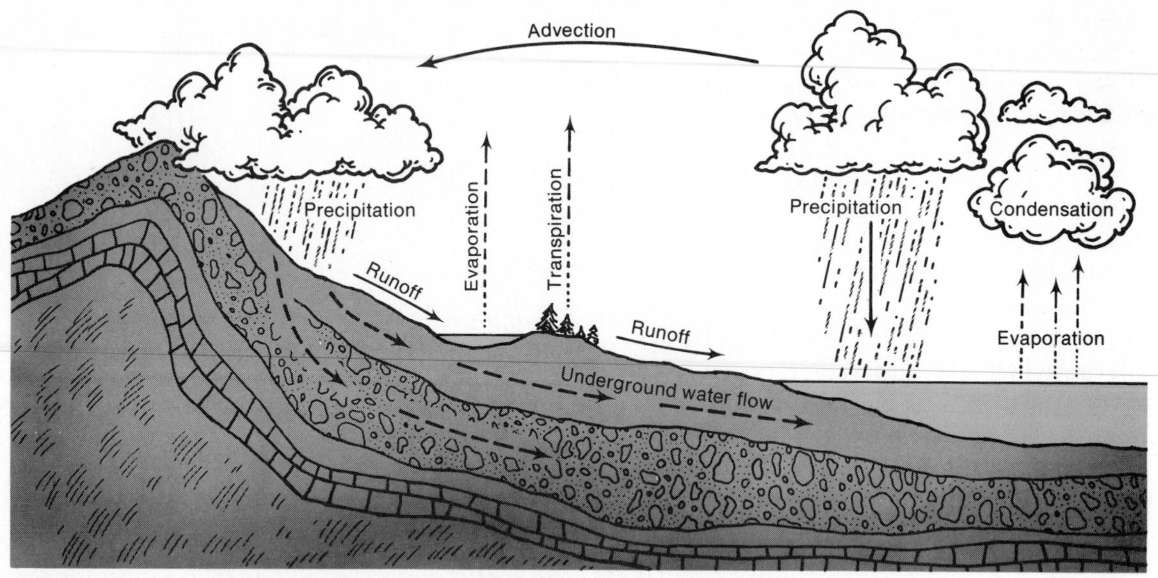

Figure 9-2. The major components of the hydrologic cycle are evaporation of moisture upward from surface to air and precipitation downward from air to surface. Other important elements of the cycle include transpiration of moisture from vegetation to atmosphere, surface and subsurface flow of water from land to sea, condensation of water vapor to form clouds from which precipitation may fall, and advection (horizontal movement) of moist air from one place to another.

balance in place. Evaporation exceeds precipitation over the oceans, whereas the opposite is true over the continents. From the data in Table 9-1 it can be seen that both evaporation and precipitation are much more common over oceans than over continents, but oceans receive less moisture from precipitation than they lose through evaporation. This is explained by the advection of moist maritime air onto land areas so that there is less moisture for precipitation over the ocean. The atmosphere is the only route by which moisture can be moved from the sea to the land, except for coastal spray and storm waves.

On and Beneath the Surface

Precipitation falling on the sea is simply incorporated immediately into the mass of the ocean; that which falls on land may experience a more complicated series of events. Rain falling on a nonvegetated surface will either collect on that surface, run off down the slope, or infiltrate the ground. If it remains on the surface, some or all of it will eventually evaporate. Most of the land surfaces of the Earth are covered with vegetation, however, which intercepts the falling precipitation. Thus in the early stages of a rainstorm, the ground beneath a tree may remain dry for some time, as the moisture collects on leaves and branches above. Before long, however, the capacity of the foliage to retain water will be exceeded, and moisture that does not evaporate will drip and flow to the ground below.

Once water reaches the ground, it evaporates, infiltrates the soil, runs off down the slope, or is ponded in some fashion. The water that sinks into the ground may be stored temporarily as soil moisture, or it may percolate further down to become part of the underground water. Much of the soil moisture eventually will be evaporated or transpired (through plants) back into the atmosphere, and much of the underground water eventually will reappear at the surface. Then, sooner or later, and in one way or another, most of the water that reaches the surface will evaporate again, and the rest will be incorporated into streams and rivers to flow into the oceans.

The total flow of water, on and beneath the surface, from the continents to the oceans, amounts to 7 percent of all circulating moisture in the global hydrologic cycle. This balances the excess of precipitation over evaporation that is received on the continents.

TABLE 9-1 Comparisons of Moisture Balance of Continents and Oceans

	Percentage of total world		
	Surface area	Precipitation	Evaporation/transpiration
Oceans	71%	77%	84%
Continents	29%	23%	16%

Duration of the Cycle

The continuous nature of the hydrologic cycle maintains our planetary water balance. Although it is a closed system and is believed to have an unvarying total capacity, there is enormous variation in the cycling of individual molecules of water. Molecules of water may be stored in oceans and deep lakes, or as glacial ice, for thousands of years without participating in the cycle other than as part of the total storage; and moisture trapped in rocks deeply buried beneath the surface of the Earth may be excluded from any activity in the cycle for literally hundreds of thousands of years. However, a small fraction of the Earth's total moisture is in almost continuous action in the dynamics of the hydrologic cycle. Water can run off over the surface and down hundreds of miles (kms) to the sea in only a few days. And moisture evaporated into the atmosphere may remain there for only a few minutes or hours before it is precipitated back to Earth. Indeed, at any given moment the atmosphere contains only a few days' potential precipitation. Thus it is clear that the global water supply includes an enormous amount of moisture, and that the hydrologic cycle has the flexibility of incorporating all of it in the long run, but only a tiny proportion of the total is actually involved at any given time.

Contemporary Moisture Inventory

As shown in Figure 9-3, the moisture of the world is found in seven types of storage, all of which are interrelated in terms of the hydrologic cycle:

1. *Oceans* contain by far the largest amount. Although they occupy only 71 percent of the Earth's surface, they contain more than 97 percent of all water.
2. *Glaciers* contain about 2 percent of all moisture, but this amounts to some three-fourths of the world's fresh water.
3. *Underground water* encompasses about 0.5 percent of the world total.
4. *Surface waters*, including rivers, lakes, and inland seas, include less than 0.2 percent of all water.
5. *Soil moisture* makes up about 0.1 percent of the total.
6. *Atmospheric moisture* is only a tiny fraction—0.0001 percent.
7. *Biological water*—that stored in plant and animal tissue—is proportionally negligible.

The Oceans

Despite the fact that most of the Earth's surface is oceanic and the greatest relief features are beneath the sea rather than on the continents, humanity's knowledge of the seas has until very recently been very limited. The ocean is a hostile environment for most air-breathing creatures, particularly so for humankind. Thus only with great care can humans venture beneath the sea's surface, and only within the last three decades or so has sophisticated equipment been available to catalogue and measure details of the maritime environment.

How Many Oceans?

From a broad viewpoint, there is but one ocean. It spreads over almost three-fourths of the Earth's surface, interrupted here and there by continents and islands. Although literally tens of thousands of bits of land protrude above the blue waters, the "world ocean" is so vast that half a dozen continent-sized portions are totally devoid of islands, without a single piece of land breaking the surface of the water.

In generally accepted usage, the world ocean is subdivided into four principal parts—the Pacific, Atlantic, Indian, and Arctic oceans. Their boundaries are not everywhere precise, and around some of their margins are partly landlocked smaller bodies of water that are called *seas*, *gulfs*, *bays*, and other related terms. Most of these smaller bodies can be considered as portions of one of

Figure 9-3. Moisture inventory of the Earth.

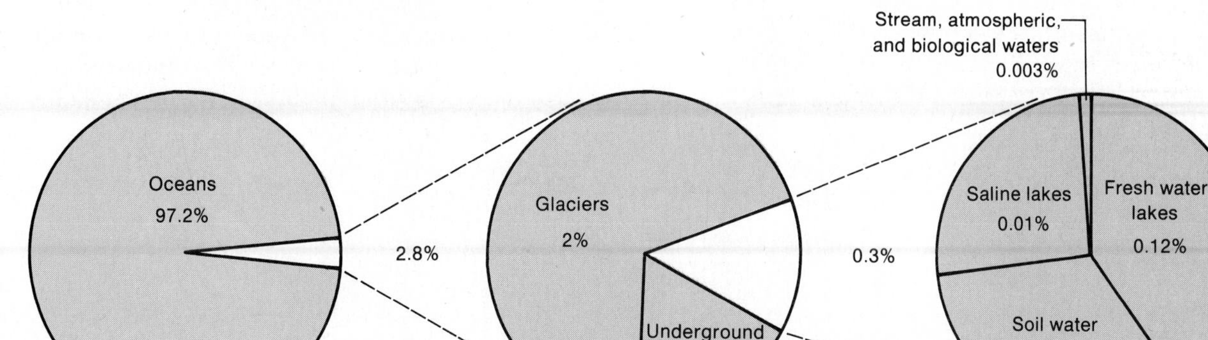

the major oceans, although a few are so narrowly connected (as the Black Sea or the Mediterranean Sea or Hudson Bay) to a named ocean as to deserve separate consideration. This matter of nomenclature is further clouded by multiple usage of the term *sea*, which is sometimes used synonymously with the general term *ocean*. A "sea" may designate a specific smaller body of water around the edge of an ocean, or occasionally it may refer to a completely isolated inland body of water. See Figure 9-4 and Table 9-2 for the location and description of the oceans and major seas of the world.

The *Pacific Ocean* is twice as large as any other body of water. Five of the seven recognized continents are on its fringes. It occupies about one-third of the total area of the Earth, more than all the world's land surfaces combined. See Figure 9-5. It contains the greatest average depth of any ocean as well as the deepest known oceanic trenches. Although it extends almost to the Arctic Circle in the north and a few degrees beyond the Antarctic Circle in the south, it is largely a tropical ocean. Its greatest girth is in equatorial regions; almost one-half of the 24,000-mile (38,500-km) length of the equator is in the Pacific. Its character often belies its tranquil name, for it houses some of the most disastrous of all storms, and most of the world's major volcanoes are either in it or around its edge.

The *Atlantic Ocean* is slightly less than half the size of the Pacific. Its latitudinal extent is about the same, but its east-west spread is only about half as great. Its average depth is also only a little less than that of the Pacific. The *Indian Ocean* is a little smaller and has a slightly less average depth than the Atlantic. Nine-tenths of its area are south of the equator. The *Arctic Ocean* is much smaller and shallower than the other three. It is connected to the Pacific by a relatively narrow passageway between Alaska and Siberia, but it has a broad and rather indefinite connection with the Atlantic between North America and Europe. Some people refer to the waters around Antarctica as the *Antarctic Ocean* or the *Great Southern Ocean*, but this distinction is not generally accepted.

Characteristics of Ocean Waters

Wherever they are found, the waters of the world ocean have many similar characteristics, but they show significant differences from place to place. The differences are particularly notable in the surface layers, down to a depth of a few hundred feet or meters. Below this level, our limited knowledge indicates considerable uniformity in a cold, dark environment.

Figure 9-4. The four principal oceans and the major seas of the world. The seas are numbered on the map: (1) South China Sea, (2) Caribbean Sea, (3) Mediterranean Sea, (4) Bering Sea, (5) Gulf of Mexico, (6) Sea of Okhotsk, (7) East China Sea, (8) Sea of Japan, (9) Hudson Bay, (10) North Sea, (11) Yellow Sea, (12) Black Sea, (13) Red Sea, and (14) Baltic Sea.

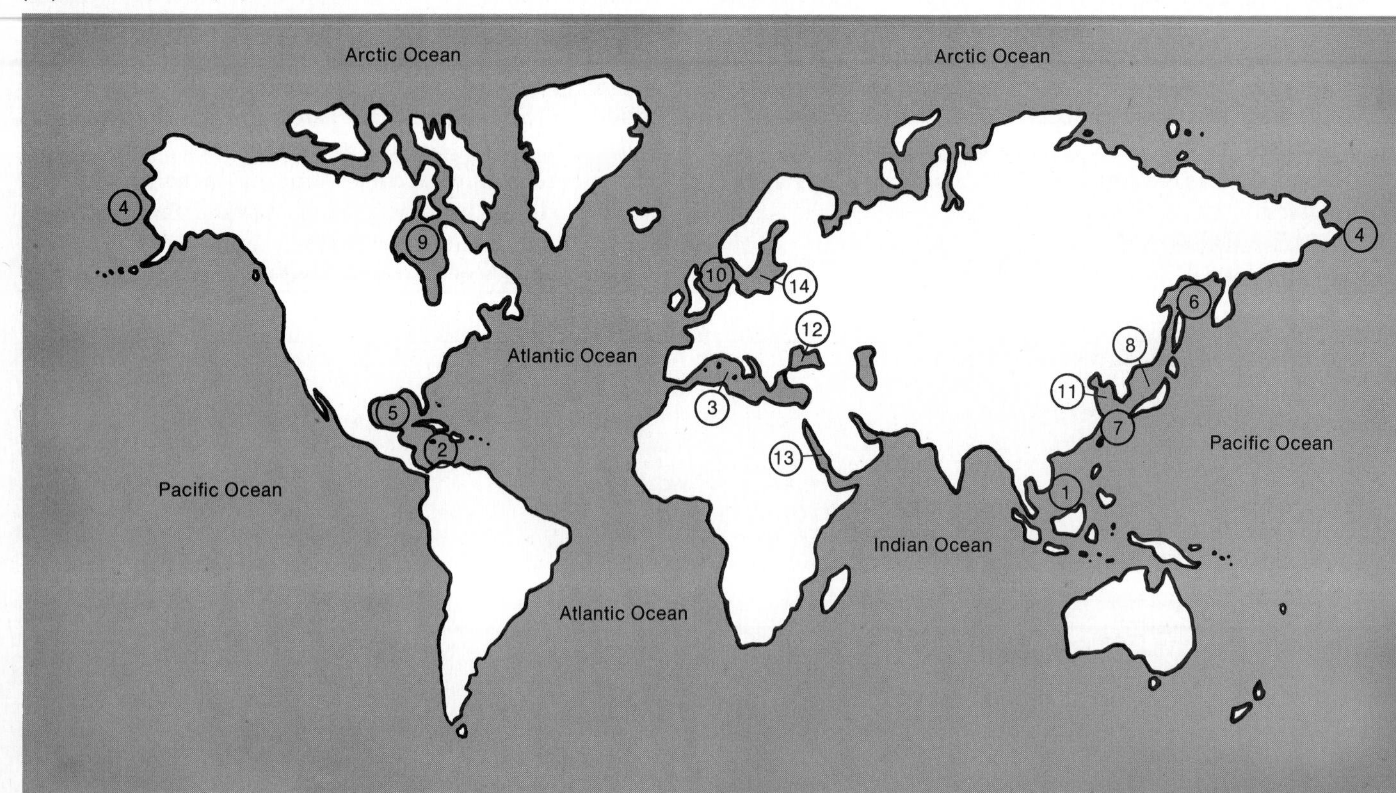

TABLE 9-2 Oceans and Major Seas of the World

Name	Approximate surface area (square miles)	(square km)	Proportion of all water (%)
Pacific Ocean	64,186,000	166,884,000	46%
Atlantic Ocean	31,862,000	82,841,000	23%
Indian Ocean	28,350,000	73,710,000	20%
Arctic Ocean	5,427,000	14,110,000	4%
South China Sea	1,150,000	2,990,000	
Caribbean Sea	971,000	2,525,000	
Mediterranean Sea	969,000	2,519,000	
Bering Sea	875,000	2,275,000	
Gulf of Mexico	600,000	1,560,000	
Sea of Okhotsk	550,000	1,430,000	
East China Sea	480,000	1,248,000	
Sea of Japan	405,000	1,053,000	
Hudson Bay	318,000	827,000	
North Sea	222,000	577,000	
Yellow Sea	220,000	572,000	
Black Sea	190,000	494,000	
Red Sea	175,000	455,000	
Baltic Sea	160,000	416,000	

Note: Proportions of all water are omitted for the 14 seas because of their minimal size in relation to the world's oceans.

Chemical Composition Seawater consists mostly of hydrogen and oxygen, of course, but its distinctiveness is due to its content of dissolved minerals, which makes up something less than 4 percent of its total bulk. Almost all known minerals are found to some extent in seawater, but by far the most important are sodium and chlorine, which form the common salt of the sea. The *salinity* of seawater is a measure of the concentration of dissolved salts, which are mostly sodium chloride but also include chlorides of magnesium, sulfur, calcium, and potassium. The average salinity of seawater is about 35 parts per thousand, or 3.5 percent of total weight. The geographical distribution of surface salinity varies in relation to the amount of evaporation and the addition of fresh water (primarily by rainfall and stream discharge). Where evaporation is great, salinity is relatively high; where the inflow of fresh water is great, salinity is relatively low. Typically the lowest salinities are found where rainfall is heavy and near the mouths of major rivers. Salinity is highest in partly landlocked seas in dry, hot regions because the evaporation rate is high and stream discharge is minimal. As a general pattern, salinity is relatively low in equatorial regions because of heavy rainfall, cloudiness, and humidity, all of which inhibit evaporation, and also because of considerable river discharge; it rises to a general maximum in the subtropics where precipitation is low and evaporation is extensive; and it decreases to a general minimum in the polar regions where evaporation is mini-

Figure 9-5. Polar perspectives of the oceans.

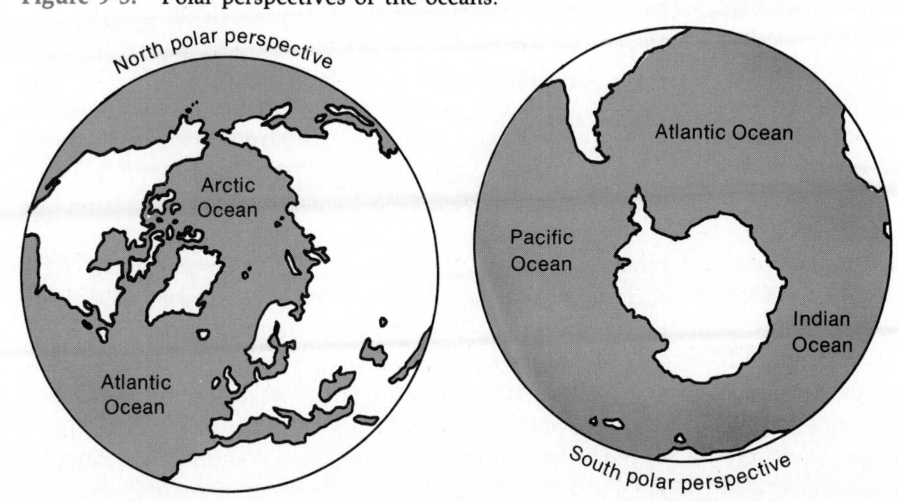

mal and there is considerable inflow of fresh water from rivers and ice caps.

Temperature As is to be expected, surface seawater temperatures generally decrease with increasing latitude. The temperature often exceeds 80° F (26½° C) in equatorial locations, and decreases to 28° F (−2° C), the average freezing point for seawater, in arctic and antarctic seas. The western sides of oceans are nearly always warmer than the eastern margins because of the movement of major ocean currents (see Chapter 5). This is due to the contrasting effects of poleward-moving warm currents on the west and equatorward-moving cool currents on the east.

Density Density is the ratio of mass to volume of a substance. Seawater density varies with temperature and degree of salinity; high salinity produces high density, whereas high temperature creates low density. Deep water has high density because of low temperature as well as the pressure of the overlying water. Surface layers of seawater tend to contract and sink in cold regions; opposite conditions prevail in warmer areas. The movement of surface currents also affects this situation, particularly by producing an upwelling of colder, denser water in some localities.

Movements of Ocean Waters

The liquidity of the ocean keeps it in continual motion. Its surface is almost always ridged with swells and waves, and below the surface there is a less conspicuous restlessness of movement. Oceanic waters can be set in motion by all sorts of external forces. The movement of almost anything over the surface—whether it is wind, a boat, or a swimmer—will activate motion in surface water. Disturbances in the Earth's crust adjacent to the ocean can also trigger significant movements. And the gravitational attraction of nearby heavenly bodies engenders the greatest movements of all. The motions of oceanic waters can be grouped under three headings: waves, currents, and tides. Waves mostly consist of a change in the shape of the ocean surface, with little actual displacement of water. Currents involve a considerable movement of water, particularly horizontally, but also vertically and obliquely. By far the greatest vertical movements of ocean waters are due to tides.

Tides On the shores of the world ocean, almost everywhere, the sea level fluctuates regularly. For about 6 hours the water rises, and then for about 6 hours it falls. These rhythmic oscillations have continued unabated, day and night, winter and summer, for eons. *Tides* are essentially bulges in the sea surface in some places that are compensated by sinks in the surface at others. Thus they are primarily vertical motions of the water. In shallow water areas around the margins of the oceans, however, the vertical oscillations may produce significant horizontal water movements as well.

Currents The world ocean contains a variety of currents that shift vast quantities of water both horizontally and vertically. Some of these are set in motion by contrasts in temperature and salinity characteristics. Surface currents are caused primarily by wind flow. All current movements are likely to be influenced by the size and shape of the particular ocean, the configuration and depth of the sea bottom, and the Coriolis effect.

Some current movement involves subsidence of surface waters downward; other vertical flows bring an upwelling of deeper water to the surface. Geographically speaking, however, the most important currents are the major horizontal flows that comprise the general circulation of the various oceans, and which were discussed in Chapter 4.

Waves To the casual observer, the most conspicuous motion of the ocean is provided by waves. Most of the sea surface is in a state of constant agitation, with wave crests and troughs bobbing up and down most of the time. Moreover, around the margin of the ocean, waves of one size or another lap, break, or pound on the shore in endless procession.

In point of fact, most of this movement is like running in place, with little forward progress involved. Waves in the open ocean are mostly just shapes, and the movement of a wave across the sea surface is a movement of form rather than of substance. Only when a wave "breaks" does any significant shifting of water take place. Waves are discussed in some detail in Chapter 18.

"Permanent" Ice

Second only to the world ocean as a storage reservoir for moisture is the solid portion of the hydrosphere—the ice of the world. Although miniscule in comparison with the amount of water in the oceans, the moisture content of ice at any given time is more than twice as large as the combined total of all other types of storage (underground water, surface waters, soil moisture, atmospheric moisture, biological water).

By far the largest portion of the ice component consists of glaciers. Approximately 10 percent of the land surface of the Earth is covered by glacial ice. Most of this is encompassed in the ice sheet that covers almost all the Antarctic continent. See Figure 9-6. A significant but much smaller ice cap covers most of the world's largest island, Greenland. Still smaller ice caps are found on a dozen or so Arctic islands, mostly in far northern Canada. Finally, there are many thousands of highland ice fields

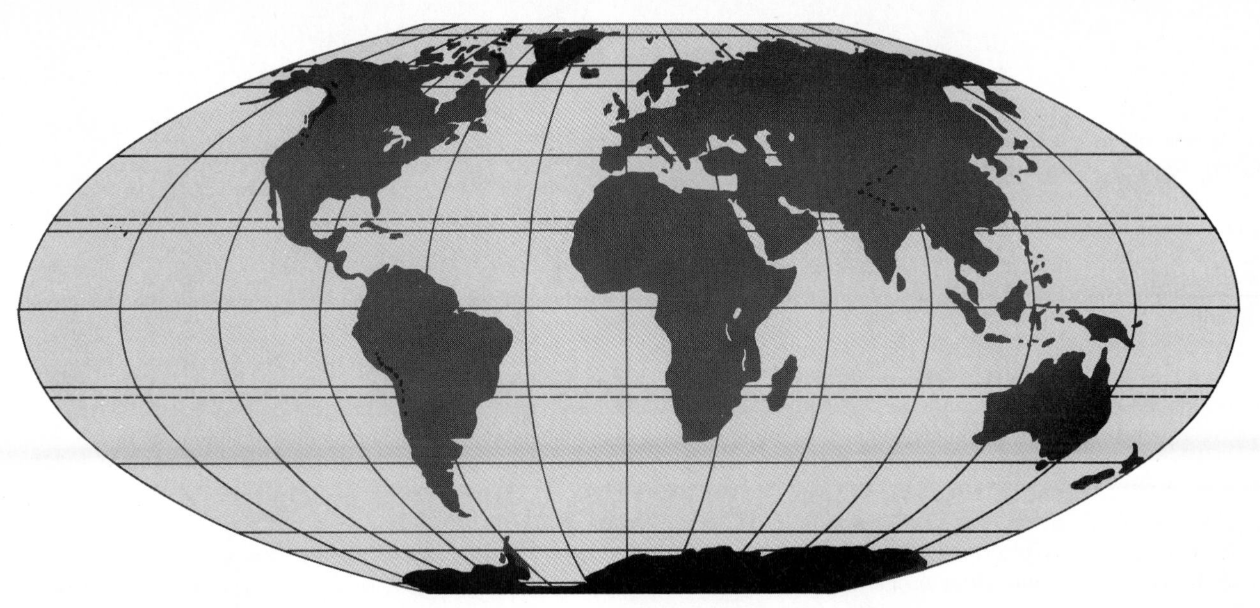

Figure 9-6. The distribution of glacial ice over the Earth is confined primarily to Antarctica and Greenland.

and alpine glaciers in the higher mountains of the world, at all latitudes and on all continents except Australia.

Oceanic ice is more limited in extent and less massive in volume than glacial ice. It occurs as ice packs, shelves, floes, and bergs. The largest "permanent" *ice pack* occupies most of the surface of the Arctic Ocean; on the other side of the globe, an ice pack fringes most of the Antarctic continent. Both of these packs become greatly enlarged during their respective winters, their areas being essentially doubled by increased freezing around their margins. An *ice shelf* is a mass of glacial ice that floats on the surface of the sea but is permanently attached to an ice cap or a glacier on land. There are a few small ice shelves in the Arctic, mostly around Greenland, but several gigantic shelves are attached to the Antarctic ice sheet, most notably the Ross ice shelf of some 40,000 square miles (100,000 km^2). *Ice floes* and *icebergs* are masses of ice that break off from larger ice bodies (ice sheets, glaciers, ice packs, and ice shelves) and float independently in the sea. See Figure 9-7. A *floe* is a large, flattish, tabular mass, whereas a *berg* is a chunkier, more rugged piece of ice. Both floes and bergs sometimes drift hundreds of miles equatorward before they melt. Some Antarctic ice floes are enormous; the largest ever observed was ten times as large as the state of Rhode Island. Despite the fact that some oceanic ice freezes directly from seawater, all forms of oceanic ice are composed entirely of frozen fresh water because the salts from the sea are never taken up in the freezing process.

A relatively small proportion of the world's ice occurs beneath the surface of the land as *ground ice*. It occurs only in areas where the temperature is continually below the freezing point, so it is restricted to high-latitude and high-altitude regions. Most permanent ground ice is characterized as *permafrost*, or permanently frozen subsoil. It is widespread in northern Canada, Alaska, and Siberia, and is found in small patches in many high mountain areas. Some ground ice is aggregated as lenses or veins of frozen water, but most of it develops as ice crystals in spaces among soil particles.

Figure 9-7. An iceberg is a chunk of independently floating ice.

Surface Waters

The surface waters of the land comprise only a tiny fraction (about 0.2 percent) of the world's total moisture supply, but from the human viewpoint, they are of incalculable value because of ready availability. Lakes, swamps, and marshes abound in many parts of the world, and all but the driest parts of the continents are seamed by rivers and streams.

Focus on Tides

Tides are set in motion by the gravitational attractions of the moon and sun. All celestial bodies exert gravitational force, but only these two are large enough and close enough to the Earth so that their influence is appreciable. The effect of this gravitational pull on the atmosphere and on the solid part of the Earth's crust is minimal, but water bodies, being fluid, are responsive to it. Tidal variation is exceedingly small in inland bodies of water. Even the largest lakes usually experience a tidal rise and fall of no more than 2 inches (5 cm). Effectively, then, tides are important only in the world ocean, and they are normally noticeable only around its edge where the water meets the land.

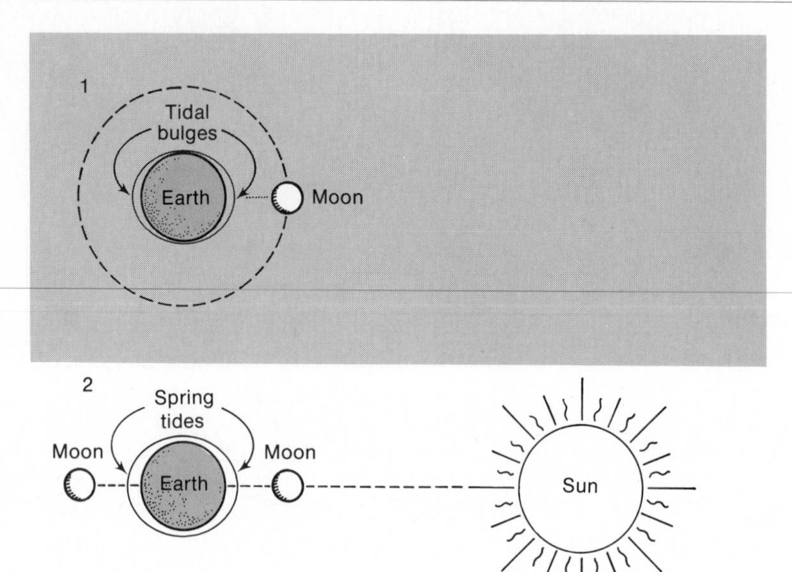

Although the sun is much larger than the moon, it is much further away from the Earth, so that the moon is more important than the sun in causing earthly tides. The sun has about 27,000,000 times as much mass as the moon, but it is also nearly 400 times as far away from the Earth. When these two factors are equated, the result is that sun-caused tides are only 44 percent as high as those created by the moon. In actuality, the gravitational influence of the two bodies combines to produce a single system of tidal variations on the Earth.

In essence, tides follow the moon in its apparent course around the Earth. There is always a bulge in the world ocean on the side of the Earth that faces the moon as the water is pulled away from the lithosphere. On the opposite side of the Earth, facing away from the moon, there is a similar bulge, due to the fact that the solid part of our planet is being pulled away from the ocean. These two bulges produce high tides on opposite sides of the Earth. At the same time, there are compensating low tides halfway between the two bulges. Schematically, this can be diagrammed like a clock: When there are high tides at 12:00 and 6:00 o'clock, there are low tides at 3:00 and 9:00 o'clock.

As the Earth rotates eastward, the tidal progression appears to move westward in an endless procession. The tides rise and fall twice in the interval between two "rising" moons, which is about 50 minutes more than a 24-hour day. The combination of the Earth's rotation and the moon's revolution around the Earth means that the Earth makes a bit more—about 12° more—than a full rotation between each rising of the moon. Thus two complete tidal cycles have a duration of about 24 hours and 50 minutes. This means that on all oceanic coastlines, there are normally two high tides and two low tides in a little more than a day.

The magnitude of tidal fluctuation is quite variable in time and place, but the sequence of the cycle is generally similar everywhere. From its lowest point

Lakes

In simplest terms, a *lake* is a body of water surrounded by land. No minimum or maximum size is attached to this definition, although the word *pond* is often used to designate a very small lake. Well over 90 percent of the surface water of our planet is contained in lakes; by contrast, less than 1 percent of all surface water is found in rivers and streams at any given moment.

With a relatively few exceptions, the origin of lakes is not related to stream activity. Most lakes are fed and drained by streams, but their genesis is usually due to other factors. Two conditions are necessary for the formation and continued existence of a lake: (1) some sort of a natural basin with a restricted outlet, and (2) sufficient input of water to keep the basin at least partly filled. A great many natural situations can create lake basins, most of them involving geologic or geomorphic processes.

Most of the world's lakes contain fresh water, but some of the largest lakes are saline. See Figure 9-8 for the location of the world's largest lakes. Indeed, more than 40 percent of the lake water of the planet is salty, although this figure is perhaps misleading because the world's biggest lake (the Caspian Sea), which is larger than the next six largest combined, is a salt lake. Any lake that has no natural drainage outlet, either as a surface stream or as a sustained subsurface flow, will become saline. Most small salt lakes and some large ones are *ephemeral,* which means that they contain water only sporadi-

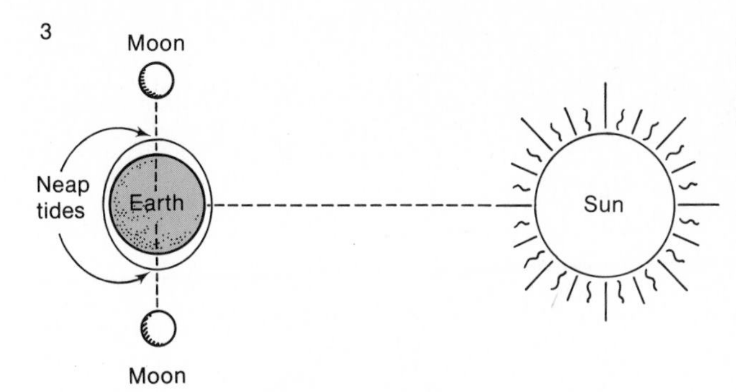

Juxtaposition of the sun, moon, and Earth accounts for variations of oceanic tides on the Earth. The three basic relationships are illustrated here: (1) Normal tidal bulges produce low tides at top and bottom, and high tides on the sides. (2) When the sun and moon are positioned along the same line, spring tides (the highest of high tides) are produced. (3) When the sun and moon are arranged vertically to one another with respect to the Earth, neap tides (the lowest of tides) result.

the water rises gradually for about 6 hours and 13 minutes, so that there is an actual movement of water toward the coast, in a *flood tide*. At the end of the flooding period, the maximum water level, *high tide*, is reached. Soon the water level begins to drop, and for the next 6 hours and 13 minutes there is a gradual movement of water away from the coast, as an *ebb tide*. When the minimum water level (*low tide*) is reached, the sequence reverses and the cycle begins again. The vertical difference in elevation between high and low tide is called the *tidal range*.

Changes in the relative positions of the Earth, moon, and sun induce periodic variations in tidal ranges. The greatest variations occur when the three bodies are positioned in a straight line, which usually occurs twice a month near the times of the full and new moons. When thus aligned, the joint gravitational pull of the sun and moon is along the same line, so that they accentuate one another. This takes place whether the moon is directly between the Earth and the sun or the Earth is directly between the moon and the sun. In either case, this is a time of maximum tidal variation, called *spring tides* (the name has nothing to do with the season called Spring). When the sun and moon are located at right angles to one another with respect to the Earth, their individual gravitational pulls are diminished because they are in markedly different directions. This results in lower than normal tidal variations, called *neap tides*. The neap tide juxtaposition generally takes place twice a month at about the time of first-quarter and third-quarter moons.

Tidal range variations are also affected, but to a lesser degree, by the moon's nearness to the Earth. The moon follows an elliptical orbit in its revolution around the Earth, the nearest point (*perigee*) being about 31,200 miles (50,000 km), or 12 percent, closer than the farthest point (*apogee*). During perigee, tidal variations are greater than during apogee.

The tidal range fluctuations described above apply all over the world at the same times of the month. There are also, however, enormous variations in tidal range along different seacoasts, due primarily to the physical configurations of coastline and sea bottom. Along most coasts there is a moderate tidal range of 5 to 10 feet (1.5 to 3 m). Some partly landlocked seas, such as the Mediterranean, have almost negligible tides. Other places, such as the northwest coast of Australia, experience enormous tides of 35 feet (10.5 m) or so. The world record tidal range is found at the upper end of the Bay of Fundy in eastern Canada, where a 50-foot (15-m) water-level fluctuation twice a day is not uncommon, and a wall of seawater (called a *tidal bore*), several inches to several feet in height, rushes up the Petitcodiac River for many miles at similar time intervals.

cally and are dry much of the time because they are in dry regions with insufficient inflow to maintain them on a permanent basis.

The water balance of most lakes is maintained by surface inflow, sometimes combined with springs and seeps below the lake surface. A few lakes are fed entirely by springs. Most freshwater lakes have but a single surface outlet, i.e., one stream that serves as a drainage channel.

Lakes are distributed very unevenly over the land. See Table 9-3. They are most common in regions that were glaciated in the relatively recent geologic past because glacial erosion and deposition have deranged the normal drainage patterns and have created innumerable basins. One has only to compare the northern and southern parts of the United States to recognize this fact; north of the Ohio and Missouri rivers the land was extensively glaciated and lakes abound, whereas south of these rivers there was no glaciation and lakes are rare. Most of Europe and the northern part of Asia demonstrate a similar relationship of past glaciation and contemporary lakes. Some parts of the world that are notable for lakes, however, were not glaciated. For example, the remarkable series of large lakes in eastern and central Africa is due to major crustal movements and volcanic action, and the many thousands of small lakes in Florida were formed by sinkhole collapse when rainwater dissolved calcium from the massive limestone bedrock.

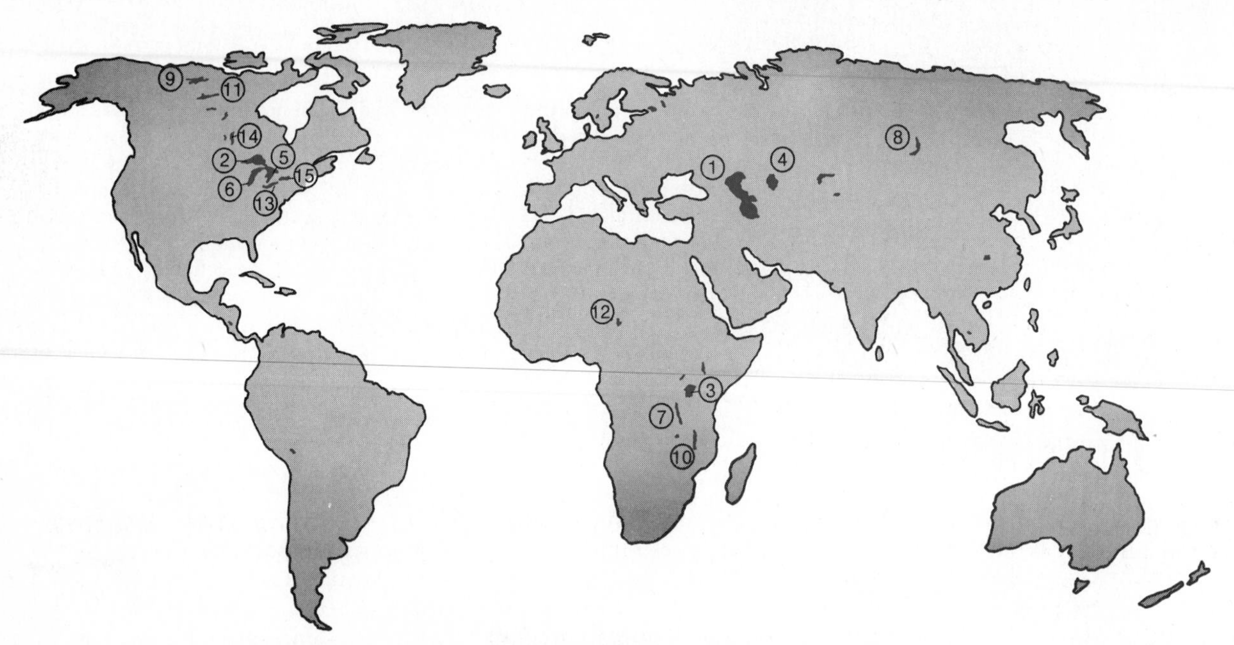

Figure 9-8. The world's largest lakes: (1) Caspian Sea, (2) Lake Superior, (3) Lake Victoria, (4) Aral Sea, (5) Lake Huron, (6) Lake Michigan, (7) Lake Tanganyika, (8) Lake Baykal, (9) Great Bear Lake, (10) Lake Malawi, (11) Great Slave Lake, (12) Lake Chad, (13) Lake Erie, (14) Lake Winnipeg, and (15) Lake Ontario.

TABLE 9-3 The World's Largest Lakes

		Area	
Name	Continent	(sq. miles)	(sq. km)
Caspian Sea	Asia	143,250	372,450
Lake Superior	North America	31,700	82,420
Lake Victoria	Africa	26,700	69,400
Aral Sea	Asia	25,700	66,800
Lake Huron	North America	23,000	59,800
Lake Michigan	North America	22,300	58,000
Lake Tanganyika	Africa	12,650	33,000
Lake Baykal	Asia	12,200	31,700
Great Bear Lake	North America	12,100	31,500
Lake Malawi	Africa	11,550	30,000
Great Slave Lake	North America	11,300	29,400

Most lakes are very temporary features in the natural landscape. Few have been in existence for more than a few thousand years, which is almost momentary in the grand scale of geologic time. Inflowing streams bring sediment to fill them up; outflowing streams cut channels progressively deeper to drain them; and as the lake becomes shallower, a continual increase in plant growth accelerates the process of infilling. Thus the destiny of most lakes is to disappear, relatively soon.

One of the most notable things that humanity has done to alter the natural landscape is to produce artificial lakes, or *reservoirs*. This has been accomplished largely by the construction of dams, ranging from small earth mounds heaped across a gully to immense concrete structures blocking the world's major rivers. Some reservoirs are so large as to be comparable in area to medium-sized natural lakes. These activities have had immense ecological and economic consequences, not all of them beneficial.

Swamps and Marshes

Closely related to lakes, but less numerous and containing a much smaller volume of water, are swamps and marshes. These are flattish surface areas that are submerged in water at least part of the time but are shallow enough to permit the growth of water-tolerant plants. See Figure 9-9. They are usually associated with flat coastal plains, broad river valleys, or recently glaciated areas. Sometimes they represent an intermediate stage in the infilling of a lake. The conceptual distinction between the terms is that a *swamp* has a plant growth that is dominantly trees, whereas a *marsh* is primarily vegetated with grasses and sedges.

Rivers and Streams

Although containing only a small proportion of the world's water at any given time, streams comprise an extremely dynamic component of the hydrologic cycle. They provide the full framework for the drainage of the land surface, moving water, sediment, and dissolved chemicals ever seaward. See Figure 9-10. The occurrence of rivers is closely, but not absolutely, related to precipitation patterns. Table 9-4 lists the longest rivers of the world. Humid lands have many streams, most of which maintain perennial flow; whereas dry lands have few streams, almost all of which are intermittent in flow. Streams will be discussed in more detail in Chapter 16.

Figure 9-9. Standing water is characteristic of both swamps and marshes. The dominant swamp vegetation is trees (as in this North Carolina scene), whereas they are virtually absent in marshes (as in this Pennsylvania setting). [USDA-SCS photos.]

Underground Water

Beneath the surface of the land is another important component of the hydrosphere—underground water. Its total quantity is far greater than that contained in all surface lakes and streams combined. Moreover, underground water is much more widely distributed than surface water. Whereas lakes and rivers are found in restricted locations, underground water is almost ubiquitous; i.e., it occurs beneath the land surface throughout the world. Its quantity is sometimes limited, its quality is sometimes poor, and its occurrence is sometimes at great depth, but almost anywhere on Earth one can dig deep enough and find water.

Almost all underground water comes originally from above. Its source is precipitation on the surface that percolates directly into the soil or seeps downward eventually from lakes and streams. Once the moisture gets underground, several different things can happen, depending largely on the nature of the soil and rocks into which it infiltrates. The situation can probably best be understood by visualizing a vertical subsurface cross section. Usually three, and often four, hydrologic zones are arranged one below another. See Figure 9-11.

Zone of Aeration The topmost band is called the *zone of aeration* or *vadose zone*. As with the other zones, its depth can be quite variable, from a few centimeters to hundreds

Figure 9-10. Streams provide a sense of motion to the natural landscape, as in this springtime scene in Idaho. [USDA-Soil Conservation Service photo.]

TABLE 9-4 The World's Longest Rivers

River	Continent	Empties into	Length (miles)	(kilometers)
Nile	Africa	Mediterranean Sea	4130	6600
Amazon	South America	Atlantic Ocean	3900	6200
Missouri-Mississippi	North America	Gulf of Mexico	3740	6000
Yangtze	Asia	East China Sea	3400	5450
Ob-Irtysh	Asia	Kara Sea	3400	5450
Hwang Ho	Asia	Yellow Sea	2900	4650
Amur	Asia	Tartar Strait	2800	4500
Zaire (Congo)	Africa	Atlantic Ocean	2700	4300
Lena	Asia	Laptev Sea	2700	4300
Mackenzie	North America	Beaufort Sea	2635	4200
Mekong	Asia	South China Sea	2600	4150
Niger	Africa	Gulf of Guinea	2600	4150
Yenisey	Asia	Kara Sea	2500	4000
Parana	South America	Atlantic Ocean	2450	3900
Volga	Europe	Caspian Sea	2300	3700

Focus on Australia's Great Artesian Basin

Australia is by far the driest of the inhabited continents, both in terms of precipitation and of surface waters (lakes and rivers). Strangely enough, however, it possesses extraordinary resources of underground water, mostly occurring as confined water. The following figure delineates the principal artesian/subartesian basins.

By far the most notable of all the world's artesian basins is the one in east-central Australia called the Great Artesian Basin. It underlies some 670,000 square miles (1,750,000 km²) in parts of three states and the Northern Territory, which is about one-fifth of the total area of the continent. The geologic structure can be likened to an enormous saucer, with sedimentary rock layers outcropping at the surface in the low mountains of northeastern Queensland, dipping gently to a position far below the surface, and then rising to the surface again on the western margin of the basin. The

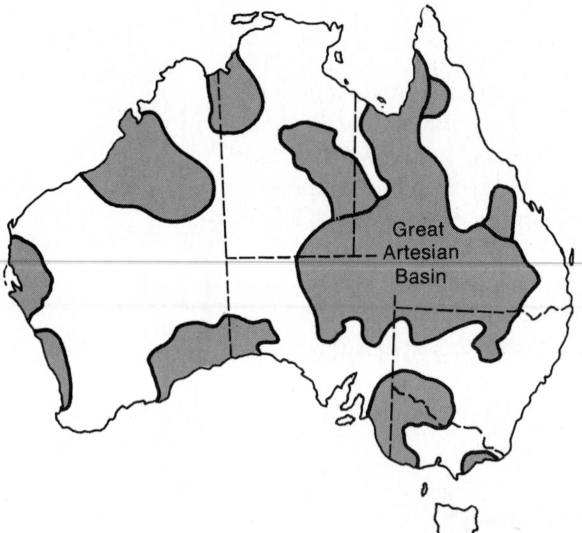

The principal artesian basins of Australia.

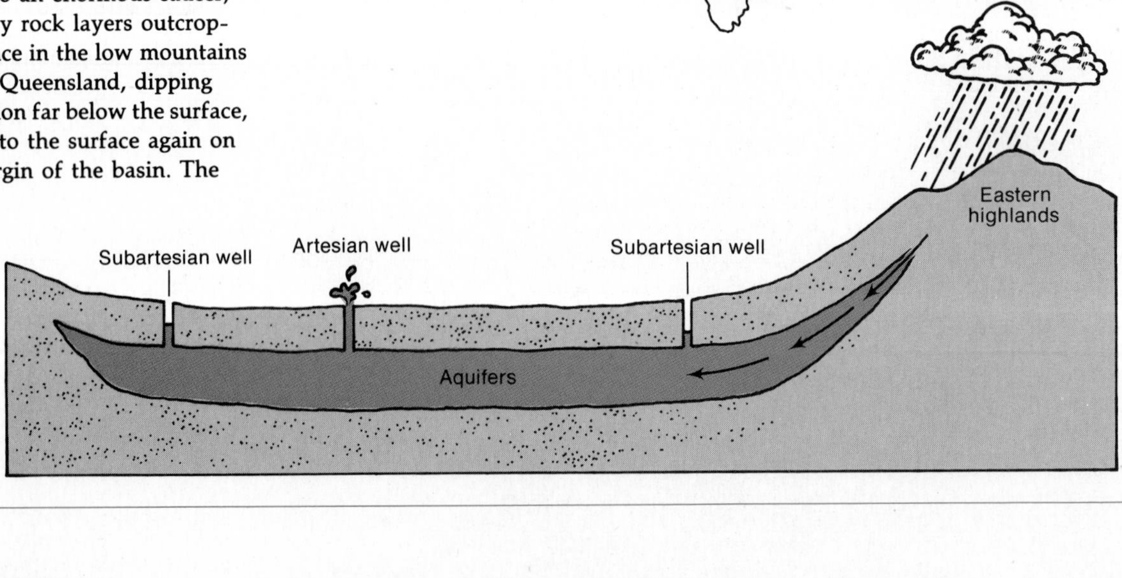

Simplified vertical cross section of Australia's Great Artesian Basin. The aquifers are supplied by rainfall in the eastern highlands, with the subsurface water slowly moving downward and westerly.

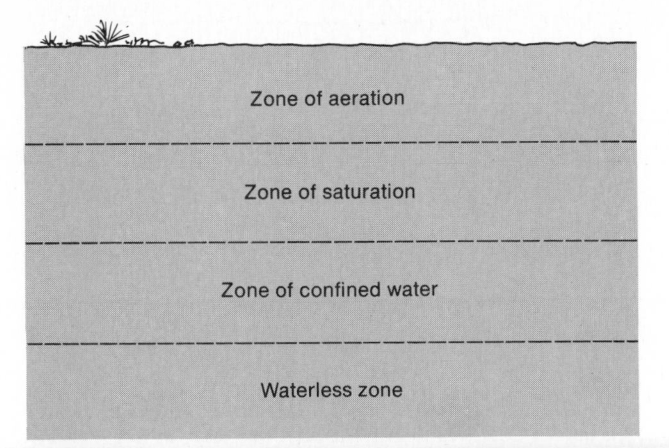

Figure 9-11. Schematic cross section to illustrate the four hydrologic zones.

of yards or meters. It contains a fluctuating amount of moisture in the pore spaces of the soil (or soil and rock). After a rain these pore spaces may be saturated with water, water that drains away rapidly; most of the time the interstices contain both air and water. Strictly speaking, this moisture is *soil water*, which will be discussed in Chapter 12.

Zone of Saturation Next below is the *zone of saturation* or *phreatic zone*, in which all pore spaces and cracks in the bedrock and regolith are fully saturated with water. The moisture in this zone is called *groundwater*; it seeps slowly through the ground following the pull of gravity and guided by rock structure. The top of the saturated zone is referred to as the *water table*; see Figure 9-12. The orientation and slope of the water table usually conform roughly to the slope of the land surface above, nearly always approaching closer to the surface in valley bottoms and being more distant from it beneath a ridge or hill. Where the water table intersects the Earth's surface, water flows out, usually in the form of a spring or a seep. A lake, swamp, marsh, or permanent stream is almost always an indication that the water table reaches the surface there. In humid regions the water table is higher than in arid regions, which means that the zone of saturation

A typical flowing *bore* in Australia's Great Artesian Basin. This is the Clifton bore in northwestern New South Wales. [TLM photo.]

three principal aquifers of sandstone alternate with beds of shale, which form aquicludes above, between, and below the sandstone layers.

Water from rainfall and streams infiltrates into the aquifers on the elevated eastern margin of the basin 1000 to 2000 feet (300 to 600 m) above sea level. (There is very little intake of water along the arid western rim of the basin.) From these eastern intake areas the water percolates down and along the aquifers very slowly (a few inches to a few yards per day) in a general southwesterly direction. The maximum depth of the aquifers is about 7000 feet (2100 m) below sea level.

Some natural discharge occurs in lower parts of the basin through upward leakage, but most of the water remained confined until it was tapped by human endeavor. The first wells (called *bores* in Australia) were drilled just over a century ago; approximately 25,000 have been completed since then, of which about 20 percent were artesian and the remainder, subartesian. In the early years there was enormous output from most of the bores, with maximum total flow achieved about 1915. The continuing excess of use over recharge has produced a steady decline in total flow since that time despite a continuing increase in the number of bores in operation.

Unfortunately, this abundance of underground water in a dry portion of the driest continent has not provided a substantial solution to the water supply scarcity. The three principal problems, in addition to declining yields, are depth, heat, and salinity:

1. The aquifers are deep and it is costly to drill far enough to reach them. The average bore depth is 1650 feet (500 m), and some have been sunk to more than 6500 feet (2000 m).
2. The deeper the well, the hotter the water, on the average. Bore bottom temperatures in excess of 200° F (93° C) have been recorded. When the water reaches the surface, sometimes it must be run in an open ditch ("bore drain") for a mile or so before it cools sufficiently for livestock to drink it.
3. The greatest problem, however, is salinity. Although the dissolved mineral content is not particularly high, it is generally too salty for human use or for serious irrigation. Most Great Artesian Basin water can be used only for watering livestock.

The Great Artesian Basin is a valuable asset to the livestock industry, making possible the stocking of perhaps 30 million sheep and 3 million cattle in a region where surface waters would not support one-twentieth that number. It does not, however, make it possible to "open up" the Outback to closer settlement; neither people nor crops can survive on its waters.

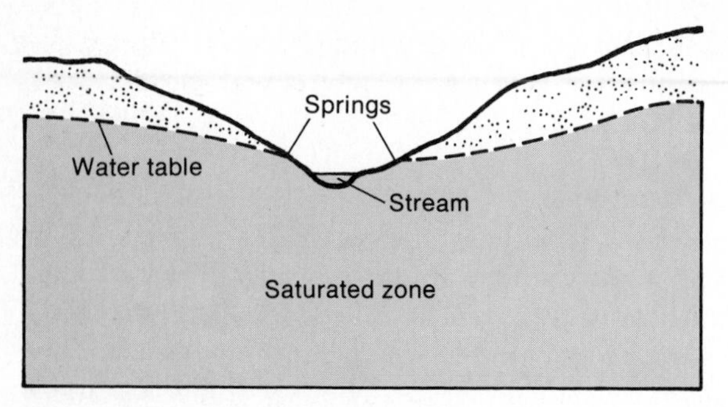

Figure 9-12. Schematic relationship of the water table to surface water. The water table is the top of a subsurface saturated zone. Wherever the water table touches the Earth's surface, water will be found, either as standing water such as lakes and marshes or as flowing water such as streams and springs.

Figure 9-13. Wherever there is an active well, the water table will be drawn downward into an inverted cone of depression.

is nearer the surface. In fact, some desert areas have no saturated zone at all.

A well dug into the zone of saturation will fill with water up to the level of the water table. When water is taken from the well faster than it can flow in from the saturated rock, the water table will sink in the immediate vicinity, in the approximate shape of an inverted cone. This striking feature is called a *cone of depression*. See Figure 9-13. If many wells are withdrawing water faster than it is being replenished naturally, the water table may be significantly depressed over a large area.

The lower limit of the zone of saturation is marked by the absence of pore spaces and therefore the absence of water. This may be caused by a single layer of rock that is so dense that it is impermeable to moisture, or it may simply be that the increasing depth has created so much pressure that no pore spaces exist in any rocks at that level.

Zone of Confined Water In many, but not most, parts of the world, a third hydrologic zone lies beneath the zone of saturation, separated from it by impermeable rock. This so-called *zone of confined water* contains one or more permeable rock layers into which water can infiltrate. Such a layer is called an *aquifer*. Often permeable layers alternate with impermeable ones so dense as to exclude water; these impermeable layers are known as *aquicludes*. Water cannot penetrate an aquifer in this deep zone by infiltration from above because of the impermeable barrier, so any water that it contains must have percolated along the aquifer from a more distant area where no aquiclude interfered. Characteristically, then, an aquifer in the confined water zone is a sloping or dipping layer that reaches to, or almost to, the surface at some location where it can absorb infiltrating water. The water works its way down the sloping aquifer from the catchment area, building up considerable hydrostatic pressure in its confined situation.

If a well is drilled from the surface down into the aquifer, which may be at considerable depth, the confining pressure will force water to rise in the well. If the pressure is sufficient to force the water to the surface without artificial pumping, the free flow that results is called an *artesian well*. See Figures 9-14 and 9-15. If the

Figure 9-14. A constantly flowing artesian well in central Texas. [USDA-Soil Conservation Service photo.]

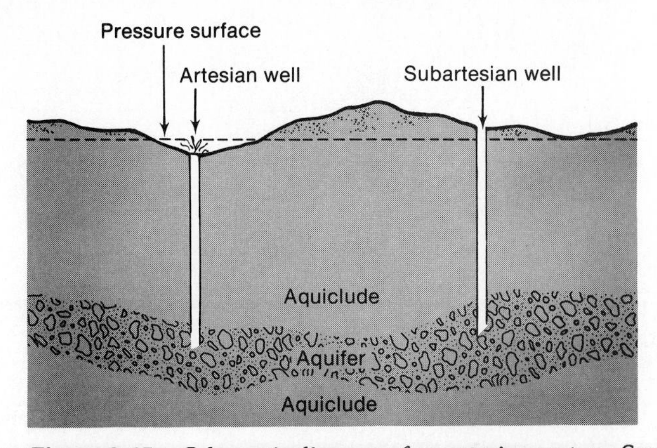

Figure 9-15. Schematic diagram of an artesian system. Surface water penetrates the aquifer in its higher portion (recharge area) and infiltrates downward. It is confined to the aquifer by impermeable strata (aquicludes) above and below. If wells are dug through the upper aquiclude into the aquifer, the confining pressure forces the water to rise in the well. In an artesian well, the pressure forces the water to the surface; in a subartesian situation, the water is forced only partway to the surface.

pressure forces the water partway up the well shaft, but it must be pumped the rest of the way to the surface, it is classed as a *subartesian well*.

Unlike the distribution of groundwater, which is closely related to precipitation, the distribution of confined water is quite erratic over the world. Confined water underlies many arid or semiarid regions that are poor in surface water or groundwater, thus providing a critical resource for these dry lands.

Waterless Zone At some depth below the surface no water can exist in pore spaces because the overlying pressure increases the density of the rock. This *waterless zone* generally begins several miles or kilometers beneath the land surface.

Review Questions

1. As a substance, water has several unique characteristics. Describe some of them.
2. Can water ever move upward, against the pull of gravity? Explain.
3. Explain the role of evaporation in the hydrologic cycle.
4. Where is most of the world's fresh water found?
5. How many oceans are there? Why is this a difficult question to answer?
6. Is the Pacific Ocean significantly different from other oceans? Explain.
7. Explain the relationship of the sun and moon to earthly tides.
8. Which are more important in the hydrologic cycle—lakes or streams? Explain.
9. Explain the relationship of *confined water* and *groundwater*.
10. Distinguish between an *artesian* well and a *subartesian* well.

Some Useful References

Chorley, R. J., *Introduction to Geographical Hydrology*. London: Methuen Publications, 1971.

———, *Water, Earth, and Man*. London: Methuen Publications, 1969.

Leopold, L. B., *Water: A Primer*. San Francisco: W. H. Freeman & Company, Publishers, 1974.

Olsen, R. E., *A Geography of Water*. Dubuque, Iowa: Wm. C. Brown Company, Publishers, 1970.

Ward, R. C., *Principles of Hydrology*, 2nd ed. New York: McGraw-Hill Book Company, 1975.

10 Cycles and Patterns in the Biosphere

Of the four principal components of our Earth system, the biosphere is the least precise. The atmosphere consists of the envelope of air that surrounds our planet, the lithosphere is the solid portion of the Earth, and the hydrosphere is composed of the various waters of the Earth; these three "spheres" are relatively distinct and easy to visualize. The biosphere, on the other hand, is an overlapping concept that impinges spatially on the other three. It consists of the incredibly numerous and diverse array of individual organisms—plant and animal—that populate our planet. Most of these organisms exist at the interface between atmosphere and lithosphere, but some live largely or entirely within the hydrosphere or the lithosphere, and others move relatively freely from one "sphere" to another. Thus it can be seen that the biosphere is more a concept of things than of space. The elements of the biosphere are very significant components of our physical environment, but only under certain circumstances, such as where there is a dense stand of trees or other plants growing together, is there is a sizable spatial aspect.

The Geographic Approach to the Study of Organisms

Even the simplest of living organisms is an extraordinarily complex entity. When a student sets out to learn about an organism, whether it is an alga or an anteater, a tulip or a turtle, he/she embarks on a complicated quest for knowledge. An organism differs in many ways from other aspects of the environment, but most significantly in that it is alive and its sustenance depends on an enormously intricate set of life processes. To learn about even a single organism is a truly herculean task. It is much easier to understand a cloud than a chrysanthemum, simpler to comprehend a moraine than a moose, less difficult to perceive a tornado than a tarantula. If our goal as a geographer was to seek a relatively complete understanding of the world's organisms, we would need at least the life span of a Methuselah. As beneficial as such knowledge would be, however, a student of geography can aim for a less ambitious attainment. With organisms, as with every other feature of the world, the geographer must focus on certain aspects rather than on a complete comprehension of the whole.

The geographic viewpoint is the viewpoint of broad understanding, whether dealing with plants and animals, or anything else. This does not mean that the individual organism is to be ignored; rather it implies that generalizations and patterns are to be sought and their significance assessed. Here, as elsewhere, the geographer is interested in distributions and interrelationships.

Biogeochemical Cycles

The web of earthly life consists of a great variety of organisms coexisting in a diversity of associations. Processes and interactions within the biosphere are exceedingly intricate. Organisms can survive in multiplicity and diversity through a bewildering complex of systemic flows—flows of energy, water, and nutrients—that nourish them. These flows are variable in different parts of the world, in different seasons of the year, and under varying local circumstances.

It is generally believed that for the last billion or so years, the atmosphere and hydrosphere of the Earth have been composed of approximately the same balance of chemical components that now persists. This implies a planetwide steady-state condition in which the various chemical elements have been maintained by cyclic passage through the tissues of plants and animals—first absorbed by the organism as sustenance and then returned to the air/water/soil through decomposition. These grand cycles, ultimately energized by solar insolation, sustain all life on our planet. They have continued unperturbed for millennia, at rates and scales almost too vast to conceptualize. In recent years, however, the rapid burgeoning of human population and the accompanying extraordinary rate of consumption of Earth resources have had a deleterious effect on every one of these cycles. None of the damage is irreparable, but the danger is increasingly apparent.

The sun is the basic energy source on which all life ultimately depends. Solar energy can ignite life processes in the biosphere only through photosynthesis, the production of organic matter by chlorophyll-containing bacteria and plants. In photosynthesis, carbon dioxide is reduced to form organic compounds and molecular oxygen.

Only a very small fraction of the solar insolation reaching the Earth is diverted into the biosphere; thus only about 0.1 percent of received solar energy is fixed in photosynthesis. More than half of that total is used immediately in the plant's own respiration; the remainder is temporarily stored, or shifted around to other portions of the plant. Eventually this remainder enters consumer food chains, either a grazing/browsing food chain or a food chain of decomposition. An increasing proportion of this energy is being diverted to the direct support of one species—*Homo sapiens*. If the biosphere is to function properly, the components of the biogeochemical flows must undergo continual recycling. This is to say that after utilization, they must be converted, at the expense of some solar energy, into a reusable form. For some elements this can be accomplished in less than a decade; for others it may require hundreds of millions of years.

The Energy Cycle

The most important of the cycles is that involving solar energy, as it is fundamental to life on Earth. Although readily absorbed by some substances, solar energy is also readily re-radiated. Thus it is difficult to store and easy to lose. Happily, most places on the Earth receive a daily renewal of the supply.

There are significant latitudinal and seasonal variations in the amount of incoming solar radiation over the Earth. Moreover, this solar energy is redistributed by the circulation of the atmosphere and the oceans. Details of these processes have been discussed in previous chapters. The biosphere is a temporary recipient of a small fraction of this solar energy, which is fixed by green plants in photosynthesis, much of it soon to be reradiated back into the atmosphere. During its brief stay in the biosphere, however, this solar energy sustains all life. Figure 10-1 illustrates energy flow in the biosphere.

The Water Cycle

The most abundant single substance in the biosphere, by far, is water. It is the medium of life processes and the source of their hydrogen. Most organisms contain considerably more water in their mass than anything else.

As discussed in the previous chapter, most of the world's water supply—some 97 percent of the total—is contained in the oceans. Another 2 percent or so is locked up in glacial ice. Less than 1 percent of the total world supply, then, is found in all other forms. At any given time the bulk of this tiny fraction of the total is underground; only a miniscule amount is found in surface lakes and streams. Considerably smaller still is the amount of moisture contained in the atmosphere as water vapor. And still less by a full order of magnitude than the amount of moisture in the atmosphere is the quantity contained in the total biomass of the world.

There are two ways in which water is encompassed in organic life: *in residence*, with its hydrogen chemically bound into plant and animal tissues; and *in transit*, as part of the transpiration/respiration stream. Water has extraordinary physical properties that endow it with a unique chemistry. Its role in life processes is unparalleled.

The water cycle in the biosphere is considered to be a closed system in which precipitation balances evapotranspiration. Over the oceans, evaporation slightly exceeds precipitation; whereas on the continents, precipitation exceeds evapotranspiration by 25 or 30 percent. More details of the hydrologic cycle were discussed in Chapter 9.

Figure 10-1. A simplistic illustration of energy flow in the biosphere.

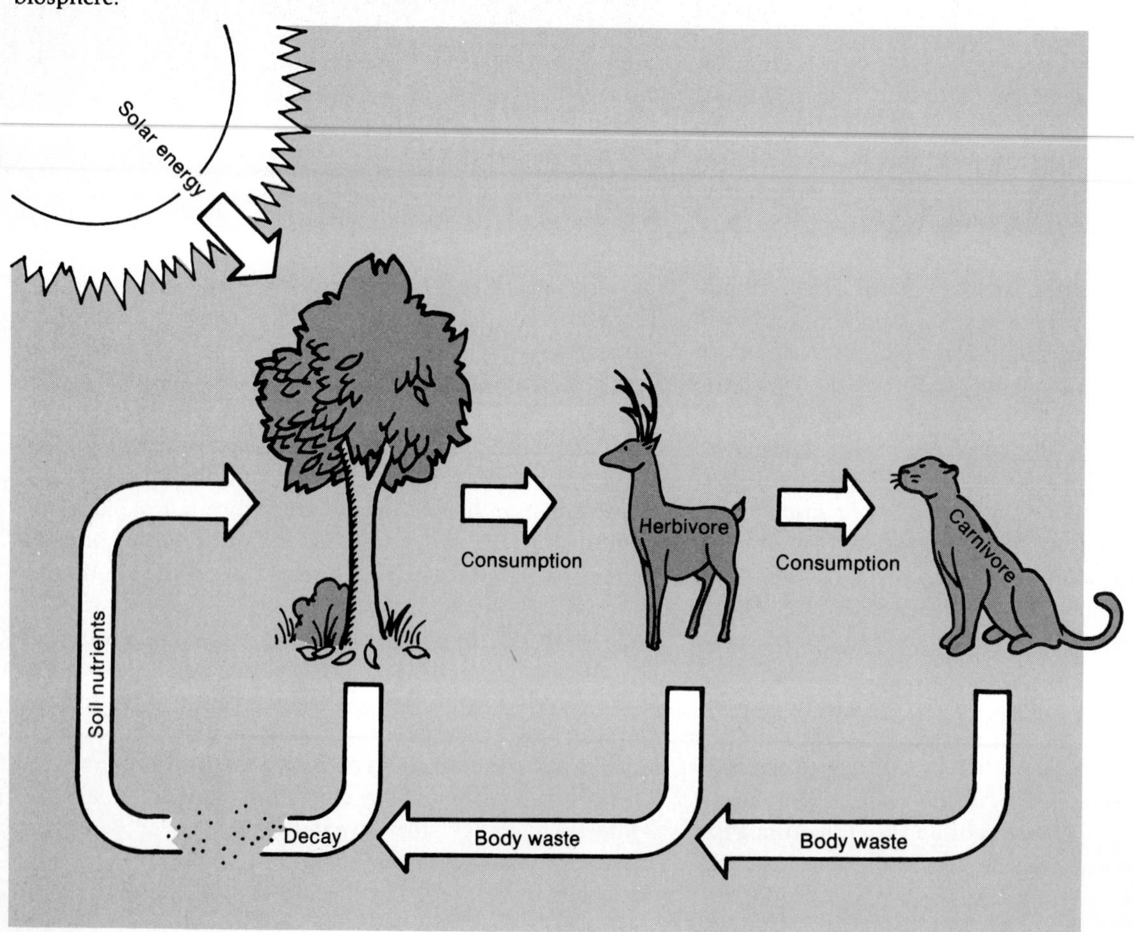

The Carbon Cycle

Carbon is one of the basic elements of life and forms a part of all living things. The biosphere contains a complex mixture of carbon compounds, more than half a million in total. These compounds are in a continuous state of creation, transformation, and decomposition.

The main carbon cycle is conversion from carbon dioxide to living matter and back to carbon dioxide. See Figure 10-2. This conversion is initiated by the fixation of carbon dioxide from the atmosphere through photosynthesis into carbohydrate compounds. Some of the carbohydrates are then consumed directly by the plant, which generates more carbon dioxide to be released through its leaves or roots. The rest becomes part of the plant tissue, which is either consumed by animals or eventually decomposed by microorganisms. Plant-eating animals convert some of the consumed carbohydrates back to carbon dioxide for release to the air by respiration; the remainder is decomposed by microorganisms after the animal dies. The carbohydrates acted upon by microorganisms are ultimately oxidized into carbon dioxide and returned to the atmosphere. A similar cycle takes place in the ocean.

The basic cycle as described above operates relatively rapidly (the time measured in years or centuries) and encompasses only a small proportion (thought to be less than 1 percent) of the total quantity of carbon on or near the Earth's crust. The overwhelming bulk of near-surface carbon has been concentrated over millions of years in sedimentary geologic deposits—such as coal, petroleum, and carbonate rocks—composed of dead organic matter that accumulated mostly on sea bottoms and was subsequently buried by other sediments. This carbon reservoir is normally incorporated into the active cycle only with geologic gradualness. In the last century and a half, however, mankind has added considerable carbon dioxide to the atmosphere by extracting and burning fossil fuels (coal, oil, gas) whose carbon had been fixed by photosynthesis many millions of years ago. The impact of this rapid acceleration of freed carbon dioxide is likely to have far-reaching effects on the biosphere.

Figure 10-2. The carbon cycle. Atmospheric carbon is "fixed" (assimilated) by plant photosynthesis into carbohydrate compounds. Through various paths these compounds eventually can be converted again to carbon dioxide and returned to the atmosphere.

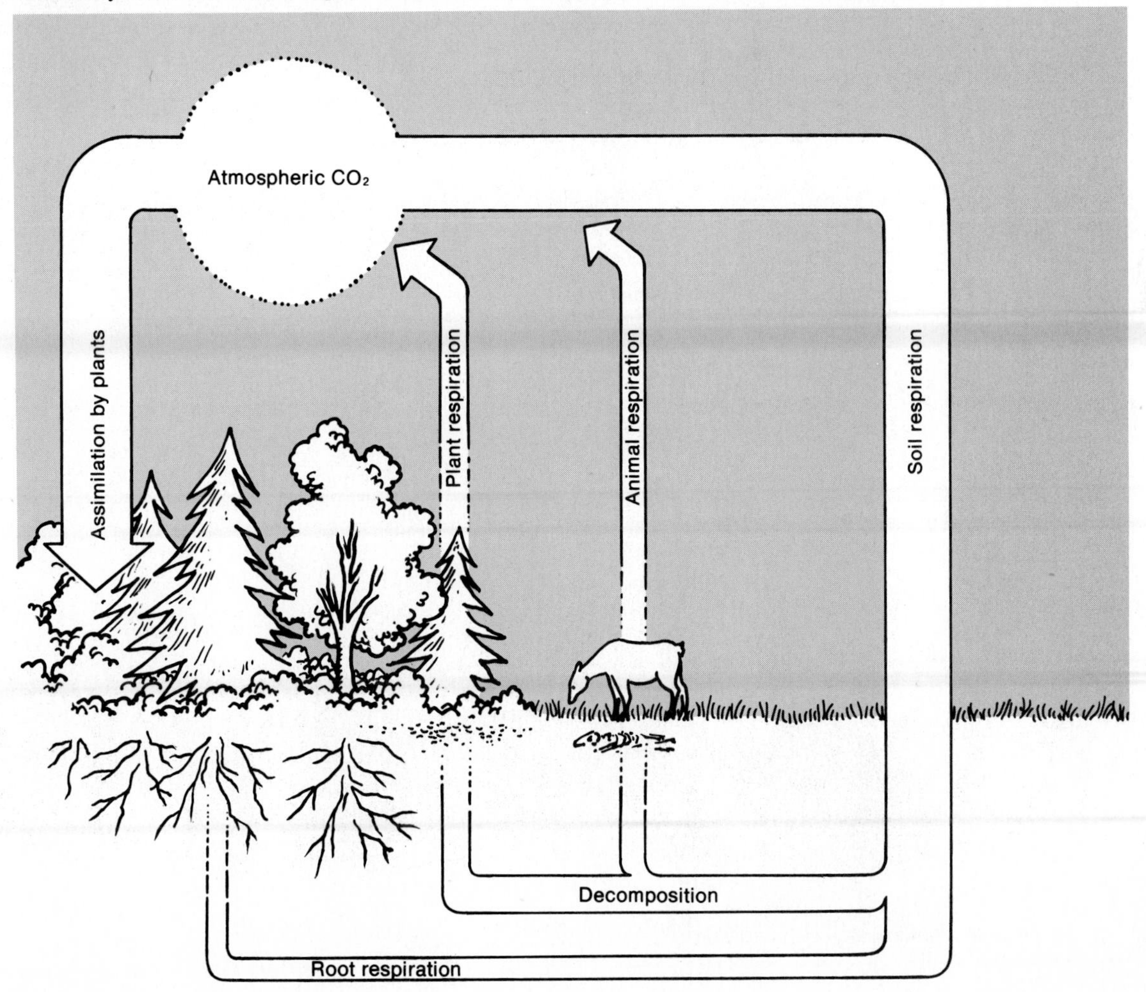

The Oxygen Cycle

Oxygen is a building block in most organic molecules and comprises a significant proportion of the atoms in living matter. The oxygen now in the atmosphere is largely or entirely of biological origin, which means that it was put there by plants. Thus primitive plants made possible the evolution of higher plants and animals by providing free oxygen for their metabolism.

The oxygen cycle (Figure 10-3) is an extremely complicated one, and will be only briefly summarized here. Oxygen occurs in a great many chemical forms and combinations and is released into the cycle in a variety of ways. Most of it is decomposed from water molecules by light energy in photosynthesis, and then added to the atmosphere by plant transpiration or animal respiration. Much is eventually recycled through the biosphere by means of photosynthesis through plant tissues again. Other interactions that play a part in the cycle include atmospheric ozone, oxygen involved in the oxidative weathering of rocks, oxygen stored in and sometimes released from carbonate rocks, and various other processes, including some (such as the burning of fossil fuels) that are human-induced.

The Nitrogen Cycle

Although nitrogen is an apparently inexhaustible component of the atmosphere and is one of the substances necessary for all forms of life, only a very few types of organisms are capable of directly utilizing it in its gaseous form. For the vast majority, nitrogen is usable only after it has been *fixed*, or activated by splitting molecular nitrogen into free nitrogen that can be incorporated into another compound. See Figure 10-4. Some nitrogen is fixed in the atmosphere by such energetic phenomena as lightning and cosmic radiation, some is fixed in the ocean by marine organisms, but the principal source of free nitrogen is action by soil microorganisms and associated plant roots on atmospheric nitrogen in air spaces in the soil.

Once atmospheric nitrogen has been fixed into an available form (nitrates), it can be assimilated by green plants, some of which are in turn eaten by animals. The animals then excrete nitrogenous wastes. These wastes, as well as the dead animal and plant material, are attacked by bacteria, releasing nitrite compounds as a further waste product. Other bacteria can convert the nitrites to nitrates, making them available again to green plants. Still other types of bacteria are capable of *denitrification*, which con-

Figure 10-3. The oxygen cycle. Free oxygen is essential for almost all forms of life. It is made available to the air through a variety of processes and is recycled in a variety of ways.

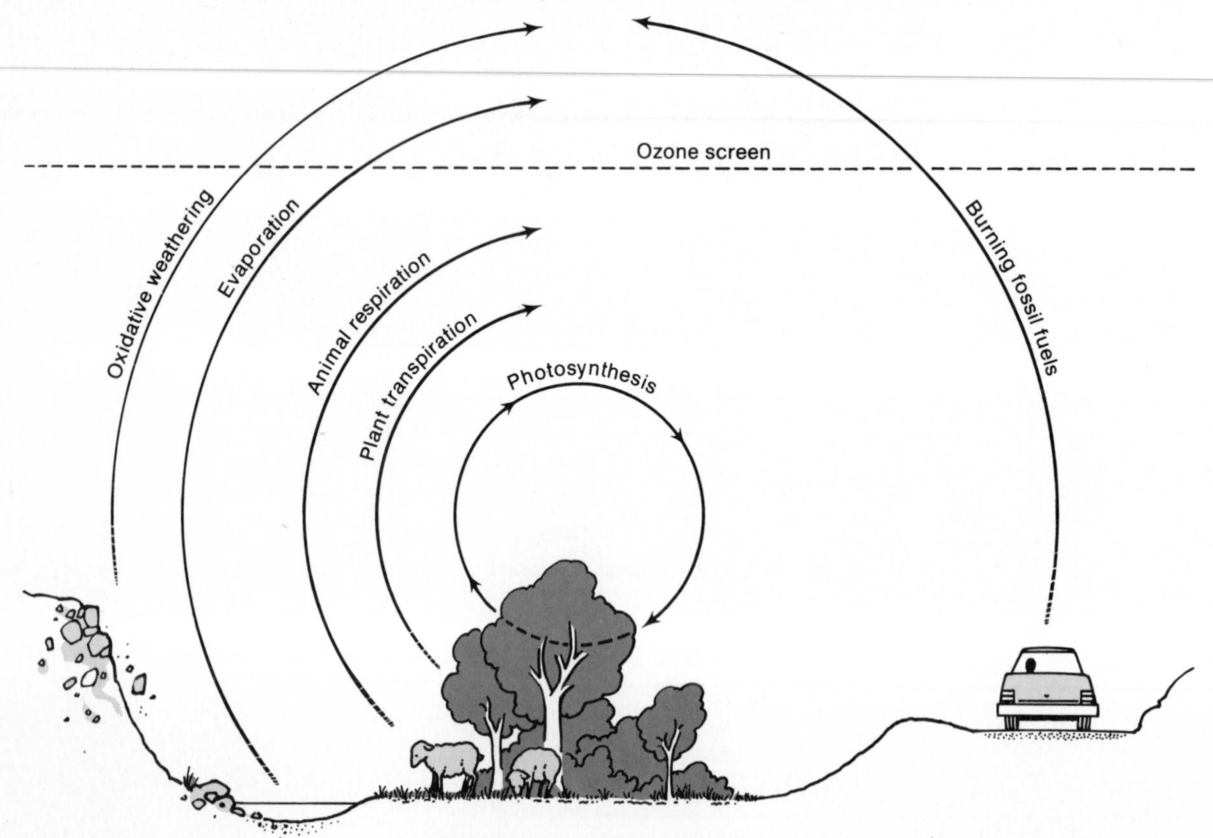

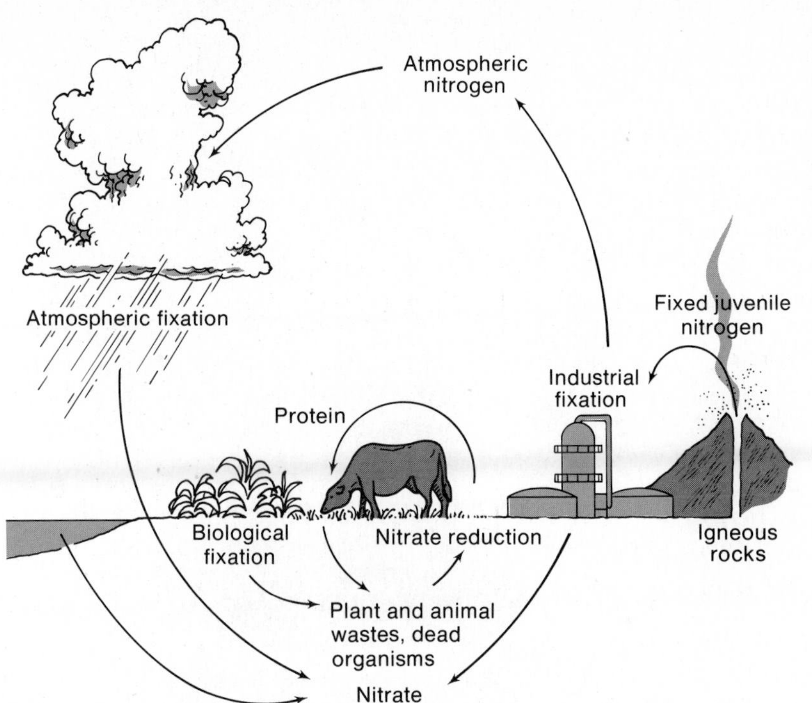

Figure 10-4. The nitrogen cycle. Atmospheric nitrogen is "fixed" (made available as a nitrate compound) in various ways so that it can be assimilated by green plants, some of which are eaten by animals. Dead plant and animal materials, as well as animal wastes, contain nitrites, which can be converted by bacteria into nitrates, and thus continue the cycle. Other bacteria denitrify the nitrate compounds, releasing free nitrogen into the air again.

verts nitrates into free nitrogen in the air. This free nitrogen is then carried by rain back to the Earth, where it enters the soil/plant portion of the cycle once more.

Human activities have produced a major modification in the natural nitrogen cycle. The synthetic manufacture of nitrogenous fertilizers and widespread introduction of nitrogen-fixing legume crops (such as alfalfa) have significantly changed the balance between fixation and denitrification. The short-term result has been an excessive accumulation of nitrogen compounds in many lakes and streams, which badly depletes the oxygen supply of the water and upsets natural ecosystems; the long-term results are still unclear.

Other Mineral Cycles

Although carbon, oxygen, and nitrogen—along with hydrogen—are the principal chemical components of the biosphere, many other minerals occur as critical nutrients (though sometimes in very small quantities) for plant and animal life. Most notable among these are phosphorous, sulfur, and calcium, but more than a dozen others are occasionally significant.

The cycles of these nutrients are variable, but they are generalized in the diagram shown in Figure 10-5. The broad pattern shows an accumulation of nutrients through atmospheric precipitation, rock weathering, leaf fall, and the dying of plants and animals; dispersal through decomposition and surface and subsurface runoff of water; and intake by means of assimilation into plant and animal tissues.

The general conclusion is that biotic nutrients are finite and limited in quantity. They are rearranged and circulated in cycles that are extremely variable from place to place. Some of them are significantly susceptible to human interference and modification.

Figure 10-5. The general nutrient cycle. A variety of minerals are in circulation through the biosphere as nutrients for organisms. They accumulate from many sources, including precipitation, weathering, and decomposition of organic matter. Mostly they are dispersed through surface and subsurface runoff and are eventually assimilated through plant roots or animal consumption.

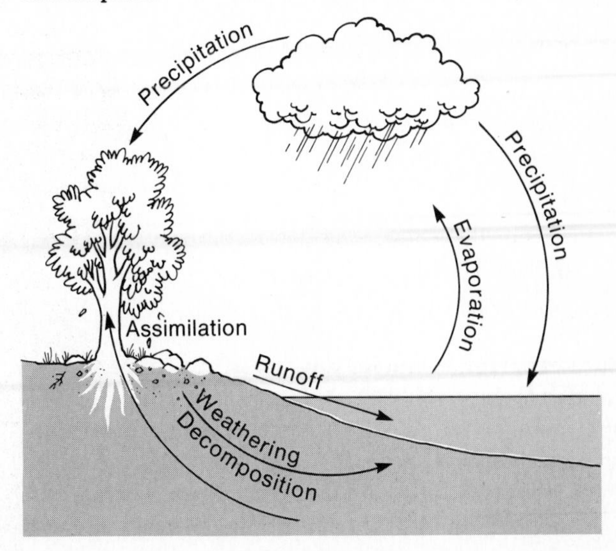

Food Chains

The unending flows of energy, water, and nutrients through the biosphere are channelled in significant part by direct passage from one organism to another in a process referred to as *food chains*. A food chain is a simple concept: One organism eats another, thereby absorbing energy and nutrients; the second organism is consumed by a third, with similar results; the third organism is eaten by a fourth; etc. In nature, however, the matter of who eats whom may be extraordinarily complex, with a bewildering number of interlaced strands.

Chain is probably a misleading word in this context because it implies an orderly linkage of equivalent units. It would be more accurate to think of each of the links as an energy transformer, which uses some of the energy originally derived by the first unit in the chain for its own sustenance, then passes the balance on to the next link in the chain.

The fundamental unit in the food chain concept consists of plants that trap solar energy through photosynthesis. The plants are then eaten by herbivorous animals, which are considered as the *primary consumers*, which in turn become food for other animals, called *secondary consumers*. There may be many levels of secondary consumers—for example, the plant-eating beetle is eaten by a frog, which is devoured by a snake, which is consumed by a hawk, etc.

A food chain can also be conceptualized as a *food pyramid*, based on the quantity and bulk of its units. See Figure 10-6. At the bottom of the pyramid is a multitude of energy-producing plants. The second level consists of a significant number of primary consumers. Above that are one or several levels of carnivorous (meat-eating) secondary consumers, each succeeding level consisting of fewer and usually larger animals. The final consumers at the top of the pyramid are usually the largest and most powerful predators in the area.

The final consumers, at the apex of the pyramid, do not constitute the final link in the food chain, however. When they die, they are fed upon by tiny (mostly microscopic) organisms, which function as decomposers, returning the nutrients to the soil to be recycled into yet another food chain.

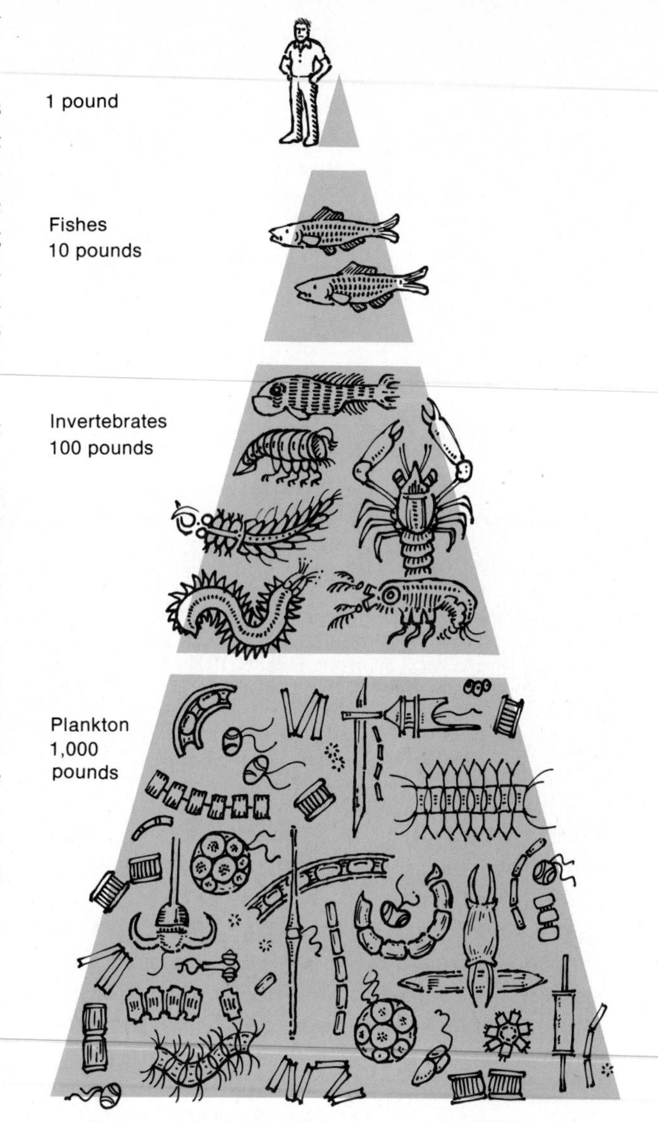

Figure 10-6. Schematic illustration of a food pyramid. It takes half a ton of plankton to provide a one-pound weight gain for a human. [Adapted from Life Nature Library/*Ecology*, drawing by Otto van Eersel (New York: Time-Life Books Inc., Publisher, 1963) p. 37. © 1963 by Time, Inc.]

The Search for a Meaningful Classification

When a geographer studies any set of phenomena, he/she attempts to work out a classification that permits the grouping of individuals in some meaningful fashion. In some cases, the geographer borrows classification schemes from specialists in other disciplines; often, however, classifications devised for other purposes are not particularly useful for geographic studies, and the geographer must develop a different one.

The systematic study of plants and animals is primarily the domain of the biologist, and many biological classifications have been devised. By far the most significant and widely used is the binomial system originally developed by Linnaeus in the eighteenth century. This system focuses primarily on the morphology of the organisms and groups them on the basis of structural similarity. Such a classification is generally useful for geographers, but it has certain shortcomings that preclude its total acceptance. The principal disadvantage of the Linnaean system for geographical use is that it is based entirely on anatomical similarities, whereas geographers are more interested in distribution patterns and habitat preferences.

It would be nice to be able to say that geographers have come up with a more appropriate classification, but such is not the case, nor is a universally accepted geographical classification of organisms ever likely to be developed. Too many subjective decisions would have to be made, making widespread agreement on any classification very unlikely.

Seeking Pertinent Patterns

Among the life forms of our planet are nearly 600,000 species of plants and more than twice that many species of animals. With such an overwhelming diversity of organisms, how can we go about studying their distributions and relationships in any meaningful manner? A logical approach is to decide on some generalizing procedures and useful groupings, and then consider the patterns that emerge.

The term *biota* refers to the total complex of plant and animal life; as such it is essentially synonymous with the term *biosphere*. The basic subdivision of biota separates *flora*, or plants, from *fauna*, or animals. In this book we will recognize a further fundamental distinction—between oceanic biota and terrestrial biota.

The inhabitants of the oceans are generally divided into three groups: plankton, nekton, and benthos.

1. *Plankton* consists of plants and animals that float about, drifting with the currents and tides. Most are microscopic in size.
2. *Nekton* is the term applied to animals that swim freely in the oceans. Most are fishes, but some significant mammals and reptiles are included.
3. *Benthos* is composed of animals and plants that live on or in the ocean bottom. The plant component exists only in shallow waters where sunlight penetrates and is mostly "seaweed" or algae. The animal component of benthos lives everywhere on the bottom, from shallowest to deepest waters. Some of the animals—snails, worms, clams, sponges, starfish, crabs, and lobsters—are free-moving, but others are fixed in position for all or nearly all of their lives, such as oysters, sea anemones, and corals.

Although the life forms of the oceans are varied and fascinating, and despite the fact that 70 percent of the Earth's surface is oceanic, in this book we will pay scant attention to oceanic biota. This is primarily because of constraints of time and space, but it is also in keeping with the idea that oceanography is too vast a subject to be given sketchy treatment in this volume.

The terrestrial biota is much more diverse than that of the oceans. It will be the focus of our interest in the remaining pages of the section devoted to the biosphere.

Ecosystems and Biomes

In our search for meaningful organizing principles for the comprehension of the biosphere, two concepts seem to be of particular value—ecosystem and biome.

Ecosystem: A Term for All Scales

The term *ecosystem* is a contraction of the phrase *ecological system*. An ecosystem is considered to include all the organisms in a given area, but it is more than simply a community of plants and animals existing together. The ecosystem concept is functional and encompasses the totality of interactions among the organisms and between the organisms and the nonliving portion of the environment in the area under consideration. The nonliving portion of the environment includes soil, rocks, water, sunlight, atmosphere, etc., but it essentially can be considered as nutrients and energy.

An ecosystem, then, is fundamentally a biological community, or an association of plants and animals, expressed in functional terms. The concept is built around the flow of energy among the various components of the system, which is the essential determinant of how a biotic community functions. See Figure 10-7.

This functional, systemic concept is very attractive as an organizing principle for the geographic study of the biosphere. It must, however, be approached with caution because of the variable scales at which it can be applied. There can be almost infinite variety in the magnitude of ecosystems that we might recognize. At one extreme, for example, we can conceive of a planetary ecosys-

Figure 10-7. The flow of energy in a simple ecosystem. Energized by the sun, grass feeds a rabbit, which is eaten by a hawk, whose decaying remains furnish nutrients for the grass.

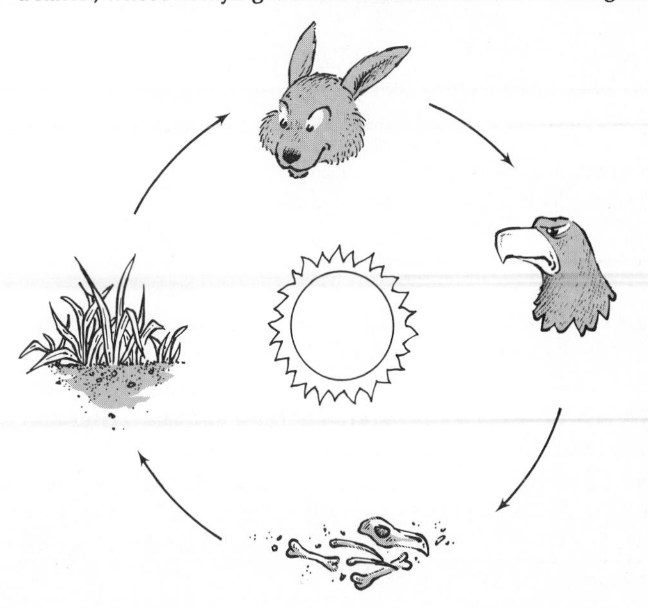

Focus on Biological Taxonomy

Taxonomy is the science of classification. As a term it was originally applied to the classification of plants and animals, although its meaning has been broadened to encompass any sort of systematic classification. Our concern here is only with biological taxonomy.

Mankind has attempted to devise meaningful classifications of plants and animals for thousands of years. One of the most useful of the early classifications was designed by Aristotle in the third century B.C., and this system was in general use for nearly twenty centuries. In the late 1700s, the Aristotelean classification was finally replaced by a much more comprehensive and systematic one developed by the Swedish naturalist Carolus Linnaeus. Linnaeus made use of ideas from other biologists, but the system is largely his own work.

The Linnaean system of classification is generic, hierarchical, comprehensive, and binomial. *Generic* means that it is based on observable characteristics of the organisms it classifies, primarily their anatomy, structures, and details of reproduction. *Hierarchical* means that the organisms are grouped on the basis of similar characteristics, with each succeedingly lower level of grouping having a larger number of similar characteristics and therefore containing fewer individuals in the group. *Comprehensive* means that all plants and animals, existing and extinct, can be encompassed within the system. *Binomial* means that every kind of plant and animal is identified by two names.

The binomial naming of organisms is highly systematized. Each type of living thing has a name with two parts. The first part, in which the first letter is capitalized, designates the *genus*, or group; the second part, which is not capitalized,

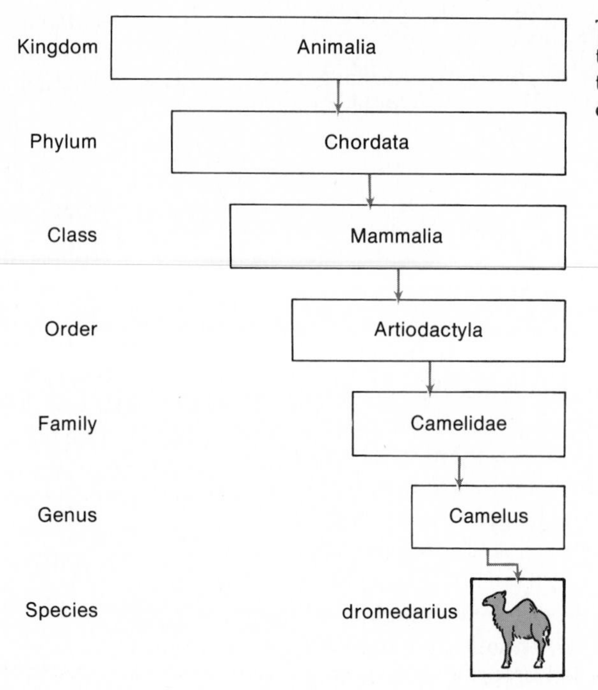

The taxonomic classification of an animal, as illustrated by the Arabian camel, or dromedary.

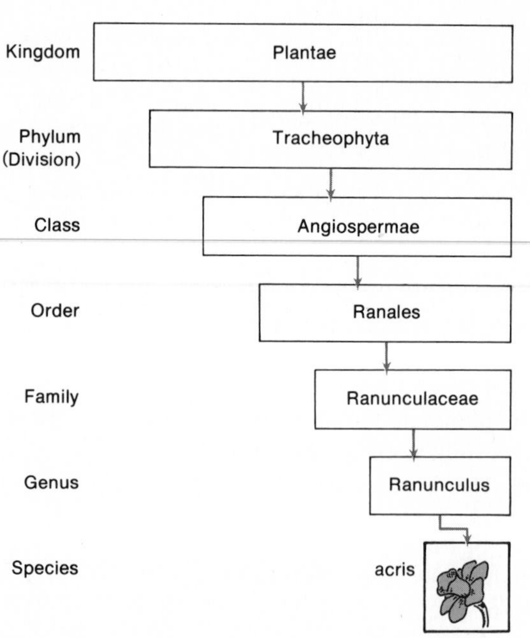

The taxonomic classification of a plant, as illustrated by the buttercup.

tem that encompasses the entire biosphere; at the other end of the scale, it would be possible to consider the ecosystem of a fallen log, or of the underside of a rock, or even of a drop of water. There is endless variety in the possibilities.

Therefore, if we are going to try to identify and understand broad distributional patterns within the biosphere, we must focus only on ecosystems that can be recognized at a useful scale.

Biome: A Scale for All Biogeographers

Among terrestrial ecosystems, the type that provides the most appropriate scale for understanding world distribution patterns is called a *biome*. A biome is a large, recognizable assemblage of plants and animals in functional interaction with its environment. It is usually identified and named on the basis of its dominant vegetation association, which normally comprises the bulk of the *biomass* (the indicates the *species*, or specific kind of organism. The combination of genus and species is referred to as the *scientific name*; it is always in Latin, although many of the words have Greek derivations.

Each type of organism, then, has a scientific name that distinguishes it from all other organisms. Although the popular name may be variable, or even indefinite, the scientific name is unvarying. Thus in different parts of the Western Hemisphere, the large native cat may be called a mountain lion, cougar, puma, panther, painter, or *león*; but its scientific name is always *Felis concolor*.

The intellectual beauty of the Linnaean system is twofold: (1) Every organism that has ever existed can be fitted into the scheme in a logical and orderly manner, for the system is capable of indefinite expansion. (2) The various hierarchical levels in the system provide a conceptual framework for understanding the relationships among different organisms or groups of organisms. This is not to say that the system is perfect. Linnaeus believed that species were unchanging entities, and his original system had no provision for variations. The concept of subspecies was a major modification of the system that was introduced subsequently to accommodate observed conditions of evolution.

Nor is the system even completely objective. Whereas its concept and general organization are accepted by scientists throughout the world, details of the classification depend on judgments and opinions made by biologists. These judgments and opinions are based on careful measurements and observations of plant and animal specimens, but there is often room for differing interpretations of the relevant data. Consequently, some details of biological taxonomy are disputed and even controversial.

Such details notwithstanding, the general system provides a magnificent framework for biological classification. The seven main levels of the system, in order from largest to smallest, are: (1) kingdom, (2) phylum, (3) class, (4) order, (5) family, (6) genus, (7) species.

Kingdom is the broadest category and contains the largest number of organisms. Until recently, the system recognized only two kingdoms: one encompassing all plants and the other including all animals. Increasingly, however, taxonomists encountered difficulty in accommodating the many varieties of one-celled and other simple microscopic organisms into such a two-kingdom system. It is now widely, but by no means universally, accepted that five kingdoms exist at the highest level of taxonomic distinction:

1. *Monera* are the simplest known organisms, consisting of one-celled bacteria and blue-green algae. Their cell walls have a different composition from those of other organisms.
2. *Protista* consist of other one-celled organisms and some simple multicelled algae, most of which were formerly classed as plants.
3. *Fungi* also were previously classified as plants, but it is now recognized that they differ in origin, direction of evolution, and primary nutrition from plants and therefore deserve separation.
4. *Plantae* include the multicelled green plants and higher algae.
5. *Animalia* consist of the multicelled animals.

Phylum is the second major level of the system. Of the two or three dozen phyla within the animal kingdom, one, *Chordata*, includes all animals with backbones, which means almost every animal more than an inch or two in length. In the plant kingdom, the term *division* is often used in place of *phylum*. Most large plants belong to the division *Tracheophyta*, or vascular plants, which have efficient internal systems for transporting water and sugars and a complex differentiation of organs into leaves, stem, and roots.

The third principal level is *class*. Among the several dozen animal classes, the most important are *Mammalia* (mammals), *Aves* (birds), *Reptilia* (reptiles), *Amphibia* (amphibians), and two classes that encompass fishes. There are somewhat fewer classes of plants, of which the most notable is *Angiospermae*, the flowering plants.

The fourth level of the classification is called *order*, and the three lower levels are *family*, *genus*, and *species*. As in all levels of the hierarchy, each succeeding lower level contains organisms that are increasingly alike. Species is the basic unit of the classification. In theory, only members of the same species are capable of breeding with one another. In practice, however, interbreeding is possible among just a few species (always within the same genus), although the offspring of such interspecific breeding are nearly always infertile (i.e., totally incapable of reproducing). In some cases, species are further subdivided into *subspecies*, also called *varieties* or *races*.

The relative diversity of living and extinct species is worthy of note. About 1,500,000 species of living organisms have been identified and described. From 3 to 10 million additional species, mostly microscopic in size, may not yet be identified. And an estimated 500 million species have become extinct in the history of the Earth. Thus more than 95 percent of all evolutionary lines have already completely disappeared.

total weight of all organisms—plant and animal) in the area under consideration, as well as being the most obvious and conspicuous visible component of the landscape.

There is no universally recognized classification of the world's biomes, but scholars commonly accept about ten major biome types. These are listed below and will be discussed in greater detail in Chapter 11:

Tropical rain forest
Tropical deciduous forest
Tropical scrub
Tropical savanna
Desert
Mediterranean woodland and shrub
Midlatitude grassland
Midlatitude deciduous forest
Boreal forest
Tundra

A biome is composed of much more than merely the plant association that gives it its name. A variety of other kinds of vegetation usually grows among, under, and occasionally over the dominant association. Diverse animal species also occupy the area. Often, significant and even predictable relations are seen between the biota (particularly the flora) of the biome and the associated climate and soil types.

As one peruses the map showing major biome types of the world (Plate IV), the arbitrary nature of regional boundaries should be noted. The major biotic communities do not occupy sharply defined areas in nature, no matter how sharp the demarcations may appear on a map. Normally the communities merge more or less imperceptibly with one another through *ecotones*—transition zones of competition in which the typical species of one community intermingle or interdigitate with those of another. In Plate IV, as in any small-scale map of biomes, the boundaries are drawn rather arbitrarily through the ecotones.

Environmental Relationships

The survival of plants and animals depends on an intimate and sometimes precarious set of relationships with other elements of the environment. The details of these relationships vary with different species, but we can generalize about many of them.

These relationships can be discussed at variable scales of generalization. However, you must remember that a generalization that is a truism at one scale may be quite invalid at another. For example, if we are considering global or continental patterns of biotic distribution, we will be concerned primarily with gross generalizations that deal with average conditions, seasonal characteristics, latitudinal extent, zonal winds, and other broad-scale factors. On the other hand, if our interest is focussed on detailed distributions in a small area such as an individual valley or a single hillside, we will be more concerned with such localized environmental factors as degree of slope, direction of exposure, and permeability of topsoil. Whatever the scale, there will nearly always be exceptions to the generalizations, and the smaller the scale, the more numerous will be the exceptions. (Remember that a small scale applies to a relatively large portion of the Earth's surface, and vice versa.) Thus in a region that is generally humid there will probably be many localized sites that are extremely dry, such as a cliff or a sand dune. And in even a very dry desert there are likely to be several places that are perennially damp, such as an oasis or a spring.

The Influence of Climate

At almost any scale the most prominent environmental constraints on biota are exerted by various climatic factors, as discussed below.

Light No plants can survive without light. It is essentially for this reason that vegetation is absent from deeper ocean areas where light does not penetrate. The basic process whereby plants produce stored chemical energy is called *photosynthesis*. A green plant takes carbon dioxide from the air and combines it with water to form energy-rich carbohydrate compounds that we know as sugars. This process is activated by light. All terrestrial environments except caves, however, receive at least some light, so darkness is not an absolute limit on terrestrial vegetation.

Another important relationship of light to life forms, both plant and animal, involves the response of an organism to the length of exposure to light in a 24-hour period, which is called *photoperiodism*. Over all the Earth, except in the immediate vicinity of the equator, the seasonal variation in the photoperiod becomes greater with increased distance from the equator. Fluctuation in the photoperiod stimulates seasonal behavior—such as flowering, leaf fall, mating, and migration—in both plants and animals.

Moisture The broad distribution patterns of the biota are governed more significantly by moisture relationships than by any other single environmental factor. A prominent trend throughout biotic evolution is the specialization or adaptation of plants and animals to excesses and especially to deficiencies in moisture availability. See Figure 10-8. The availability of water is not wholly dependent on the climatic component—precipitation versus evaporation—but it is very largely determined by these atmospheric dynamics.

Temperature The temperature of the air and the soil is also important to biotic distribution patterns. In very hot and in very cold climates, fewer species of both plants and animals are able to survive than in areas of more moderate temperatures. This is an especially critical factor in cold regions. Plants, in particular, have a limited tolerance for low temperatures because they are continually exposed to the weather, and they experience tissue damage and other physical disruption when their cellular water freezes. Animals in some instances are able to avoid the bitterest cold by moving around to seek shelter. Even so, the cold weather areas of high latitudes and high altitudes have a limited variety of both animals and plants.

Wind The influence of wind on biotic distributions is more limited than that of the other climatic factors. Where winds are persistent, however, they often serve as a constraint. See Figure 10-9. The principal negative effect of wind is that it causes excessive drying, thus emphasizing moisture deficiency. In cold regions, also, wind escalates the loss of body heat from animals, severely intensifying the coldness. On the positive side, wind sometimes aids in the dispersal of biota by carrying pollen, seeds, lightweight organisms, and flying creatures.

Figure 10-8. Even the most stressful environments often contain distinctive and conspicuous plants. The largest of the cacti—organ pipe on the left and saguaro on the right—prosper in the parched Sonoran desert along the border that separates Arizona from Mexico. [TLM photo.]

Figure 10-9. Some plants are remarkably adaptable to environmental stress. In this timberline scene from northern Colorado, there is no question about the prevailing direction of wind. Persistent wind from the right has so dessicated these subalpine firs that branches are only able to survive if they grow directly toward the left, which places the trunk of the tree between them and the wind. [TLM photo.]

Figure 10-10. Wildfires are commonplace in many parts of the world. This ground fire in a eucalyptus woodland in northeastern New South Wales (Australia) is almost an annual occurrence under natural conditions. [TLM photo.]

Edaphic Influences

Soil characteristics, known as *edaphic* factors, are also influential in biotic distributions. These factors are direct and immediate in their relationship to flora, but they are usually indirect in their effect on fauna. The soil is a major component of the habitat of vegetation, and its characteristics are particularly deterministic with regard to rooting capabilities and nutrient supply. Especially significant are soil texture, soil structure, humus content, chemical composition, and the relative abundance of soil organisms.

Topographic Influences

Topographic influences on biotic distributions are variable. In terms of global patterns, the general topographic characteristics comprise the most important factor. For example, a plains region has a very different assemblage of plants and animals from that of a mountainous region. At a more localized scale, the factors of slope and drainage are likely to be significant. This involves primarily the steepness of the slope, its orientation with regard to sunlight, and the relative ease with which water flows over the surface or sinks into the ground.

Wildfire

Most environmental factors that affect the distribution of plants and animals are passive, and their influences are slow and gradual. Occasionally, however, abrupt and catastrophic events also play a significant role. These include floods, earthquakes, landslides, insect infestations, droughts, and other dramatic occurrences. By far the most important is wildfire. See Figure 10-10. In almost all portions of the continents, except for the perenially wet regions where fire simply cannot start, and the perenially dry regions where there is an insufficiency of combustible vegetation, uncontrolled natural fires have occurred with surprising frequency. Fires generally result in complete or partial devastation of the flora and the killing or driving away of all or most of the fauna. These results, of course, are only temporary; sooner or later, vegetation sprouts and animals return. At least in the short run, however, the composition of the biota is changed, and if the fires occur with sufficient frequency, the change may be more than temporary.

Although wildfire is a catastrophic event, it does not necessarily create a catastrophe. Indeed, the occurrence of a wildfire can be very helpful to the sprouting of certain plants and the maintenance of certain floristic associations. In some cases, grasslands are sustained by relatively frequent natural fires, which inhibit the encroachment of tree seedlings. Moreover, many plant species, particularly certain trees such as the California redwood and the southern yellow pine, scatter their seeds only after the heat of a fire has caused the cones or other types of seedpods to open.

Predictable Correlations

Great intellectual satisfaction can be derived from the considerable order and logic seen in the distribution of the biotic components of the environment. This is generally true at the species level, but it is much more apparent in the context of groups of communities of species, and particularly at the broad scale of biomes. The point is that there are recognizable and, in many cases, predictable patterns of biotic distributions, based on environmental relationships.

In terms of broad distribution patterns, climate and vegetation have a particularly close correlation. To illustrate this principle at its most fundamental level, we might consider the distribution of forests over the world. Where, under natural conditions, do forests grow? There are many facets to the answer but by far the most important is the availability of moisture. A relative abundance of moisture during the growing season usually implies that trees will flourish and that a forest is likely to be the dominant plant association. See Figure 10-11. This dependence on moisture is primarily because trees, unlike other plants, must have a mechanism for transporting mineral nutrients a relatively great distance from the place of acquisition (roots) to the place of need (leaves). Such transport can only take place in a dilute solution; therefore much water is needed by trees throughout the growing season. Other plant forms can flourish in areas of relatively high precipitation, of course, but they rarely become dominant because they are shaded out by trees. The broad generalization, then, is that trees in particular and forests in general are usually found wherever there is a relative abundance of precipitation. Low-growing plants dominate the landscape only in areas where trees cannot flourish, which means mostly in regions of relatively sparse precipitation.

Figure 10-11. Wherever precipitation is adequate, trees are likely to grow in close proximity, so that a forest is the dominant vegetation association. This fall scene is in a deciduous forest in Michigan. [Courtesy Michigan Department of Natural Resources.]

The relationship between climate and faunal distributions is less pervasive and obvious, but it is still important. If we were to try to understand, for example, the global distribution of fur-bearing animals, we would find that there are climatic correlations, but that they are not so distinct. Where do fur-bearers live? See Figure 10-12. There are several hundred species of mammals whose skin is covered with a fine, soft, thick, hairy coat that is referred to as *fur*. They range in size from tiny mice and moles to the largest of bears. An examination of their ranges would show three generalized habitat preferences:

1. Many species live in locations where winters are long and cold, either because of latitude or altitude.
2. A number of species live in aquatic environments.
3. The remainder are widely scattered over the continents, occupying a considerable diversity of habitats.

From a quick look at the evidence, we can reach the tentative conclusion that many of the fur-bearing species, including all those with the heaviest, thickest fur, live in regions where cold temperatures are common. The climatic correlation in this case is partial and indistinct.

Figure 10-12. Most fur-bearing animals, such as this raccoon, live in cool or cold habitats. [Courtesy Michigan Department of Natural Resources.]

In this chapter we have been considering broad relationships between climatic factors and gross biotic distributions. Clearly, the climatic influence on general vegetation distributions is stronger than that of any other environmental factor. In terms of faunal distribution, climate is also of considerable significance, but the most important factor is the vegetation pattern. Other environmental elements are also influential in biotic distributions, but they are much less important than climate for plants and vegetation for animals, with regard to gross patterns.

Review Questions

1. Describe the basic characteristics of the carbon cycle.
2. Most of the carbon at or near the Earth's surface is not involved in any short-term cycling. Explain this situation.
3. Explain the differences between *nitrogen fixation* and *denitrification*.
4. What is the relationship between a *food chain* and a *food pyramid*?
5. The Linnean system of classification is described as being *generic, hierarchical, comprehensive,* and *binomial*. Explain the terminology in relation to the classification.
6. Explain the difference between *ecosystem* and *ecotone*. Between *biome* and *biomass*.
7. Explain how both *photosynthesis* and *photoperiodism* are dependent on sunlight.
8. What are the beneficial effects of wildfire?

Some Useful References

DARLINGTON, P. J., *Zoogeography*. New York: John Wiley & Sons, Inc., 1957.

HENRY, S. M., ed., *Symbiosis: Its Physiological and Biochemical Significance*. New York: Academic Press, Inc., 1966.

HUTCHINSON, G. E., *The Ecological Theater and the Evolutionary Play*. New Haven, Conn.: Yale University Press, 1965.

PHILLIPSON, J., *Ecological Energetics*. New York: St. Martin's Press, 1966.

Scientific American Editors, *The Biosphere: A Scientific American Book*. San Francisco: W. H. Freeman & Company, Publishers, 1970.

WHITTAKER, R. H., *Communities and Ecosystems*, 2nd ed. New York: Macmillan Publishing Co., Inc., 1975.

11 Terrestrial Flora and Fauna

The most basic studies of organisms made by geographers are usually concerned with an analysis of distributions. What is the range or habitat of certain species or groups of plants/animals? What are the reasons behind these distribution patterns? What is the significance of such distributions?

The Consideration of Natural Distributions

At the most fundamental level, the natural distribution of any species or group of organisms is determined by the balance between four conditions: evolutionary development, migration, reproductive success, and extirpation.

Evolutionary Development

All species have developed through the vagaries of the evolutionary process. The Darwinian theory of *natural selection*, sometimes referred to as "the survival of the fittest," explains the origin of any species as a normal process of descent, with variation, from parent forms. The progeny best adapted to survive in the struggle for existence will survive, whereas those less well adapted will perish. This long, slow, and essentially endless process accounts for the development of all organisms.

Thus in our search for understanding the distribution of any species, or group of species, we begin with a consideration of where it first evolved. Each originated somewhere. In some cases, there was a very localized beginning; in others, similar evolutionary development took place at several scattered localities. Unfortunately, the evidence is not always clear in this regard.

As an example of extremes, we might consider the contrast in apparent origin of two important groups of plants—acacias and eucalypts. See Figures 11-1 and 11-2. Acacias comprise an extensive genus of shrubs and low-growing trees that is represented by numerous species in relatively low-latitude portions of every continent that extends into the tropics or subtropics. Eucalypts, on the other hand, are a genus of trees native only to Australia and a few adjacent islands. Acacias apparently evolved in a number of different localities, but all evidence points to the origination of eucalypts on only a single small continent.

Consequently, the first piece in our puzzle of trying to understand the present distribution of organisms is to determine where the species originally evolved.

Migration

Throughout the millennia of the Earth's history, there has always been significant movement of organisms from one place to another. Animals possess active mechanisms for locomotion—legs, wings, fins, etc.—and their possibilities for migration are obvious. Plants are also mobile, however. Although most individual plants become rooted, and therefore fixed in location for most of their life, there is much opportunity for passive migration, particularly in the seed stage. Wind, water, and animals are the principal natural mechanisms for the dispersal of seeds.

The contemporary distribution pattern of many organisms, then, is often the result of natural migration or dispersal from an original center(s) of development. Among thousands of examples that could be used to illustrate this process are the following:

1. The cattle egret (*Bubulcus ibis*) apparently originated in southern Asia. During the last few centuries, it has spread from that continent into other warm areas of the world, particularly to Africa. In recent decades its range has expanded more dramatically. At least as early as the nineteenth century, some cattle egrets crossed the Atlantic Ocean from West Africa to Brazil, but they were unable to find suitable ecological conditions and thus did not become established. The twentieth century introduction of extensive cattle raising in tropical South America, however, apparently provided the missing ingredient, and egrets quickly adapted to the newly suitable habitat. See Figure 11-3. Their descendants spread northward throughout the subtropics and are now common inhabitants of the Gulf coastal plain in southeastern United States, are well established in California, and even occur in southern Canada. Also within this century, cattle egrets have dispersed at the other end of their "normal" range to enter northern Australia.

2. The coconut palm (*Cocos nucifera*) is believed to have originated in southeastern Asia and adjacent Melanesian islands. It is now, however, extraordinarily widespread along the coastlines of tropical continents and islands, worldwide. Most of this dispersal apparently has come about because coconuts, the large hard-shelled seeds of the plant, can float in the ocean for months or years without losing their fertility. Thus they have been washed up on beaches throughout the world, where they can colonize successfully if environmental conditions are right. This natural dispersion was significantly augmented by human help in the Atlantic region, particularly by the deliberate transport of coconuts from the Indian and Pacific ocean areas to the West Indies.

Thus the second piece in our contemporary distribution puzzle is to unwind the patterns of migration or dispersal.

Figure 11-1 A eucalyptus forest scene in eastern New South Wales (Australia), near Eden. [TLM photo.]

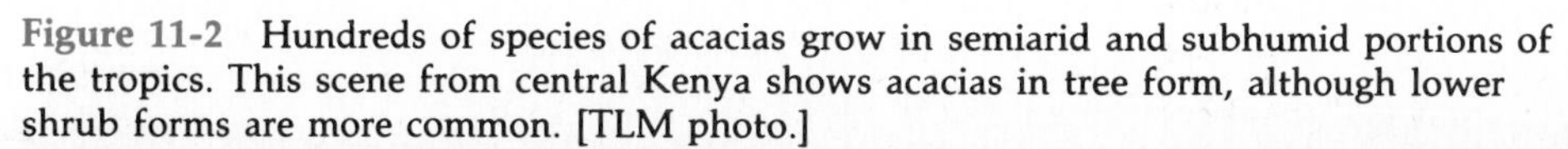

Figure 11-2 Hundreds of species of acacias grow in semiarid and subhumid portions of the tropics. This scene from central Kenya shows acacias in tree form, although lower shrub forms are more common. [TLM photo.]

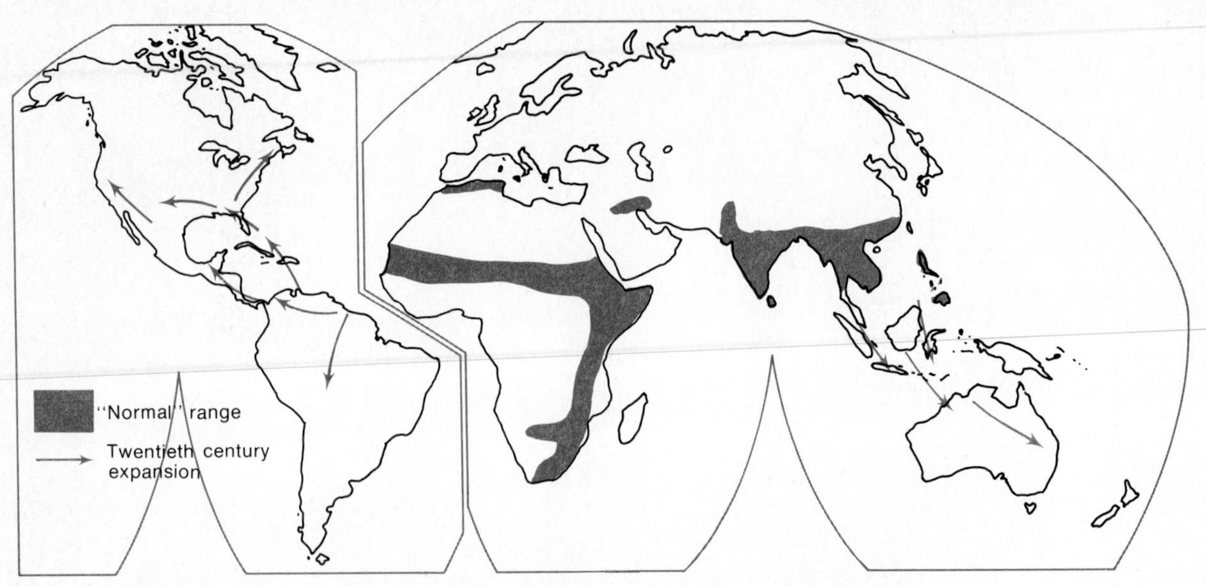

Figure 11-3 Natural dispersal of an organism. During the twentieth century, cattle egrets expanded from their African/Asian range into new habitats in the Americas and Australia.

Reproductive Success

A key factor in the sustenance of any biotic population is its reproductive success. A breeding group of any quantity, from a single pair to a species of worldwide extent, is always in a reproductive contest for continuing existence with other breeding populations of the same and/or different species. Poor reproductive success can come about for a number of reasons—heavy predation, climatic change, failure of food supply, etc. Changing environmental conditions are also likely to "favor" one group over another. Thus, reproductive success is usually the limiting factor that allows one competing population to flourish while another languishes, and comprises the third piece of the puzzle.

Extirpation

As migration expands the range of a species, so range also can be diminished by the dying out of some or all of the population of the organism. The history of the biosphere is replete with examples of such range diminution, varying from minor adjustments in a small area to extinction over the entire planet. The point is that evolution is a continuing process; no species is likely to be a permanent inhabitant of the Earth, and during the period of its ascendency, there is apt to be a great deal of distributional variation within a species, part of which is caused by local extinctions.

One of the simplest and most localized examples of this process is represented by *plant succession*, in which one type of vegetation is replaced naturally by another. Plant succession is a normal occurrence in a host of situations; a very common one involves the infilling of a lake. See Figure 11-4. As the lake gradually fills with sediments,

Figure 11-4 A simple case of plant succession, as illustrated by the infilling of a small lake. Successional colonization by different plant associations involves a change from lake to marsh to meadow to forest.

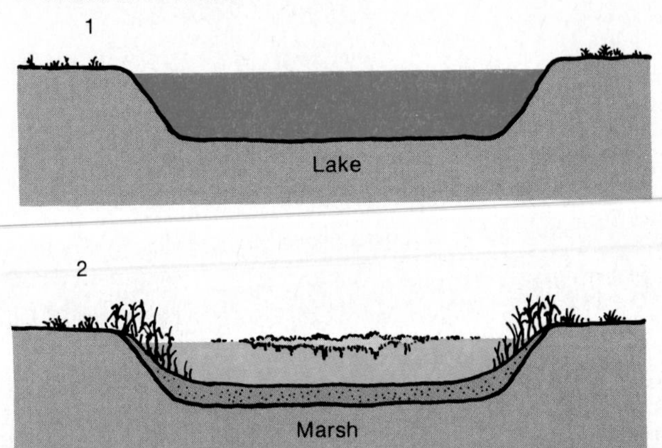

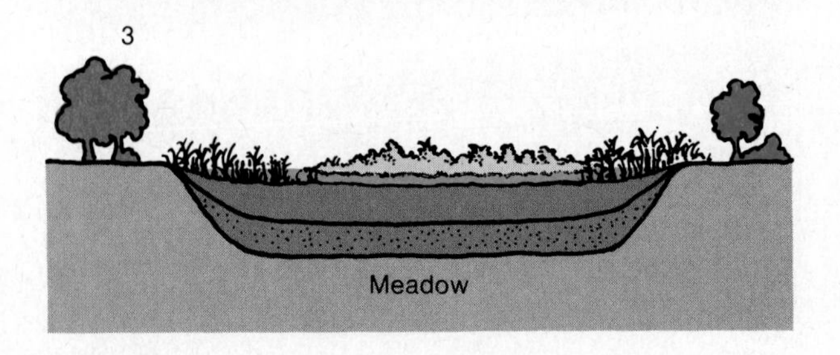

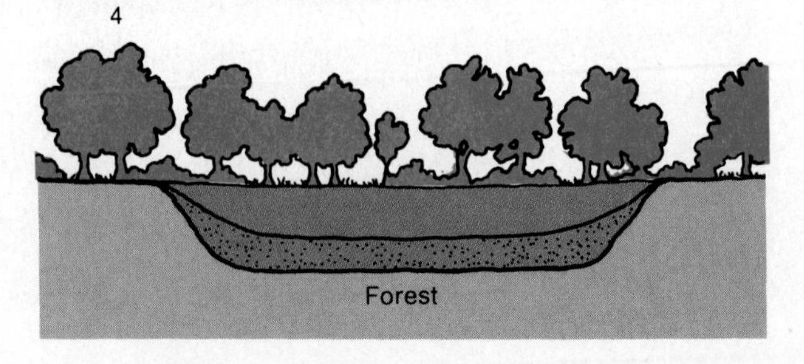

the aquatic plants of the lake bottom are slowly choked out, at the same time that the sedges and reeds and mosses of the shallow waters of the lake edge become more numerous and extensive. Continued infilling of sediments and organic debris further diminishes the aquatic habitat and allows for the increasing encroachment of low-growing land plants such as grasses and shrubs. As the process continues, trees move in to colonize the site, completing the transition from lake to marsh to meadow to forest. This simple sequence, which may require hundreds or even thousands of years, involves a series of local plant extirpations that is the inevitable result of plant succession, wherever it may occur. A similar series of local animal extirpations would accompany this plant succession because of the significant habitat changes; thus lake animals are replaced by marsh animals, which in turn are replaced by meadow and forest animals.

The biotic changes encompassed within plant succession are not to be confused with actual extinction. Extinction is permanent, but species succession is not. See Figure 11-5. Although a particular plant species may not be growing at a given time in a given situation, it may reappear quickly if environmental conditions change and if there is an available seed source (either from seeds that have lain dormant for a long period in that locale or from adjacent areas from which dispersal can take place).

At the other end of the scale is the extreme case of extinction in which a species, or group of species, is extirpated over the entire world, thus being eliminated forever from the landscape. This has taken place many times in the history of life on Earth, but the most dramatic example is probably the disappearance of the dinosaurs. For many millions of years those gigantic reptiles were the dominant life forms of our planet, yet in a relatively short period of geologic time they were all wiped out. The reasons behind their extermination are imperfectly understood, but the fact remains that there have been many such natural extirpations of entire species in the history of the world.

The disappearance of a species, then, can be a very localized event, or it can take place on a worldwide scale. Regardless of its magnitude, however, extinction is the fourth significant factor in the search to understand the puzzle of contemporary distribution patterns of plants and animals.

Figure 11-5 Plant succession is a continual fact of life in the biosphere. This scene from northern Idaho is representative. Nearly a century ago this area was devastated by a forest fire. The taller trees are tamaracks, which have become established as a subclimax association. Beneath them grow cedars, which eventually will over-top them and shade them out, becoming the climax vegetation of this area. [TLM photo.]

Terrestrial Flora

The natural vegetation of the land surfaces of the Earth is of particular interest to the geographer for three reasons:

1. Over much of the Earth, the terrestrial flora is the most significant visual component of the landscape. Plants often grow in such profusion that they hide or mask all other elements of the environment. Topography, soils, animal life, and even water surfaces are often obscured or obliterated by their presence. Only in areas of rugged terrain, unusually harsh climate, or significant human activities are plants not likely to dominate the landscape.
2. Vegetation is a sensitive indicator of other environmental attributes. Floristic characteristics typically reflect subtle variations in sunlight, temperature, precipitation, evaporation, drainage, slope, soil conditions, and other natural parameters. Moreover, the influence of vegetation on soil, animal life, and microclimatic characteristics is frequently pervasive.
3. Vegetation often has a prominent and tangible influence on human settlement and activities. In some cases, it is a barrier or hindrance to human endeavor; in other instances, it provides an important resource to be exploited or developed.

Characteristics of Plants

Although it might seem simple to distinguish between plants and animals, such is not always the case, particularly where simple organisms are concerned. Nevertheless, certain distinctions between flora and fauna generally pertain. For example, the ability to ingest solid food into its body through a mouth and then discharge waste products after digestion is typical of most animals but not of plants. Furthermore, most animals are capable of self-locomotion during all or part of their lives, whereas plants cannot do this. On the other hand, most plants contain green coloring matter (chlorophyll), which can use solar energy in food production, but this is rarely found in animals.

Despite the seemingly fragile appearance of many plants, most varieties are remarkably hardy. Although greatly exposed to the elements and without any ability to seek shelter, plants are capable of surviving the harshest of environmental circumstances. They survive, and often flourish, in the wettest, driest, hottest, coldest, and windiest places on Earth. Much of their survival potential is based on a subsurface root system that is capable of sustaining life despite whatever may happen to the above-surface portion of the organism. The survival capability of a species is also dependent in part on its reproductive mechanism. Plants that endure seasonal climatic fluctuations from year to year are called *perennials*, whereas those that perish during times of climatic stress but leave behind a reservoir of seeds to germinate during the next favorable period are called *annuals*.

Plant life varies remarkably in form, from microscopic algae to gigantic trees. Most plants, however, have common characteristic features—roots to gather nutrients and moisture, stems and/or branches for support and for transportation of the nutrients, and leaves to absorb and convert solar energy for sustenance. See Figure 11-6.

Environmental Adaptations

Despite the relative hardiness of most plants, there are definite tolerance limits that govern their survival, distribution, and dispersal. During hundreds of millions of years of development, plants have evolved a variety of protective mechanisms to shield against harsh environmental conditions and to enlarge those tolerance limits. The mechanisms are largely physiological, involving changes in the structure of plants in order to increase their survival capabilities under conditions of environmental stress. Some of the most prominent of these evolutionary adaptations are discussed below.

Xerophytic Adaptations The descriptive term that is applied to plants that are structurally adapted to with-

Figure 11-6 Characteristic plant features are roots, stems, branches, and leaves.

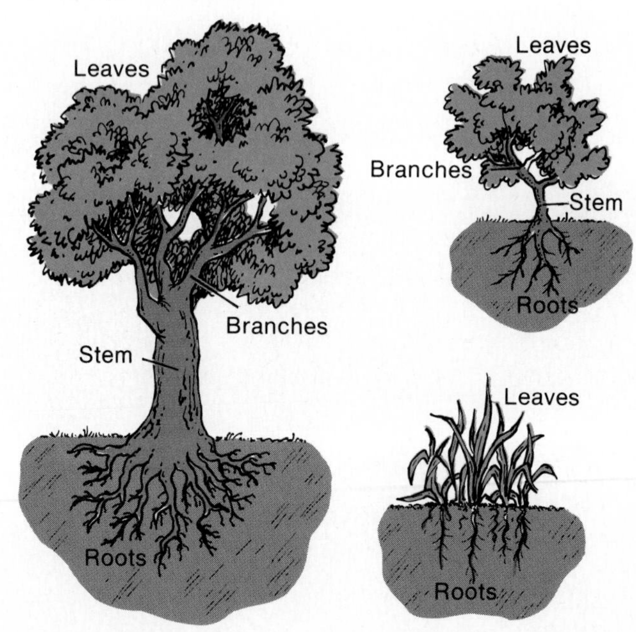

stand protracted dry conditions is *xerophytic*. Xerophytic adaptations can be grouped into four general types:

1. Roots are modified in shape or size to enable them to seek widely for moisture. Sometimes tap roots are extended to extraordinary lengths to reach subterranean moisture. On the other hand, root modification may involve the growth of a large number of thin hairlike rootlets to penetrate the tiny pore spaces in the soil.
2. Stems are sometimes modified into fleshy, spongy structures that can store moisture. Plants with such fleshy stems are called *succulents*; most cacti are prominent examples.
3. Leaf modification takes many forms, all of which are designed to decrease the amount of water that is transpired to the atmosphere from the leaf surface. This is sometimes accomplished by developing a leaf surface that is hard and waxy to inhibit water loss, or white and shiny to reflect insolation and thus reduce evaporation. Still more effective is for the plant to have either tiny leaves or no leaves at all. In many types of dryland shrubs, for example, leaves have been replaced by thorns, from which there is virtually no transpiration.
4. Perhaps the most remarkable floristic adaptation to aridity is not structural but involves the plant's reproductive cycle. Many desert plants are capable of lying dormant for years at a time without perishing. When rain eventually arrives, they can promptly initiate and pass through an entire annual cycle of germination, flowering, fruiting, and seed dispersal in only a few days, then lapsing into dormancy again if the drought resumes.

Hygrophytic Adaptations A surplus of moisture is not nearly as disastrous to most plants as a moisture deficiency. Some plants, however, are particularly suited to growing in a wet terrestrial environment, such as a marsh, swamp, bog, or shallow pond. Such vegetation is said to be *hygrophytic* or *hydrophytic* (synonymous terms). The evolutionary adaptations of hygrophytes are much less specialized than those of xerophytes. Hygrophytes are likely to have extensive root systems to anchor them in the soft ground, and hygrophytic trees often develop a widened, flaring trunk near the ground to provide better support. Many hygrophytic shrubs and lesser plants that grow in standing or moving water have weak, pliable stems that can withstand the ebb and flow of currents rather than standing erect against them; the buoyancy of the water, rather than the stem, provides support for the plant.

Other Kinds of Adaptations Various kinds of plants have evolved mechanisms to accommodate many other aspects of environmental stress. Some are adjusted to long periods of high temperatures or protracted intervals of cold weather. Some are adapted to unusual concentrations of minerals in the soil, particularly salts. Some are adapted to fire, actually being able to germinate only when great heat causes their seedpods to open. There are many other examples of specialized evolutionary adaptations that could be cited.

The Critical Role of Competition

As important as climatic, edaphic, and other environmental characteristics may be, it is necessary to realize that a particular vegetative type or association will not necessarily occupy an area due to these conditions alone. Plants are just as competitive with one another as are animals or used-car salesmen. Of the dozens of species that might be suitable for an area, only one or a few are likely to survive. This is not to say that all plants are mutually competitive; indeed, thousands of ecologic niches can be occupied, and many of these in no way impinge on one another. However, the basic fact is that most plants draw their nutrients from the same soil and their energy from the same sun, and what one plant obtains cannot be used by another.

Climate and the other environmental elements are essentially permissive factors; competition will decide which species or communities will actually flourish. The taller plants usually have an advantage—at least eventually—in the competition if for no other reason than that they can overshadow their competitors. Almost all plants require sunlight to survive, and a taller plant that shades out a shorter one will win the competition. Largely for this reason, forests dominate the flora wherever the environment is not too restrictive. Shrubs and herbaceous plants can be the dominant plant cover only on a temporary basis or in areas where they are not shaded out by trees.

Floristic Terminology

Certain details of natural vegetation terminology should be summarized prior to a consideration of floristic associations.

Taxonomic Summation An important biological distinction can be made between "higher" plants and "lower" plants. Lower plants include the following major groups:

1. *Bryophytes* are mosses and liverworts. Presumably they have never in geologic history been very important among plant communities, except in localized situations.
2. *Pteridophytes* are ferns, horsetails, and club mosses, which are spore-bearing plants. During much of geologic history, great forests of tree ferns, giant horsetails, and tall club mosses dominated continental vegetation. They are less important today.

The so-called higher plants reproduce by means of seeds and are encompassed within two broad groups:

1. *The gymnosperms* ("naked seeds") carry their seeds in cones, and when the cones open, the seeds fall out. This is the more primitive of the two groups. This group was more important in the geologic past; today the only large surviving gymnosperms are cone-bearing trees such as pines. See Figure 11-7.
2. *Angiosperms* ("vessel seeds") have seeds that are encased in some sort of a protective body, such as a fruit, a nut, or a seed pod. Most higher plants—including trees, shrubs, grasses, crops, weeds, garden flowers—are angiosperms. Along with a few conifers, they have dominated the vegetation of the planet for the last 50 or 60 million years.

Descriptive Terms Several other terms are commonly used to describe types or associations of vegetation. Their definitions are not always precise, but their meanings generally are clear. Important terms and term distinctions include the following:

1. The fundamental distinction among higher plants is often on the basis of the composition of the stem or trunk. *Woody plants* have stems composed of hard fibrous material, whereas *herbaceous* plants have soft stems. Woody plants are mostly trees and shrubs; herbaceous plants are mostly grasses, forbs, and lichens.
2. The presence or absence of a seasonal leaf fall is an important distinguishing characteristic for trees. An *evergreen* tree is one that sheds its leaves on a sporadic or successive basis, but at any given time appears to be fully leaved; in other words, it has no apparent seasonal variation in its complement of leaves. A *deciduous* tree, on the contrary, is one that experiences an annual period in which all leaves die and usually fall from the tree, due either to a cold season or a dry season.
3. Trees are also often described in terms of their leaf shapes. *Broadleaf* trees have leaves that are flat and expansive in shape, whereas *needle-leaf* trees are adorned with thin slivers of tough, leathery, waxy needles rather than typical leaves. Almost all needle-leaf trees are evergreen, and the great majority of all broadleaf trees are deciduous, except in the rainy tropics where everything is evergreen.
4. *Hardwood* and *softwood* are two of the most unsatisfactory terms in biogeography, but they are widely used in everyday parlance, so we must not ignore them. Hardwoods are angiosperm trees that are usually broad-leaved and deciduous. Their wood has a relatively complicated structure, but it is not always hard. Softwoods are gymnosperms; nearly all are needle-leaved evergreens. Their wood has a simple cellular structure, but it is not always soft.

Figure 11-7 Although the terminology is somewhat confusing, there are conspicuous differences between hardwood (*angiosperm*) and softwood (*gymnosperm*) trees. The most obvious difference is in general appearance.

Spatial Groupings of Plants

As a geographer endeavors to comprehend the floristic characteristics of the environment, attention is sometimes directed to individual plants, but even more commonly to the spatial groupings of plants—whether in small communities, large associations, or extensive biomes. Over most land surfaces the natural vegetation occurs in considerable variety; but regardless of species diversity, the plant association usually can be described or classified on the basis of its dominant members or its dominant appearance, or both.

Critical Concepts Before considering the major vegetational associations, we should remember that the floristic pattern of the Earth is impermanent. The plant cover that exists at any given time and place may be in a state of constant change or may be relatively stable for millennia before experiencing significant changes. But, sooner or later, change is inevitable. Sometimes the change is slow and orderly, as when there is a long-term trend toward different climatic conditions over a broad area, or with

Focus on Biotic Rejuvenation of Krakatau

One of the most dramatic single natural events in recent history was the final explosive eruption of the volcano Krakatau. This mountain occupied the small island of the same name, located in the East Indies between the large islands of Java and Sumatra, about 25 miles (40 km) from each. After a long and erratic eruptive history, Krakatau finally blew its top in 1883, in a spectacular explosion that sent 6 cubic miles (38 km³) of rock into the air. The explosion created a vast ocean wave that killed 36,000 people on adjacent islands, produced volcanic dust that colored sunsets around the world for a year, and obliterated every trace of life on the remaining remnant of the island.

Our attention here, however, is not focussed on the apocalyptic explosion, but rather on the biotic aftermath. Prior to the explosion, the lower portions of the island had been covered with a tropical rainforest, and the upper slopes had supported a similarly dense forest of various mountain species, a biome identical to that on nearby Java and Sumatra. After the explosion, the sterilized island remnant was lifeless and covered with a thick coating of volcanic ash, a situation analogous to the primeval rock from which the Earth was formed in the dim geologic past. Nearby, however, were other islands that could provide ample biotic resources for the recolonization of devastated Krakatau. Scientists were thus provided with an unparalleled opportunity to witness the speedy repopulation of a landmass under ideal environmental conditions.

The first reinvaders were blue-green algae, which became established within the first few months after the totally inorganic surface had cooled. In short order, other forms of life began to appear—blown by the wind, floating on the ocean, carried by birds. Spiders, for example, were found on the island within the first year, and within three years at least two dozen species of herbaceous plants had become established in the newly developing soil.

In less than two decades, most of the lower portions of the island had developed a grass cover and had taken on the appearance of a savanna; the upper slopes were largely vegetated with ferns. Seeds of the wild sugar cane, carried in by the wind, had formed some extensive stands, and young coconut trees, nurtured from coconuts that had drifted in on the waves, dotted the seashore. Twenty-five years after the explosion, biologists were able to identify more than 260 species of animals, mostly insects, on the island.

As time went on, the relatively rapid pace of plant succession continued. Woody shrubs replaced the grasses, and they, in turn, gave way to trees. The first trees were drought-resistant varieties quite atypical of rainforest growth; but as their development continued, more and more of the surface of the island was shaded. Active weathering in this wet tropical climate, along with the accumulation of more and more organic matter on the surface, brought into existence a true soil, providing a more stable base for tree growth.

Eventually, nearly all of the surface of the island was shaded from the direct rays of the sun by the forest canopy. This provided a sustaining environment for the seeds of true rainforest trees that were brought in by wind or birds, allowing them to germinate, grow, and finally over-top and shade out many of the original trees.

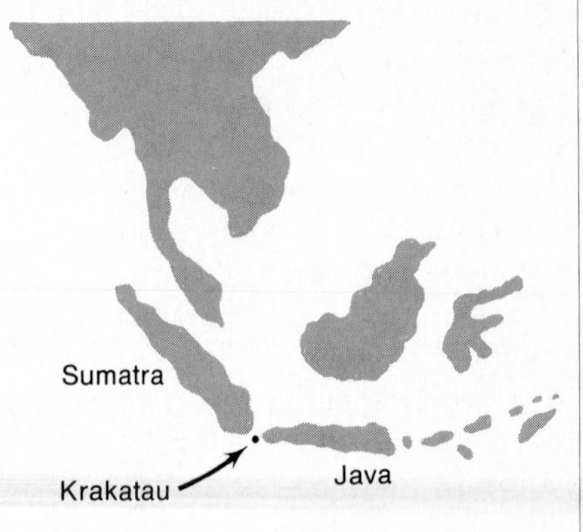

The location of Krakatau.

Now, just after the centennial anniversary of Krakatau's big bang, the floristic landscape of the island is visually very similar to its appearance before the explosion. A forest that is essentially a tropical rainforest covers most of the land, providing clear evidence that under optimum conditions, it is possible for vegetation succession to take place with considerable rapidity. However, the flora of Krakatau today is considerably poorer than it was in 1883 because many species that were there before the explosion have not yet found their way back from neighboring islands. The equilibrium condition of a climax vegetation is not yet reestablished. Such development can be anticipated, but at a slower pace than the major reinvasions of recent years.

Similarly, Krakatau's fauna has not yet reached its pre-explosion diversity. More than 1,300 species of animals have been identified, including a variety of reptiles, birds, and bats; however, several hundred more species of animals were present in the years preceding the catastrophe. The passage of time undoubtedly will find many of them reestablished.

---

the simple plant succession that accompanies the natural infilling of a marsh. Upon occasion, however, the change may be abrupt and chaotic, as in the case of a wildfire.

The idea of continual change in the floristic cover of the land is particularly well displayed in the concept of plant succession, which was discussed briefly in Chapter 10. Succession implies a sequential change in the vegetation, which nearly always proceeds from simpler to more complex plant communities. Accompanying the floral change, inevitably, are changes in other aspects of the local ecosystem—fauna, soils, microclimate, water balance, etc. Plant succession is not a random process, although it may follow a great many different avenues and may involve an extraordinary diversity of species.

At some time in the Earth's history, all parts of the present landmasses were newly "created," and therefore unvegetated. When such new land is first available for terrestrial vegetation, it is occupied temporarily by some plant community, which will give way to another and then another and another. Many complicated changes can be expected until some sort of floristic stability is attained. Each succeeding community alters the local environment, particularly the microclimate and the surface soil, making possible the establishment of the next community. The general sequential trend is toward taller plants, more complex patterns, and greater stability in species composition. The longer that plant succession continues, the more slowly does change take place, because the more advanced communities usually contain species of greater longevity.

These successional changes may be spread over a period of centuries—sometimes many centuries—until eventually a plant association of constant composition comes into being. This stable association, developing at the end of a long succession of changes, is generally referred to as the *climax vegetation*. The implication of this term is that the dominant plants of a climax community have demonstrated that, of all possibilities for that particular situation, they can compete the most successfully. Thus they represent the optimal floristic cover for that environmental context. The climax vegetation presumably will persist, unchanged, for an indefinite period until the next environmental disturbance. The climax vegetation is, then, a community in equilibrium with prevailing environmental conditions. When these conditions change, the climax stage will be disturbed and a renewed successional sequence will be initiated.

Identification and Mapping of Groupings The geographer's attempt to recognize spatial associations of plants faces some significant difficulties. Plant communities that are similar in appearance and in environmental relationships can occur in widely separated localities and are likely to contain totally different species compositions. At the other extreme, exceedingly different plant communities can often be detected within a very small area. As the geographer tries to identify patterns and recognize relationships, generalization invariably is needed. This generalization must accommodate gradations, transition zones (ecotones), interdigitations, and other kinds of irregularities in the extent of plant associations. When associations are portrayed on maps, their boundaries in nearly all cases represent approximations.

Another special problem facing the student who would learn about the Earth's environment is that in many areas the natural vegetation has been completely removed or replaced through human interference. No other element of the physical landscape has been so extensively metamorphosed by mankind as the Earth's flora. Many of the human-induced changes are obvious—forests have been cut, crops have been planted, pastures have been seeded, urban areas have been paved—but others are much more subtle. Over extensive areas of the Earth's surface the existence of a climax vegetation is the exception rather than the rule. Most world maps that purport to show natural vegetation ignore the factor of human interference and are actually maps of "theoretical" or "potential" natural vegetation, in which the map-maker provides assumptions concerning what the natural vegetation would be if it had not been modified by human activity. Thus maps of natural vegetation at a world scale, as presented in this book, must be used with care. The concepts of climax vegetation and of major plant associations are very useful, but they have their limitations.

The Major Floristic Associations There are many ways to classify and identify plant associations. For broad geographical purposes, emphasis is usually placed on the structure and appearance of the dominant plants. The major associations that are generally recognized include the following:

1. *Forests* consist of an assemblage of trees growing closely together so that their individual leaf canopies generally overlap. See Figure 11-8. This means that the ground is largely in shade, which usually precludes the development of much undergrowth. Forests require considerable annual precipitation and can survive in widely varying temperature regimes. Except where moisture is inadequate or the growing season very short, forests are likely to become the climax vegetation association.
2. *Woodlands* are tree-dominated plant associations in which the trees are spaced more widely apart and do not have interlacing canopies. Undergrowth may be dense or sparse, but it is not inhibited by lack of sunlight. Woodlands usually are found on the dry side of forests.
3. *Shrublands* are plant associations that are dominated by relatively short woody plants, generally called *shrubs* or *bushes*. Shrubs take a variety of forms, but most have several stems branching near the ground and a leafy foliage that begins very close to ground level. Trees and grasses may be interspersed with the shrubs but are less prominent in the landscape. Shrub lands have a wide latitudinal range, but they are generally restricted to semiarid or arid locales.
4. *Grasslands* may contain scattered trees and shrubs, but the landscape is dominated by grasses and forbs (broadleaf herbaceous plants). Prominent types of grassland include *savanna* (low-latitude grassland characterized by tall forms), *prairie* (tall grassland in the midlatitudes), and *steppe* (short grasses and bunchgrasses of the midlatitudes). Grasslands are associated with semiarid and subhumid climates.
5. *Deserts* are typified by widely scattered plants, with much bare ground interspersed. See Figure 11-9. *Desert*

Figure 11-8 A dense coniferous forest on Vancouver Island, British Columbia. The stream is the Somass River. [TLM photo.]

Figure 11-9 The desert of southern Arizona has much bare ground surface, but there is also a considerable growth of xerophytic vegetation, of which various species of cactus (such as prickly pear in lower right, ocotillo in lower left, and giant saguaro) are particularly prominent. [TLM photo.]

is actually a climatic term, and desert areas may have a great variety of vegetation, including grasses, succulent herbs, shrubs, and straggly trees. Some extensive desert areas are comprised of loose sand, bare rock, or extensive gravel, with virtually no plant growth.

6. *Tundra* consists of a complex mix of very low-growing plants, including grasses, forbs, dwarf shrubs, mosses, and lichens, but no trees. Tundra occurs only in the perenially cold climates of high latitudes or high altitudes.
7. *Wetlands* have a much more limited extent than the associations described above. See Figure 11-10. They are characterized by shallow standing water all or most of the year, but with vegetation rising above the water level. The most widely distributed wetlands are *swamps* (with trees as the dominant plant forms) and *marshes* (with grasses and other herbaceous plants dominant).

Vertical Zonation The mountainous areas of the world usually have a very distinct pattern of vertical zonation in vegetation patterns. See Figure 11-11. Significant altitudinal changes in short horizontal distances cause various plant associations to exist in relatively narrow bands or zones on the mountain slopes. This altitudinal zonation is largely due to the effects of altitude on the climatic elements of temperature and precipitation. The essential implication is that altitude compensates for latitude; i.e., to travel from sea level to the top of a tall tropical peak is roughly equivalent environmentally to a horizontal journey from the equator to the shores of the Arctic Ocean. This relationship of altitude and latitude is shown most clearly by the variation in elevation of the upper treeline. For example, trees cease to grow at an elevation of about 16,000 feet (5000 m) in the equatorial Andes of South America; at 20° north latitude in central Mexico, the treeline is at about 14,000 feet (4250 m); at 40° north latitude in Colorado, it is at about 11,000 feet (3350 m); at 60° north latitude in western Canada, it occurs at about 2500 feet (900 m); and at 70° north latitude in northern Canada, the treeline is at sea level.

Treeline variation represents only one facet of the broader design of vertical zonation in vegetation patterns. See Figure 11-12. All vegetation zones are displaced downward with increasing distance from the equator. This principle accounts for significant vegetational complexity in all mountainous areas.

Local Variations In the preceding sections we have been concerned largely with a consideration of extensive vegetation associations. But there are also significant local variations from the general patterns in response to a wide variety of local environmental conditions. Most such variations are limited in extent. A few, however, are broader and more predictable in application, such as the two examples cited below:

1. *Exposure*. The orientation of a particular slope to weather elements is often a critical determinant of vegetation composition. Exposure has many aspects, but one of the most pervasive is simply the angle at which sunlight strikes the slope. If the sun's rays arrive at a relatively direct angle, they are much more effective

Figure 11-10 Wetlands are sometimes tree-covered, as in this Louisiana swamp, and sometimes dominated by lesser vegetation, as in this Vermont marsh that is dotted with muskrat houses. [USDA-SCS photos.]

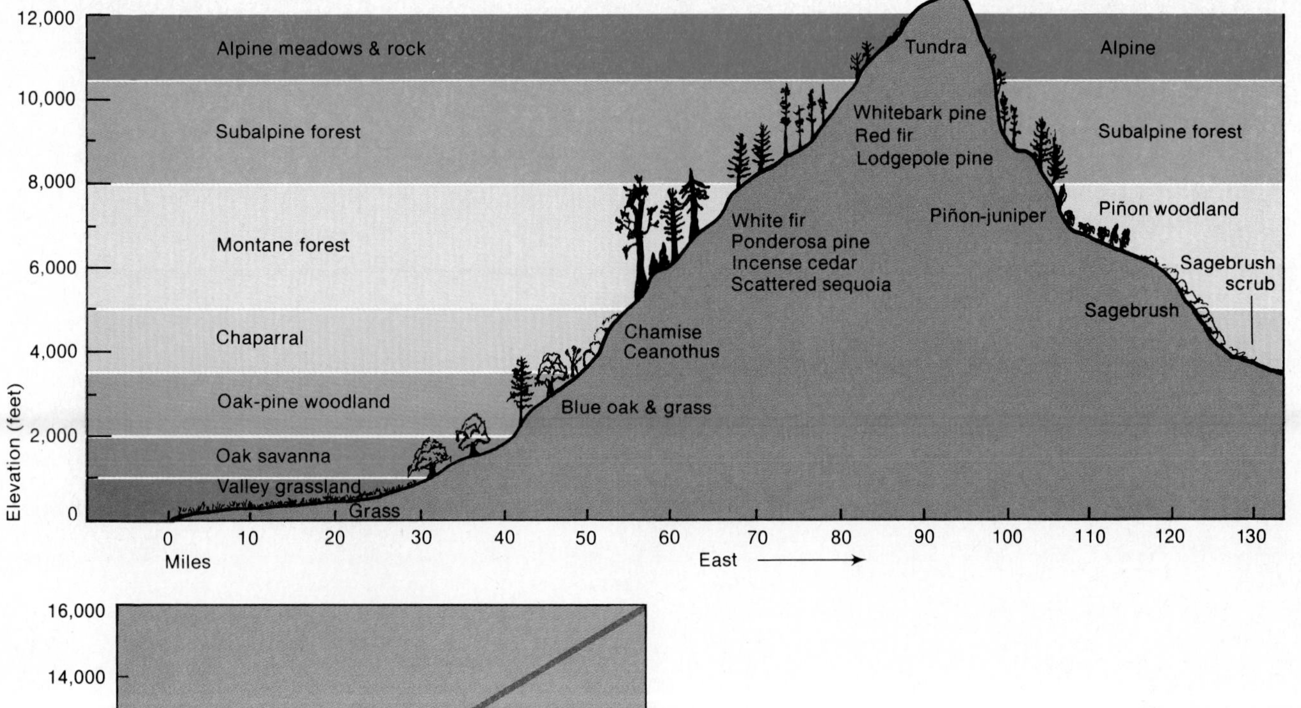

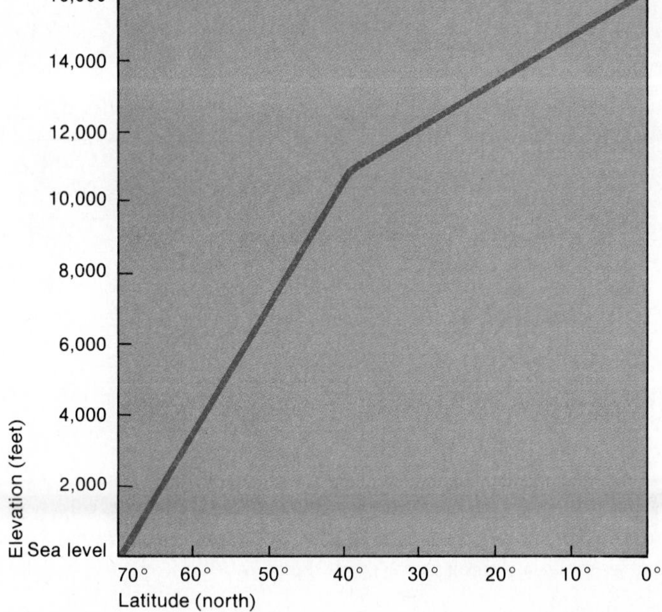

Figure 11-11 Two illustrations of the vertical zonation concept. The first shows a west-east profile of California's Sierra Nevada Range, indicating the principal vegetation associations that occur at different altitudes on the western (wet) and eastern (dry) side of the mountains. The graph shows the approximate variation of the tree-line altitude by degrees of latitude in the Northern Hemisphere.

in heating the ground and thus in evaporating available moisture. Such a sun slope (called *adret*) is relatively hot and dry, and its vegetation will not only be sparser and smaller but is likely to have a different species composition from adjacent slopes with different exposure. The opposite condition is a *ubac* slope, which is oriented so that sunlight strikes it at a low angle and is thus much less effective in heating and evaporating. This produces more luxuriant vegetation of a richer diversity.

2. *Valley-Bottom Location*. Most major vegetation associations reveal a significantly different floral composition in the bottomlands of the principal valleys, sometimes restricted to immediate stream-side locations and sometimes extending more broadly over the valley floor. The difference is primarily a reflection of the perennial availability of subsurface moisture near the stream and is manifested in a more diversified and more luxuriant flora. This type of vegetation is particularly prominent in relatively dry regions, where stream courses may be lined with trees even though no other trees are to be found in the landscape. Such anomalous stream-side growth is called *riparian* vegetation. See Figure 11-13.

Figure 11-12 Even from a distance, the "banding" of vegetation along the mountainside is obvious. From the floor of the San Luis Valley (near Alamosa, Colorado) to the top of Sierra Blanca in the Sangre de Cristo Mountains, at least five vertical zones of plant growth are easily recognizable. [TLM photo.]

Figure 11-13 Riparian vegetation is particularly prominent in deserts. This stream course in northwestern New South Wales (Australia) would be inconspicuous if it weren't for the trees growing along it. [TLM photo.]

Terrestrial Fauna

Skip 215-221

Animals occur in much greater diversity than plants over the Earth. As objects of geographical study, however, they have been relegated to a place of lesser importance for at least two reasons:

1. Animals are much less prominent in the landscape. Apart from extremely localized situations—such as waterfowl flocking on a lake, an insect plague attacking a crop, caribou migrating across the tundra, or a school of fish in a feeding frenzy—animals tend to be secretive and inconspicuous, whereas the vegetation is not only fixed in position but also serves as a relatively complete ground cover wherever it has not been removed by human interference.
2. Environmental interrelationships are much less clearly evidenced by animals than by plants. This is due in part to the inconspicuousness of wildlife, which renders it more difficult to study; but it is also due to the fact that animals are mobile and therefore more able to adjust to environmental variability.

This is not to say that fauna is inconsequential for students of geography. Under certain circumstances, wildlife is a prominent element of physical geography; and in some regions of the world, it is an important resource for human use and/or a significant hindrance to human activity. Moreover, it is increasingly clear that animals are sometimes more sensitive indicators than plants of the health of a particular ecosystem.

Characteristics of Animals

The diversity of forms of animal life is not realized by most people. We commonly think of animals as being relatively large and conspicuous creatures that run across the land or scurry through the trees, seeking to avoid contact with mankind. In actuality, the term *animal* encompasses not only the larger, more complex forms but also hundreds of thousands of species of smaller and simpler organisms that may be inconspicuous or even invisible.

The variety of animal life is so great that it is difficult to find many unifying characteristics. The contrast, for example, between an enormous elephant and a microscopic protozoan is so extreme as to make their kinship appear ludicrous. Animals really have only two universal characteristics, and even these are so highly modified in some cases as to be almost unrecognizable:

1. Animals are *motile*, which means that they are capable of spontaneous or self-generated movement.
2. Animals are largely dependent for sustenance upon eating plants or other animals; they are incapable of manufacturing their food from air, water, and sunlight, which plants can do.

Environmental Adaptations

As with plants, animals have evolved slowly and diversely through eons of time. The process of evolution has made it possible for animals to diverge remarkably in making adjustments to differing environments. Just about every existing environmental extreme has been met by some (or many) evolutionary adaptations that make it feasible for some (or many) species of animals to survive and even flourish.

Physiological Adaptations The majority of animal adaptations to environmental diversity have been physiological in nature, which is to say that these adaptations have involved anatomical and/or metabolic changes in the organism. A classic example is the size of fox ears. Ears are prime conduits for body heat in furred animals, as they provide a relatively bare surface for its loss. Arctic foxes have unusually small ears, which minimizes heat loss; desert foxes, on the other hand, possess remarkably large ears, which are a great advantage during the blistering heat of desert summers. The catalogue of similar adaptations is almost endless: webbing between toes to make swimming easier, dense fur to keep out the cold, broad feet that won't sink in soft snow, increase in size and number of sweat glands to aid in evaporative cooling, and a host of others.

Behavioral Adaptations An important advantage that animals have over plants, in terms of adjustment to environmental stress, is that the former can move about and therefore can modify their behavior in order to minimize the stressful conditions. Animals can seek shelter from heat or cold or flood or fire; they can travel relatively long distances in search of relief from drought or famine; they can shift from daytime (diurnal) to nighttime (nocturnal) activities to minimize water loss during hot seasons; and some can go into a state of dormancy to avoid periods of inclement weather. Such techniques as *migration* (periodic movement from one region to another), *hibernation* (spending winter in a dormant condition), see Figure 11-14, and *estivation* (spending a dry/hot period in a torpid state) are behavioral adaptations that are employed regularly by many species of animals.

Reproductive Adaptations Harsh environmental conditions are particularly destructive to the newly born. As partial compensation for this factor, many species have evolved specialized reproductive cycles or have developed modified techniques of baby care. During lengthy periods of bad weather, for example, some species delay mating or postpone nest building. If fertilization has already taken place, some animal reproductive cycles are capable of almost indefinite delay, resulting in a protracted egg or larval stage or even of total suspension of development

Figure 11-14 A hibernating (such as this chipmunk) or estivating animal can avoid stressful cold or dry seasons by remaining dormant and thus avoiding the need for food and water. [Leonard Lee Rue, III, Photo Researchers, Inc.]

The desert is a classic illustration of a stressful environment for biotic life. A pervasive scarcity of moisture in all forms is exacerbated by high temperatures and strong winds that accelerate water loss from organisms through evaporation. Deserts often appear to be lifeless lands, with only a scattering of sparse vegetation and no obvious evidence of animals. It is not hard to believe that any animal that can live in the desert must be equipped with one or more clever survival techniques, nor is it too far-fetched to imagine that desert animals might, indeed, be nonexistent.

The truth of the matter is that desert animals are usually inconspicuous, but they are far from being nonexistent. During daylight, particularly in the hot season, desert fauna is likely to verge on total invisibility; thus most of the animals are more conspicuous during cooler periods, i.e., at night and in winter. In locations where water is permanent or prolonged in occurrence—water holes, oases, exotic streams—the faunal diversity may be astounding. And all species are favored when the rare rains arrive; reproduction is rapid for almost all forms of life.

The simplest and most obvious adaptation to life in an arid land is for the animals to remain near a perpetual source of water. Thus the permanent streams and enduring springs of the desert attract a resident faunal population that is often rich and diverse. Birds are more dependent on an open source of water than any other nonaquatic animals, so they often remain near such features throughout the day, or at least visit the water with some frequency. Many other kinds of animals also congregate near open water, though usually not as overtly as birds. Even in areas where open water is not available, it is rare for the desert to be uniformly inhospitable. There are usually some small pockets of localized favorable habitat that permit remnant populations of animals to survive until more propitious conditions allow them to recolonize large areas or to move on.

Often animals cannot remain near permanent water in a vast arid region, so the next best thing is to follow the rains in nomadic fashion. Many species demonstrate remarkable instincts for knowing where precipitation has occurred and will travel tens or even hundreds of miles to take advantage of locally improved conditions. This trait is again most prominently displayed by birds, which have greater powers of mobility than other animals. A high proportion of desert bird species can be classed as nomads, in contrast to the large proportion of sedentary and migratory species in other parts of the world. Some of the larger mammals also have rain-following habits. Some African antelopes and Australian kangaroos have been observed to double in numbers almost overnight after a good rain, a feat impossible except by rapid, large-scale movement from one area to another.

Apart from mobility, a number of other behavioral techniques have been adopted by various kinds of animals to ensure survival in the desert. A significant method used by smaller creatures is to spend a great deal of time underground, beneath the level of dessicating heat. Most desert rodents and reptiles live in underground burrows, and many lesser creatures do the same. Desert frogs, while not numerous, are noted burrowers; some have been found as deep as 6 feet (2 meters) underground. Freshwater crayfish and crabs often survive long dry spells by burying themselves. Most desert ants and termites live in underground nests.

Two other dessication-resisting habits—estivation and nocturnality—are often, but not always, combined with burrowing. Desert frogs usually estivate in their burrows, and some add to their protection by wrapping themselves in cocoons. Many land mollusks are also estivators, although they are unable to burrow and spend their resting time in rock crevices or attached to roots. Estivation precludes the need for water, of course, and thus allows the estivator to exist for months or even years during a total drought.

The adoption of a nocturnal activity pattern is not necessarily a reflection of climatic stress. Many animals the world over are active at night rather than during daylight in order to evade predators, or to accommodate preferred feeding habits, or for some other reason. In arid regions, however, nocturnal habits provide the added advantage of reducing heat stress and water loss. Most desert animals except birds, ungulates, and many flying insects are almost completely nocturnal. There is clearly much more wildlife activity at night than in the daytime.

Various other specialized behaviors have been adopted by desert animals to cope with dessication. A prime example is the stance adopted by several dozens of species of lizards during the hottest part of the day. They characteristically face the sun with their bodies parallel to the sun's rays, thus exposing the bare minimum of body surface to the radiant heat.

Of all animals, however, the dromedary, or one-humped camel (*Camelus dromedarius*), has developed the most remarkable series of adjustments to the desert environment, allowing it to survive and prosper under conditions of aridity that would be fatal to almost all other large animals. The principal adaptations are outlined below:

1. *Anatomical adaptations*. The summer coat of the dromedary is relatively light-colored and shiny, reflecting solar radiation. Although the hair is short, it lies flat and serves as an effective barrier against heat gain from the environment. The upper lip of the dromedary is deeply cleft, with a groove that extends to this cleft from each nostril, presumably so that any moisture from the nostrils can be caught in the mouth rather than being wasted. The nostrils consist of horizontal slits that can be closed tightly to keep out blowing dust and sand. The eyes are set beneath shaggy, beetling brows, which help to shield them from the sun's glare; the eyes are further protected from blowing sand by a complex double set of eyelids. The feet are broad and distended into pads that provide insulation from hot ground, and their firm but elastic tissue gives excellent traction on either sand or stony pebbles.
2. *Physiological adaptations*. The heat balance of dromedaries is distinguished by a remarkable daily fluctuation of body temperature during hot weather.

Figure 11-2-1 Dromedaries are superbly adapted to desert living. [Courtesy John Isaac, United Nations.]

Mankind and most other large animals have body temperatures that fluctuate by only 2° to 4° F (1° to 2° C) on hot days, whereas the range for dromedaries is 12° F (7° C). This means that the dromedary sweats little except during the very hottest hours, while a human in the same environment perspires almost from sunrise to sunset, the resultant evaporative cooling irretrievably using up body water. Dromedaries are also extremely economical in water use through producing only a small volume of urine and voiding very little water in their feces. Moreover, they are able to maintain bloodstream moisture during extreme heat stress (by a mechanism not clearly understood) nearly three times more efficiently than any other large mammal, which means that they can tolerate extreme dehydration without their body temperature rising to a fatal level. The remarkable efficiency of their heat and water budgets allows them to go without drinking for long periods (although they cannot "store" water in their hump, as was once believed), and when they do begin to drink, they are capable of rapid and complete rehydration in only a short time.

Some smaller desert animals have even more miserly water budgets than dromedaries. Indeed, a few species of rodents, most notably kangaroo rats, can exist from birth to death without ever taking a drink, surviving exclusively on moisture ingested with their food.

Perhaps the most astounding of all faunal adaptations to arid conditions is one that is overtly very similar to a vegetational adaptation—the ability to delay reproductive processes over long dry periods until more favorable conditions occur, at which time rapid breeding or birth can allow remarkable population regeneration. Australian desert kangaroos, for example, are capable of "delayed implantation," in which a fertilized blastocyst can remain in an inactive state of development in the uterus during a period of difficult living, and then spontaneously resume normal development after conditions have improved. Desert birds may experience enormous die-offs when waters dry up during a prolonged drought, but when rains finally come, the survivors may begin nest construction within a week and ovulation within a fortnight. Moreover, clutch size (the number of eggs laid) may increase, as may the number of clutches produced.

Invertebrates, too, can take advantage of a favorable weather change to proliferate their numbers extraordinarily. Many of them survive the dry period in an egg or larval stage that is extremely resistant to dessication; when the drought breaks, the egg hatches or the larva develops into an active stage that may not be drought-resistant at all. Development continues with great rapidity until the adult form can breed and more drought-resistant eggs can be laid. This is a commonplace circumstance for brine shrimp, crayfish, grasshoppers, locusts, flies, mosquitos, and various other desert arthropods; cases have been known in which more than a quarter of a century passed between successive generations of these tiny creatures in which the adult life cycle lasts for only a few weeks!

of the blastocyst (the early stage of the embryo) until the weather improves. If the young have already been born, they sometimes remain longer than usual in nest or den or pouch, and there may be a longer period of feeding of the young by the adults. When good weather finally returns, some species are capable of hastened estrus (the period of heightened sexual receptivity by the female), nest building, den preparation, etc., and the progeny produced may be in greater than normal numbers.

Competition Among Animals

Competition among animals is even more intense than that among plants because it involves not only indirect rivalry for space and resources but also the direct antagonism of predation. See Figure 11-15. Animals with similar dietary habits compete for food and occasionally for territory. Animals in the same area also sometimes compete for water. And animals of the same species often compete for territory and for mates. Across this matrix of ecological rivalry is spread a prominent veneer of predator/prey relations, which represents one of the three dominant activities of most animals, the other two being food-getting and procreation.

Many animals live together in social groups of varying sizes, generally referred to as *herds* or *flocks* or *colonies*. This is a common, but by no means universal, behavioral characteristic among animals of the same species and sometimes encompasses several different species in a communal relationship. Within such groupings there may be a certain amount of cooperation among unrelated animals, but competition for both space and resources is likely to be prominent as well.

Complicating corollaries to the concept of competition are presented by the relationships called *symbiosis* and *parasitism*. *Symbiosis* involves a mutually beneficial relationship between two species, as exemplified by the tickbirds that invariably accompany rhinoceroses; the former aids the latter by removing insects of various kinds that infest the crevices and orifices of its skin, and the birds are in turn benefited by having a readily available supply of food. *Parasites* are organisms of one species that infest the body of a creature of another species, obtaining their nutriment from the host, which is usually weakened and sometimes killed by the actions of the parasite.

Competition among animals is a major part of the general struggle for existence that characterizes natural relationships in the biosphere. Individual animals are concerned either largely or entirely with their own survival (and sometimes for that of their mates), in response to normal primeval instincts. In some species, this concern is broadened to include their own young, although such maternal (and, much more rarely, paternal) instinct is by no means universal among animals. Still fewer species show individual concern for the group, as represented in colonies of ants or troops of baboons. For the most part, however, animal survival is a matter of every creature for itself, with no individual "helping" another, and no individual deliberately destroying another apart from normal predatory activities.

Kinds of Animals

The vast majority of all species of animals are so tiny and/or secretive as to be either invisible or extremely inconspicuous. Their size and habits, however, are not valid indicators of their significance for geographic study. Very minute and seemingly inconsequential organisms sometimes play exaggerated roles—as vectors of disease, hosts of parasites, sources of infection, providers of scarce nutrients, or some other prominent function—in the bio-

Figure 11-15 Predation is a normal and pervasive relationship among animals. This Michigan coyote has caught a rabbit. [Courtesy Michigan Department of Natural Resources, Lansing.]

sphere, thereby attracting much more attention than their size and numbers might otherwise warrant. For example, no geographic assessment of Africa, however cursory, can afford to ignore the presence and distribution of the small tsetse fly and the tiny protozoan called *Trypanosoma*, which together are responsible for the transmission of Trypanosomiasis ("sleeping sickness"), a widespread and deadly disease for humans and livestock over much of the continent.

Zoological classification is much too detailed for use in most geographical studies, in part because of the vast numbers of tiny species that are identifiable. Therefore, presented below is a brief summation of the principal kinds of animals that might be recognized in a general study of physical geography. See also Figure 11-16.

Invertebrates Animals without backbones are called *invertebrates*. More than 90 percent of all animal species are encompassed within this broad grouping, which is sometimes referred to as "lower animals." Invertebrates include worms, sponges, mollusks, various marine animals, and a vast host of creatures of microscopic or near-microscopic size. Very prominent among invertebrates are the *arthropods*, which include insects, spiders, centipedes, millipedes, and crustaceans (lobsters, etc.).

Vertebrates Vertebrates have backbones that protect the main nerve (or spinal) cord. They are commonly known as "higher animals." Geographers generally follow biologists in recognizing five principal groups of vertebrates:

1. *Fishes* are aquatic animals and the only vertebrates that can breathe under water (a few species are also capable of breathing directly from air). Most fishes inhabit either fresh water or salt water, but some species are capable of living in both environments, and several species are *anadromous*, which means that they spawn in freshwater streams but live most of their lives in the ocean.
2. *Amphibians* are semiaquatic animals. In the larval stage, they are fully aquatic and breathe through gills; as adults, they are air-breathers by means of lungs and through their glandular skin. Most amphibians are either frogs or salamanders.
3. *Reptiles* are a bit further up the evolutionary scale. Whereas fishes must live in water, and amphibians require water for the first part of their life, most reptiles are totally land-based. Ninety-five percent of all reptile species are either snakes or lizards. See Figure 11-17. The remainder are mostly turtles and crocodilians. As with amphibians, reptiles are cold-blooded, which

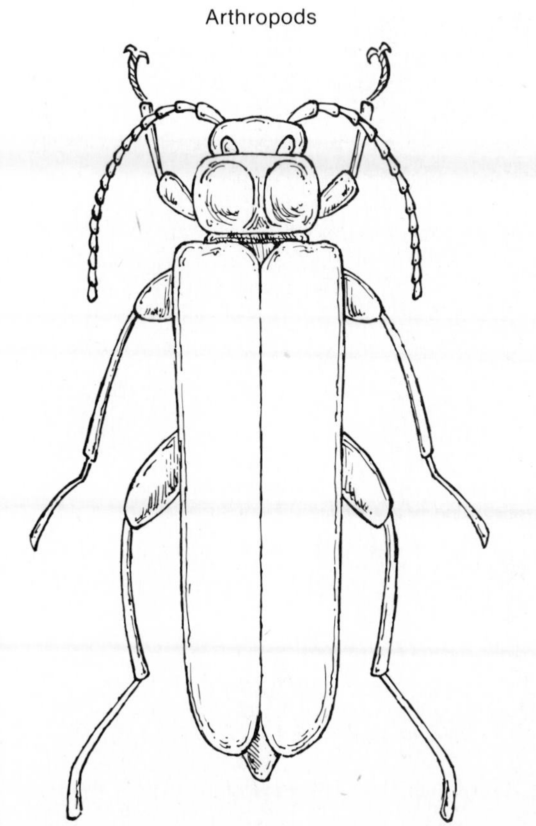

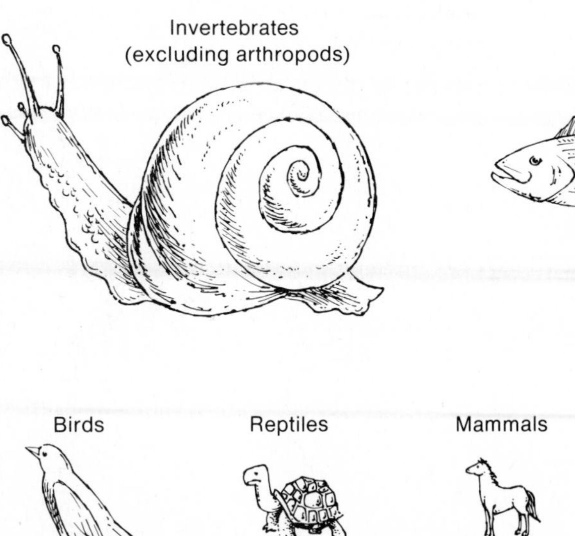

Figure 11-16 Relative abundance of different types of animals. The size of the animal portrayed is indicative of the comparative number of species in that group.

Figure 11-17 Lizards are perhaps the most widespread of all reptiles. This is a blue-tongue skink in the northern part of western Australia. [TLM photo.]

means that they cannot exist in the cooler parts of the Earth and become inactive in winter wherever they live.

4. *Birds* are believed to have evolved from reptiles; indeed, they have so many reptilian characteristics that they have been called "feathered reptiles." There are over 9000 species of birds, all of which reproduce by means of eggs. Birds are so adaptable that some species can live almost anywhere on the Earth's surface.
5. *Mammals* are the highest form of animal life; i.e., they have evolved into organisms of the greatest complexity and most diverse capabilities. Mammals are distinguished from all other animals by several internal characteristics as well as by two prominent external features; that is, only mammals produce milk with which they feed their young, and only mammals possess true hair (some mammals have very little hair, but no creatures other than mammals have any hair). Mammals are also notable for being warm-blooded, an attribute that they share only with birds. Thus, the body temperature of mammals and birds stays about the same under any climatic conditions, which enables them to live in almost all parts of the world.

The great majority of all mammals are *placentals*, which means that their young grow and develop in the mother's body, nourished by an organ known as the *placenta*, which forms a vital connecting link with the mother's bloodstream. A small group of mammals (about 135 species) are *marsupials*, whose females have pouches in which the young, which are born in a very undeveloped condition, live for several weeks or months after birth. The most primitive of all mammals are the *monotremes*, of which only half a dozen species exist; they are egg-laying mammals. See Figure 11-18.

The 5000 known species of mammals can be grouped in a variety of ways. Prominent among the major categories are *cetaceans* (whales, dolphins, etc.), *carnivores* (flesh-eaters such as dogs, cats, bears, and seals), *rodents* (gnawing animals), *lagomorphs* (rabbits and hares), *ungulates* (hoofed mammals), and *primates* (humans and humanlike mammals such as apes and monkeys).

Zoogeographic Regions

The distribution of animals over the world is much more complex and irregular than that of plants, primarily because animals are more mobile and capable of more rapid dispersal. Therefore the patterns of faunal distribution are more complicated and less predictable than those of plants. As with plants, however, the broad distributions of animals are reflective of the general distribution of energy and of food diversity. Thus the richest faunal assemblages are found in the permissive environment of the humid tropics, and the dry lands (deserts) and cold lands (polar and high-mountain regions) have the sparsest representations of both species and individuals.

When considering the global patterns of animal geography, most attention is usually paid to the distribution of terrestrial vertebrates, with other animals being given only casual notice. The classical definition of world zoogeographical regions is credited to the British naturalist

Figure 11-18 The platypus is one of only two kinds of monotremes, or egg-laying mammals. It is found only in aquatic environments in eastern Australia. [Jerry Cooke, Photo Researchers, Inc.]

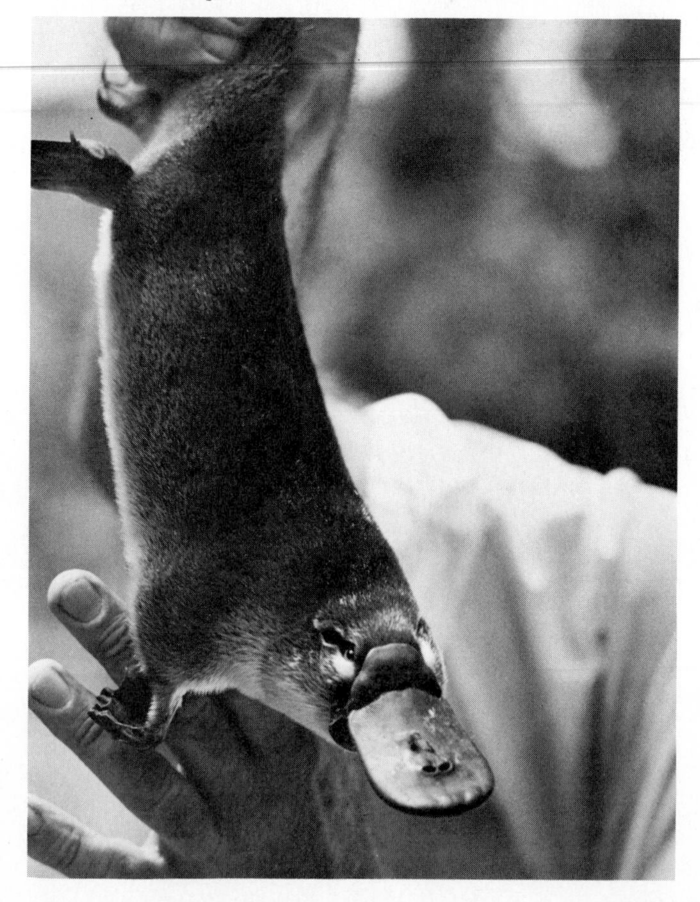

A. W. Wallace in the nineteenth century, whose scheme is based importantly on the work of P. L. Sclater. This system has been modified in detail, but its basic tenets are still generally accepted by zoogeographers. As shown in Figure 11-19, six major and three lesser zoogeographical regions are recognized. Each of these regions will be described below. It should be understood that this or any system of faunal regions represents average conditions and cannot portray some common pattern in which different groups of animals fit precisely. It is simply a composite of many diverse distributions of contemporary fauna.

The Ethiopian or Paleotropic Region includes Africa south of the Sahara as well as the southern edge of the Arabian peninsula. It is primarily a tropical and subtropical region and has a rich and diverse fauna. It is separated from other regions by an oceanic barrier on three sides and a broad desert on the fourth. Its freshwater fish fauna is very distinctive, and it contains many mammals that have no living relatives outside Africa. It has, however, many faunal affinities with the Oriental Region and (especially as regards birds) with the Palearctic Region.

The Oriental Region encompasses tropical Asia and major associated islands in the East Indies and the Philippines. It is separated from the rest of Asia by significant mountain ranges. Its faunal assemblage is generally similar to that of the Ethiopian Region, with somewhat less diversity. It has some endemic groups (found nowhere else) and a few relationships with the Palearctic and Australian regions.

The Palearctic Region includes the rest of Asia, all of Europe, and most of North Africa. Its fauna as a whole is much poorer than that of the two regions previously discussed, which is presumably a function of its location in higher latitudes with a more rigorous climate. It has some affinities with all three bordering regions, particularly with the Nearctic.

The Nearctic Region consists of the nontropical portions of North America. Its faunal assemblage is relatively poor and is largely a transitional mixture of Palearctic and tropical American groups. It has few important groups of its own except for freshwater fishes. The considerable similarities between Palearctic and Nearctic fauna have persuaded some zoologists to group them into a single superregion, the *Holarctic*, which had a land connection in the recent geologic past (the Bering land bridge), across which considerable faunal dispersal took place.

The Neotropic Region encompasses all of South America and the tropical portion of North America. Its fauna is rich and distinctive, which reflects both a variety of habitats and a considerable degree of isolation from other regions. Neotropic faunal evolution often followed a path different from that in other regions.

The Australian Region is restricted to the continent of Australia and some adjacent islands, particularly New Guinea. Its fauna is by far the most distinctive of any major region, primarily due to its lengthy isolation from other principal landmasses and the resultant unusual trends in evolution. Within the region there are many significant differences between the fauna of Australia and that of New Guinea.

The Madagascar Region, restricted to the island of that name, has a fauna very different from that of nearby Africa. It is dominated by a relic assemblage of unusual forms in which primitive primates (lemurs) are notable.

The New Zealand Region has a unique fauna dominated by birds, with a remarkable proportion of flightless types. It has almost no terrestrial vertebrates (no mammals; only a few reptiles and amphibians).

The Pacific Islands Region includes a great many far-flung islands, mostly quite small. Its faunal assemblage is very limited.

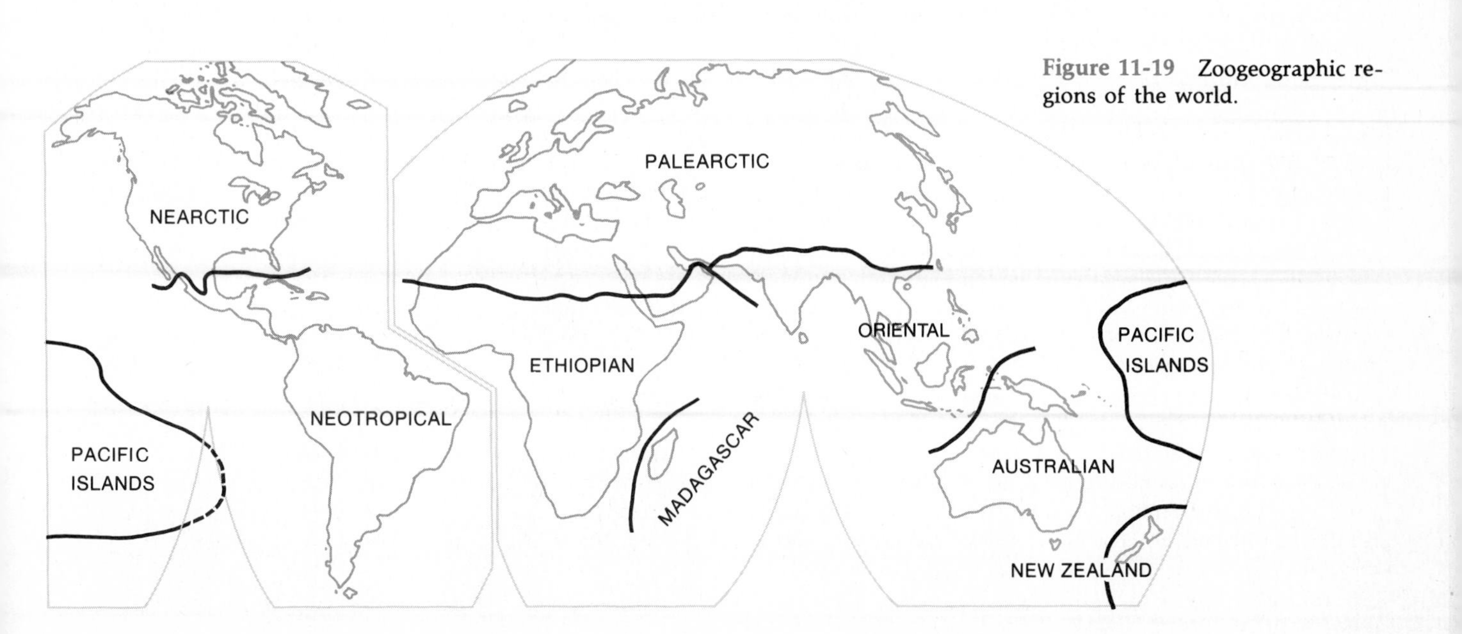

Figure 11-19 Zoogeographic regions of the world.

Focus on Wallace's Line: A Zoogeographic Boundary Zone

Boundaries between adjacent zoogeographical regions are, for the most part, conceptual rather than actual. That is to say, they represent generalized limits instead of precise delimitations. Most such boundaries are marked by oceans or by major complexes of topographic barriers. In every case, however, some species (sometimes many species) range on both sides of the boundary.

Perhaps the most interesting and certainly one of the most distinctive of these zoogeographical boundaries is represented by *Wallace's Line*, identified by Alfred Russell Wallace as the border between the Oriental and Australian regions. Wallace was a naturalist who worked extensively in the East Indies and South America, and he independently developed a theory of evolution at about the same time as Charles Darwin. Concern about the impending publication of Wallace's ideas stimulated the cautious Darwin into publishing his long-considered classic, *The Origin of Species*.

The faunal complements of the Oriental and Australian regions are remarkably different. The Oriental mammals are entirely placental, and very few of them have ranges that extend to Australia; the Australian mammals are dominated by marsupials, none of which are found in Asia. Similarly, there is no overlap between freshwater fishes of the two regions. Amphibians and reptiles of the two regions are shared in relatively few cases. Only among some groups of birds and invertebrates is there much similarity between Australia and Asia. The islands of the East Indies comprise a geologically unstable area of land and sea that has effectively isolated Australia from infiltration by most Oriental species throughout recent geologic history, allowing a very different evolutionary development.

Wallace delimited the boundary between the Oriental and Australian regions as a precise line separating various islands. He was particularly struck by the difference in birds between the small islands of Bali and Lombok, which are less than 20 miles (32 km) apart. His line is laid between these two islands and extends north to separate Sulawesi from Celebes, then curves northeasterly to exclude all of the Philippines from the Australian Region. Some scholars have supported this precision; one ornithologist even noted that the birds of Bali and Lombok are more different than those of England and Japan.

Other zoologists, while agreeing in principle to the conceptualization of the boundary, have questioned its precision. For example, more than 70 percent of the birds on Lombok are Oriental, although the dominant species are Australian on Lombok and Oriental on Bali. The ichthytologist M. Weber has proposed an alternate boundary, called *Weber's Line*, which is some distance to the west of Wallace's Line, as a more appropriate separation of the two major zoogeographic realms; this contention is based primarily on the statistical superiority of Oriental vertebrate species throughout the Lesser Sunda Islands. And still other zoologists prefer a boundary that is just west of New Guinea, considering that this border represents the western margin of pure New Guinean fauna.

In separating the Oriental and Australian regions, then, scholars have suggested three clear-cut boundaries for what is obviously not a clear-cut margin. Here, as with most zoogeographic boundaries, there is a transition zone. Indeed, the transition zone itself has been given a distinctive name, *Wallacea*, to apply to the area between New Guinea and Wallace's Line. Within Wallacea, the vertebrate fauna changes progressively from west to east, from predominantly Oriental to predominantly Australian. There is, however, a greater penetration of Oriental species than Australian species into Wallacea. The facts are clear; only the interpretation is variable: boundary or transition zone?

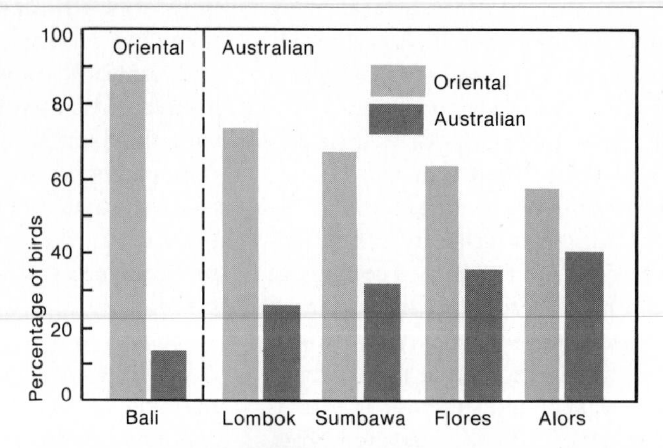

Wallacea is the name applied to a transition zone between the Oriental and Australian realms, especially as far as bird speciation is concerned. The length of the bars in this graph corresponds to the proportional numbers of bird species from the two realms for each of the indicated islands. [Adapted from Life Nature Library/*Ecology*, drawing by Eric Gluckman (New York: Time-Life Books, Inc., Publisher, 1963), p. 12. © 1963 by Time, Inc.]

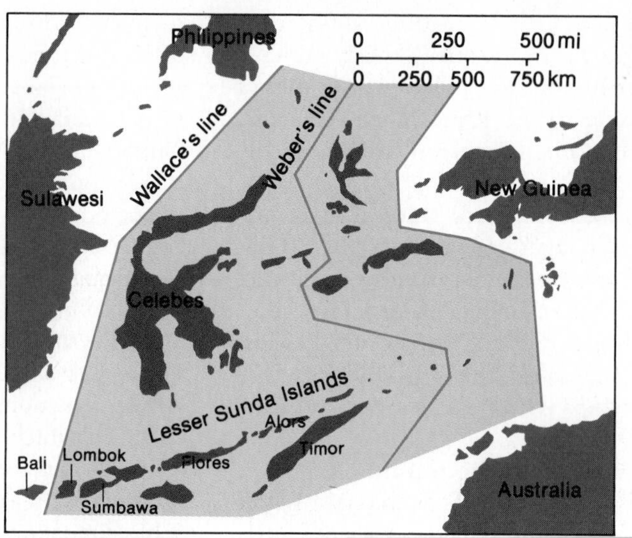

The three zigzag north-south lines represent (1) Wallace's Line (the westernmost), (2) Weber's Line (in the center), and (3) the western boundary of the Australian zoogeographic region (the easternmost).

The Major Biomes

As a summation of the geography of the biosphere, it is logical to identify and describe the major terrestrial ecosystems, or biomes, of the world. Most biomes are named for their dominant vegetation association, but the biome concept also encompasses fauna as well as interrelationships with soil, climate, and topography.

Tropical Rainforest

The rainforest of the low latitudes, also called a *selva*, is probably the most complex of all terrestrial ecosystems. It contains a bewildering variety of trees growing in close conjunction. See Figure 11-20. Mostly they are tall, high-crowned, broadleaf species that never experience a seasonal leaf fall because the concept of seasons is unknown in this environment of continual warmth and moistness. The selva has a layered structure; the second layer usually forms a complete canopy of interlaced branches that provides continuous shade to the forest floor. Bursting through the canopy to form the top layer are the forest giants, tall trees that sporadically rise to great heights above the general level. Beneath the canopy is an erratic third layer of lower trees that is able to survive in the shade. Sometimes still more layers of increasingly shade-tolerant trees grow at lower levels.

Undergrowth is normally sparse in the selva because the lack of light precludes the survival of most green plants. Only where there are gaps in the canopy, as alongside a river, does light reach the ground, resulting in the dense undergrowth associated with a "jungle." Most of such epiphytes like orchids and bromeliads hang from or perch on the tree trunks and branches.

The interior of the selva, then, is a region of heavy shade, high humidity, windless air, continual warmth, and an aroma of mold and decomposition. As plant litter accumulates on the forest floor, it is acted upon very rapidly by plant and animal decomposers, which find optimal conditions for their activities. The upper layers of the forest are areas of high productivity, and there is a much greater concentration of nutrients in the vegetation than in the soil. Indeed, most selva soil is surprisingly infertile.

Rainforest fauna is largely *arboreal* (tree-dwelling) because the principal food sources are in the canopy rather than on the ground. Large animals are generally scarce on the forest floor, although there are vast numbers of invertebrates. The animal life of this biome is characterized by creepers, crawlers, climbers, and flyers—monkeys, arboreal rodents, birds, tree snakes and lizards, and multitudes of invertebrates.

Figure 11-20 Schematic diagram of tropical rain-forest tree forms. A complete canopy of interlacing branches is conspicuous, with some smaller trees whose tops are well below the canopy, and some individual trees that reach above the canopy level.

Tropical Deciduous Forest

There is structural similarity between the selva and the tropical deciduous forest, but several important differences are usually obvious. In the tropical deciduous forest, the canopy is less dense, the trees are somewhat shorter, and there are fewer layers in the forest structure, all of which represents a response to less total or periodic precipitation. Due to a pronounced dry period that lasts for several weeks or months, many of the trees shed their leaves at the same time, allowing the penetration of light to the forest floor. This produces an under-story of lesser plants that often grows in such density as to produce classical "jungle" conditions. The diversity of tree species is not as great in this biome as in the selva, but there is a greater variety of shrubs and other lesser plants.

The faunal assemblage of the tropical deciduous forest is generally similar to that of the rainforest. Although there are more ground-level vertebrates than in the selva, arboreal species are particularly conspicuous in both biomes.

Tropical Scrub

Extensive areas in the tropics and subtropics are dominated by a vegetation association that consists of low-growing, scraggly trees and tall bushes, usually with an extensive under-story of grasses. The trees range from 10 to 30 feet (3 to 9 m) in height. Their density is quite variable, sometimes growing in close proximity to one another but often spaced much more openly. Species diversity is much less than in the tropical forest biomes; frequently just a few species will comprise the bulk of the taller growth over vast areas. In the more tropical and wetter portions, most of the trees and shrubs are evergreen; elsewhere most species are deciduous. In some areas a high proportion of the shrubs are thorny. See Figure 11-21.

The fauna of tropical scrub regions is notably different from that of the biomes previously discussed. There is a moderately rich assemblage of ground-dwelling mammals and reptiles, and of birds and insects.

Tropical Savanna

Savanna lands are dominated by a plant cover that consists primarily of grasses, usually relatively tall grasses. Sometimes the grasses form a complete sod, but sometimes there is bare ground among dispersed tufts of grass in a "bunchgrass" pattern. Some savanna areas are virtually without shrubs or trees, but in most cases, a wide scattering of both types dots the grass-covered terrain. This mixture of plant forms is often referred to as *parkland* or *park savanna*.

The savanna biome has a very pronounced seasonal rhythm. During the wet season, the grass grows tall, green, and luxuriant. At the onset of the dry season, the grass begins to wither, and before long the above-ground portion is dead and brown. At this time, too, many of the trees and shrubs shed their leaves. The third "season" is the time of wildfires. The accumulation of dry grass provides abundant fuel, and most parts of the savannas experience natural burning every year or so. The recurrent grassfires are salutary for the ecosystem, as they burn away the unpalatable portion of the grass without causing significant damage to shrubs and trees. When the rains of the next wet season arrive, the grasses spring into growth with renewed vigor.

Figure 11-21 A thorn scrub scene in northern Namibia. [TLM photo.]

Figure 11-22 The plains of Africa are famous for their immense herds of ungulates. In some localities these conditions still persist. The antelope here are topi, and the location is the northern Serengeti of Kenya. [TLM photo.]

The faunal complement of the savanna lands is variable from continent to continent. The African savannas are the premier "big game" lands of the world, with an unmatched richness of large animals, particularly ungulates (Figure 11-22) and carnivores, but also including a remarkable diversity of other faunal forms. The Latin American savannas, on the other hand, have only a sparse population of large wildlife, whereas the Asian and Australian areas are intermediate between these two extremes.

Desert

In previous chapters we have noted a general decrease in precipitation as one moves away from the equator in the low latitudes. This moisture progression is matched by a gradation of dominant vegetation associations from the selva biome of the equator to the desert biome of the subtropics. The desert biome also occurs extensively in midlatitude locations in Asia, North America, and South America.

Desert vegetation is surprisingly variable. It consists largely of drought-resisting plants with structural modifications that allow them to conserve moisture, and of drought-evading plants capable of hasty reproduction during brief rainy times. The plant cover of desert areas is usually sparse, with considerable bare ground dotted by a scattering of individual plants. See Figure 11-23. Typically the plants are shrubs, which occur in considerable variety, each with its own mechanisms to combat the stress of limited moisture availability. Grasses and other herbaceous plants are widespread but sparse in desert areas. Despite the dryness, trees can be found sporadically in the desert, especially in Australia.

Figure 11-23 Most desert areas have a great deal of unvegetated ground, but few are as barren as this alluvial fan in California's Death Valley. [National Park Service photo.]

Animal life is exceedingly inconspicuous in most desert areas, leading to the erroneous idea that animals are nonexistent. In actuality, most deserts have a moderately diverse faunal assemblage, although the variety of large mammals is limited. A large proportion of desert animals avoid the principal periods of dessicating heat (daylight in general and the hot season in particular) by resting in burrows or crevices during the day and prowling at night. Various other behavioral and physiological adaptations to environmental stress have been discussed earlier in this chapter.

Generally speaking, life in the desert biome is characterized by an appearance of inert stillness. In favorable times (at night, and particularly after rains) and in favored places (around water holes and oases), however, there is a great increase in biotic activities, and sometimes the total biomass is of remarkable proportions.

Mediterranean Woodland and Shrub

In six widely scattered and relatively small areas of the midlatitudes, all of which experience the pronounced dry summer/wet winter precipitation regime of the mediterranean climatic type, is found a biome in which the dominant vegetation associations are physically very similar but taxonomically quite varied. This mediterranean biome is mostly dominated by a dense growth of woody shrubs, which is known as a *chaparral* (Figure 11-24) in North America but has other names in other areas. A second significant plant association of mediterranean regions is an open grassy woodland, in which the ground is almost completely grass-covered, but it has a considerable scattering of trees as well.

The plant species are variable from region to region. Oaks (*Quercus*) of various kinds are by far the most significant genus in the Northern Hemisphere mediterranean lands, sometimes occurring as prominent medium-sized trees but also appearing as a more stunted, shrubby growth. In all areas, the trees and shrubs are primarily broadleaf evergreens. Their leaves are mostly small and have a leathery texture or waxy coating, which inhibits water loss during the long dry season. Moreover, most plants have deep roots.

Summer fires are relatively common in this biome, and many of the plants are adapted to rapid recovery after a wildfire has swept over the area. Indeed, some species have seeds that are released for germination only after the heat of a fire has caused their seedpods to open. Part of the seasonal rhythm of this biome is that winter floods sometimes follow summer fires, as slopes left unprotected by the burning away of grass and lower shrubs

Figure 11-24 The California chaparral is characterized by a dense growth of woody shrubs, usually with considerable bare ground beneath the spreading arms of the plants. [TLM photo.]

are susceptible to abrupt erosive runoff if the winter rains arrive before the vegetation has a chance to resprout.

The fauna of this biome is not particularly distinctive. Seed-eating, burrowing rodents are common, as are some bird and reptile groups. There is a general overlap of animals between this biome and those adjacent.

Midlatitude Grassland

Vast areas of relatively continuous grassland occur widely in the midlatitudes of all continents. This is a general response to the lack of sufficient precipitation to support larger plant forms, or to the frequency of fires (both natural and human-induced) that prevents the growth of tree or shrub seedlings. In the wetter areas, the grasses grow tall, and the term *prairie* is often applied. In drier regions, the grasses are shorter in stature; such growth is often referred to as *steppe*. See Figure 11-25. Sometimes a continuous sod is missing, and the grasses grow in discrete tufts as bunchgrass or tussock grass.

Most of the grass species are perennials, lying dormant during the winter and sprouting anew during the following summer. Trees are mostly restricted to riparian situations, whereas shrubs and bushes occur sporadically on rocky sites. Grass fires are fairly common in summer, which helps to explain the relative scarcity of shrubs. The woody plants cannot tolerate fires and generally can survive only on dry slopes where there is little grass cover to sustain a fire.

These grasslands have provided extensive pastures for grazing animals, which originally occurred in large numbers but of relatively few species. The larger herbivores often were migratory prior to human settlement. Many of the smaller animals spend all or part of their lives underground, where they find some protection from heat, cold, and fire.

Midlatitude Deciduous Forest

Extensive areas on all Northern Hemisphere continents, as well as more limited tracts in the Southern Hemisphere, were originally covered with a forest of largely broadleaf deciduous trees. Except in hilly country, a large proportion of this forest has been cleared for agriculture and other types of human use, so that very little of the original natural vegetation remains.

The forest is characterized by a fairly dense growth of tall broadleaf trees with interwoven branches that provide a complete canopy in summer. Some smaller trees and shrubs exist at lower levels, but for the most part, the forest floor is relatively barren of undergrowth. In winter the appearance of the forest changes dramatically due to the seasonal fall of leaves.

Tree species vary considerably from region to region, although most are broad-leaved and deciduous. See Figure 11-26. The principal exception is in eastern Australia, where the forest is composed almost entirely of varieties of eucalyptus, which are broadleaf evergreens. Northern Hemisphere regions have a northward gradational mixture with needle-leaf evergreen species. An unusual situation in southeastern United States finds extensive stands of pines (which are needle-leaf evergreens) rather than deciduous species occupying most of the well-drained sites above the valley bottoms.

This biome generally has the richest assemblage of fauna to be found in the midlatitudes, although it does not have the diversity to match that of most tropical biomes. It has (or had) a considerable variety of birds and mammals, and in some areas, reptiles and amphibians are well represented. Summer brings a diverse and active population of insects and other arthropods. All animal life is less numerous (partly due to migrations and hibernation) and less conspicuous in winter.

Figure 11-25 Short-grass steppe in the Great Plains of the United States. [Courtesy Department of Economic Development, State of Nebraska, Lincoln.]

Figure 11-26 Much of northeastern United States was originally covered with a deciduous forest. The forest persists today mostly in hilly or mountainous areas. This scene is in West Virginia. [USDA photo.]

Boreal Forest

One of the most extensive components of the biosphere is the boreal forest, or *taiga*, which occupies a vast expanse of northern North America and Eurasia. See Figure 11-27. This great northern forest contains perhaps the simplest assemblage of plants of any biome. Most of the trees are coniferous (cone-bearing), nearly all of which are needle-leaf evergreens, with the important exception of the tamarack or larch, which drops its needles in winter. The variety of species is limited; most are pines, firs, and spruces, extending broadly in homogenous stands. In some places, the coniferous cover is interrupted by areas of deciduous trees. These deciduous stands are also of limited variety (mostly birch, poplar, and aspen), and often represent a subclimax situation following a forest fire.

The trees grow taller and more densely near the southern margins of the taiga, where the summer growing season is longer and warmer. Near the northern margins, the trees are spindly, short, and more openly spaced. Undergrowth is normally not dense beneath the forest canopy, but a layer of deciduous shrubs sometimes grows in profusion. The ground is usually covered with a complete growth of mosses and lichens, with some grasses in the south, and with a considerable accumulation of decaying needles over all.

Poor drainage is typical in summer, due partially to permanently frozen subsoil, which prevents downward percolation of water, and partially to the derangement of normal surface drainage by the action of glaciers during the recent ice age. Thus bogs and muskeg swamps are numerous, and the ground generally has a spongy feel to it in summer. During the long winters, of course, all is frozen and icy.

Figure 11-27 A typical view along the southern margin of the North American boreal forest in Minnesota. The trees are small, close-growing, and of uniform species composition. [Courtesy U.S. Forest Service.]

The immensity of the boreal forest gives an impression of biotic productivity, but such is not the case. Harsh climate, floristic homogeneity, and slow plant growth produce only a limited food supply for animals. Species diversity of the fauna is limited, although the number of individuals of some species is astounding. With relatively few animal species in such a vast biome, populations sometimes fluctuate enormously within the space of only a year or so. Mammals are represented prominently by fur-bearers and by a few species of ungulates. Birds are numerous and fairly diverse in summer, but nearly all migrate to milder latitudes in winter. Insects are totally absent in winter but are superabundant during the brief summer.

Tundra

The tundra is essentially a cold desert in which moisture is scarce and summers are so short and cool that trees are unable to survive. See Figure 11-28. The plant cover consists of a considerable mixture of species, many of them in dwarf forms. Included are grasses, mosses, lichens, flowering herbs, and a scattering of low shrubs. These plants often occur in a dense, but ground-hugging, arrangement, although some places have a more sporadic cover with considerable bare ground interspersed. The plants complete their annual cycles hastily during the brief summer, when the ground is often moist and waterlogged because of inadequate surface drainage and particularly inadequate subsurface drainage.

Animal life is dominated by birds and insects during the summer. Extraordinary numbers of birds flock to the tundra for summer nesting, migrating southward as winter approaches. Mosquitos, flies, and other insects proliferate astoundingly during the short warm season, laying eggs that can survive the bitter winter. Other forms of animal life are scarcer—a few species of mammals and freshwater fishes but almost no reptiles or amphibians.

Figure 11-28 A tundra scene on Ellesmere Island in the Northwest Territories of Canada. [NFB Phototheque Onf ©. Photo by Scott Miller & David Hiscocks.]

Biomes and Climate

One of the truly striking relationships in physical geography is the notable correlation between the gross global pattern of distribution of biomes and that of major climatic types. This correlation is readily apparent when one compares the maps of Plates II and IV. The interrelationships of climate, flora, and fauna are exceedingly complex, but a reasonable generalization is to note that broad vegetation patterns are generally dependent on broad climatic patterns, and that the distribution of animal life is significantly influenced by the climatic and vegetational milieu. An understanding of the world pattern of biomes requires consideration of many factors, but clearly a comprehension of the climatic pattern is the starting point.

Human Modification of Natural Distribution Patterns

Thus far in our discussion of the biosphere, our attention has been focussed on "natural" conditions, i.e., events and processes that take place in nature without the aid or interference of human activities. Such natural processes have been ongoing for millennia, and their effects on the distribution patterns of plants and animals normally have been very slow and gradual. The pristine environment, uninfluenced by mankind, experiences its share of abrupt and dramatic events, to be sure, but environmental changes generally proceed at a very leisurely pace. When *Homo sapiens* appear, however, the tempo changes dramatically.

Mankind is capable of exerting extraordinary influences on the distribution of plants and animals. Not only is the magnitude of the changes likely to be great, but also the speed with which they are effected is sometimes exceedingly rapid. In broadest perspective, humanity exerts three kinds of direct influences on biotic distributions—physical removal of organisms, habitat modification, and artificial translocation of organisms.

Habitat Modification

Although the eradication of an organism is of immediate catastrophic significance to that individual, the modification of habitat is likely to be of much greater importance to an entire species. This is another activity in which mankind excels. The soil environment is changed by farming, grazing, engineering, and construction practices; the atmospheric environment is degraded by the introduction of impurities of various kinds; the waters of the Earth are impounded, diverted, and polluted—all of these deeds influence the native plants and animals in the affected areas. Of even greater importance to the animal life, normally, is any human-induced modification of the natural vegetation.

Artificial Translocation of Organisms

Mankind is capable of elaborate rearrangement of the natural complement of plants and animals in almost every part of the world. This is shown most clearly with domesticated species—crops, livestock, pets. There is now, for example, more corn than native grasses in Iowa, more cattle than native gazelles in the Sudan, more canaries than native thrushes in Detroit.

In our study of physical geography, however, we are not concerned with domesticated conditions; our interest is in the natural or "wild" state. Even so, mankind has accounted for many introductions of wild plants and animals into "new" habitats that they did not naturally occupy; such organisms are called *exotics* in their new homelands. In some cases, the introduction of exotic species was deliberate. A few examples among a great many that could be cited include the taking of prickly-pear cactus from Arizona to Australia, crested wheat grass from Russia to Kansas, European boar from Germany to Tennessee, pronghorn antelope from Oregon to Hawaii, and red fox from England to New Zealand.

Physical Removal of Organisms

One of mankind's most successful skills is in eliminating other living things. As human population increases and spreads over the globe, there is often a wholesale removal of native plants and animals to make way for the severely modified landscape that is thought to be necessary for civilization. The natural plant and animal inhabitants are cut down, plowed up, paved over, burned, poisoned, shot, trapped, and otherwise eradicated in actions that may well have far-reaching effects on overall distribution patterns.

Frequently, however, the introduction of an exotic species was an accidental result of human carelessness. The European flea, for example, has become one of the most widespread creatures on the face of the Earth due to the fact that it has been an unseen accompaniment to human migrations all over the world. Similarly, the English sparrow and European brown rat have been inadvertently introduced to all inhabited continents by traveling as stowaways on ships.

One other type of human-induced translocation of animals involves the deliberate release or accidental escape of livestock to become established as a "wild" (properly termed *feral*) population. This has happened in many parts of the world, most notably in North America and Australia.

It can be recognized, then, that an understanding of contemporary distribution patterns of plants and animals requires consideration of not only the natural factors—evolutionary development, migration, reproductive success, extinction—but also the influence of human activities.

Distributional Summation: Some Patterns Examined

The patterns of distribution of flora and fauna are sometimes governed by natural factors that are easy to discern, but sometimes the relationships are very complicated and obscure. In addition to recent and contemporary environmental influences, one must weigh the historic trends of evolution and dispersion, as well as the more recent modifications induced by human activities. To illustrate the range of this complexity, let us consider three case studies.

1. The Distribution of Selva: Environmental Simplicity

On any map showing the world distribution of major plant associations, or biomes, one of the conspicuous units that is always recognized is the *selva*, or tropical rainforest. The selva is a distinctive assemblage of tropical vegetation that is dominated by a great variety of tall, high-crowned trees. The species composition of the rainforest varies widely from place to place, but its general physical aspect is similar wherever it occurs.

Although the selva is found in many parts of the tropical continents and islands, it has a disjunct distribution, meaning an irregular and fragmented distribution. In general, it is found in the equatorial zone, but not everywhere in the equatorial zone. Additionally, it occurs along certain coastal stretches located several degrees of latitude away from the equator. A vast extent of selva exists in northern South America (primarily within the watershed of the Amazon River), central Africa (mostly within the watershed of the Zaire River), and the East Indies, with more limited patches in Central America,

Focus on the Changing Distribution of Camels

As an example of the complexity of the problem of unravelling the story behind the contemporary distribution pattern of an organism, we might consider the camel. The camel is a large, conspicuous, hardy creature that has left an extensive fossil record, a record that is of great help in piecing together the outline of its evolutionary development and historical distribution patterns. Many of the details are unknown, but the general framework of the story is reasonably clear.

Paleontological evidence indicates that the cameloids evolved originally in North America. Fossil remains show that the family *Camelidae* and its immediate ancestors were confined to the North American continent during the entirety of their complicated evolutionary history, which had a duration of perhaps 40 million years. The earliest known ancestral camels were no larger than big rabbits, but continued evolution produced later American camels of very large size.

From their region of origin, the cameloids dispersed southward into South America and northwestward across the Bering land bridge into Asia. These later migrants spread westward across the dry lands of the Old World until they reached as far as Rumania in eastern Europe and all the way to the Atlantic Ocean coastline in North Africa. At least two dozen fossil genera have been discovered and identified, but by one of nature's ironic vagaries, the entire family became extinct in North America, its hearthland of evolution, presumably during early stages of the most recent Ice Age (the Pleistocene epoch).

At present, the family *Camelidae* contains only three living genera. The Old World genus is *Camelus*, comprising two species, *C. dromedarius* and *C. bactrianus*; and in South America are found the genus *Lama*, with three species, and the genus *Vicugna*, with but a single species.

The four South American species of cameloids today occupy a semidesert habitat that ranges from sea level to the Andean high country at elevations of 16,000 feet (5000 m) or so. The guanaco (*L. guanicoe*) and the vicuña (*V. vicugna*) are wild forms, but the llama (*L. peruana*) and the alpaca (*L. pacos*) have been known only in domestication during recorded history.

The Bactrian or two-humped camel (*C. bactrianus*) is native to central Asia and originally ranged as far west as Iran and European Russia. It is adapted to cold and rocky areas, in contrast to the preference for warm and sandy regions demonstrated by dromedaries. It became widely domesticated throughout central Asia.

The Arabian or one-humped camel (*C. dromedarius*) had a natural range that is imperfectly known. It is believed to have been restricted primarily to the Arabian peninsula and the margins of the Sahara Desert in North Africa. Wild dromedaries have been nonexistent,

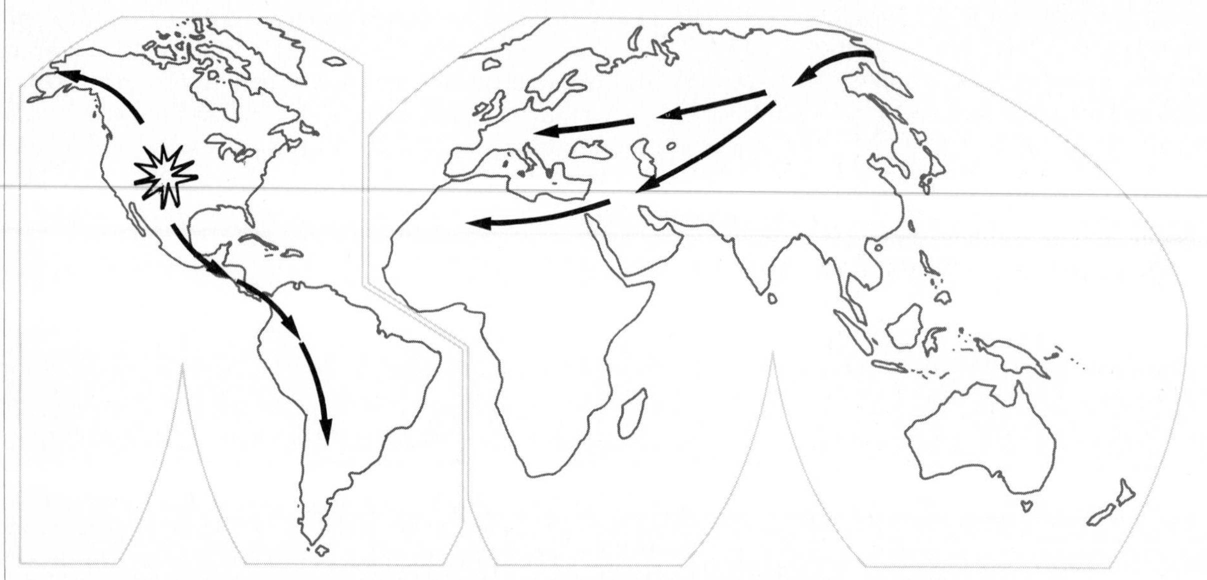

Camels evolved originally in central North America, from which area they dispersed widely.

Colombia, West Africa, Madagascar, Southeast Asia, and northeastern Australia. See Figure 11-29.

A general explanation of this distribution pattern is simple. With very limited exceptions, the selva occurs wherever relatively abundant precipitation and uniformly warm temperatures occur throughout the year. Such conditions are found only in portions of the equatorial zone and along certain windward coasts within the tropics.

If we had detailed distribution maps of the various regions of tropical rainforest for several recent years, we would notice a continuing diminution of its areal extent. In almost all selva regions, the rainforest is being reduced by deliberate human actions. The trees are being cut for timber in some areas, but much more significant is the removal of forest to make way for farming, grazing, and such construction projects as Brazil's Trans-Amazon however, for at least two thousand years, the wild stock being absorbed as camel-using people spread throughout the dry realm of southwestern Asia and North Africa.

By the middle of the nineteenth century, the distribution of wild cameloids could be summarized as follows:

1. There were probably no wild dromedaries anywhere in the world.
2. Several thousand wild Bactrian camels ranged widely through interior China and Mongolia, in central Asia.
3. The wild cameloids of South America, the guanaco and the vicuña, existed in the thousands, and perhaps in the tens of thousands, in Bolivia, Peru, Argentina, and Chile.

During the latter part of the nineteenth century, domesticated dromedaries were imported as work stock into three widely separated dryland regions—southwestern United States, southwestern Africa, and inland Australia. In all three of these locations, the dromedaries eventually became superfluous as working livestock, and several dozens or several hundreds of the animals were simply turned loose to shift for themselves. In both the United States and southern Africa, the camels were unsuccessful in surviving. For reasons that are not perfectly clear (but probably due to deliberate killing by the aboriginal inhabitants of these areas), the stray dromedaries soon died out in these areas. In Australia, however, the camels found a happy home. They took readily to a feral existence, and significant populations of "wild" camels built up throughout much of central Australia, reaching a population total that may have exceeded 30,000 by the late 1930s.

During the last half-century, both the range and the numbers of wild cameloids have been significantly reduced, partly because of deliberate killing but particularly due to the diminution of available range through expanded human settlement and associated activities. A contemporary map of the distribution of free-ranging cameloids would show three widely separated situations:

1. Several thousand wild guanacos and vicuñas range through the high country of Bolivia and southeastern Peru, as well as in western and southern Argentina.
2. A few thousand wild Bactrian camels are widely scattered in northeastern China and Mongolia.
3. Perhaps 15,000 to 20,000 feral dromedaries occupy extensive areas in central and western Australia.

This brief survey of the fluctuations in camel distribution exemplifies the complicated background that characterizes the history of most organisms. For some groups, the story is simpler and better understood, but for others, it is more complex and less well known.

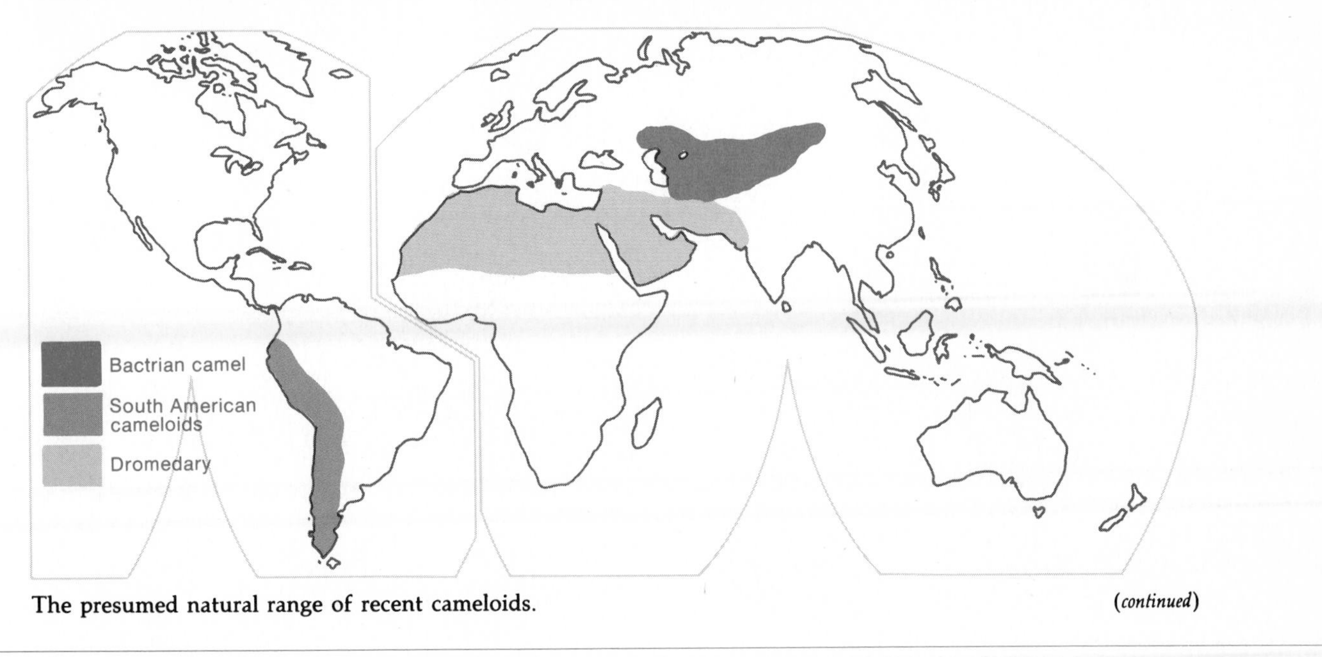

The presumed natural range of recent cameloids.

(*continued*)

Highway. This widespread elimination of a critical plant association provides the potential for wholesale extinction of a substantial number of biotic species (plant and animal) even before many of them have been "discovered" by science.

Whatever the ultimate fate of the selva, it has a gross distribution pattern that is easily comprehended. The pattern is determined primarily by two climatic elements—temperature and precipitation—with increasingly significant modifications resulting from human activities.

2. The Distribution of Antelope: Continental Diversity

The word *antelope* is a collective term that refers to several dozen species of hoofed animals, nearly all of which have

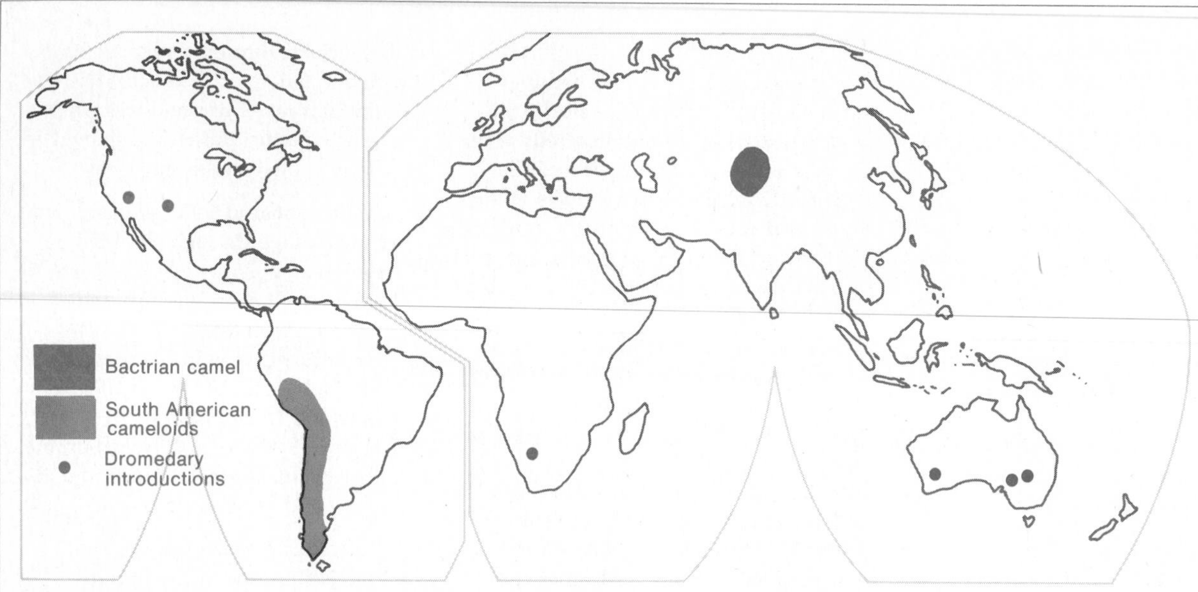

The range of wild cameloids in the midnineteenth century.

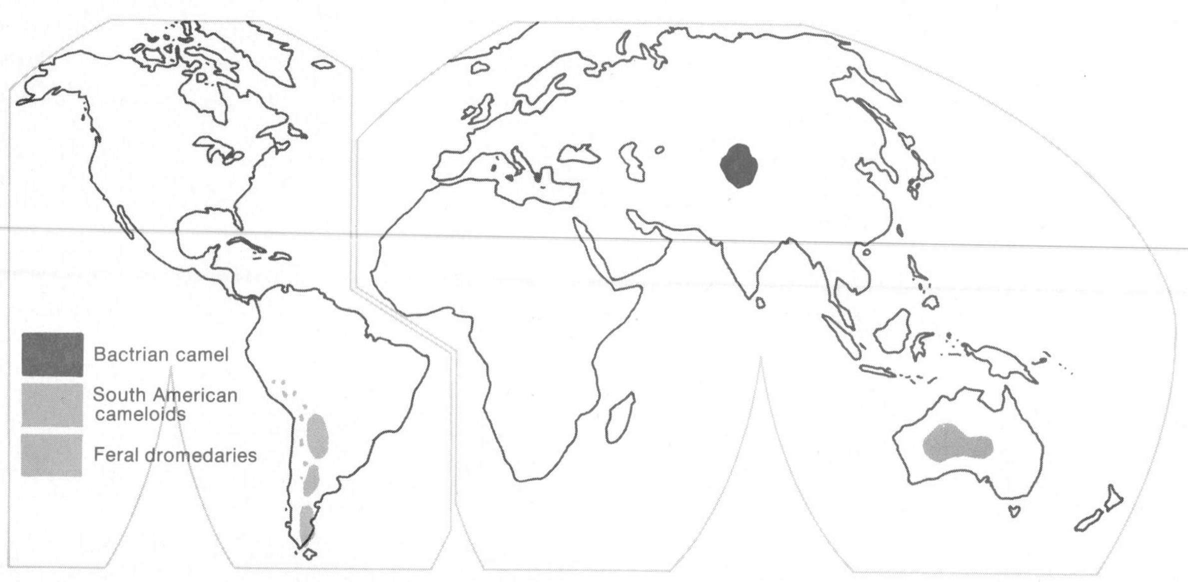

The contemporary distribution of wild and feral cameloids.

permanent, hollow, unbranched horns. All antelope subsist on vegetation of some sort; most are gregarious in behavior (i.e., they usually live in groups or herds); and there is a tremendous range in size from the tiny dikdik, which is the size of a jack rabbit, to the giant eland, which is as big as a large horse.

Most antelope live in Africa, and almost all of the rest inhabit Asia. See Figure 11-30. About 75 different species range over nearly all parts of Africa; and another 15 species are endemic to Asia, where their native habitat encompasses much of the southern and central parts of that continent. Europe has been a marginal home for antelope, with only one species ranging across European Russia into Poland. No antelope are native to the Western Hemisphere, although one North American species, the pronghorn, is very similar to the true antelope, its characteristics and behavior being almost indistinguishable from those of many African and Asian antelope.

In many parts of the world, natural biotic distribution patterns have been rearranged through deliberate and accidental human efforts to translocate plants and animals. When an exotic (foreign) species is released in a new area, the results are usually one of two extremes. Either it dies out in a short time because of environmental hazards and/or competitive/predatory pressures or the introduced species finds both a benign climate and an unfilled ecologic niche and is able to flourish extraordinarily. When the latter situation occurs, the results are occasionally salutary, but in many cases, they are unsatisfactory and sometimes absolutely disastrous.

The world abounds with bad examples that could be cited—sparrows and starlings, rabbits and pigs, mongooses and mynas, lantana and prickly pear, mesquite and broomweed, Australia and New Zealand, Hawaii and Mauritius; the list of things and places is virtually endless. An argument could be made, however, that the recent and contemporary biotic history of Florida represents the worst example of all, made all the more frightening because the cumulative impact will almost surely be much worse in the near future than it is already.

A Florida drainage canal completely choked with water hyacinth. [USDA photo.]

Florida is a very attractive place for human settlement. Particularly in the last decade or so, it has experienced one of the highest rates of in-migration of people ever known anywhere in the world. The major inducement is climate. A mild winter more than compensates for a hot summer, particularly when the latter can be ameliorated by artificial air conditioning. Florida's subtropical climate is also very suitable for many plants and animals. The tropical regions of the world contain an incredible variety of biota. Only a small fraction of this total is native to Florida, but a great many alien tropical species are capable of survival and proliferation in the Sunshine State if they can but gain a foothold.

Mankind has been energetically and capriciously willing to help in providing that foothold. In the last few years, Florida has become the major world center for the animal-import industry, and it is almost as important as a focus for plant imports as well. In an average year, more than 60 million fish, 2 million reptiles, half a million birds, and 150,000 mammals are imported into the United States for eventual sale to pet shops, zoos, research institutes, etc.; more than half of these creatures are brought, at least initially, to Florida, where the warm, moist climate and abundant water resources have encouraged the establishment of a great concentration of firms that engage in the breeding and wholesaling of animals and plants. It is inevitable that many of these organisms escape from confinement and try their luck at survival in the wild. Others are deliberately turned loose—by fishers who want new quarry, pet owners who are tired of their pets, gardeners who would like a new shrub in the backyard. Still others are brought in inadvertently, in the holds of freighters or the baggage of travelers. Through these and similar events, which have been accelerating year by year, Florida has become what has been termed a "biological cesspool" of introduced life forms. More than fifty species of exotic animals, not counting invertebrates, have now taken up residence in Florida, and exotic plants are almost too numerous to count.

Not all the introduced species have established viable, breeding populations in the wild, but many of them have, and some occur in plague proportions. Exotics are most likely to prosper in a new environment when the natural ecosystems of the host region are unstable. Florida has experienced massive disruption of its ecosystems in recent years due to the explosive human population increment and the associated accelerated changes in land use. Particularly contributory has been the modification of drainage systems in this flattish state of normally expansive water surfaces (lakes, swamps, marshes, everglades). A great deal of artificial drainage has taken place, primarily by the construction of drainage canals, canals that provide routes of access for easy dispersal of aquatic organisms.

In summary, the Florida environment has been significantly altered by human activities, which makes for unstable or unhealthy local and regional ecosystems, thus providing opportunities for invasions by exotics. In a healthy ecosystem, all the resources are

(*continued*)

being more or less fully utilized, and there is little chance for exotics to gain an ecologic niche (unless the exotic is unusually vigorous). In contemporary Florida, however, human interference has destabilized the natural ecosystems, and human-induced introductions provide a large and steady source of exotic plants and animals on a year-round basis.

Dozens of species of exotic plants have become widespread, and almost all of them are continually expanding their ranges. Prominent among them is the melaleuca tree from Australia. Seeds from these "paperbark" trees were deliberately sown by airplane in the 1930s in the hope of developing a timber industry in the swamplands of southwestern Florida. Only recently have they begun to spread extensively, in response to increased artificial drainage and to wildfires. They prosper and flourish in disturbed land much better than the native species do. The spread of melaleuca has changed swamp to forest, radically altering the entire regional ecosystem, but the lumber potential has turned out to be negligible.

Much more extensive and deleterious in impact has been the spread of exotic aquatic weeds, which now infest more than half a million acres (200,000 ha) of Florida's waters. The two most significant are the water hyacinth and the hydrilla:

1. The Amazon water hyacinth grows incredibly fast. A single plant can double its mass in two weeks under ideal conditions. These plants grow so thickly as to impede boat traffic and shade out other flora, sometimes to the extent that the water receives virtually no oxygen, and biological deserts are created. They are very difficult to cut out, and even when this is done, the decay of displaced portions puts more nutrients into the water, which promotes even lusher growth.
2. The hydrilla is a native of tropical Africa and Southeast Asia that was brought to Florida as an aquarium plant but has now spread vastly in the wild. It has the form of long green tentacles, growing an inch (2.5 cm) a day, which can become intertwined and form dense mats that will stop an outboard motor propeller dead. It overwhelms native plants and can even thrive in deep water where there is almost total darkness. Tiny pieces, when broken off, can regenerate into a new plant, so it is easily spread by birds, boat propellers, and other things that move from lake to lake. It has already clogged more than 100,000 acres (40,000 ha) of waterways in Florida and has become established in almost all other southeastern states.

The walking catfish is a clumsy walker, but it covers more ground than any other fish. The shoes in this photo do not belong to the fish! [Charles Trainer, Rapho/Photo Researchers, Inc.]

Exotic animals are less overwhelming in their occurrence, but many species are already well established and some are spreading rapidly. A sampling of nonaquatic species includes Mexican armadillos, Indian rhesus monkeys, Australian parakeets, Cuban lizards, Central American jaguarundi cats, South American giant toads, Great Plains jack rabbits, Amazonian parrots, and many others.

Exotic fish are much more numerous than their animal counterparts and pose even more serious problems, primarily through fierce competition with native species. Most of the drainage systems of Florida are interconnected by numerous irrigation and drainage canals; thus any introduced freshwater species now has access to most of the stream systems of the state. Moreover, Florida has a great abundance of springs that have stable water temperatures throughout the year, providing havens for many tropical fish that would otherwise find winter water temperatures below their tolerance limits.

South American acaras are already the dominant canal fish throughout southern Florida; and African tilapias are the most numerous fish in many lakes in the central part of the state. The greatest present and potential threat, however, is the so-called walking catfish from Southeast Asia, which numbers in the millions throughout the state and is considered to be "out of control," with "no practical method of eradication." These catfish eat insect larvae until the insects are gone; then they eat other fish. They are overwhelming rivals of almost all the native fish, eventually reducing the entire freshwater community to a single species—the walking catfish. They are significantly hardier than other fish because if they don't like the local waters, or if the waters dry up, they can simply hike across land, breathing directly from the air, until they find a new lake or stream. And they are proliferating spectacularly.

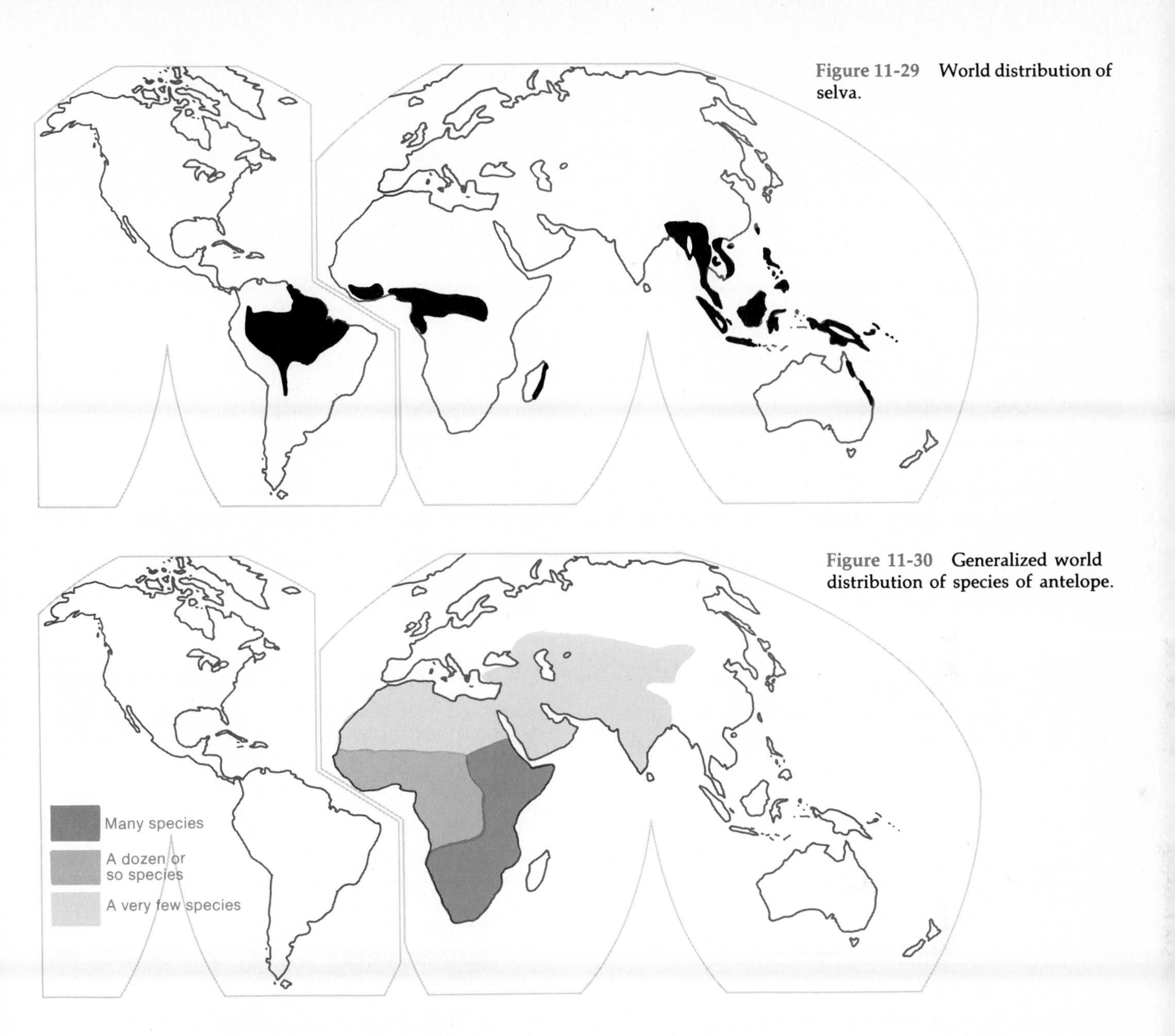

Figure 11-29 World distribution of selva.

Figure 11-30 Generalized world distribution of species of antelope.

The vast majority of all antelope are found in relatively similar habitats—flattish country in which the dominant vegetation is either grass or scrubby brush. There is clearly a close correlation between the natural ranges of antelope and the extent of these major types of vegetation. Only a handful of species—less than 5 percent of the total—are at home in forests or mountains or swamps. Thus the environmental relationship between faunal distribution and floral association is very close, and the topographic factor is also important.

In the case of these particular animals, however, much more is involved than a straightforward environmental relationship. There is extraordinary disparity in antelope distribution from continent to continent. Almost all parts of Africa have, or recently had, a numerous population of antelope; particularly in the grasslands and scrub country, they numbered in the millions. In Asia, their range has been widespread, but the number of species and individuals is much less than in Africa. The other continents are virtually without antelope: a marginal range of one species in eastern Europe; a single related species, although not a true antelope, in the grasslands of North America; and none at all in South America or Australia despite the extensive potential habitat on both continents.

We must examine the history of antelope evolution to explain such continental disparity. Although the full evolutionary story is still obscure, nearly all antelope species seem to have evolved in Africa and southern Asia, and with a few exceptions, have not dispersed extensively from their original center(s) of development. Apparently, a significant constraint on the contemporary distribution pattern of antelope has been the impact of human activities. Because most antelope are relatively large, attractive, gregarious, and meaty, they have always been appealing quarries for hunters, and in some areas their population has been reduced to insignificance through hunting pressure. The spread of human population and associated competing land use (particularly for grazing and farming), however, has been an even more important limiting factor. The range available to these animals, particularly in Asia, has been reduced severely, and the skyrocketing human population in Asia and Africa promises even greater reductions in the near future.

An understanding of the irregular distribution pattern of antelope, then, requires consideration of the complex balance among three factors—evolutionary development, habitat preference, and human influences.

3. The Biota of Australia: A Unique Case

We have seen that there is a fairly expectable distribution of major plant associations on all the continents, related primarily to latitudinal position and nearness to the coast. This "normal" pattern generally fits in Australia as it does in other continents, but the actual taxonomic components of the associations often are quite different in Australia than elsewhere.

The most notable vegetation distinction is that nearly all the native trees in Australia—more than 90 percent of the total—are members of a single genus, *Eucalyptus*. Moreover, the eucalypts, of which there are more than 400 species, are native to no other continent but Australia. The shrubs and bushes of Australia are also dominated by a single genus, *Acacia*, which encompasses about half of the intermediate level (between grasses and trees) flora of the continent. Australia has several unusual grasses as well, but their distinctiveness is less pronounced.

These unusual floristic developments seem to be largely the result of isolation. During long periods in the geologic past, the Earth's climate was more equable than it is today, with the result that all continents experienced the evolution of relatively similar plant groups. During the last 3 million years or so, however, continental climatic trends have been more disparate, with greater variations and extremes. This has engendered specialized evolutionary development among the isolated flora of Australia.

If the flora of Australia is unusual, the fauna is absolutely unique; its assemblage of terrestrial animal life is completely without parallel in other parts of the world. This, too, is primarily the result of isolation. Through a chain of varied geological events and biological repercussions, the Australian continent functioned for millions of years as a faunal asylum, where rare and vulnerable species were able to flourish in relative isolation from the competitive and predatory pressures that influenced animal evolution in other parts of the world. The results of this isolation are bizarre, especially with regard to the highest forms of animal life—the mammals.

Unlike all other continents, the Australian fauna is dominated by a single primitive mammalian order, the marsupials (Figure 11-31), an order that has long disappeared from most other parts of the world. Australia also provides the only continental home for an even more primitive group, the monotremes. More remarkable, perhaps, is what is lacking in Australia—placental mammals, so notable elsewhere, are limited and inconspicuous. Australia is completely without representatives from such common groups as cats, dogs, bears, monkeys, hoofed animals, and many others. Partially because of the specialized character of the native mammals, feral livestock have become established in the wild in much greater profusion in Australia than elsewhere on Earth; there are nine feral species that exist in numbers that aggregate to several millions. In addition to the remarkable nature of the mammalian fauna, there are also unusual aspects to the bird, reptile, and invertebrate populations of Australia.

The biota of Australia, then, has many singular aspects, due primarily to the circumstances of isolated evolution, with significant recent human-induced modifications.

Figure 11-31 The kangaroo is the classic example of the unusual marsupial fauna of Australia. [Courtesy Australian Tourist Commission.]

Review Questions

1. Explain the concept of *plant succession*.
2. What are the characteristics that distinguish between plants and animals?
3. Describe some typical *xerophytic* adaptations of vegetation.
4. List a tree that is an example of each of the following: *gymnosperm*; *angiosperm*; *coniferous*; *deciduous*; *evergreen*; *hardwood*; *softwood*.
5. Explain the concept of *climax vegetation*.
6. Why do geographers study plants more than they study animals?
7. Distinguish between the three types of animal adaptations (physiological, behavioral, reproductive) to the environment.
8. List and briefly describe the principal groups of vertebrates.
9. Select one zoogeographic region and describe its location and its principal characteristics.
10. Discuss the distribution and characteristics of the *selva*.

Some Useful References

BENNETT, C. F., JR., *Man and Earth's Ecosystems*. New York: John Wiley & Sons, Inc., 1975.

DARLINGTON, P. J., *Biogeography of the Southern End of the World*. New York: McGraw-Hill Book Company, 1975.

DE LAUBENFELS, D. J., *A Geography of Plants and Animals*. Dubuque, Iowa: Wm. C. Brown Company, Publishers, 1970.

MARGULIS, L., AND K. V. SCHWARTZ, *Phyla of the Five Kingdoms*. San Francisco: W. H. Freeman & Company, Publishers, 1981.

NEWBIGIN, M. I., *Plant and Animal Geography*. London: Methuen & Co. Ltd., 1968.

RILEY, D., AND A. YOUNG, *World Vegetation*. Cambridge, England: Cambridge University Press, 1966.

SHELFORD, V. E., *The Ecology of North America*. Urbana, Ill.: University of Illinois Press, 1963.

VANKAT, J. L., *The Natural Vegetation of North America: An Introduction*. New York: John Wiley & Sons, Inc., 1979.

12 Soils

The final major component of our study of physical geography is the lithosphere. *Lithosphere* refers to the crust of the Earth, the solid fundament that undergirds mankind's earthly habitat. The lithosphere is just as complex as the atmosphere, biosphere, or hydrosphere, but it contrasts with these three other realms in its enormity and particularly in its seeming stability.

A human can easily observe clouds forming, flowers blooming, wildlife migrating, and rivers flowing, but the dynamics of the lithosphere, with a few spectacular exceptions, operate with such incredible slowness that the crust of the Earth often appears changeless. "The everlasting hills" is a phrase that is considered by most laymen to be a literal expression aptly describing the permanence of the Earth's topography, whereas in reality it is an exaggerated hyperbole that fails to recognize the remarkable alterations that can take place over a lengthy time span.

A few lithospheric processes—such as earthquakes and volcanic eruptions—occur abruptly, but the vast majority operate so slowly as to be unrecognizable by the casual observer. Thus the student of topographic change must search for clues to the ponderous processes that shape the land—evidence from the past that helps to interpret the landscape of the present and to predict the landscape of the future. Our goal in the remaining chapters of this book is to try to understand the contemporary topography of the Earth's surface and to explain the processes that have caused it to be as it is.

Soil and Regolith

As the sun is the source of energy for all life on Earth, so is the soil the essential medium in which all terrestrial life is nurtured. Almost all land plants sprout from this precious medium that is spread so thinly across the continental surfaces, with an average worldwide depth of only about 6 inches (15 cm).

Despite the implication of the well-known simile, "as common as dirt," soil is one of the most complex features produced in nature. Soil is an infinitely varying mixture of weathered mineral particles, decaying organic matter, living organisms, gases, and liquid solutions. It is a dynamic mixture involving a slow but continuous disintegration of once-solid rock and decomposition of once-alive organisms. It is a locale of complicated chemical reactions involving a rich variety of ions and molecules in solution and colloidal particles in suspension.

Preeminently, however, soil is a zone of plant growth. Although the concept of soil almost defies definition, it can be conceptualized as a relatively thin surface layer of mineral matter that normally contains a considerable amount of organic material and is capable of supporting living plants. It occupies that part of the outer skin of the Earth that extends from the surface down to the maximum depth to which living organisms penetrate, which means basically the area occupied by plant roots. Soil is significantly characterized by its ability to produce and store plant nutrients, an ability that is made possible by the interactions of such diverse factors as water, air, sunlight, rocks, plants, and animals.

Although thinly distributed over the surface of the land, soil functions as a fundamental interface where atmosphere, lithosphere, hydrosphere, and biosphere meet. The bulk of most soils is inorganic, so soil is usually classed as part of the lithosphere, but its relationship to the other three "spheres" is both intimate and complex.

The development of soil is initiated by the physical and chemical disintegration of rock that is exposed to the atmosphere and the action of water percolating down from the surface. This disintegration is called *weathering*, the details of which will be discussed at some length in Chapter 15. The basic result of weathering is the weakening and breakdown of solid rock, the fragmentation of coherent rock masses, the making of little rocks from big ones. The principal product of weathering is a layer of loose inorganic material, called *regolith*, which overlies the unfragmented rock below. See Figure 12-1. Normally the regolith has a crude gradation of particle sizes, with the largest and least fragmented pieces at the bottom, immediately adjacent to the consolidated bedrock, and a progressive diminution of the fragments upward.

Above the regolith, usually but not invariably, is the soil itself. It is composed largely of unconsolidated

Figure 12-1. Schematic vertical cross section from surface to bedrock, showing the relationship of soil and regolith.

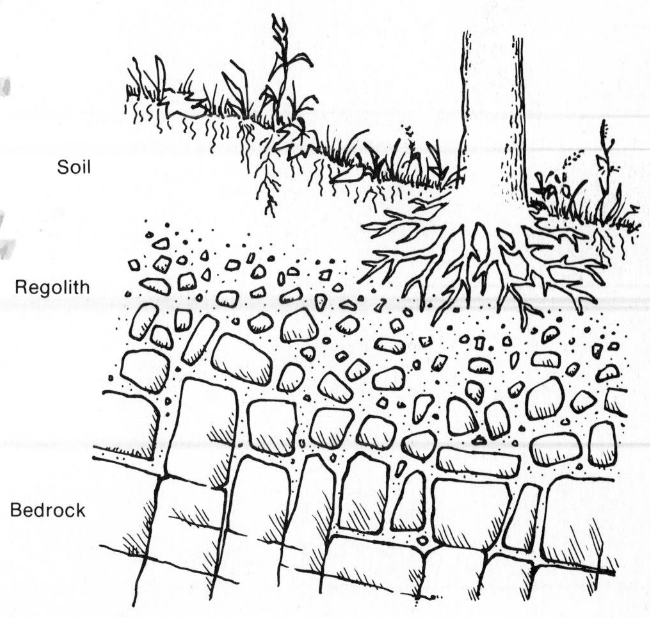

finely fragmented mineral particles, the ultimate product of weathering. It normally also contains an abundance of living plant roots, a variety of dead and rotting plant parts in varying stages of decomposition, an unbelievable quantity of microscopic plants and animals both living and dead, and a variable amount of air and water. Soil is not the end product of a process, but rather a stage in a never-ending continuum of physical/chemical/biotic activities, which, at any given time and place, represents a dynamic life layer that has a unique set of physical and chemical properties. See Figure 12-2.

Soil-Forming Factors

Three kinds of variables interact in producing any particular soil: (1) the chemical composition of the rock that provides most of the bulk material, (2) the environmental conditions under which the rock is converted into soil, and (3) the length of time that the rock has experienced that environment. These variables influence the five principal soil-forming factors, which will be discussed below. See also Figure 12-3.

The Geologic Factor

Almost all soil is composed largely of weathered fragments of rock. The source of the fragments is the *parent material*, which may consist of solid bedrock or of loose sediments that have been transported from elsewhere by the action of water, wind, or ice. Whatever the parent material, it is sooner or later disintegrated and decomposed at and near the surface, thus providing the raw material for soil formation.

The nature of the parent material often has a prominent influence on the characteristics of the soil that develops from it, and particularly in the early stages of soil formation, it sometimes overwhelmingly dominates all other factors. The chemical composition of the parent material obviously will be reflected in the resulting soil; for example, a bedrock containing much calcium is likely to produce a calcium-rich soil, or a parent material lacking in calcium can be expected to yield a calcium-deficient

Figure 12-2. Soil develops through complex interactions of physical and biological processes.

Figure 12-3. Five "factors" are interactive in soil formation. The geologic fundament and the topographic slope are acted upon by climatic and biologic agencies over a time continuum to produce a natural soil.

soil. Physical characteristics of the parent material also may be influential in soil development, particularly in terms of texture and structure of the soil. A bedrock that weathers into relatively large-sized particles, such as a sandstone, normally will produce a coarse-textured soil, one that is easily penetrated by air and water to some depth. On the other hand, a shale parent material weathers into very minute fragments, yielding a texture that has a minimum of pore spaces for air and water penetration.

The influence of parent material on the characteristics of the resulting soil becomes blurred by time. "Young" soils are likely to be very reflective of the rocks or sediments from which they were derived. With the passage of time, however, other soil-forming factors become increasingly important, and the significance of the parent material diminishes. Eventually the influence of the parent material may be completely obliterated, so that it is impossible to ascertain the nature of the rock from which the soil evolved.

It is evident, then, that the geologic factor provides the bulk of the material from which soil is developed, but it is a passive factor whose relative significance fades with time.

The Climatic Factor

One of the very active factors in soil formation is climate. Temperature and moisture are the climatic variables of greatest significance. As a basic generalization, both the chemical and biological processes in the soil are usually accelerated by high temperatures and abundant moisture and are slowed down by low temperatures and lack of moisture. High temperatures tend to speed up the chemical weathering of parent material, the decay activities of soil microorganisms, and the rate of plant growth. Similarly, an abundance of water percolating through the soil and regolith facilitates deeper and more rapid weathering and enhances plant growth. One predictable result is that soils tend to be deepest in warm, humid regions and shallowest in cold, dry regions. However, many other variable characteristics are strongly influenced by temperature and moisture conditions and not always in predictable fashion.

It is difficult to overemphasize the role of moisture moving through the soil. The flow is mostly downward following gravitational pull, but it is sometimes lateral in response to drainage opportunities, and in specialized circumstances, the movement is upward. Wherever and however water moves, it always carries along dissolved chemicals in solution and usually also tiny particles of matter. Thus moving water is ever engaged in rearranging the chemical and physical components of the soil, as well as contributing to the variety and availability of plant nutrients.

In terms of general soil characteristics, climate is likely to be the single most influential of the soil-forming factors. This is to say that temperature and moisture conditions, in the long run, tend to be the principal determinants of regional soil characteristics, providing a greater imprint than parent materials or any other single factor. This generalization has many exceptions and when soils are considered on a local scale, climate is likely to be less prominent as a determinant.

The Topographic Factor

Another passive factor in soil formation is provided by the topography, manifested primarily by aspects of slope and drainage. Wherever soil develops, its vertical extent undergoes continual, if usually very slow, change. This change comes about through a lowering of both the bottom and the top of the soil layer. The bottom of the soil slowly progresses downward as weathering penetrates deeper into the regolith and parent material and as the plant roots are extended to greater depths. At the same time, the surface of the soil is being lowered by sporadic removal of its uppermost layer through normal (also called *geologic*) erosion. This erosion is accomplished through the physical removal of individual soil particles by running water, wind, and gravity.

Where the land is flat, soil tends to develop at the bottom more rapidly than it is removed at the top by normal erosion. This does not mean that the downward development is speedy, rather that surface erosion is extraordinarily slow. Thus, the deepest soils are usually on flat land. Where slopes are relatively steep, on the other hand, surface erosion progresses more rapidly than soil deepening, with the result that such soils are nearly always thin and immaturely developed. See Figure 12-4.

Most soils are well drained, so that moisture relationships are relatively unremarkable in their development. Some, however, have inefficient natural drainage, which imparts significantly different characteristics. For example, a waterlogged soil tends to contain a high proportion of organic matter, and the biological and chemical processes that require free oxygen are impeded. Most ill-drained soils are in valley bottoms or in some other flat locale because soil drainage is usually related to the slope of the land. In some cases, however, such subsurface factors as *permeability* (a soil characteristic in which there are interconnected pore spaces through which water can move) and the presence or absence of impermeable layers are more influential than the slope of the surface.

The Biologic Factor

Plants and animals comprise another of the active factors in soil formation. This biologic factor in particular gives life to the soil and makes it more than just dirt. Every soil contains a quantity (sometimes an enormous quantity) of living organisms, and every soil incorporates an amount (sometimes a vast amount) of dead and decaying organic matter.

Vegetation of various kinds grows in the soil and performs certain vital functions. Plant roots work their way down and around, providing passageways for drainage and aeration, as well as being the vital link between soil nutrients and the growing plants. Detached plant parts (leaves, flowers, stems, trunks, etc.) and entire dead plants are frequently added to the soil as they accumulate and decay.

Many kinds of animals contribute to soil development as well. Even such large surface-dwelling creatures as elephants and bison affect soil formation by compaction with their hooves, rolling in the dirt, grazing the vegetation, and dropping excreta. Many smaller animals such as gophers, mice, worms, ants, termites, and others spend most or all of their lives in the soil layer, tunneling here and there, moving soil particles upward and downward, and providing passageways for water and air to penetrate. See Figure 12-5. The mixing and plowing effect of soil fauna is sometimes remarkably extensive. Earthworms, in particular, are noted for the beneficial effect of their earth-moving on plant growth. Ants, worms, and all other land animals fertilize the soil with their waste products and contribute their carcasses for eventual decomposition and incorporation into the soil when they die.

Another important component of the biologic factor is represented by microorganisms, both plant and animal, that occur in uncountable billions. Algae, fungi, protozoans, actinomycetes, and other minuscule organisms all play a role in soil development, but bacteria probably make the greatest contribution overall. This is because certain types of bacteria are responsible for the decomposition and decay of dead plant and animal material and the consequent release of nutrients into the soil in a form usable by plants.

Figure 12-4. Slope is a critical determinant of soil depth. On flat land, soil normally develops more deeply with the passage of time; whereas on relatively steep slopes, the soil is removed rapidly by erosion and thus remains shallow.

The Chronologic Factor

For soil to develop on a newly exposed land surface requires time, which may be quite variable depending on the nature of the exposed parent material and the characteristics of the environment. Soil-forming processes are generally very slow, and centuries, or tens of centuries, may be required for an inch or two of soil to evolve on a newly exposed surface. A warm, moist environment is inducive to soil development, and the relevant processes operate optimally under such conditions. Normally of much greater importance, however, are the attributes of the parent material. For example, soil can develop from some kinds of sediments relatively quickly because the breakdown of rock into small particles has already been accomplished. A soil that originates on a bedrock surface, on the other hand, must undergo a much longer period of development because the solid rock disintegrates very slowly.

Most soil develops with geologic slowness—so slowly that changes are almost imperceptible within a human life span. It is possible, however, for a soil to be degraded, either through the physical removal associated with accelerated erosion or through depletion of nutrients, in only a few years. In the grand scale of geologic time, then, soil can be formed and reformed, but in the dimension of human time, it is a nonrenewable resource.

Figure 12-5. Grass-eating termites are very important components of the soil fauna in many subtropical dry lands. In addition to their above-ground "ant-hill" homes, they burrow widely underground, accomplishing much mixing and cultivation of the soil. This scene is in northern Queensland, Australia. [TLM photo.]

Soil Components

Soil is made up of a variety of natural components existing together in myriad combinations. They can be classified, however, into just a few groups of substances, as discussed below.

Inorganic Materials

The bulk of most soils consists of mineral matter, mostly in small but macroscopic particles. Inorganic material also occurs as microscopic clay particles and as dissolved minerals in solution. About half the volume of an "average" soil is composed of granular mineral matter in small sizes called *sand* and *silt*. These particles may consist of a great variety of minerals, depending on the nature of the parent material from which they were derived, and are simply fragments of the wasting rock. Most common are bits of quartz, which are composed of silica (SiO_2) and appear in the soil as very resistant grains of sand. Other prominent minerals include some of the feldspars and micas.

The smallest particles in the soil are *clay*, which is a new mineral combination found only in the soil and not in the parent material from which the soil was derived. Clay has properties significantly different from those of the larger (sand and silt) fragments. Clay particles are colloidal in size, which means that they are larger than molecules but too small to be seen with the naked eye. Clay has an important influence on the chemical activity in the soil because many chemical reactions occur at the surfaces of soil particles, and clay particles are so tiny and numerous that they provide the greatest expanse of surface area. Clay particles occur as thin laminar sheets that resemble minute flakes and are composed primarily of silica and of oxides of aluminum and iron. Water moves easily between these sheets, and dissolved substances in the water are attracted to and held by the clay sheets. Since the sheets are negatively charged, they attract positively charged ions (*cations*). Many essential plant nutrients occur in soil solutions as cations, with the result that clay is an important reservoir for plant nutrients, just as it is for soil water.

Organic Matter

Although organic matter generally comprises less than 5 percent of total soil volume, it has an enormous influence on soil characteristics and plays a fundamental role in the biochemical processes that make soil an effective medium of plant growth. Some of the organic matter consists of living organisms; some is comprised of dead but undecomposed plant parts and animal carcasses; some is in the form of a stable decomposition product called humus; and some is in an intermediate stage of decomposition.

The living organisms of the soil are varied and incredibly numerous. Apart from the roots of higher plants, evidence of this may be inconspicuous or invisible, but most soils are virtually seething with life. A single acre (0.4 ha) may contain a million earthworms, and the total number of organisms in an ounce (28 g) of soil is likely

The abundance and variety of animal life connected with the soil are quite surprising. Such organisms vary in size from the gigantic to the microscopic, and in numbers from a few per acre to billions per gram. The organic life of the soil ranges from almost impalpable protozoans existing in colloidal slime to larger animals that may accidentally alter certain soil characteristics. Of all the creatures, however, it is probable that the earthworm is the most important to soil formation and development.

The cultivating and mixing activities of earthworms are of great value in improving the structure, increasing the fertility, lessening the danger of accelerated erosion, and deepening the profile of the soil. The distinctive evidence of this value is that the presence of many well-nourished earthworms is almost always a sign of productive, or potentially productive, soil.

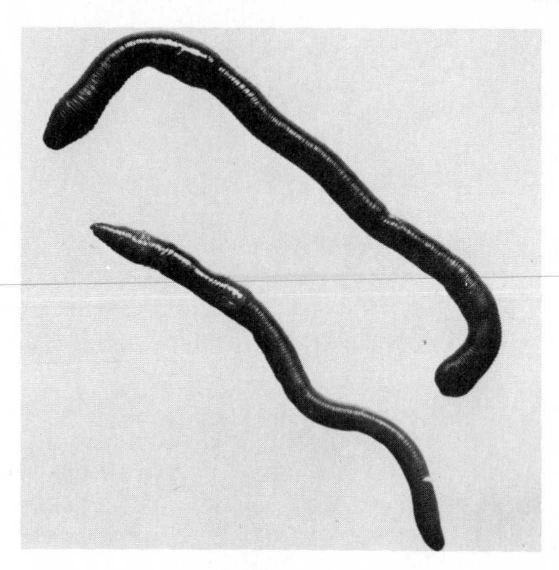

A close view of two of humans best friends—earthworms. [USDA photo.]

Several dozen varieties of segmented worms are often referred to as *earthworms*, but by far the most widespread and important is the "common earthworm," *Lumbricus terrestris*. These organisms are not widely distributed over the world; they are generally restricted to humid regions that have temperate or cool temperature regimes. Although they are common in eastern North America and most of Europe, in other parts of the world their occurrence is sometimes patchy and localized. Their preferred habitat is a heavy, compact soil with moderate acidity. They are unable to live in soils with a high alkaline content; if the soil has a pH value of less than 6, earthworms are almost certain to be absent. Where conditions are favorable, they may number in the hundreds of thousands per acre; in such areas they may comprise as much as 80 percent by weight of the total soil fauna. Indeed, in some prime New Zealand pasture land, the weight of earthworms beneath the surface has been found to amount to as much as the total weight of sheep on the surface!

Almost every activity undertaken by earthworms has some sort of direct effect on the soil. They move about beneath the surface by means of small tunnels that they construct as they progress. The tunnels, however, are not "dug" in the ordinary sense of the word. Rather, the earthworm pushes its way through the soft earth and literally eats its way through firmer soil. It obtains nourish-

to exceed 100,000,000,000. Microorganisms far exceed higher forms of life, both in total numbers and in cumulative mass. They are active in rearranging and aerating the soil and in yielding waste products that are links in the chain of nutrient cycling. Some make major contributions to the decay and decomposition of dead organic matter, and others are active in nitrogen fixation (making nitrogen available for plant use).

Many higher plants shed their leaves, twigs, and stalks on a regular or sporadic basis. Most of these dead plant parts accumulate right at the surface of the soil, where they are referred to collectively as *litter*. The eventual fate of most litter is decomposition, in which the solid parts are broken down into their essential chemical elements, which are then either absorbed into the soil or washed away. In cold, dry areas, litter may remain undecomposed for a very long time; where the climate is warm and moist, however, decomposition may take place almost as rapidly as litter accumulates.

After most of the residues have been decomposed, the remaining dark-colored (brown or black), gelatinous, chemically stable fraction of organic matter is referred to as *humus*. Humus represents a new chemical structure composed of materials liberated from the decaying bodies and parts of animals and plants. It tends to loosen the structure and lessen the density of the soil, thereby improving its physical capacity for plant growth. Moreover, humus plays a similar role to that of clay in that it is a catalyst for chemical reactions and a reservoir for plant nutrients and soil water.

Soil Air

Nearly half the bulk of an average soil is made up of pore spaces. These spaces provide a labyrinth of interconnecting passageways, called *interstices*, among the soil particles. This labyrinth permits the penetration of air and water into the soil. On the average, the pore spaces are about half filled with air and half with water, but at any given time and place, the amounts of air and water are quite variable, the quantity of one varying inversely with that of the other. See Figures 12-6 and 12-7. After a rain, most pore spaces will be filled with water, and little air is found. After a long dry spell, the amount of ment from the organic matter that is ingested, but it also takes into its alimentary canal large amounts of inorganic material, which is subsequently excreted in the form of *casts*. These casts consist of earthy matter bound together with decomposed organic residues, and they contribute considerable fertility to the soil and furnish a suitable environment for many microorganisms. Both the chemical and physical nature of the casts are important to the soil, whether on the surface or beneath it. The deposition of casts on the surface, which is particularly noticeable after a rain or heavy dew, involves the transportation of great quantities of subsurface material upward. Earthworms also drag leaves and other organic matter down into their tunnels.

At least seven beneficial functions have been attributed to earthworms:

1. Their innumerable tunnels facilitate drainage and aeration and the deepening of the soil profile.
2. The continual movement of the creatures beneath the surface tends to bring about the formation of a crumby structure, which is generally favorable for plant growth.
3. They rearrange material in the soil, especially by bringing deeper matter to the surface, where it can be weathered more rapidly. Where earthworms are numerous, they may deposit as much as 25 tons (24 t) of casts per acre (0.4 ha) on the surface in a year.
4. The soil is further mixed by material being carried and washed downward into their holes from the surface. This is notably in the form of leaf litter dragged downward by the worms, which fertilizes the subsoil.
5. The digestive actions and tunneling of earthworms form aggregate soil particles that increase porosity and resist the impact of raindrops, helping to deter erosion.
6. Nutrients in the soil are increased by the addition of casts, which have been shown to be 5 times richer in available nitrogen, 7 times richer in available phosphates, and 11 times richer in available potash than the surrounding soil.
7. Nitrification is also promoted by the presence of earthworms, due to increased aeration, alkaline fluids in their digestive tracts, and the decomposition of earthworm carcasses.

The mere presence of earthworms does not guarantee that a soil will be highly productive, as there may be other kinds of inhibiting factors such as a high water table. Nevertheless, the functions listed above show that an earthworm-infested soil has a higher potential productivity than similar soils lacking earthworms. In various controlled experiments, the addition of earthworms to wormless soil has enhanced plant productivity by several hundred percent.

In many parts of the world, of course, earthworms are lacking. They are, for example, almost totally absent from arid and semiarid regions. In these dry lands, some of the earthworm's soil-enhancing functions are carried out by ants and earth-dwelling termites, but much less effectively.

The precise importance of earthworms to soil development is difficult to assess, but there can be no question that soil productivity is significantly increased by their activities. Thus the impact of the earthworm on agricultural output is beyond measure. Charles Darwin conducted a classic study of earthworm activity a century ago and reached the conclusion that they may have played a more important role in the history of the world than any other type of animal. This is perhaps too sweeping an assertion, although many farmers are probably willing at least to modify the old aphorism concerning dogs to state, "Man's best friend is the earthworm."

water may be extremely depleted, with air occupying most of the interstices.

Air in the soil has significantly different characteristics from that in the atmosphere above. This air is found in openings generally lined with a film of water; and since it exists in such close contact with water and is not exposed to moving air currents, it is highly saturated. Soil air is also very rich in carbon dioxide and poor in oxygen because plant roots and soil organisms remove oxygen from, and respire carbon dioxide into, the pore spaces from which the carbon dioxide escapes (slowly) into the atmosphere.

Soil Water

Moisture is an important component of all soils and exists in widely varying quantities that depend on climatic conditions, surface vegetation, and the physical characteristics of the soil. *Soil water* differs from *groundwater*, which was discussed in Chapter 9.

Water comes into the soil largely by infiltration of rainfall and snowmelt from above. Some is also added

Figure 12-6. For plant growth, a soil ideally will have a volumetric composition that is about half solid material and about half pore spaces. Most of the solids will be mineral; a small amount will be organic. The pore spaces, on the average, will be about half filled with air and about half with water.

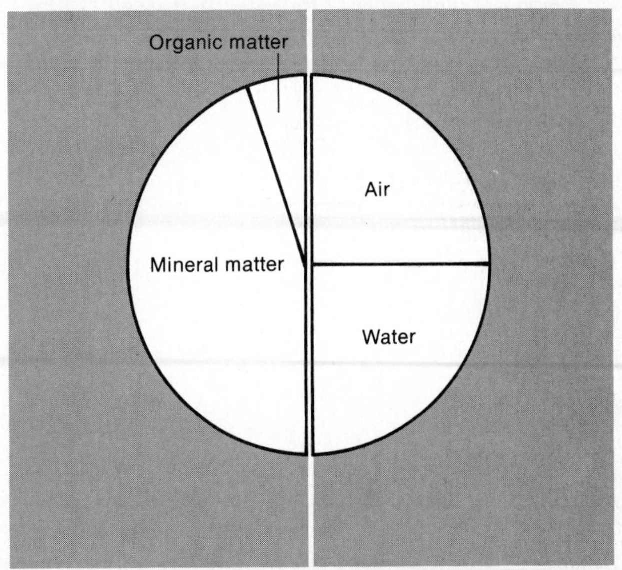

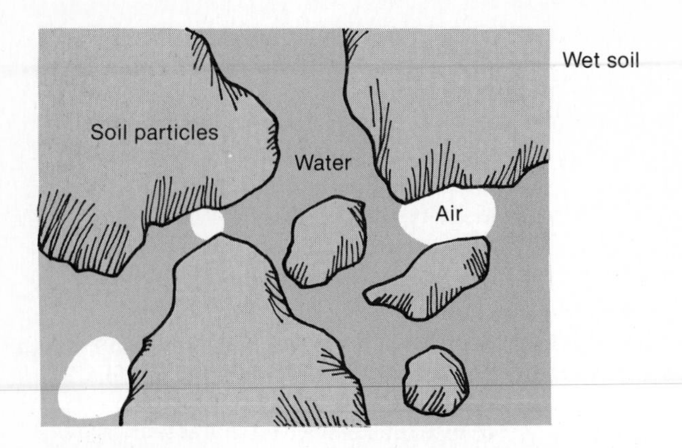

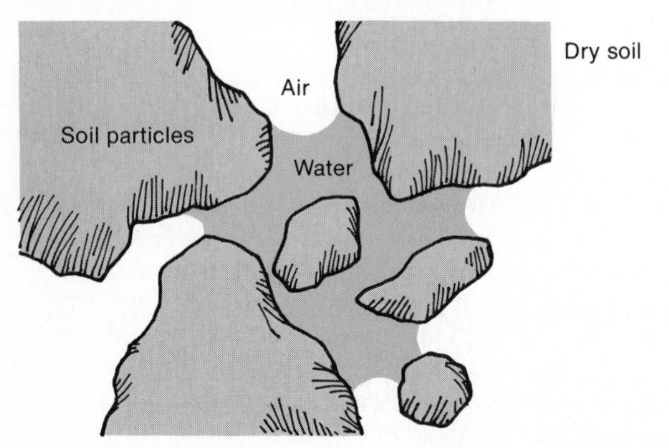

Figure 12-7. Pore spaces within the soil are filled with a varying balance of water and air. The interstices of wet soil contain much water and little air; in dry soil there is much air and little water.

from below when groundwater is pulled upward by *capillary action*. Once water has penetrated the soil, it envelops in a film of water each solid particle that it contacts, and it fills, or partially fills, the pore spaces. Water can be lost from the soil by percolation down into the groundwater, by upward capillary movement to the surface where it evaporates, or by plant use (*transpiration*).

Water performs a number of important functions in the soil. First of all, it is an effective solvent, dissolving essential soil nutrients and making them available to plant roots. It is also required for many of the chemical reactions of clay and for the actions of microorganisms that produce humus. In addition, it can have considerable influence on physical characteristics of the soil by moving particles around.

Four forms of soil moisture are recognized, as discussed below. See also Figure 12-8.

Gravitational Water Gravitational water is temporary in occurrence in that it results from prolonged infiltration from above (usually due to prolonged precipitation) and is pulled downward through the interstices toward the groundwater zone below by gravitational attraction. It is thus resident in the soil for a relatively short time and is not very effective in supplying plants because it drains away rapidly once the external supply ceases.

Gravitational water, however, accomplishes significant functions during its passage through the soil. It picks up fine particles of soil from the upper layers and carries them downward, in a process called *eluviation*. These particles are eventually deposited at a lower level, which is known as *illuviation*. This work of gravitational water is a mixing process that makes the topsoil coarser and more open in texture and the subsoil denser and more compact. Gravitational water also is involved in chemical action, for it dissolves soluble materials and carries them downward in solution, to be partly redeposited at lower levels. This process, called *leaching*, tends to deplete the topsoil of soluble nutrients.

Figure 12-8. The four forms of soil moisture are displayed here in sequence from least available (on the left) to most available (on the right). [After Donald Steila, *The Geography of Soils: Formation, Distribution and Management* (Englewood Cliffs, N.J.: Prentice-Hall, Inc., 1976), p. 45. By permission of Prentice-Hall.]

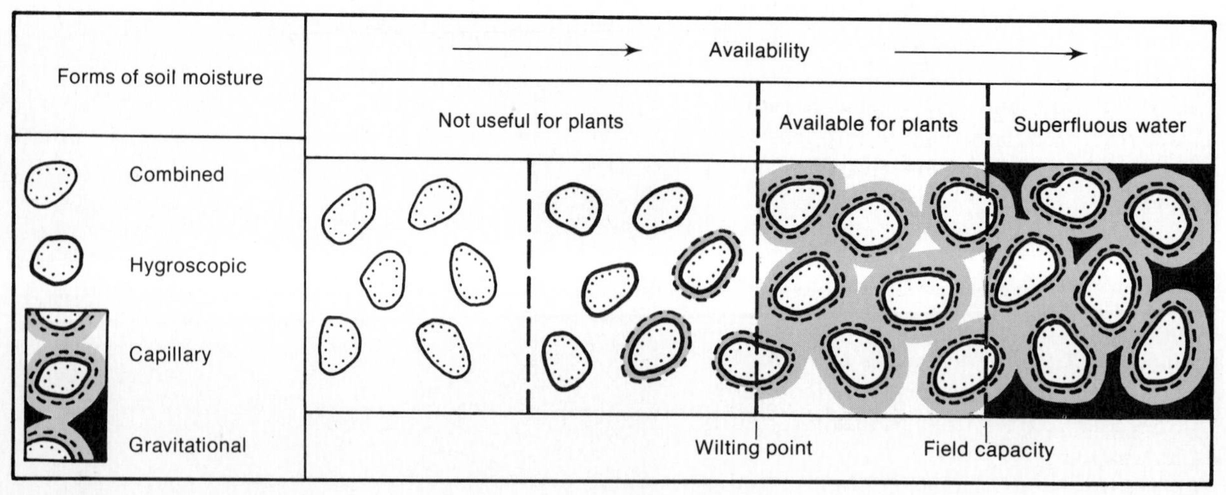

Capillary Water Capillary water remains after gravitational water has drained away. It consists of moisture held at the surface of soil particles by surface tension, which is the attraction of water molecules for each other (the same property that causes water to form rounded droplets rather than dispersing in a thin film). Capillary water is by far the principal source of moisture for plant use. Its surface tensions are sufficiently great that it does not respond to gravitational pull. Rather it is free to move about equally well in all directions—downward, upward, or laterally—in response to capillary tension. It tends to move from wetter areas toward drier ones, which accounts for the upward movement of capillary water when no gravitational water is percolating downward.

Hygroscopic Water Hygroscopic water consists of a microscopically thin film of moisture bound rigidly to all soil particles by *adhesion*, the attraction of water molecules to solid surfaces. It adheres so tightly to the particles that it is normally unavailable for plant use.

Combined Water Combined water is least available of all. It is held in chemical combination with various soil minerals, and is only freed if the chemical is altered.

For plant usage, capillary water is the most important. Gravitational water is largely superfluous in this regard. After gravitational water has drained away, the remaining volume of water represents the *field capacity* of the soil. If drought conditions prevail and the capillary water is all used or evaporated, plants are no longer able to extract moisture from the soil, and the *wilting point* is reached. Thus the amount of soil moisture available for plant use is essentially the difference between the field capacity and the wilting point.

Soil Properties

As one looks at, feels, smells, tastes, and otherwise examines soils, various physical and chemical characteristics appear useful in describing, differentiating, and classify-

Focus on the Soil-Water Balance

Moisture is added to the soil by infiltration of rainfall or snowmelt from the surface. Soil moisture is diminished largely through evapotranspiration. The dynamic relationship between these two processes is referred to as the *soil-water balance*. This balance is influenced by a variety of factors, including soil and vegetation characteristics, but it is primarily determined by temperature and humidity conditions.

At any given time and place there is likely to be either a surplus or a deficit of water in the soil. Such a condition is only temporary, and it varies in response to changing weather conditions, particularly those related to seasonal changes. Generally speaking, warm weather causes increased evapotranspiration, which diminishes the soil water supply; whereas cool weather slows evapotranspiration, allowing more moisture to be retained in the soil.

In a hypothetical Northern Hemisphere midlatitude location, the beginning of the calendar year (January) is a time of soil-water surplus because low temperatures inhibit evaporation and there is little or no transpiration from plants. The soil is likely to be at or near field capacity at this time as the result of a slow accumulation of moisture and little use. With the arrival of spring, temperatures are on the rise and plant growth accelerates, so that both evaporation and transpiration are considerably enhanced. The soil-water balance tips from a surplus to a deficit condition. The water deficit builds to a peak in middle or late summer, as temperatures reach their greatest heights and the water need of plants is at its maximum. In late summer and fall, as air temperature decreases and plant growth slackens, evapotranspiration diminishes rapidly. At this time, the soil-water balances shifts once again to a surplus condition, which continues through the winter. Then, the cycle begins again.

The accompanying diagram illustrates the annual sequence just described. Such a variation of the soil-water balance through time is called a *soil-water budget*.

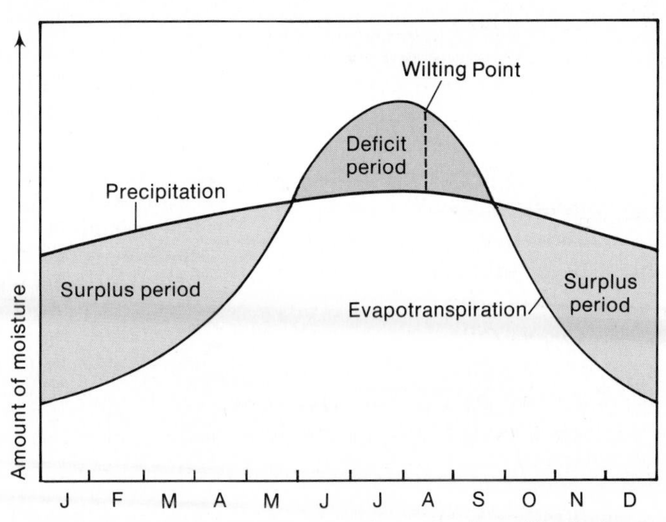

A hypothetical soil-water budget for a Northern Hemisphere midlatitude location.

ing them. Some soil properties are easily recognized, but most can be ascertained only by precision measurement.

Color

The most conspicuous property of a soil is usually its color, but color is by no means the most definitive property. Soil color can provide clues about its nature and capabilities, but the clues are sometimes misleading. Soil scientists recognize 175 different gradations of color in their descriptions, based on the widely accepted Munsell color chart. The standard colors are generally shades of black, brown, red, yellow, gray, and white. Soil color occasionally reflects the color of the unstained mineral grains, but in most cases, it is imparted by stains on the surface of the particles, due to metallic oxides or organic matter.

Black or dark brown colors usually indicate a considerable humus content; the blacker the soil, the more humus it contains. This is a strong hint of inherent high fertility because humus is an important catalyst in releasing nutrients to plants. Dark pigment is not invariably a sign of fertility, however, because it may be due to other factors, such as poor drainage or high carbonate alkalinity.

Reddish and yellowish colors generally indicate iron oxide stains on the outside of soil particles. These colors are most common in tropical and subtropical regions where many minerals are leached away by gravitational water, leaving insoluble iron compounds behind. In such situations, a red color bespeaks good drainage, and a yellowish hue suggests poor drainage. Red soils are also common in desert and semidesert environments where the color is carried over intact from reddish parent materials, rather than representing a surface stain.

Light-colored soils—gray or white—may develop in varying environments. In humid areas, a light color implies so much leaching that even the iron has been removed; but in dry climates, it indicates an accumulation of white alkali or other salts.

Texture

All soils are composed of myriad particles of varying sizes, ranging from microscopic bits to large stones, although smaller particles usually predominate. Rolling a sample of soil about between the fingers can provide a "feel" for the principal particle sizes. Table 12-1 shows the standard classification of particle sizes, in which the size groups are called *separates*. See also Figure 12-9.

The three principal separates are sand, silt, and clay. The larger particles—sand and silt—are fragments of the wasting parent material and are mostly the grains of minerals found commonly in rocks, especially quartz, feldspars, and micas. These coarser particles comprise the inert materials of the soil mass, its skeletal framework.

TABLE 12-1 Standard U.S. Classification of Soil Particle Sizes

Separate	Diameter
Gravel	Greater than 0.08 in. (2.0 mm)
Very coarse sand	0.04–0.08 in. (1.0–2.0 mm)
Coarse sand	0.02–0.04 in. (0.5–1.0 mm)
Medium sand	0.01–0.02 in. (0.25–0.5 mm)
Fine sand	0.004–0.01 in. (0.1–0.25 mm)
Very fine sand	0.002–0.004 in. (0.05–0.1 mm)
Coarse silt	0.0008–0.002 in. (0.02–0.05 mm)
Medium silt	0.00024–0.0008 in. (0.006–0.02 mm)
Fine silt	0.00008–0.00024 in. (0.002–0.006 mm)
Clay	Less than 0.00008 in. (0.002 mm)

Only the clay particles are significantly reactive; i.e., they can take part in the intricate chemical activities that occur in the soil. Clay represents new mineral structure built up from the dissolved matter that is liberated from weathered rock.

No soil is made up of particles of uniform size, so the texture of any soil represents the proportional occurrence of the various separates. The texture triangle (Figure 12-10) shows the standard classification of soil texture based on the percentage of each separate by weight. Near the center of the triangle is the term *loam*, which is the name given to a texture in which none of the separates dominates the other two. This fairly even texture mix is generally the most productive for plants.

Figure 12-9. Comparative sizes of soil separates. This magnified example shows the size relationship of sand, silt, and clay. For scale, a portion of the head of a pin is shown in lower right. [After Robert A. Muller and Theodore M. Oberlander, *Physical Geography Today: A Portrait of a Planet*, 2nd ed. (New York: Random House, Inc., 1978), p. 260. © 1978 by Random House, Inc. By permission of Random House.]

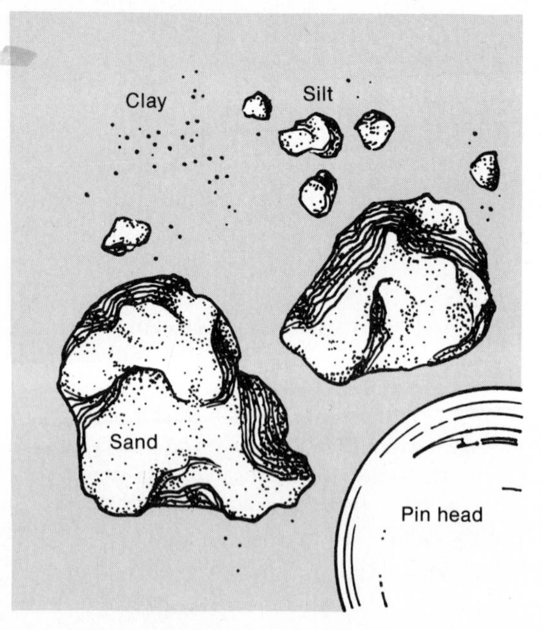

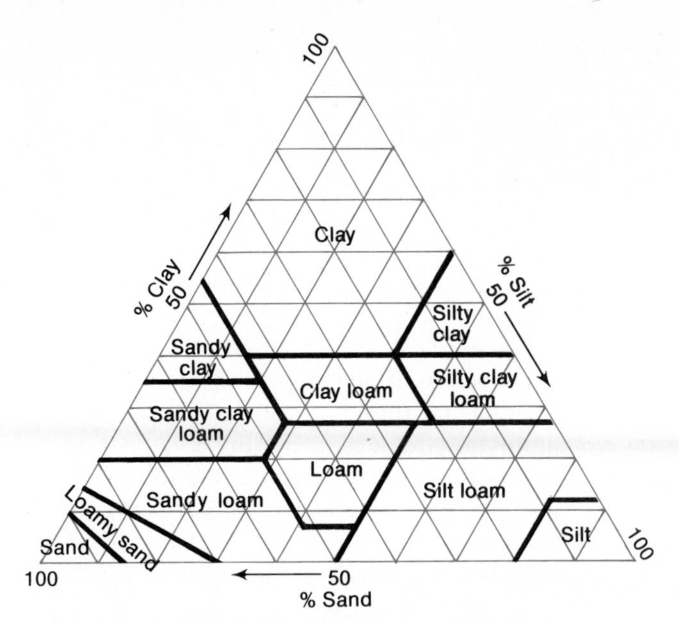

Figure 12-10. The standard texture triangle.

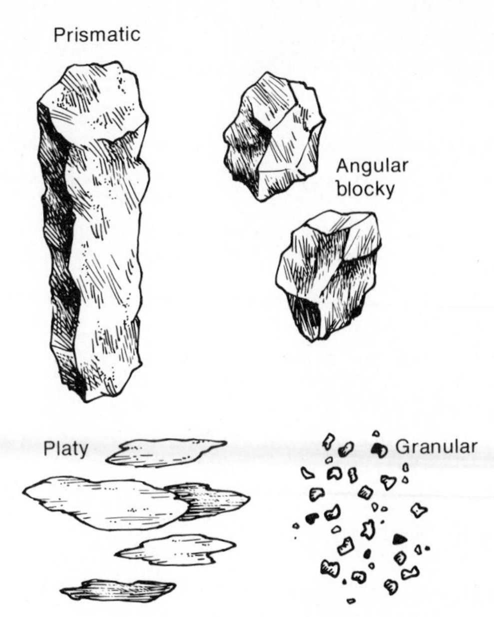

Figure 12-11. Commonly recognized structural types (greatly magnified).

Structure

The individual particles of most soils tend to aggregate into larger masses or clumps, called *peds*, which determine the structure of the soil. The size, shape, and stability of such aggregates have a marked influence on the ease of movement of water, air, and organisms (including plant roots) in the soil, and consequently on soil fertility.

Some soils, particularly those composed largely of sand, do not develop a true structure, which is to say that the individual grains do not aggregate into peds. Silt and clay particles readily aggregate in most instances. Other things being equal, aggregation is usually greatest in moist soils and least in dry ones. There is no generally accepted classification of structure types, although Figure 12-11 shows some of the more common types of aggregations. Aeration and drainage are usually facilitated by aggregates of intermediate size; both massive and fine structures tend to inhibit these processes.

Structure is an important determinant of a soil's porosity and permeability. *Porosity* refers to the amount of pore space between the soil particles and between the peds. As such, it is a measure of the capacity of the soil to hold water and air. *Permeability* is the ability of the soil to allow water to pass through it. Although these two concepts appear to be related, the relationship is not simple; that is, the most porous materials are not necessarily the most permeable. Clay, for example, is the most porous separate, but it is the least permeable because it soaks up water rather than allowing it to pass through.

Colloids

The colloidal complex of soil is one of its most vital characteristics. Colloids are composed of microscopic particles smaller than about 0.1 micrometer in size. They may be both organic and inorganic, and they represent the chemically active soil particles. When mixed with water, the tiny particles remain suspended indefinitely as a homogenous, murky solution. Colloids function as a bank in which plant nutrients can be stored for future use, to be withdrawn when needed. Some, but not all, have remarkable storage capacities. Colloids are also major determinants of a soil's water-holding capacity. They function as a virtual sponge, soaking up water, whereas the inert, coarse soil particles can maintain only a surface film of water. Inorganic colloids consist of clay in thin, crystalline, platelike forms created by the chemical alteration of larger particles. Organic colloids represent decomposed organic matter in the form of humus. Both varieties are able to attract and hold cations in great quantities. Their cation exchange capacity is critical in providing nutrients for plant growth. In general, it is valid to state that the greater the colloidal content of a soil, the greater its inherent fertility.

Acidity/Alkalinity

Nearly all nutrients are provided to plants in solution since plants (and animals) are only able to absorb nutrients when dissolved in a liquid. An overly alkaline soil solution is inefficient in dissolving minerals and releasing their nutrients. On the other hand, if the solution is highly acidic, the nutrients are likely to be dissolved and leached away too rapidly for plant roots to absorb them. The optimum situation, then, is for the soil solution to be neutral, neither too alkaline nor too acidic.

The chemist's symbol for the measure of acidity/alkalinity of a solution is pH, which is based on the relative concentration of active hydrogen ions. The scale

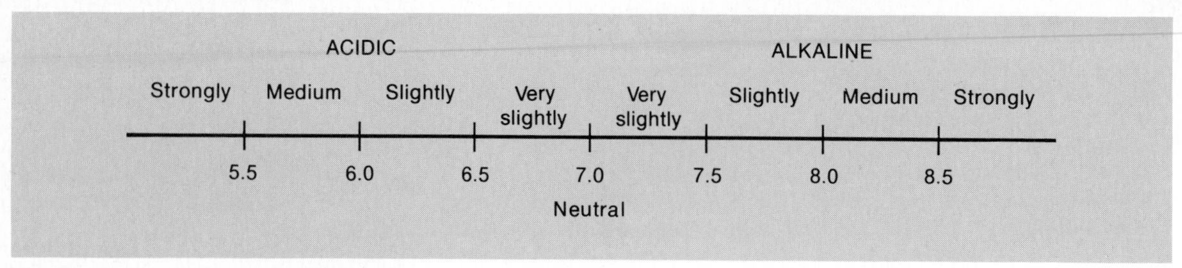

Figure 12-12. The standard pH scale.

ranges from 0 to 14; the lower end represents acidic conditions, higher numbers indicate alkaline conditions. Neutral conditions are represented by a value of 7, which is most suitable for the great majority of plants as well as being the optimum situation for the activities of most microorganisms. See Figure 12-12.

Soil Profiles

The development of any soil is expressed in two dimensions—depth and time—which have a close relationship with one another. The weathering of parent material, the addition of organic matter, and various chemical and biological reactions all combine to create new soil. As these soil-forming processes continue to operate—unless they are overbalanced by an even faster rate of soil loss through erosion—the soil becomes continually deeper. Along with deepening usually comes an increasing vertical variation in soil characteristics. Such properties as texture, structure, color, porosity, density, organic content, pH, colloidal complex, and others begin to show differences with depth. These differences are a clear indication of increasing time; they only develop gradually with age. There is no straight-line relationship between depth and age; some soils deepen and develop much more rapidly than others. However, any soil at least several inches deep that shows significant vertical variation in physical and chemical characteristics must have been in the process of development for many centuries.

The vertical variation of soil properties is not random but rather an ordered layering with depth. Soil tends to have more or less distinctly recognizable layers, called *horizons*, each with differing characteristics. The horizons are positioned approximately parallel to the land surface, one above the other, normally separated by a transition zone rather than a sharp line. A vertical cross section (as might be seen in a road cut or the side of a trench dug in a field) from the Earth's surface down through the soil layers into the parent material beneath is referred to as a *soil profile*.

Time is a critical passive factor in profile development, but the vital active factor is surface water. If there is no surface water, from rainfall or snowmelt or some other source, to infiltrate or percolate down through the soil, there can be no profile development. Descending water carries material from the surface downward, and from the topsoil into the subsoil, by eluviation and leaching. Material from above is mostly deposited in zones a few inches or a few feet lower down. In the usual pattern, topsoil becomes a somewhat depleted horizon through eluviation and leaching, and subsoil develops as a layer of accumulation due to illuviation.

Figure 12-13 presents an idealized sketch of a well-developed soil profile, in which five horizons are differentiated. These five horizons are summarized below:

O—The O horizon is the immediate surface layer in which organic matter, both fresh and decaying, makes up most of the volume. It is sometimes designated as the A_0 horizon.

A—This is a mineral horizon that also contains considerable organic matter. It is usually dark in color. It represents a zone of eluviation from which certain constituents (especially clay, iron, and aluminum) have been removed, leaving a relative concentration

Figure 12-13. Idealized diagram of a soil profile. The true soil, or solum, consists of the O, A, and B horizons.

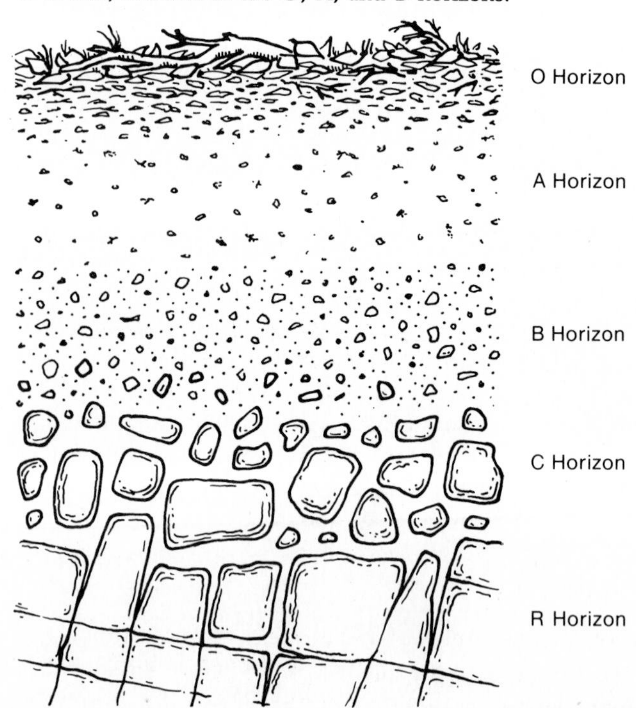

of others (such as quartz and other resistant minerals in larger particle size).

B—The B horizon is a mineral horizon of illuviation where most of the materials removed from above have been deposited. Usually of heavier texture, greater density, and relatively greater clay content than the A horizon.

C—The C horizon is unconsolidated parent material (regolith) beyond the reach of plant roots and most soil-forming processes except weathering. It is lacking in organic matter.

R—This horizon is consolidated bedrock.

The true soil, or *solum*, includes only the top three horizons: O, the organic surface layer; A, the topsoil; and B, the subsoil. Soil scientists sometimes recognize a number of variations of this basic profile pattern, especially by subdividing the major horizons on the basis of relatively minor vertical differences. The A horizon, for example, may have subdivisions labeled A_1, A_2, A_3, etc.

A "normal" profile with all horizons is typical of a humid area on well-drained but gentle slopes in an environment that has been undisturbed for a relatively long period of time. In many parts of the world, however, such idealized conditions do not pertain, and the soil profile may be "abnormal"—having one horizon particularly well developed, or one horizon missing altogether, or a fossil horizon formed under a different past climate, or an accumulation of a hardpan, or surface layers removed through accelerated erosion, or some other variation. Moreover, many soils are too young to have evolved a normal profile. A soil containing only an A horizon atop partially altered parent material (C horizon) is said to be *immature*. The formation of an illuvial B horizon is normally an indication of a *mature* soil.

The profile is such an important indicator of the characteristics and capabilities of a soil that it is the principal diagnostic factor in soil classification. The almost infinite variety of soils in the world usually are grouped and classified on the basis of differences exhibited in their profiles.

Pedogenic Regimes

Soil-forming factors have a great variety of characteristics and thus can coexist and interact in almost limitless variations to produce soils of all descriptions. Although the world pattern that results is exceedingly complex and only imperfectly known, nevertheless, soils can be classified into various groups and types on the basis of specific characteristics or associations or relationships.

Fundamental to an understanding of soil classification and distribution is the realization that only a handful of major pedogenic (i.e., soil-forming) regimes exist. These regimes can be thought of as environmental settings in which certain physical/chemical/biological processes prevail. The major regimes are distinguished largely on the basis of climate, primarily as reflected in temperature and moisture availability (see Figure 12-14), and secondarily on the basis of vegetation cover.

In regions where there is normally a surplus of moisture—which is to say that annual precipitation exceeds annual evapotranspiration—water movement in the soil is predominantly downward, and leaching is a prominent process. In such areas where temperatures are relatively high throughout the year, *laterization* is the dominant regime; where winters are long and cold, *podzolization* predominates; and where the soil is saturated most of the time due to poor drainage, *gleization* is notable. Drier regions have a moisture deficit because potential evapotranspiration is greater than precipitation, so the principal soil moisture movement is upward (through capillarity) and leaching is limited. *Calcification* and *salinization* are the principal pedogenic regimes under these conditions.

Although these five major regimes produce relatively distinctive soils, there is a multitude of local variations and transitional situations. However, the major regimes are the most useful at the broadest, or global, scale of generalization.

Laterization

The processes associated with laterization are typical of the warm, moist regions of the world. A significant annual moisture surplus is a requisite condition for laterization. This soil type is most prominent, then, in the humid tropics and subtropics, including regions dominated by forest, shrub, and savanna vegetation.

Figure 12-14. Schematic graph to show the relative temperature and moisture relationships of the five principal pedogenic regimes.

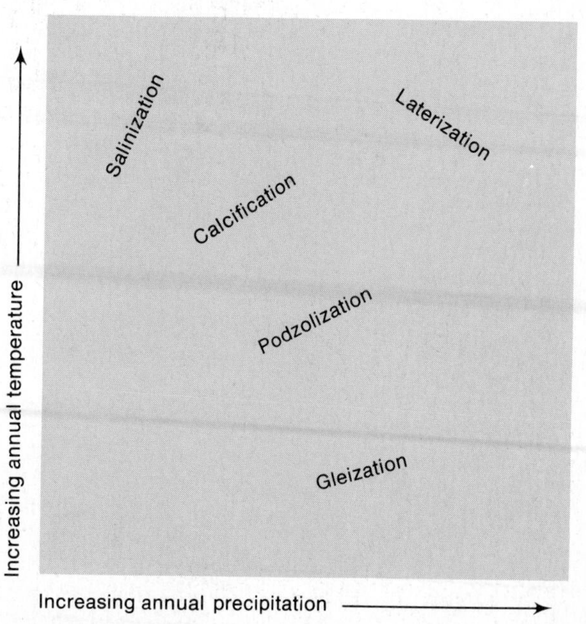

A laterization regime is characterized by rapid weathering of parent material, dissolution of nearly all minerals, and the speedy decomposition of organic matter. Probably the most distinctive feature of laterization is the leaching away of silica, the most common constituent of most rock and soil and highly resistant to being dissolved. That silica is removed in this fashion indicates the extreme effectiveness of chemical weathering and leaching under this regime. Most other minerals are also leached out rapidly, leaving behind primarily iron and aluminum oxides and barren grains of quartz sand. This residue normally imparts a reddish color to the resulting soil. See Figure 12-15. The A horizon is highly eluviated and leached, whereas the B horizon has a considerable concentration of illuviated materials.

Although the natural vegetation of these regions is normally luxuriant, plant litter is rapidly decomposed, and little humus is incorporated into the soil. This is due to the superabundance of soil microorganisms and the favorable environment for their activities. Even so, plant nutrients are not totally removed by leaching because the natural vegetation, particularly in a forest, quickly absorbs many of the nutrients in solution. If the vegetation is relatively undisturbed by human activities, this regime has the most rapid of nutrient cycles, and the soil is not totally impoverished by the speed of mineral decomposition and leaching. Where the forest is cleared for agriculture or some other human purpose, however, most base nutrients are likely to be lost from the cycle because the tree roots that would bring them up are gone. The soil, then, rapidly becomes impoverished, and hard crusts of iron and aluminum compounds are likely to form.

The general term applied to soils produced by laterization is *latosols*. These soils sometimes develop to great depths (tens of feet or several meters) because of the strong weathering activities and the fact that laterization continues year-round in these benign climates. Most latosols have little to offer as agricultural soils, for the reasons noted above, but laterization often produces such concentrations of iron and aluminum oxides that mining them can be profitable.

Podzolization

Podzolization also occurs in regions with a positive moisture balance and involves considerable leaching, but beyond those two characteristics, it bears little similarity to laterization. It is found primarily in areas where the vegetation has limited nutrient requirements and where the plant litter is itself acidic. These conditions are most prominent in middle- and high-latitude locales with a coniferous forest cover. Thus podzolization is largely a Northern Hemisphere phenomenon because there is not much land in the higher middle latitudes south of the equator. The typical location for podzolization is under a boreal forest in subarctic climates, which is found only in the Northern Hemisphere.

Figure 12-15. Laterization produces deeply weathered, reddish soils, as in this scene from the island of Viti Levu in Fiji. [TLM photo.]

In these cool regions, chemical weathering is slow, but mechanical weathering from frost action is relatively rapid during the unfrozen part of the year. Moreover, much of the land was bulldozed by Pleistocene glaciers, leaving an abundance of broken rock debris at the surface. The underlying bedrock (source of the surface glacial deposits) consists mostly of ancient crystalline rocks rich in quartz and aluminum silicates and is poor in base minerals important in plant nutrition. The boreal forests, dominated by conifers, require little in the way of soil nutrients, and their litter returns little in the way of nutrient minerals when it decays. The plant litter is largely needles and twigs, which accumulate on the surface of the soil and decompose slowly. Microorganisms do not thrive in this environment, so humus production is retarded. Moisture is relatively abundant in summer, so that leaching of whatever nutrient bases are present, along with iron/aluminum oxides and colloidal clays, from the topsoil is relatively complete.

Podzolization, then, produces soils that are shallow, acidic, and with a fairly distinctive profile. There is usually an O horizon at the surface consisting of largely undecomposed plant litter. The A horizon is eluviated to a silty or sandy texture and is so leached as to appear bleached. It usually has an ashy, light gray color (*podzol* is derived from a Russian word meaning "ashes"), which is imparted by its high content of silica. The illuviated B horizon is a receptacle for the iron/aluminum oxides and clay minerals leached from above and has a sharply contrasting darker color (sometimes with an orange or yellow tinge). Soil fertility is generally low, and a crumbly structure makes the soil very susceptible to accelerated erosion if the vegetation cover is disturbed, whether by human activities or by such natural agencies as wildfire.

Gleization

Gleization is a regime that is restricted to waterlogged areas, normally in a cool climate. Although occasionally widespread, it is generally much more limited in occurrence than the regimes previously discussed. Poor drainage that produces a waterlogged environment can be associated with flat land, but it can also result from a topographic depression, a high water table, or various other conditions. In North America, gleization is particularly prominent in areas around the Great Lakes where recent glacial deposition has interrupted preglacial drainage patterns.

The general term for soils produced by gleization is simply *gley soils*. They characteristically have a dark-colored, highly organic A horizon, where decomposition proceeds slowly because bacteria are inhibited by the lack of oxygen in a waterlogged situation. This slow decay yields organic acids that cannot oxidize iron to produce reddish colors, and the pH is invariably low. The B horizon in a true gley soil is poorly developed, but where waterlogging is more limited, a distinctive B horizon may show various colors.

Gley soils are too acidic and oxygen-poor to be productive for anything but water-tolerant vegetation. If drained artificially, however, and fertilized with lime to counteract the acid, their fertility can be greatly enhanced.

Calcification

In semiarid and arid climates where precipitation is less than potential evapotranspiration, leaching is either absent or transitory. Natural vegetation in such areas consists of grasses or shrubs. See Figure 12-16. Calcification is the dominant pedogenic process in these regions, as typified by the prairies of North America, the steppes of Eurasia, and the savannas and steppes of the subtropics.

Both eluviation and leaching are restricted by the absence of percolating water, so materials that would be carried downward in other regimes become concentrated in the soil. Moreover, there is considerable upward movement of water by capillary action in dry periods. Calcium carbonate ($CaCO_3$) is the most important chemical compound active in a calcification regime. It is carried downward by limited leaching after a rain and often is concentrated in the B horizon to form a dense layer (called a *hardpan*), then brought upward by capillary water and by grass roots, and finally returned to the soil when the grass dies. Little clay is formed because of the limited amount of chemical weathering. Organic colloidal material, however, is often present in considerable quantity.

Where calcification takes place under undisturbed grassland, the resulting soils are likely to have remarkable agricultural productivity. Humus from decaying grass parts yields abundant organic colloidal material, imparting a dark color to the soil and contributing to a structure that is favorable for the retention of both nutrients and soil moisture. Grass roots tend to bring calcium up from the B horizon sufficiently to inhibit or delay the formation of calcic hardpans. Where shrubs are the dominant vegetation, roots are fewer but deeper, so that nutrients are brought up from deeper layers, and litter accumulates at the surface with less humus being incorporated into the soil. In true deserts, the soils tend to be shallower and sandier, calcic hardpans may form near the surface, little organic matter accumulates either on or in the soil, and the soils are not very different from the parent material.

Salinization

In arid and semiarid regions, it is fairly common to find areas with inadequate drainage, particularly in enclosed valleys and basins. Moisture is drawn upward by intense evaporation. The evaporating water leaves behind various salts in or on the surface of the soil, sometimes in such

Figure 12-16. Native grassland (to the right of the fence) on the high plains of southern Colorado, with the Sangre de Cristo Range in the background. Wheatgrass and alfalfa have been planted to the left of the fence. [Courtesy USDA-Soil Conservation Service.]

quantity as to impart a brilliant white surface color to the land. These salts, which are mostly chlorides and sulfates of calcium and sodium, are toxic to most plants and soil organisms, and the resulting soil is able to support very little biotic life apart from a few salt-tolerant grasses and shrubs.

Soils developed under such a salinization regime sometimes can be made productive through careful water management. Irrigation water can provide a bloom to the land, but artificial drainage is equally necessary or the salt accumulation aspects of salinization will be intensified. Indeed, human-induced salinization has ruined good agricultural land in various parts of the world many times in the past.

Soil Classification

One of the most significant goals that can be achieved by scholarly studies is the development of classification systems. If phenomena can be meaningfully classified (i.e., arranged in logical groups, orders, and hierarchies), it becomes easier to remember them and to understand the relationships among them. Our consideration of physical geography thus far has included various classifications. In no other aspect of the subdiscipline, however, is the matter of classification more complicated than with soil. The principal reason for this complexity is that so many variables are involved, in regard to factors that influence soil formation as well as characteristics that develop in the soil.

Earlier Classifications

Unlike the classification of most other natural phenomena, the bases of soil classification have changed significantly over the last century, largely in response to increasing knowledge about the soil itself. Early global soil classifications were mostly on the basis of relationships with the parent material, which was then considered to be the primary factor in determining soil characteristics. New approaches introduced in the late 1800s, particularly by Russian soil scientists, focussed on other aspects of soil genesis, especially the interrelated roles of climate and vegetation. Various worldwide classifications based on these pedogenetic aspects have been devised, most of them simply being modifications of previous systems.

In 1938, the U.S. Department of Agriculture gave its official blessing to a classification system that became widely accepted in the United States and in much of the rest of the world, again based largely on soil/climate/vegetation relationships. This system was generally popular with geographers because it was a *genetic* classification, which means that it was based on soil-forming conditions and processes. Thus the relationship between the soil and other aspects of the environment was an integral part of the system, and the distribution of major soil categories could be fruitfully compared with the distribution of other major environmental complexes, particularly climate and vegetation.

The Seventh Approximation

During recent years, however, soil scientists have become increasingly disenchanted with this 1938 classification because it did not seem to be either sufficiently precise or comprehensive enough to encompass the incredible variety of soils that actually exist and to place them in proper niches in a hierarchy. Accordingly, a new system was developed slowly and laboriously by the Soil Survey Staff of the U.S. Department of Agriculture during the 1950s and 1960s. This *U.S. Comprehensive Soil Classification System* is widely referred to simply as the *Seventh Approximation* because six tentative versions of the plan had previously been circulated among soil scientists for testing and reactions. The Seventh Approximation has now largely superseded the 1938 system among soil scientists in the United States and is increasingly being accepted (sometimes with modifications) in other countries. The acceptance is slow in some countries (and by some scientists in the United States) because it is such a radical departure from previous classification systems and because the terminology is confusing, at least initially. Even so, it is likely to become increasingly acceptable internationally, and equally likely to continue to undergo revisions.

The basic characteristic that sets the Seventh Approximation apart from previous systems is that it is *generic* in nature, which means that it is organized systematically on the basis of observable soil characteristics. The focus is on the existing properties of a soil rather than on the environment, genesis, or properties it would possess under virgin conditions. The logic of such a system is theoretically impeccable. A soil type has certain properties that can be observed, measured, and at least partly quantified. This makes it possible to determine in what ways, and how much, one soil differs from another.

Like other logical generic systems, the Seventh Approximation is a hierarchical system, which means that it has several levels of generalization, with each higher level encompassing several members of the level immediately below it. There are only a few similarities among all the members of the highest-level category, but the number of similarities increases with each step downward in the hierarchy, so that in the lowest-level category, all members have mostly the same properties.

At the highest level (the smallest scale of generalization) of the Seventh Approximation is the soil *order*, of which only ten are recognized worldwide. The soil orders, and many of the lower-level categories as well, are distinguished from one another largely on the basis of certain diagnostic properties, which are often expressed in combination to form *diagnostic horizons*. The two basic types of diagnostic horizon are the *epipedon* (based on the Greek word *epi*, meaning "over" or "upon"), which is essentially the A horizon or the combined O/A horizon, and the *subsurface horizon*, which is roughly equivalent to the B horizon. (Note that all A and B horizons are not necessarily diagnostic, so the terms and concepts are not synonymous.)

Soil orders are subdivided into *suborders*, of which about 50 are recognized in the United States. The third level consists of *great groups*, which number about 225 in the United States. Successively lower levels in the classification are *subgroups*, *families*, and *series*. This logical system permits the user to work at any level of the classification. About 12,000 soil series have been identified in the United States to date, and the list will undoubtedly be expanded in the future. For the purpose of comprehending general world distribution patterns, however, we need only the order or suborder level.

The relationship between the Seventh Approximation categories and those of previous soil classification systems is vague because of the different criteria of classification. At the lowest level—the series—there is considerable agreement between the Seventh Approximation and the 1938 USDA system, but at other levels they diverge due to their different theoretical foundations. Most geographers have been slow to accept the Seventh Approximation system because they were generally comfortable with the conspicuous correlations that were apparent between high-level categories (called "great soil groups") of the 1938 classification and global patterns of such other environmental elements as climate and vegetation. In other words, the genetic bias of the 1938 system seemed well suited for broad-scale geographic uses, whereas the generic approach of the Seventh Approximation yields categories that have less clear relationships with other environmental components. This is not to say that there are no relationships between Seventh Approximation categories and pedogenetic factors, but rather that the relationships are not as apparent nor as easy to demonstrate in terms of global or continental patterns of distribution.

Nevertheless, the Seventh Approximation system has superseded all previous classifications, at least in the

Focus on the Tongue-Twisting Terminology of the Seventh Approximation System

The U.S. Comprehensive Soil Classification System, or Seventh Approximation (1962), utilizes an entirely new nomenclature "invented" for the purpose. This nomenclature is a set of *synthetic* terminology, which means that groups of basic syllables are rearranged in multiple combinations to produce names for the various soil types. Only a few dozen syllables are combined to create the several hundred names in the four upper levels of the hierarchy. The beauty of each term thus created is that it is highly descriptive of one or more properties of the soil so named.

The awkwardness of the terminology is threefold:

1. Most of the terms are "new" and have never before appeared in print or in conversation. Thus they look strange, and the words do not easily roll off one's tongue. It is almost a small new language.
2. Many of the words are difficult to write and great care must be taken in the spelling of seemingly bizarre combinations of letters.
3. Many of the syllables have such similarity of sound that precise clarity of pronounciation is difficult, although this is essential for the resulting words to serve their purpose.

These difficulties are readily apparent and have not served to endear the Seventh Approximation to new or casual users.

Nevertheless, the new terminology has a sound theoretical base, and once the user has an acquaintance with the vocabulary, every syllable of every word displays a factual meaning. Almost all of the syllables are derived from Greek and Latin roots, in contrast to the English and Russian terms used in previous systems. In some cases, the appropriateness of the classical derivatives may be open to question, but the uniformity and logic of the terminology extend throughout the system. The infrequent or casual user of the system will experience much greater difficulty with the nomenclature than the person who works with it in more detail.

The top level in the system's hierarchy is the soil *order*. The names of orders are made up of three or four syllables, the last of which is always *sol* (from the Latin word *solum*, or "soil"). The next-to-last syllable consists of a single linking vowel (*i* or *o*). The syllable or two that begin the word contain the formative element of the name and imply a distinctive

Name Derivations of Soil Orders

Name of order	Root	Derivation	Formative element for suborders	Pronunciation as in
Entisol	ent	"Recent"	ent	Enter
Vertisol	vert	Latin, *verto*: "to turn"	ert	Vertical
Inceptisol	incept	Latin, *inceptum*: "beginning"	ept	Inept
Aridisol	arid	Latin, *aridus*: "dry"	id	Arid
Mollisol	moll	Latin, *mollis*: "soft"	oll	Doll
Spodosol	spod	Greek, *spodos*: "wood ash"	od	Trod
Alfisol	alf	Aluminum, *al*, and iron, *fe*	alf	Alfalfa
Ultisol	ult	Latin, *ultimus*: "ultimate or last"	ult	Ultimate
Oxisol	ox	"Oxide"	ox	Ox
Histosol	hist	Greek, *histos*: "tissue"	ist	History

Name Derivations of Soil Suborders

Root	Derivation	Connotation	Example of suborder name
alb	Latin, *albus*: "white"	Presence of a bleached eluvial horizon	Alboll
and	Japanese, *ando*: a volcanic soil	Derived from pyroclastic material	Andept
aqu	Latin, *aqua*: "water"	Associated with wetness	Aquent
ar	Latin, *arare*: "to plow"	Horizons are mixed	Arent
arg	Latin, *argilla*: "white clay"	Presence of a horizon containing illuvial clay	Argid
bor	Greek, *boreas*: "northern"	Associated with cool conditions	Boroll
ferr	Latin, *ferrum*: "iron"	Presence of iron	Ferrod
fibr	Latin, *fibra*: "fiber"	Presence of undecomposed organic matter	Fibrist
fluv	Latin, *fluvius*: "river"	Associated with floodplains	Fluvent
fol	Latin, *folia*: "leaf"	Mass of leaves	Folist
hem	Greek, *hemi*: "half"	Intermediate stage of decomposition	Hemist
hum	Latin, *humus*: "earth"	Presence of organic matter	Humult
ochr	Greek, *ochros*: "pale"	Presence of a light-colored surface horizon	Ochrept

United States, and geographers must accommodate to it and must learn to appreciate its obvious merits. One method of easing the transition from previous systems to the Seventh Approximation is by pointing out equivalent categories in the two systems. Thus geographers explaining world soil distribution patterns often list great soil groups (from the 1938 classification) that are approximately equivalent to orders or suborders of the Seventh Approximation system. Unfortunately, the search for environmental relationships sometimes blurs the searcher's judgment, with the result that equivalence may be perceived when it does not actually occur. Thus the reader is warned that the equivalents between categories in the two systems listed in Table 12-2 are only approximate.

Name Derivations of Soil Suborders (*continued*)

orth	Greek, *orthos*: "true"	Most common or typical group	Orthent
plag	German, *plaggen*: "sod"	Presence of a human-induced surface horizon	Plaggept
psamm	Greek, *psammos*: "sand"	Sandy texture	Psamment
rend	Polish, *rendzina*: a type of soil	Significant calcareous content	Rendoll
sapr	Greek, *sapros*: "rotten"	The most decomposed stage	Saprist
torr	Latin, *torridus*: "hot and dry"	Usually dry	Torrox
trop	Greek, *tropikos*: "of the solstice"	Continually warm	Tropert
ud	Latin, *udus*: "humid"	Of humid climates	Udoll
umbr	Latin, *umbra*: "shade"	Presence of a dark-colored surface horizon	Umbrept
ust	Latin, *ustus*: "burnt"	Of dry climates	Ustert
xer	Greek, *xeros*: "dry"	Annual dry season	Xeralf

Samples of Name Derivations for Great Groups

Root	Derivation	Connotation	Example of great group name
calc	Latin, *calcis*: "lime"	Presence of calcic horizon	Calciorthid
ferr	Latin, *ferrum*: "iron"	Presence of iron	Ferrudolf
natr	Latin, *natrium*: "sodium"	Presence of a natric horizon	Natraboll
pale	Greek, *paleos*: "old"	An old development	Paleargid
plinth	Greek, *plinthos*: "brick"	Presence of plinthite	Plenthoxeralf
quartz	German, *quarz*: "quartz"	High quartz content	Quartzipsamment
verm	Latin, *vermes*: "worm"	Notable presence of worms	Vermudoll

Example of Complete Hierarchical Nomenclature of a Soil Series

Order	Entisol
Suborder	Aquent
Great Group	Cryaquent
Subgroup	Sphagnic Cryaquent
Family	skeletal, mixed, acid, Sphagnic Cryaquent
Series	Aberdeen

characteristic of the soil order. Thus the names of the ten soil orders contain ten syllables that serve as formative elements for the order names, as well as appearing in all names in the next three levels of the classificational hierarchy.

The *suborder* is the second level in the hierarchy. All suborder names contain two syllables. The second syllable identifies the order to which the suborder belongs; for example, *ent* for the order Entisol. The first syllable indicates a distinctive characteristic of the suborder. For example, *aqu* is derived from the Latin word for "water," so *Aquent* is the suborder within the Entisols that exhibits evidence of wetness. The names of the four-dozen suborders are constructed from about two-dozen root elements, as shown in the table of name derivations on pages 256 and 257.

The third level of the hierarchy contains *great groups*. Their names are constructed by grafting one or more syllables, as prefixes, to the name of the appropriate suborder. Hence, a Cryaquent is a cold soil that is a member of the Aquent suborder of the Entisol order, because *cry* comes from the Greek word for "coldness." The formative elements (prefix syllable or syllables) of great group names are derived from about 50 root words, samples of which are presented in the second table on this page.

The next level is called *subgroup*, of which more than 1000 are recognized in the United States. Each subgroup name consists of two words, the second of which is the same as that of the relevant great group. The first name is derived from one or more of the formative elements introduced at higher levels in the hierarchy, with a handful of exceptions. Thus a *Sphagnic Cryaquent* is of the order Entisol, the suborder Aquent, the great group Cryaquent, and it contains sphagnum moss (derived from the Greek word *sphagnos* for "bog.")

The *family* is the penultimate level in the hierarchy. It is not given a proper name, but is simply described by one or more lower-case adjectives, as "a skeletal, mixed, acid family."

Finally, at the lowest level of the hierarchy, there is the *series*. Series are named for geographic locations, as "Miami" or "Aberdeen."

In summation, neither the system not its terminology is simple, but both are logical and relatively comprehensive and can contribute significantly to our understanding of the geography of soils.

The Mapping Question

A map is a basic tool of geographic study, and one of the fundamental problems that confronts any geographic inquiry is how the phenomena under study should be mapped. There are precise techniques for mapping features that are located at specific points or phenomena that are spread uniformly over an area. However, if the object of study is a generalized abstraction, its depiction on a map is necessarily imprecise, and choice of the mapping technique becomes more subjective. The higher levels of the soil classification are generalized abstractions. They do not represent phenomena that actually exist, but are generalizations of average or typical conditions

TABLE 12-2 Approximate Equivalents in U.S. Soil Classification Systems

Seventh approximation order	1938 System	
	Mostly included	Significantly included
Entisols	Lithosols Regosols	Alluvial soils Low-humic gley soils
Vertisols	Grumusols	
Inceptisols	Ando soils Brown forest soils Half-bog soils Sols bruns acides Tundra soils	Alluvial soils Brown podzolic soils Latosols Low-humic gley soils
Aridisols	Desert soils Red desert soils Sierozems Solonchak	Brown soils Calcisols Reddish brown soils Solonetz soils
Mollisols	Brunizems Chernozems Chestnut soils Reddish prairie soils Rendzina soils	Alluvial soils Brown soils Calcisols Humic gley soils Reddish chestnut soils Solonetz soils
Spodosols	Brown podzolic soils Ground-water podzols Podzol soils	
Alfisols	Gray-brown podzolic soils Gray wooded soils Planosols	Degraded chernozems Noncalcic brown soils Reddish brown soils Reddish chestnut soils Solonetz soils
Ultisols	Ground-water laterite soils Humic latosols Reddish brown lateritic soils Red-yellow podzolic soils	Latosols
Oxisols	Laterite soils	Latosols
Histosols	Bog soils	

over broad areas. Thus the selection of an appropriate mapping technique can significantly influence our understanding of the situation.

Most soil maps use the same timeworn technique of areal expression. If one soil type (at whatever level of generalization is being studied) is more common in an area than any other, that area will be classified by the prevailing type and will be colored or shaded appropriately. Such a map is effective in indicating the principal type of soil in each region and is useful in portraying the general distribution of the major soil types. Plate 5 and Figure 12-17 show the national distribution of soil orders in this fashion, which is commensurate with other maps (i.e., climate and vegetation) in this book.

Maps of this type are compiled through generalization of data, and the smaller the scale, the greater the generalization that is necessary. Thus more intricate patterns can be shown on the larger scale map of the United States (Figure 12-17) than on the smaller scale world map (Plate V). In either case, the map is only as good as its generalizations are meticulous.

Soil maps according to the Seventh Approximation often appear visually confusing and seem to lack understandable patterns, particularly when compared with similar maps based on earlier classifications. This confusion arises, at least in part, because the new system has a different logical framework, which perhaps calls for a different mapping technique. An alternative method of cartographic expression has been proposed by geographer Philip Gersmehl in an attempt to portray the distribution patterns more clearly. His technique is to plot the proportional occurrence of particular soil orders in certain-size areas (he chose 1000 km² as the mapping unit) by means of point symbols (dots) on the map. This approach, as applied to the nine soil orders that are widespread in the United States, is demonstrated in Figures 12-18 through 12-26. Its great advantage is in providing a clearer picture of the distribution of a particular order; its princi-

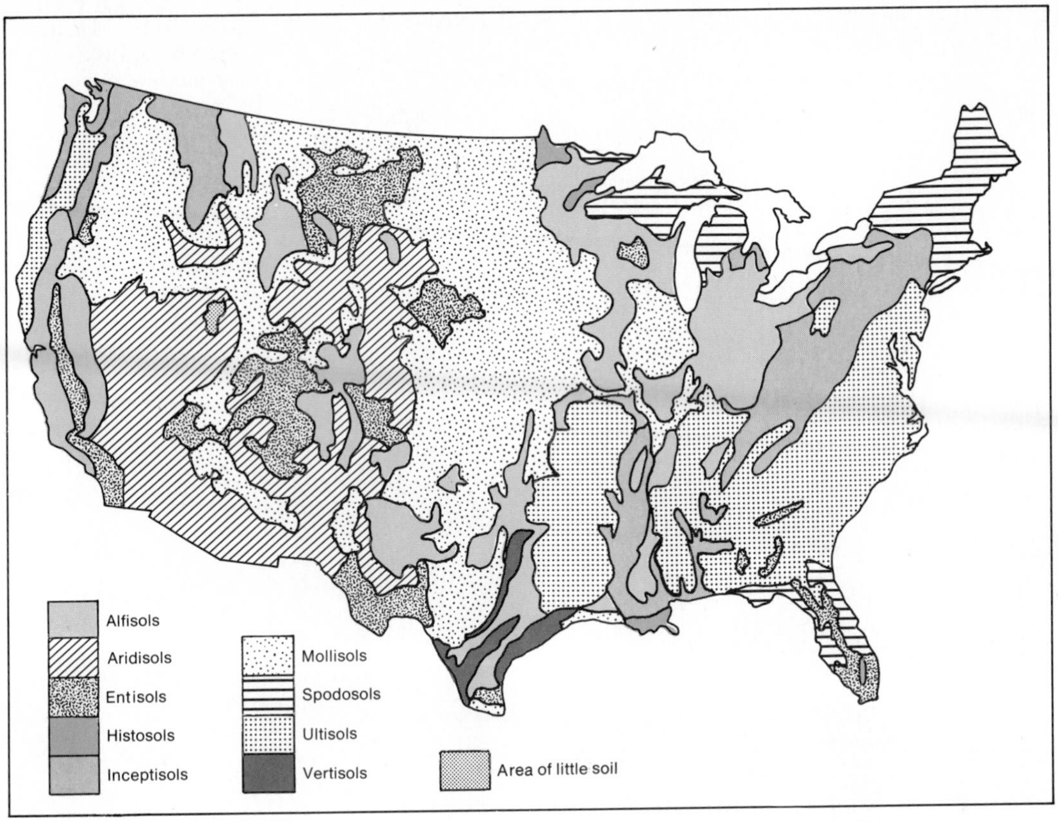

Alfisols
Aridisols
Entisols
Histosols
Inceptisols
Mollisols
Spodosols
Ultisols
Vertisols
Area of little soil

Figure 12-17. Generalized soil map of the conterminous states.

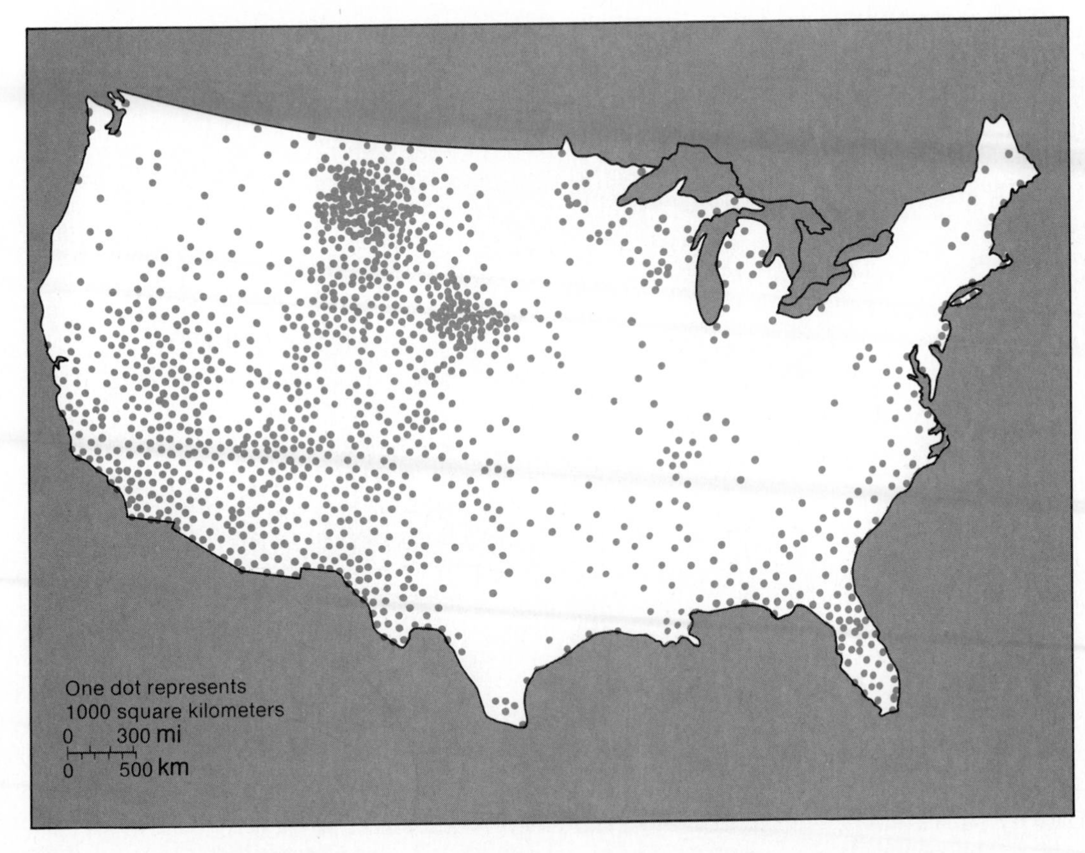

Figure 12-18. Distribution of Entisols in the conterminous states. [After Philip J. Gersmehl, "Soil Taxonomy and Mapping," *Annals of the Association of American Geographers,* 67 (September 1977), 423. By permission of the Association of American Geographers.]

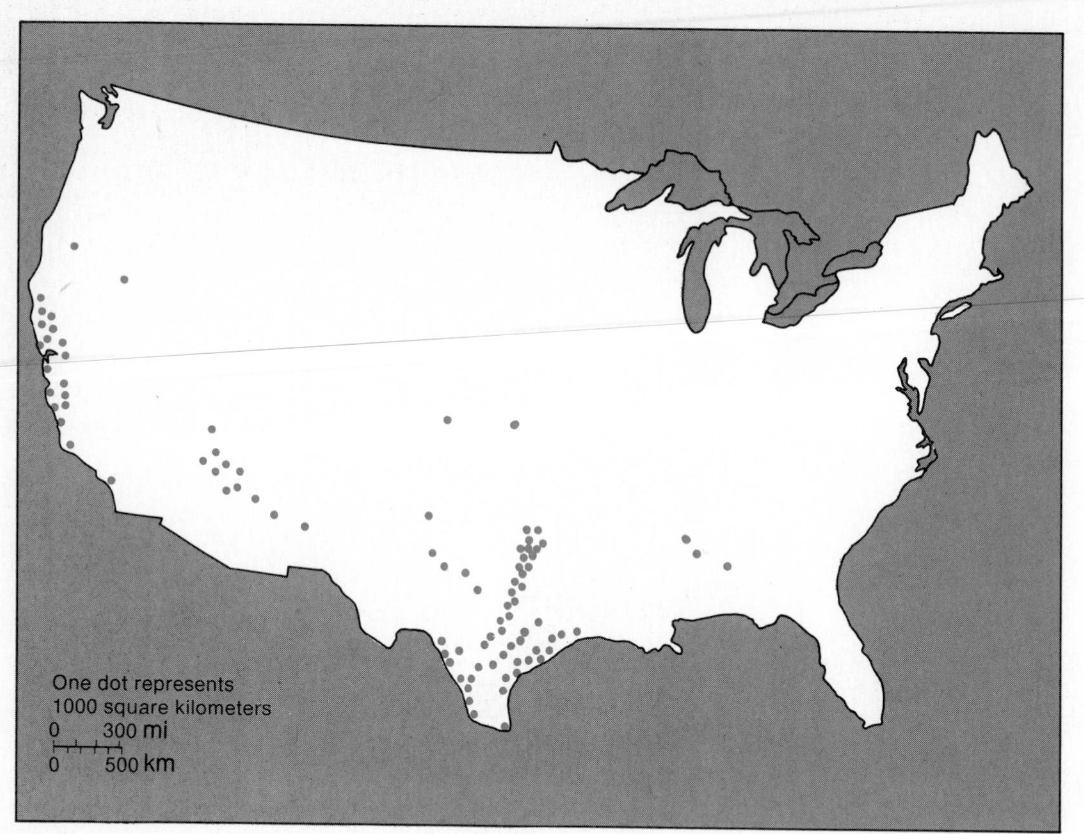

Figure 12-19. Distribution of Vertisols in the conterminous states. [After Phillip J. Gersmehl, "Soil Taxonomy and Mapping," *Annals of the Association of American Geographers,* 67 (September 1977), 424. By permission of the Association of American Geographers.]

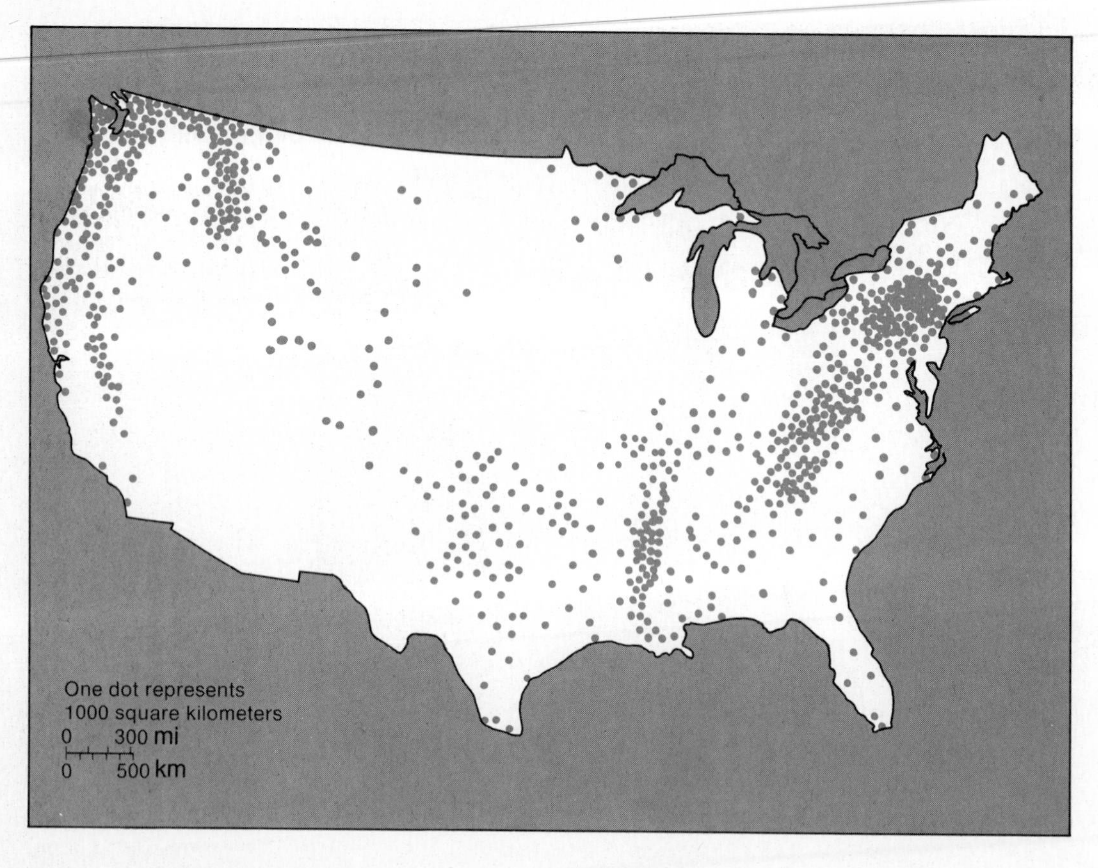

Figure 12-20. Distribution of Inceptisols in the conterminous states. [After Philip J. Gersmehl, "Soil Taxonomy and Mapping," *Annals of the Association of American Geographers,* 67 (September 1977), 423. By permission of the Association of American Geographers.]

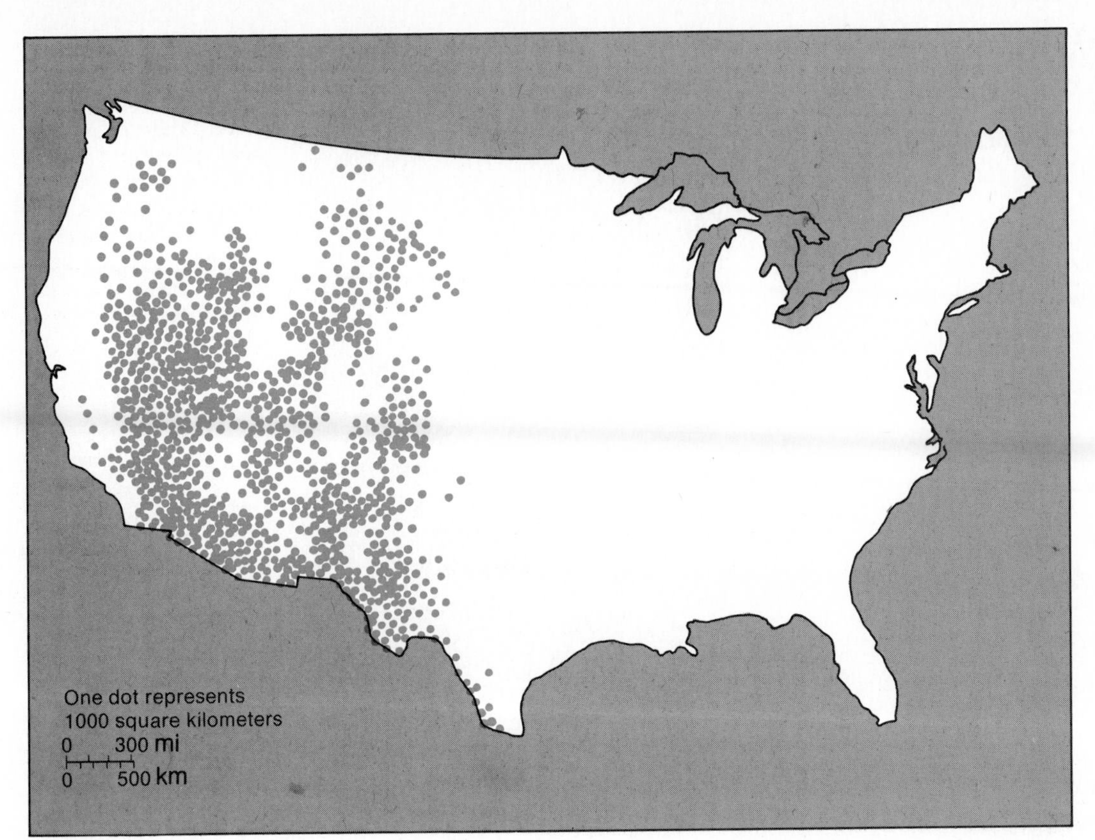

Figure 12-21. Distribution of Aridisols in the conterminous states. [After Philip J. Gersmehl, "Soil Taxonomy and Mapping," *Annals of the Association of American Geographers,* 67 (September 1977), 424. By permission of the Association of American Geographers.]

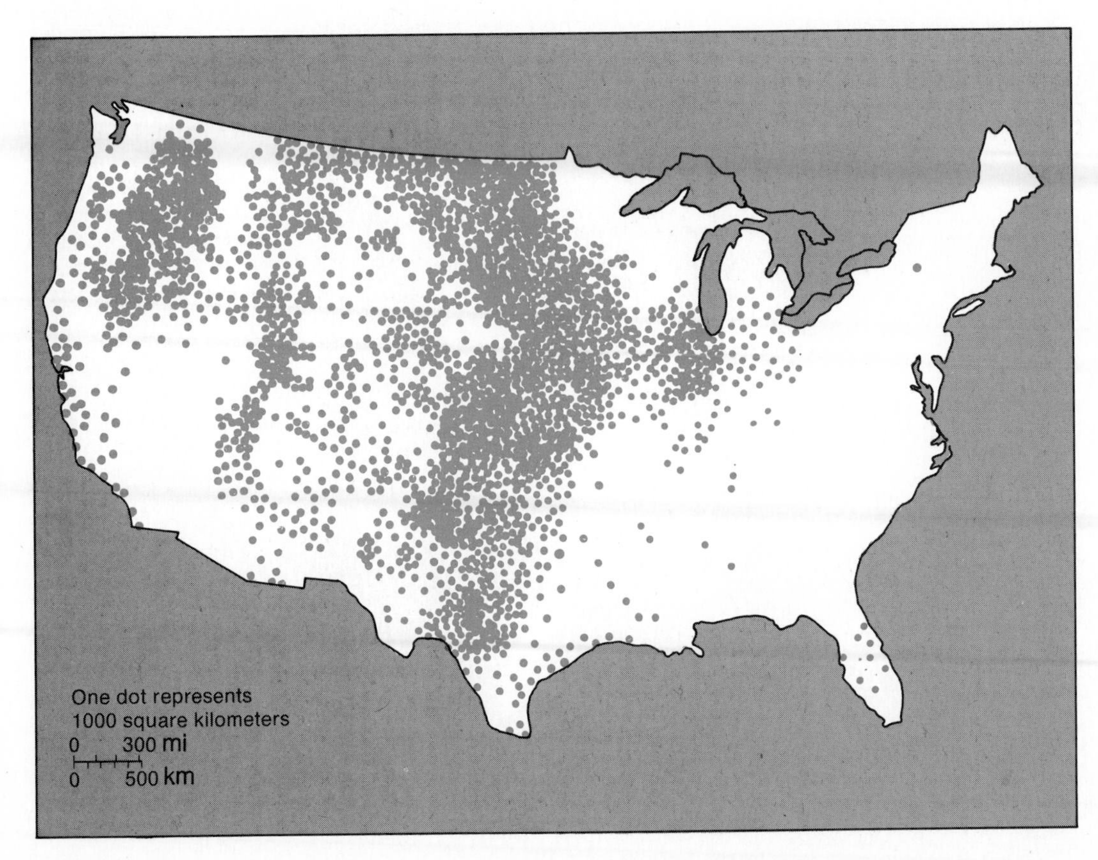

Figure 12-22. Distribution of Mollisols in the conterminous states. [After Philip J. Gersmehl, "Soil Taxonomy and Mapping," *Annals of the Association of American Geographers,* 67 (September 1977), 425. By permission of the Association of American Geographers.]

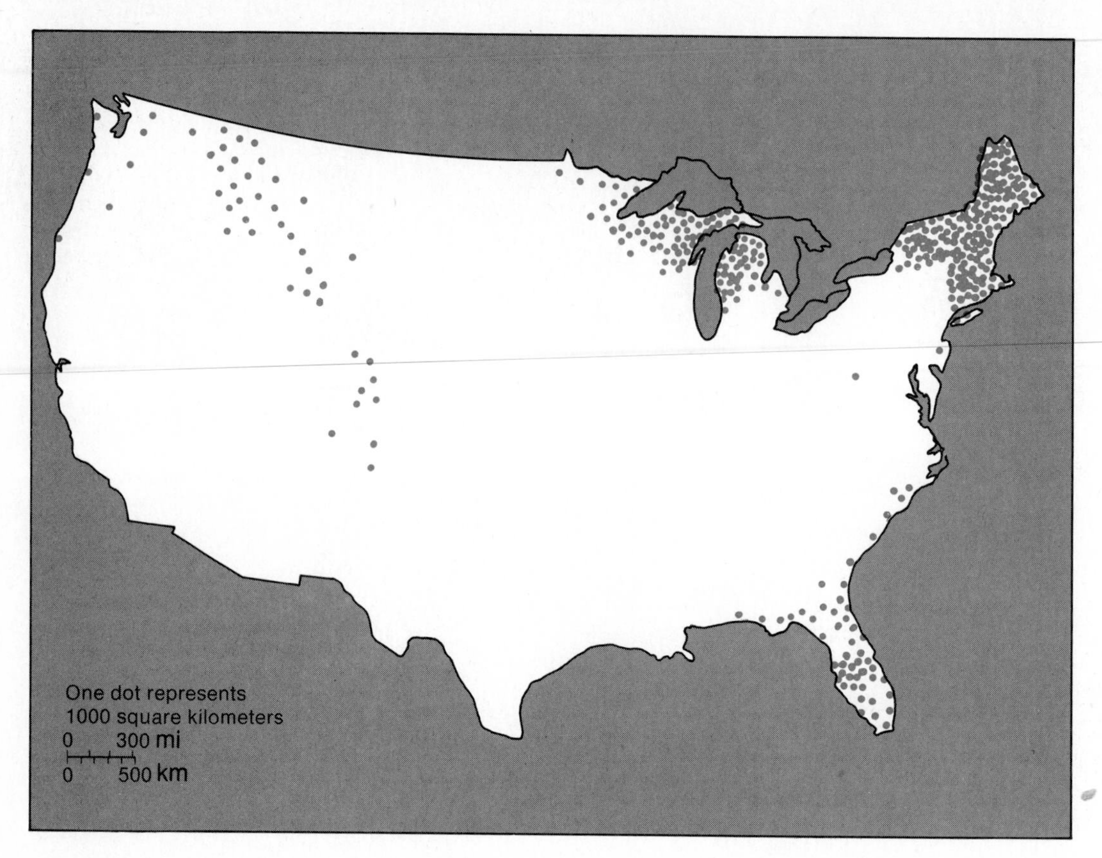

Figure 12-23. Distribution of Spodosols in the conterminous states. [After Philip J. Gersmehl, "Soil Taxonomy and Mapping," *Annals of the Association of American Geographers,* 67 (September 1977), 425. By permission of the Association of American Geographers.]

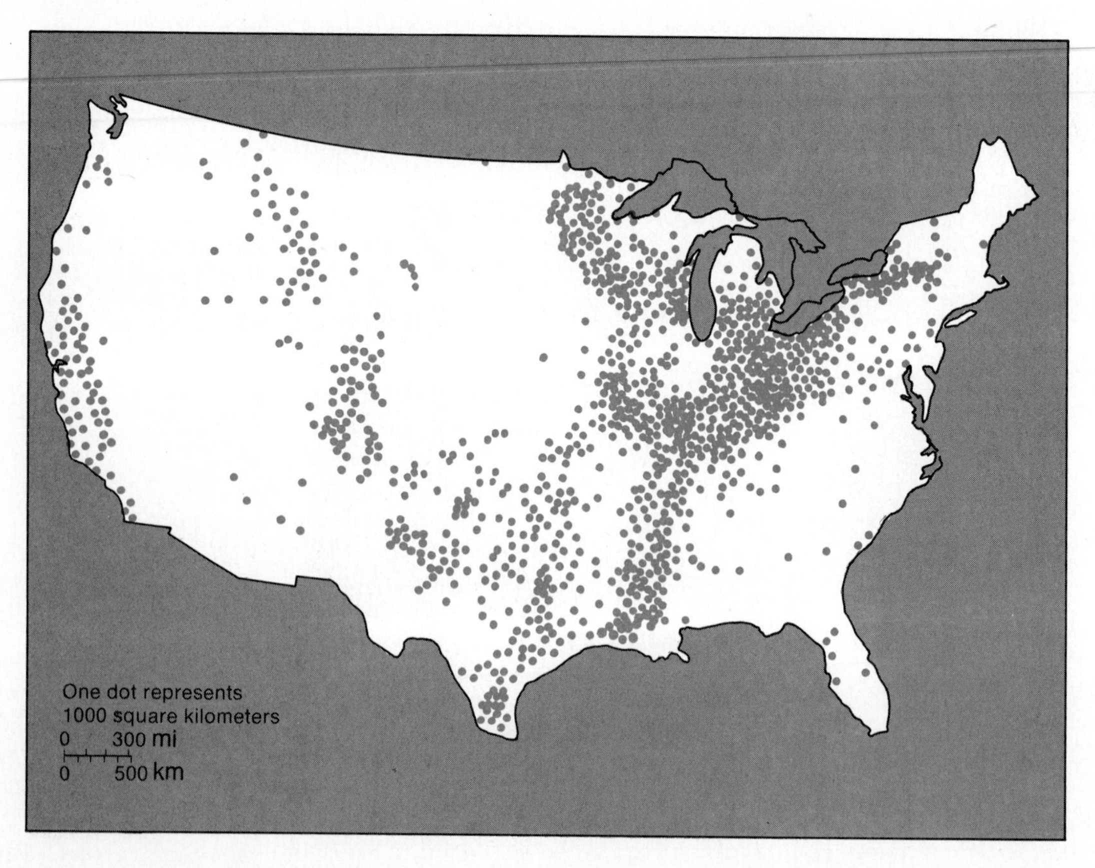

Figure 12-24. Distribution of Alfisols in the conterminous states. [After Philip J. Gersmehl, "Soil Taxonomy and Mapping," *Annals of the Association of American Geographers,* 67 (September 1977), 426. By permission of the Association of American Geographers.]

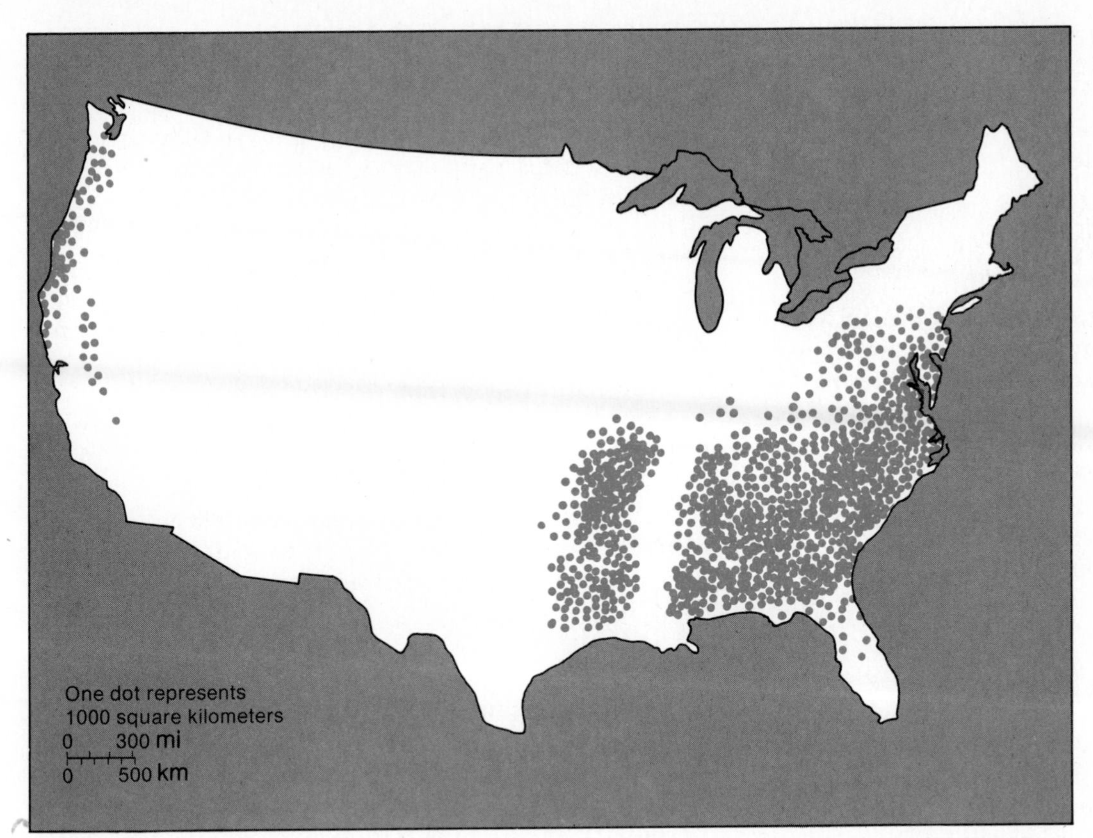

Figure 12-25. Distribution of Ultisols in the conterminous states. [After Philip J. Gersmehl, "Soil Taxonomy and Mapping," *Annals of the Association of American Geographers,* 67 (September 1977), 426. By permission of the Association of American Geographers.]

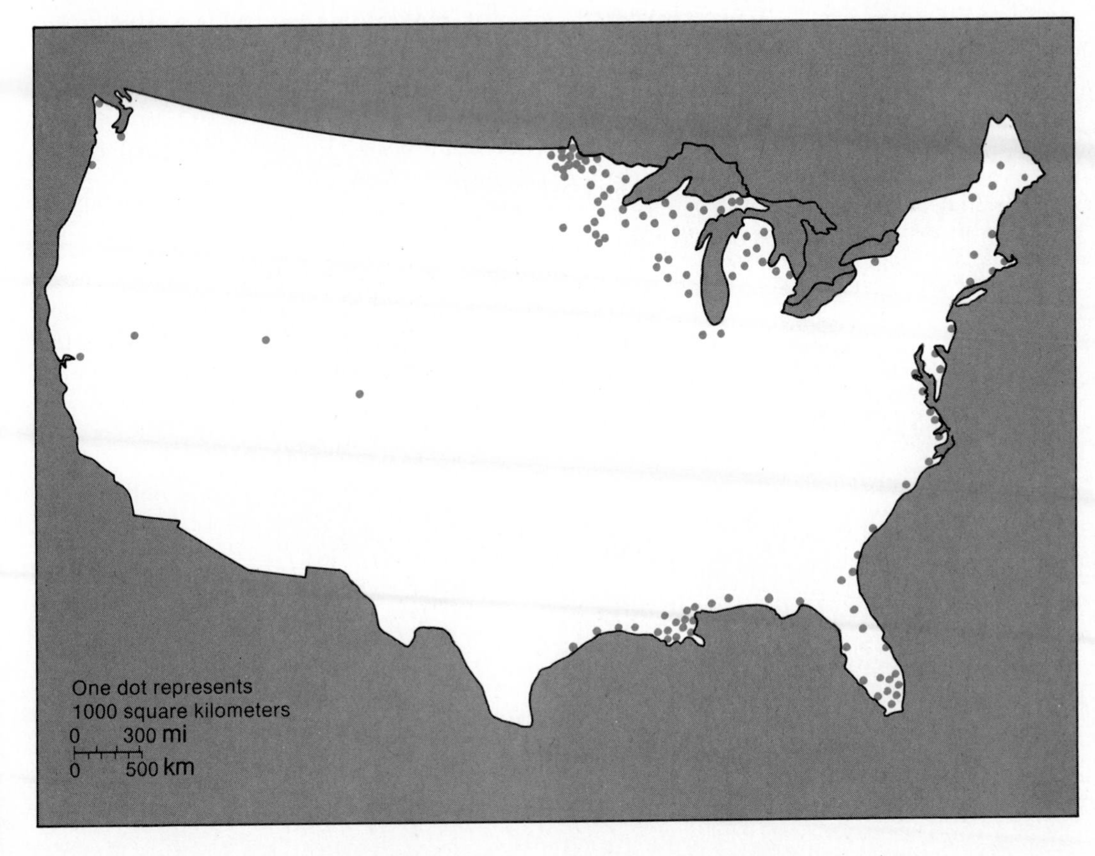

Figure 12-26. Distribution of Histosols in the conterminous staes. [After Philip J. Gersmehl, "Soil Taxonomy and Mapping," *Annals of the Association of American Geographers,* 67 (September 1977), 427. By permission of the Association of American Geographers.]

pal disadvantage lies in the difficulty of absorbing discrete patterns from several different maps for a unified perspective.

There is no simple solution to the mapping of complex data on a small scale.

Global Distribution of Major Soils

Of the ten orders of soils recognized in the U.S. Comprehensive Soil Classification System, nine are arranged in a hierarchy in which each succeeding order represents an increased degree of weathering, particularly as expressed by mineral alteration and profile development. See Figure 12-27. The tenth order, Histosols, is essentially an organic soil that lies outside the concept of this hierarchy. We will discuss each order in turn, with limited reference to suborders.

Entisols

Entisols are the least well developed of all soils. They have experienced little mineral alteration and are virtually without pedogenic horizons. Their undeveloped state is usually a function of time (the very name of the order connotes recency); most Entisols are surface deposits that have not been in place long enough for pedogenetic factors to have had much effect, such as floodplains, rapidly eroding slopes, or desert sandhills. Some Entisols, however, are very old, and the lack of horizon development is due to a mineral content that does not alter readily, or a very cold climate, or some other factor totally unrelated to time. The distribution of Entisols is therefore very widespread and cannot be specifically correlated with particular moisture or temperature conditions or with certain types of vegetation or parent materials. See Figure 12-28. In the United States, Entisols are most prominent in the dry lands of the West but are found in most other parts of the country as well. Entisols are commonly thin and/or sandy and have limited productivity, although those developed on recent alluvial deposits tend to be quite fertile. See Figure 12-29.

Figure 12-27. The relationship between soil orders and degree of weathering. The relative height of the band for each order is proportional to the approximate worldwide areal extent of that order. Nine of the ten soil orders fit this hierarchy; the base of the pyramid represents histosols and nonsoil surfaces, which have no relationship to degree of weathering.

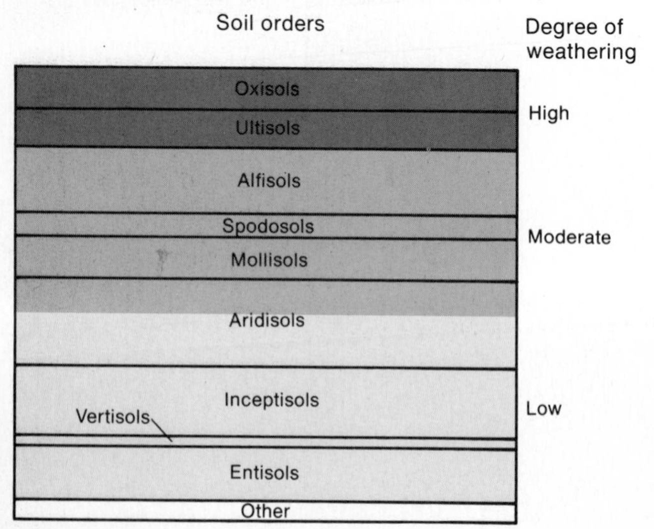

There are five suborders of Entisols. *Aquents* occupy wet environments where the soil is more or less continuously saturated with water; they may be found in any temperature region, from very cold to tropical. *Arents* lack horizons because of human interference, particularly due to large agricultural or engineering machinery. *Fluvents* form on recent water-deposited sediments that have satisfactory drainage. *Orthents* develop on recent erosional surfaces. *Psamments* occur in sandy situations, where the sand is either shifting or stabilized by vegetation.

Vertisols

Vertisols comprise a specialized type of soil with a large quantity of clay that becomes a dominant factor in the soil's development. The clay content of Vertisols is described as "swelling" or "cracking" clay. This clay-type soil has an exceptional capacity for absorbing water: When moistened, it swells and expands; as it dries, deep wide cracks are produced. These cracks are sometimes an inch (2.5 cm) wide and as much as a yard (1 m) deep. Some surface material falls into the cracks, and more is washed in when it rains. When the soil is wetted again, more swelling takes place and the cracks close. This alternation of wetting and drying, expansion and contraction, produces a churning effect that mixes the soil constituents (the name *Vertisol* connotes an inverted condition), inhibits the development of horizons, and may even cause minor irregularities in the surface of the land. See Figure 12-30.

An alternating wet and dry climate is needed for Vertisols to be produced because the sequence of swelling and contraction is necessary for its formation. Thus the wet/dry regime of tropical and subtropical savannas provides an ideal climate, but there must also be the proper

Figure 12-28. Generalized world distribution of Entisols.

Figure 12-29. Profile of an Entisol in Thailand. As is characteristic of Entisols, differentiating horizons are difficult to discern. [Courtesy Hari Eswaran, USDA-Soil Conservation Service.]

parent material to yield the clay minerals. Consequently, Vertisols are widespread in distribution but are very limited in extent. See Figure 12-31. The principal occurrences are in eastern Australia, India, and East Africa. They are uncommon in the United States, although prominent in some parts of Texas and California.

The fertility of Vertisols is relatively high, as they tend to be rich in nutrient bases. They are difficult to till, however, because of their sticky plasticity, so they are often left uncultivated.

The four principal suborders of Vertisols are distinguished largely on the frequency of "cracking," which is based on the climatic regime. *Torrerts* are in arid regions where the cracks remain open most of the time. *Uderts* are found in humid areas where cracking is irregular. *Usterts* are associated with monsoonal climates and have a more complicated cracking pattern. *Xererts* occur in mediterranean climates and have cracks that open and close regularly once each year.

Figure 12-30. A Vertisol profile from the central valley of California. Many cracks typically are found in Vertisols. [Courtesy USDA-Soil Conservation Service.]

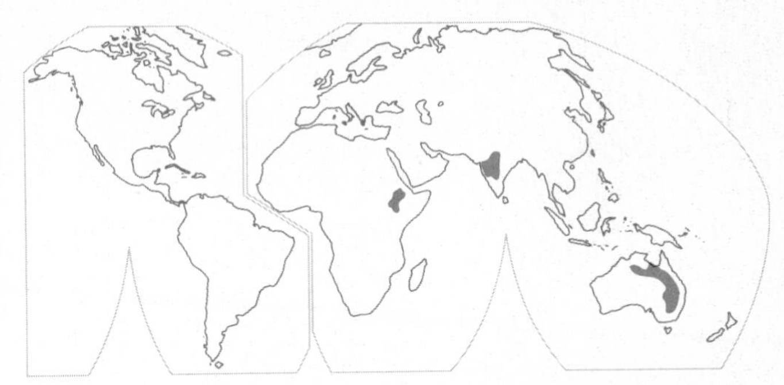

Figure 12-31. Generalized world distribution of Vertisols.

Inceptisols

Another immature order of soils is the Inceptisols. Their distinctive characteristics are relatively faint, not yet prominent enough to produce diagnostic horizons. See Figure 12-32. If the Entisols can be called youthful soils, the Inceptisols might be classed as adolescent. They are primarily eluvial soils and lack illuvial layers.

Like Entisols, they are widespread over the world in differing environments. See Figure 12-33. Also like Entisols, they include a variety of fairly dissimilar soils whose common characteristic is lack of maturity. They are most common in tundra and mountain areas but are also notable in older valley floodplains. Their world distri-

Figure 12-32. Profile of a New Zealand Inceptisol with a distinctive B horizon of white pebbly material. [Courtesy Hari Eswaran, USDA-Soil Conservation Service.]

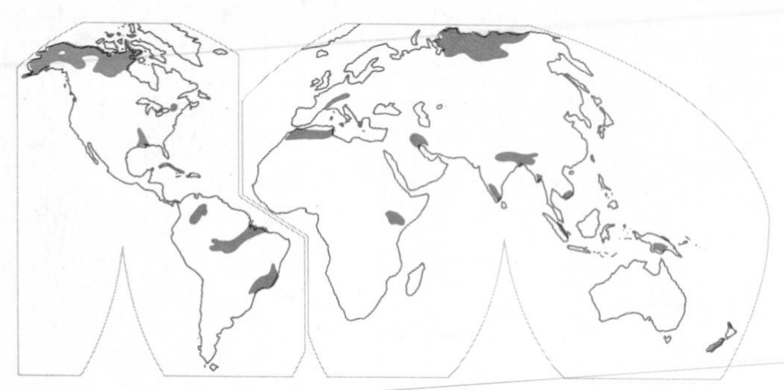

Figure 12-33. Generalized world distribution of Inceptisols.

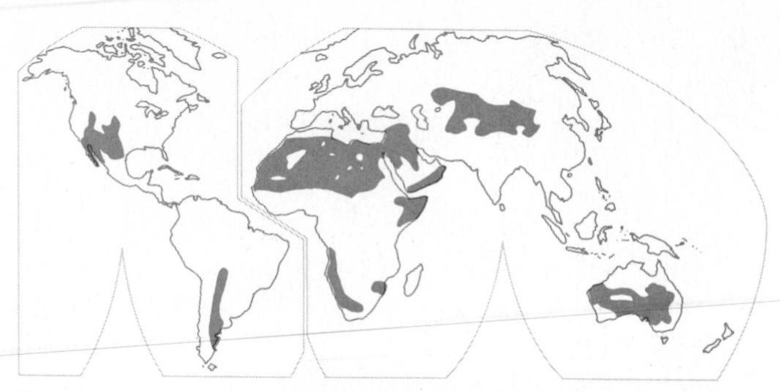

Figure 12-34. Generalized world distribution ofAridisols.

bution pattern is very irregular. This is also true in the United States, where they are most typical of the Appalachian Mountains, the Pacific Northwest, and the lower Mississippi Valley.

The six suborders of Inceptisols—*Andepts*, *Aquepts*, *Ochrepts*, *Plaggepts*, *Troperts*, and *Umbrepts*—have relatively complicated distinguishing characteristics.

Aridisols

Nearly one-fifth of the land surface of the Earth is covered with Aridisols, the most extensive spread of any soil order. See Figure 12-34. They are preeminently soils of the dry lands, occupying environments that do not have enough water to remove soluble minerals from the soil. Thus their distribution pattern is largely correlated with that of desert and semidesert climate.

Aridisols are typified by a thin profile, greatly lacking in organic matter and typically sandy in texture—characteristics that are clearly associated with a dry climate and a scarcity of penetrating moisture. See Figure 12-35. The epipedon is almost invariably light in color. There are various kinds of diagnostic subsurface horizons, nearly always with a distinctly alkaline reaction. Most Aridisols are unproductive, particularly because of lack of moisture; if irrigated, however, some display remarkable fertility. The threat of salt accumulation is ever present.

Two suborders are generally recognized on the basis of degree of weathering. *Argids* have a distinctive subsurface horizon with clay accumulation, whereas *Orthids* do not.

Figure 12-35. The typical sandy profile of an Aridisol; in this case, from New Mexico. [Courtesy USDA-Soil Conservation Service.]

Mollisols

The distinctive characteristic of Mollisols is the presence of a mollic epipedon, which is a mineral surface horizon that is dark, thick, contains abundant humus and base nutrients, and retains a soft character (rather than becoming hard and crusty) when it dries out. See Figure 12-36. Mollisols can be thought of as transition soils that evolve in regions not dominated by either humid or arid conditions. They are typical of the midlatitude grasslands, and are thus most common in central Eurasia, the North American Great Plains, and the Pampa of Argentina. See Figure 12-37.

The grassland environment generally maintains a rich clay/humus content in the soil. The dense, fibrous mass of grass roots permeates uniformly through the epipedon and to a lesser extent into the subsurface layers. There is an almost continuous process of death and decay of plant parts, which produces nutrient-rich humus that is in demand by the living grass. Thus the nutrient cycle is an active one.

Mollisols on the whole are probably the most productive of the soil orders. They are generally derived from loose parent material, rather than from bedrock, and tend to have favorable structure and texture for cultivation. They are not overly leached, so nutrients are generally retained within reach of plant roots. Moreover, they comprise a favored habitat for earthworms, which contribute to softening and mixing the soil.

The seven recognized suborders of mollisols—*Albolls*, *Aquolls*, *Borolls*, *Rendolls*, *Udolls*, *Ustolls*, and *Xerolls*—are distinguished largely, but not entirely, on the basis of relative wetness/dryness.

Figure 12-36. A South Dakota Mollisol with a typical mollic epipedon, a surface horizon that is dark and replete with humus. [Courtesy USDA-Soil Conservation Service.]

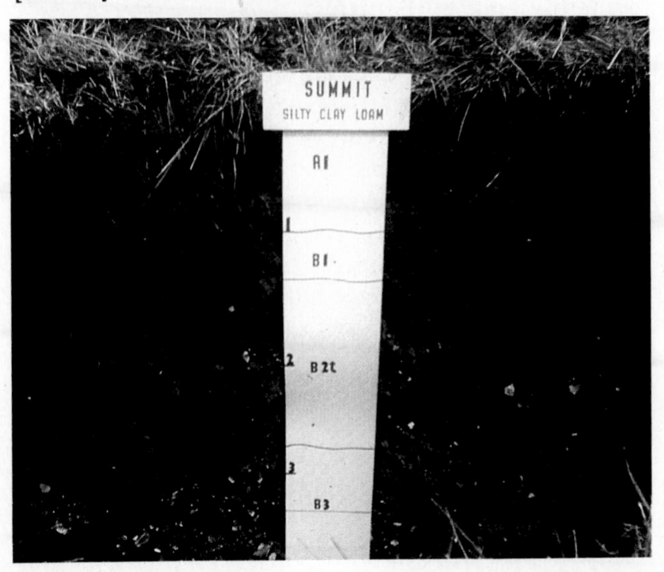

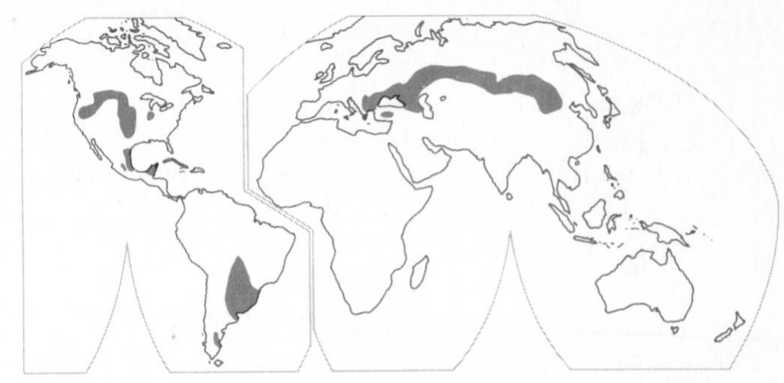

Figure 12-37. Generalized world distribution of Mollisols.

Spodosols

The key diagnostic feature of a Spodosol is the occurrence of a spodic subsurface horizon, an illuvial layer where organic matter and aluminum accumulate, and which has a dark, sometimes reddish, color. The overlying A horizon is light colored and heavily leached. See Figure 12-38.

Figure 12-38. The profile of a Spodosol often shows a light-colored A horizon overlying a reddish B horizon. This example is from Quebec. [Courtesy USDA-Soil Conservation Service.]

Above that is usually an O horizon of organic litter. Such a soil is a typical result of podzolization; the podzol soils of the old classification are equivalent.

Spodosols are notoriously infertile. They have been leached of useful nutrients and are acid throughout. They do not retain moisture well and are lacking in humus and often in clay.

Spodosols are most widespread in areas of coniferous forest where there is a subarctic climate. See Figure 12-39. Other types of soils (such as Alfisols, Histosols, and Inceptisols) also occupy these regions, however, and Spodosols are sometimes found in quite different environments, such as poorly drained portions of Florida.

Of the four suborders of Spodosols, most widespread are the *Orthods*, which represent the typical Spodosols. *Aquods*, *Ferrods*, and *Humods* are differentiated particularly on the basis of the amount of iron in the spodic horizon.

Alfisols

Alfisols are the most wide-ranging of the maturely developed soils, occurring extensively in the low- and middle-latitude portions of the continents. They are found in a variety of temperature and moisture regimes and under diverse vegetation associations. By and large they tend to be associated with transitional environments and are less characteristic of regions that are particularly hot or cold or wet or dry. Their global distribution, as shown in Figure 12-40, is extremely varied. They are also widespread in the United States, with particular concentrations in the Midwest.

Figure 12-39. Generalized world distribution of Spodosols.

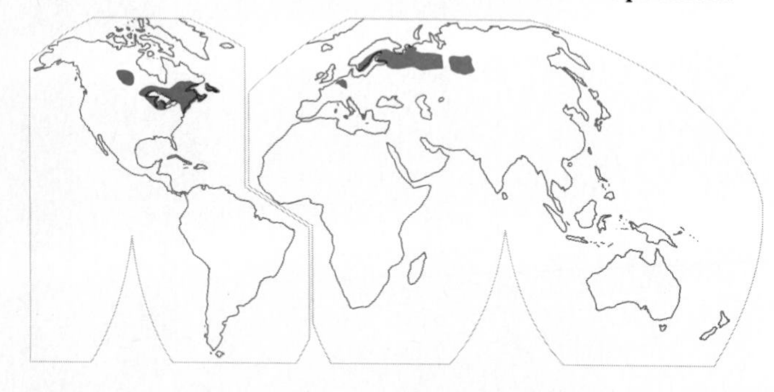

Figure 12-40. Generalized world distribution of Alfisols.

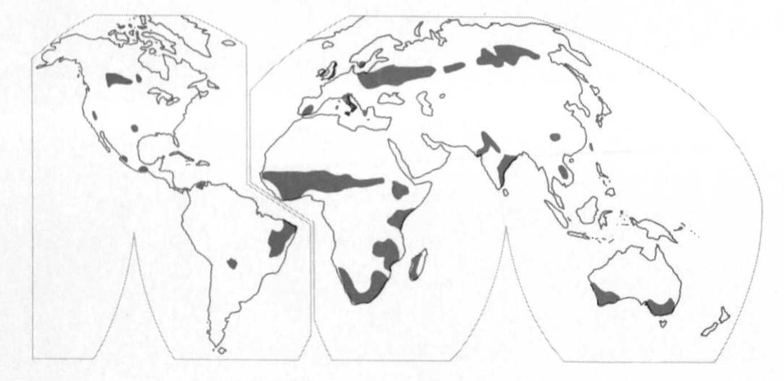

These soils are distinguished by a subsurface clay horizon and a medium to generous supply of plant nutrients and water. The epipedon (surface horizon) is ochric (light colored; see Figure 12-41), but beyond that, it has no characteristics that are particularly diagnostic and can be considered as an ordinary eluviated horizon. The relatively moderate conditions under which Alfisols develop tend to produce balanced soils that are reasonably fertile. As an order, Alfisols rank second only to Mollisols in agricultural productivity.

Alfisols have five suborders: *Aqualfs* have characteristics associated with wetness. *Boralfs* are associated with cold boreal forest lands. *Udalfs* are brownish or reddish soils of moist midlatitude regions. *Ustalfs* are similar in color but subtropical in location and usually have a hard surface layer in the dry season. *Xeralfs* are found in regions of mediterranean climate and are characterized by a massive, hard surface horizon in the dry season.

Ultisols

Ultisols are roughly similar to Alfisols except that they are more thoroughly weathered and more completely leached of nutrient bases. They have experienced greater mineral alteration than any other soil in the midlatitudes, although they also occur in the low latitudes. Many pedologists believe that the ultimate fate of Alfisols is to degenerate into Ultisols.

Typically, Ultisols are reddish in color from the significant proportion of iron and aluminum in the A horizon. Usually they have a fairly distinct layer of subsurface clay accumulation. The principal properties of Ultisols have been imparted by a great deal of weathering and leaching. See Figure 12-42. Indeed, the connotation of the name (derived from the Latin, *ultimos*) is that it represents the ultimate stage of weathering among soils in the conterminous United States. The result is a fairly deep soil that is acidic, lacking in humus, and has a relatively low fertility due to the lack of bases.

Ultisols have a fairly simple world distribution pattern. See Figure 12-43. They are mostly confined to humid subtropical climates and to some relatively youthful tropical land surfaces. In the United States, they are restricted largely to the southeastern quarter of the country.

Five suborders of Ultisols—*Aquults*, *Humults*, *Udults*, *Ustults*, and *Xerults*—are recognized. The distinction among them is largely on the basis of temperature and moisture regimes and their influence on the epipedon.

Oxisols

The most thoroughly weathered and leached of all soils are the Oxisols, which invariably display a high degree of mineral alteration and profile development. They occur mostly on ancient landscapes in the humid tropics, particularly in Brazil and in equatorial Africa, and to a lesser

Figure 12-41. An Alfisol from southern Texas. The reddish surface horizon in this profile is characteristic of Udalfs and Ustalfs of drier localities. [Courtesy USDA-Soil Conservation Service.]

Figure 12-42. A tropical Ultisol, from Thailand. It is reddish throughout its profile, indicative of much leaching and weathering. [Courtesy USDA-Soil Conservation Service.]

Figure 12-43. Generalized world distribution of Ultisols.

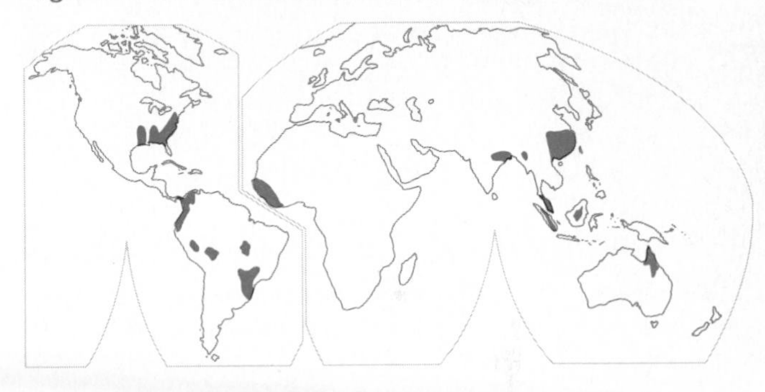

extent in southeast Asia. See Figure 12-44. They are often in a spotty distribution pattern, mixed with less-developed Entisols, Vertisols, and Ultisols. Oxisols are totally absent from the United States, except for Hawaii, where they are common.

Oxisols are essentially the products of laterization; they represent soils that were called *Latosols* in the older classification systems. Oxisols have evolved, then, in warm, moist climates, although some are now found in drier regions, which is an indication of climatic change since the soils developed. The diagnostic horizon for Oxisols is a subsurface dominated by oxides of iron and aluminum and with a minimal supply of nutrient bases (this is called an *oxic horizon*). These are deep soils but not inherently fertile. See Figure 12-45. The natural vegetation is efficient in cycling the limited nutrient supply, but if the flora is cleared (to attempt agriculture, for example), the nutrients are rapidly leached out, and the soil becomes impoverished.

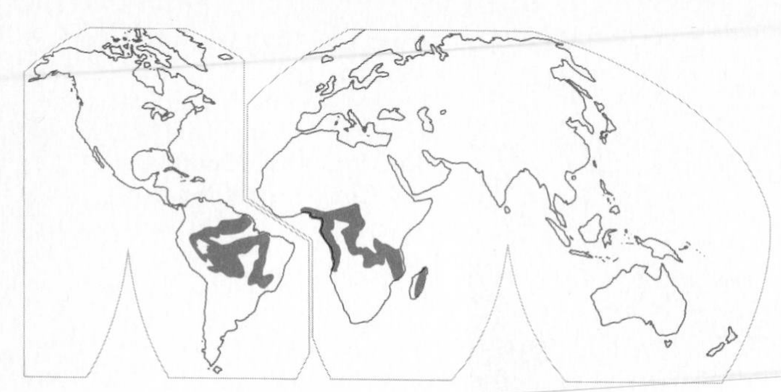

Figure 12-44. Generalized world distribution of Oxisols.

Figure 12-45. Oxisols are tropical soils of considerable impoverishment. The reddish color in this profile from Hawaii indicates considerable leaching. [Courtesy USDA-Soil Conservation Service.]

The five suborders of Oxisols—*Aquox*, *Humox*, *Orthos*, *Torros*, and *Ustox*—are distinguished from one another primarily by the effects that varying amounts and seasonality of rainfall have on the profile.

Histosols

Least important among the soil orders are the Histosols, which occupy only a small fraction of the Earth's land surface, a much smaller area than any other order. These are organic, rather than mineral, soils and invariably are saturated with water all or most of the time. They may occur in any waterlogged environment but are most characteristic in middle- and high-latitude regions that experienced Pleistocene glaciation. In the United States, they are most common around the Great Lakes, but they also occur in southern Florida and Louisiana. Nowhere, however, is their occurrence extensive.

Some Histosols are composed largely of undecayed or only partly decayed plant material, whereas others consist of a thoroughly decomposed mass of muck. See Figure 12-46. The lack of oxygen in the waterlogged soil slows down the rate of bacterial action, and the soil becomes deeper mostly by growing upward, that is, by more organic material being added from above.

Histosols are usually black and acidic and are naturally fertile only for water-tolerant plants. If drained, they can be very productive agriculturally for a short while. Before long, however, they are likely to dry and shrink and oxidize, which leads to compaction, susceptibility to wind erosion, and danger of fire.

The four suborders of Histosols—*Fibrists*, *Folists*, *Hemists*, and *Saprists*—are recognized in terms of the degree of decomposition of the plant material.

Figure 12-46. The dark color of this Histosol profile is typical. Histosols are organic soils. This example is from Brazil. [Courtesy USDA-Soil Conservation Service.]

Distribution of Soils in the United States

The proportional extent of the various soils in the United States is quite different from that of the world as a whole. This difference is due to many factors, most important being that the United States is essentially a midlatitude country, and it lacks significant expanses of area in the low and high latitudes. Table 12-3 provides a comparison of the proportional areas occupied by the ten soil orders, nationally and globally, as well as indicating the proportional areas of suborders that are significant in the United States. The statistics are generalized estimates prepared by the Soil Conservation Service of the U.S. Department of Agriculture and should not be considered as definitive.

Mollisols are much more common in the United States than in the world as a whole; they are the most prevalent soil order throughout the Great Plains and in much of the West. Also significantly more abundant in the United States are Inceptisols and Ultisols. Almost totally lacking from this country are Oxisols, an important world soil order mostly restricted to tropical areas. Other major orders that are proportionally less extensive in the United States are Aridisols and Entisols.

TABLE 12-3 Approximate Proportional Extent of Soil Orders and Suborders

		Percentage of land area occupied		
Order	Suborder	United States		World
Alfisols		13.4		14.7
	Aqualfs		1.0	
	Boralfs		3.0	
	Udalfs		5.9	
	Ustalfs		2.6	
	Xeralfs		0.9	
Aridisols		11.5		19.2
	Argids		8.6	
	Orthids		2.9	
Entisols		7.9		12.5
	Aquents		0.2	
	Fluvents		0.3	
	Orthents		5.2	
Histosols		0.5		0.8
	Fibrists		0.2	
	Hemists		0.2	
	Saprists		0.1	
Inceptisols		18.2		15.8
	Andepts		1.9	
	Aquepts		11.4	
	Ochrepts		4.3	
	Umbrepts		0.7	
Mollisols		24.6		9.0
	Aquolls		1.3	
	Borolls		4.9	
	Udolls		4.7	
	Ustolls		8.8	
	Xerolls		4.8	
Oxisols		—		9.2
Spodosols		5.1		5.4
	Aquods		0.7	
	Orthods		4.4	
Ultisols		12.9		8.5
	Aquults		1.1	
	Humults		0.8	
	Udults		10.0	
	Xerults		1.0	
Vertisols		1.0		2.1
	Uderts		0.4	
	Usterts		0.6	

Review Questions

1. What is the relationship of *weathering* to *regolith*?
2. List and briefly describe the five principal soil-forming factors.
3. Why are earthworms generally considered to be "beneficial" to humans?
4. Explain the importance of clay as a constituent of soil.
5. Describe the four forms of soil moisture.
6. What can you learn about a soil from its color?
7. What is a *soil profile*?
8. How does the U.S. Comprehensive Soil Classification (Seventh Approximation) differ from previous soil classifications?
9. Why is it so difficult to portray soil distribution with reasonable accuracy on a small-scale map?
10. Select one of the soil orders and describe its distribution and characteristics.

Some Useful References

BRADY, N. C., *The Nature and Properties of Soils*. New York: Macmillan Publishing Co., Inc., 1974.

BRIDGES, E. M., *World Soils*, 2nd ed. Cambridge, England: Cambridge University Press, 1978.

BUOL, S. W., F. D. HOLE, AND R. J. MCCRACKEN, *Soil Genesis and Soil Classification*. Ames, Iowa: Iowa State University Press, 1973.

FOTH, H. D., *Fundamentals of Soil Science*. New York: John Wiley & Sons, Inc., 1978.

PITTY, A. F., *Geography and Soil Properties*. London: Methuen & Co. Ltd., 1978.

Soil Survey Staff, *Soil Taxonomy*. Soil Conservation Service, U.S. Department of Agriculture, Agriculture Handbook No. 436. Washington, D.C.: U.S. Govt. Printing Office, 1975.

STEILA, D., *The Geography of Soils*. Englewood Cliffs, N.J.: Prentice-Hall, Inc., 1976.

13 Introduction to Landform Study

Our Massive Earth

From the human viewpoint, our planet has an almost incalculable enormity. We have noted in Chapter 1 that the Earth has a diameter of 8000 miles (12,800 km) and a circumference of 25,000 miles (40,000 km), measurements that are well beyond our normal scale of living and thinking. To comprehend the nature of such a massive body is a colossal task.

As students of physical geography, however, our endeavor is greatly simplified because we can largely ignore the interior of the Earth and concentrate our attention on the surface. Two factors make this possible: (1) The characteristics of the Earth's interior are very imperfectly understood by anyone, and (2) the focus of geographic inquiry is primarily on mankind's zone of habitation.

The Unknown Interior

Our knowledge of the interior of the Earth is scanty and is based entirely on indirect evidence. No human activity has explored more than a minute fraction of the vastness beneath the surface. No person has penetrated as much as one-thousandth of the radial distance from the surface to the center of the Earth; the deepest existing mine shaft extends a mere 2.4 miles (3.8 km) into the Earth. Nor have direct probes extended much deeper; the deepest known drill holes from which sample cores have been brought up have penetrated only a modest 7 miles (11 km) into the Earth. Thus it is clear that direct evidence of the nature and composition of the Earth's interior is based only on very superficial sampling, mere scratches on the surface.

Even so, a considerable body of inferential knowledge concerning the interior of our planet has been amassed by geophysical means, primarily by monitoring shock waves that are transmitted through the Earth from earthquakes or from man-made explosions. Analysis of such seismic waves, augmented by related data on the Earth's magnetism and gravitational attraction, has enabled the conceptualization of various models of the Earth's structure.

Figure 13-1 is one rendition of this inferred structure. Four generalized shells, arranged concentrically, have been postulated. The outermost shell, called the *crust*, is by far the thinnest, estimated to extend from as little as 5 miles (8 km) to as much as 50 miles (80 km) beneath the surface; it is shallowest under the oceans and deepest under the continents. The *mantle* is another apparently rigid, dense, solid shell extending to a depth estimated at 1800 miles (2900 km). Beneath the mantle is the *outer core*, thought to be molten (liquid), with a radius of perhaps 1500 miles (2400 km). The innermost portion is referred to as the *inner core*, a supposedly rigid mass with a radius of about 600 miles (960 km).

This generalized model of the Earth's interior is inexact and probably inadequate. The fragility of our supposed understanding of the depths was dramatically revealed in the 1960s when the notion of *continental drift*, propounded in the early 1900s but held in disdain by most scientists for half a century, was revived and expanded. Recent seismic and magnetic evidence makes it clear that the drift concept is valid, and its elaboration as the theory of *plate tectonics* is now almost universally accepted by Earth scientists. The mechanics of plate tectonics will be discussed subsequently, but the point here is that although our understanding of the crust and upper mantle has fundamentally changed in the last two decades, it is still by no means complete, which only emphasizes the vastness of our ignorance about the nature of the deeper interior of our planet.

Geographers Focus on the Surface

Even if we possessed relatively precise knowledge of the Earth's interior, however, it would not hold our attention, *as geographers*, for long. The focus of geographical study

Figure 13-1. The presumed vertical structure of the Earth's interior.

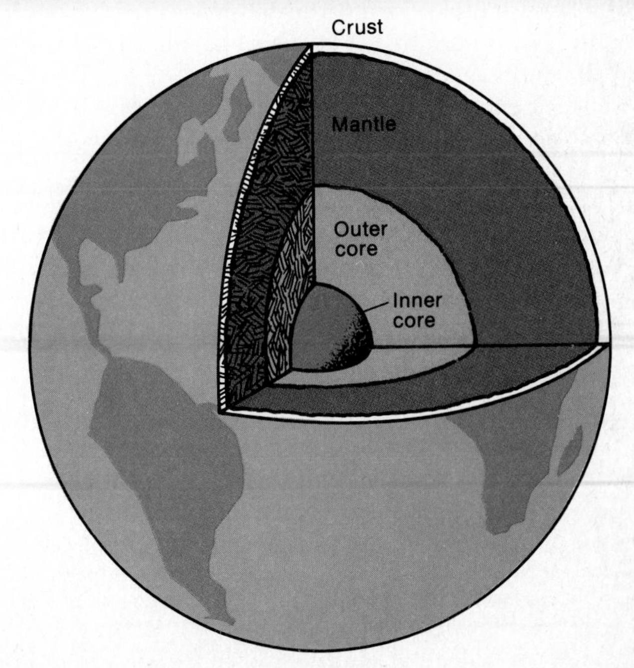

is in the zone of human habitation, that part of the environment having some human interaction. Mankind's interaction with the Earth's interior has been very superficial. In the same way that the geographer's concern with the upper atmosphere is limited largely to its effect on weather and climate in mankind's domain below, so the geographer's concern with the lower lithosphere (and deeper portions of the Earth) is restricted primarily to its influence on topography and other environmental elements in mankind's domain above. As geographers we want to know about the Earth's interior only as it helps us to comprehend the nature and characteristics of the surface.

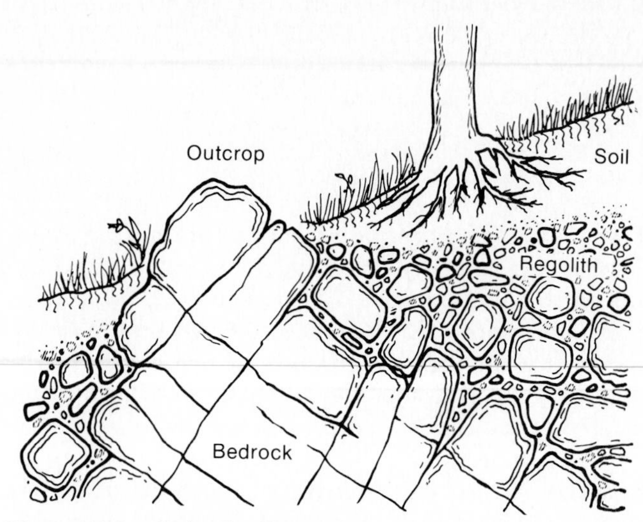

Figure 13-2. Bedrock relationships.

Composition of the Crust

As with all forms of matter, chemical elements occur in varying combinations to form the materials of the lithosphere. About 90 of these basic chemical substances are found in the Earth's crust, occasionally as discrete elements but usually bonded with one or more other elements to form compounds. These naturally formed compounds and elements of the lithosphere are called *minerals*. See Table 13-1. There are more than 2000 different minerals, most of which have homogeneous and unvarying chemical compositions as well as distinctive physical characteristics.

Within the Earth is an unknown amount of molten mineral matter called *magma*. At or near the surface, however, almost all the lithosphere is a solid, generally known as *rock* and composed of aggregated mineral particles that occur in bewildering variety and complexity. Solid rock sometimes is found right at the surface as an *outcrop*; but over most of the earth's land area, the *bedrock* is covered by a layer of broken and partly decomposed rock particles referred to as regolith. See Figure 13-2. Soil, when present, is above the regolith.

The enormous variety of rocks can be systematically classified in a logical fashion. However, for a topographic study, a detailed knowledge of *petrology* (the characteristics of different kinds of rocks) is unnecessary, although an understanding of the major rock types and their basic attributes can be an important asset. All rocks can be categorized into one of three fundamental types, depending on their genesis, or mode of origin, as discussed below. See also Figure 13-3.

TABLE 13-1 Necessary Features of a Mineral

1. It must be found in nature.
2. It must be made up totally of inorganic substances (never alive).
3. It must have the same chemical composition wherever found.
4. Its atoms must be arranged in a regular pattern and form solid units called *crystals*.

Igneous Rocks

In the beginning, all rocks on the Earth were igneous. The initial formation of our planet involved the cooling and solidification of magma, which produced the first solid material of the Earth—igneous rocks. The word *igneous* signifies a fiery inception; all igneous rocks have been formed by magmatic cooling, which has continued to occur throughout the history of our planet. There are within the Earth forces of tremendous strength that cause magma to rise toward the surface, where it cools and solidifies, becoming part of the solid crust.

There are a great many different kinds of igneous rocks, and their characteristics are quite variable. Their principal shared trait is that they are crystalline in structure, which means that as the magma cools, the various minerals assume the form of hard crystals. If the cooling is slow, as happens far beneath the Earth's surface where literally thousands of years may be required for full cooling, the crystals can develop to large size, giving the rock a very coarse-grained appearance. When rapid cooling occurs, as on the surface where full solidification may be complete within hours, the crystals may be so small as to be invisible without microscopic inspection.

An important distinction can be made among different igneous rocks on the basis of their differing conditions of solidification. *Extrusive* rocks were spewed out onto the Earth's surface while still molten, solidifying quickly in the open air. Their most familiar form is the spectacular ejection of lava from volcanoes, but quieter instances of extrusion, such as the massive oozing of magma from fissures in the Earth, have also been widespread in the Earth's history. During extrusion, much of the volatile matter in the magma escapes as gases prior to crystallization, and the rapid cooling may produce glassy texture for all or part of the resulting rock. Of the many kinds of extrusive rocks, the most common is a fine-grained (often speckled) rock called *basalt*, which is particularly widespread in the ocean basins. See Figure 13-4.

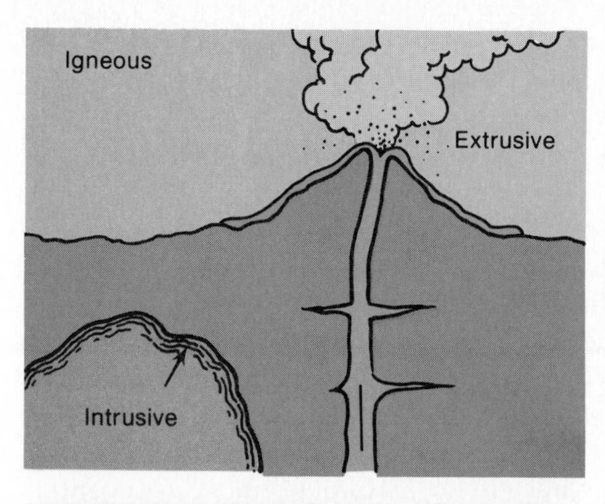

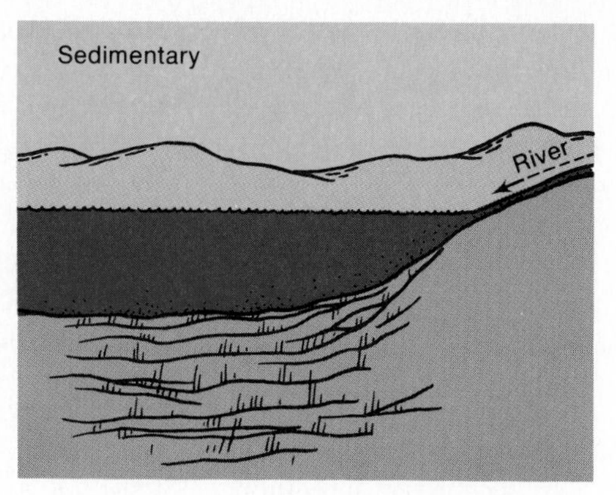

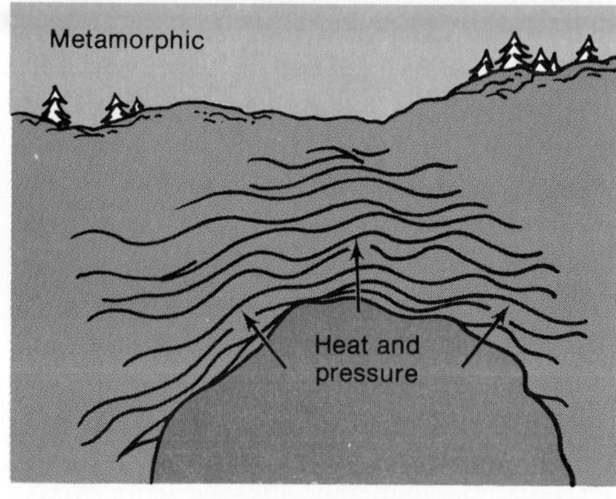

Figure 13-3. There are three basic types of rocks: (1) Igneous rocks are formed by the cooling of magma; (2) sedimentary rocks result from consolidation of deposited particles; (3) metamorphic rocks are produced when heat and pressure act upon preexisting rocks.

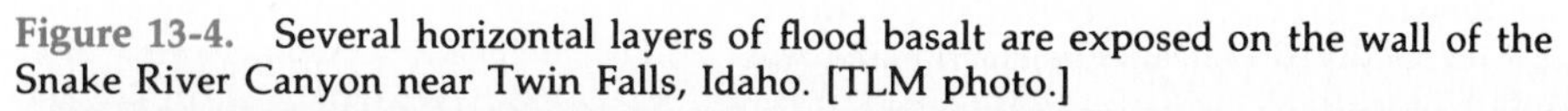

Figure 13-4. Several horizontal layers of flood basalt are exposed on the wall of the Snake River Canyon near Twin Falls, Idaho. [TLM photo.]

Intrusive rocks cool and solidify beneath the Earth's surface, where surrounding nonmagmatic material serves as insulation, which greatly retards the rate of cooling. Although originally buried, they may subsequently become important to topographic development by being pushed upward to the surface or by being exposed by erosion. The most common and well-known intrusive rock is *granite*. See Figure 13-5.

Sedimentary Rocks

The original crust of the cooling Earth consisted of igneous rocks; all other types are of secondary derivation. External processes, mechanical and chemical, operating on these and subsequently formed rocks, have resulted in their decay and disintegration. This produces fragmented mineral material, some of which is removed by water, wind, ice, gravity, or a combination of these agencies. Much of this material is transported by water moving in rivers or streams as *sediment*, e.g., sand and mud. Eventually the sediment is deposited somewhere in a quiet body of water, particularly on the floor of an ocean. Over a long period of time, sedimentary deposits can build to a remarkable thickness—many thousands of feet. The sheer weight of this massive overburden exerts an enormous pressure, which causes adhesion and interlocking of the individual particles. In addition, chemical cementation normally takes place. Various cementing agents—especially silica, calcium carbonate, and iron oxide—precipitate from the water into the pore spaces. This combination of pressure and cementation consolidates and transforms the sediments into *sedimentary rock*.

During sedimentation, materials are sorted roughly by size; the finer particles are carried further than the heavier particles. Other variations in the composition of the sediments are due to processes and rates of deposition, changes in climatic conditions, ocean current movements, and other factors. Consequently, most sedimentary deposits are built up in more or less distinct horizontal layers, called *strata*, which vary in thickness and composition. See Figure 13-6. The resulting parallel structure, or *stratification*, is a characteristic feature of most sedimentary rocks. Although originally deposited and formed in horizontal orientation, the strata may later be uplifted, tilted, and deformed by pressures from within the Earth. See Figure 13-7.

Sedimentary rocks are generally classified on the basis of mode of formation: mechanical, chemical, or organic. The mechanically accumulated sedimentary rocks (also called *detrital* or *clastic*) are composed of fragments of preexisting rocks, in the form of boulders, gravel, sand, silt, or clay; by far the most common are sandstone (composed of sand grains) and shale (composed of silt and clay particles). Chemically accumulated sedimentary rocks are usually formed by the deposition or precipitation of soluble materials or sometimes by more complicated chemical reactions. Calcium carbonate is a common component of such rocks, and limestone is the most wide-

Figure 13-5. Massive outcrops of granite at Sylvan Lake in the Black Hills of South Dakota. Granite is an intrusive igneous rock that solidified beneath the surface and was subsequently exposed by erosion. [TLM photo.]

Figure 13-6. Nearly horizontal strata of interbedded limestone and shale in a road cut near Lyons, Colorado. [TLM photo.]

Figure 13-7. Sedimentary layers that have been uplifted and tilted into vertical orientation on Mt. Angeles in the Olympic Mountains of Washington. [TLM photo.]

spread result. Organically accumulated sedimentary rocks are formed from massive accumulated remains of dead plants or animals. Limestone can also be formed in this fashion from skeletal remains of coral and other lime-secreting sea animals.

There is considerable overlap among the three modes of formation described above, with the result that not only are there many different kinds of sedimentary rocks but there also are many gradations among them. Taken together, however, the vast majority of all sedimentary rocks are either limestone, sandstone, or shale.

Metamorphic Rocks

Metamorphic rocks were originally something else but have been drastically changed by massive forces of heat and/or pressure working on them from within the Earth. The metamorphic result is often quite different from the

One of the most important aspects for students of the landscape to be aware of is the protracted nature of topographic development. The terrain features that appear on the Earth's surface at any given time represent only a momentary condition in an ongoing evolution that began in the remote past and will continue into the indefinite future. To exemplify this somewhat abstract concept in practical terms, let us consider a simplified version of the geomorphic history of a part of the Rocky Mountains in Colorado, called the Front Range. The historical narrative will be related as a continued story in five episodes, with each episode appearing in this book at the point where there is systematic discussion of the pertinent terrain-forming or terrain-shaping process.

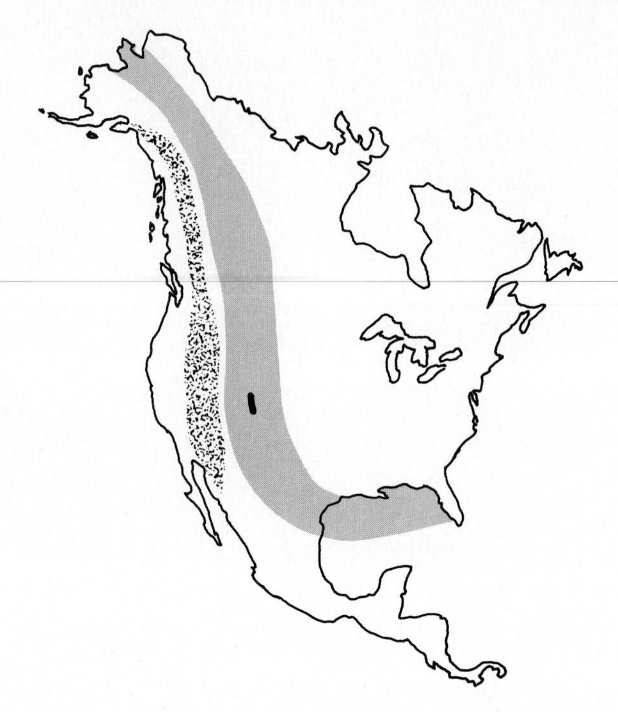

An extensive geosynclinal sea occupied much of western North America during part of the Mesozoic period. Sediments from ancestral western mountains were deposited to great depth in this sea. The approximate location of the contemporary Front Range is indicated on the map.

The Front Range—Episode One: Sedimentation

The terrain of north-central Colorado today is characterized by rugged mountains and steep slopes. In the past, however, this portion of the Earth's surface has been very different. The formation of the Rocky Mountains began in the Mesozoic period, some 150 to 200 million years ago. At that time, a vast sea occupied the area where the Front Range now stands. Millions of years of erosion brought immense amounts of alluvial material (gravel, sand, mud, silt) into this sea from nearby uplands; these materials accumulated on the sea bottom to remarkable depths. The weight of overlying sediments packed and squeezed the lower layers into compacted rock, and various chemicals cemented the particles more firmly together. Sedimentary rocks of several kinds, particularly sandstones, shales and conglomerates, were formed to a thickness of about 10,000 feet (3000 m).

(To be continued. Episode Two begins on page 316.)

original rock; the rocks are changed in structure, texture, composition, and general appearance. See Figure 13-8. Metamorphism is virtually a "cooking" process that partially melts the rock, causing its mineral components to be recrystallized and rearranged. The usual result is a banding, or *foliation*, that gives a wavy-layered appearance, although if the original rock was dominated by a single mineral (as sandstone or limestone), such foliation does not normally develop.

Some rocks, when metamorphosed, change in predictable fashion. Thus, limestone usually becomes marble, sandstone normally is changed into quartzite, and shale often turns into slate. In many cases, however, the metamorphosis is so great that it is difficult to ascertain the nature of the original rock. By far the most common metamorphic rocks are schist (in which the foliations are very narrow) and gneiss (broad foliations), both of which represent a high degree of metamorphism.

The lithosphere as a whole has a very uneven distribution of the three principal rock types. Figure 13-9 is a diagram of the relationships among the three types.

Figure 13-8. This bedrock exposure in northeastern California, near Alturas, shows darker basalt overlying light-colored tufa. The basalt was extruded onto the tufa in molten form, and it "cooked" the rock immediately below. The great heat metamorphosed the upper portion of the tufa. Visual evidence of the metamorphosis is seen in the darker color of the upper tufa. [TLM photo.]

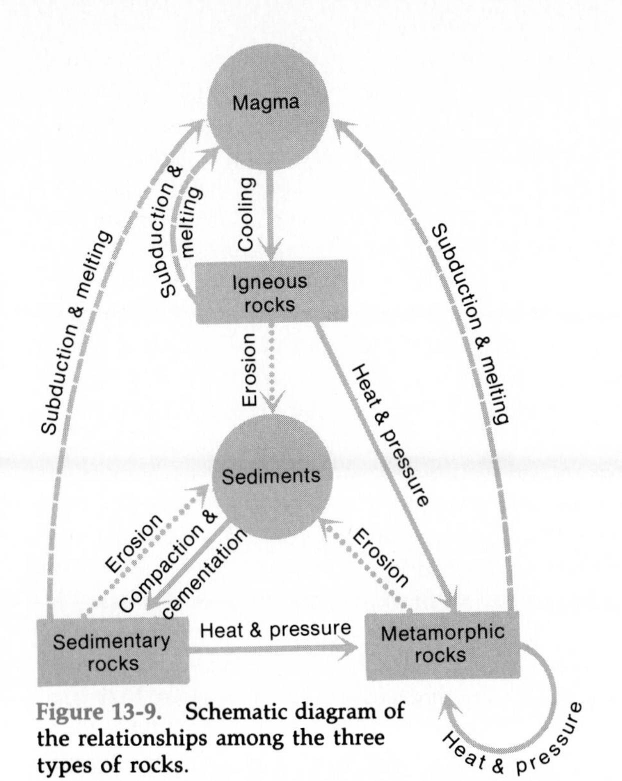

Figure 13-9. Schematic diagram of the relationships among the three types of rocks.

Sedimentary rocks comprise the most common bedrock (perhaps as much as 75 percent) of the surface of the continents, although they constitute only a very small proportion of the total crust. Igneous rocks apparently make up the bulk of the crust, but the volume of metamorphic rocks (which are relatively minor at the Earth's surface) is conceivably even greater because an enormous amount of metamorphosis has taken place beneath the crustal surface. See also Table 13-2, which shows the relative abundance of crustal components.

TABLE 13-2 Relative Abundance of Crustal Components

There are about 90 chemical elements, but nearly 75 percent of the weight of the Earth's crust is provided by only 2 of them.

There are about 2,000 minerals, but more than 60 percent of the weight of the Earth's crust is provided by only 4 of them.

There are about 100 principal rock types, but 80 percent of the volume of the Earth's crust is provided by only 7 of them.

Some Critical Concepts

Before we begin a geographic analysis of the lithosphere, a few critical concepts and basic items of terminology are essential to our study.

Basic Terms

Our attention is directed primarily to *topography*, which is the surface configuration of the Earth. A *landform* is an individual topographic feature, of whatever size; thus the term could refer to such a minor feature as a cliff or a sand dune, and equally to a major element of the landscape such as a peninsula or mountain range. When expressed in the plural—*landforms*—the term is less restrictive and is generally considered to be synonymous with *topography*. Our endeavor in this section of the book is *geomorphology*, or the study of the characteristics, origin, and development of landforms.

One other term of frequent usage is *relief*, which refers to the difference in elevation between the highest and lowest points in an area—the vertical variation from mountain top to valley bottom. See Figure 13-10. It can be used at any scale. Thus the maximum world relief is approximately 13 miles (21 km), which is the difference in elevation between the top of Mt. Everest and the bottom of the Mariana Trench. On the other hand, the *local relief* of any small area is simply the altitudinal difference between the highest and lowest points in that area.

Figure 13-10. Relief is shown in spectacular fashion at Devil's Tower in northeastern Wyoming. From the base to the top of this volcanic tower there is a total vertical change in elevation of 865 feet. [TLM photo.]

Uniformitarianism

Fundamental to any logical understanding of topographic development is acceptance of the doctrine of uniformitarianism. This doctrine holds that "the present is the key to the past." This means that the processes that formed the topography of the past are the same ones that have shaped contemporary topography; they are still functioning in the same fashion and, barring unforeseen cataclysm, they will be responsible for the topography of the future. The processes involved are not temporary, and with only a few exceptions, are not abrupt. They are mostly permanent and slow-acting. The development of landforms is a virtually endless event, with the topography at any given time simply representing a temporary balance in a continuum of change.

Geologic Time

Probably the most mind-boggling concept in all physical geography is the vastness of geologic time. In our puny human scale of time, we deal with such brief intervals as hours, months, and centuries, which does nothing to prepare us for the scale of the Earth's history. The colossal sweep of geologic time encompasses epochs of millions and hundreds of millions of years, which are periods that are extremely difficult for the human mind to encompass. See Figure 13-11 for a diagram of the relative duration of geologic time intervals.

To state that the Earth is thought to have existed for about 4.6 billion years, or that the Age of Dinosaurs persisted for some 160 million years, or that the Rocky Mountains were initially uplifted approximately 65 million years ago is to enumerate temporal expanses of almost unfathomable scope. To reduce such numbers into a manageable frame of reference, we might envision the entire history of the Earth as being compressed into a single calendar year. On such a scale, the first primitive forms of one-celled life would appear in late March, and these primitive algae and bacteria would have the world to themselves until mid-November, when the first multicelled organisms finally would begin to evolve. Mammals would not arrive until late in the second week of December. Humankind would appear on the scene on December 31 at about 11:45 P.M. The age of written history would cover only the last two seconds of the year—just the last two ticks of the clock!

Only with such an extraordinary time scale as this can one give credence to the doctrine of uniformitarianism, or accept that the Grand Canyon is a youthful feature that was carved by that relatively small river that can be seen deep in its inner gorge, or believe that Africa and South America have literally drifted 2000 miles (3200 km) apart. Indeed, the remainder of this book can only be considered as fanciful fiction unless one can rely on the concept of geologic time. The geomorphic processes generally operate with excruciating slowness, but the vastness of geologic history provides a suitable time frame for their accomplishments.

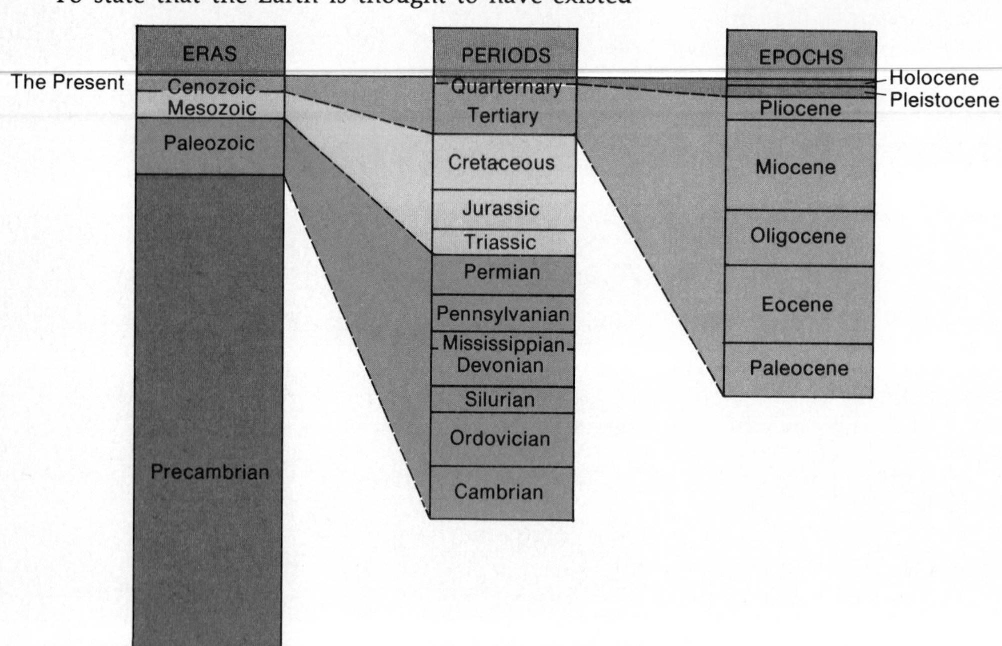

Figure 13-11. The relative duration of geologic time intervals. The height of the column is proportional to the length of the various units of geologic time. Geologic history is divided into four major units, called *eras*. The eras are subdivided into *periods*, and the more recent periods are further subdivided into *epochs*. The Precambrian era was seven or eight times longer than the other three eras combined.

The Study of Landforms

To assert that the geographer's task in studying landforms is simplified by being only marginally concerned with the interior of the Earth does not, by any stretch of the imagination, mean that it is simple. Although our focus of attention is on the surface, that surface is vast, complex, and often obscured. Even without considering the 70 percent of our planet that is covered with oceanic waters, we must realize that more than 58 million square miles (150 million km²) of land are scattered over seven continents and innumerable islands. This area encompasses the widest possible latitudinal range and experiences the full diversity of environmental conditions. Moreover, much of the surface is obscured from our view by the presence of vegetation, soil, or the works of mankind. We must try to penetrate those obstructions, observe the characteristics of the lithospheric surface, and encompass the immensity and diversity of a worldwide landscape. This is far from a simple task.

To organize our thinking for such a complex endeavor, we can isolate certain basic elements for an analytic approach:

1. *Structure* refers to the nature, arrangement, and orientation of the materials comprising the observed feature(s). Essentially this is the geologic underpinning of the landform. Is it composed of bedrock? If so, what kind? What are the physical and chemical characteristics? What is the configuration? If not, what is the nature and orientation of the material? See Figure 13-12.
2. *Process* considers the actions that have combined to produce the landform. A variety of forces—usually geologic, hydrologic, atmospheric, and biotic—are at work in shaping the features of the lithospheric surface, and their interaction is critical to the formation of the feature(s). See Figure 13-13.
3. *Slope* is the fundamental aspect of shape for any landform. The angular relationship of the surface is essentially a reflection of the contemporary balance among the various components of structure and process. The inclinations and lengths of the slopes provide important details for both description and analysis of the feature(s). See Figure 13-14.
4. *Drainage* refers to the movement of water (from rainfall and snowmelt) over the surface or down into the soil and bedrock. Although moving water is an outstanding

Figure 13-12. The structure of these abrupt sandstone cliffs in southeastern Utah is easy to see. [TLM photo.]

Figure 13-13. The spectacular terrain of Glacier National Park was produced by a variety of interacting forces, but the contemporary shape of the topography is particularly due to the process of glaciation. This view is across St. Mary Lake toward Mt. Jackson. [TLM photo.]

Figure 13-14. Slope is a conspicuous visual element of the landscape in desert areas because the land surface is not disguised with much vegetation. These rocky hills (the Macdonnell Ranges) in central Australia show many different slope angles. [TLM photo.]

force under the "process" heading above, the ramifications of slope wash, streamflow, stream patterns, and other aspects of drainage are so significant that the general topic of drainage should be considered as a basic element in landform analysis.

Once these basic elements have been recognized and identified, the geographer is prepared to proceed with an analytical consideration of the topography by providing answers to the fundamental questions that are at the heart of any geographic inquiry:

What? The *form* of the feature(s).
Where? The *distribution* and *pattern* of the landform assemblage.
Why? An explanation of *origin* and *development*.
So what? The *significance* of the topography in relation to other elements of the environment and to human life and activities.

Internal and External Geomorphic Processes

The topography of the Earth's surface exists in infinite variety, apparently in much greater diversity than on any other known planet. This variety reflects the complexity of interactions between process and structure—the multiplicity of shapes and forms that result as the geomorphic processes exert their inexorable effects on the geologic fundament.

These processes are relatively few in number, but they are extremely varied in their nature and operation. Basically they are either internal or external. The *internal* processes operate from within the Earth, energized by massive and generally unpredictable forces that are imperfectly understood and that apparently operate without regard to any surface or atmospheric influences. They result in *tectonic* activity, or crustal movements of various kinds. In general, they are constructive, uplifting, building forces that tend to increase the relief of the land surface.

In contrast, the *external* processes are largely subaerial; i.e., they operate at the base of the atmosphere and draw their energy mostly from sources above the lithosphere, in the atmosphere or the oceans. Unlike internal processes, however, external processes are well understood and their behavior is often predictable. Moreover, their behavior may be significantly influenced by the characteristics of the preexisting topography, particularly its shape and the nature of the surface materials. The external processes may be thought of generally as wearing down or destructive forces that eventually tend to diminish topographic irregularities and decrease the relief of the Earth's surface.

These two groups of processes, internal and external, thus work in more or less direct opposition to one another. Their battleground is the surface of the Earth, the interface between lithosphere and atmosphere, where this remarkable struggle has persisted for literally billions of years and may continue endlessly into the future.

In succeeding chapters we will consider these various processes—their nature, dynamics, and effects—in some detail; but it may be useful here to summarize the processes so that they can be glimpsed in totality, prior to individual treatment. It should be noted, however, that our classification is imperfect; some items are clearly separate and discrete, whereas others overlap with each other. This outline, then, represents a simple, logical way to approach an understanding of the processes, but is not necessarily the only or ultimate framework for comprehension of the complex interactions.

A. Internal processes
   1. Massive crustal rearrangement
   2. Vulcanism
      (a) Extrusive
      (b) Intrusive
   3. Diastrophism
      (a) Broad warping
      (b) Folding
      (c) Faulting
B. External processes
   1. Weathering
   2. Mass movements
   3. Erosion/Deposition
      (a) Fluvial (running water)
      (b) Aeolian (wind)
      (c) Glacial (moving ice)
      (d) Solution (ground water)
      (e) Waves and currents (ocean/lake)

The Question of Scale

In a systematic study of the geomorphic processes, two general topics should be kept in mind—*scale* and *pattern*. The question of scale is fundamental in geography. Regardless of the subject of geographic inquiry, the recognizable features and associations are likely to vary considerably depending on the scale of observation. In simplest terms, this means that in a close-up view, one recognizes different aspects of the landscape than would be seen from a more distant view. Comprehension of immense variations in size and perspective are essential to a proper understanding of the geomorphic lineaments of our planet.

Five different *orders of relief* can be recognized on the surface of the lithosphere, as described below. See also Figure 13-15.

As a specific example of the complexity and significance of scale, let us focus our attention on a particular place on the Earth's surface and view it from different perspectives. The place is located in north-central Colorado within the boundaries of Rocky Mountain National Park, some 8 miles due west of the town of Estes Park. It encompasses a small valley called Horseshoe Park, through which flows a clear mountain stream named Fall River and adjacent to which is the steep slope of Bighorn Mountain.

1. To illustrate the largest scale of ordinary human experience, we will hike northward from the center of Horseshoe Park up the side of Bighorn Mountain. At this level of observation, the first topographic feature of note is a smooth stretch of Fall River that we must cross. We will walk over a small *sand bar* at the south edge of the river, wade for a few steps in the *river*, step up an 18-inch (46-cm) *bank* onto the *mountainside*, noting the dry bed of a small intermittent *pond* on our left in passing. After 20 minutes or so of steep uphill scrambling, we will reach a rugged granite *outcrop* (locally called Hazel Cone or Poop Point), which presents us with an almost vertical *cliff* face to climb.
2. At a significantly different scale of observation, we might travel for 20 minutes by car in this same area. The road through Horseshoe Park is part of U.S. Highway 34, which is called the Trail Ridge Road in this portion of Rocky Mountain National Park. After 20 minutes, we will have reached a magnificent viewpoint high on the mountain to the southwest of Horseshoe Park. From this vantage point, our view of the country through which we hiked is significantly enlarged. We can no longer recognize the sand bar, the bank, or the dry pond, and even the rugged cliff of Poop Point appears as little more than a pimple on the vast slope of Bighorn Mountain. Instead, we see that Fall River is a broadly *meandering stream* in a flat *valley* and that Bighorn Mountain is an impressive *peak* rising high above.
3. Our third observation of this area might take place through the window of a commercial airplane flying at a height of perhaps 39,000 feet (12 km) on its regular run between Omaha and Salt Lake City. From this elevation, Fall River is nearly invisible, and only careful observation would reveal Horseshoe Park at all. Bighorn Mountain itself is now merged indistinguishably as part of a *mountain range*, the Mummy Range, which itself is seen as a minor offshoot of a much larger and more impressive *mountain system* called the Front Range.
4. A fourth level of observation would be available to us if we could hitch a ride on an orbiting satellite some 100 miles (160 km) above the Earth. Our brief glimpse of northern Colorado would probably be inadequate to distinguish the Mummy Range, and even the 250-mile (400-km) long Front Range would appear only as a component of the mighty Rocky Mountain *cordillera*, which extends

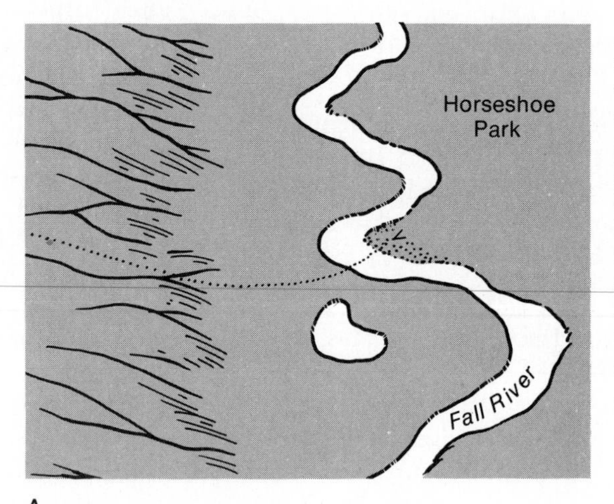

An experience with change of scale: (A) A close view of Horseshoe Park from the west. The dotted line shows our route. (B) Looking down on Horseshoe Park from Trail Ridge Road. (C) An aerial view of the Mummy Range and part of the Front Range. (D) A high-altitude look at Colorado. (E) North America as seen from a distant spacecraft.

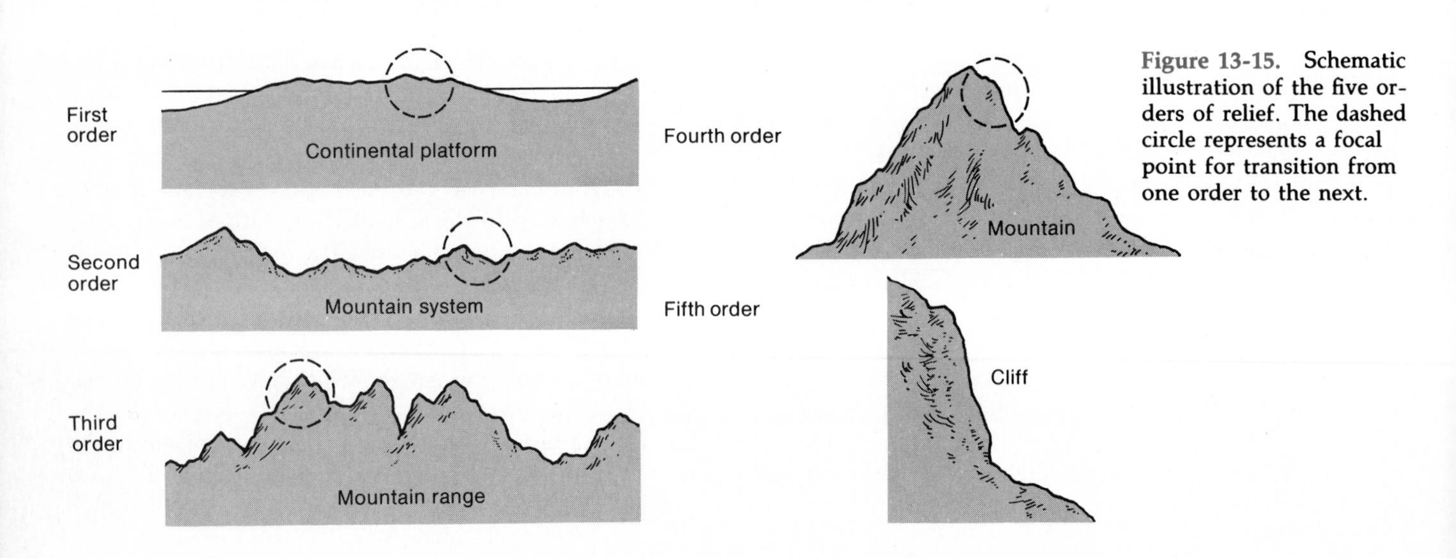

Figure 13-15. Schematic illustration of the five orders of relief. The dashed circle represents a focal point for transition from one order to the next.

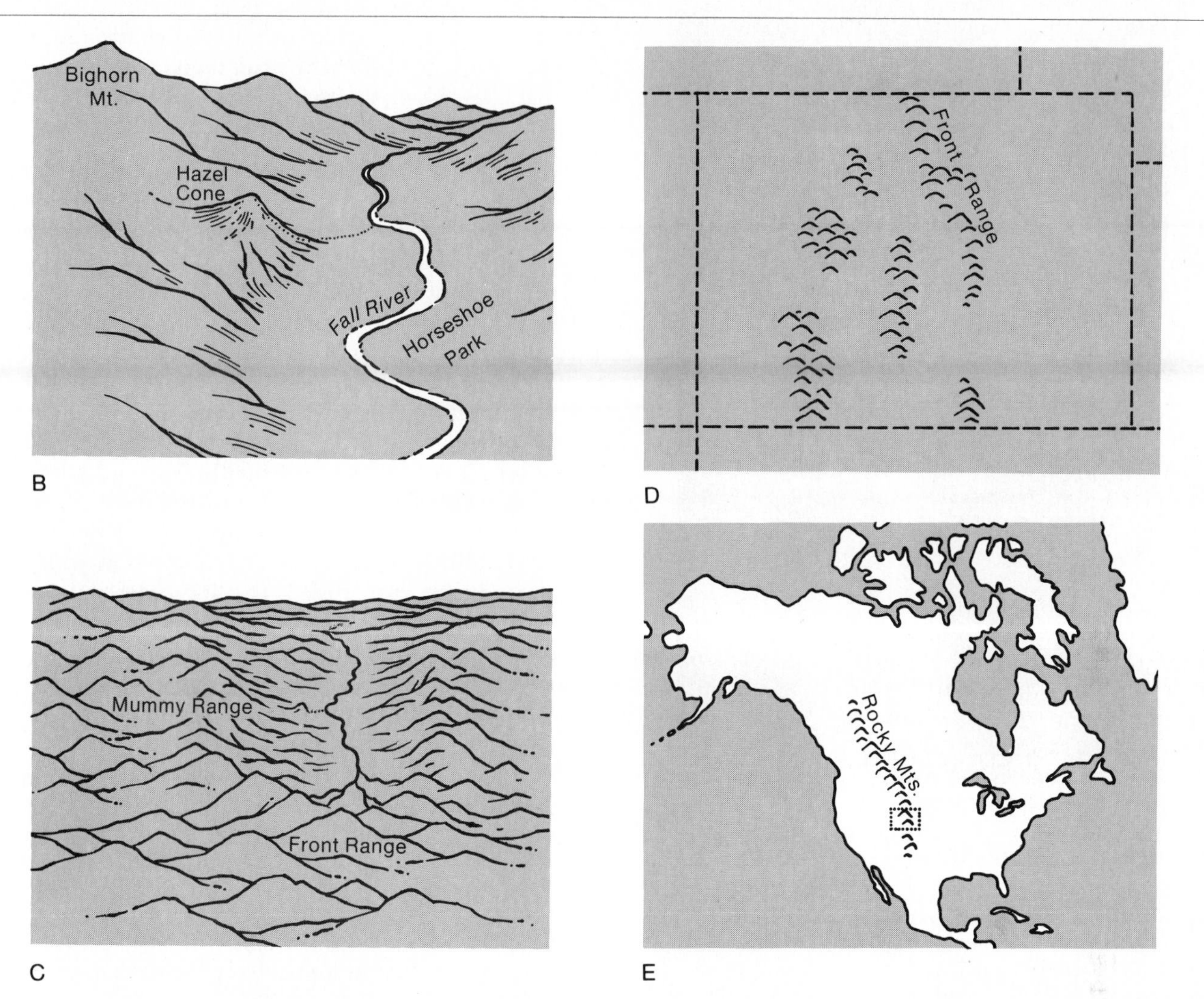

from New Mexico to northern Canada.

5. A final viewpoint possible to humans could come from a spacecraft rocketing its way toward some distant heavenly body. Looking back in the direction of Horseshoe Park from a near-space position, one might possibly recognize the Rocky Mountains, but the only conspicuous feature in this small-scale view would be the North American *continent*.

1. *First-order relief* represents the small-scale end of the spectrum, which means that the features are the largest that can be recognized: *continental platforms* and *ocean basins*. Although the shoreline at sea level appears as a conspicuous demarcation between land and water, it is not the accepted boundary between the platforms and the basins. Each continent has a margin that is submerged, called the *continental shelf*. This shelf is a terrace of varying width, in some places only a few miles wide and in other localities extending seaward for several hundred miles. At its outer edge, usually at a depth of about 600 feet (200 m) below present sea level, the slope pitches more steeply and abruptly into the ocean basins. See the view of first-order relief features seen from *Apollo 16*, in Figure 13-16.
2. *Second-order relief* consists of major mountain systems (such as the Andes or the Rockies) and other extensive surface formations of sub-continental extent (such as the Mississippi lowland or the Canadian shield). Second-order relief features (like those of all other orders) may be found in ocean basins as well as on continental platforms, most conspicuously in the form of the great undersea mountain ranges that are usually referred to as *ridges*.
3. *Third-order relief* encompasses specific landform complexes of lesser extent and generally of smaller size than those of the second order, with no precise separation between the two. Typical third-order features include discrete mountain ranges, groups of hills, and large river valleys. See Figure 13-17.

Figure 13-16. Only first-order relief features can be clearly identified in this view of North America from a departing lunar mission. [*Apollo 16* photo, courtesy of NASA.]

4. *Fourth-order relief* comprises the sculptural details of the third-order features, including such individual landforms as a mountain, mesa, or hill. See Figure 13-18.
5. *Fifth-order relief* consists of the small individual features that may be part of the fourth-order relief, such as a sand bar, cliff, or waterfall. See Figure 13-19.

Although these orders of relief are generally discrete and observably different from one another, they nevertheless represent somewhat arbitrary categories, and there is some overlap among them.

An important correlation with this classification is the degree of permanency represented by the various orders. Although no features of the Earth's crust are permanent, the lower-numbered orders include features that are usually more long-lived than those of the higher-numbered orders. Thus a second-order feature normally will exist for a much longer time than a third-order feature, and so forth.

As an analogy, one might compare the orders of relief with a setting for theatrical presentations in which geomorphic dramas are featured. The first-order features represent the foundation for construction, a stable and long-lasting undergirding that may outlast a sequence of buildings erected on it. The second-order features represent a theater built on the foundation, which will be in use for a long time and in which will be presented many different dramas. The third-order features represent

Figure 13-17. Various third-order relief features (mostly mountain ranges and valleys) are identifiable in this high-altitude view of northern Pakistan. [*Apollo 7* photo, courtesy of NASA.]

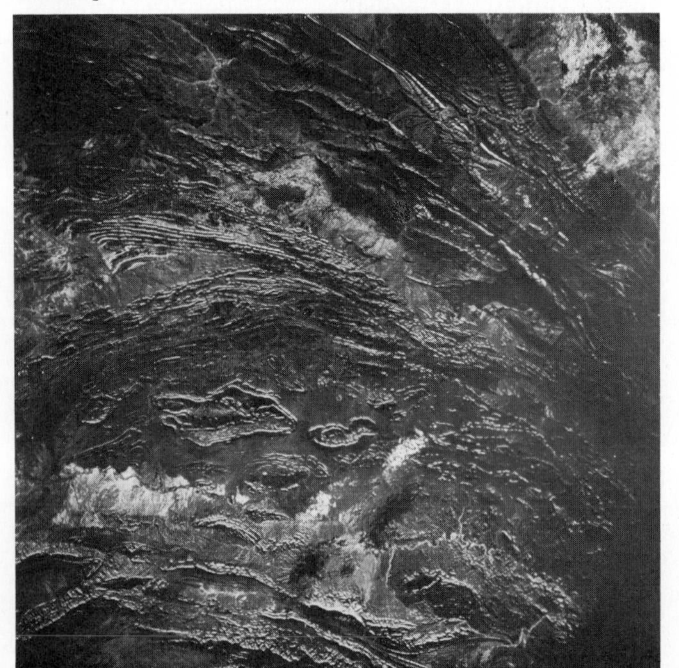

Figure 13-18. Fourth-order relief features are comprised of individual landforms, such as Mitten Butte in the Monument Valley of southeastern Utah. [TLM photo.]

Figure 13-19. The fifth order of relief is represented by a waterfall. This is Millstream Falls in northern Queensland, Australia. [TLM photo.]

the stage installed in the theater, which will be used for a number of presentations, but which will be changed several times during the life of the theater itself. The fourth-order features represent the scenery, which will be reconstructed with each succeeding drama. The fifth-order features represent the props, which will be revised several times during the course of a single play.

The Pursuit of Pattern

A prime desideratum in any geographic study is to detect patterns in the areal distribution of phenomena. If features have a disordered and apparently haphazard distribution, it is more difficult to comprehend the processes and relations that are involved; however, if there is some perceptible pattern to their distribution—some more or less predictable spatial arrangement—it becomes simpler for us to understand both the reasons for the distribution pattern and the interrelationships that pertain.

In previous portions of this book we have been concerned with geographic elements and complexes that exhibit distribution patterns of some predictability. For example, one can say with some assurance that at about 50° latitude on the west coasts of continents there will be a certain type of climate, a particular association of natural vegetation is likely to occur, the native animal life will have some expectable components, and soil-forming processes will be roughly predictable. These relatively orderly arrangements have been helpful, and perhaps even comforting, in our studies thus far.

Now, alas, we enter into a major portion of physical geography in which orderly patterns of distribution are much more difficult to discern. There are, to be sure, a few aspects of predictability; for example, one can anticipate that in desert areas, certain geomorphic processes will be more conspicuous than others and certain landform features are likely to be found. Overall, however, the global distribution of topography is very disordered and irregular. The pursuit of a predictable pattern is, for the most part, a thankless endeavor, and to comprehend the world distribution of landforms requires an unfortunate degree of sheer memorization.

Largely for this reason, the geomorphology portion of this book concentrates less on distribution and more on process. Comprehension of the dynamics of topographic development is more important to an understanding of systematic physical geography than any amount of detailed study of landform distribution. Thus, in succeeding chapters we will concentrate in turn on the principal internal and external processes that shape the landforms of the continents.

Review Questions

1. Why are geographers more interested in the Earth's surface than its interior?
2. Distinguish among *sedimentary*, *igneous*, and *metamorphic* rocks.
3. Distinguish between *intrusive* and *extrusive* rocks.
4. What is the difference between *topography* and *relief*?
5. Distinguish between *internal* and *external* processes that shape the Earth's surface.

Some Useful References

BRUNSDEN, D., and J. DOORNKAMP, eds., *The Unquiet Landscape*. Bloomington, Ind.: Indiana University Press, 1974.

EICHER, D. L., *Geologic Time*, 2nd ed. Englewood Cliffs, N.J.: Prentice-Hall, Inc., 1976.

FLINT, R. F., *The Earth and Its History*. New York: W. W. Norton, & Co., Inc., 1973.

GARNER, H. F., *The Origin of Landscape*. New York: Oxford University Press, 1974.

HAMBLIN, W. K., *The Earth's Dynamic Systems*. Minneapolis, Minn.: Burgess Publishing Company, 1978.

PITTY, A. F., *The Nature of Geomorphology*. London: Methuen & Co. Ltd., 1982.

14 The Internal Processes

The forces that operate within the Earth are vast, complex, and largely mysterious. Geophysical and geological research increasingly generates more data to improve our understanding, but the energetics and dynamics of the internal processes of the planet are as yet only partially comprehended.

Massive forces beneath and within the lithosphere actively reshape the configuration of the crustal surface. The crust is buckled and bent; the land is raised and lowered; the rocks are fractured and folded; solid material is melted; and molten material is solidified. These actions have been ongoing for billions of years and are fundamentally responsible for the gross shape of the lithospheric surface at any given time. They do not always act independently and separately, but in this chapter we will attempt to isolate them for analysis.

Massive Crustal Rearrangement

Until recent years, most Earth scientists assumed that the planet's crust was rigid, with continents and ocean basins fixed in position and significantly modified only by changes in sea level and periods of mountain building. The uneven shapes and irregular distribution of the continents were a matter of puzzlement, but it was generally accepted that the present arrangement was emplaced in some ancient age when the Earth's crust cooled from its original molten state. At that time the pattern of the first-order relief features was set.

Plasticity

The "rigid Earth" theory was seriously called into question in recent years by a variety of discoveries and hypotheses. Prominent among these has been the recognition that the igneous rocks of the upper crust apparently occur in two layers, well differentiated in several characteristics, especially density. See Figure 14-1.

The lower of the two layers is thought to be continuous, underlying both the ocean basins and the continents. It is nearer the surface under the basins and deeper beneath the continental masses. This layer is called the *sima*, named for its two most prominent mineral compounds, *si*lica and *ma*gnesium. The sima is chemically basic, dark in color, relatively young in age, basaltic in nature, and relatively dense (at least 3.0 g/cm^3).

The upper layer is believed to be discontinuous, apparently underlying only the continental masses, where it sits as immense bodies of rock embedded in the underlying sima. It is called the *sial*, named for its common constituents of *si*lica and *al*uminum. It is lighter in color and chemically more acidic than the sima. It is generally granitic in nature and notably less dense (averaging about 2.8 g/cm^3).

The general crustal structure, then, appears to consist of continental masses of sial "floating" in a foundation of sima. Such a structure casts serious doubts on the theory of rigidity and introduces a concept of *plasticity* in regard to the movement of continental masses.

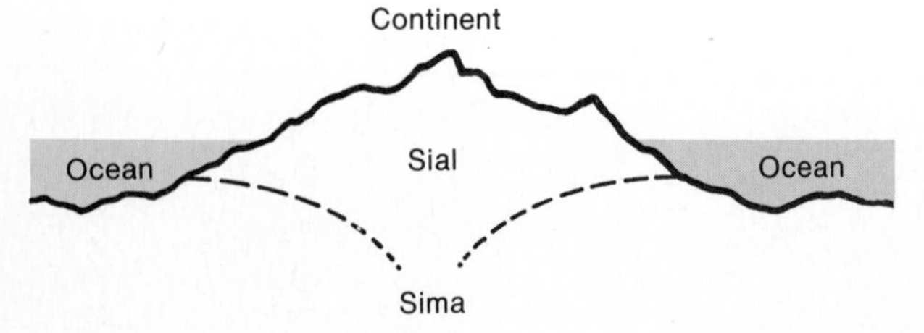

Figure 14-1. Igneous rocks on the upper crust of the Earth can be categorized as belonging either to sial or sima. Sial is an upper layer associated with the continents, whereas sima underlies both ocean basins and continental sial.

Isostasy

Somewhat related to the concept of plasticity is the principle of *isostasy*, or the maintenance of the hydrostatic equilibrium of the Earth's crust. In simplest terms, the addition of a significant amount of weight onto a portion of the crust will cause it to sink, whereas the removal of a massive weight will allow the crust to rise, in a sort of balancing reaction.

The detailed dynamics of isostasy are not clear. For example, the depth of the adjustment is unknown; it may involve only an immediate surface section, or it may include an entire sialic continental mass, or it may extend clear through the sima to the base of the lithosphere. Moreover, the areal extent of the isostatic adjustment sometimes covers only a few hundred square miles and sometimes extends over half a continent. Also unclear is the immediacy of the isostatic response: In some cases, the raising or lowering seems to take place in direct accordance with the removal or addition of weight; but in others, there is a lengthy lag before the adjustment begins. These points of confusion notwithstanding, however, the eventual topographic results are obvious—the raising or lowering of some portion of the surface.

Isostatic reactions can be set in motion by a variety of causes. See Figure 14-2. The crust may be depressed, for example, by the deposition of a significant amount of sediment on the continental shelf, or by the accumulation of a great weight of glacial ice on a landmass, or even by the weight of water trapped behind a large dam. Conversely, the crust may "rebound" to a higher elevation because of the removal of a significant amount of material by erosion, deglaciation (the melting of an ice sheet), or drainage of a large body of water.

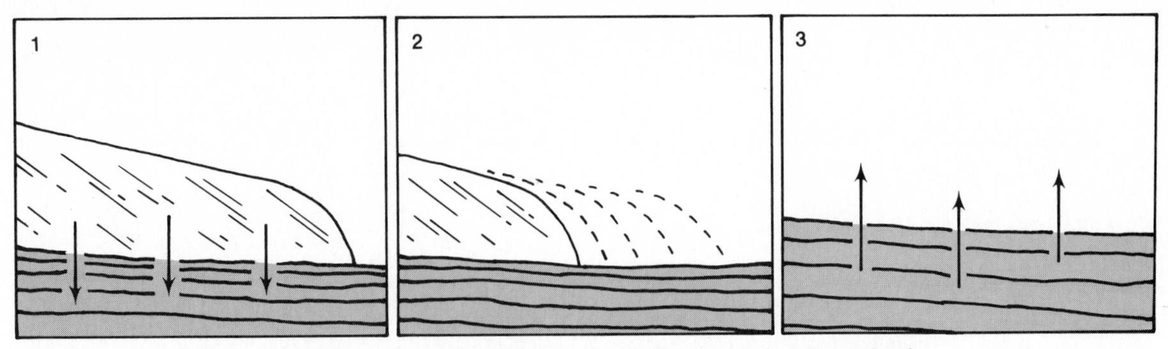

Figure 14-2. Schematic example of isostatic adjustment: (1) During a glacial epoch, the heavy weight of the accumulated ice depresses the crustal surface. (2) Deglaciation removes the weighty overburden. (3) The surface rises, or "rebounds," after the weight is removed.

Continental Drift

The concepts of plasticity and isostasy have provided portions of a framework for the potential understanding of a much more massive and dramatic restructuring of the crust. The theory of continental drift proposes that the present continents were originally connected as one or two large landmasses, landmasses that have broken up and literally drifted apart over the last several tens of millions of years. The drifting continues today so that the contemporary position of the continents is by no means their ultimate one.

The idea of a single supercontinent from which large fragments separated has been around for a long time. Various scientists—naturalists, physicists, astronomers, geologists, botanists, geographers—from a number of countries—Britain, France, Italy, Germany, Austria, South Africa, the United States—have enunciated their views in this regard, based on fragmentary evidence and deduction, since the days of Sir Francis Bacon in the early 1600s. However, the idea was generally unacceptable to the scientific community at large, until fairly recently.

During the second and third decades of the twentieth century, the notion of continental drift was revived, most notably by the German meteorologist Alfred Wegener, who put together the first relatively comprehensive theory to describe and partially explain the phenomenon. Wegener postulated a massive supercontinent, which he called "Pangaea," as existing about 200 million years ago, which broke up into several large sections that have continually moved away from one another and now comprise the contemporary continents. See Figure 14-3. Wegener assiduously accumulated a mass of evidence to support his hypothesis, most notably the remarkable number of close affinities on either side of the Atlantic Ocean. The coastlines of the subequatorial portions of Africa and South America fit together with jigsaw-puzzlelike precision, and there are other important elements of congruity in the coastline shapes on either side of the North Atlantic as well. Moreover, the petrological and paleontological records on either side of the Atlantic show many distributions that would be continuous if the ocean did not intervene, for example, the same or similar rocks, structures, and fossils. The dramatic evidence (see Figure 14-4) of this former transatlantic connection, along with data brought forward from other areas by Wegener, attracted much attention and generated much controversy. Some Southern Hemisphere geologists, particularly in South Africa, responded with enthusiasm. They gathered supportive evidence in the form of ancient glacial deposits and fossilized plant remains that exhibited similar distributions in southern landmasses (South America, South Africa, Madagascar, India, Australia, and more recently discovered in Antarctica) that are now widely separated, evidence that could be most logically explained by continental drift.

The general response to the Wegenerian hypothesis, however, was one of disbelief. Despite the vast number of distributional coincidences, most scientists felt that two difficulties made the theory improbable if not impossible: (1) The Earth's crust was believed to be too rigid to permit such large-scale motions; and (2) no suitable mechanism could be conceived of that would provide enough energy to displace such large landmasses for a long journey. The fatal flaw in Wegener's theory was its inability to explain the mechanism of the breakup and movement. His explanation, which he agreed was

Figure 14-3. The presumed arrangement of the Pangaea supercontinent as of about 200 million years ago.

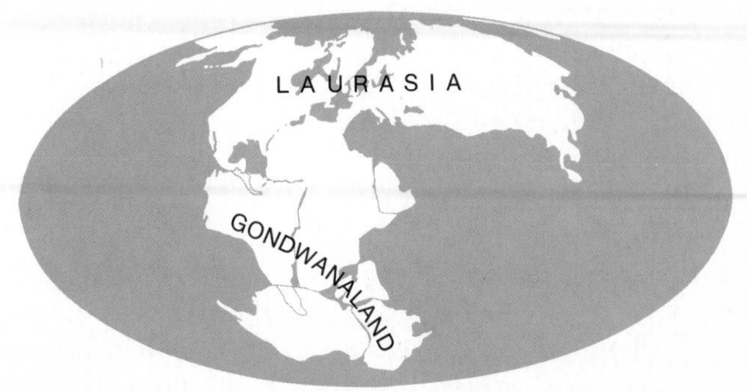

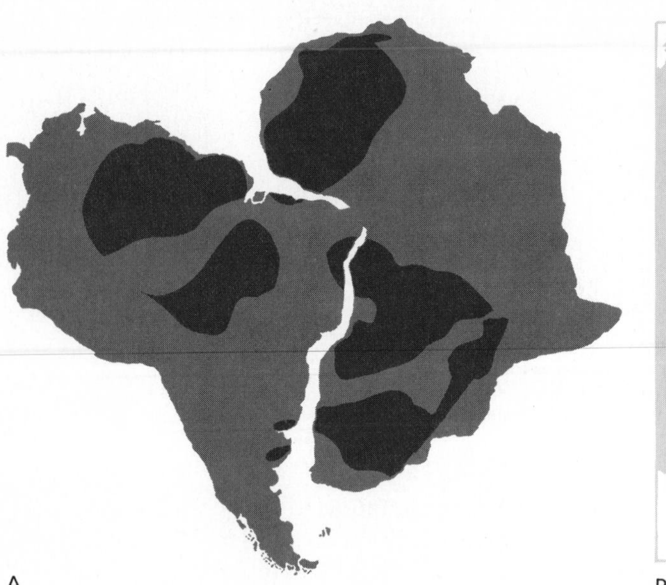

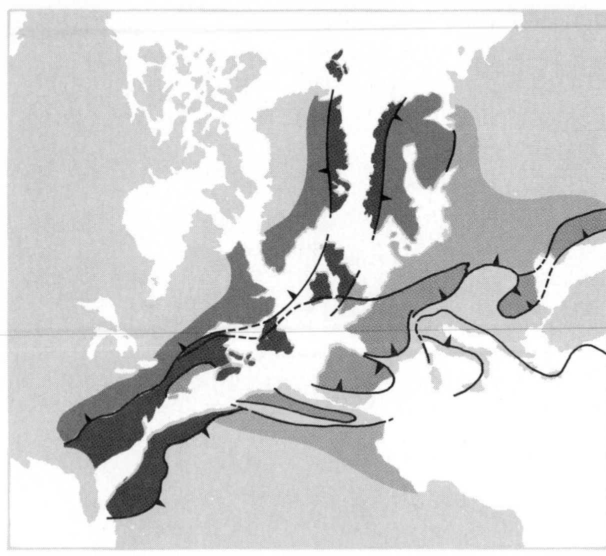

A B

Figure 14-4. Some of the most decisive evidence for continental drift comes from the transatlantic connection. The matching of the coasts of Africa and South America is particularly persuasive, whereas the "fit" across the North Atlantic is less clear-cut. (A) In the first diagram, the ancient continental blocks of Africa and South America are shown in darker hue, with the younger areas (mostly zones of sedimentary deposition and volcanic activity) shown in lighter tone. (B) Across the North Atlantic are four zones of similar geological activity, which have a complex relationship but are very difficult to correlate unless Europe, North America, and North Africa were once together.

probably inadequate, was based on gravitational pull and centrifugal force.

Several other scholars attempted to formulate explanations of the necessary impelling force. Most dramatic was that of the American geologist F. B. Taylor, who suggested that some catastrophic event caused a large chunk of the Earth's crust to break loose and be propelled into space, creating a great gap where the Pacific Ocean is now, and thus initiating the actions of continental drift. This was an exciting but unconvincing hypothesis, and it did little to strengthen the reputation of the continental drift theory. Indeed, most Earth scientists ignored or even debunked the idea of continental drift for the better part of half a century after Wegener's theory was presented.

Plate Tectonics

The questionable validity of continental drift notwithstanding, continuing research has revealed more and more about crustal mechanics. Geologists, geophysicists, seismologists, oceanographers, and physicists have accumulated a large body of data on the characteristics of the ocean floor and the nature of the underlying simatic crust. It began to become apparent that the process of *convection*, normally associated with fairly rapid flows of liquid or gas, was at work in the interior of the Earth. A very sluggish thermal convective system appears to operate within the Earth, bringing deep-seated molten material slowly to the surface, and, in turn, pulling remelted crustal rocks into the depths. See Figure 14-5.

Systematic depth soundings have proved that running across the floors of all the oceans for some 40,000 miles (64,000 km) is a continuous system of large ridges.

Figure 14-5. Schematic illustration of subsurface convection. Molten material reaches the surface at the midocean ridges and moves laterally across the ocean floor to be subducted downward at the edge of the plate, where it is melted and recycled into the system.

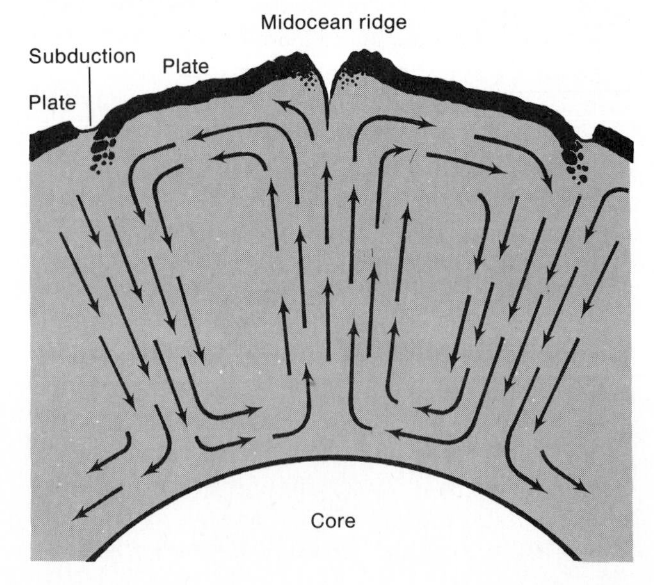

These ridges are mostly located at some distance from the continents, often in the very middle of the oceans. The Mid-Atlantic Ridge (see Figure 14-6) is the most prominent example of this phenomenon. Moreover, deep trenches occur at many places in the ocean floors, often around the margins of the ocean basins.

In the early 1960s, a new theory was propounded, most notably by the American oceanographer Harry Hess, which has come to be known as *sea-floor spreading*. See Figure 14-7. It is believed that oceanic ridges were formed by rising currents of deep-seated material, often accompanying volcanic eruptions, which spread laterally to form new ocean floors. Thus new crustal material appears at the ridges and slowly moves outward, at a rate of perhaps $\frac{1}{2}$ inch (1 cm) per year. At other places in the ocean basins, usually associated with trenches at the margins, older ocean floor material descends into the interior, in a process called *subduction*, where it is presumably melted and recycled into the convective system.

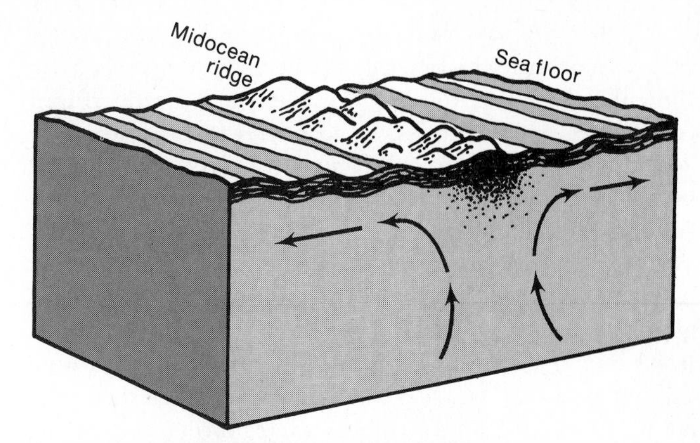

Figure 14-7. Sea-floor spreading involves the rise of rock material from within the Earth and its lateral movement away from the midocean ridges. This extremely gradual process is accompanied by a pattern of relatively symmetrical stripes that shows an alternating orientation of magnetic attraction.

The validity of the idea of sea-floor spreading has been confirmed by two sets of evidence—paleomagnetism and core sampling. When any rock containing iron is formed, it is magnetized so that the iron grains become oriented toward the direction of the magnetic pole. This then becomes a permanent record of the polarity of the Earth's magnetic field at the time the rock was solidified. During the last hundred million years of the Earth's history, the Earth's magnetic field is known to have reversed itself, with the north and south magnetic poles changing places, more than 170 times. Thus, if the sea floor has spread laterally by the addition of new material at the oceanic ridges, there should be a relatively symmetrical pattern on either side of the ridges, reflecting the changes in past magnetic orientation (*paleomagnetism*). Such has been shown to be the case, with parallel stripes of magnetic orientation facing in opposite directions alternately (Figure 14-7).

Final confirmation of sea-floor spreading was obtained from holes drilled into the sea floor by a research ship. Several hundred *core samples* from the sea-bottom sediments have been analyzed, from which the depths and ages of the sedimentary layers have been determined. Almost invariably the thickness and age of the sediments increase with their distance from the oceanic ridges, indi-

Figure 14-6. The principal midocean ridges form a continuous worldwide system, mostly far removed from any present continent. The principal oceanic trenches, on the other hand, are mostly situated close to continental margins.

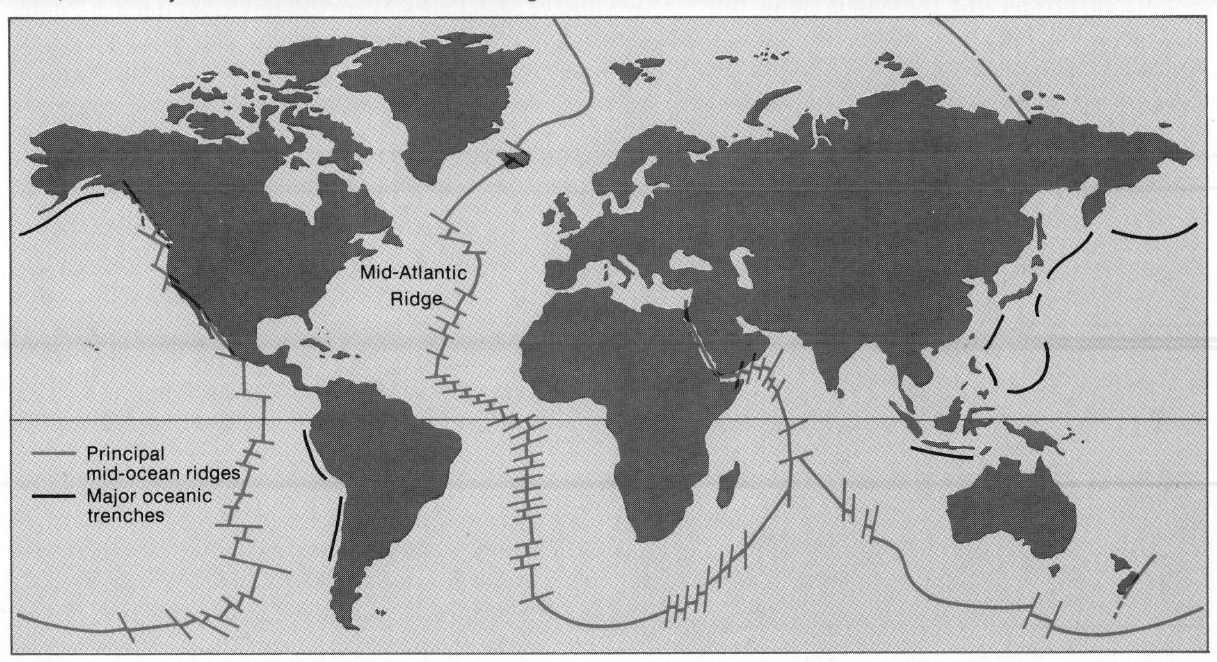

cating that the sediments furthest from the ridges have been in existence the longest. Conversely, sediments near the ridges are thinner and younger, and at the ridges themselves, the material is almost all igneous, with little accumulation of sediment. Thus the sea floors can be likened to gigantic conveyor belts, moving ever outward from the oceanic ridges.

On the basis of these details and a variety of other related evidence, the concept of plate tectonics is now generally accepted. According to this theory, the upper portion of the lithosphere is believed to consist of a mosaic of rigid plates, with an estimated average depth of about 60 miles (100 km); these plates are embedded in a somewhat plastic layer of undetermined extent called the *asthenosphere*. The plates vary considerably in their areal extent; some appear to be almost hemispheric in size, whereas others are much smaller. The actual number of plates, and some of their boundaries, is as yet unclear. Six major plates, and perhaps an equal number of smaller ones, are postulated. Most of the plates consist of both continental (sial) and oceanic (sima) crust. They move ever so slowly over the asthenosphere, the movements frequently bringing two plates together on a collision course.

At present, three kinds of plate boundaries are recognized. See Figure 14-8. At a *divergent* boundary, normally represented by an oceanic ridge, new sea-floor material is being added, and plates are moving away from one another. Such locations are often associated with shallow earthquakes and volcanic activity. At a *convergent* boundary, plates moving in opposite directions meet, and the result of the collision normally is a vast crumpling of the edges as one plate slides under the other. This subduction is a slow but nevertheless catastrophic event in which the crustal material of the submerging plate is gradually incorporated, presumably through melting, into the superheated depths below. The subduction zone is usually an area of crustal instability and is characterized by the occurrence of many deep-seated earthquakes and

Figure 14-8. A schematic representation of the three kinds of plate boundaries: divergent, convergent, transcurrent.

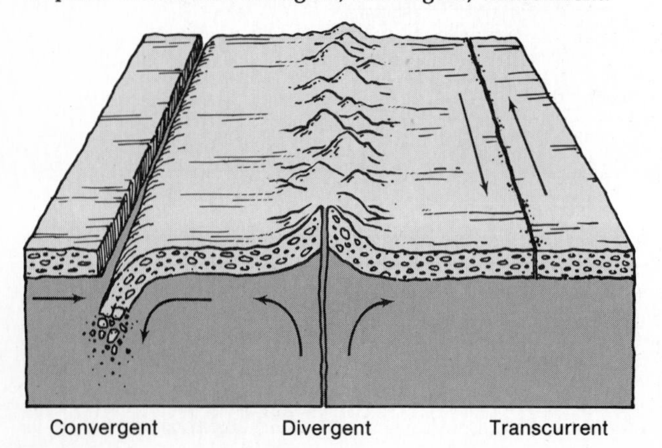

considerable volcanic activity. Convergent boundaries are typically located around the edges of ocean basins and are often associated with abyssal trenches. The third boundary type is called *transcurrent*, and involves two plates slipping past one another laterally, in a typical fault structure. Transcurrent boundaries seem to be much less common than the other types, but they, too, are associated with earthquakes.

The concept of plate tectonics provides us with a grandiose framework for understanding the massive crustal rearrangement that apparently has taken place during the relatively recent history of the Earth. The single original supercontinent, Pangaea, as visualized by Wegener, is now generally accepted as having indeed existed. It is further thought that Pangaea subsequently was broken apart into two massive pieces, Laurasia in the Northern Hemisphere and Gondwanaland in the Southern Hemisphere. See Figure 14-9. The various crustal plates, with their attached continents and parts of continents, became separated and drifted in various directions (Figure 14-10), their divergence often associated with sea-floor spreading and their convergence frequently involving collision, subduction, and mountain building.

Most, but not all, of the major mountain areas of the world are found in zones of plate convergence, and the origin of the mountains can easily be conceptualized as resulting from the collision of plates. For example, the Himalayas and associated ranges of central Asia, the world's highest mountains, apparently were formed primarily by the crumpling of the edges of two plates in a convergent boundary zone. See Figure 14-11. The Indian subcontinent was originally a part of the Gondwanaland (Southern Hemisphere) supercontinent, but it broke away and drifted northward, eventually to collide with the Eurasian landmass, the resulting upheaval producing massive ranges and knots of mountains. The mighty Andes Mountains, along the entire western edge of South America, were probably formed in a somewhat similar fashion in the collision of the so-called East Pacific and American plates.

Many aspects and implications of plate tectonics are still unknown or unclear. For example, in a convergent boundary zone, why does subduction occur in some places and mountain upheaval in others? Why are some plates so much larger than others? What determines the zones of crustal weakness where plate boundaries occur? Why did the cycle of plate movement, from Pangaea to the present, begin so late (the last 200 million years or so) in the Earth's history? What is the ultimate cause of plate movement? These and a host of other puzzles await solutions.

The present state of our knowledge about plate tectonics, however, is ample to provide a firm basis for understanding the gross patterns of most of the world's first- and second-order relief features—the size, shape, and distribution of the continents and ocean basins and

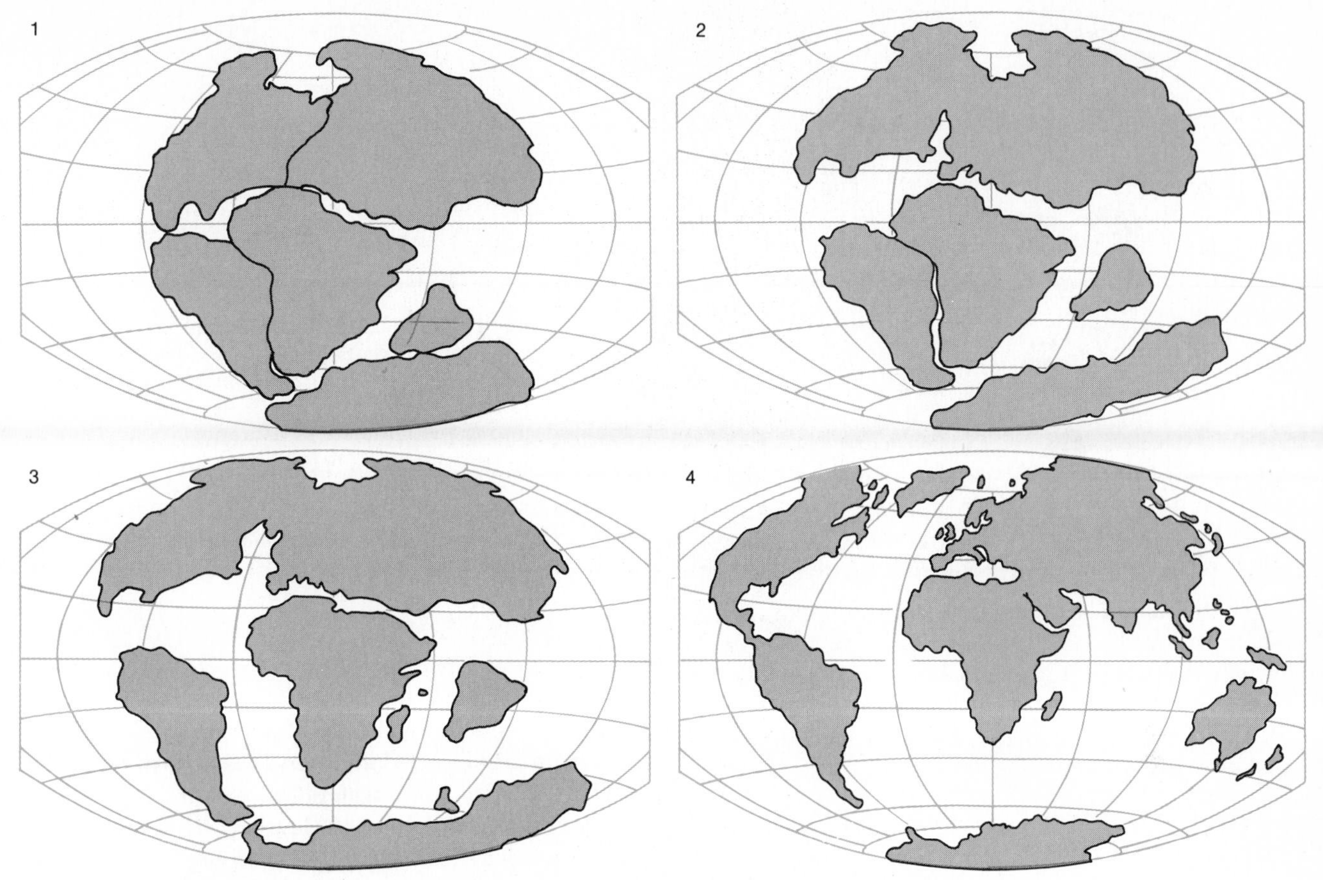

Figure 14-9. Highly generalized illustration of how Pangaea might have broken into separate continental masses. Each diagram represents a progression of about 65 million years from the preceding one, starting about 200 million years ago and concluding with the contemporary arrangement.

Figure 14-10. The major crustal plates and their generalized directions of drift.

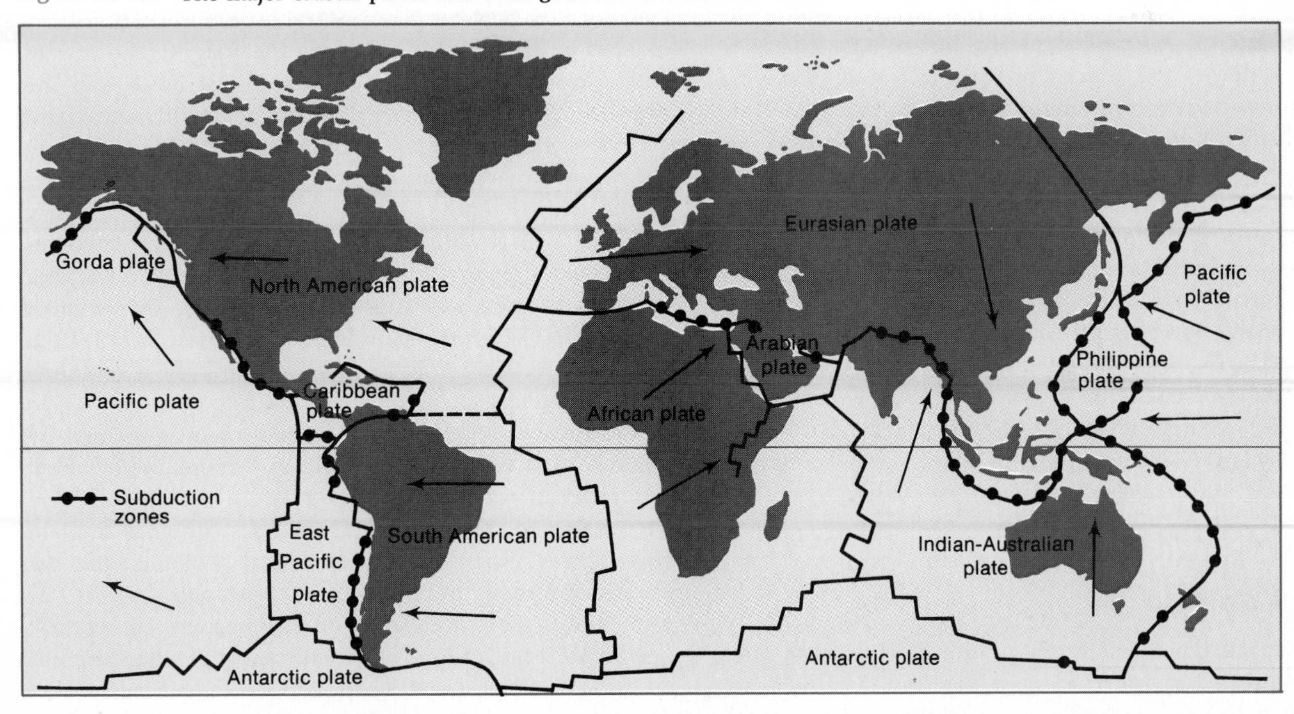

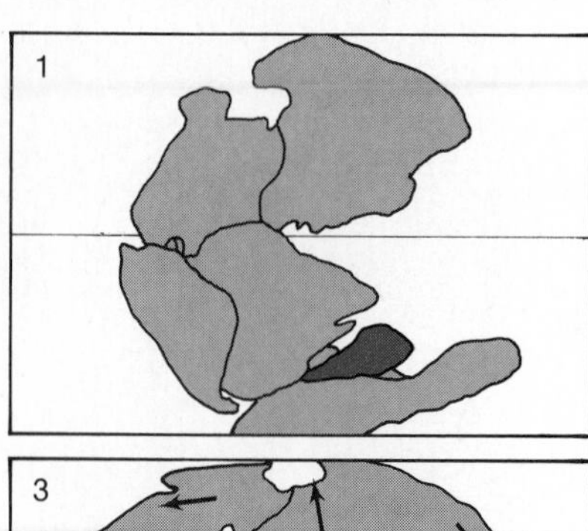

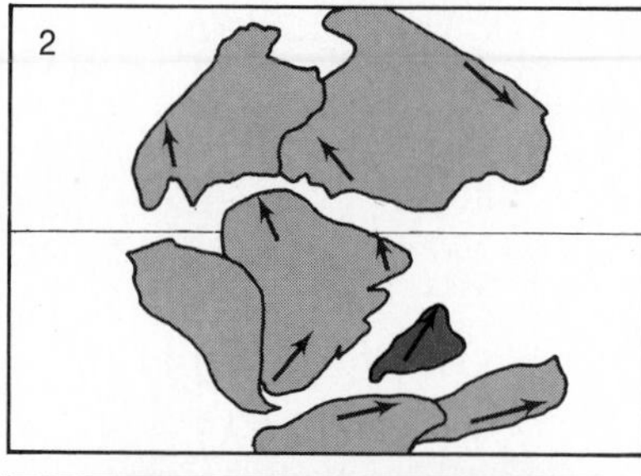

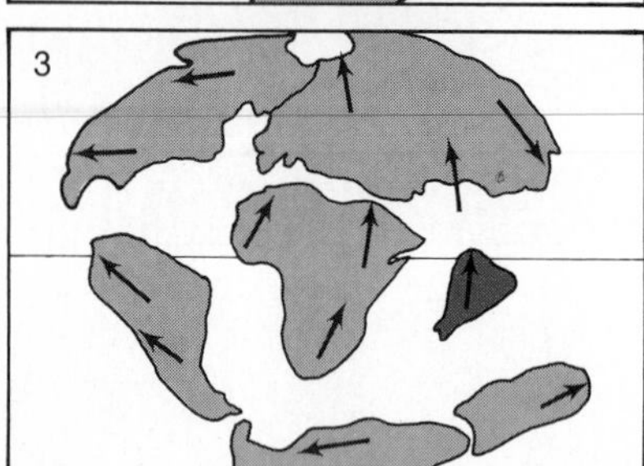

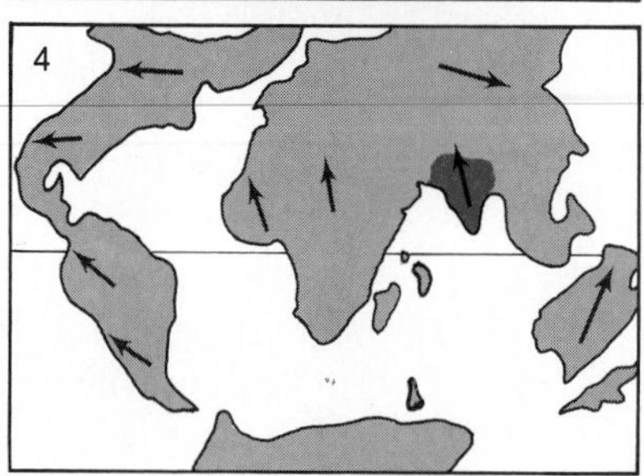

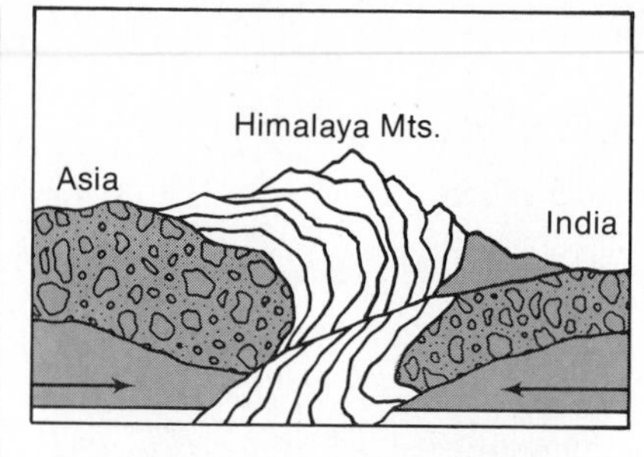

Figure 14-11. The landmass that we now know as India broke loose from Gondwanaland some 200 million years ago and drifted north to crash into the Asian plate of Laurasia; shown in steps (1) to (4). This collision of massive plates apparently was the basic cause of the upheaval that created the world's highest mountains, the Himalayas.

many of the mountain cordilleras. For a comprehension of larger-scale or more detailed topographic features, we must turn our attention to less spectacular, but no less fundamental, internal processes, which are often directly associated with the movements of the lithospheric plates.

Diastrophism

Diastrophism is a general term that refers to the deformation of the earth's crust. It involves various kinds of movements of crustal material and implies that the material is solid (i.e., rock) and not in a molten state. The rocks may be bent or broken in a variety of ways, in response to great pressures exerted from below or within the crust itself. Some of these vast pressures clearly result from movements of the crustal plates, but others appear to be unrelated to the actions of plate tectonics. Some are caused by the rise of molten material from below, either intrusively or extrusively, but most seem to have no causal relationship with magma. Whatever the causes of diastrophic pressures, the results are often conspicuous in the landscape. Diastrophic movements are particularly obvious where sedimentary rocks are involved because all sedimentary strata are initially deposited in a horizontal or near-horizontal attitude, and if they now appear in a bent, broken, or nonhorizontal orientation, they have clearly experienced some sort of diastrophic deformation.

For purposes of description and analysis, diastrophic movements can be separated into three types—*broad warping*, *folding*, and *faulting*—although the actual separation in nature is not always so discrete and clear-cut.

Broad Warping

Relatively extensive portions of the Earth's crust have been subjected to uplift or depression innumerable times in the history of our planet. Areas that are now well above sea level are often deeply covered with marine sediments, indicating that they were once in a submarine environment and consequently at a lower elevation than their present location. Conversely, many areas that are now covered by oceanic waters (for example, the North Sea) were once well above sea level.

The causes of broad crustal warping are multifarious. In many cases, isostasy is involved—downwarping resulting from some addition of weight on the surface, or upwarping due to a removal of weight. Frequently, however, there is no apparent relationship to an isostatic adjustment, and some other explanation must be sought.

When broad warping takes place in an ocean basin or in the interior of a continent, it may be relatively inconspicuous and perhaps is recognizable only through instrumented measurements. However, the warping of a coastal area causes an immediate change in sea level, and the resulting topographic development is readily apparent.

A notable example of the effects of broad warping can be seen by comparing the eastern and western coasts of the United States. Although the geomorphic history of these two coastlines is fairly complex and involves more than simple warping, the gross effects of warping are both obvious and strongly contrasting. The Atlantic coast is highly irregular, indented, embayed, and fretted by long estuaries. It is a classic example of a coastline of submergence, in which the land has been sinking in relation to sea level. The flat coastal plain is seamed by a series of broad, shallow valleys, through which flow a number of rivers on their way from the Appalachian Mountains to the Atlantic Ocean. As the coastal plain was warped downward in the recent geologic past, the sea has invaded the valleys, and the lower reaches of each one is now drowned and has become an estuary. See Figure 14-12. Opposite conditions prevail on the Pacific coast, which has been upwarped in the recent past and is clearly a coastline of emergence. No coastal plain

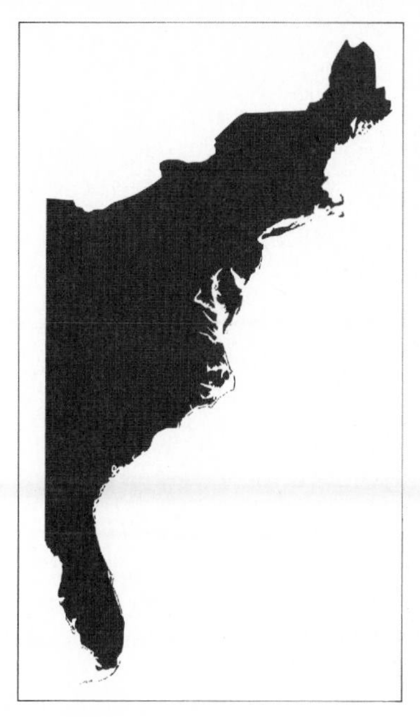

Figure 14-12. The Atlantic coast shows "drowned" characteristics (right), indicative of downwarping, which do not appear on the Pacific coast (left).

is present, the topography being hilly or mountainous right down to the coast. The present coastline is regular and smooth, with few embayments and no estuaries of any size at all. Topographic evidence of the emergent upwarping is manifested in a series of old beach lines, wave-cut benches, and wave-eroded cliffs that now occur well above the present sea level.

Folding

When crustal rocks are subjected to certain forces, particularly to lateral compression, they are often deformed by being bent, in a process called *folding*. The notion of folding is sometimes difficult for people to conceptualize. Our common experience is that rocks are hard and brittle, and if subjected to stresses, they might be expected to break; bending is harder to visualize. In nature, however, when great pressure is applied for long periods, particularly in an enclosed, buried, subterranean environment, the result is often a slow plastic deformation that can produce folded structures of incredible complexity. See Figure 14-13. Folding can occur in any kind of rock, but it is obviously most recognizable in sedimentary strata.

The process of folding can take place at almost any scale. Some folds can be measured in no more than inches or centimeters, whereas others can develop over such broad areas that crest and trough are tens of miles apart.

The configuration of the folds can be equally variable. In some cases, the folding is simple and symmetrical; elsewhere it may be of extraordinary complexity and totally without symmetry. Indeed, the most severe crumpling of a dishrag can be duplicated in rocks by the actions of diastrophic folding. Moreover, the structure may become even more complex by breakage of the rock (faulting), which the stresses may engender in addition to the complicated folding.

Structural geologists recognize many different kinds of folds, with a lengthy nomenclature to classify them. For introductory physical geography, however, only a few

Figure 14-13. The sedimentary layers in this road cut near Kiowa in central Montana have been tightly folded. [TLM photo.]

If folds are large enough, and if they are positioned at or near the Earth's surface, they often are responsible for the gross appearance of conspicuous topographic features. The actual relief and detailed configuration of the topography in nearly all cases will be determined by external processes, primarily various forms of erosion (see subsequent chapters), which act upon the fold structure of the bedrock. Some typical examples of folding are demonstrated here.

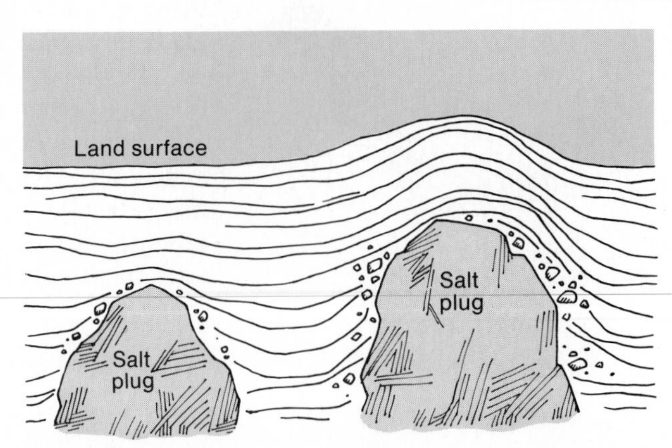

Schematic cross section of a portion of the Texas coastal plain. Both salt plugs interrupt the horizontal sedimentary strata, but only the one on the right exerted sufficient pressure to cause doming at the surface.

Doming One of the simplest types of folds is a dome. Domes can be produced in a variety of ways but are generally the result of some sort of rock being pushed up into a sequence of relatively horizontal sedimentary strata. The coastal plain of Texas and Louisiana, for example, is underlain by a series of sedimentary beds that are almost horizontal, dipping gently downward to the south; the surface expression of this structure is an exceedingly flat plain, with very little local relief. The only discordant elements in this otherwise very simple structural/topographic pattern that dominates the landscape for more than 40,000 square miles (102,000 km^2) are several hundred large plugs of rock salt, which have been intruded upward from below. The principal result of these intrusions is a breaking through of the lower sedimentary strata and a doming of the beds that are immediately above the salt plugs. Although a dome structure has been formed above each of the salt plugs, in the great majority of instances the effect has not been massive enough to cause an actual uplift at the surface. In only a handful of cases is the dome structure close enough to the surface to produce a topographic feature. Where this has occurred, a small hill rises a few tens of feet above the otherwise featureless plain.

Parallel Folds When sedimentary strata are subjected to lateral compression, a series of simple parallel folds is often formed, with an alternating sequence of anticlines and synclines. If the folding takes place near enough to the surface, or if it is subsequently exposed by uplift and/or erosion, the topographic result is likely to be a series of long,
narrow ridges and valleys in a parallel arrangement. The simplest relation between structure and topography, and one that often occurs in nature, finds the upfolded anticlines producing ridges and the downfolded synclines forming valleys.

A converse relationship may eventuate, however, with valleys developing on the anticlines and ridges on the synclines. This inverted topography is most easily explained by the effects of tension and compression on the folded strata. Where a bed is arched over an upfold, tension cracks can form and provide easy footholds for erosional forces to remove materials and incise downward into the underlying strata. Conversely, the compression that acts upon the downfolded beds increases their density and therefore their resistance to erosion. Thus, over a long period of time, the upfolds may be eroded away faster than the downfolds, producing anticlinal valleys and synclinal ridges.

terms are necessary. See Figure 14-14. Most important are the simple folds, which are widespread and common in areas of sedimentary bedrock. A *monocline* is a one-sided fold connecting horizontal or gently inclined strata. A simple symmetrical upfold is called an *anticline*, and a simple downfold is referred to as a *syncline*. Also relatively common is an upfold that has been pushed so vigorously from one side that it becomes oversteepened enough to have a reverse orientation on the other side; such a structure is referred to as an *overturned fold*. If the pressure is enough to break the oversteepened limb and cause a shearing movement, the result is an *overthrust fold*.

The landforms that result from folding are even more varied than the folds themselves because other processes are involved in the final shaping of the surface. Most folding takes place in a constricted subterranean

Figure 14-14. The basic types of folds.

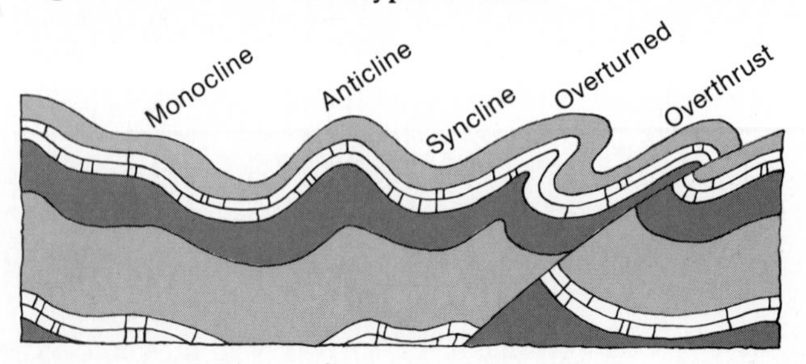

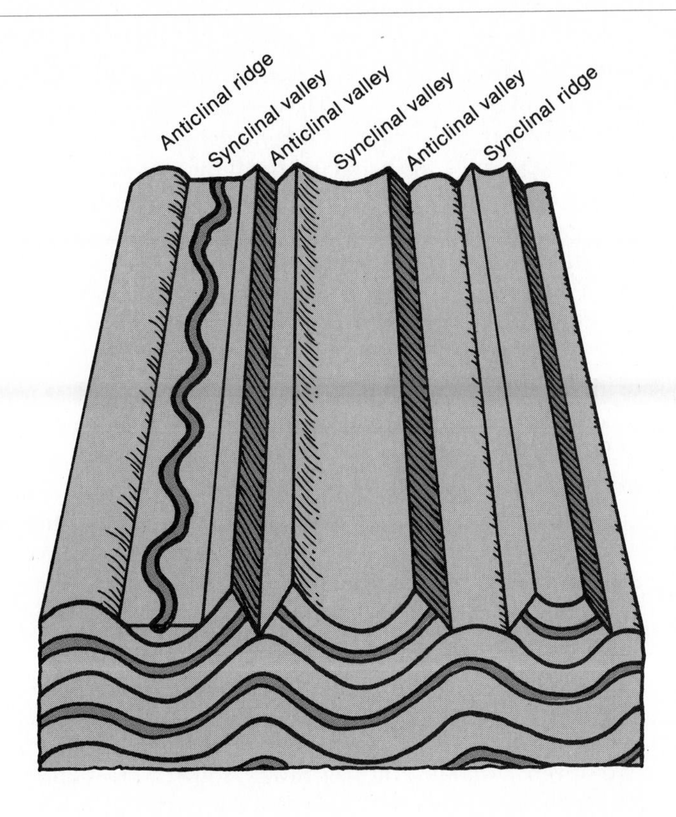

Schematic vertical cross section through the ridge and valley section of the Appalachians. Both anticlines and synclines produce ridges and valleys.

Generalized cross section through the Swiss Alps, showing the enormous complexity of fold structures.

Both of these sets of circumstances, as well as many variations, can be found in the so-called ridge and valley section of the Appalachian Mountains, an area that is world-famous for its remarkably parallel sequence of mountains and valleys developed on folds. The ridge and valley section extends for about 1000 miles (1600 km) in a northeast-southwest direction across parts of nine states, with a width that varies from 25 to 75 miles (40 to 120 km). Structural/topographic relations within this vast area are varied, but by far the most characteristic are the two patterns described above: anticlinal ridges and synclinal valleys or anticlinal valleys and synclinal ridges.

Complex Folds The simplicity of doming and the repetition of parallel folding are relatively easy to envisage, but sometimes there are more complicated configurations of folding that produce much less understandable and less predictable topography. The geometry of complex folding can be of infinite variability, with extremes of crumpling and overturning. The resultant topographic features can be equally variable, depending, as always, on the actions of the external processes.

One example of such a situation is portrayed in the figure above, which displays some of the remarkable compressional folding that has taken place in the Swiss Alps and the generalized topographic profile that has resulted.

environment; when the structures are exposed at the surface, erosional processes modify them in a great variety of ways.

Faulting

Another prominent result of the various stresses in the crust is the breaking apart of rock material. When rock is forcefully broken with accompanying displacement (i.e., an actual movement of the crust on one or both sides of the break), the action is called *faulting*. See Figure 14-15. The movement can be vertical or horizontal or a combination of both. Faulting usually takes place along zones of weakness in the crust; such an area is referred to as a *fault zone* or *fault plane*, and the intersection of that zone with the Earth's surface is called a *fault line*. See Figure 14-16.

Movement along a fault zone is sometimes very slow, but it may also occur as a sudden slippage. A single slippage may result in a displacement of only a centimeter or so, but in some cases the movement may be as much as 20 or 30 feet (6 to 9 m). Successive slippages may be years or even centuries apart, but the cumulative displacement over millions of years, encompassing both sudden slippages and slow movements, could conceivably amount to enormous distances—as much as hundreds of

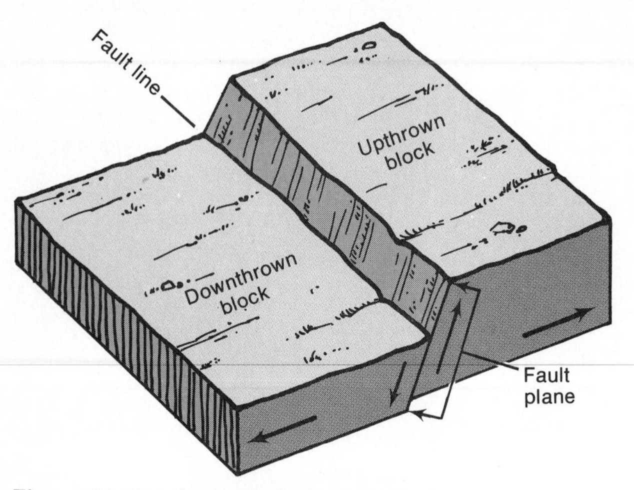

Figure 14-15. A simple fault structure.

miles horizontally and tens of miles vertically. In all known cases, however, the magnitude of total fault displacement is considerably less than these extreme figures.

Usually, but not exclusively, associated with faults are abrupt movements of the crust known as *earthquakes*. These movements vary from mild, imperceptible tremors to wild shakings that may persist for many seconds. Although usually of very limited importance in the shaping of landforms, earthquakes sometimes cause much destruction and suffering for mankind.

The depth of fault actions is unknown, but major faults seem to penetrate many miles into the Earth's crust.

Figure 14-16. A prominent fault line runs through the town of Hollister, California. Dramatic evidence of its activity is shown by the offset in curb, sidewalk, and stone wall. [TLM photo.]

Indeed, the deeper fault zones apparently serve as conduits to allow both water and heat to approach the surface from the depths. Frequently springs are found along fault lines, sometimes with hot water gushing forth. Volcanic activity is also associated with some fault zones, as magma forces its way upward in the zone of weakness.

Fault lines are often marked by other prominent topographic features. Most notable are *fault scarps*, which are steep cliffs that represent the edge of a vertically displaced block. Some fault scarps are as much as 2 miles (3.2 km) in height and extend for more than 100 miles (160 km) in virtually a straight line. The abruptness of their rise, the steepness of their slope, and the linearity of their orientation combine to make some fault scarps extremely spectacular features in the landscape, as exemplified by the east face of California's Sierra Nevada or the west face of Utah's Wasatch Range. Other prominent fault-line topographic features include linear erosional valleys, carved in the weakened rock along the fault zone; displaced stream courses, associated with lateral movement of the fault blocks; and *sag ponds*, caused by the collection of water from springs and/or runoff into sunken ground resulting from the jostling of the Earth in the area of fault movement.

Although structural geologists recognize more than two dozen different kinds of faults, they can be generalized into four principal types on the basis of direction and angle of movement. See Figure 14-17. Two types involve displacement that is mostly vertical, a third encompasses only horizontal movement, and the fourth variety includes both horizontal and vertical offsets.

1. A *normal fault* results from tension stresses in the crust. It produces a very steeply inclined fault plane, with the block of land on one side being pushed up, or

Figure 14-17. The principal types of faults.

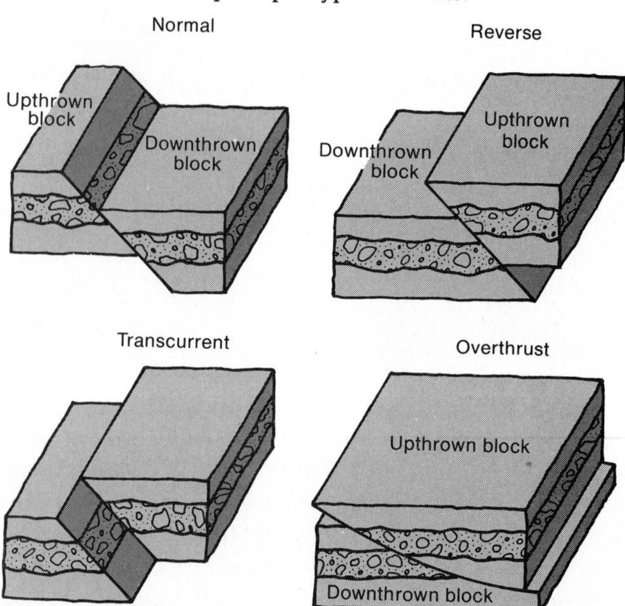

Focus on Earthquakes

Although many of the processes that operate in nature are slow and either benign or inoffensive to humanity, a few—such as tornados, hurricanes, volcanic eruptions, and earthquakes—are abrupt and capable of creating an enormous amount of destruction in a very short time. As shapers of terrain, earthquakes are of minor significance, but as potential producers of instantaneous havoc, they are overwhelming.

An earthquake is essentially a vibration in the crust that is produced by shock waves from a sudden displacement along a fault. The fault movement amounts to an abrupt release of energy from a long, slow accumulation of strain, which is an ongoing process in crustal deformation. The faulting may take place right at the surface, but it usually originates at considerable depth, extending downward as much as 400 miles (640 km) beneath the surface. The pent-up energy that is released moves through the lithosphere in several kinds of waves from the center of motion (called the *focus*). These *seismic waves* are transmitted outward in widening circles, almost like ripples produced when a rock is thrown into a pond, gradually losing momentum with increasing distance from the focus. The strongest shocks and greatest crustal vibration are usually felt directly above the focus on the surface, which is referred to as the *epicenter* of the earthquake.

Epicenter

Focus

Schematic diagram to show the relationship of focus, epicenter, and seismic waves of an earthquake. The waves are indicated by the concentric circles.

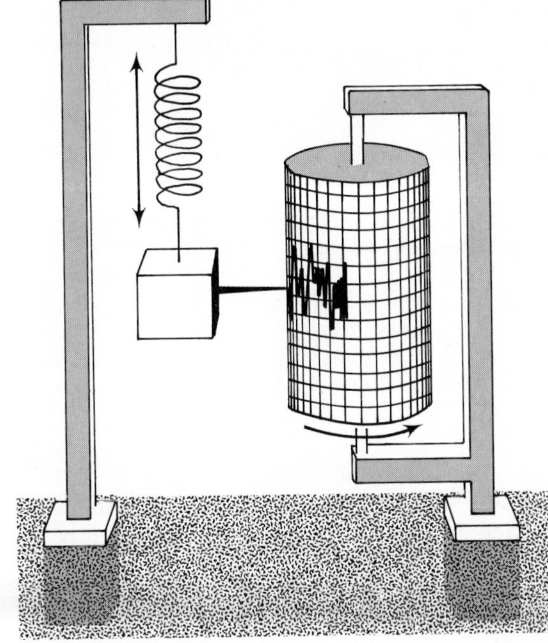

A simplified seismograph. The two posts, which are anchored in bedrock, pick up vibrations in the crust. The pendulum, suspended by a wire coil from one of the posts, traces the movement of the other post on the rotating drum.

Seismographs around the world record the arrival times and forces of seismic waves from earthquakes. Comparing records from different seismograph stations allows the focus of a quake to be pinpointed with great precision and its strength to be determined. Several different scales are used to indicate the violence of an earthquake, but by far the most widely used is the Richter scale of earthquake magnitudes, devised by California seismologist Charles F. Richter in 1935 to describe the amount of energy released in a single quake. The scale is logarithmic, so each successively higher number represents an energy release that is ten times greater than the preceding number. The scale numbers range from 0 to 9 but theoretically have no upper limit. Any earthquake with a Richter number of 8 or above is considered to be catastrophic, and lesser-numbered shocks can cause immense damage under certain conditions. The most violent known earthquakes have recorded 8.9 on the Richter scale. In comparison, the famous San Francisco quake of 1906 reached an estimated 8.3, the southern Alaska earthquake of 1964 recorded 8.5, and the San Fernando Valley shock of 1971 attained only 6.6.

upthrown, with relation to the *downthrown* block on the other side. A prominent fault scarp usually is formed.

2. A *reverse fault* is produced from compression, with the upthrown block rising steeply above the downthrown block, so that the fault scarp would be severely oversteepened if erosion did not act to smooth the slope somewhat. Landslides normally accompany reverse faulting.
3. In a *transcurrent* or *strike-slip fault*, the movement is entirely horizontal, produced by shearing, with the adjacent blocks being displaced laterally with respect to each other. This action, of course, does not produce a scarp, and the fault line is topographically inconspicuous except where stream courses are displaced or sag ponds develop.
4. More complicated in structure and more impressive

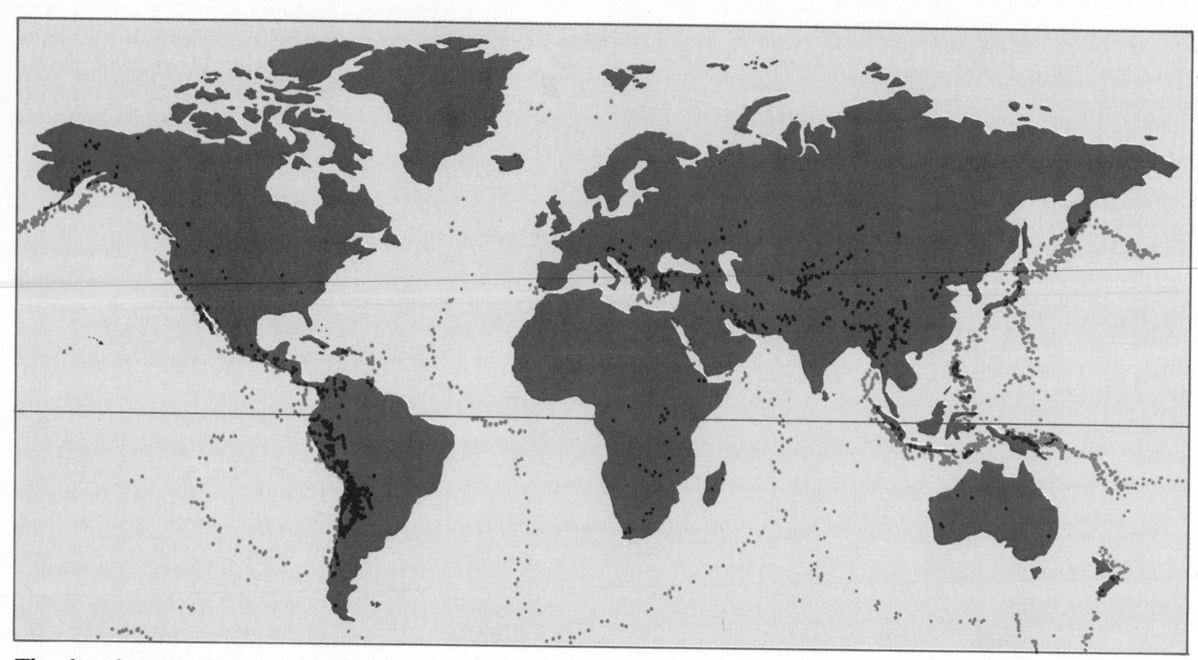

The distribution of epicenters for all earthquakes of at least 5.5 magnitude from 1963 through 1977. Their relationship to midocean ridges and oceanic trenches is striking.

The effect of an earthquake on the topography is distinct from, although obviously related to, actual movements along a fault line. The most notable earthquake-caused terrain modifications are usually landslides, which may be triggered off in profusion in hilly or mountainous terrain. The landslides themselves sometimes produce significant secondary effects such as blocking streams and thereby creating instant new lakes.

Another kind of hazard associated with earthquakes involves water movements in lakes and oceans. The abrupt crustal vibrations can set great waves in motion in lakes and reservoirs, causing them to overflow shorelines or dams in the same fashion that water can be sloshed out of a dishpan by shaking it. Much more significant, however, are great seismic sea waves, or *tsunamis*, which are sometimes generated by seafloor movement associated with undersea earthquakes. These waves, sometimes occurring in a sequential train, move with great rapidity across the ocean. They are all but imperceptible in deep water; but when they reach shallow coastal waters, they sometimes build up to several feet or even tens of feet in height and may crash on the shoreline with devastating effect (where they are often incorrectly called *tidal waves*).

Although the effect of earthquakes on topography may be limited, it is sometimes cataclysmic for humanity. Most quakes are so slight as to occur without recognition, sometimes being mistaken as the rumbling of a passing truck. More severe earthquakes, however, can cause devastating damage within a few seconds. Rubble can be shaken off buildings, entire buildings may collapse, gas mains can be broken, igniting dangerous fires, and villages and even cities may be destroyed. Structural damage sometimes amounts to hundreds of millions of dollars, and loss of life can be counted in the thousands. Indeed, at least five earthquakes in history have killed more than 100,000 people each.

In any given year, literally tens of thousands of earthquakes occur somewhere in the crust, most of them followed by aftershocks. These aftershocks may number in the hundreds after a single quake and may continue for several days with diminishing intensity.

Earthquakes may occur anywhere, even in the middle of apparently very stable continental areas. Most, however, take place in association with the boundary zones of the great crustal plates, particularly along the midocean ridges and in the subduction areas of ocean margins. The greatest concentrations of earthquake epicenters are found around the rim of the Pacific Ocean.

Despite their awesome potential and occasionally devastating results, earthquakes represent only minor adjustments in the infinite continuum of events and process interactions that combine to maintain tectonic equilibrium in the Earth's crust—a scientific fact that may be of little comfort to a person who lives on the edge of the Pacific Ocean!

in their dynamics are *overthrust faults*, in which compression forces the upthrown block to override the downthrown block at a relatively low angle, sometimes for many miles. Overthrusting occurs frequently in mountain building, resulting in unusual geological relationships such as older strata being piled on top of younger rocks.

Faulting sometimes occurs as an isolated event, but it usually accompanies more complex diastrophic action.

Where the Earth's crust experiences stresses sufficient to cause faulting, other kinds of diastrophic movement (folding, warping) may also occur. Moreover, such stresses are frequently associated with zones of weakness in the crust, and molten magma sometimes finds its way to the surface through these zones. Thus, a simple individual fault may occur in isolation, but much more common are multiple faults, or a mixture of faulting and folding, or an even more complex association of diastrophism and vulcanism and ultimately the forces of plate tectonics.

The surface of the Earth is subjected to enormous stresses by the movement of the crustal plates and by the associated rise of magma into the divergent zones, and particularly by the crumpling and subduction that occur when plates collide or slide over one another. The tension and compression involved in these processes are instrumental in causing many of the lesser diastrophic movements described previously. Thus an appreciation of plate tectonics helps us to understand the formation of not only first-order relief features but also many of the second-, third-, and fourth-order landforms as well.

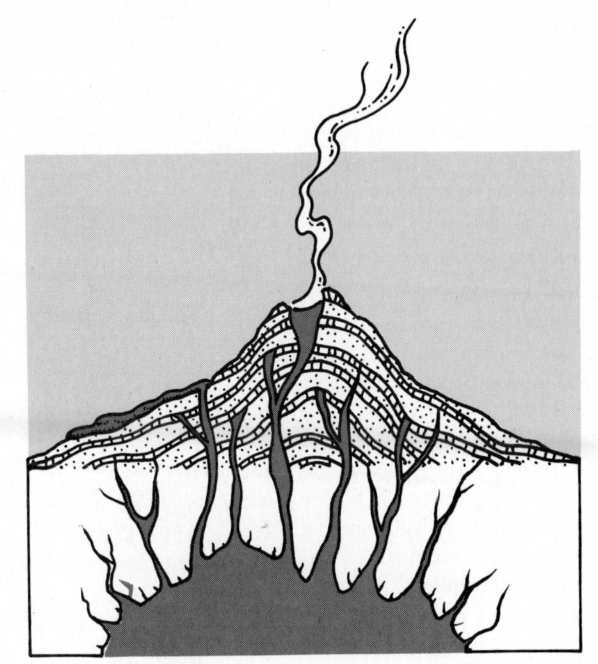

Figure 14-18. Idealized diagram of a volcano.

Vulcanism

Vulcanism is a general term that refers to the movement of magma from the interior of the Earth to or near the surface. It sometimes consists of explosive volcanic eruptions that are among the most spectacular and terrifying events in all nature, but it also involves much more quiescent phenomena, including the slow solidification of molten material below the surface.

We have noted previously the distinction between extrusive and intrusive igneous rocks; a similar differentiation can be made between extrusive and intrusive vulcanism. If the magma is expelled onto the Earth's surface while still in a molten condition, the activity is *extrusive* and represents true vulcanism. On the other hand, if the molten material solidifies within the crust, it is an *intrusive* activity. A further distinction is sometimes made to identify *plutonic* activity as involving deep-seated magma that solidifies far below the surface.

Extrusive Vulcanism

Molten magma extruded onto the surface of the Earth, where it cools and solidifies, is designated as *lava*. See Figure 14-18 for an idealized diagram of a volcano erupting. The ejection of lava into the open air is sometimes volatile and explosive, devastating the area for miles around; in other cases, it is gentle and quiet, affecting the landscape more gradually. All instances, however, cause clear-cut alteration of the landscape because the fiery lava is an inexorable force until it cools, even if it is expelled only slowly and in small quantity.

The explosive eruption of a volcano is an awesome spectacle. In addition to an outward flow of liquid lava, such solid matter as rock fragments, solidified lava blobs, ashes, and dust (collectively called *pyroclastic material*), as well as gas and steam, may be hurled upward in prodigious quantities. In some cases, the volcano literally explodes, disintegrating in an enormous self-destructive blast. The supreme example within historic times, perhaps, was the final eruption of the volcano Krakatau, which occupied a small island in the East Indies. See Figure 14-19. When it exploded in 1883, the noise was heard 1500 miles (2400 km) away in Australia, and 6 cubic miles ($20\frac{1}{2}$ km^3) of material were blasted into the air. The volcanic island itself completely disappeared, leaving only open sea where it had been. The tsunamis that were generated drowned more than 30,000 people, and sunsets in various parts of the world were colored by fine volcanic dust for many months afterwards.

The nature of a volcanic eruption apparently is determined largely by the chemistry of the magma that feeds it, although the relative strength of the surficial crust and the degree of confining pressure to which the magma is subjected may also be important. The chemical relations are complex, but the critical component seems to be the relative amount of silica (SiO_2) in the magma. A high content of silica usually indicates a somewhat cooler magma in which a portion of the heavier minerals has already crystallized and a considerable amount of gas has already separated. Some of this gas is trapped in bubbles and pockets in the magma under great pressure. As the magma approaches the surface, the confining pressure is diminished, and the pent-up gases are released

Certain conspicuous fault associations recur frequently in various parts of the world, providing a gross configuration to the terrain that is very prominent in the landscape.

Fault-Block Mountains Under certain conditions of crustal stress, a surface block may be severely faulted and upthrown on one side without any faulting or uplift on the other. When this happens, the block is tilted asymmetrically, producing a steep slope along the fault scarp and a relatively gentle slope on the other side of the block. The classic example of a fault-block range is California's Sierra Nevada, which is an immense block nearly 400 miles (640 km) north-south and about 60 miles (96 km) east-west. The spectacular east face of the Sierra Nevada represents a fault scarp that has a vertical relief of about 2 miles (3.2 km) in a horizontal distance of only about 12 miles (19 km). In contrast, the general slope of the western flank of the range, from crest to hinge line, has a vertical dimension of about 12,000 feet (3600 m), spread over a horizontal distance of nearly 50 miles (80 km). The topographic shape of the range has of course been modified by other processes, but its general configuration has been determined specifically by block faulting.

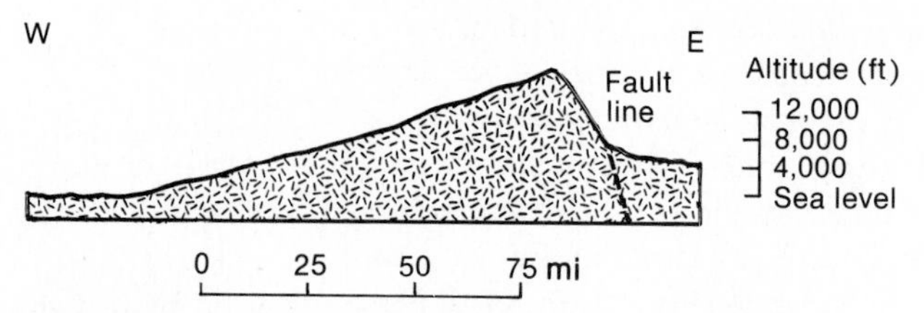

The western slope of the Sierra Nevada is long and grossly gentle, whereas the eastern slope is short and steep. This is the result of enormous block faulting on the eastern side.

Mt. Rundle is a fault-block mountain in the Canadian Rockies. The abrupt northeastern face, seen here, represents the fault scarp. The far (hidden) side of the mountain slopes much more gently down to the hinge line. [TLM photo.]

Horst Another frequent occurrence in nature is the uplift of a block of land between two parallel faults, which produces a structure called a *horst*. The same result can be achieved if the land on both sides has been downthrown. In either case, the horst is an elevated block in comparison with the surrounding land. It may take the form of a plateau or a mountain mass with two steep, straight sides.

Graben At the other extreme is a graben, which is a block of land bounded by parallel faults in which the block has been downthrown, producing a distinctive structural valley with a straight, steep-sided fault scarp on either side. A world-famous graben is Death Valley in California, a prominent trough that extends for about 50 miles (80 km) between fault scarps of the Panamint Mountains to the west and the Black Mountains to the east, which are only about 5 miles (8 km) apart.

Grabens and horsts often occur side by side in juxtaposition with one another, sometimes in multiplicity where there is a repetition of parallel faulting. The basin and range country of the western interior of the United States is a vast sequence of horsts and grabens, mixed with fault-block ranges and some more complex structures, encompassing most of Nevada and portions of surrounding states. The following diagram presents a simplified cross section at approximately the 39th parallel of latitude, extending from the massive fault-block range of the Sierra Nevada to the almost as impressive west-facing fault scarp of Utah's Wasatch Range.

Rift Valleys Downfaulted graben structures occasionally extend for extraordinary distances as linear structural valleys enclosed between typically steep fault scarps. Such lengthy trenches are called *rift valleys*, and they comprise some of the Earth's most notable structural lineaments. Most conspicuous is the series of trenches that stretches north-south across much of East Africa, with extensions northward into the Near East.

This mighty rift system extends for almost 4000 miles (6400 km), nearly one-sixth of the circumference of the Earth. Although a single major longitudinal trench persists for almost the entire length of the structural trend, several offshoot trenches complicate the pattern considerably, particularly near the equator where two lengthy north-south rifts are 400 miles (640 km) apart. Different portions of the trenches vary significantly in their characteristics—for example, the floor of the rift is more than

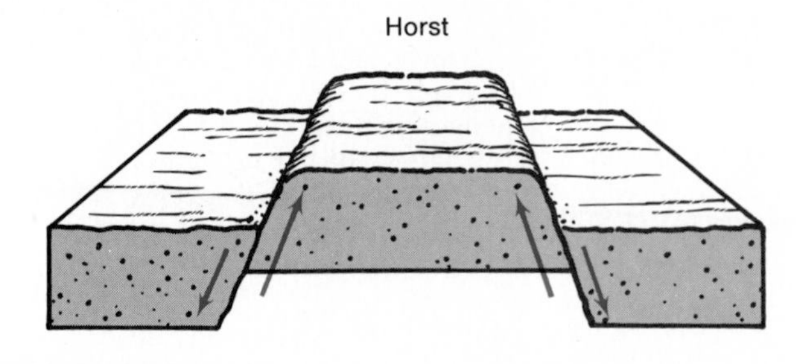

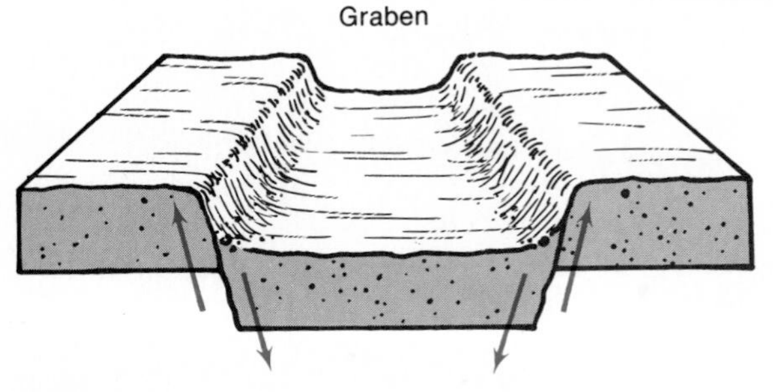

Horst and graben.

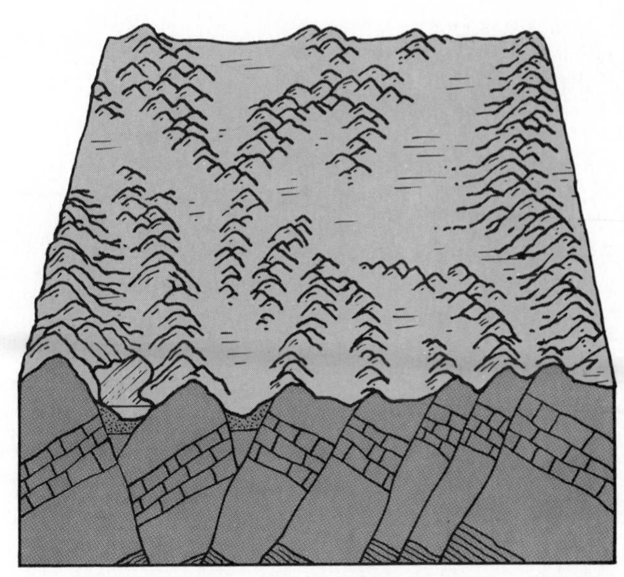

A simplified cross section of the basin and range country of Nevada and Utah. Many fault-block mountain ranges and linear valleys occur in roughly parallel arrangement in this region.

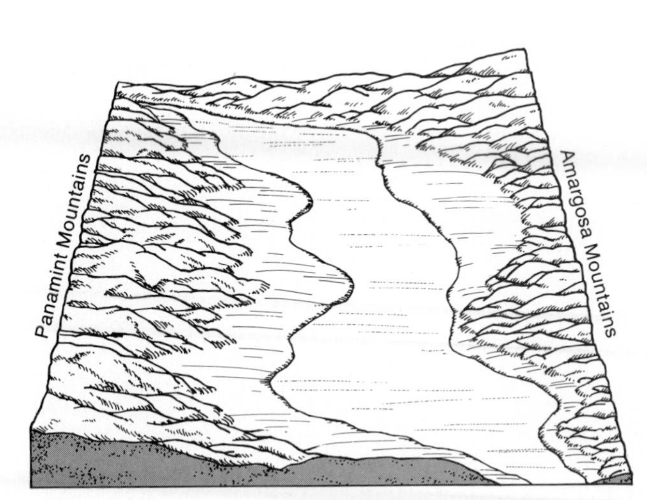

Schematic diagram of Death Valley from the south.

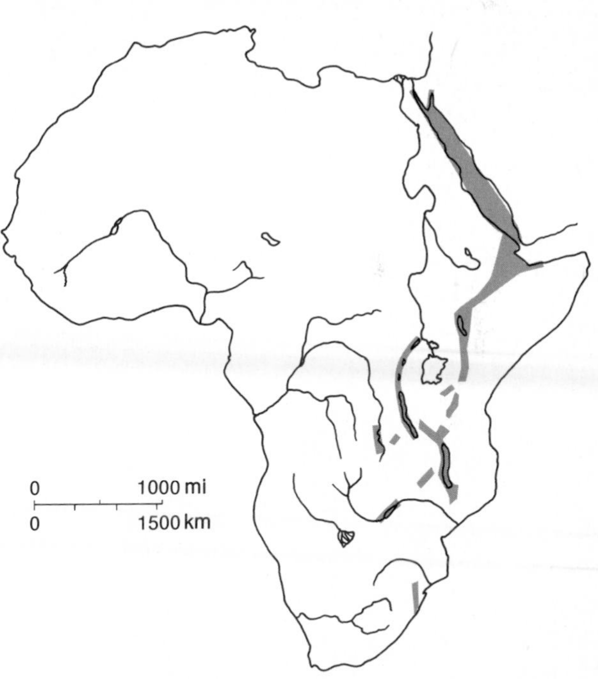

The great rift valley systems of East Africa.

2000 feet (600 m) below sea level in some places and more than 5000 feet (1500 m) above sea level in others—but almost everywhere the sides of the valleys are bounded by steep fault scarps that rise abruptly for hundreds or thousands of feet above the floor. The rift valleys are from 20 to 80 miles (32 to 128 km) in width and encompass many of the major water bodies of that part of the world, such as the Dead Sea, Red Sea, Lake Turkana, Lake Mobutu (Albert), Lake Tanganyika, and Lake Malawi. There is considerable debate as to the actual mechanism of formation of these structural troughs, but it may well be related to incipient ocean-floor spreading associated with the movement of lithospheric plates. Faulting, however impelled, has clearly formed this remarkable system of rifts, and sporadic earthquakes and volcanic activity indicate that the Earth's movements that produced the trenches have not yet ceased.

North America also has a major linear structural trench that shows the

(continued)

An *Apollo 7* photo on a clear day shows rift valleys in dramatic fashion. Egypt's Sinai Peninsula occupies the center of the photograph, its edges marked by the Gulf of Suez to the west and the Gulf of Aqaba to the east. North of the latter, the rift valley trend is emphasized by the appearance of the Dead Sea. [Courtesy of NASA.]

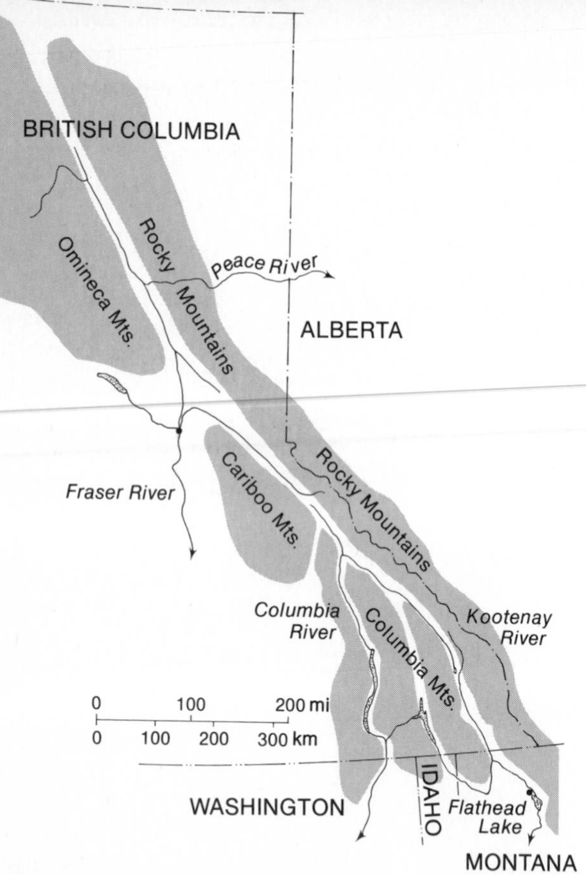

The Rocky Mountain Trench is a remarkable linear trough in western North America.

characteristics of a rift valley, although a full understanding of its origin has not yet been worked out. The Rocky Mountain Trench extends for more than 1000 miles (1600 km) from Montana's Flathead Lake in a virtually straight line north-northwest across British Columbia almost to the border of the Yukon Territory, comprising the longest and most persistent gash in the surface of the continent. In comparison with its length, the trench is extremely narrow, varying only from 2 to 10 miles (3.2 to 16 km). Although the floor of this remarkable valley is generally flat, isolated hilly areas and low ridges separate the drainage basins of six different rivers that flow longitudinally along portions of the trench, often in opposite directions. Unlike the East African rifts, no large natural lakes (with the exception of Flathead Lake) occupy the trench. The remarkable linearity of the valley and the regularity of its box-like sides are clear evidence that faulting played a major role in its formation.

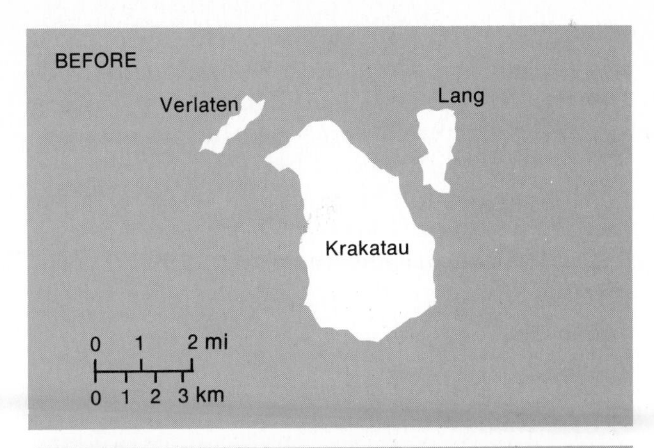

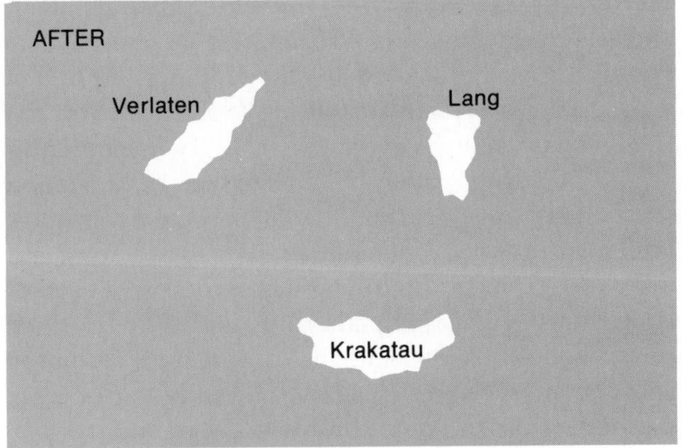

Figure 14-19. Krakatau, before and after its cataclysmic explosion. Verlaten Island became larger; Lang Island remained about the same size; Krakatau Island was mostly blown away.

Figure 14-20. Various kinds of volcanoes are identified by scientists. The most common types are shield, composite, plug dome, and cinder cone.

in explosive fashion. The initial explosion eases the pressure within the magma, but it may trigger a lengthy chain reaction of further gas formation and more explosions lasting for days. If, on the other hand, the magma is chemically more basic, with a lesser proportion of silica, it is likely to be considerably hotter, and most of the magmatic gases will still be dissolved, producing a much more fluid mixture. The resulting eruption may yield a great outpouring of lava, quietly and without explosions. See Figure 14-20.

The eruptive style of a volcano can change with time, but it usually does not. If the initial eruptions from a particular crater are explosive, future eruptions will probably also be of an explosive nature. Most other aspects of volcanic activity are much less predictable, however, particularly the timing of eruptions. In the vast majority of cases, a volcanic eruption will begin with little or no warning. This is a particularly critical factor for people living in the vicinity because the interval between eruptions may amount to years, decades, or even centuries, and the populace may be totally unheeding of potential danger. Indeed, a volcano that has not erupted for several generations may suddenly be galvanized into action.

In terms of recency of activity, volcanoes are described as being active, dormant, or extinct. An *active* volcano has had relatively frequent activity. A *dormant* volcano has been quiet for some time but has erupted within recorded history and is considered to be potentially active. A volcano that has not been known to erupt at any time since its discovery is considered to be *extinct*. Within the conterminous 48 states of the United States there was only one volcano (prior to the 1980 eruption of Mt. St. Helens) that was classified as active, Mt. Lassen in California, which last erupted in 1917 but still occasionally produces gas and steam. A few old volcanoes, notably California's Mt. Shasta and Washington's Mt. Baker and Mt. Rainier, show signs of potential activity but have not erupted in recorded time; and there are hundreds of totally extinct volcanoes, primarily in the west coast states. Both Alaska and Hawaii have many volcanoes in all three classes, including a few that are very active.

All volcanoes are "relatively" temporary features in the landscape. Some may have an active life of only a few years, whereas others are sporadically active for thousands of centuries. At the other end of the scale, new volcanoes are spawned from time to time. One of the most notable recent examples is Parícutin, which, abruptly and without warning, began to erupt and grow out of a corn field in central Mexico in 1943. It was active frequently for a few years and then sporadically for a few more, eventually reaching a height of 1500 feet (455 m) above the surrounding land during its last eruption in 1952. It is now considered to be dormant. Even more recent was the birth of Surtsey, which rose out of the sea as a new island off the coast of Iceland in a series of spectacular eruptions beginning in 1964.

Although volcanoes are very conspicuous landscape features associated with extrusive vulcanism, many of

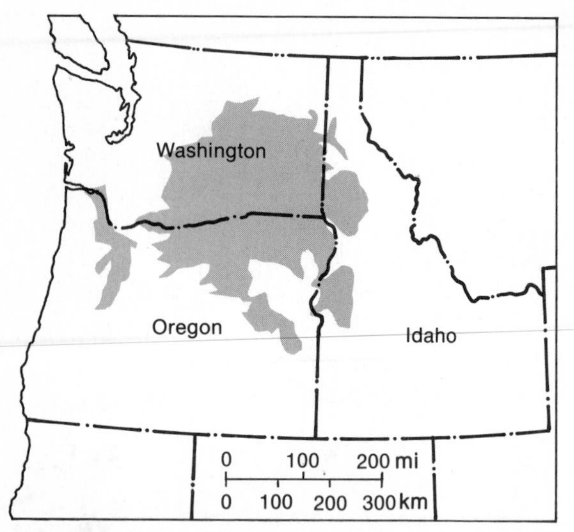

Figure 14-21. The extent of the Columbia Plateau flood basalts of the Pacific Northwest.

the world's most extensive lava flows were not extruded from true volcanoes but rather issued quietly from great fissures that developed in the crustal surface. The mechanism that produces these great outpourings of lava is not clear; but apparently it involves both the midocean ridges, where upwelling magma fills the gaps between diverging plates, and more localized "hot spots," where magma is expelled through cracks in the plate itself. The lava that flows out of these fissures is nearly always basaltic and frequently comes forth in great volume. The term *flood basalt* is applied to the vast accumulations of lava that build up, layer upon layer, sometimes covering tens of thousands of square miles to depths of many hundreds of feet. A prominent example of flood basalts in the United States is the Columbia Plateau, which covers some 50,000 square miles (130,000 km^2) in Washington, Oregon, and Idaho. See Figure 14-21. Larger outpourings are evidenced on other continents, most notably the Deccan Plateau of India, and even more extensive flood basalts probably occupy much of the ocean floors. Over the world as a whole, more lava has issued quietly from fissures than from the combined outpourings of all volcanoes.

As is the case with all internal processes, extrusive vulcanism can produce an almost infinite variety of terrain features. Four particular types of landforms, however, are distinctively associated with volcanic activity, as discussed below.

Lava Flows Whether originating from a volcanic crater or a crustal fissure, a lava flow spreads outward over the crust at an attitude that is approximately parallel with the surface over which it is flowing, and this parallelism is maintained as the lava cools and solidifies. Although some viscous flows cling to relatively steep slopes, the vast majority of all flows eventually solidify in a horizontal orientation that grossly resembles the stratification of sedimentary rock, particularly if several flows have accumulated on top of one another. The topographic expression of a lava outpouring, then, whether single or multiple, is often a flattish plain or plateau. The stratified nature of sequential flows may be conspicuously exposed by subsequent erosion, as streams usually incise very steep-sided gullies or gorges into lava flows, and the layered sequence of flows can be seen in cross section on the side of a gorge. The character of the actual surface of the flow varies with the nature of the lava and the extent of subsequent erosion, but as a general rule the surface of relatively recent lava flows tends to be extremely irregular and fragmented.

Volcanic Peaks Most volcanic peaks take the form of cones with a symmetrical profile. See Figure 14-22. Some, whose ejected material is acidic, may be very high and have relatively steep slopes, whereas others, composed

Figure 14-22. The most conspicuous volcanic peak in the conterminous United States is Mt. Rainier. It has not erupted for several thousand years, but is not considered to be extinct. [TLM photo.]

of basic rocks, have very low-angled slopes. A common denominator of nearly all volcanic peaks is a *crater*, normally set conspicuously at the very apex of the cone. Frequently, smaller subsidiary cones develop around the base of a principal peak, or on its sides, or even in the crater itself. See Figure 14-23.

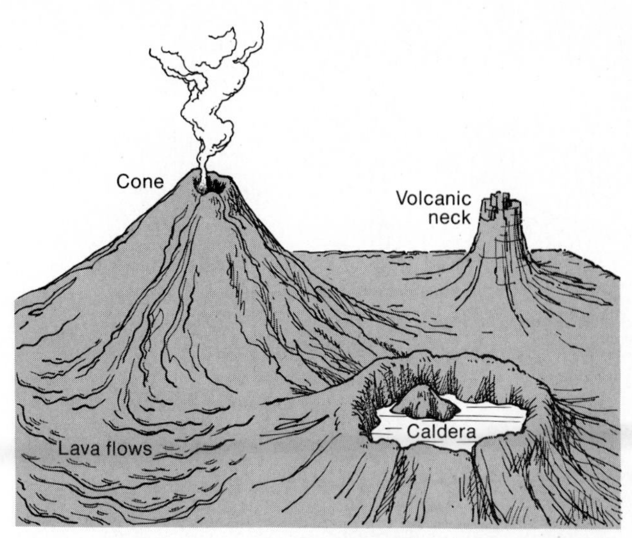

Figure 14-23. Some prominent volcanic features.

Calderas Uncommon in occurrence but spectacular in result is the formation of a caldera, which is produced when a volcano literally "blows its top" in a violent explosion. The entire summit area may be destroyed in such an explosion, some of it blown outward in fragments and the remainder subsequently collapsing into the subsurface magma reservoir that is partially voided by the explosion. The resultant landform is a vast circular pit that usually fills partly with water to form a deep lake. North America's most famous caldera is that occupied by Oregon's Crater Lake. See Figure 14-24. Less than 7000 years ago Mt. Mazama was a volcanic cone that reached an estimated altitude of 12,000 feet (3640 m) above sea level. See Figure 14-25. Its final cataclysmic eruption removed (by explosion and collapse) the upper 4000 feet (1820 m) or so of the peak and produced a caldera whose bottom is 4000 feet (1820 m) below the crest of the remaining caldera rim. Half this depth has filled with water, creating one of the deepest lakes in North America. After the final explosion, a subsidiary volcanic cone built up from the bottom of the caldera and now breaks the surface of the lake as Wizard Island.

Volcanic Necks More limited still, but very prominent where it does occur, is a *volcanic neck*, a small, sharp spire that rises abruptly above the surrounding land. It represents the pipe or throat of an old volcano, filled with solidified lava after its final eruption. Eons of erosion have removed the less resistant material that made up the cone, leaving the harder, lava-choked neck as a conspicuous residual remnant.

Figure 14-24. Crater Lake is a classic example of an enormous caldera, formed by the explosive collapse of a huge volcano. Wizard Island, the small volcanic cone in the lake, was built up subsequently. [U.S. Forest Service photo.]

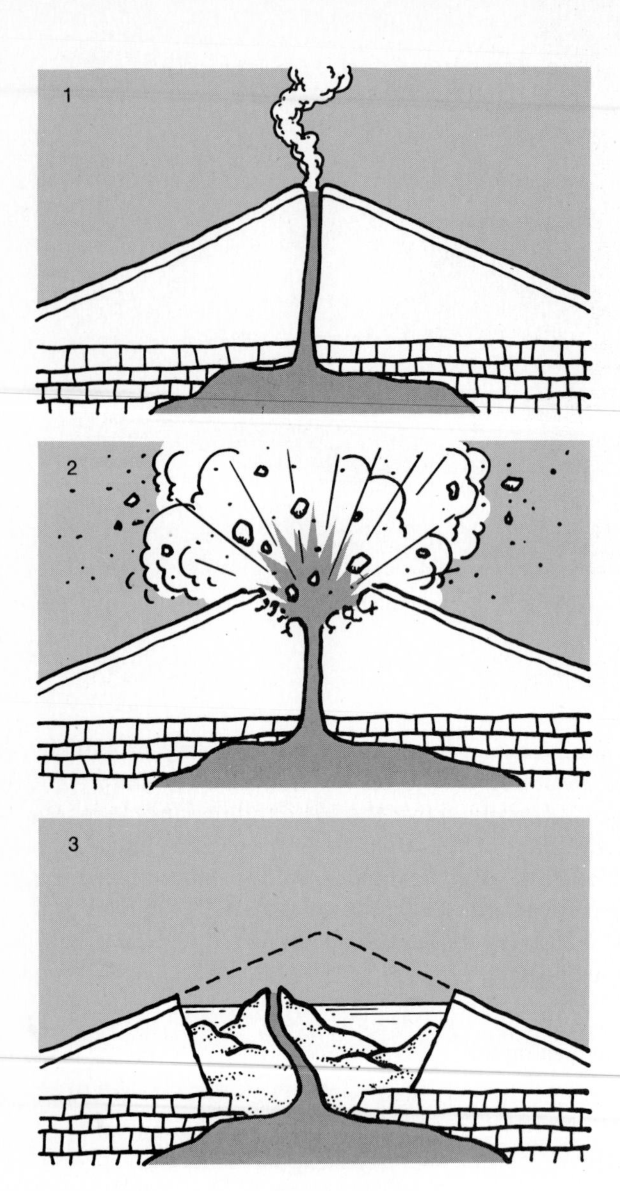

Figure 14-25. A simplified version of the formation of Crater Lake: (1) Mt. Mazama was a prominent active volcano. (2) Some 7,000 years ago, it experienced a cataclysmic explosive eruption, followed by collapse of much of the crater and its walls. (3) It has become a lake-filled caldera, with a smaller cone appearing as an island.

Areas of extrusive vulcanism are widespread over the world, but their distribution is highly irregular, with many volcanic features in some regions and none in vast sections of the globe. Volcanic activity is primarily associated with the boundary zones of crustal plates. At a divergent plate boundary, magma wells up from the interior both by eruption from active volcanoes and by flooding out of fissures. At convergent plate boundaries, where subduction is taking place, volcanoes are often formed in association with the turbulent descent and melting of crustal material and the crustal crumpling resulting from plate collision.

Figure 14-26 shows the distribution of major active volcanoes in the world. It is readily apparent that the most notable area of vulcanism is around the margin of the Pacific Ocean, often referred to as the Pacific Ring of Fire or the Andesite Line (because the lava issuing from the volcanoes of the Pacific rim consists primarily of andesite, a distinctive mineral association). Most of the world's prominent volcanoes, both active and inactive, are associated with this Pacific ring. Another prominent volcanic zone is along the line of collision between the Eurasian Plate of the Northern Hemisphere and several northward-moving plates from the Southern Hemisphere. This zone of activity can be traced eastward through the Mediterranean Basin, across Turkey and Iran, through the Himalayas, and down through Indonesia to meet the Pacific Ring of Fire in New Guinea.

Somewhat less conspicuous on the map are the mid-ocean ridges of the Atlantic and Pacific. This is misleading because these are zones of very active vulcanism, but much of the activity is hidden beneath the sea and often involves flood basalts rather than explosive eruptions. The sporadic and scattered occurrence of volcanoes elsewhere in the world usually is a reflection either of minor plate boundaries or of "hot spots" in the crust where magma has broken through a plate surface.

Intrusive Vulcanism

When magma solidifies below the Earth's surface, it produces igneous rock. If this rock is pushed upward into the crust either before or after solidification, it is called an *igneous intrusion*. Most such intrusions have no effect on the surface landscape, but sometimes the igneous mass is raised high enough to deform the overlying material and change the shape of the surface. In many cases, the intrusion itself is exposed at the surface through uplift and/or erosion. When intrusions are thus exposed to the external processes, they often become conspicuous because they are usually resistant to erosion, and with the passage of time, they stand up relatively higher than the surrounding land.

Intrusions come in all shapes and sizes, and their composition may vary considerably. Moreover, their relationship to overlying or surrounding rock is quite variable. The intrusive process is usually a disturbing one for preexisting rock. The rock may be pushed aside or upward by the intrusion; it may be dissolved and incorporated into the hot magma; or it may experience contact metamorphism from being exposed to the heat and pressure of the rising intrusion. In rare cases, however, there may be little or no disturbance, as with a small intrusion that is inserted between preexisting sedimentary beds with little deformation or metamorphism.

Although there can be an almost infinite variety in the forms assumed by igneous intrusions, most can be broadly encompassed within a classification that contains only a half-dozen types, as described below. See also Figure 14-27.

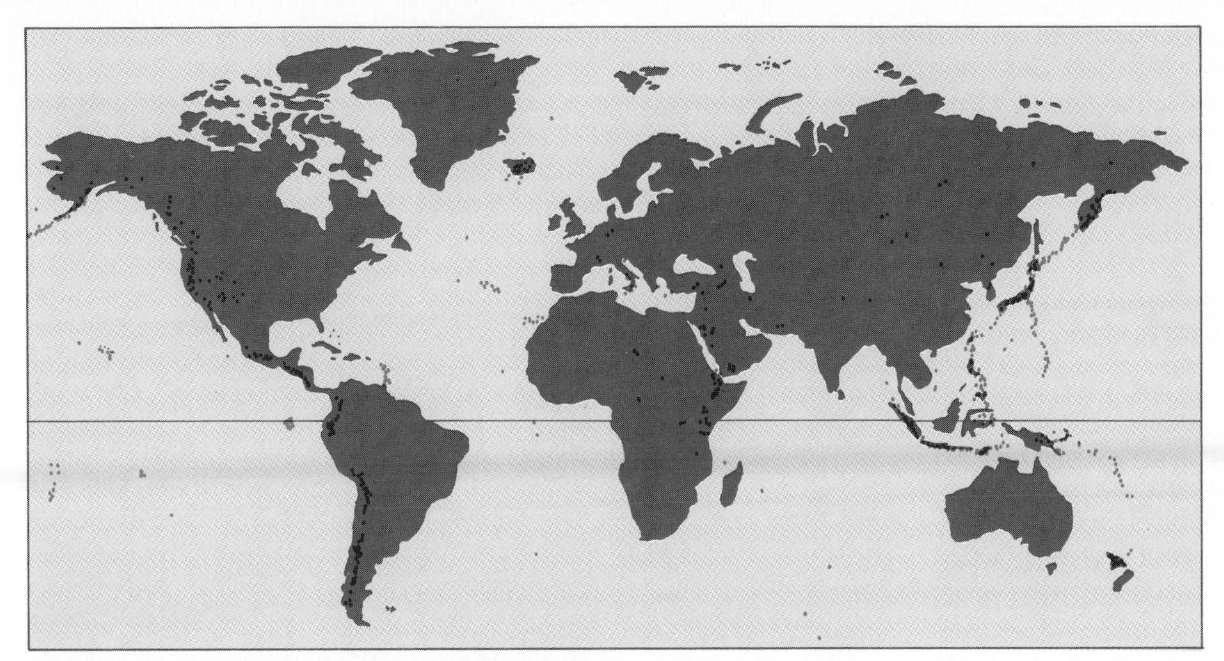

Figure 14-26. Distribution of volcanoes known to have erupted at some time in the past.

1. *Batholith*. By far the largest and most amorphous of intrusion types is the batholith, which is a subterranean igneous body of indefinite depth and enormous size. Batholiths often form the core of major mountain ranges, their intrusive uplift being a fundamental part of the mountain-building process. Such notable ranges as the Sierra Nevada in California, Idaho's Sawtooth Mountains, and Colorado's Front Range (which includes Pike's Peak) were created at least partially by the uplift of massive batholiths; almost all the bedrock exposed in the high country of these ranges consists of batholithic granite, which was originally covered by an extensive overburden of other rocks that has since been stripped away by erosion.

Figure 14-27. Some typical forms of igneous intrusions.

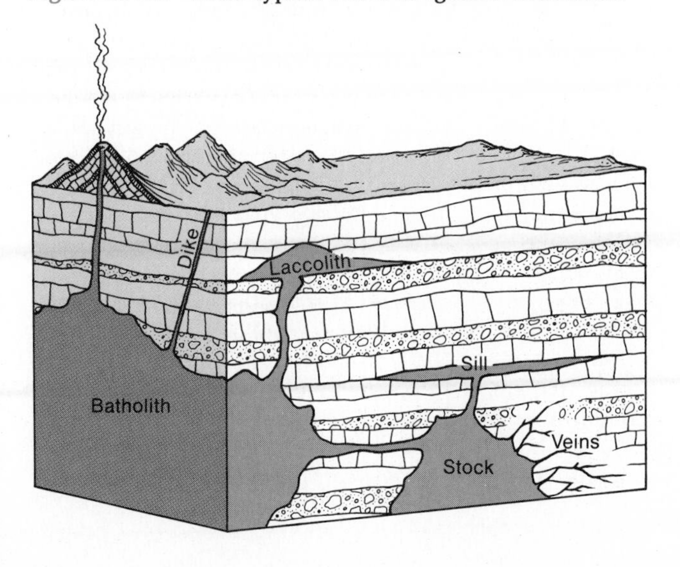

2. *Stock*. Similar in nature to a batholith but much smaller in size is a stock. It is also amorphous in shape and indefinite in depth, but it has a surface area of only a few tens of square miles at most. Many stocks apparently are connected to batholiths as offshoots.
3. *Laccolith*. A specialized form of intrusion is the laccolith, which is produced when slow-flowing, viscous magma is forced between horizontal layers of preexisting rock. Although continuing to be fed from some subterranean source, the magma resists flowing and instead builds up into a mushroom-shaped mass that domes the overlying strata. If this is near enough to the Earth's surface, a rounded hill will rise, like a blister, above the surrounding area. Many laccoliths are small, but some are so large as to form the cores of hills or mountains in much the same fashion as batholiths. The Black Hills of South Dakota, for example, have a laccolithic core, as do several of the geologically famous mountain groups (Henry, Abajo, LaSal) in southern Utah.
4. *Dike*. Probably the most widespread of all intrusive forms is the dike. A dike consists of a vertical or nearly vertical sheet of magma that is thrust upward into preexisting rock, sometimes forcing its way into vertical fractures and sometimes melting its way upward. See Figure 14-28. Dikes are notable because they are vertical, quite narrow (a few centimeters to a few yards wide), and usually quite resistant to erosion. As with most igneous intrusions, their depth is indeterminate; indeed, they often serve as conduits through which deep-seated magma can reach the surface. In some cases, they are quite lengthy features, extending for miles or even tens of miles in one direction. When exposed at the surface by erosion, dikes commonly

In the spring and summer of 1980, the attention of the populace of the United States was gripped as never before by a volcano. In the south-central part of the state of Washington, the splendid peak of Mt. St. Helens, capped with a "permanent" crown of ice and snow for more than a century, suddenly began extrusive activity. Starting on March 27th, the long dormant volcano entered a period of sporadic eruption that devastated the surrounding area, temporarily paralyzing much human activity within hundreds of miles and becoming the principal topic of conversation across the country for weeks.

Before the massive eruptions of 1980, Mt. St. Helens had a classic conical shape. [Courtesy U.S. Geological Survey, Department of the Interior.]

This was an unprecedented situation in the United States. Through the years, the media had carried occasional references to eruptions on the Big Island of Hawaii or in Alaska's Valley of Ten Thousand Smokes, but these were peripheral to the mainstream of American consciousness and were generally looked upon as nothing more than exotic tourist attractions. *Real* volcanic activity, producing destruction and suffering, was a phenomenon most Americans associated with remote and underdeveloped lands such as Java or Costa Rica, or with ancient historical events such as the burial of Pompeii or the explosion of Krakatau. Yet Mt. St. Helens is remote neither in space nor time. The destruction it has created so far is enormous, and it simplication for the near future is even more impressive.

The gross details of Mt. St. Helens' behavior in 1980 are awesome enough. After 123 years of dormancy, the unstable crust abruptly became turbulent. The initial earthquake of the active period was recorded on March 20th, followed in the succeeding week by several hundred more, generally in the range between 2.0 and 4.5 on the Richter scale. The actual eruptions began on March 27th and continued sporadically for the next 51 days, consisting almost entirely of expelling steam, smoke, and ash, and creating a prominent crater at the top of the peak. Then came the first and (to date) most devastating of the explosive blasts—on May 18, 1980.

Without warning an earthquake unhinged the entire north slope of the mountain, uncapping bottled-up gases and magma with the force of 500 Hiroshima-sized atomic bombs. A major vertical eruption was within the range of expectations, but this explosion extended laterally as well. Thus a column of steam and ash was projected 70,000 feet (21,300 m) above the mountain, through three layers of rain clouds, and at the same time, 150 square miles (390 km²) on the north side of the volcano were totally devastated by the sidewise blast. A cubic mile of mountain, 12 percent of its total bulk, was simply pulverized and blown away, the elevation of the peak being instantly reduced from 9677 feet (2950 m) to 8400 feet (2560 m).

Spirit Lake, one of the most beautiful bodies of water in America, disappeared in an avalanche of boiling mud and rock, as did 25 smaller lakes. Millions of trees were scattered like matchsticks over an 8-by-15-mile (13-by-24-km) area north of the mountain by a "glowing avalanche" of superheated gas and fiery lava fragments moving at more than 100 mph (160 km/hr). A churning mudflow, riding on a sled of air at up to 200 mph (320 km/hr), carried millions of tons of debris into the Toutle River, where a raging mud flood knocked out every bridge for 30 miles (48 km), overwhelmed the Cowlitz River valley, and by the next day had clogged the Columbia River shipping channel to less than half its normal depth of 40 feet (12 m). The toll was 70 dead and missing people, 11 million fish were estimated to have perished, and the final body count of larger animals and birds exceeded $1\frac{1}{2}$ million.

Westerly winds carried most of the ash to the east, resulting in virtual blackouts for such nearby cities as Yakima, absolutely stopping all airplane and most highway traffic for several hundred miles in the downstream flow, leaving a 2-inch (5-cm) blanket of ash on the ground as far away as western Idaho (260 miles or 418 km), and circumnavigating the globe in only 17 days. And this was the result of only the first explosion!

Later in 1980, Mt. St. Helens experienced four other major eruptions, several minor eruptions, and more than 5000 earthquakes. The mountaintop has been significantly lowered and totally reshaped. Its upper slopes remain devastated, yet lower down, life has rejuvenated. Spiders were spinning webs

The new shape of Mt. St. Helens—late 1980. [Courtesy of USDA-Soil Conservation Service.]

in the devastated area within a week of the greatest eruption. New growth of ferns and skunk cabbages and even trees was sprouting before the end of the summer.

From a scientific standpoint, Mt. St. Helens presents an unparalleled opportunity for volcano watching. More scientists have more instruments perched at more vantage points than ever before in history. Yet clearly the most important lesson is for the public. The saga of Mt. St. Helens in 1980 may appear to be an unusual, brief, and isolated sequence of events. But such is not the case.

In lithospheric terms, Mt. St. Helens is a young and not particularly notable volcanic peak that is part of a mountain range (the Cascades) that includes 14 other volcanoes of equal or greater significance and extends from northern California to southern British Columbia. Eight of those peaks have erupted in the past two centuries, the most recent (prior to Mt. St. Helens) being Mt. Lassen, which blew more than 170 times in the decade following 1913. Mt. St. Helens itself has a complex geological history, involving a young volcano built on the eroded remnants of an older one. Most of the visible part of the present mountain has actually been formed within the last 1000 years. Although there are Indian legends of earlier activity, the only eye-witness accounts of eruptions involve sporadic events from 1835 until 1857, after which dormancy set in until 1980.

Debris, ash, and mudflows
Area of total devastation from blast
Major flood areas
Cowlitz River
Toutle River
Longview
Mt. St. Helens
Columbia River
Seattle
Portland
0 5 mi
0 5 10 km

The immediate results of the May 18, 1980 eruption of Mt. St. Helens.

Apart from Mt. Lassen, at its extreme southern end, the Cascade Range during the twentieth century has been noted for its serene and majestic beauty rather than its potential turbulence. It is, however, clearly a part of the Pacific Ring of Fire, and the evidence of past volcanic activity is ubiquitous. In 1975, for the

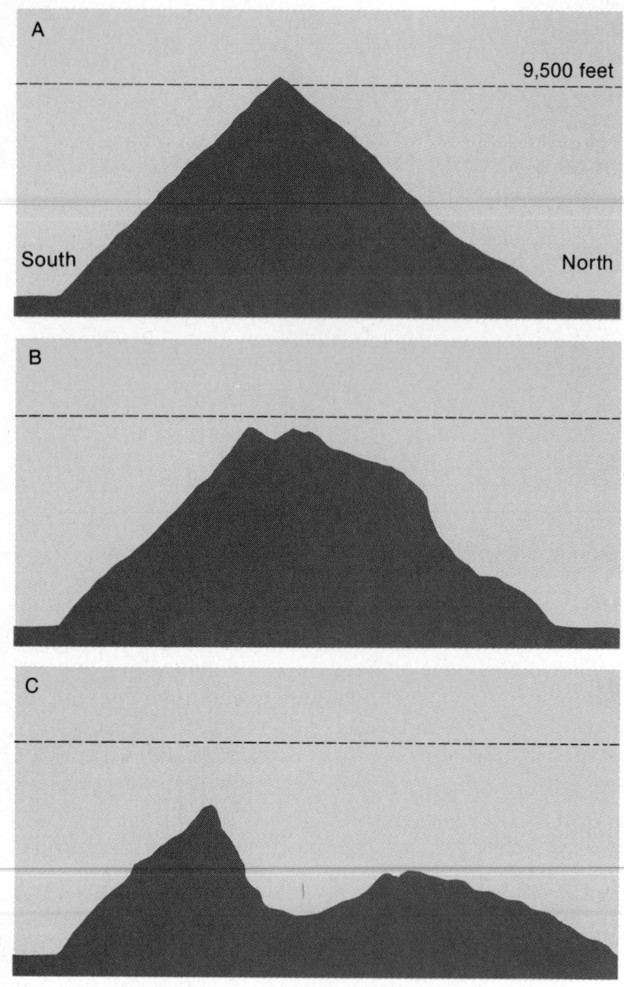

The changed profile of Mt. St. Helens: (A) as of March 1, 1980; (B) as of May 17, 1980; (C) as of May 19, 1980.

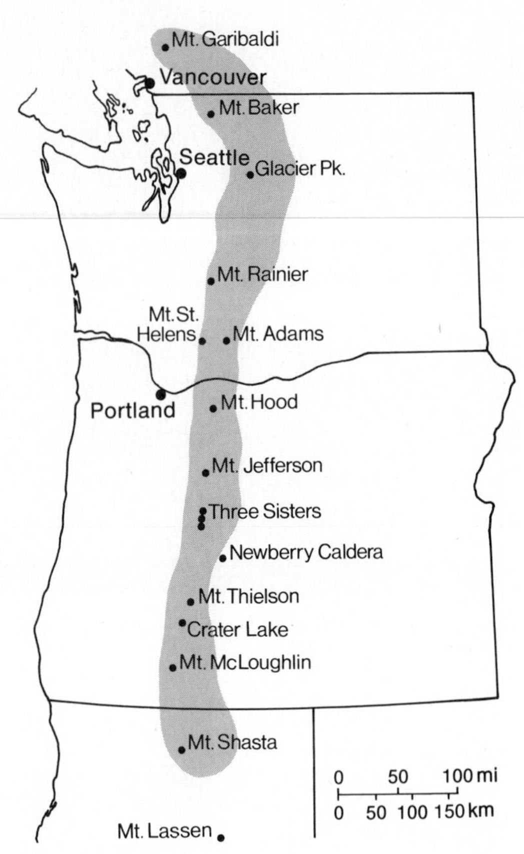

Prominent volcanoes (active, dormant, and extinct) of the Cascade Range.

first time since Mt. Lassen became quiescent in 1921, sulfurous activity appeared in the Cascades: Mt. Baker began spouting steam from various vents, its first fumings for 125 years. Much scientific interest was focussed on this phenomenon, but it was thoroughly upstaged by Mt. St. Helens five years later.

What predictive conclusions can be gleaned from all this?

Will Mt. St. Helens now subside into dormancy again? Possible, but not probable. Once an eruptive period begins, such volcanoes tend to be active sporadically for years, if not decades. The citizens of Yakima, and even Portland, have no reason to dispose of their ash-filtering face masks at this point.

Will a different hot spot become important? Don't bet against it. Mt. Baker has already demonstrated its anxiety. The internal plumbing of the Cascade Range is almost a total mystery. Mt. Baker might be the next to erupt; beware, Vancouver! Or, since Mt. Rainier is situated halfway between Mt. Baker and Mt. St. Helens, maybe Seattle should be the nervous city?

Perhaps the most meaningful conclusion is that our specific predictive knowledge of volcanoes is extraordinarily limited. Mt. St. Helens will undoubtedly be a good teacher, but there is a great deal more to learn.

Figure 14-28. An igneous dike was intruded into softer bedrock. Subsequent erosion has left the resistant dike standing as an abrupt natural wall. This scene is in southern Colorado, near LaVeta. [TLM photo.]

form sheer-sided walls that rise a few feet or yards above the surrounding terrain, standing out in stark relief due to their density and hardness. Dikes often are found in association with volcanoes, occurring as radial walls extending outward from the volcano like spokes of a wheel. Notable examples of radial dike development can be seen around Shiprock in northwestern New Mexico and around the Spanish Peaks in south-central Colorado. See Figure 14-29.

5. *Sill*. A sill is also a long, thin intrusive body, but its orientation is determined by the structure of the preexisting rocks. It is formed when magma is forced between strata that are already in place; the result is often a horizontal sheet between horizontal sedimentary layers. Sills are much less common than dikes, and their landscape expression is usually inconspicuous except where one serves as a cap rock protecting softer rocks underneath, which can produce a steep-walled mesa, butte, or cliff.
6. *Vein*. Least prominent among igneous intrusions, but widespread in occurrence, are thin veins of magma that may occur individually or in profusion. They are commonly formed by the forcing of molten material into small fractures in the preexisting rocks, but they can also result from melting by an upward surge of magma. Intrusive veins may take very irregular shapes, but normally they have a generally vertical orientation.

Figure 14-29. Shiprock, in northwestern New Mexico, as sketched from the north. Three prominent radial dikes can be seen extending outward from this old volcanic plug.

The Complexities of Crustal Configuration

In this chapter we have considered each of the principal internal processes in turn, which is a helpful way to systematize knowledge. This presents an artificial and misleading picture, however, because in nature these processes are multiple and interrelated, and the resulting complexity is sometimes bewildering. To attempt a more balanced assessment, let us consider a highly simplified statement of the origin of the gross contemporary topographic features of a small part of the Earth's surface—a mountainous section of northwestern Montana that encompasses the spectacular scenery of Glacier National Park. This area, now part of the northern Rocky Mountains, was below sea level for many millions of years. Most of the rocks in the region were formed in the dim geologic past—the Precambrian era—when much of the area now occupied by the Rocky Mountain cordillera consisted of a large, shallow, ocean-filled trough known as a *geosyncline*. Muds and sands were washed into this Precambrian sea for millennia, and a vast thickness of sedimentary strata, as much as 20,000 feet (6000 m), built up. These limestones, shales, and sandstones accumulated as six distinct formations, each with a conspicuous color variation, known collectively as the Belt Series.

Occasionally during this lengthy epoch of sedimentary accumulation, igneous activity added variety to the rocky fundament. Most notable was a vast outpouring of flow basalt that issued from fissures in the ocean floor and was extruded in the form of a submarine lava flow with a thickness varying between 50 and 275 feet (15 and 83 m). Further sediments were then deposited on top of the lava flow.

Igneous intrusions were also injected from time to time, including one large sill and a number of dikes. The igneous rocks, both the flow basalt and the various intrusions, initiated contact metamorphism, whereby the tremendous heat of the igneous material converted some of the adjacent sedimentaries into metamorphic rocks (mostly changing limestone to marble and sandstone to quartzite).

The Front Range—Episode Two: Batholithic Intrusion and Uplift (continued from page 278)

At the close of the Mesozoic era, some 65 million years ago, the "growth" of the Rocky Mountains began. The uplift occurred very slowly but was of magnificent proportions. For the Front Range, it took the form of a huge anticlinal fold or great arch in which a massive batholith of ancient (Precambrian) crystalline rocks, composed largely of granite but including a considerable amount of schist, was pushed upward from below. This caused the thick series of sediments that had been deposited in the Mesozoic sea to arch over the batholith. As the uplift progressed, the increasing elevation of the sedimentary strata speeded up streamflow and thus accelerated the erosive action. Consequently, erosion removed the uparching rocks at approximately half the rate of their uplift. The eventual result was that all of the sediments were stripped away from the crest of the anticline, leaving nothing but crystalline rocks in the high country of the Front Range. Thousands of feet of rock were removed; if no erosion had taken place, the Front Range would now be approximately twice its present height. The accompanying figure shows the structural relationship between batholith and upfolded sediments. The crystalline core of the range is bordered on either side by the truncated edges of upturned sedimentary strata that were left dipping steeply (to east and west) away from the axis of the great arch.

(To be continued. Episode Three begins on page 367.)

The intrusion of a granitic batholith arched the sedimentary rocks into a great anticlinal fold. The massive uplift was accompanied by extensive erosion so that the sediments were stripped away from the anticlinal crest, leaving only crystalline rocks in the high country of the Front Range.

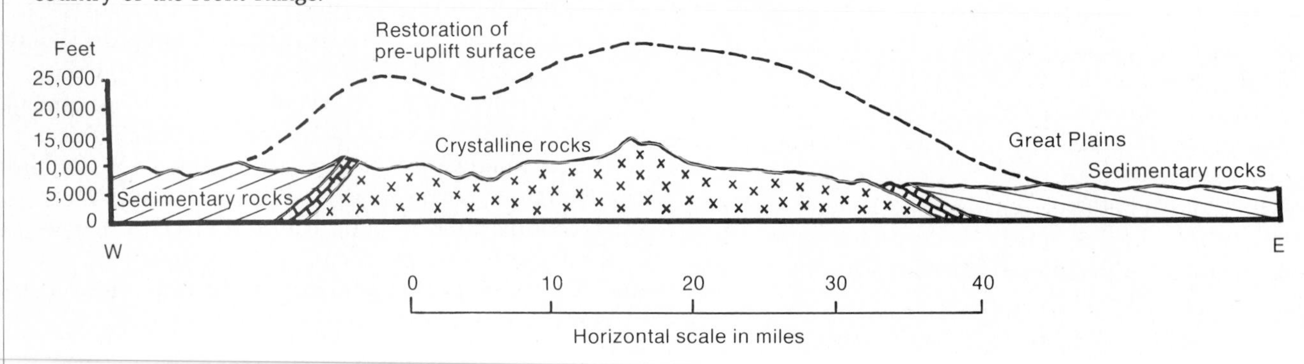

After a long gap in the geologic record, during which the Rocky Mountain region was mostly above sea level, the great geosyncline once again sank below the ocean during the latter part of the Mesozoic era (Cretaceous period), and another thick series of sediments was deposited. This was followed by a period of mountain building that is so significant in western North America that it has been named the Rocky Mountain Revolution. In the Glacier Park area, the rocks of the geosyncline were compressed and uplifted, converting the site of the former sea into a mountainous region.

Along with uplift came extreme lateral pressure from the west, convoluting the gently downfolded strata into a prominent anticline. Continuing pressure then overturned the anticline toward the east, with a lengthening western limb and a truncating eastern limb. This additional strain on the rock and the persistent crustal pressure eventually caused a vast rupture and faulting. The entire block was then pushed eastward by one of the greatest thrust faults known, the Lewis Overthrust. See Figure 14-30. This remarkable fault forced the Precambrian sedimentaries out over the Cretaceous strata that underlie the plains to the east, by as much as 20 miles (32 km). The plane of the thrust fault was only slightly above the horizontal, nowhere exceeding a dip of 10°. This had the peculiar effect of placing older rock layers on top of much younger strata, in an overturned relation to the normal sequence of deposition. The terrain thus produced is referred to as "mountains without roots." Chief Mountain is world-famous as a rootless mountain because of its conspicuous location as an erosional outlier east of the main range.

Other episodes of faulting also occurred, although none were as massive as the Lewis Overthrust. Most notable among these subsidiary movements was a normal fault with an almost vertical displacement of several thousand feet that took place along the western edge of the overthrust geosynclinal block, creating a prominent fault scarp along that margin of the mountains.

The contemporary topography of the Glacier National Park area was molded primarily by the complicated series of events presented in simplified fashion here. Included was a sample of almost every facet of the internal processes at work—sedimentation, intrusive and extrusive igneous activity, metamorphism, simple and complex folding, and simple and complex faulting. Details of the terrain and the specific landforms that now exist, however, were carved from these gross lineaments by the

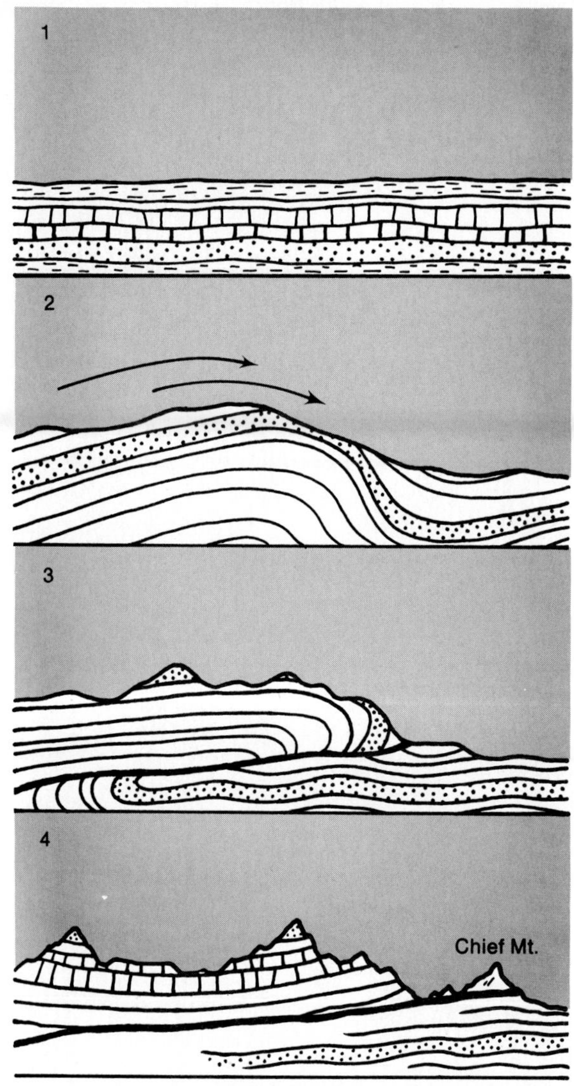

Figure 14-30. Sequential development of the Lewis Overthrust: (1) initial sea-bottom deposition of sediments; (2) uplift and folding; (3) continued pressure from the west causes overturning of the fold and faulting along the eastern limb of the anticline; (4) subsequent erosion produces the present topography, with Chief Mountain as a residual outlier to the east of the range.

external processes—the action of weathering, gravity, water, and ice—which will be discussed in subsequent chapters.

Review Questions

1. Explain why the "rigid Earth" theory is no longer acceptable.
2. What is the relationship between *plasticity* and *isostasy*?
3. Explain how the presence of the Mid-Atlantic Ridge supports the theory of plate tectonics.
4. Explain how sea-floor spreading supports the theory of plate tectonics.
5. Distinguish among the three kinds of plate boundaries.
6. Explain the difference between *folding* and *faulting*.
7. Why is there such a concentration of both earthquakes and volcanoes around the margin of the Pacific Ocean?
8. What lessons have been learned from the recent eruptions of Mt. St. Helens?
9. Describe the major forms of igneous intrusions.
10. How is it possible for such a hard, brittle substance as a rock to be bent without breaking?

Some Useful References

BOLT, B. A., *Earthquakes: A Primer*. San Francisco: W. H. Freeman & Company, Publishers, 1978.

BULLARD, F. M., *Volcanoes of the Earth*. Austin, Texas: University of Texas Press, 1976.

DECKER, R., AND B. DECKER, *Volcanoes*. San Francisco: W. H. Freeman & Company, Publishers, 1981.

Scientific American Editors, *Continents Adrift and Continents Aground*: A Scientific American Book. San Francisco: W. H. Freeman & Company, Publishers, 1976.

———, *Earthquakes and Volcanoes: A Scientific American Book*. San Francisco: W. H. Freeman & Company, Publishers, 1980.

SHEETS, P. D., AND D. GRAYSON, *Volcanic Activity and Human Ecology*. New York: Academic Press, Inc., 1979.

UYEDA, S., *The New View of the Earth: Moving Continents and Moving Oceans*. San Francisco: W. H. Freeman & Company, Publishers, 1978.

15 Preliminaries to Erosion: Weathering and Mass Wasting

If the internal processes of landscape formation are overwhelming, the external processes are inexorable. During all the time that continents are drifting, that the crust is bending and breaking, that volcanoes are erupting, and that intrusions are forming—during all the time that these grandiose events are grossly revising the configuration of the Earth's surface—another suite of natural forces is simultaneously at work. These are the external forces, often mundane and even miniscule in contrast to those described in the preceding chapter. Yet their eventual cumulative effect is awesome; they are capable of wearing down anything that the internal forces can erect. No rock is too resistant, no mountain is too massive to withstand their unrelenting puissance. Ultimately, the external forces will prevail. The specific shape of the Earth's surface will be sculptured by these external forces. The detailed configuration of peaks, slopes, valleys, and plains will be molded by the work of gravity, water, wind, and ice, or some combination thereof.

The total effect of these actions—disintegration, wearing away, and removal of rock material—is generally encompassed by the term *denudation*, which implies an overall lowering of the surface of the continents. See Figure 15-1. Denudation is accomplished by the interaction of various agencies and forces, but for analytical purposes we can consider three types of activities—weathering, mass wasting, and erosion. *Weathering* encompasses the breakdown of rock into smaller components by atmospheric and biotic agencies. *Mass wasting* involves the downslope movement of broken rock material due to gravity, sometimes lubricated by water. *Erosion* consists of more massive and generally more distant removal of fragmented rock material.

Weathering

The first step in the shaping of the Earth's surface by external processes is accomplished by weathering, that is, the mechanical distintegration and/or chemical decomposition that destroys the coherence of bedrock and begins to fragment rock masses into progressively smaller components. Only after a certain amount of fragmentation can the erosive processes remove and redistribute the rock materials.

Wherever bedrock is exposed at the surface, it undergoes recognizable changes. It often has a different appearance, in terms of color or texture. Most significant from a topographic standpoint, the exposed surface rock is likely to be looser and less coherent than the underlying rock. Blocks or chips may be so loose that they can be detached with little effort. Sometimes pieces are so "rotten" that they can be crumbled by finger pressure. Slightly deeper in the bedrock, there is firmer, more solid rock, although along cracks or crevices, "softening" may extend to considerable depths.

All rock that is exposed at the surface will be weathered. See Figure 15-2. Moreover, in some cases, the weathering may reach as much as several thousand feet beneath the surface. This is made possible by open spaces within

Figure 15-1 Denudation is accomplished by a combination of weathering, mass wasting, and erosion.

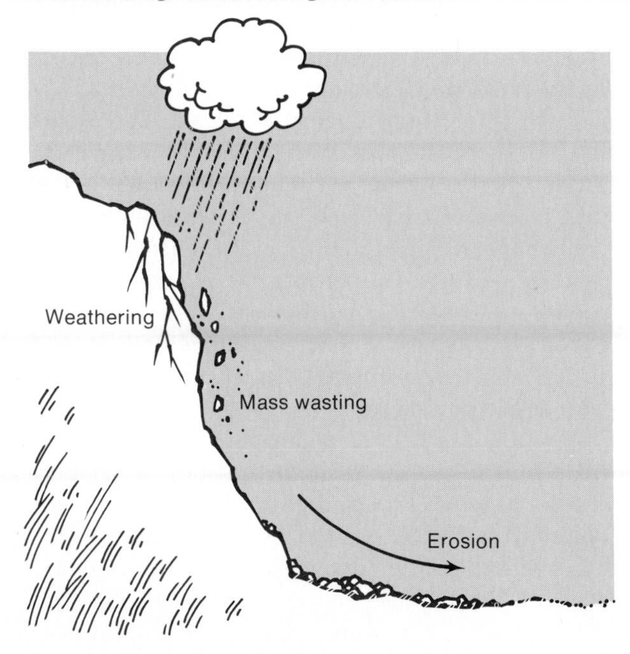

Figure 15-2 A bare granite surface in Yosemite National Park is totally exposed to weathering. Cracks and fissures in the granite allow the weathering to pentrate well below the surface. [U.S. Department of the Interior, National Park Service photo.]

the rock bodies and even within the mineral grains. Such openings are surprisingly numerous near the surface, but at depths of more than a few kilometers or miles, the confining pressures are so great that openings are virtually nonexistent. Subsurface weathering is initiated along these openings, which can be penetrated by agents from above (water, air, roots). As time passes, the weathering effects spread from the immediate vicinity of the openings into the denser rock beyond.

Openings within the surface and near-surface bedrock are frequently submicroscopic in size, but they may also be large enough to be conspicuous and are sometimes huge. In any case, they occur in vast numbers and provide avenues along which the weathering agents can attack the bedrock and break it apart. Broadly speaking, five types of openings are common:

1. *Microscopic open spaces* occur in profusion, but they are generally so tiny that they provide very limited access to weathering agents. They may consist of spaces between crystals of igneous or metamorphic rocks, pores between grains of sedimentary rocks, or minute fractures within or alongside mineral grains in any kind of rock.
2. *Joints* are the most common structural features of the rocks of the lithosphere. They are cracks that develop due to stress, but the rocks show no appreciable movement parallel to the walls of the joint. These cracks are innumerable in all rock masses, serving to divide them into blocks of various sizes. Because of their ubiquity, joints are the most important of all rock openings in facilitating weathering.
3. *Faults* allow even easier penetration of weathering agents into subsurface areas because not only fracturing but also displacement is involved. Faults, however, are much less common than joints.
4. *Lava vesicles* are holes of various sizes, usually small, that develop in cooling lava when gas is unable to escape during solidification.
5. *Solution cavities* often are formed in calcareous rocks (particularly limestone) as the soluble minerals are dissolved and carried away by percolating water. Most such cavities are small, but sometimes huge holes and even massive caverns will be created by the removal of large amounts of solubles.

The agents of weathering are relatively few in number, but they are complex in their interactions. Most, as implied by the term *weathering*, are atmospheric. Because it is in gaseous form, the atmosphere is able to penetrate readily into all cracks and crevices that extend down into the bedrock. From a chemical standpoint, oxygen, carbon dioxide, and water vapor are the three atmospheric components of greatest importance in rock weathering. Temperature changes are also of considerable significance. Most notable, however, is moisture in its liquid form, which can penetrate downward effectively into openings in the bedrock. Biotic agents also contribute to weathering, in part through the burrowing activities of animals and the rooting effects of plants, but especially by producing chemical substances that attack the rock more directly.

The total effect of these various agents is complicated and is influenced by a variety of factors—the nature and structure of the bedrock, the abundance and size of openings in the rock, the surface configuration, prevailing climatic conditions, the vegetative cover, and the variety and abundance of burrowing animals. For analytic purposes, however, it is convenient to recognize two major categories of weathering—physical and chemical. We will discuss them in turn, while realizing that they usually act in concert.

Physical Weathering

Physical weathering involves the mechanical disintegration of rock material without any change in its chemical composition. In essence, big rocks are made into little rocks by various stresses that cause fracturing of the rock into smaller fragments. Most physical weathering occurs at or very near the surface, but under certain conditions, it may be accomplished at considerable depth.

Probably the most important single agent of physical weathering is the freeze/thaw action of water in open spaces in rock. When water freezes, it decreases in density and expands in volume. The expansion sometimes amounts to almost one-tenth of its original volume. Moreover, the upper surface of the water freezes first, and the formation of ice continues progressively downward, which means that the principal force of expansion is exerted against the confining rock rather than upward. It is easy to see why a frozen water pipe or automobile radiator can be ruptured under such pressure; the same principle applies in fracturing rocks. Even the strongest of rocks cannot withstand the frequent alternation of freezing and thawing. Repetition is the key to understanding the inexorable force of *frost shattering* or *frost wedging*, as this phenomenon is usually termed. See Figure 15-3. Regardless of the size of the opening in the rock, if it contains water, when the temperature falls below 32° F (0° C), ice will form, wedging its way downward. When the temperature rises above 32°, the ice will melt, and the water will sink further into the slightly enlarged cavity. With renewed freezing, the wedging is repeated. Such a freeze/thaw pattern may be repeated millions of times through the eons of earth history, providing what is literally an irresistible force.

Frost wedging in large openings may produce large boulders, whereas that occurring in small openings may granulate the rock into sand and dust particles, with every size gradation in between. The physical characteristics of the rock are important determinants of the rate and magnitude of breakdown, as are the temperature and moisture variations. The process is most effective where

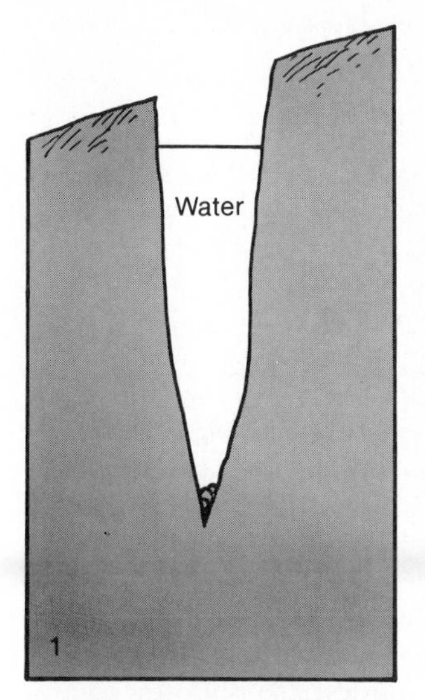

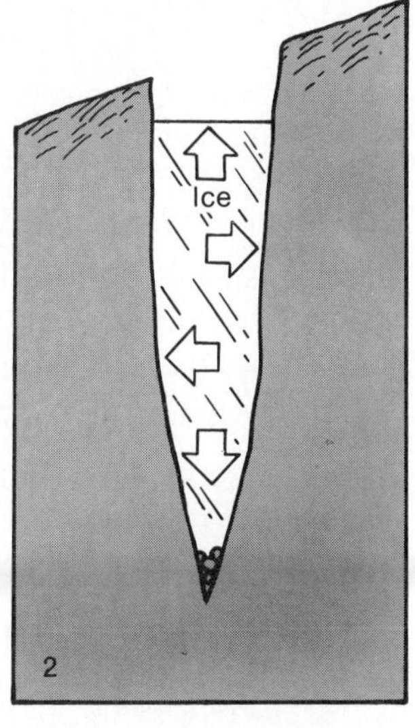

Figure 15-3 Schematic illustration of frost wedging. When water in a rock crack freezes, the ice expansion exerts a force that can deepen and widen the crack, especially if the process is repeated many times.

freezing is frequent and intense—in high latitudes, in midlatitudes during winters, and at high altitudes. It is most conspicuous above the tree line of mountainous areas (see Figure 15-4), where broken blocks of rock are likely to be found in profusion everywhere except on slopes that are too steep to allow them to lie without sliding downhill. Only in areas of both low latitude and low altitude is frost wedging completely unknown, due to the lack of freezing temperatures.

Related to frost wedging, but much less significant in its total effect, is *salt wedging*, caused by the crystallization of salts from evaporating water. In areas of dry climate, water is often drawn upward in rock openings by capillary action (a subject discussed previously in connection with soil formation). This water nearly always carries dissolved mineral salts. When the water evaporates, as it commonly does, the salts are left behind as tiny crystals. With time, the crystals grow and expand, prying apart the rock and weakening its internal structure, much in the fashion previously described for freezing water, although less intensely. In more humid areas, salt wedging is inconsequential because the soluble salts are flushed away by percolating groundwater.

Temperature changes not accompanied by freeze/thaw conditions also accomplish physical weathering, but much more gradually than the processes just described. The fluctuation of temperature from day to night and from summer to winter can cause minute changes in the volume of most mineral particles, forcing expansion when heated and contraction when cooled. This volumetric variation weakens the coherence of the mineral grains and tends to break them apart. Millions of repetitions are normally required for much weakening or fracturing to occur, although the intense heat of forest or brush fires can speed up the process. This factor is most significant in arid areas and near mountain summits where direct solar radiation is intense during the day and radiational cooling is prominent at night.

Chemical changes (to be discussed in more detail in the next section) also contribute to physical disintegration of rock masses. Various chemical actions can cause an increase in volume of the affected mineral grains. This swelling sets up strains that weaken the coherence of the rock and can actually cause fractures to occur.

Some biotic activities are also direct contributors to physical weathering. Most notable is the penetration of growing plant roots into cracks and crevices, which exerts an expansive force that widens the openings. See Figure 15-5. This factor is especially conspicuous where trees grow out of joint or fault planes, with their large

Figure 15-4 Frost wedging is an especially pervasive force on mountaintops above the treeline, as in this scene from Squaretop Peak in the Wind River Range of Wyoming. Frost-shattered rock litters the surface. [Courtesy Doug Virtue.]

Focus on Jointing

Joints and faults are basically similar structures. Both represent a physical response to stress, involving the fracturing of the rock. The principal difference between them is that faulting encompasses relative displacement of the walls parallel with the fracture, whereas jointing does not. Some faults, of course, involve extremely massive displacement, but the only movement associated with jointing is very slight separation of the adjacent blocks at right angles to the walls. Another usual distinction is that faults generally are individual or occur only in small numbers, whereas joints normally are multitudinous. A further difference is that faults sometimes appear as major landscape features, extending for tens or even hundreds of miles, whereas an individual joint is normally a minor structure with an extent of only a few inches or feet.

Almost all lithospheric bedrock is jointed, sometimes resulting from contractive cooling of molten material, sometimes caused by contraction of sedimentary strata as they dry, and sometimes due to diastrophic tensions in the Earth's crust. Most joints are relatively small features, often invisible to the naked eye; in other cases, they consist of obvious cracks that are traceable for many feet or meters horizontally.

At the surface of the Earth, the separation between blocks on either side of a joint may be conspicuous because weathering emphasizes the fracture. Below the surface, however, the visible separation is minimal, with the blocks fitting very closely together and the rock masses generally appearing to be solid and unbroken.

Joints are relatively common in most rock, but they are clearly more abundant in some places than in others. Where numerous, they are usually arranged in *sets*, each of which is a series of fractures of approximately parallel orientation. Frequently two prominent sets of joints intersect almost at right angles; such a combination constitutes a *joint system*. A well-developed joint system, particularly in sedimentary rock with prominent natural bedding planes, can divide stratified rock into a remarkably regular series of close-fitting blocks. Generally speaking, jointing is more regularly patterned, and the resulting blocks are more sharply defined in fine-grained rocks than in those of coarser crystals or particles.

In some places, large joints, or sets of joints, extend for long distances and through a considerable thickness of rocks; these are termed *master joints*. Much more numerous are minor fractures, which are sometimes limited to a single stratum. Master joints characteristically play a deterministic role in topographic development by functioning as a plane of weakness, a plane that is more susceptible to weathering and erosion than the

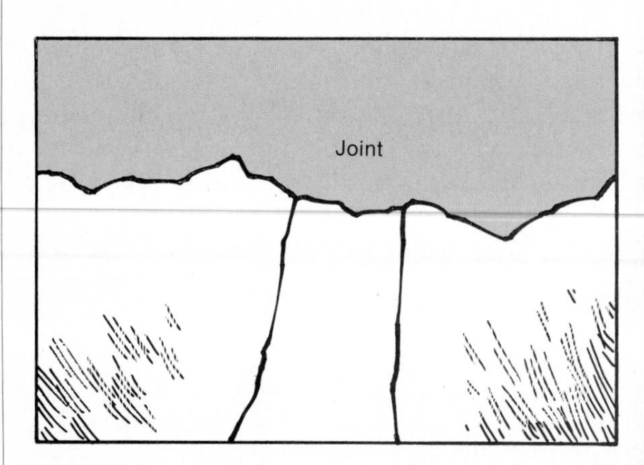

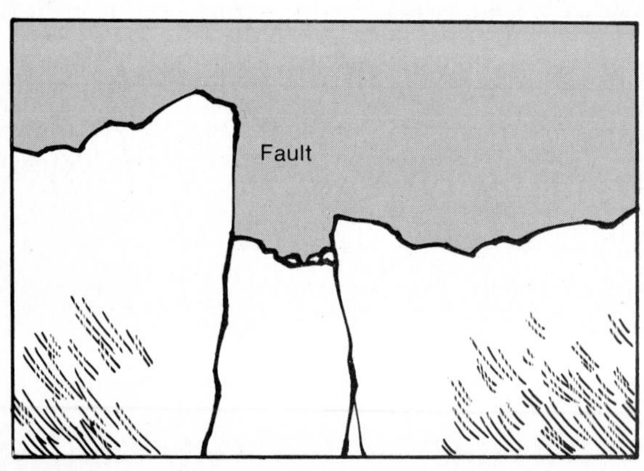

The essential difference between a joint and a fault is that the former involves no relative displacement on either side of the crack.

Extreme frequency of jointing is displayed in this basalt formation in Devil's Postpile National Monument in California. The cooling of the molten basalt produced the parallel joint sets. [National Park Service photo.]

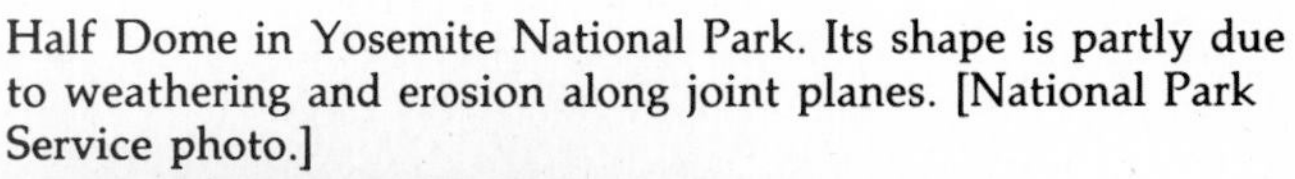
Half Dome in Yosemite National Park. Its shape is partly due to weathering and erosion along joint planes. [National Park Service photo.]

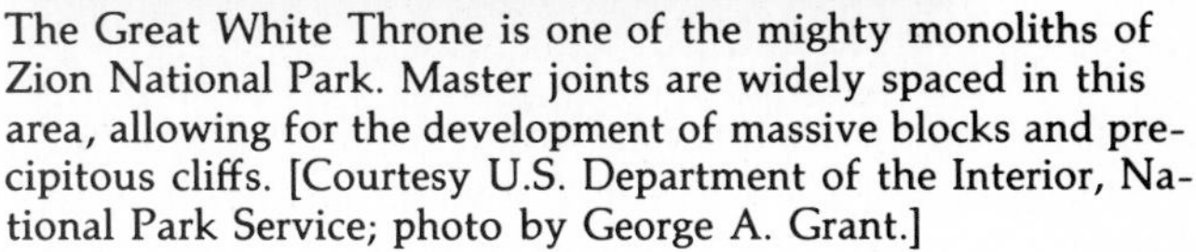
The Great White Throne is one of the mighty monoliths of Zion National Park. Master joints are widely spaced in this area, allowing for the development of massive blocks and precipitous cliffs. [Courtesy U.S. Department of the Interior, National Park Service; photo by George A. Grant.]

The closely spaced joint systems of Bryce Canyon contribute to intricate sculpturing by weathering and erosion. [National Park Service photo.]

rock around it. Thus the location of large features of the landscape, such as valleys and cliffs, may be influenced by the position of master joints.

As examples of the influence of jointing on topographic evolution, let us consider three well-known landscape features in western United States:

1. Half Dome is a gigantic granite monolith that towers above Yosemite Valley in California's Sierra Nevada. The shape of the dome has been partly determined by weathering and erosion along two principal joint sets. The sheer northern face is controlled by a set of vertical master joints. The rounded back slopes of the dome were sculptured by the splitting off of thin concentric shells of granite, made vulnerable by a set of curved minor joints.
2. Zion National Park in southern Utah is dominated by a number of massive sandstone monoliths that rise sheerly for more than 2000 feet (600 m) above the flattish floor of Zion Canyon. The canyon itself was eroded by a stream that follows a set of master joints that extends through the entire thickness of the sandstone formation. Several tributary canyons join the main canyon at approximately right angles, thereby separating the monoliths from one another by deep narrow gorges with almost vertical sides, following a second set of joints. Thus almost all the stupendous cliffs in and around Zion Canyon have developed in response to the joint systems. In this area, the master joints are widely spaced, which accounts for the extraordinary massiveness of the individual sandstone blocks.
3. In Bryce Canyon National Park, less than 100 miles (160 km) from Zion Canyon, the joint systems are much more closely spaced. Consequently, a very intricate pattern of badlands topography has been developed by weathering and erosion. No massive features occur; rather, the cliffs and scarps and pinnacles are etched in delicate detail.

Jointing is a matter of considerable practical importance in any human activity involving bedrock excavation. In quarrying, mining, tunneling, or foundation excavation, the presence of jointing is likely to facilitate rock removal. Without the natural fractures, every rock fragment would have to be drilled or blasted or scraped loose from the bedrock mass. On the other hand, jointing can be detrimental to some types of rock work by loosening rock unwantedly. In tunnels and mines, a solid roof rock is desired, and the looseness associated with jointing may create hazardous conditions.

Figure 15-5 Roots can be incredibly persistent in their penetration of rock crevices, contributing to the weathering of the rock mass. In this case, tree roots are invading a marble bedrock outcrop in a park in New York City. [Mary M. Thatcher, Photo Researchers, Inc.]

roots showing amazing tenacity and persistence as wedging devices. Furthermore, the activities of burrowing animals sometimes are factors in rock disintegration. The total effect of these biotic actions is probably significant, but it is difficult to assess because it is obscured by subsequent chemical weathering that immediately begins to function in the openings created or enlarged by the activities of plants and animals.

When acting alone, physical weathering functions to break up rock masses into ever smaller pieces, producing boulders, cobbles, pebbles, sand, silt, and dust. In most situations, this mechanical disintegration is accompanied by chemical weathering, which produces a different kind of change in the rock materials.

Chemical Weathering

Chemical weathering involves the decomposition of rock by the alteration of rock-forming minerals. Almost all minerals are subject to chemical alteration when exposed in nature to atmospheric and biotic agents. Some, such as quartz, are extremely resistant to chemical change, but many others are very susceptible to alteration. Most rocks are composed of various minerals in combination, and few cannot be significantly affected by chemical weathering because the alteration of even a single significant mineral constituent can lead to the eventual disintegration of the entire rock mass.

Virtually all chemical weathering requires the presence of moisture to some degree. Thus an abundance of water enhances the effectiveness of chemical weathering, and chemical processes operate more rapidly in humid climates than in arid areas. Moreover, chemical reactions are more rapid under high temperature conditions than in cooler regions. Consequently, chemical weathering is most efficient and conspicuous in warm, humid climates. In cold, dry lands, there is less chemical weathering (although the removal of dissolved salts is sometimes notable), and the influence of physical weathering is dominant.

Some of the chemical reactions that affect rocks are very complex, but others are simple and predictable. The principal reacting agents are the commonplace ones of water, carbon dioxide, and oxygen. Most significant among the processes are oxidation, hydrolysis, and carbonation. These processes often take place more or less simultaneously, largely because they all involve the presence of water, which invariably contains dissolved atmospheric gases. Water percolating into the ground acts as a weak acid due to the presence of these gases and of decay products from the local vegetation; this increases its capability to engender chemical reactions.

Focus on Some Important Examples of Chemical Weathering

Chemical weathering involves a great variety of reactions, many of which are highly individualized on the basis of the chemical components of the rock. Two of the most common and significant processes are illustrated below.

Oxidation Silicates or carbonates of iron or manganese combine with water and oxygen to produce iron or manganese oxides (Fe_3O_4, FeO, FeS_2), which may be attacked and altered. An example is:

$$4FeO + 3H_2O + O_2 \longrightarrow 2Fe_2O_3 \cdot 3H_2O$$

ferrous oxide + water + oxygen ⟶ ferric oxide (bound with water)

The ferric oxide often appears as rust, a relatively soft red or yellow or brown form of iron.

Carbonation Carbonic acid is produced in several ways in the lithosphere, always involving the combining of water and carbon dioxide:

$$H_2O + CO_2 \longrightarrow H_2CO_3$$

water + carbon dioxide ⟶ carbonic acid

Carbonic acid combines readily with calcium carbonate to produce calcium bicarbonate, a salt that is easily soluble in water:

$$CaCO_3 + H_2CO_3 \longrightarrow Ca(HCO_3)_2$$

calcium carbonate + carbonic acid ⟶ calcium bicarbonate

When the dissolved oxygen in water comes into contact with certain rock minerals, *oxidation* is initiated. This is the chemical union of oxygen atoms with atoms from various metallic elements to form new products, which are usually more voluminous, softer, and more easily eroded than the original compounds. When iron-bearing minerals are affected by oxygen, iron oxide is produced; this is probably the most common oxidation effect in the lithosphere. This phenomenon is called *rusting*, and the prevalence of rusty red stains on the surface of many rocks attests to its widespread occurrence. Similar effects are produced by the oxidation of aluminum. Since iron and aluminum are very common in the Earth's crust, a reddish brown color is seen in many rocks and soils, particularly in tropical areas where oxidation is the most notable chemical weathering process. The implication of rusting for weathering is significant because the oxides usually are softer and more easily removed than the original iron-containing minerals.

Hydrolysis is a chemical union of water with another substance to produce a new compound that also is nearly always softer and weaker than the original. Igneous rocks are particularly susceptible to hydrolysis because their silicate minerals combine readily with water. Hydrolysis invariably increases the volume of the mineral, which can contribute to mechanical disintegration, in addition to the chemical change that produces a weaker mineral. In tropical areas where water frequently percolates to considerable depth, hydrolysis often occurs far below the surface.

Focus on Exfoliation

One of the most striking of all weathering processes is *exfoliation*, in which curved layers peel off the bedrock in sheets. Curved and concentric sets of joints, usually minor and inconspicuous, develop in the bedrock, and parallel shells of rock break away in succession from the original rock mass, somewhat analogous to the separation of layers of an onion. The sheets that split off are sometimes very thin, only a millimeter or so in width. In other cases, however, they are much more massive and may be several feet thick.

Exfoliation apparently occurs only in granite and related intrusive rocks, but the exact mechanism is not fully understood. It probably involves both chemical and physical weathering, but mechanical disintegration appears to be much more significant than chemical alteration. Fracturing resulting from temperature variations may be an important component of the mechanism, although exfoliation is also found in areas where temperature fluctuations are not marked. Volumetric changes in minerals, which set up strains in the rock, may also be involved. Such changes are most notably produced by *hydration*, in which water molecules become attached to other substances, which causes a wetting and swelling of the original substance without any change in its chemical composition. This has a weakening effect on the rock mass and is usually sufficient to produce some fracturing.

One other likely mechanism for exfoliation has been hypothesized and widely accepted, that is, cracking from the relief at removal of an overlying weight—a process called *unloading*. The intrusive bedrock originally may have been deeply buried beneath a heavy overburden. When the overlying material is stripped away by erosion, the release of pressure allows expansion in the granite. The outer layers are unable to contain the expanding mass, and the expansion can be absorbed only by cracking along the sets of joints.

Whatever the causes of exfoliation, the results are conspicuous. The outer shells of rock peel away from the main mass, progressively exposing lower layers. If the rock mass is a large one, such as Half Dome or one of the other granite monoliths overlooking Yosemite Valley, its surface configuration will consist of imperfect curves punctuated by several partially fractured shells of the surface layers, and the mass itself is referred to as an *exfoliation dome*. Exfoliation may also occur on detached boulders, of whatever size, which usually results in more perfectly rounded shapes, with each layer of shelling revealing a smaller spherical mass.

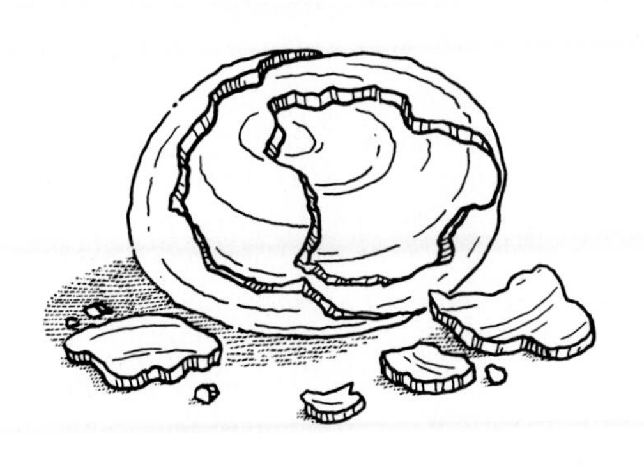

Examples of exfoliation at different scales: an individual boulder and a massive hill.

Carbonation is a process by which the carbon dioxide in water reacts with carbonate rocks such as limestone and dolomite to produce a very soluble product (calcium bicarbonate), which can readily be removed by runoff or percolation, and which can also be deposited in crystalline form if the water is evaporated.

These and other less common processes of chemical weathering are continually at work at and beneath the surface of the Earth. Most affected rocks change their physical appearance. Their coherence is weakened, and at the surface, loose particles are produced that are quite unlike the parent material. Beneath the surface, the rock holds together but in a chemically altered condition. The major eventual products of chemical weathering are clays, very common constituents of the surface, which are produced solely by rock decomposition through chemical alteration.

Mass Wasting

The denudation of the Earth's surface is accomplished by a continuum of action that can be more or less clearly divided into two, and often three, stages. The initial stage consists of weathering, in which the bedrock is broken up in place and thus made available for movement. Weathering is nearly always a prelude to the transfer of rock materials; only rarely is it bypassed by such events as direct gouging of bedrock by a moving glacier or the chipping of rock by windblown sand. The final stage is erosion, the large-scale and often distant removal of fragmented rock materials. The ultimate destiny of all weathered material is to be carried away by erosion, although at any given time, of course, a great mass of unremoved weathered material is still in place all over the continental surfaces. In between the initial weathering and the ultimate erosion, there is usually an intermediate stage in which weathered material is moved a relatively short distance downslope under the direct influence of gravity. This is called *mass wasting*, also sometimes referred to as *mass movement* or *gravity transfer*. Mass wasting is often circumvented by the direct action of erosive agents upon weathered material, but it normally serves as the second step in a three-step denudation sequence.

Gravity is the energizing force in mass wasting. Throughout our planet, gravity is inescapable; everywhere it pulls objects toward the center of the Earth. Where the land is flat, the influence of gravity on topographic development is minimal; but even on gentle slopes, minute effects are likely to be significant in the long run, and on steep slopes, the results are often immediate and conspicuous. Any loosened material will be impelled downslope by gravity—in some cases, falling abruptly or rolling rapidly; in other instances, flowing or creeping with imperceptible gradualness.

The materials involved in these movements are as varied as the products of weathering themselves. Gigantic boulders respond to the pull of gravity in much the same fashion as do particles of dust, although the larger the object, the more immediate and pronounced the effect. Of particular importance, however, is the implication of "mass" in mass wasting. Large units of material—fragmented rock, regolith, soil—are often moved.

All rock materials, from individual fragments to cohesive layers of soil, will lie at rest on a slope if undisturbed, unless the slope is so steep that the angular relationship becomes overbalanced. The steepest angle that can be assumed by loose fragments on a slope without downslope movement is called the *angle of repose*. The angle of repose represents a fine balance between the pull of gravity and the cohesion and friction of the rock material. The precise angle varies with the nature and internal cohesion of the material. If additional material accumulates on a slope that is near the angle of repose, it may upset the balance (because the added weight overbalances the friction) and may cause all or part of the material to slide downward.

If water is added to the rock material through rainfall, snowmelt, or subsurface flow, the mobility is usually increased, particularly if the fragments are of small size. Water is a lubricating medium, and it diminishes friction among the particles so that they can slide past one another more readily. Water also adds to the buoyancy and weight of the mass, which makes for a lower angle of repose and adds momentum once movement is under way. It is clear, then, that mass movement will be particularly likely during and after heavy rains.

Another facilitator of mass movement is the presence of clay. As noted in the section on soils, clays readily absorb water, and this factor combines with their extremely fine-grained texture to produce a substance that is very slippery and mobile. Masses of material associated with or resting on top of a clay stratum can often be set in motion by rainfall or an earthquake shock, even on very gentle slopes. Indeed, some clay formations are called "quick clays" because they spontaneously change from a relatively solid mass into a near-liquid condition as the result of a sudden disturbance or shock.

In subarctic regions and in high-latitude locations, mass movements are often initiated and continued through the heaving action of ice derived from ground water. The presence of thawed and saturated ground in summer overlying permanently frozen subsoil (permafrost) contributes further to downslope movements in such regions. Some geomorphologists assert that mass movement of debris is the single most important means of transport of weathered material in the subarctic.

Although some types of mass wasting are rapid and conspicuous, others are slow and gradual. The principles involved are generally similar, but the extent of the activity and particularly the rate of movement are quite variable. In our consideration here we will proceed from the most rapid to the slowest, discussing the characteristics of each as if they were discrete, although in nature the various types often overlap.

Fall

The simplest and most obvious form of mass wasting involves direct falling of pieces of rock downslope. When a rock fragment is loosened by weathering on a very steep slope, it may simply be dislodged and fall, roll, or bounce down to the bottom of that segment of slope. This is a very characteristic event in mountainous areas, particularly due to frost action. Normally the fragments do not travel far before they become lodged, although the lodging may be unstable and temporary. Pieces of rock, of whatever size, that fall in this fashion are referred to collectively as *talus* or *scree*.

Sometimes the fragments accumulate relatively uniformly along the base of the slope, in which case the resultant landform is called a *talus slope* or *talus apron*. More characteristically, however, the dislodged rocks collect in sloping, cone-shaped heaps called *talus cones*. See Figure 15-6. This pattern is commonplace because most steep bedrock slopes and cliffs are seamed by vertical ravines and gullies that funnel the falling rock fragments into piles directly beneath the ravines, usually producing a series of talus cones side by side along the base of the slope or cliff.

Some of the falling fragments, especially the larger ones with their greater momentum, tumble and roll to the base of the cone. Most of the new talus, however, comes to rest at the upper end of the cone. The cone thereby grows up the mountainside, sometimes reaching upward many hundreds of feet (dozens of meters). The angle of repose for talus is very high, generally about 35° and sometimes as great as 40°. The slope of a talus accumulation is gently concave upward, with the steepest angle near the apex of the cone. New material is frequently added to the top of the cone, where the fragments invariably are in delicate equilibrium, and each new piece that bombards down from above may cause a readjustment, with considerable downhill sliding. The freshest blocks, then, tend to be at the upper end of the cone. There is also, however, a rough sorting of fragments according to size, with the larger pieces rolling further downslope and the smaller bits lodging higher up. With the passage of time, all talus slowly migrates downslope, encouraged by the freeze/thaw action of water in the many open spaces of the cones and aprons.

Figure 15-6 Talus cones often accumulate at the base of steep slopes.

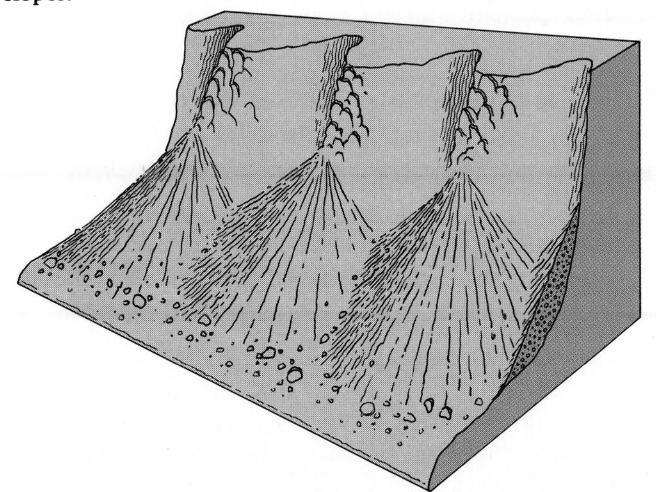

Slide

In rugged terrain, landslides carry large masses of rock and earth bodily downslope, abruptly and often catastrophically. The event may last only a few seconds or minutes, but it is often followed by smaller subsequent slides from above or alongside the original. A landslide is essentially an instantaneous collapse of a slope and does not necessarily involve the lubricating effects of water or clay. In other words, the sliding material represents a rigid mass that is suddenly displaced, rather than encompassing any sort of plastic flow. The presence of water may contribute to the action; many slides are triggered by rains that add weight to already overloaded slopes. Landslides may be activated by other stimuli as well, most notably by earth tremors. Slides are also sometimes initiated simply by lateral erosion of a stream that undercuts its bank and thus oversteepens the slope above.

Some slides move only unconsolidated regolith, but many of the larger slides also encompass masses of bedrock that may be detached along joint planes and fracture zones. The rapid downslope surge invariably involves the violent disintegration of much of the material, regardless of the size of the blocks originally dislodged.

Landslide action is not only abrupt but also rapid. Precise measurement of the rate of movement is obviously impossible, but eye-witness accounts have included sweep-second timing that affirms speeds of more than 100 miles (160 km) per hour, and occasionally more than twice this rate. Thunderous noise accompanies the slide, and blasts of air are created that can strip leaves, twigs, and even branches from nearby trees.

As most landslides occur in steep, mountainous terrain, the great mass of material (displaced volume is sometimes measured in cubic miles) that roars downslope may choke the valley at the bottom of the slope. Moreover, the momentum of the slide may push material several hundred feet *up* the slope on the other side of the valley. One characteristic result is the creation of a natural dam across the width of the valley, blocking the valley-bottom stream and producing a new lake, which becomes ever larger until it either over-tops the dam or cuts a path through it.

Large landslides are irresistible forces, obliterating and/or burying everything in their paths. The spectacular and destructive nature of these happenings makes them a matter of great concern for mankind, on a par with hurricanes, tornadoes, tsunamis, volcanic eruptions, and earthquakes. The catastrophic results for human life and property are mostly concentrated in the immediate path of the slide, although subsequent flooding from naturally dammed streams may wreak more extensive havoc.

Countless examples of catastrophic landslides come from many parts of the world, particularly in Asia, along the Andes Mountains in South America, and in various parts of Europe. Presented here are descriptions of four of the more disastrous landslides that have occurred in the United States and Canada in the twentieth century.

The Frank (Alberta) slide. The mountain slope at the upper left broke loose and roared down into the valley, partly burying the town of Frank. The river was temporarily blocked by the slide, although it soon reestablished its course.

Frank, Alberta In the spring of 1903, an enormous mass of limestone abruptly broke loose from the side of Turtle Mountain, which towers some 3000 feet (910 m) above the valley of the Crow's Nest River in the eastern foothills of the Rocky Mountains in southwestern Alberta. The geology of the area was structurally unstable because during the mountain-building epoch, massive limestone had been thrust-faulted over weaker rocks, and the lower part of the limestone had been shattered by the movements along the fault zone, producing an area of weakness beneath more cohesive rock. Moreover, a set of large joints was inclined downward toward the valley.

Although earthquakes in the area two years before had created no obvious problems, they may have contributed to the instability of the slope. The landslide occurred without warning, and it moved some 40 million cubic yards (30 million m^3) of bedrock and overburden down the side of Turtle Mountain. The leading edge of the slide advanced more than 2 miles (3.2 km) across the valley and climbed some 400 feet (120 m) up the opposite slope. The small coal-mining town of Frank, in the valley bottom at the northern edge of the slide, was partly buried by the debris, with the loss of 70 lives in approximately 2 minutes.

Gros Ventre, Wyoming In the foothills of the Gros Ventre Mountains south of Yellowstone Park in northwestern Wyoming, a large landslide broke loose in June of 1925. It roared across the valley of the Gros Ventre River and climbed more than 300 feet (90 m) up the opposite slope. No people were in the area at the time of the slide; thus the only fatalities were six head of cattle. However, the debris formed a natural dam that blocked the river and created a lake some 5 miles (8 km) long. Two years later, the lake waters suddenly breached the dam and a wall of water and debris flooded down the valley below, washing away the village of Kelly and drowning seven people.

Madison Valley, Montana Just before midnight in the late summer of 1959, the geologically unstable area of Yellowstone Park was rocked by a major earthquake (7.8 on the Richter scale). Of the many striking geomorphic results of the shock, by far the most conspicuous and tragic took place in the narrow valley of the Madison River a few miles west of the national park in Montana. Hebgen Lake, a 7-mile (11-km) long reservoir in the valley, experienced a series of *seiches* (abrupt "tidal wave" actions in which sudden oscillation of the surface causes a surge of water to flow across the lake), and an enormous quantity of water was hurled over the dam as if sloshed out of a dishpan, and it swept down the valley as a flash flood. This wall of water sped some 7 miles (11 km) downstream, almost to the point where the constricted canyon opened out into rolling wheat and grazing land. Just as the surge approached the canyon mouth, a resistant buttress of sedimentary rock (dolomite) on the steep south slope gave way, releasing an enormous mass of unstable schist that had been retained above and behind it. Some 43 million cubic yards (33 million m^3) of debris cascaded down the slope, filled the valley bottom, spurted up the opposite hillside, and sprawled up and down the canyon for $1\frac{1}{2}$ miles (2.4 km). A Forest Service campground, crowded with sleeping vacationists, was partly buried by the slide.

The landslide and the flash flood both reached the canyon mouth at about the same time. The slide dammed the river and forced the surging water back into the unburied portion of the campground. Thus the campers who had escaped being crushed were further endangered by the force of the ricocheting water and then the inexorable flooding as a new lake formed behind the natural dam. The final death toll from this tragic episode will never be known. Nine bodies were found, and indirect evidence (mostly reports from families and friends of people thought to be in the area at the time) indicated that at least 19 other people were buried in the slide.

An immediate result of the cataclysm was the creation of a continually enlarging lake, called Quake Lake, upstream from the blocking debris. Not only did this impoundment produce pressures that might rupture the natural dam, but its upstream end would eventually lap against the foundations of Hebgen Dam, quite possibly undermining it. The downvalley area, which

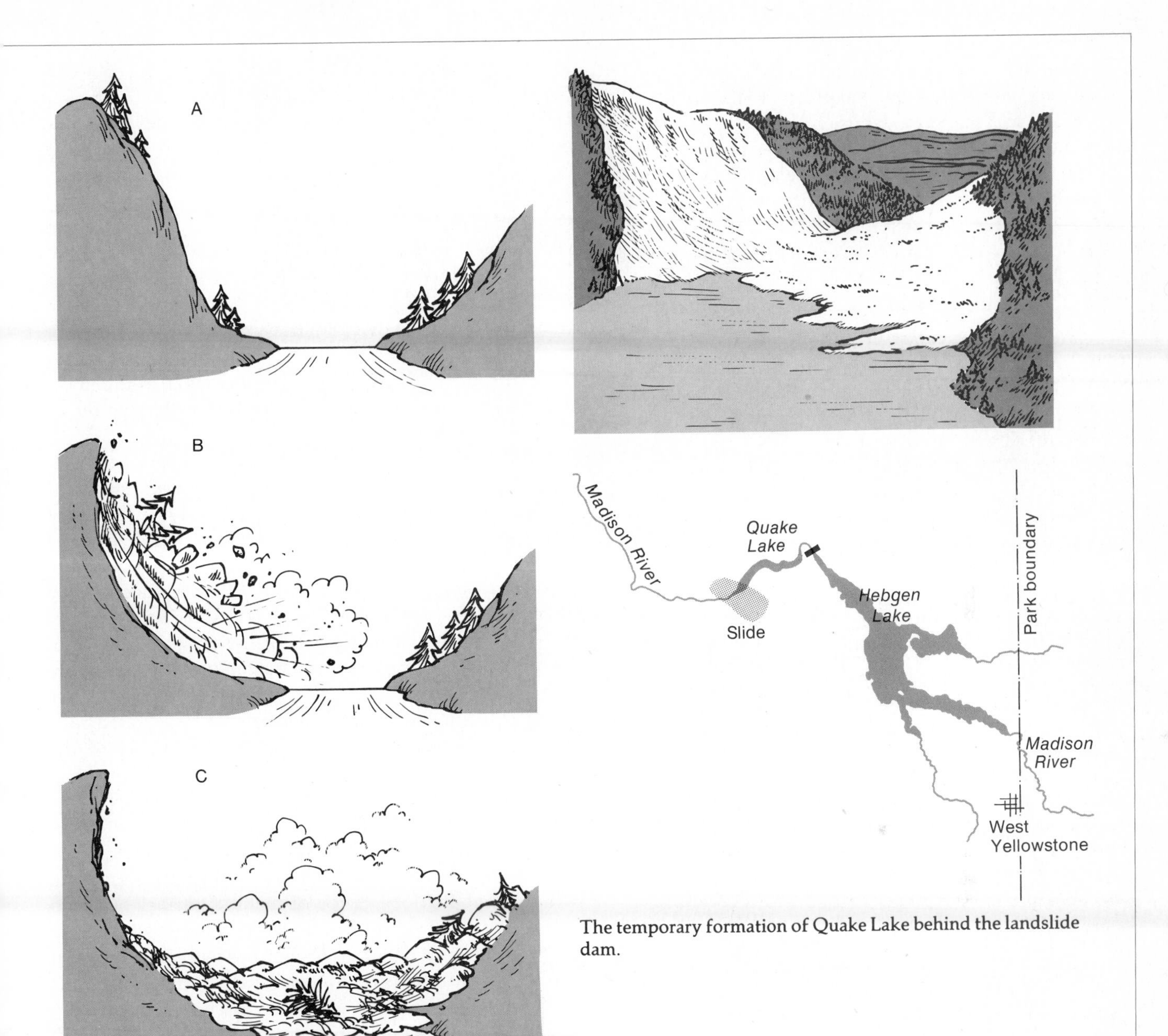

The slope collapse and landslide that dammed the valley of the Madison River. This three-stage sketch looks downvalley.

The temporary formation of Quake Lake behind the landslide dam.

included three towns, thus faced the twin threat of abrupt flooding from the broken natural and Hebgen dams, and memories of the postlandslide Gros Ventre tragedy were much in the minds of officials. Consequently, nearly $2 million were spent in the next ten weeks to construct a deep channel across the top of the slide, allowing the Madison River to resume its normal flow regime.

Hope, British Columbia In midwinter of 1965, a mass of bedrock and surficial sediments estimated to contain 15 million cubic yards (11.4 million m³) slid 3500 feet (1067 m) down the southwestern slope of Johnson Peak in the Cascade Mountains some 11 miles (18 km) east of Hope, British Columbia. In the valley bottom, the debris overran or displaced large glacial deposits, remnants of an earlier slide, a small lake, and several acres of forest. The more mobile materials advanced 500 feet (152 m) up the opposite slope, and the frontal blast of air shook snow from trees and deposited sand and dust for several hundred feet beyond. Debris accumulated to a depth of 250 feet (76 m), burying four people who were traveling along the valley-bottom highway at the time.

The immediate topographic result of a landslide is three fold:

1. On the hill where the slide originated, there is a deep and extensive scar, usually exposing a mixture of open bedrock and scattered debris.
2. In the valley bottom where the slide material comes to rest, there is a massive pile of highly irregular debris, usually in the form of a broad ridge or low-lying cone. Its surface consists of a hummocky jumble of totally unsorted material, ranging from immense boulders to fine dust.
3. On the upvalley side of the debris, a lake may form.

An extremely common form of mass wasting is a *slump*, which is considered to be in the slide category. See Figure 15-7. Slumping involves slope collapse with a backward rotation. The upper portion of the moving material tilts down and back, and the lower portion moves up and out. The top of the slump is usually marked by a crescent-shaped scarp face, sometimes with a steplike arrangement of smaller scarps and terraces below it. The bottom of the slump has the appearance of a typical earthflow (see page 330), with a bulging lobe of saturated debris protruding downslope or into the valley bottom.

Figure 15-7 Comparison of earthflow and slump.

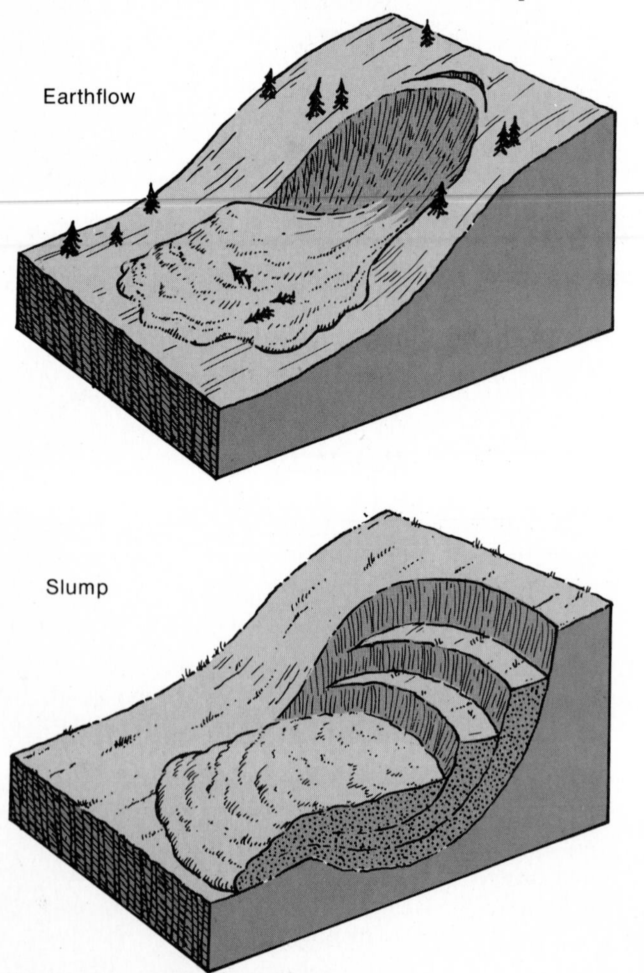

Flow

A much less spectacular form of mass wasting, but one that is conspicuous where it occurs, involves *flow* movements, in which a sector of a slope becomes unstable, normally due to the addition of water, and slips gently downhill. In some cases, the flow is fairly rapid, but normally it is a very gradual and sluggish movement. Usually the center of the mass moves more rapidly than the base and sides, which are retarded by friction.

Many flows are relatively small, often encompassing an area of only a few square yards (square meters). More characteristically, however, they cover several tens or hundreds of acres. Normally, they are relatively shallow phenomena, including only soil and regolith in the mass, but under certain conditions, a considerable amount of bedrock may be involved.

As with other forms of mass movement, gravity is the impelling force. Water is nearly always an important catalyst to the movement; the surface materials become unstable with the added weight of water, and their cohesion is diminished by waterlogging so that they are more responsive to the pull of gravity. The presence of clay also promotes flow, as clay minerals become very slippery when lubricated with any sort of moisture.

The most common of the flow movements is simply termed an *earthflow*, in which a portion of a slope moves a limited distance downhill, normally in a saturated condition during or after a heavy rain. At the top of the area from which the movement occurred there is usually a distinct interruption in the surface of the preflow slope; this may take the form of one or more irregular cracks or fissures, or a prominent oversteepened scarp face. The flow effect is most conspicuous in the lower portion, where a bulging lobe or toe of material pushes out onto the valley floor. This type of slope failure is relatively common on hillsides that are not densely vegetated and often results in blocked transportation lines (roads, railways) in valley bottoms. Property damage is occasionally extensive, but the rate of movement is usually so sluggish that there is no threat to life.

Similar in name but different in dynamics to the earthflow is the *mudflow*. It originates on slopes in arid and semiarid country when a heavy rain follows a long dry spell, producing a cascading runoff that is too voluminous to be absorbed into the soil. Fine debris is picked up from the hillsides by the runoff and is concentrated in the valley bottoms, where it flows downvalley as a viscous mass. The leading edge of the mudflow continues to accumulate load, becoming increasingly stiff and retarding the flow movement of the more liquid upstream portions, so that the entire mudflow moves haltingly down the valley or canyon. When such a mudflow reaches the mouth of the canyon and abruptly leaves its confining walls for the more open slopes of the piedmont zone, the pent-up liquid behind the glutinous leading edge breaks through with a rush, spreading the slimy mass of muddy debris into a wide sheet. Mudflows often pick up large pieces of rock debris, including huge boulders, and carry them along as part of their load. In some cases, the large pieces are so numerous that the term *debris flow* is used in preference to mudflow.

An important distinction between an earthflow and a mudflow is that the latter consists of movements over the slope and down established drainage channels, whereas the former involves a slope collapse and has no relationship to the drainage network. Moreover, mudflows are normally much more rapid, with a rate of movement intermediate between the sluggish surge of an earthflow and the rapid flow of a stream of water. Mudflows are potentially (and actually) more dangerous to humanity than earthflows because of the more rapid movement, the larger quantity of debris involved, and the fact that the mudflow often discharges abruptly across a desert piedmont zone, which is likely to be a favored area for human settlement and intensive agriculture.

A much more limited and specialized type of flow movement is represented by *rock glaciers*. In some rugged mountain areas, talus accumulates in great masses in favorable locations, and these masses may move slowly but distinctly downslope under their own weight. The movement is in discernible "streams," and their general resemblance to valley glaciers gives rise to their name. Their flow is primarily due to gravitative transfer, aided by freeze/thaw temperature changes, and may be largely independent of any lubricating effect of water. Rock glaciers are normally found on relatively steep slopes, but they sometimes extend for thousands of feet (hundreds of meters) down valleys and even out onto adjacent flatter areas.

Creep

The slowest and least perceptible form of mass wasting is called *creep*. It consists of a very gradual downhill movement of soil and regolith that is so unobtrusive that normally it can be recognized only by indirect evidence. Generally the entire slope is involved, not just a portion of it. Creep is such a pervasive phenomenon, however, that it occurs all over the world on sloping land. Although most notable on steep, lightly vegetated slopes, it also occurs on gentle slopes with a dense plant cover. Wherever weathered materials are available for movement on land that is not flat, creep is a persistent form of mass wasting.

The creeping process is universal. Infinite numbers of tiny bits of lithospheric material, as well as many larger pieces, march slowly and sporadically downslope from the places where they were produced by weathering or deposited after previous erosional action. When free water is present in the surficial material, creep is usually accelerated because the lubricating effect allows individual particles to move more easily and because water adds to the weight of the mass.

Soil creep is caused by the interaction of various factors. The most significant mechanism involves the alternation of freeze/thaw and wet/dry conditions. When soil particles are frozen or moistened, they tend to be displaced upward at a vertical angle to the slope. After thawing or drying, however, the particles settle downward, not directly into their original position but rather are pulled slightly downslope by gravity. See Figure 15-8. With countless repetitions, this process can result in downhill movement of the entire slope.

Many other agents also contribute to creep. Indeed, any activity that produces a disturbance of soil and regolith on a sloping surface is a contributor because gravity affects every rearrangement of particles, attracting them downslope. For example, burrowing animals pile most of their excavated material downslope, and subsequent infilling of burrows is mostly by material from the upslope side. As plant roots grow, they also tend to displace particles downslope. Animals that walk on the surface exert a downslope movement as well. Even the shaking of earthquakes or thunder produces disturbances that stimulate creep.

Wherever it occurs, creep is a very slow process, but its rate of movement is faster under some circum-

Figure 15-8 The movement of a typical particle in a freeze/thaw situation. Freezing lifts the particle at a vertical angle to the slope (from A to B); upon thawing, the particle settles slightly downslope (from B to C).

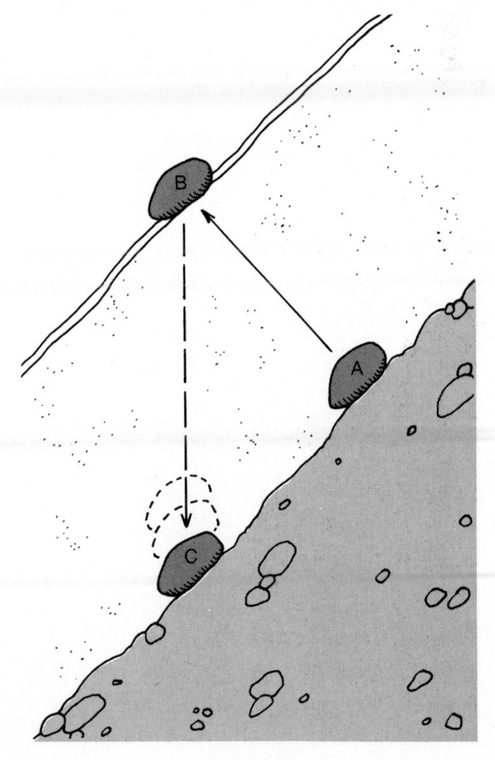

Figure 15-9 Visual evidence of soil creep may be shown by the displacement or bending of trees, fences, utility poles, or retaining walls.

stances than others. The principal variables are slope angle, vegetative cover, and moisture supply. Creep operates faster on steep slopes than on gentle ones, for obvious reasons. Deep-rooted and dense-growing vegetation inhibits creep because of the binding effect of the roots on the soil. Also, as mentioned previously, saturated slopes generally experience faster rates of creep than dry ones. Although extreme rates of movement of up to 2 inches (5 cm) per year have been reported, much more common would be a rate of just a fraction of a centimeter in a year.

Whatever the rate of creep, it is much too slow for the human eye to recognize, and the results in the soil itself are all but invisible. Creep is usually comprehended only because certain works of man are affected sufficiently to provide visual evidence. Most commonly this involves fence posts and utility poles being tilted downhill. See Figure 15-9. Retaining walls may be broken or displaced, and even roadbeds may be disturbed. Trees also sometimes provide visual evidence of creep by growing at odd angles or by having a downslope convex bend in their trunks resulting from the conflict between the tree's natural tendency to grow vertically and the downhill displacement of the creeping soil.

Unlike the other forms of mass wasting discussed previously, creep produces few distinctive landforms. Rather it induces an imperceptible diminishing of slope angles and gradual lowering of hilltops—i.e., a widespread but minute smoothing of the land surface—as the surface material slowly creeps downslope.

Under certain conditions, and usually on steep grassy slopes, creep can produce a complicated terracing effect that resembles a network of faint trails. These are called *terracettes*. Large animals, particularly ungulates (cattle, sheep, deer, etc.), tend to walk along the terracettes, accentuating their outlines until the entire hillside is covered with a maze of small pathlike terraces.

A special form of creep that produces a distinctive surface appearance is *solifluction* (meaning "soil flowage"), a process that is largely restricted to tundra landscape beyond the tree line either at high latitudes or high altitudes. See Figure 15-10. During the summer, the near-surface portion of the ground (called the *active layer*) thaws, but the meltwater cannot percolate deeper because of

Figure 15-10 Solifluction in the high country. The surface has slipped gradually downslope about 200 feet (600 m) in some places here on Trail Ridge in north-central Colorado. [TLM photo.]

the permanently frozen subsoil (permafrost) below. The spaces between the soil particles become saturated, and the heavy surface material sags slowly downslope. Movement is erratic and irregular, with lobes overlapping one another in a haphazard fish-scale pattern. The lobes move only a few inches or centimeters per year, but they remain very obvious in the landscape, in part because of the scarcity of vegetation. Where solifluction occurs, drainage channels are usually scarce because water flow during the short summer is mostly laterally through the soil rather than across the surface.

The effects of weathering and mass wasting on the lithospheric surface are sometimes distinctive and obvious, but more often they are unobtrusive and inconspicuous. The role of weathering and mass wasting in the continuing drama of denudation is fundamental but preliminary. They function to fracture, disaggregate, and loosen rock material, often moving the resulting debris a short distance as well. This work is preparatory to the principal denudation activities, which are accomplished by the agents of erosion, and which serve as the prime foci of our attention in the remaining chapters of this book.

Review Questions

1. Distinguish between *weathering*, *mass wasting*, and *erosion*.
2. Explain how it is possible for weathering to take place beneath the surface of a rock.
3. What is the difference between a *joint* and a *fault*?
4. Explain the mechanics of *frost wedging*.
5. Why is chemical weathering more effective in humid than in arid climates?
6. What is the relationship of *oxidation* to *rusting*?
7. What are some of the suggested causes of *exfoliation*?
8. How does the *angle of repose* affect mass wasting?
9. What is the function of clay in mass wasting?
10. What is the difference between *earthflow* and *mudflow*?

Some Useful References

CARROLL, C., *Rock Weathering*. New York: Plenum Publishing Corporation, 1970.

OLLIER, C. D., *Weathering*. London: Longman, 1969.

SHARPE, C. F. S., *Landslides and Related Phenomena*. New York: Columbia University Press, 1938.

ZARUBA, Q., and U. MENCL, *Landslides and Their Control*. Amsterdam, N.Y.: American Elsevier Publishing Co., 1969.

16 The Fluvial Process

The shaping of the lithospheric surface is accomplished by a variety of agents working in concert. In Chapter 14 we considered the gamut of internal forces; in Chapter 15 we detailed the actions of weathering and mass wasting; now in chapters 16 through 19 we will concentrate on those external forces that are prominent as agents of erosion and deposition. These include wind, moving ice, coastal waters, and subsurface waters. However, by far the most important of the external forces involved in denudation is running water moving over the surfaces of the continents.

The work accomplished by running water is probably more significant in shaping landforms than that of all other external forces combined. This is not because running water is necessarily more forceful than the other agents. Indeed, such forces as moving ice and pounding waves often apply much greater amounts of energy than can be mustered by surface runoff or streamflow. The key to the dominance of running water as a shaper of continental landforms is in its ubiquity; almost all parts of all continents except Antarctica experience occasional, frequent, or continuous movement of surface water. Other external forces, with the exception of wind, which is trivial in its power as a terrain sculptor, are much more limited in their occurrence, and thus in their effectiveness, than running water.

Running water is referred to as a *fluvial process* in this context. This encompasses both the general movement of surface water down the slope of the land surface, called *overland flow*, and the more precisely channelled movement of water along a valley bottom, or *streamflow*. On all continental surfaces not covered by "permanent" glaciers or snowfields, almost without exception the fluvial process dominates as a shaper of surface forms. This is true even in desert areas where rains are rare and surface streams are virtually nonexistent.

In this chapter we will examine many of the details of the fluvial process, seeking an understanding of the mechanics of flow, the characteristics of surface drainage systems, the mechanism of fluvial erosion/transportation/deposition, and the nature of the landforms that are produced. In the following chapter (Chapter 17), we will focus our attention on the arid environments where fluvial action is more limited; in those regions the fluvial process is significantly modified, and the resulting landforms often have different characteristics.

Some Fundamental Concepts

Prior to a consideration of the specifics of the fluvial process and its topographic results, a few basic concepts must be clarified.

Valleys and Interfluves

In a fundamental sense, the surfaces of the continents can be considered to consist of two topographic elements—valleys and interfluves. The distinction between the two is not always obvious in nature, but the conceptual demarcation is clear. In this content, all the continental surfaces fit in one or the other cateogry—valleys or interfluves.

A *valley*, in the broad meaning of the term, encompasses that portion of the total terrain in which a drainage system is clearly established. See Figure 16-1. It includes the *valley bottom*, which is the lower and flatter area that is partially or totally occupied by the channel of a stream, as well as the *valley sides*, which are the somewhat steeper (sometimes much steeper) slopes that rise above the valley bottom on either side. In some situations, the valley bottom is narrow and elongated with a limited areal extent; in other cases, it may be extraordinarily broad and extensive. The valley sides, likewise, may be exceedingly variable, with slopes that are steep or gentle and an extent that is limited or expansive. The outermost, or upper, limit of a "valley" in this context is not always readily apparent on the land, but it can be clearly conceptualized as a "lip" or "rim" at the top of the valley sides above which drainage channels are either absent or indistinct. See Figure 16-2.

An *interfluve* is the higher land above the valley sides that separates adjacent valleys. Interfluve means "between valleys." Some interfluves consist of ridgetops or mountain crests with precipitous slopes, but others are simply broad and flattish divides between drainage systems. Conceptually, all parts of the terrain not in a valley constitute a portion of an interfluve.

These simplistic definitions are not always applica-

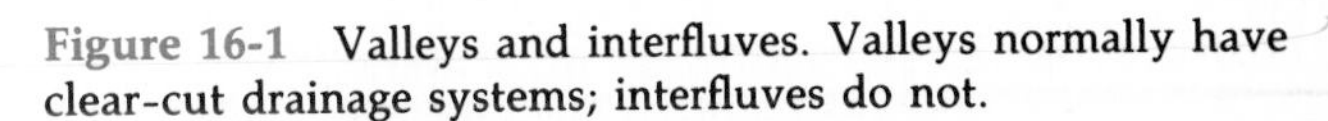

Figure 16-1 Valleys and interfluves. Valleys normally have clear-cut drainage systems; interfluves do not.

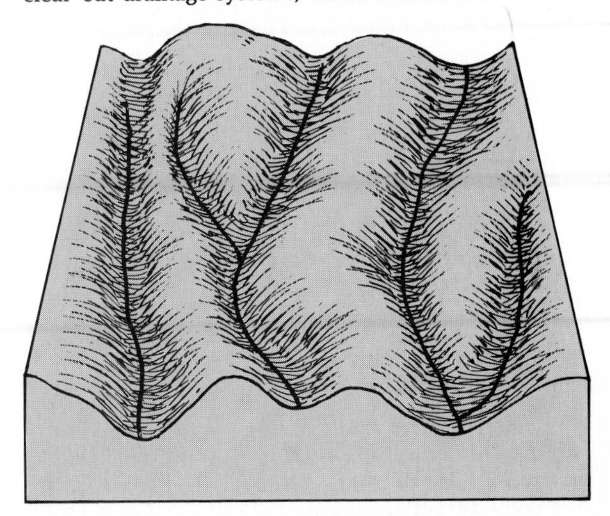

Figure 16-2 The distinction between interfluve and valley is seen clearly here. The grass-covered area is an interfluve where runoff is unchannelled. When water trickles over the upper edge, or "lip" of the valley, it rapidly becomes channelled into the established drainage system. [TLM photo.]

ble in nature. Some terrain elements defy distinct classification. Swamps and marshes, for example, may be on interfluves but are usually in valleys, although with no clearly established drainage system. Such exceptional cases, however, should not inhibit our acceptance of the valley/interfluve concept.

Erosion and Deposition

Fundamental to any consideration of denudation in general and the shaping of landforms in particular is an understanding of the relationship between erosion and deposition. All external forces accomplish both activities; they remove fragments of bedrock, regolith, and soil from their original positions (*erosion*), and they relocate them somewhere further downslope (*deposition*). The fluvial process, which is our concern in this chapter, produces one set of residual landforms by its erosional activities and quite different set of landforms by its depositional activities.

Erosion by Overland Flow The initial opportunity for fluvial erosion occurs when rain starts to fall. Unless the impact of rain is absorbed by vegetation or some other protective covering, the direct collision of raindrops with the ground is strong enough to blast fine soil particles upward and outward, shifting them a few millimeters laterally. On sloping ground, most particles experience a net downhill movement by this *splash erosion*. See Figure 16-3. In the first few minutes of a rain, much of the water infiltrates into the soil, and there is little runoff. During heavy or continued rain, however, particularly if the land is sloping and there is a sparse vegetative cover, infiltration is greatly diminished, and most of the water proceeds downslope as overland flow. The water

Figure 16-3 The awesome impact of a raindrop produces splash erosion. [USDA-Soil Conservation Service photo.]

flows across the surface as a thin sheet, transporting material already loosened by splash erosion, in a process termed *sheet erosion*. As the overland flow increases in volume, the resulting turbulence tends to disaggregate the sheet flow into multitudinous tiny channels called *rills*. This more concentrated flow pattern loosens additional material and scores the slope with numerous parallel seams; this sequence of events is termed *rill erosion*. If the process continues, the rills begin to coalesce into fewer and larger channels called gullies, and *gully erosion* becomes recognizable. See Figure 16-4. As the gullies increase in size, they tend to become incorporated into the drainage system of the adjacent valley, and the sequential change from overland flow to streamflow is accomplished.

Overland flow, splash erosion, and sheet erosion are normal conditions in nature. Rill erosion and gully erosion are less common but sometimes occur under natural conditions, particularly in areas of steep slopes, sparse vegetation, and heavy rains. Human activities (especially vegetation removal) often speed up these natural processes, resulting in greatly increased rates of overland flow and *accelerated erosion*.

Erosion by Streamflow Once the surface flow is channelled into a stream, its ability to erode is greatly increased by the enlarged volume of water. Erosion is accomplished in part by the direct hydraulic power of the moving water, which has an impact and a dragging effect on material at the bottom and sides of the stream and can thus excavate and move considerable quantities of unconsolidated debris. Banks can also be undermined by streamflow, particularly at times of high water, dumping more loose material into the water to be swept downstream.

The erosive capability of streamflow is also significantly enhanced by the abrasive tools that it picks up and carries along with it. All sizes of rock fragments, from silt to boulders, exert a chipping and grinding effect as they are swirled or bounced or rolled downstream by the moving water. These "tools" break off more fragments from the bottom and sides of the channel, and they impact against one another, becoming both smaller in size and rounder in shape from the wear and tear of the frequent collisions as they are moved downstream. See Figure 16-5. The eventual result of this *abrasion* is to reduce almost all stream-carried debris to very small silt particles.

Figure 16-4 Intense gullying on a volcanic slope in central Mexico. [K. Segerstrom, U.S. Geological Survey photo.]

Figure 16-5 Rock particles transported by streams are inevitably rounded by frequent collision in their downstream path. This scene from the upper reaches of the Nisqually River in Washington shows relatively large boulders that have not yet been transported very far, but they are already quite round in shape. [TLM photo.]

A certain amount of chemical action also accompanies streamflow, and some of the same processes discussed under chemical weathering—particularly solution action and hydrolysis—also help erode the stream channel by *corrosion*.

The erosive effectiveness of streamflow varies enormously from one situation to another. It is determined primarily by the *velocity* and *turbulence* of the flow on one hand, and the *resistance* of the bedrock on the other. Velocity, in turn, is governed by the *gradient* of the stream bed (the steeper the gradient, the faster the flow) and by the *volume* of flow (more water normally means a higher velocity of flow). The degree of *turbulence* is determined in part by the velocity (faster flows are more turbulent) and in part by the roughness of the stream channel (an irregular channel surface increases turbulence).

The pattern of erosion in the stream bed and on its sides is often irregular because of swirls and eddies in the streamflow and the resulting erratic paths of the abrasive materials being carried along. This is particularly true where steep stream gradients engender considerable turbulence in the flow. As a result, troughs, chutes, plunge pools, circular potholes, and other irregular excavations occur in the bedrock of the stream channel, caused by the grinding and boring and erratic movements of the stream's abrasive tools.

Transportation Any water moving downslope, whether part of overland flow or streamflow, is capable of transporting rock material along with it, as part of its *load*. At any given time and place, the load carried by overland flow is likely to be small in comparison with what a stream can transport. In total, however, the amount of material transported by overland flow is incredibly large. Eventually, most of this great mass of material reaches the valley bottoms, where it is added to the rock debris eroded by streams to constitute the *stream load*.

Streams effectively sort the debris that they transport; the finer, lighter material moves more rapidly than the coarser and heavier material. Essentially, the stream load contains three fractions, as discussed below. See also Figure 16-6.

1. Some minerals, largely salts, are dissolved in the water itself and are carried invisibly in solution as the *dissolved load*.
2. Very fine particles of clay and silt are carried in suspension, moving along with the flow of water without ever touching the stream bed. These tiny particles have a very slow settling velocity, even in still water (fine clay may require as much as a year to sink 100 feet in perfectly quiet water), so in any downstream flow this *suspended load* will continue to move.
3. Sand, gravel, and larger rock fragments constitute the *bedload*. The smaller particles are moved along with the general streamflow in a series of jumps or bounces, by *saltation*. Coarser pieces are shifted by *traction*, rolling or sliding along the stream bed.

The total load capacity of a stream varies tremendously from time to time, adapting to fluctuations in volume and velocity of flow. It is difficult to overemphasize the significance of the greatly expanded capacity of a stream to transport material at flood time. As a general rule, however, it can be noted that the greatest bulk of material is transported as suspended load, and that the dissolved load usually exceeds the bedload in total weight.

Deposition Whatever is picked up must eventually be set down, which means, from the standpoint of topographic development, that erosion is inevitably followed

Figure 16-6 A stream moves its load in various ways. The dissolved and suspended fractions are carried in the general flow. The bedload is moved by saltation and traction.

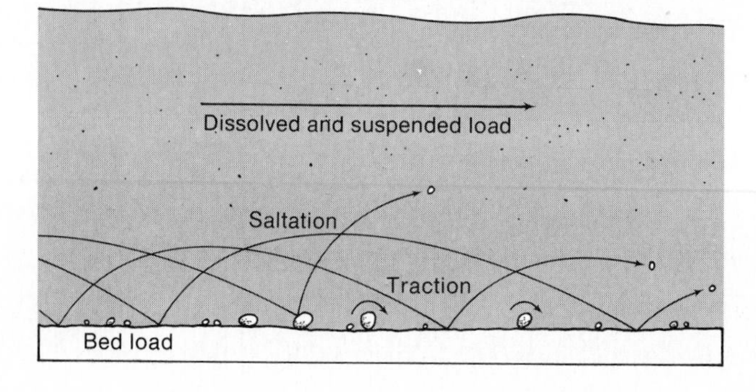

by deposition. Moving water, whether overland flow or streamflow, carries its load downslope or downvalley toward an ultimate destination either in an ocean or in some basin of interior drainage. The movement of the load is normally accomplished in spasmodic fashion; debris is transported some distance, then dropped, only to be picked up later and carried further along. Water moving fast and in large quantity can carry its debris a great distance, but sooner or later deposition will take place in response to a diminution of either velocity or volume of flow.

Diminished flow is often the result of a change in gradient, but it may also be due to slowing caused by widening or bending of the channel. The former is represented by deposition at canyon mouths where streams issue from mountainous or hilly areas; the latter is typified by deposits on floodplains or on the inside of meander curves. Eventually, however, most waterborne debris is dumped by a moving body of water (a stream or river) into a quiet body of water (an ocean or lake), where the most massive deposits are developed.

The general term applied to stream-deposited debris is *alluvium*. It is characterized by a sorting of particles on the basis of size. Finer bits of silt, for example, can be carried greater distances by a smaller flow of water than can larger particles of gravel. Thus the separation of different sizes of alluvial material is sometimes crude and indistinct, but it can also be clear-cut and precise.

Time and the River

Most of us have, on one or more occasions, stood on a bridge or on the brink of a canyon looking at a small stream far below and wondered how that meager flow could have carved the massive gorge that we see spread beneath us. On such occasions, we might be tempted to discount the doctrine of uniformitarianism and consider some brief and catastrophic origin as a more logical linking of what we can see with what we can imagine. Such reasoning, however, is false, for the uniformitarian concept easily encompasses tiny streams in extensive valleys by invoking two significant geomorphic principles:

1. The extraordinary length of geologic time, and
2. The remarkably effective work of floods.

We have previously discussed the incredible expanse of time that is involved in the evolution of topography. This vast temporal sweep provides the opportunity for countless repetitions of an action, and the repetitive movement of even a tiny flow of water can wear away the strongest rock.

Of perhaps equal importance is the fact that water does not always occur in tiny flows. Most streams have very erratic regimes; i.e., they show great fluctuations in their *discharge*, or volume of flow. Most of the world's streams carry a relatively small amount of water during most of the year and a relatively large volume during floodtime. In other words, their "normal" flow involves considerably less water than their flood flow. For some streams, flood flow occurs for several weeks or even months; but for the vast majority, the duration of high-water flow is much more restricted. In many cases, the "wet season" may encompass only one or two days out of the year. Yet the amount of denudation that can be accomplished during high-water flow is supremely greater than that done during periods of normal discharge. See Figure 16-7. The epic work of streams—the carving of great valleys, the forming of vast floodplains—is primarily accomplished by flood flows.

Figure 16-7 Much of the erosive work of streams is accomplished during flood flow. These two photos of the Potomac River are taken from the same bridge, a few months apart. [R. S. Sigafoos, U.S. Geological Survey photo.]

Slope Modification

Slope is the basic element of landforms. In the final analysis, topographic study focusses on the shapes and lengths of slopes and how these change through time. Slopes come in all shapes and sizes: steep, gentle, concave, convex, irregular, small, extensive, etc. Their inclinations and extents are modified by mass wasting, erosion, and deposition. Most slopes, however, are erosional in origin, which is to say that erosional processes are primarily responsible for their orientation and arrangement. Streamflow tends to incise the stream bed vertically, usually lengthening and steepening slopes. Mass wasting and overland flow transfer rock debris downhill and into valley-bottom streams to be carried away, which tends to make the slopes less steep as the valley is widened. In the long run, the slopes are ever-changing in shape and length, and our understanding of the topography is predicated largely on how well we can comprehend the nature, extent, and causes of these changes.

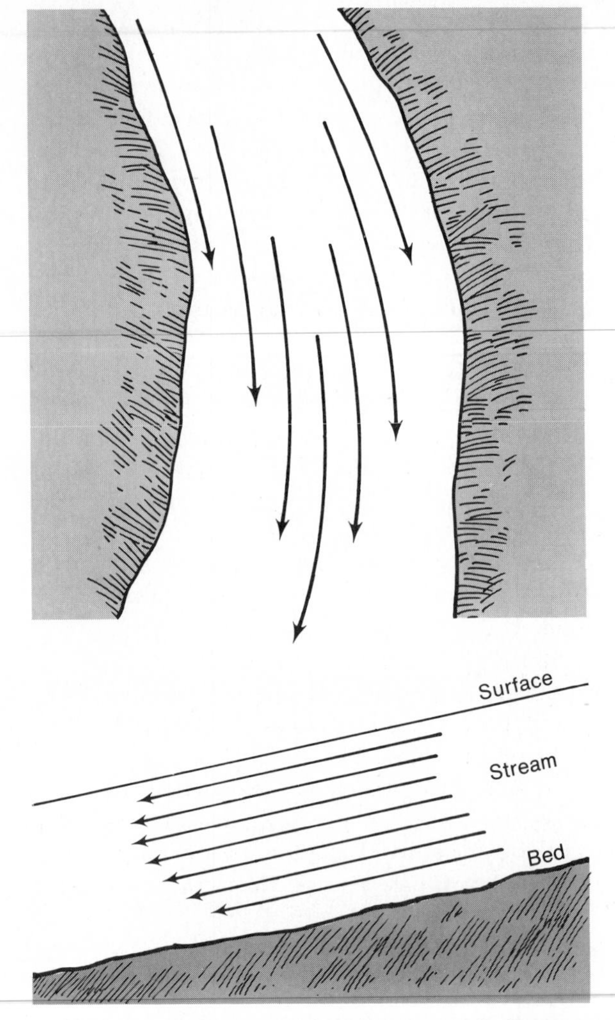

Figure 16-8 Friction retards velocity near the bottom and sides of a stream. The most rapid flow is near the center and slightly below the surface of the water, as shown in these hypothetical surface and cross-section views.

Stream Channels

Overland flow or surface runoff is a relatively simple process. It is affected by many factors, such as the nature of the rainfall, the vegetative cover, the character of the surface, and the shape of the slope, but its general characteristics are straightforward and relatively easily understood. Streamflow, on the other hand, is much more complicated, in part because streams represent not only a process of denudation but also an element of topography—an active force as well as an object of study.

Channel Flow

A basic characteristic of streamflow, and one that further distinguishes it from the randomness of overland flow, is that it is normally confined to channels, which gives it a three-dimensional nature with scope for considerable complexity. In any channel with even a slight gradient, gravitational pull overcomes frictional inertia to move the water down-channel. Except under very unusual circumstances, however, this movement is not straight and smooth and regular. Rather it tends to be unsystematic and irregular, with many directional components involved other than a simple down-channel flow, and with varying velocities in different parts of the channel.

A principal cause of this irregularity of flow is the retarding effect of friction along the bottom and sides of the channel, which causes the water to move slowest there and fastest in the center of the stream. See Figure 16-8. The amount of friction is determined by the width and depth of the channel and the roughness of its surface. A narrow, shallow channel with a rough bottom will have a much greater retarding effect on streamflow than a wide, deep, and smooth-bottomed one. One effect of frictional retardation is to use up much of the stream's energy, decreasing the amount that is available for erosion and transportation.

Turbulence

Another partial result of friction is the development of *turbulence* in the streamflow. Turbulence is also generated by internal shearing stresses between currents within the flow and by surface irregularities in the channel bed and banks. Stream velocity is also contributory to the development of turbulence; faster streams are more turbulent than slow-moving ones. The nature of turbulent flow is such that the general downstream movement is interrupted by continuous irregularities in direction and speed, producing momentary currents that can move in any direction, including upward. See Figure 16-9. Eddies and whirlpools are conspicuous examples of turbulence, as is the roiling whitewater of rapids. Even streams that

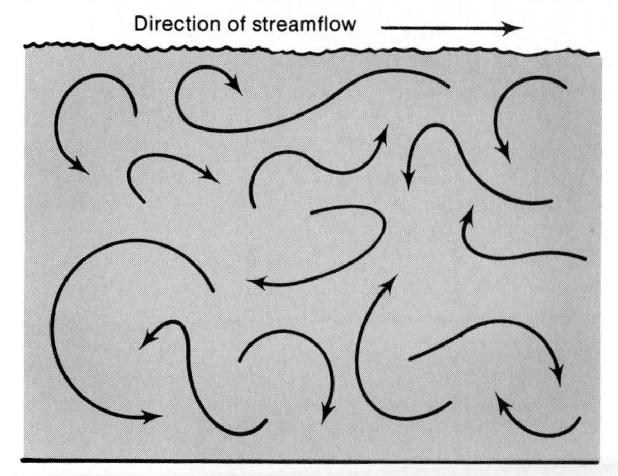

Figure 16-9 Turbulent flow is characterized by movement in all directions, which allows sediment to be picked up and carried within the flow.

appear very placid and smooth on the surface, however, are often turbulent at lower levels.

Turbulent flow creates a great deal of frictional stress as the numerous internal currents interfere with one another. This dissipates much of the stream's energy, decreasing the amount that is available to erode the channel and transport sediment. On the other hand, turbulence contributes to erosion by creating differential flow patterns that pry and lift rock material from the streambed. See Figure 16-10.

Channel Changes

Nearly every stream continually rearranges its sediment by *scouring* and *filling*, in response to variations in velocity and volume of flow. During high-water periods of fast, voluminous flow, the stream is able to scour its bottom by detaching particles from its bed and shifting most or all sediment downstream. Conversely, during low-water periods, the flow is slowed and sediment is more likely to settle to the bottom, which results in filling or aggrading the channel, thereby actually raising its bed. Thus in the course of time, a stream can alter the shape of its channel by fluctuations between scouring and filling. See Figure 16-11.

The irregularity of streamflow is manifested in nature in various ways, but perhaps most conspicuously by the variation in channel patterns. If the streamflow were smooth and regular, one might expect that stream channels would be straight and direct. However, few natural stream channels are straight and uniform for any appreciable distance. Instead, they wind about to a greater or lesser extent, sometimes developing remarkable sinuosity. In some instances, this is a response to the underlying geologic structure, but even in areas of perfectly uniform structure, such as a massive granitic batholith or horizontal sedimentary strata or a vast alluvial plain, stream channels normally exhibit irregular curving patterns. The mechanics of this winding habit are not fully

Figure 16-10 A turbulent river, such as the Thompson in central British Columbia, can accomplish considerable erosion in a relatively short time. [TLM photo.]

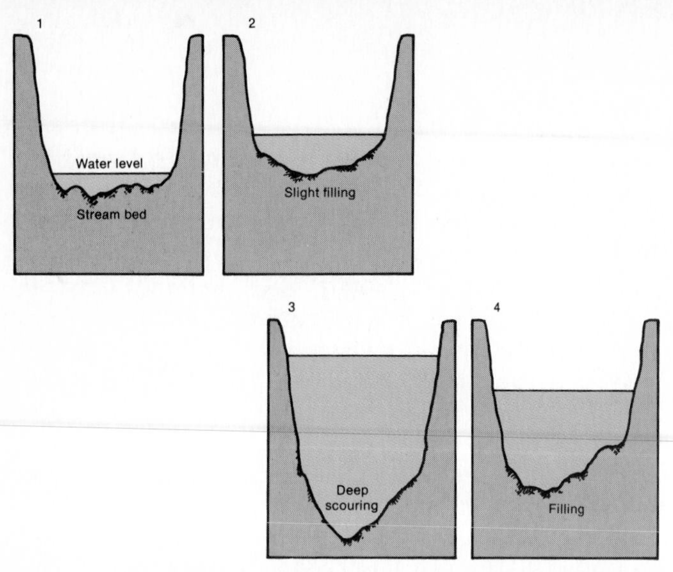

Figure 16-11 Schematic illustration of changing channel depth and shape during a flood: (1) Low water flow, prior to flooding. (2) As the volume of streamflow increases, the bed is raised slightly by filling. (3) Flood flow significantly deepens the channel by scouring. (4) As the flood recedes, considerable filling raises the channel bed again.

understood by geomorphologists, and are beyond our concern here, although we will discuss some aspects later in this chapter.

Two types of channel patterns are common and conspicuous: meandering channels and braided channels. See Figure 16-12. *Meandering streams* exhibit an extraordinarily intricate pattern of smooth curves in which the lateral and up-valley components are often more extensive than the down-valley components. *Braided streams* consist of a multiplicity of interwoven and interconnected shallow channels separated by low islands of sand, gravel, and other loose debris.

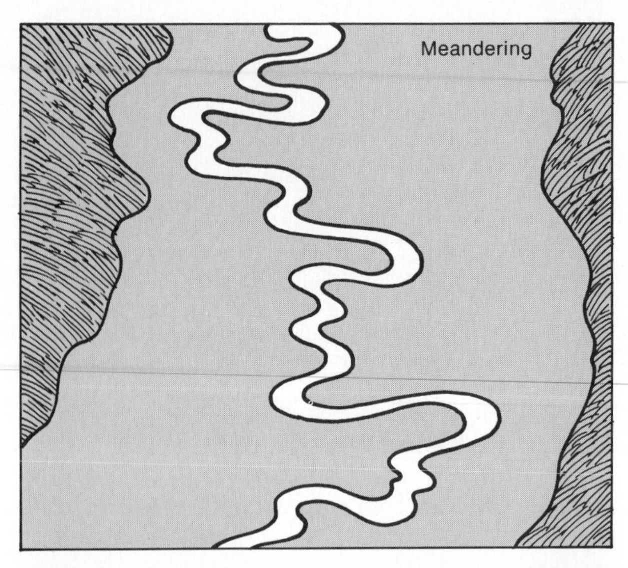

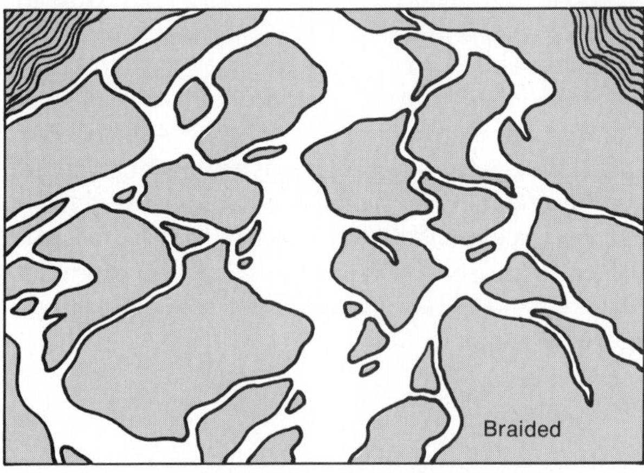

Figure 16-12 A meandering stream normally has a single principal channel that follows a very circuitous path; a braided stream has many interconnected channels.

Stream Systems

Streams and their valleys comprise a very prominent aspect of the physical landscape. Over most of the continental surfaces, they dissect the land in a myriad of patterns, each developed in response to the same universal laws, yet each with a uniqueness of its own. However, there are some important threads of similarity, and the study of individual streams can produce additional generalities to aid in our understanding of topographic development.

Drainage Basins

A *drainage basin*, also sometimes called a *watershed*, is an area that contributes overland flow and ground water to a specific stream. In other words, it consists of that stream's valley bottom, valley sides, and those portions of the surrounding interfluves that drain toward the valley. Conceptually, the drainage basin is bounded in all directions except down-valley by a *drainage divide*, which is the line of separation between runoff that descends in the direction of the drainage basin in question and that which goes toward an adjacent basin. See Figure

Figure 16-13 The drainage basins of adjacent streams are separated by drainage divides, which are located on the crests of the surrounding interfluves. In this drawing, the drainage divide for stream B is shown by a dashed line.

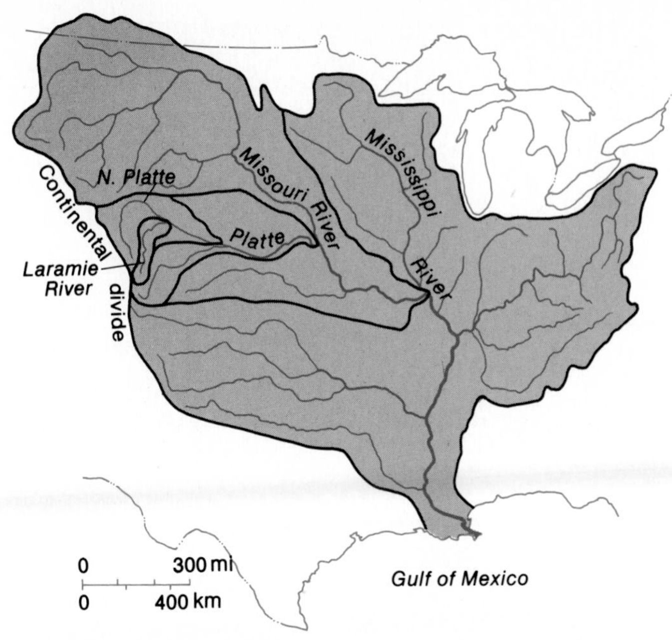

Figure 16-14 A nested hierarchy of drainage basins: The Laramie River flows into the North Platte. The North Platte drains into the Platte. The Platte is a tributary of the Missouri. The Missouri flows into the Mississippi. The Mississippi empties into the Gulf of Mexico. Rivers are shown in solid lines; the margins of the relevant drainage basins are shown in dashed lines.

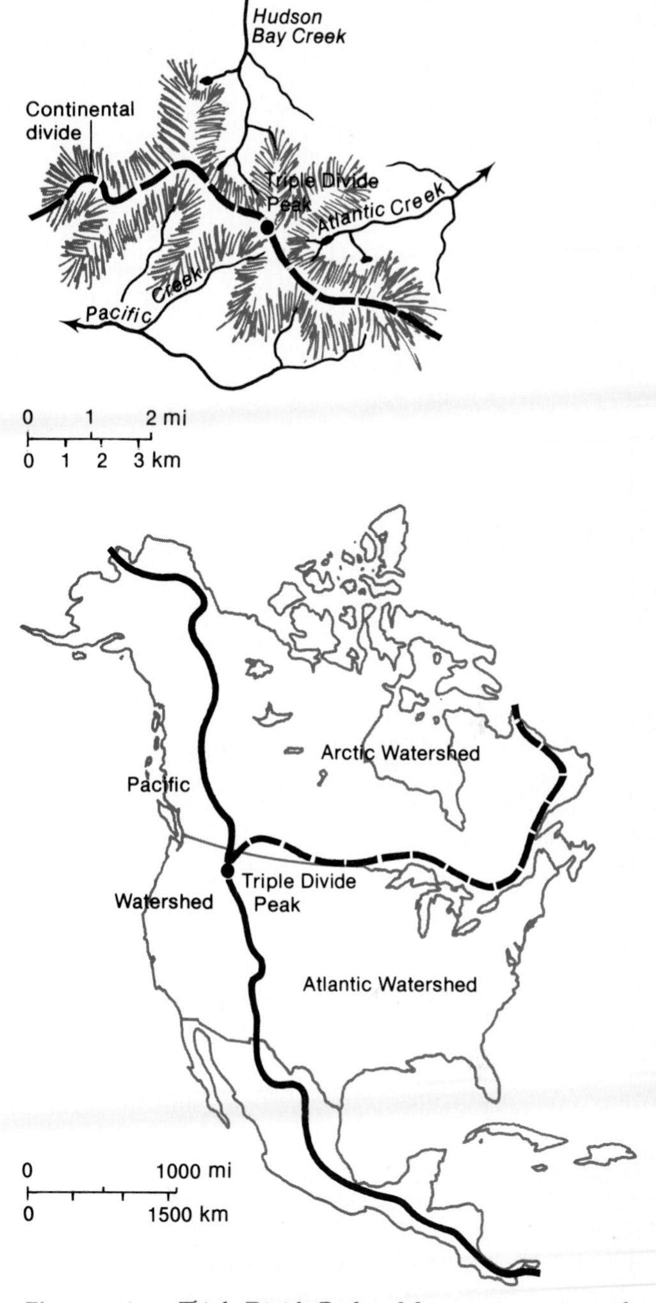

Figure 16-15 Triple Divide Peak in Montana is a unique place. Nowhere else in North America do the three continental drainage divides (Pacific, Atlantic, Arctic) meet at a single point.

16-13. Thus every stream of every size has its own particular drainage basin, but for practical purposes, the term is often reserved for major streams. The drainage basin of a principal river encompasses the smaller drainage basins of all its tributaries; consequently, larger basins include a hierarchy of smaller tributary basins. See Figures 16-14 and 16-15. The complex of streams within a drainage basin is referred to as the *drainage net.*

Stream Orders

In every drainage net, small streams join or come together to form successively larger ones. Little streams join bigger streams, and small valleys join more extensive ones. This relationship, although variable in detail, holds true for drainage basins of any size or extent. This systematic characteristic makes it possible to recognize a natural organization within a drainage net, and the concept of stream order has been devised to describe the arrangement. See Figure 16-16. A first-order stream is the smallest unit in the system and is thus conceptualized as a stream without tributaries because it represents the smallest tributary in the net. Where two first-order streams unite, a second-order stream is formed. At the confluence of two second-order streams, a third-order stream begins, and this uniting principle applies through successively higher orders in the hierarchy. The joining of a lower-order stream with a higher-order stream does not increase the order below that junction; for example, the confluence of a first-order stream with a second-order stream does not produce a third-order stream. A third-order stream is formed only by the joining of two second-order streams.

The concept of stream order is more than simply a numbers game, as several significant relationships are involved. See Figure 16-17. In a well-developed drainage system, for example, one can predict with some certainty that first-order streams and valleys will be more numerous than all others combined, and that each succeedingly higher order will be represented by significantly fewer streams. Other predictable relationships include that: (1)

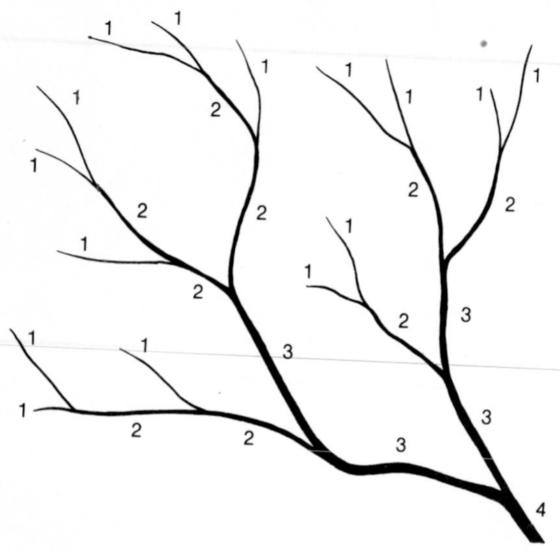

Figure 16-16 Stream orders. The branching components of a stream and its tributaries can be classified into a hierarchy of segments, ranging from smallest to largest (in this case, from 1 to 4).

the average length of streams increases regularly with increased order, (2) the average watershed area drained by streams increases regularly with increased order, and (3) the average gradient of streams decreases with increased order. Although all these facts and relationships are significant, perhaps the most important principle here is the realization that most drainage nets represent remarkably systematic and efficient natural developments.

Structural Relations

Despite the regularity of some aspects of drainage systems, many factors impinge on stream development, imparting particular characteristics that are sometimes predictable and sometimes not. Perhaps the most important of these impinging factors is the geologic/topographic structure over which or through which the stream must make its way and carve its valley. Each stream faces particular structural obstacles and impediments as it seeks the "path of least resistance" in its descending course to the sea. These obstacles sometimes are simple and easily accommodated, as in areas of flattish terrain and soft rocks; but in some cases, the rocks are resistant and the relief is formidable, necessitating both more time and greater complexity in the establishment of drainage channels.

There are hundreds of millions of streams in the world, and each channel (and valley) has its own particular characteristics. Nevertheless, the extensive replication of situations enables us to generalize about conditions and to classify streams on the basis of these generalizations. Streams may be classified by several widely accepted structural categories; this enables us to use brief terms to describe important developmental relationships. Chief among these categories are the following:

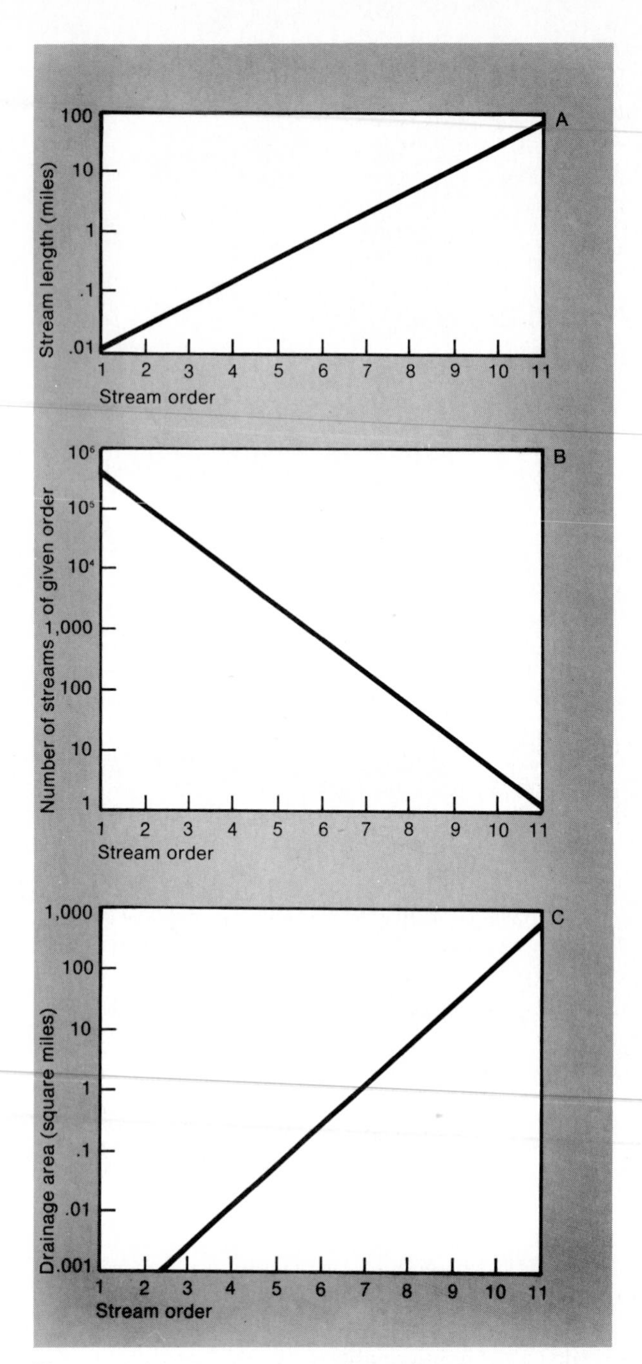

Figure 16-17 Some typical relationships between stream order and other factors: (a) First-order streams are the shortest. (b) First-order streams are the most numerous. (c) First-order streams have the smallest drainage areas.

Consequent streams The simplest and most common relationship between underlying structure and channel development is one in which the stream follows the initial slope of the land. Such a consequent stream normally is the first to develop on newly uplifted land, and many streams remain consequent throughout their evolutionary development. See Figure 16-18.

Subsequent streams Streams that develop along zones of structural weakness are termed *subsequent streams*. They

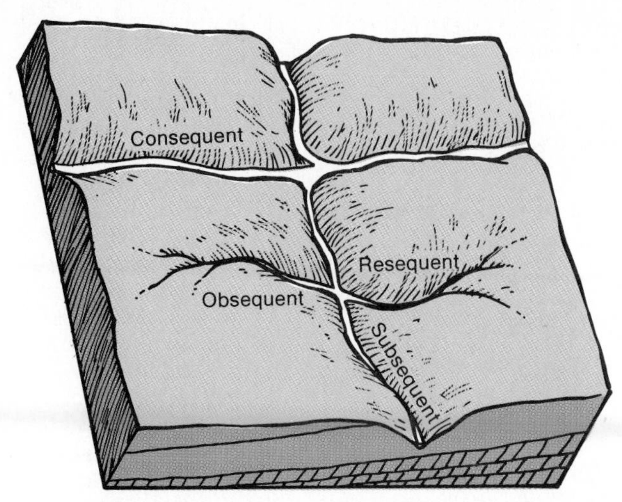

Figure 16-18 Schematic relationship of stream and structure. The consequent stream flows right to left across the diagram.

may excavate their channels along an outcrop of weak bedrock, or perhaps follow a fault zone or a master joint. Subsequent streams often trend at right angles to other drainage channels.

Obsequent streams Obsequent streams commonly flow in the opposite direction of consequents; they are often short tributaries of subsequents. Typically they develop on the downslope valley side of a subsequent.

Resequent streams Resequent streams flow in the same direction as consequents, but they develop on slopes that were formed later. They are found typically on the upslope sides of subsequent valleys.

Antecedent streams Sometimes an established drainage system will be interrupted by an uplift of land that is so slow that the stream is able to maintain its previously established course by downward erosion. In many cases, this results in a deep gorge carved through hills or mountains. As the stream antedates the existence of the uplift, it is called an *antecedent stream.* See Figure 16-19.

Superimposed streams A superimposed drainage system is one originally established upon a higher sequence of land that has been entirely or largely eroded away, so that the original drainage pattern becomes incised into an underlying sequence of rocks of quite different structure. The result may be a drainage system that bears no relation at all to the present surface structure.

Many other categories of stream types are based on structural relationships, but those identified above are the most widely recognized. In addition, many streams fit into no clearly defined structural category.

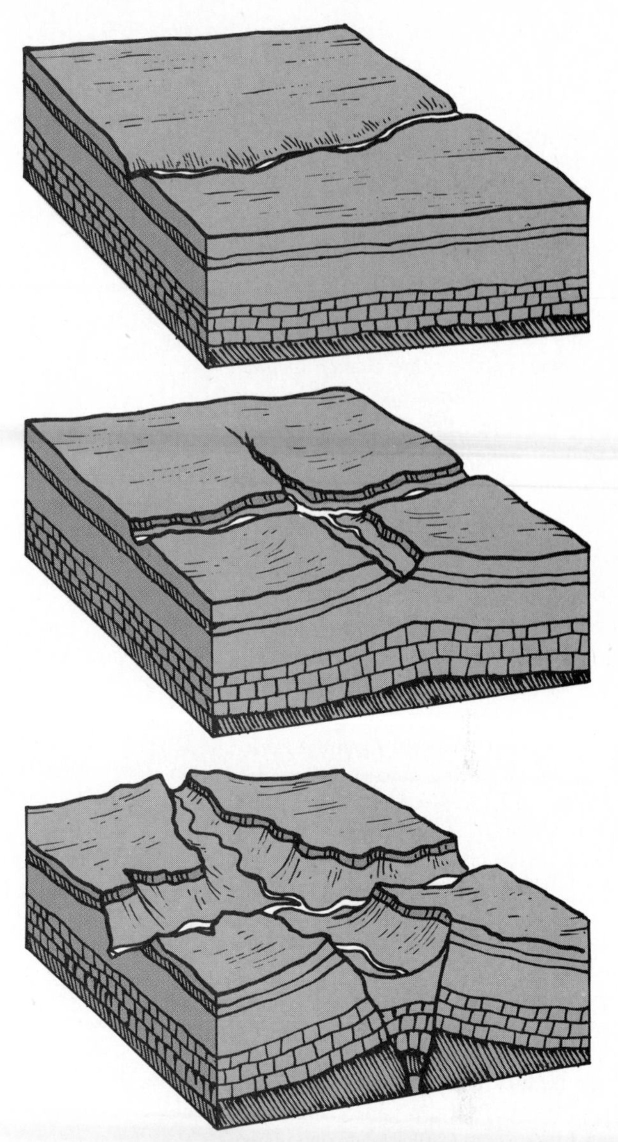

Figure 16-19 Hypothetical development of an antecedent stream. In the top diagram, the stream has an established course, flowing from right to left. In the middle diagram, anticlinal uplift is taking place, but the stream is powerful enough to maintain its original course. In the bottom diagram, uplift and faulting have increased the relief considerably, but the stream has been able to maintain its original course by downcutting.

Permanence of Flow

One tends to think of a river as being a permanent body of water, but in actual fact many of the world's streams do not flow year-round. In humid regions, the large streams and most tributaries are *permanent*, or *perennial*; but in less well-watered parts of the world, many of the major streams and most tributaries carry water only part of the time, during "the wet season" or during and immediately after rains. These impermanent flows are called *intermittent* or *ephemeral* streams. In desert areas, virtually all streams may be intermittent (see Figure 16-20), with

Most streams respond directly and conspicuously to structural controls, which is to say that their courses are distinctly guided and shaped by the nature and arrangement of the underlying rock. Many factors impinge on the determination of where channelled flow will occur, but the most significant is usually the differential resistance of the bedrock.

Although streams change in pattern and flow characteristics, they may persist through eons of time, outlasting mountain ranges and other topographic assemblages that are more temporary occupants of continental surfaces. Thus sometimes a direct relationship can be seen between the location of a stream and the contemporary structure of the land over which it flows, although sometimes it is necessary to delve into the geomorphic history of a region before the location of a drainage channel can be comprehended. We will examine here two widely differing patterns of regional drainage, as examples of the variable influence of structural controls on the location of streams.

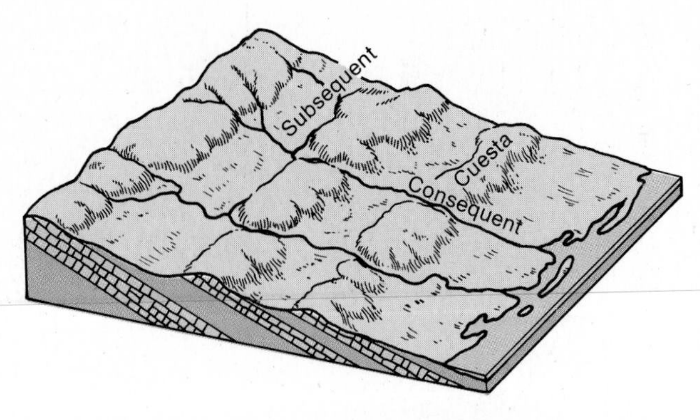

Schematic diagram of a portion of the belted coastal plain of Texas. The more resistant layers of the gently dipping sedimentary strata produce linear cuestas roughly parallel with the coastline. The long consequent streams cut through the cuestas. Tributary subsequent streams join them at right-angle junctions.

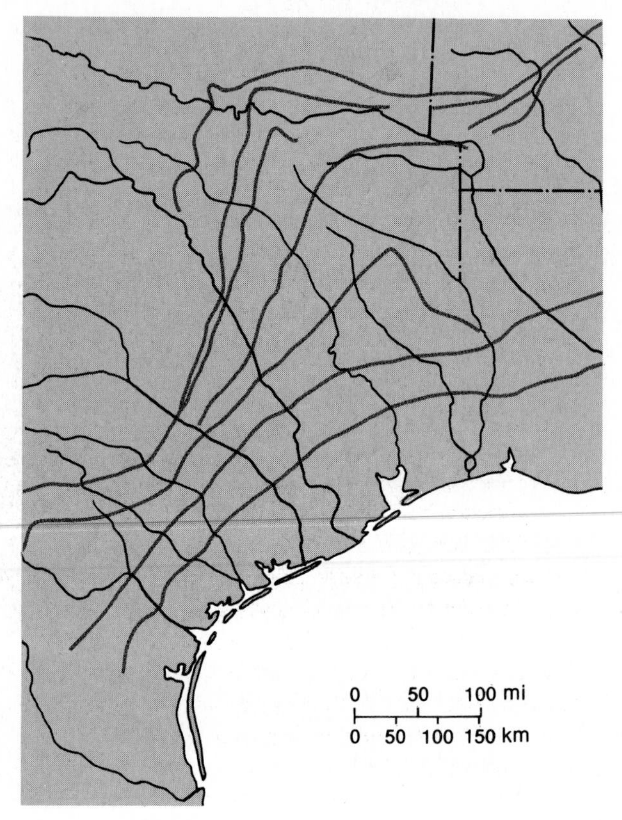

The belted coastal plain of Texas. A half-dozen major cuestas roughly parallel the Gulf coastline.

The Belted Coastal Plain of Texas

One of the simplest patterns of structural control, on a regional basis, can be seen along the coastal plain of Texas and parts of adjacent states. The geologic structure here consists of an extensive series of thick sedimentary beds that have a monoclinal dip gently southeastward toward the Gulf of Mexico. The flattish surface of the coastal plain is inclined in the same direction but at a lower angle than that of the sedimentary strata, so that progressively older rocks are exposed with increasing distance from the coast.

No two geological formations are exactly equal in their resistance to weathering and erosion, with the result that the most resistant beds protrude slightly above the general level of the coastal plain as low ridges extending in the form of "belts" roughly parallel to the coastline. These ridges have long gentle slopes on their seaward sides, with much shorter, steeper slopes on the landward margins, producing landforms called *cuestas*, some of which persist for hundreds of miles.

The initial drainage that developed on this coastal plain consisted of a number of long, relatively straight, consequent streams that flowed sluggishly in generally parallel arrangement toward the Gulf. As the resistant rocks of the cuestas became more obtrusive through the general wasting of the land, short subsequent tributary streams became established on the softer rocks immediately inland from each cuesta. As is character-

the notable exception of those that flow into the desert, bringing their water from somewhere else (these are called *exotic* streams). In Australia, for example, which is the driest of the inhabited continents, some 90 percent of all streams are intermittent. Many channels in desert areas have no water flowing in them for years at a time.

Even in humid regions, many first- and second-order streams are only intermittent in flow. These are generally

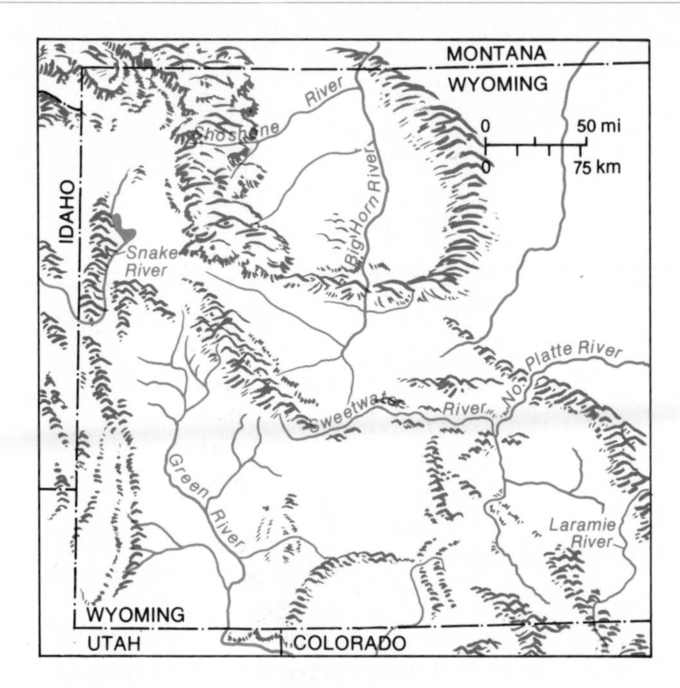

The rivers of Wyoming drain in all directions, and most of them have cut gorges directly through various mountain ranges in doing so.

istic of subsequent streams, these join the consequents at right angles. Obsequent and resequent tributaries have also developed, but the slopes are so gentle that these streams are usually short and insignificant.

The Curious Drainage Anomalies of Wyoming In contrast to the fairly simple and direct structural relationships discussed above is the remarkable complex of stream courses followed by the major rivers that drain the state of Wyoming. Almost without fail, the principal rivers of that state appear to fly in the face of fluvial logic by cutting through ranges of mountains and hills rather than flowing around them. On the accompanying map, the leading (but not the only) examples of this phenomenon are indicated. The easterly flowing Laramie, Sweetwater, and North Platte; the northerly flowing Big Horn and Shoshone; the westerly flowing Snake; and the southerly flowing Green rivers all plunge through mountains in deep and precipitous gorges (twice for the Big Horn and three times for the North Platte), although it would appear that much easier paths would be available for them in some other direction.

Why would rivers cut courses through mountains rather than going around them? How could the drainage system be so out of harmony with the principal topographic features? The answer, of course, is not capricious; rivers are not creatures of whim but follow immutable physical principles as they pursue what is always the path of least resistance in their search for the sea. The explanation for the seemingly anomalous courses of Wyoming's rivers is to be found in the region's geomorphic history, which is only now being slowly unravelled.

The earliest, and simplest, explanation for the rivers' apparent disregard of topographic barriers invoked the principle of *antecedence*; i.e., it was assumed that the mountains were uplifted after the rivers were already established, and the uplifts were sufficiently slow to permit the rivers to maintain their courses by downcutting. Further study, however, determined that the mountains were generally older than first thought and the intervening basins had previously been more deeply filled with sediments than at present. Thus *superimposition* became the favored theory, with the view that previously established rivers flowed at an older and higher level, eventually cutting down into the older rocks of the buried mountains and incising their courses without regard to the mountainous structures.

However, more detailed study indicated a still more complicated origin for the gorges, that of *anteposition*, in which the stream courses were superimposed while the mountains were low and then deepened in antecedent fashion when uplift was subsequently resumed. In this explanation, uplift impedes or dams the drainage for awhile, but eventual overflow permits accelerated downcutting. The effect of this condition is superimposition upstream from the mountain barrier and antecedence downstream from it. In this case, then, structural control can be seen to be only partial, indirect, and incomplete.

short streams with relatively steep gradients and small watersheds. If rain is not frequent, or snowmelt not continuously available, these low-order streams simply run out of water. Higher-order streams in the same regions are likely to have permanent flow because of their larger drainage areas and because previous rainfall or snowmelt that sank into the ground can emerge in the valleys as groundwater runoff long after the rains have ceased.

Figure 16-20 Most desert streams are intermittent, carrying water only rarely. This sandy stream bed is located in northwestern New South Wales, Australia. [TLM photo.]

Patterns of Streams

Thus far our attention has been focussed primarily on individual streams. If we broaden our scale of analysis, however, we can recognize that stream systems frequently occur in a fairly ordered arrangement, forming more or less conspicuous patterns of drainage in the landscape. These patterns develop largely in response to underlying structure and the slope of the land. The geologic/topographic structure often can be deduced from the drainage pattern, and conversely, the drainage pattern can often be predicted from the structure.

The most common drainage pattern over the world is a treelike, branching one that is called *dendritic*. It consists of a random merging of streams, with tributaries joining larger streams irregularly, but always at acute angles (less than 90°). The pattern roughly resembles branches of a tree or veins of a leaf. Dendritic patterns characteristically are found in areas of relatively uniform geologic structure; i.e., the underlying material is of relatively uniform resistance to erosion. In regard to pattern, the structural relationship is negative, which is to say that the underlying structure does not control the evolution of the drainage pattern.

Other recognizable patterns of drainage are more closely related to structural controls. See Figure 16-21. A *trellis* pattern is usually developed on alternating bands

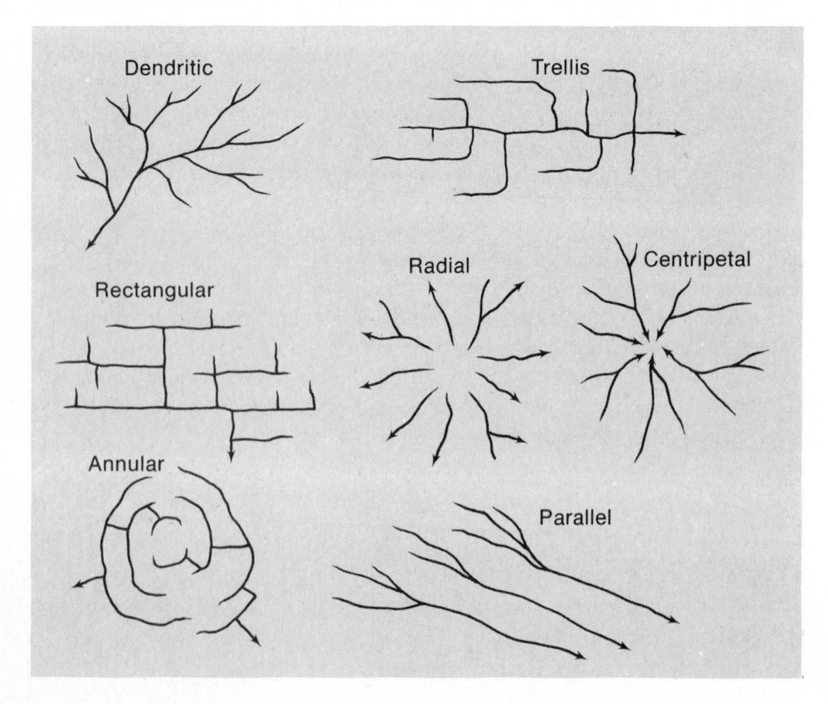

Figure 16-21 Schematic illustration of some common drainage patterns.

of hard and soft strata, with long parallel subsequent streams linked by short right-angled segments and joined by short obsequent and resequent tributaries. Related to the trellis in appearance but developed on a different structure is the *rectangular* pattern, in which the streams are essentially all subsequents that follow sets of faults and/or joints, with prominent right-angular relationships. This pattern is most commonly developed on granitic bedrock with strong joint systems. In a *radial* pattern, consequent streams flow outward in all directions from an uplifted dome or a volcanic cone. A basin structure can cause streams to converge toward the center in a *centripetal* pattern. More complex is an *annular* pattern, which can develop either on a dome or in a basin where dissection has exposed alternating concentric bands of hard and soft rock. Here principal streams follow curving courses on the softer material, occasionally breaking through the harder layers in short, right-angled segments and being joined by short obsequent and resequent tributaries. A *parallel* pattern can emerge in areas of pronounced regional slope, particularly if the gradient is gentle, with long consequent streams flowing roughly parallel to one another.

These are the most common and most recognizable patterns of regional drainage. In some cases, they are found in only limited areas, but they often occur over very extensive portions of the land. However, in many parts of the world, none of these patterns is found. The drainage system may be so irregular or complex that no particular pattern is apparent.

The Shaping and Reshaping of Valleys

The work of running water in shaping the terrain is accomplished in part by overland flow on the interfluves, but of much greater magnitude is the denudation that is associated with streamflow in the valleys. The shaping of valleys and their almost continual modification through time produces a changing sequence of landforms, a changing balance between valleys and interfluves, and an ongoing dynamism in the configuration of most parts of the continental surfaces.

A stream excavates its own valley by eroding its channel bed, which is a major role in denudation. If only downcutting were involved, the resulting valley would be a narrow, steep-sided gorge. This sometimes occurs, but usually other factors emphasize valley widening as well. In any case, a lower limit to downcutting is imposed by sea level and is expressed in the concept of base level.

Base level is an imaginary surface extending underneath the continents from sea level at the coasts. The ultimate destination of water carried in natural drainage systems is the sea, which means that fluvial downcutting below sea level is impossible, with a few unusual exceptions. The imaginary surface of base level, however, is not simply a horizontal extension of sea level; inland it is gently inclined at a gradient that would allow streams to maintain some flow under natural conditions. Sea level, then, is the *ultimate* base level, or lower limit of downcutting.

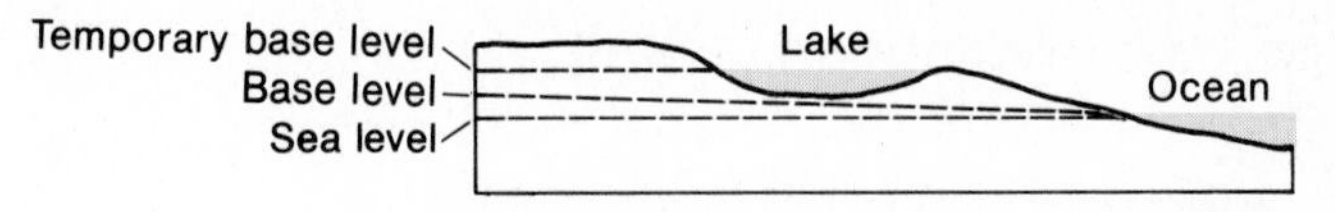

Figure 16-22. Comparison of sea level, base level, and temporary base level.

Also included in this concept is the idea of *local* or *temporary* base levels, which consist of lower limits of downcutting imposed on particular streams or sections of streams by specific structural or drainage conditions. See Figure 16-22. For example, no tributary can cut deeper than its level of confluence with the higher-order stream that it joins, so the level of their point of junction comprises a local and temporary base level for the tributary. Similarly, a lake normally serves as temporary base level for all streams that flow into it. Some valleys have been downfaulted to elevations below sea level (e.g., Death Valley in California), which produces a temporary base level that is actually lower than the ultimate base level. Over an extremely long period of time with no structural changes, it can be imagined that all local and temporary base levels will eventually be eliminated, and thus all streams will eventually be governed by ultimate base level.

The concept of base level is useful because it describes the ultimate conditions toward which streams work in constructing their own gradients or stream-bed slopes. However, there are many other factors involved in the total shaping of valleys by fluvial processes. For example, a stream can deepen its valley, but it can also widen or lengthen its valley. We will now turn our attention to the mechanics by which streams can accomplish these three ways of modifying the shapes of their valleys.

Valley Deepening

The process of valley deepening is simple and straightforward. Wherever it has either a relatively rapid velocity or a relatively large volume of flow, a stream will expend most of its energy in downcutting. This involves the hydraulic power of the moving water, the prying and lifting capabilities of turbulent flow, and the abrasive effect of the stream's bedload as it rolls and slides and bounces along the channels. Downcutting shows up most prominently in the upper reaches of a stream, where the gradient is usually steep and the valley narrow. The general effect of downcutting is to produce a deep valley with relatively steep sides and a V-shaped, cross-sectional profile.

Conspicuous regional drainage patterns occur in all parts of the world. They are easily recognizable on maps, thus providing significant clues to the regional geologic/topographic structure. Some prominent examples are discussed below.

Dendritic Pattern Dendritic patterns are more numerous than all others combined and can generally be found almost anywhere. Shown here is a small drainage basin in Wyoming.

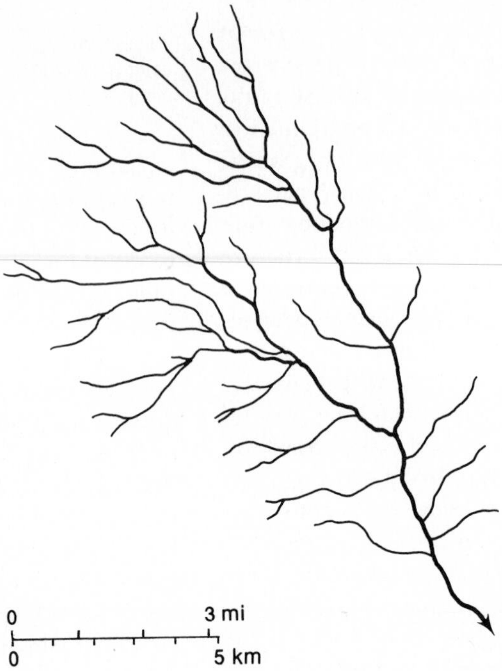

A dendritic drainage pattern, from the Pat O'Hara Mountain (Wyoming) topographic quadrangle.

Trellis Pattern Two regions of the United States are particularly noted for their trellis drainage patterns. Most extensive is the so-called ridge and valley section of the Appalachian Mountains, which extends in a northeast-southwest trend for more than 800 miles from southeastern New York across six intervening states to northern Alabama. The drainage pattern has developed in response to a tightly folded complex of ancient (Paleozoic) sedimentary strata, upon which has developed a world-famous series of roughly parallel ridges and valleys. The repetitious topographic development results primarily from differential erosion, with the configuration based on differences in the resistance of the bedrock. Parallel streams flow in the valleys between the ridges, with short right-angled connections here and there cutting through the ridges. Shown here is an area in Tennessee.

A similar structure in the Ouachita Mountains of Arkansas and Oklahoma has produced another extensive area of trellising. The marked contrast between trellis and dendritic patterns is shown dramatically by the map of the principal streams of West Virginia. The fold structures of the eastern part of the state produce trellising, whereas the nearly horizontal sedimentaries of the remainder of the state are characterized by dendritic patterns.

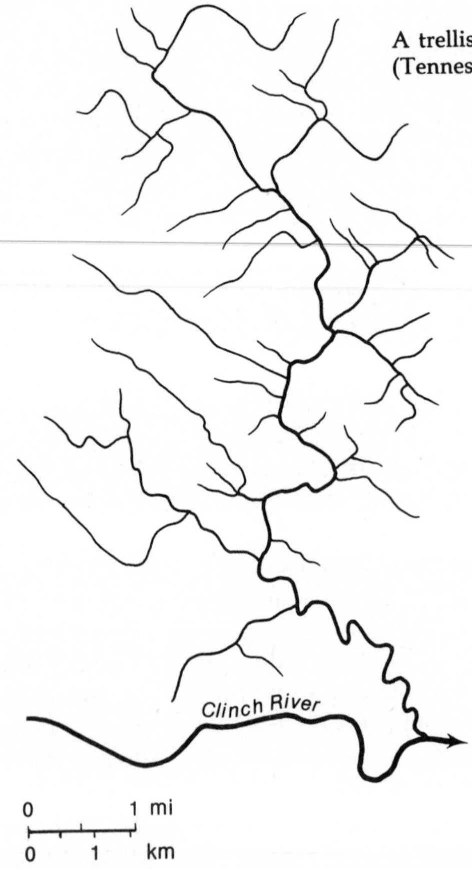

A trellis drainage pattern, from the Norris (Tennessee) topographic quadrangle.

Radial Pattern A radial pattern is usually found when consequent streams descend from some sort of concentric uplift, such as an isolated volcano. This example is from Mount Egmont on the North Island of New Zealand.

Centripetal Pattern A centripetal pattern is essentially the opposite of the radial; it is usually associated with the convergence of streams into a basin. Occasionally, however, centripetal drainage develops on a much grander scale. Shown here is the northeastern part of Australia, where rivers converge from hundreds of miles away toward the Gulf of Carpentaria, a structural basin that has been partly inundated by the sea.

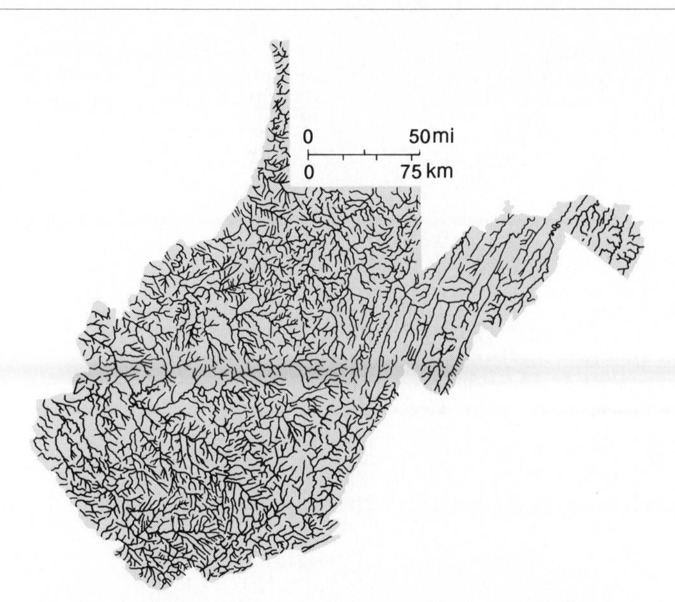

Drainage pattern contrasts in West Virginia. The trellis systems in the east are a response to parallel folding; the dendritic drainages in the west have developed because there are no prominent structural controls.

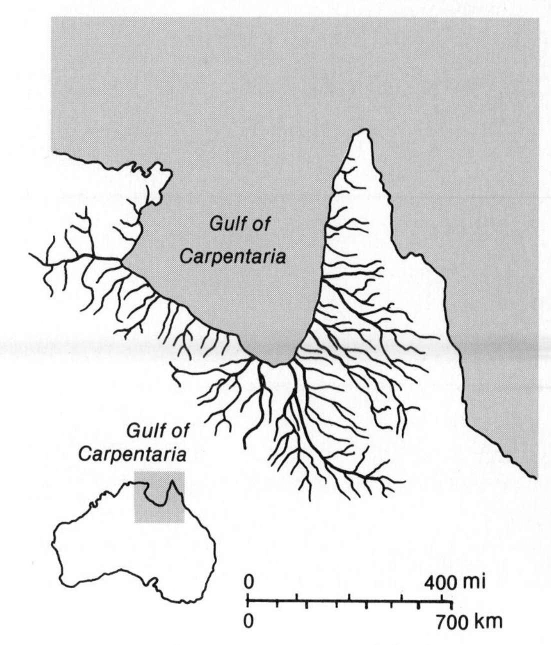

The centripetal drainage pattern of the region around the Gulf of Carpentaria in northeastern Australia.

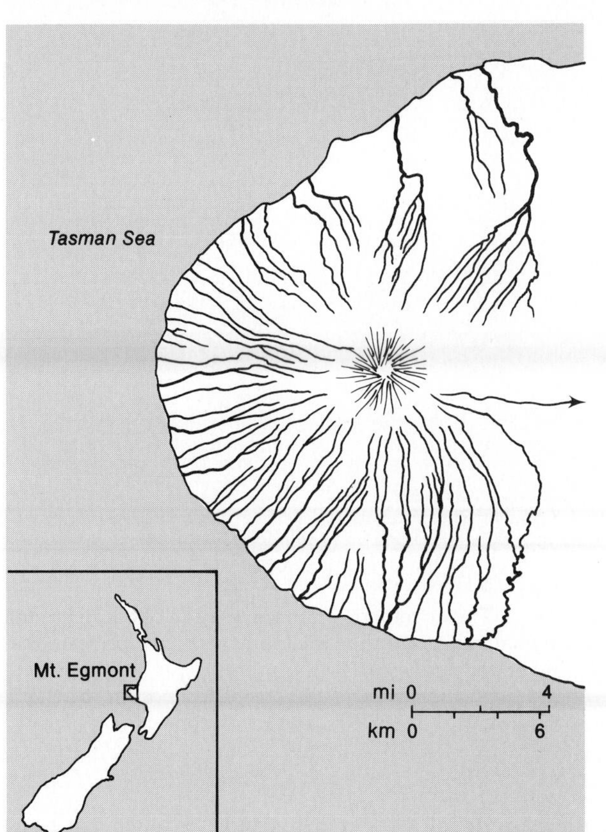

The extraordinary radial drainage pattern of Mt. Egmont in New Zealand.

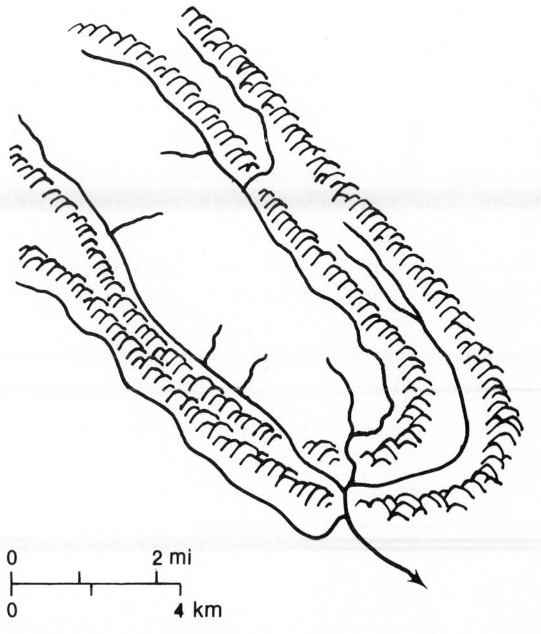

An annular drainage pattern, from the Maverick Spring (Wyoming) topographic quadrangle.

Annular Pattern The Maverick Spring Dome of Wyoming portrays a prominent example of an annular drainage pattern. This uplifted dome of ancient crystalline rocks was pushed up through a sedimentary overlay and has been deeply eroded, thus exposing crystallines in the higher part of the hills, with upturned more or less concentric sedimentary ridges (called *hogbacks*) around the margin. Major streams follow the curving valleys of the softer layers, with shorter segments breaking through the hogbacks at right angles.

Figure 16-23 A small waterfall in McDonald Creek, Glacier National Park, Montana. Knickpoints such as this usually are created by more resistant bedrock. [TLM photo.]

Waterfalls, rapids, and cascades are often found in valleys where downcutting is prominent. See Figure 16-23. They represent steeper sections of the channel gradient and cause faster and more turbulent flow, which in turn intensifies the erosive action. These irregularities in the channel profile are collectively termed *knickpoints*. They may originate in various ways but commonly are due to abrupt changes in the resistance of the underlying bedrock. The more resistant material inhibits downcutting, and as the water plunges over the waterfall or rapids with accelerated vigor, it tends to scour the channel above and along the knickpoint and fill the channel immediately downstream. This intensified action eventually wears away the harder material, but it is often a slow procedure in which the knickpoint migrates upstream with a successively lower profile until it finally disappears, and the channel gradient is smoothed. See Figure 16-24. The principle of upstream migration is important to a comprehension of fluvial erosion because it illustrates dramatically the manner in which the shaping of a valley often develops first in the lower reaches and then proceeds progressively upstream, even though the water obviously flows in the opposite direction.

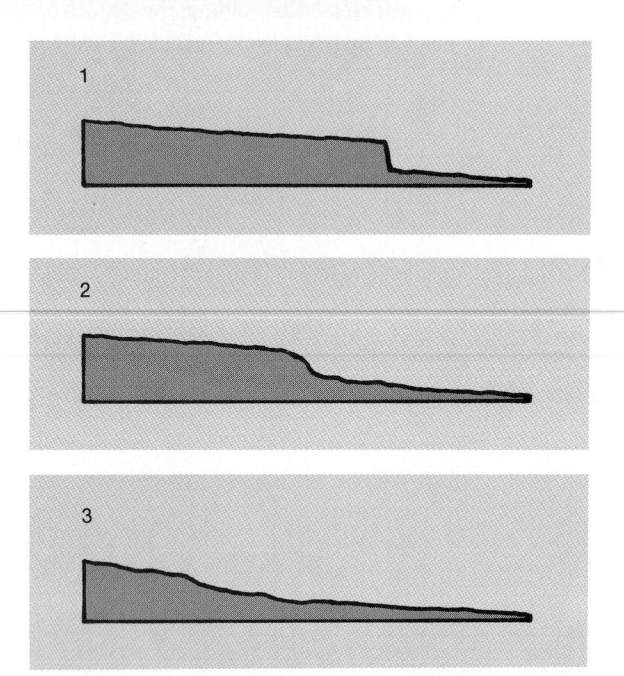

Figure 16-24 The development and migration of a knickpoint. In this schematic diagram the knickpoint retreats upstream, becoming lower and less steep in the process, due to intensified erosion and deposition around the steepened portion of the channel. Eventually (3) the knickpoint disappears completely.

Valley Widening

Where a stream gradient is steep and the channel is well above the local base level, downcutting is the dominant activity, and valley widening is likely to be slow. Even at this stage, however, some widening will take place as the combined action of weathering, mass wasting, and overland flow removes material from the valley sides. In the valley bottom, downcutting will diminish with the passage of time and eventually will virtually cease as the stream develops a gentle profile. The stream's energy is then increasingly diverted into a meandering, side-to-side flow pattern, the reasons for which are not yet fully understood. As the stream begins to waffle from side to side, *lateral erosion* is initiated. In essence, this means that the principal current of the stream swings laterally from one bank to the other, eroding where the velocity is greatest and depositing where it is least. The water

The most famous North American waterfalls is Niagara Falls. It is actually relatively low (about 175 feet or 57 m) and on a short river (about 35 miles or 56 km long), but it is a very impressive spectacle because an enormous amount of water (about 200,000 cubic feet or 17,000 cubic meters per second) cascades over it. Moreover, it is situated in the midst of a major drainage basin (the Great Lakes–St. Lawrence River system) and sits astride an international boundary. From the standpoint of physical geography, however, Niagara Falls is particularly significant because it demonstrates the immutable principle of knickpoint retreat on a massive scale and in a spectacular setting that is so easily available and so readily interpreted to the visiting public.

The entire natural drainage of the Great Lakes works its way toward the Atlantic Ocean through short connecting rivers from lake to lake. The Niagara River forms the connecting link between Lake Erie and Lake Ontario. As the contemporary drainage system became established in this area, following the last retreat of Pleistocene ice sheets, Lake Ontario was about 150 feet higher than its present level, and the Niagara River had no falls. However, an easterly outlet, the Mohawk Valley, developed for Lake Ontario, and the lake drained down to approximately its present level, exposing a prominent escarpment directly across the course of the river. The Niagara Escarpment is formed by a massive bed of resistant limestone that dips gently toward Lake Erie and is underlain by similarly dipping but softer strata of shale, sandstone, and limestone.

As the river pours over the escarpment, the swirling water undermines the hard limestone by erosion of the weaker beds beneath, leaving a lip of resistant rock projecting without support. Through the years, the lip has collapsed, with block after block of limestone tumbling into the gorge below. After each collapse, rapid undermining takes place again, leading to further collapse. In this fashion the falls has gradually retreated upstream, moving southward adistance of about 7 miles (11 km) from its original position along the trend of the escarpment, the retreat being marked by a deep gorge.

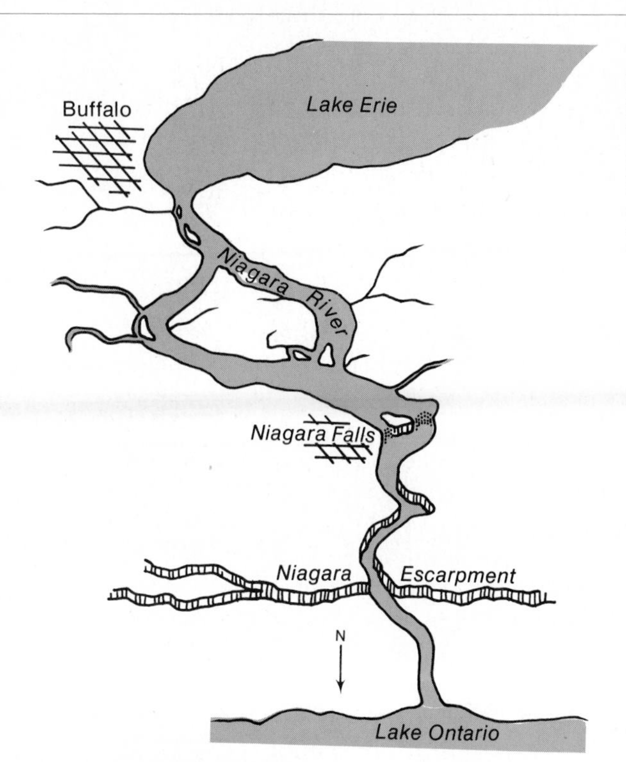

The situation of Niagara Falls. The falls was originally located where the Niagara River crosses the Niagara Escarpment, but it has now retreated upstream to its present location.

A cross section of Niagara Falls. The falls is "held up" by the massive layer of resistant limestone at the top. The weaker underlying strata are more easily eroded, which undermines the limestone, causing blocks to break off.

At present, the falls has been split around an island in the river, with most of the water descending over Horseshoe Falls, in Canada, and a much smaller amount spilling over the American Falls. The average rate of retreat of Horseshoe Falls is now about 4 feet (1.3 m) per year, whereas the smaller volume of flow over the American side causes an average retreat of only about 1 foot (30 cm) every two years. As the falls migrates upstream, it becomes progressively lower, which is the experience of almost all retreating knickpoints. When it has migrated 2 miles (3.2 km) further, in perhaps 6000 years, its height will be reduced to about 100 feet (30 m), and it will be entirely within the hard limestone layer. This will terminate the undercutting process, greatly slow the retreat, and gradually convert the falls into rapids. Barring crustal movement, however, even the rapids will eventually disappear. The ultimate destiny, then, for our most notable waterfalls is the same as that for any knickpoint—obliteration.

moves fastest on the outsides of curves, and there it undercuts the bank, whereas on the inside of a bend, there is likely to be the accumulation of a sandbar.

The current often shifts its position, so that the undercutting is not concentrated in just a few locations. Rather, over a long period of time, most or all parts of the valley sides will experience it. The undercutting encourages the slumping of material into the stream as the process of lateral erosion continues. See Figure 16-25. Throughout this period of widening of the valley floor, mass movement and overland flow continually aid in the wearing down of the valley sides, and similar activities along tributary streams also contribute to the general widening of the main valley.

The frequent shifting of stream meanders produces an increasingly broader, flattish valley floor, which is largely or completely covered with deposits of alluvium. At any given time, a stream is likely to occupy only a small portion of the flatland, although during periods of flood flow, the entire floor may be flooded, and the valley bottom is properly termed a *floodplain*. See Figure 16-26. The outer edge of the floodplain, on either side, is usually bounded by a clear-cut break in slope, marking the outer limit of lateral erosion and undercutting where the flat terrain abruptly changes to a relatively steep slope, or a line of *bluffs*. This process of valley widening and floodplain development can extend to an almost indefinite dimension; the floodplains of many of the world's largest rivers are so broad that a person standing on the bluffs at one side cannot see the bluffs on the other.

Figure 16-25 Valley widening is accomplished primarily by lateral erosion. The stream current shifts from side to side, eroding on the outside of curves and depositing on the inside. Continuation of this pattern produces an increasingly meandering channel. [After Robert E. Gabler et al., *Essentials of Physical Geography*, 2nd ed. (Philadelphia: Saunders College Publishing, 1982) p. 427. © 1982 by CBS College Publishing. By permission of CBS College Publishing.]

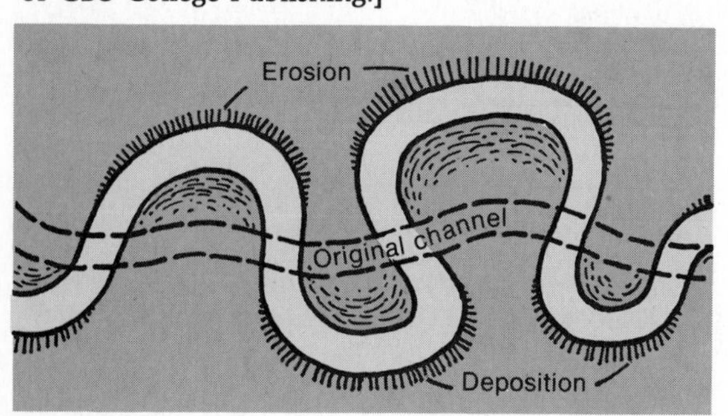

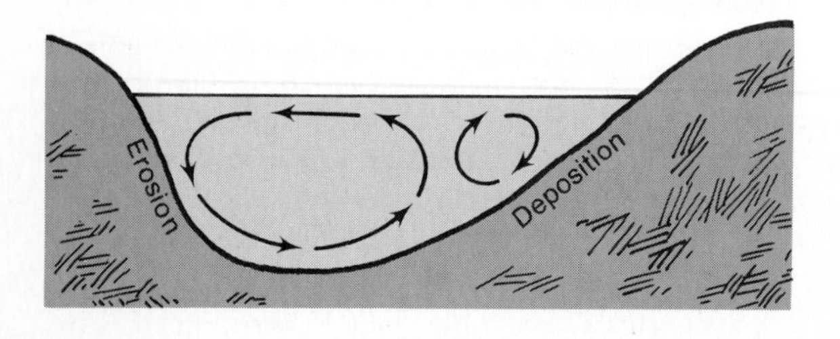

Figure 16-26 Looking east across the floodplain of the Illinois River near the town of Hardin. Our viewpoint is from the bluff on the western margin of the floodplain. In the distance we can see the bluff that marks the eastern edge of the floodplain, some 10 miles (16 km) away. [TLM photo.]

Valley Lengthening

The third dimension in the shaping of stream valleys is lengthening or extension. Whereas the concept of valley deepening is easy to grasp and the idea of valley widening is not too hard to comprehend, the notion of valley lengthening is sometimes difficult to conceptualize. Actually a stream may lengthen its valley in two quite different ways: (1) by *headward erosion* at the upper end or (2) by *delta formation* at the lower end. Both are commonplace occurrences, but the former is much more widespread and significant.

Headward Erosion No concept is more fundamental to an understanding of fluvial processes than that of headward erosion because it is at the root of comprehension of how rills, gullies, and valleys can be initiated and how they can be extended. At the outset we should recall a precept stated earlier in the chapter, that essentially all of the land surfaces of the Earth can be divided between interfluves and valleys. Overland flow descends from the interfluves into the valleys. The upper perimeter of a valley is the line where the somewhat gentler slope of an interfluve changes to the steeper slope of a valley side. Sheet flow from the interfluve surface drops more or less abruptly over this slope break, which tends to undercut the lip or rim of the perimeter, weakening it and often causing the collapse of a small amount of material. See Figure 16-27. This action can be aided and accelerated if the slope of the valley side has been oversteepened and made unstable by the incision and lowering of the stream in the headwater area.

The result of this action is a net decrease in the area of the interfluve and a commensurate increase in

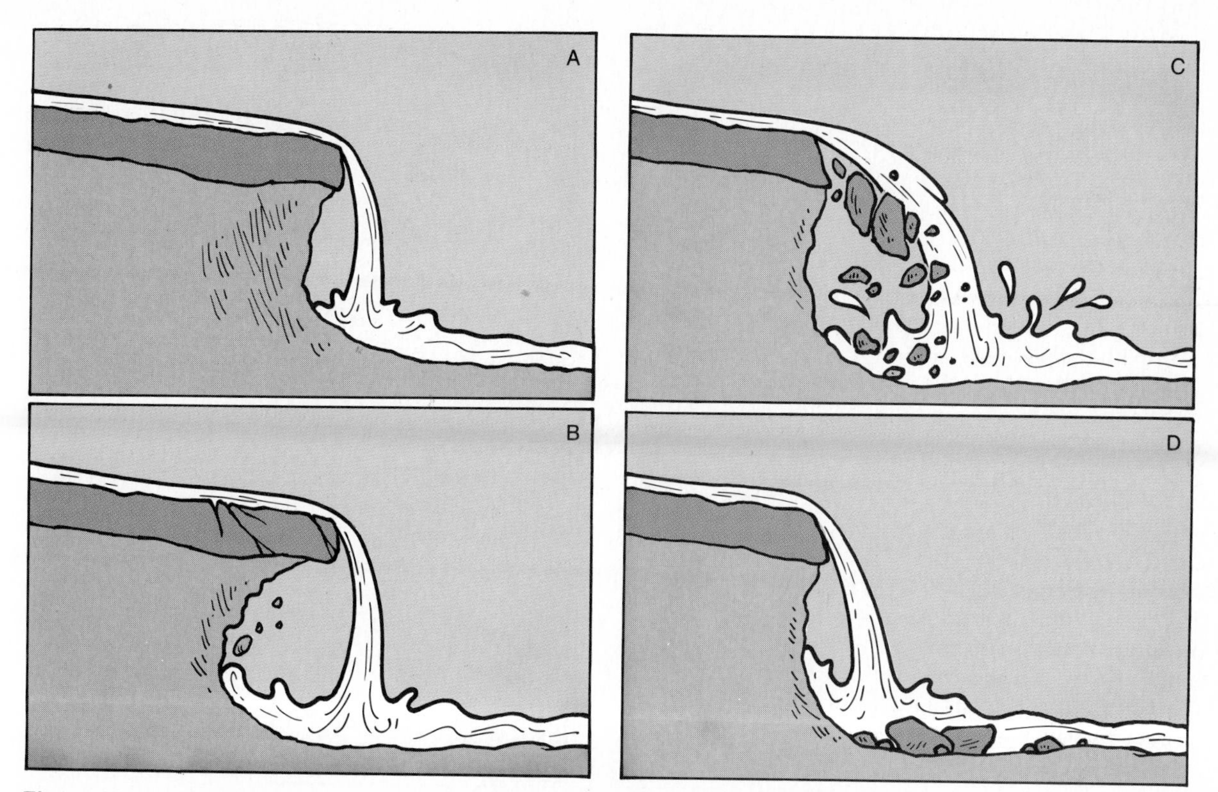

Figure 16-27 Headward erosion is accomplished by sheetflow pouring over the lip of a valley from an interfluve surface (A). The flow undercuts the lip (B) and causes collapse (C). This extends the valley headward at the expense of the interfluve (D). The height of the "waterfall" shown in this diagram is so small as to be measured in centimeters.

the area of the valley. As the overland flow of the interfluve becomes part of the channelled flow of the valley, there is a minute but distinct extension of rills and gullies into the drainage divide of the interfluve, comprising a headward extension of the valley. See Figure 16-28. Although miniscule as an individual event, when multiplied by a thousand gullies and a million years, this action can encompass the lengthening of a valley by tens of miles and the expansion of a drainage basin by hundreds of square miles. Thus the valley lengthens at the expense of the interfluve.

Figure 16-28 This Texas panhandle scene illustrates headward erosion. The grassy flat in the background, on which are found the cattle and the pond, represents an interfluve that is being cut into by headward erosion of the gorgelike valley in the foreground. [USDA-Soil Conservation Service photo.]

Delta Formation At the opposite, or seaward end, a valley can also be lengthened, in this case by deposition rather than by erosion. When rivers flow into quiet bodies of water, such as lakes and particularly oceans, their velocity is dissipated and their load deposited. Most of this debris is dropped right at the mouth of the river in a landform that has been termed for more than 2500 years a *delta*, after a fancied resemblance to the fourth letter of the Greek alphabet, Δ. The classic triangular shape is maintained by some deltas, but it is severely modified in others in response to the balance between river-deposited sediments on one hand and the removal of sediments by ocean waves and currents on the other. At some river mouths, ocean movements are so vigorous that no delta is formed at all.

The principle of headward erosion is illustrated most conspicuously and dramatically when a portion of the drainage basin of one stream is diverted into that of another by natural processes. This event, called *stream capture* or *stream piracy*, is relatively uncommon in nature, but evidence of its occurrence can be found in many places, and its implications are sufficiently significant that we should consider it in some detail.

As a hypothetical example, let us consider two streams flowing across a coastal plain, as shown in the accompanying diagram. Their valleys are separated by an interfluve, which for this example can be thought of as an undulating area of low relief. Stream A is shorter than stream B, but is also more powerful, and its valley is aligned so that headward extension will project it in the direction of stream B's valley. In the normal course of events, rainfall and snowmelt will accumulate on the interfluve and move down as overland flow into the valleys. The overland flow will become concentrated into rills and then gullies, and the gullies will become tiny tributaries to the valleys.

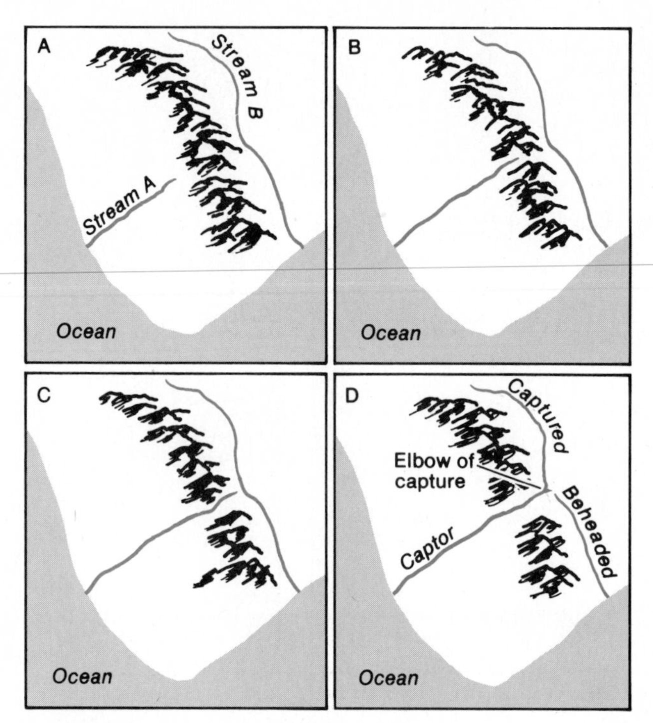

A hypothetical stream-capture sequence. Stream A is stronger than stream B. Its valley is extended by headward erosion until it captures and beheads stream B.

Due to its power and the orientation of its course, stream A will lengthen its valley headward at the expense of the interfluve. As the valley of stream A becomes larger and the interfluve becomes smaller, the drainage divide between the two valleys is shifted further toward stream B. As the process continues, the headwaters of stream A eventually will extend completely into the valley bottom of stream B, and the flow of water from the upper reaches of stream B will then be diverted into stream A. It is then that stream capture has taken place.

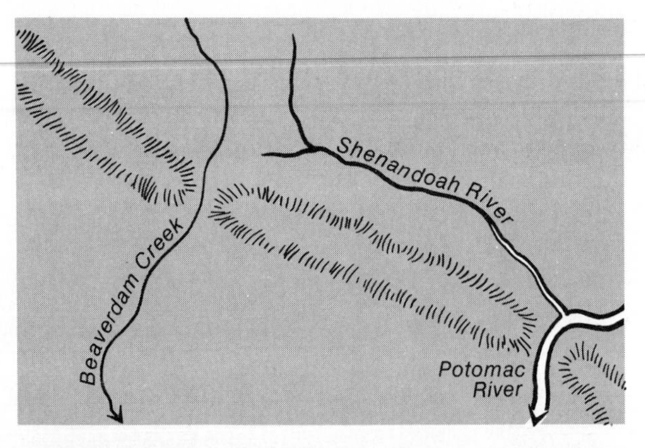

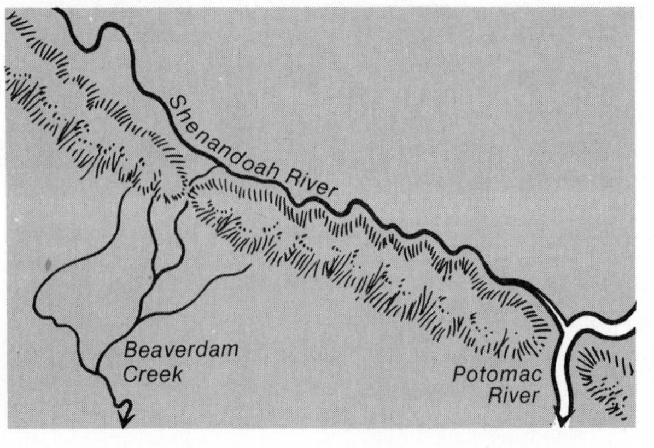

A sequence of stream capture in Virginia. Headward erosion by the Shenandoah River beheaded Beaverdam Creek and captured the flow from its upper reaches.

Identifying terminology has been devised for the various stream elements produced by such piracy. Stream A is called the *captor stream*; the lower part of stream B is the *beheaded stream*; the upper part of stream B is referred to as the *captured stream*; and the abrupt bend in the stream channel where the capture took place is called an *elbow of capture*.

Stream capture often occurs where parallel folds produce a trellis drainage pattern with subsequent and consequent streams. A typical example is shown in the accompanying figure, which portrays

The Delaware River flows from the distance through a prominent water gap in the Kittatinny Range. Pennsylvania is on the left; New Jersey on the right. [G. W. Stose, U.S. Geological Survey photo.]

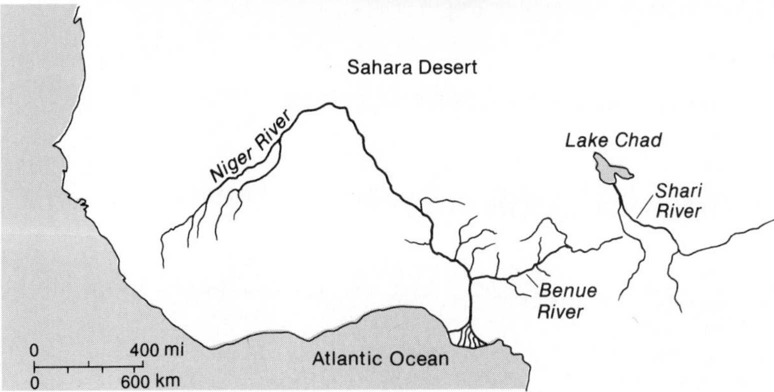

Stream capture, actual and anticipated. The upper course of the Niger River was once part of a stream that flowed into what is now the Sahara Desert. It was captured by headward erosion of the ancestral Niger River. At present, headward erosion on the Benue River gives promise of capturing the Shari River just a few tens of centuries, or hundreds of centuries, in the future.

the capture of the upper portion of Beaverdam Creek by the Shenandoah River in Virginia. Beaverdam Creek was a consequent stream that had cut its way through the Blue Ridge in a narrow gap as the linear belt of hills was uplifted. Such a narrow notch in which a stream flows through a ridge is called a *water gap*. Stream capture, however, beheaded Beaverdam Creek, as the young Shenandoah River, a subsequent stream, extended its valley by headward erosion. Due to the previous alignment of Beaverdam Creek, no elbow of capture was produced by this stream piracy, but the diversion of streamflow away from the water gap left it high and dry as the adjacent streams continued their downcutting. Such an abandoned water gap is referred to as a *wind gap*.

Stream capture on a grandiose scale can be detected by looking at a map of West Africa. The mighty Niger River has its headwaters relatively near the Atlantic Ocean, but it flows inward rather than seaward. After flowing for nearly 1000 miles (1600 km) northeasterly, it makes an abrupt right-angled turn toward the southeast, where it continues for another 1000 miles (1600 km) before finally emptying into the Atlantic. The drainage history of this region is that at some time in the past the upper reaches of what is now the Niger flowed on to the northeast into a great inland lake or sea in what is now the central Sahara Desert. This river, however, was beheaded by the ancestral Niger, producing a great elbow of capture, and leaving the beheaded stream to wither and dry up as the climate became more arid.

The map of Africa provides us with still another major point of interest concerning stream capture, but in this case the capture has not yet taken place. The Shari River of central Africa flows northwesterly into Lake Chad. Its lower course has a very flat gradient, and the stream flows sluggishly without any power for downcutting. West of the Shari is an active and powerfully downcutting river in Nigeria, the Benue, a major tributary of the Niger. Some of the tributary headwaters of the Benue originate in a flat, swampy interfluve only a short distance from the floodplain of the Shari. Since the Benue is more active than the Shari and its alignment is such that headward erosion cuts directly into the Shari drainage basin, the Benue is likely to behead the Shari before our very eyes, so to speak, provided we can wait a few thousand years.

These examples of stream piracy are interesting and significant, but the real importance of our consideration of stream capture is in helping to understand the process of headward erosion and to appreciate the enormous role it plays in shaping the terrain of the Earth.

A stream of any size that debouches into a sea or lake can produce a delta if conditions are right. Most large streams that flow into lakes do indeed form deltas. Rivers that enter oceans, however, often are unable to develop deltas because currents, tides, and waves sweep the sediments away before they can accumulate sufficiently. The entire coastline of the United States, for example, has only a handful of deltas of any size, despite hundreds of major streams that drain into the bordering seas.

Where large rivers do form deltas, the land is frequently of particular importance to mankind because the extensive areas of flat, fertile soil can support dense agricultural populations. Moreover, some of the world's major cities—such as Calcutta, Shanghai, Rotterdam, Alexandria, New Orleans, Bangkok, Rangoon, and Marseilles—are on or adjacent to deltas because of the prosperous hinterland and the opportunity to function as a link between ocean and river traffic. Some of the most notable deltas are described briefly below.

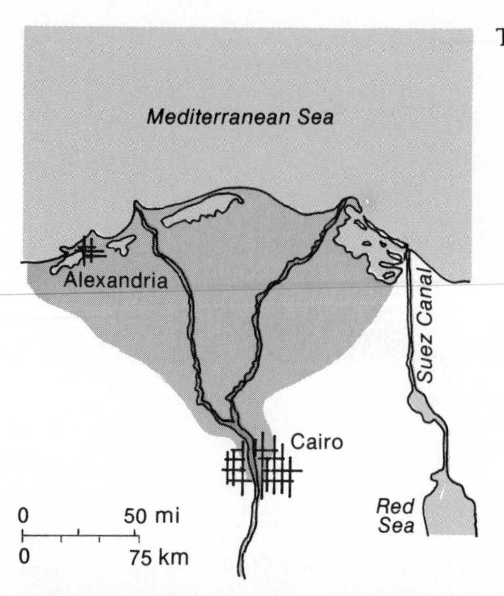

The delta of the Nile River.

The extent of the Nile delta is shown clearly in this high-altitude (*Gemini IV*) image because almost its entire area is devoted to irrigated farming. The Mediterranean Sea is on the left of the photograph, which looks southwesterly. In the middle distance is the Sinai Peninsula, which projects southward between the Gulf of Suez and the Gulf of Aqaba. [NASA photo.]

The Nile The most famous of all deltas is that of the Nile River, in part because its classic configuration gave rise to the term *delta*, and in part because its vast area of fertile land supports perhaps the densest mass of rural population anywhere in the world—some 25,000,000 people occupy the 9650 square miles (24,700 km²) of the delta, for an average density of 2600 people per square mile (1,000/km²). The Nile River rises in central Africa and flows more than 4000 miles (6400 km), largely through deserts, until it reaches the Mediterranean Sea. About 120 miles (192 km) south of the Mediterranean coast is the apex of the delta, which widens northward to form a rough triangle 150 miles (240 km) wide on the coastal side. The Nile splits into two main channels as it crosses the delta, with numerous smaller distributaries, and a network of canals has been constructed to carry irrigation water to all parts of the delta. Egypt's primate city, Cairo (population approximately 4 million), is situated at the apex of the delta, and its second city, Alexandria (population approximately $1\frac{1}{2}$ million), is on the coast near the western corner of the delta. For thousands of years the summer floodwaters of the Nile overflowed its banks (an average rise of 23 feet or 7 m at Cairo), depositing rich silt over the delta and causing it to grow out into the Mediterranean at an average rate of about 10 feet (3 m) per year. In order to control the floods and better manage the distribution of irrigation water, the massive Aswan Dam was constructed 500 miles (800 km) upstream during the 1960s. Water is now much better controlled, but great disadvantages have also resulted, most notably the loss of most of the annual addition of silt to the top-

The delta of the Mississippi River.

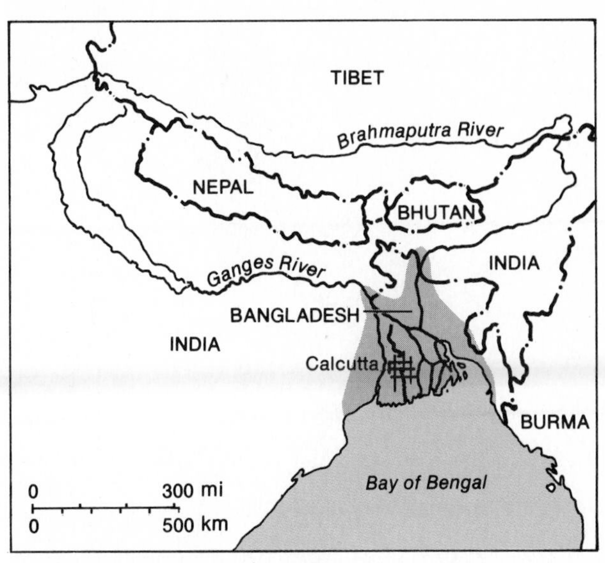

The Ganges-Brahmaputra delta.

soil of the delta. Only time will tell whether the dam is a boon or a bane to the 90 percent of Egypt's population who live on the delta.

The Mississippi The delta of the Mississippi River is by far the largest in continental United States and is one of only two deltas of any size along our entire Atlantic/Gulf coastline. It differs in many respects from that of the Nile. Originally a lengthy bay of the Gulf extended inland to what is now the southern tip of the state of Illinois, some 650 miles (1,040 km) upstream from the present mouth of the river. The delta built slowly southward from that point, and by now most of it has been converted into a broad river floodplain, although the term *delta* is still often used colloquially to refer to the entire Mississippi Valley south of Cairo, Illinois.

The lower end or "true" delta occupies most of Louisiana southeast of New Orleans and is largely a development of the last four centuries. Although the main channelled flow of the river has changed often in the past, most of the water apparently has always been carried by a single major channel. Significant distributaries today are concentrated in the southeastern corner of the delta, in addition to a maze of small natural channels, several large bayous that were main channels in the past, and a number of manufactured canals. The seaward margin of the delta is shaped more by the river and less by ocean waves and currents than is the case with most deltas.

As the sedimentary deposits have accumulated, their weight has caused tectonic subsidence of the ocean floor, producing a very irregular interfingering of land and water, with many bays, lakes, and marshes as characteristic features of the delta. As the land sinks under its load of sediment, only the crests of the natural levees alongside the distributary channels remain above sea level. This gives a very distinctive shape to the delta; the long, straight channels branch outward like the toes or claws of a bird, and the entire configuration is called a *bird's-foot delta*.

As with most deltas, this one has experienced numerous floods and major channel changes in the past. Now that people have settled in these low-lying lands (although nothing like the density of settlement in the Nile delta), strenuous efforts have been made to control flooding, primarily by building artificial levees on top of the natural levees and by constructing floodways to siphon off excess floodwater. New Orleans is the major city of the delta and one of the great ports of North America; its population is about 1 million, considerably less than either of the principal cities of the Nile delta.

The Ganges/Brahmaputra Two of the mightiest rivers of southern Asia come together to form a single massive delta at the head of the Bay of Bengal. Both rivers originate in the Himalayas, with the Ganges draining much of northern India and the Brahmaputra passing through Tibet and northeastern India prior to uniting with the Ganges in Bangladesh. Although the delta is large (roughly twice the size of the contemporary Mississippi delta), it has been building seaward at a very slow rate in recent centuries. Also, some tilting of the land has caused the Ganges to abandon a great many distributaries in the western part of the delta in favor of joining the Brahmaputra system to the east.

Still, nearly 5000 miles (8000 km) of navigable waterways crisscross the delta. The coast experiences relatively high tides, averaging about 10 feet (or 3 m), and it is devasted from time to time by tropical cyclones sweeping northward from the Bay of Bengal. Vast numbers of people occupy these low-lying lands. The largest cities of the two countries, Calcutta in India and Dacca in Bangladesh, are built largely on delta lands, and the total population of the 30,000-square-mile (78,000 km^2) delta exceeds 70 million.

The Niger Africa's largest delta has been deposited in southern Nigeria by

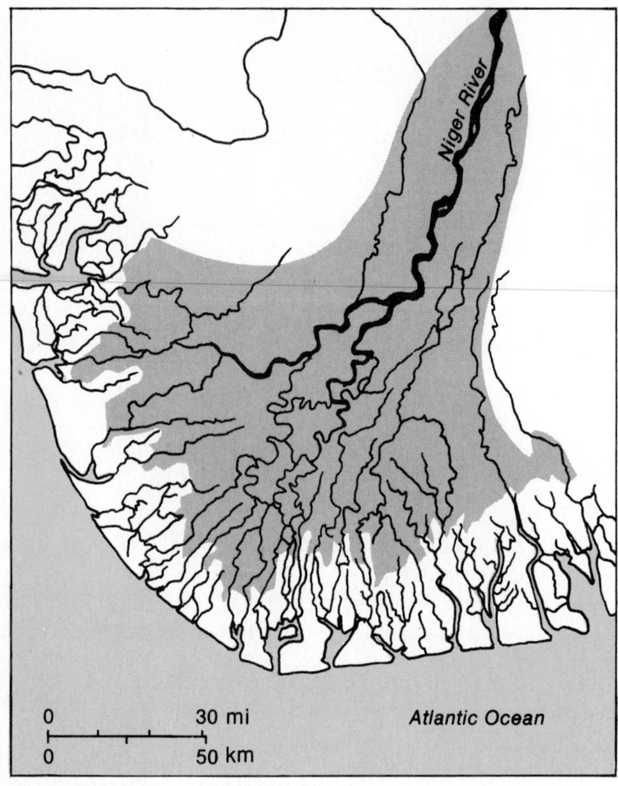

The delta of the Niger River.

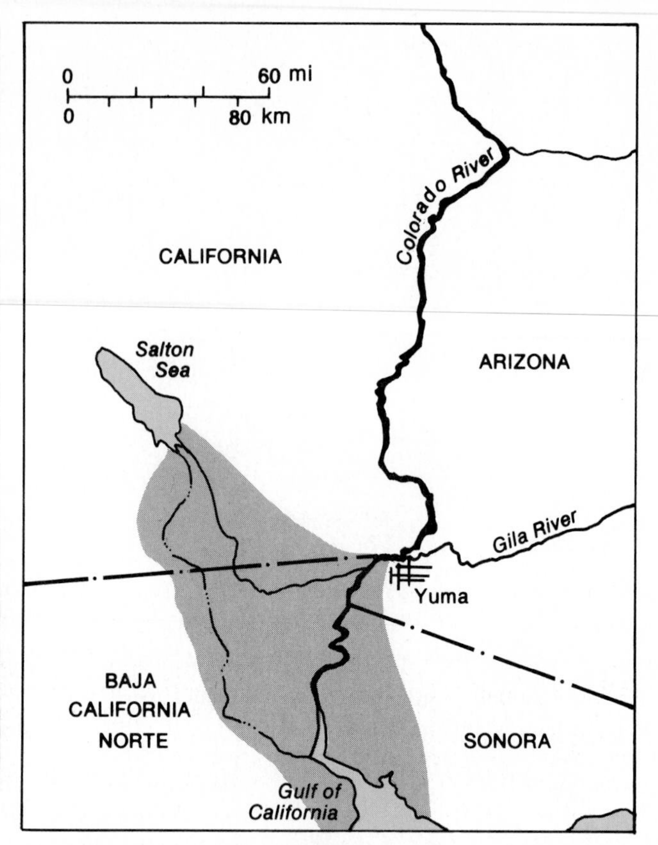

The delta of the Colorado River.

the Niger River. It has 14,000 square miles (35,840 km²) of flat land and a coastal margin that is considered to be an almost perfect example of an *arcuate delta* (curved like a bow). This smooth perimeter apparently results from the balance between a relatively small sediment load brought down by the river and a heavy and persistent ocean swell striking the coast from the south. The delta surface has a great many distributaries diffusing radially from the apex of the lowland, an extensive zone of mangrove swamps toward the outer portion of the delta, and a narrow strip of sandy beach ridges along the ocean front. This is one of the most thoroughly flooded deltas in the world, due in part to the heavy local rainfall (more than 100 inches or 250 cm per year) and in part to the annual flood that sweeps down the Niger River from its far-flung drainage basin. Much of the delta is under water for weeks at a time, with the result that the population density of the Niger delta is considerably less than that of most other major deltas.

The Colorado Although carrying much less water than the rivers described above, the Colorado River has always transported an enormous amount of sediment. Consequently, it has built a gigantic delta that spreads across the international boundary at the head of the Gulf of California (Sea of Cortez). The river originally discharged into the Gulf at the present location of the city of Yuma, some 60 miles (96 km) above its present mouth and about 150 miles (240 km) southeast of the upper end of the Gulf at that time. In relatively short order, the river built a large delta that extended across the width of the Gulf and created a natural dam that shut off the northern end from the open sea. As a result, the Salton Trough, north of the delta, lost its water through evaporation in the hot, arid climate. (Only in this century did human activities and an unusual flood bring water to the present-day Salton Sea, but that is another story.) The present delta of the Colorado River retains no distinctive outline because of this peculiar historical evolution, but its surface is exceedingly flat and marked by erratic distributary channels in characteristic deltaic fashion. Delta expansion is almost totally inactive today because most of the river's water is diverted for municipal or agricultural use before it reaches the delta; thus neither water nor sediment reaches the delta in significant quantity.

The Earth has many other important deltas, including those of the Yangtze, the Rhine, the Yukon, the Po, and the Indus, to name but a few. An equally imposing list of rivers—such as the Amazon, the Zaire (Congo), the St. Lawrence, and the Columbia—have built no deltas, primarily because active ocean currents sweep away the sediments they bring to the sea.

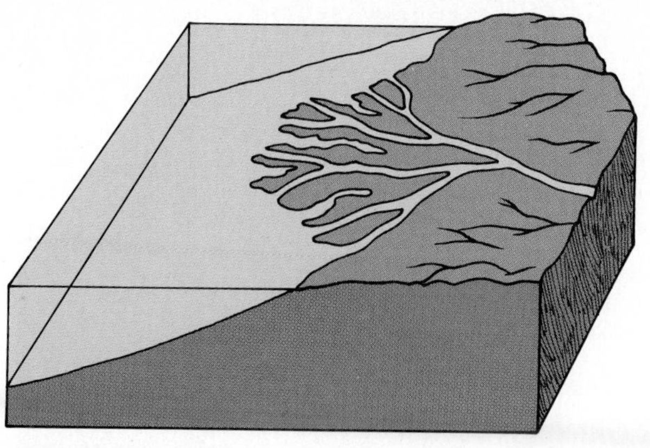

Figure 16-29 A delta often forms where a stream flows into an ocean or a lake.

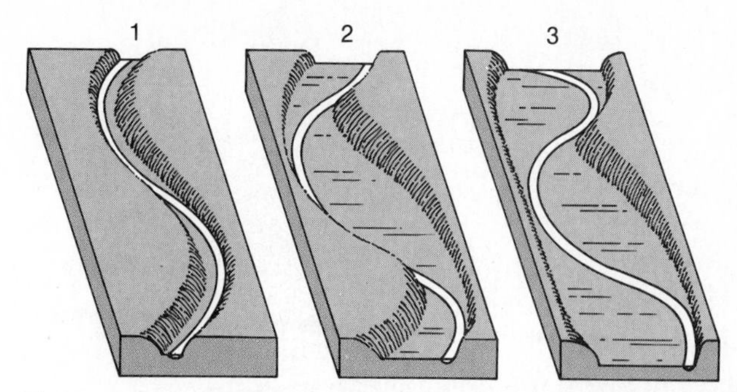

Figure 16-30 A stream widens its valley by lateral erosion, producing an increasingly broader valley floor by undercutting the bluffs on the sides of the valley. The flattish valley floor develops into a floodplain.

Generally, however, the formation of a delta is prominent and clear-cut. See Figure 16-29. The stream slows down and drops much of its load, which partially blocks the channel and forces the stream to seek another path. This path, in turn, is likely to become clogged, and the pattern is repeated. The result is that deltas usually consist of a maze of roughly parallel channels, called *distributaries*, through which the water flows slowly toward the sea. Continued deposition builds up the surface of the delta so that it is at least partially exposed above sea level. Rich alluvial sediments and an abundance of water favor the establishment of vegetation, which provides a more secure base for further expansion of the delta. In this fashion the stream valley is extended downstream.

Deposition in Valleys

Thus far we have been emphasizing the prominence of erosion in the formation and shaping of valleys. Deposition, too, has a role to play, although a more limited one in terms of overall valley configuration. Certain distinctive landforms, however, are due entirely to deposition.

Fluvial deposits may include all sizes of rock debris, but smaller particles constitute by far the greatest bulk of the total. This is primarily because of the continuing attrition suffered by any size particle that is transported downstream. The frequent impact of particles against one another and against the stream bed and sides makes them smaller and smaller and more and more numerous. The constant battering and buffeting eventually reduces the boulders, cobbles, and pebbles to sand, silt, and clay. It is no accident that most of the load carried by almost all streams in their lower courses consists of these smaller fractions. The term *alluvium* can be applied to any stream-deposited sedimentary material, but in fact it mostly refers to sand, silt, and clay.

Although the principle of fluvial deposition is very clear—deposition takes place wherever the stream velocity is inadequate to transport the load—fluctuations in flow and variations in turbulence are such that alluvial deposits can occur almost anywhere in a valley bottom. They may be found on the stream bottom, on the sides, in the center, in plunge pools, at the base of knickpoints, in overflow areas, and in a variety of other locations. Under some circumstances, particularly after a period of flood flow, alluvium may accumulate on the stream bed to such an extent that its elevation is actually raised in a process called *aggradation*, a relatively common occurrence.

The most prominent of all depositional landscapes is the *floodplain*, made possible by the lateral erosion of a meandering stream, which produces a broad flattish valley floor. See Figure 16-30. This valley floor is inundated by overflow from the stream channel during floodtime, which occurs either regularly or sporadically. The floods leave broad and sometimes deep deposits of alluvium over the entire floor, which then comprises the floodplain.

The most conspicuous feature of a floodplain is the meandering channel of the river, which frequently changes its course with the vagaries of flow in such flat terrain. See Figure 16-31. Meanders often develop narrow necks that are easily cut through by the stream, leaving abandoned *cutoff meanders*, which initially hold water as *oxbow lakes*, but gradually fill with sediment and vegetation to become *oxbow swamps*, and eventually retain their identity only as *meander scars*.

Although all of a floodplain consists of very flat terrain, the land is slightly higher along the edges of the stream channel. As the stream overflows at floodtime, the current is abruptly slowed when it leaves the normal channel by friction with the floodplain surface. This initial slowdown causes the principal deposition to take place along the margins of the main channel, producing *natural levees* on each side of the stream, which are normally the best-drained portions of the floodplain. The natural

Figure 16-31 On a flat floodplain, a stream may meander very circuitously. This is the King River near Wyndham, Western Australia. [TLM photo.]

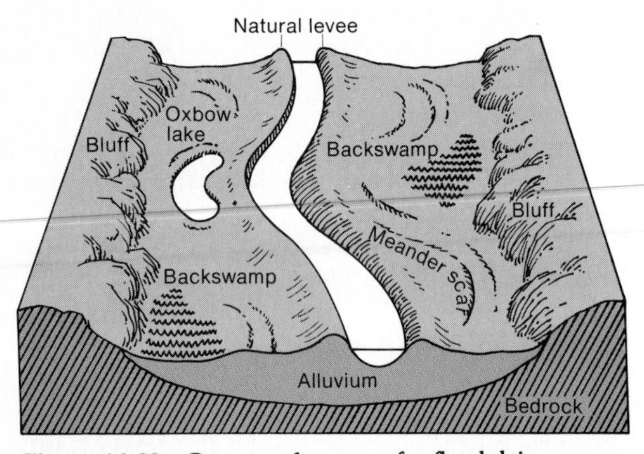

Figure 16-32 Common features of a floodplain.

levees merge outwardly and almost imperceptibly with the less well-drained and lower portions of the floodplain, generally referred to as the *backswamps*. See Figure 16-32.

Theories of Landform Development

The complexity of the various internal and external processes operating on a multiplicity of Earth materials at and near the lithospheric surface produces an infinite and bewildering variety of slopes, shapes, individual landforms, and general landscape types. The desirability of systematizing and organizing this vast array of facts and relationships into a coherent body of knowledge has been the goal of many students of geomorphology. If principles can be recognized, theories and models can be formulated. The theories and models can then be tested, and out of this procedure we can hope to gain a broader and more comprehensive understanding of the features of the terrain and of the processes that have produced them. Many scholars have contributed to theoretical considerations, and several comprehensive theories of terrain evolution have been devised. Three of these constructs seem to be the most useful for our purposes and will be discussed below.

The Geomorphic Cycle Concept

The first, and in many ways the most important, conceptual model of landscape development was propounded by William Morris Davis, an American geographer/geomorphologist who wrote and lectured extensively on his theory in the 1890s and early 1900s. He called it the *geographical cycle*, but many of his students considered the adjective to be too generalized, so it came to be known as the *cycle of erosion*. This term, in turn, was considered to be too restrictive because weathering and deposition were also involved, so the Davisian theory is now usually referred to as the *geomorphic cycle*.

Davis envisioned a circular sequence of terrain evolution in which a relatively flat surface (for example, a coastal plain) was uplifted from a lower to a higher elevation, where it was incised by fluvial erosion into a landscape of slopes and valleys and then eventually thoroughly denuded until it was once again a flat surface at low elevation. See Figure 16-33. He likened this sequential

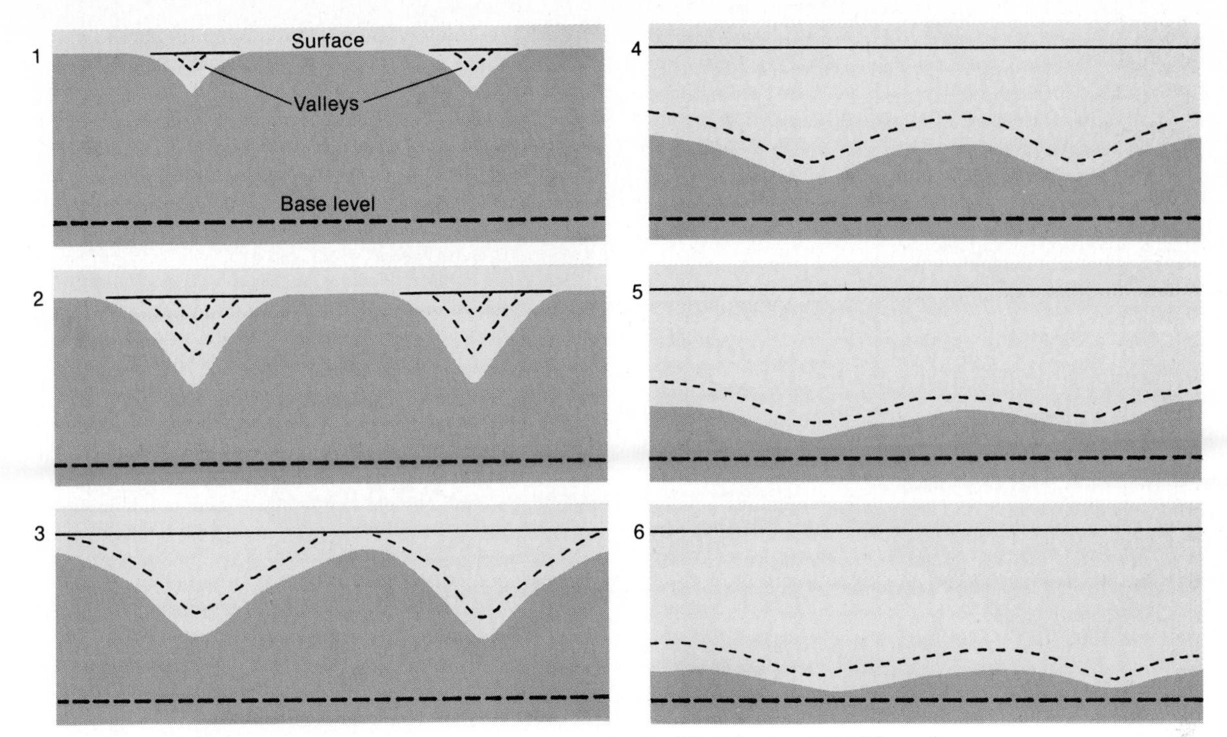

Figure 16-33 Hypothetical sequence of surface change in the Davisian geomorphic cycle, from original surface (1), through youth (2), maturity (3 and 4), and old age (5), to peneplain (6).

denudation to the life cycle of an organism, recognizing "stages" of development that he called *youth*, *maturity*, and *old age*. Each stage brought a characteristic and recognizable pattern of landform development. See Figure 16-34.

Initial Surface Davis postulated that the initial surface was uplifted relatively rapidly, so that erosion had little time to act upon it until the uplift was complete. Thus the original surface consisted of relatively flat land at an altitude considerably above sea level. He further assumed no significant subsequent crustal movement or deformation for the duration of the cycle, and a stable base level during all stages.

Youth During the youthful stage of development, consequent streams become established, and a drainage net begins to take shape. These streams then incise themselves downward to form deep, narrow, steep-sided, V-shaped valleys. These rapidly flowing streams have irregular gradients, marked by waterfalls and rapids. During this stage, most of the initial surface would be maintained as broad, flattish interfluves, largely unaffected by stream erosion, and encompassing shallow lakes and swamps because of the incomplete drainage system.

Maturity In the mature stage, the main streams approach an equilibrium condition, having worn away the falls and rapids and developed smooth profiles. Vertical erosion ceases in the main valleys, the streams begin to meander, and floodplain development is initiated. The drainage net has become much more extensive, with many more numerous and longer streams and tributaries. The interfluves are thoroughly dissected in this stage, their lakes and swamps have been drained by headward erosion, and their remnants exist only as narrow drainage divides between valleys. Whereas youth was characterized by the presence of a vast area of initial surface, maturity is marked by its absence.

Old Age With the passage of a vast amount of time, erosion reduces the entire landscape to near baselevel. Sloping land is virtually absent, and the entire region is dominated by vast and extensive floodplains over which a few major streams meander broadly and slowly.

Peneplain The end product of the geomorphic cycle is a flat and relatively featureless landscape with minimal relief. Davis called this a *peneplain* ("almost a plain"). He envisaged that there would be occasional remnants of exceptionally resistant rock rising slightly above the peneplain surface; such erosional remnants are dubbed *monadnocks*, after the example of a mountain of this name in New Hampshire. This completes the geomorphic cycle, with the land having again assumed its hypothetical preuplift form.

Rejuvenation Davis recognized that the extraordinarily long period of time without crustal deformation required for this entire cycle to be accomplished was unlikely in nature. So his theory also encompassed the concept of

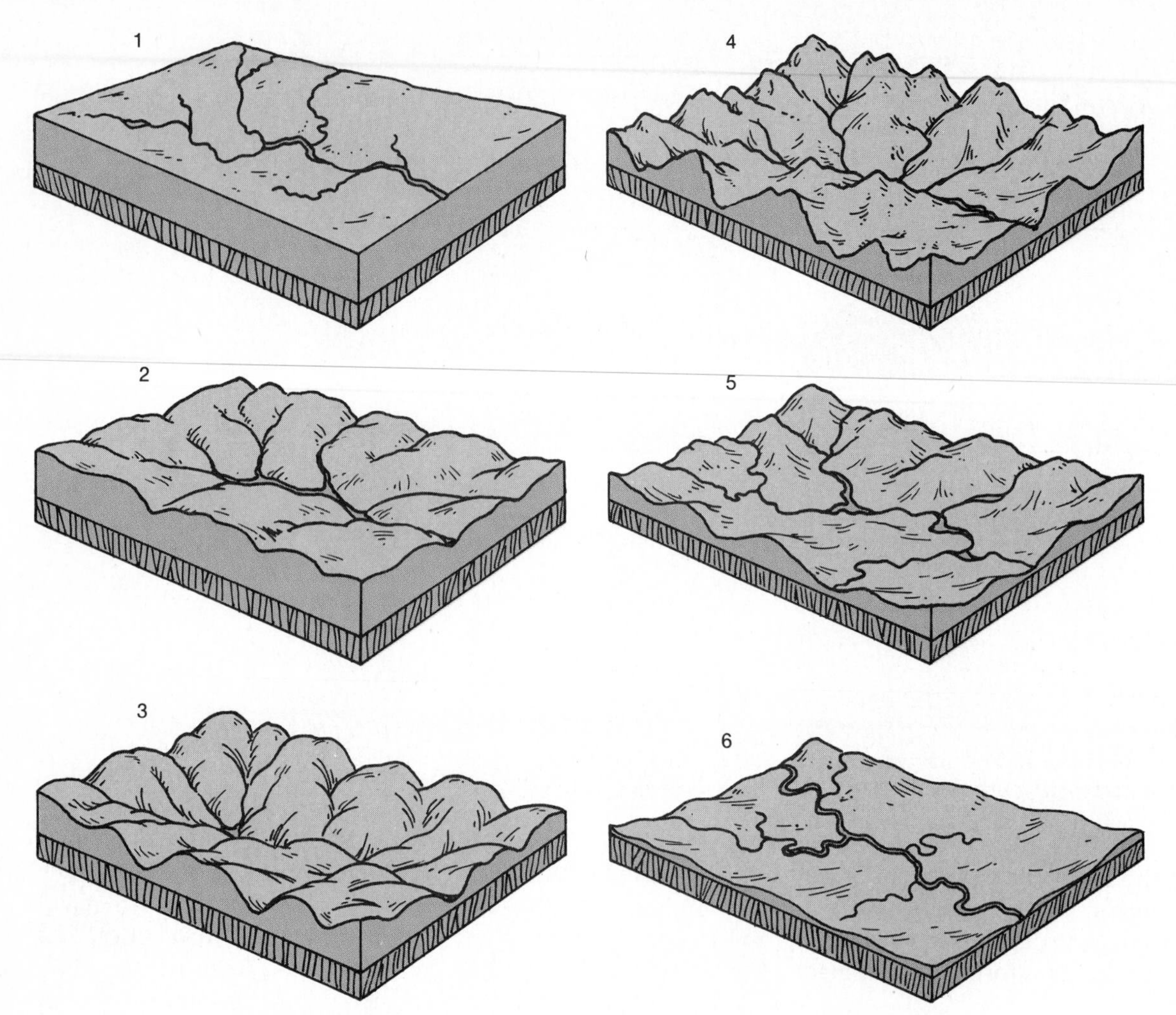

Figure 16-34 Hypothetical sequence of development in the Davisian geomorphic cycle in a humid area, from initial surface (1), through youth (2), and maturity (3) and (4), to old age (5), and peneplain (6).

rejuvenation, whereby regional uplift could raise the land and interrupt the cycle at any stage. This would reenergize the system, initiating a renewed period of downcutting and restarting the cycle again.

Topographic Analysis Davis stressed that any topographic landscape could be comprehended by analyzing the three variables of structure, process, and stage. *Structure*, as we have noted, refers to the type and arrangement of the underlying rocks and surface materials; *process* is concerned with the internal and external forces that shape the landforms; and *stage* is essentially a consideration of the length of time during which the processes have worked on the structures.

Crustal Change and Slope Development

Davis was both a prolific writer and a persuasive teacher, and his cyclical theory had a profound influence on geomorphic study for many decades, especially in the United States. However, even from the early days of this century, there were strong dissenters. Various other geomorphologists recognized imperfections in some of his assumptions and called into question some of his conclusions. For example, apparently no examples of intact peneplains exist; remnants of peneplain surfaces are recognized in some areas, but nowhere does an actual peneplain occur. A more important difficulty with the cyclic concept concerns the Davisian idea that little erosion takes place while the initial surface is being uplifted, a notion unacceptable to most geomorphologists. Moreover, the causal interplay between uplift and erosion is open to varying interpretations. Finally, there are serious doubts about the entire concept of sequential development; for example, does youth "normally" precede maturity, and is maturity "normally" followed by old age? The biologic analogy may be more misleading than helpful.

The sequential aspect of the geomorphic cycle concept is very appealing because it provides an orderly, evolutionary, and predictable train of development. Moreover, many areas in nature have the appearance of youthful, mature, or old-age topography. However, no

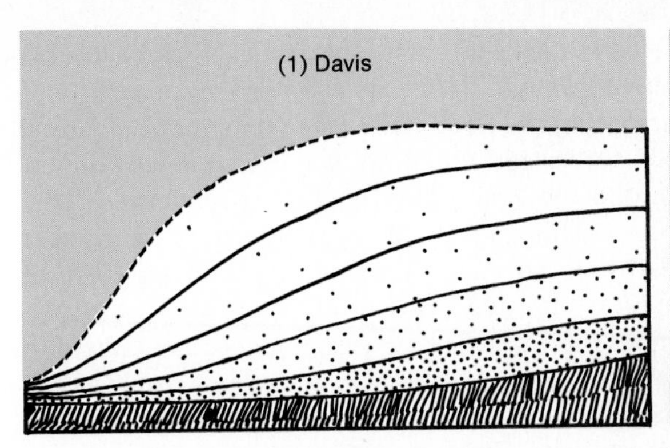

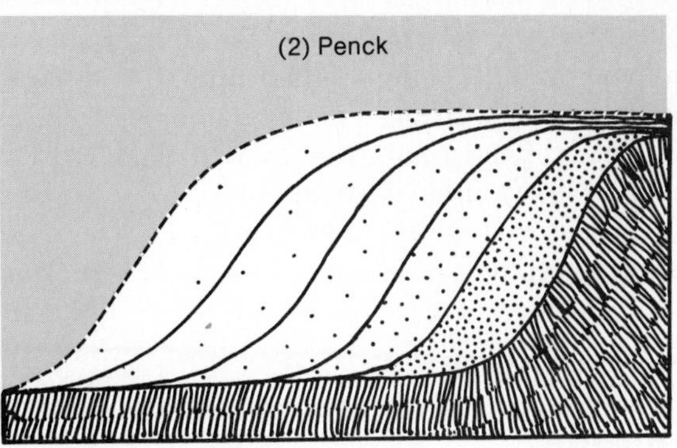

Figure 16-35 Schematic sequence of slope retreat in the Davis (1) and Penck (2) models. The Davis concept visualized a continually diminishing angle of slope, whereas Penck theorized parallel retreat that retained approximately the same slope angle.

proof is found that one stage commonly precedes another in regular fashion, and even in a single valley, the terrain characteristic of the various stages is often jumbled. For landform analysis, it is probably better to use these terms (*youth*, *maturity*, *old age*) as descriptive summaries of regional topography rather than as distinct implications of sequential development.

In the Davisian cycle, drainage divides are seen to waste away in a steady and predictable pattern. As the slope retreats, it becomes less steep and more rounded, always maintaining a convex form. In nature, however, some slopes are convex, but others are straight, and still others are concave. Walther Penck, a young German geomorphologist and a prominent early critic of Davis, pointed out in the 1920s that slopes assume various shapes as they erode. Penck stressed that uplift stimulated erosion immediately, and that slope form was significantly influenced by the rate of uplift or other crustal deformation. He argued that steep slopes, particularly, maintained a constant angle as they were eroded, retaining their steepness as they diminished in a sort of "parallel retreat," rather than being worn down at a continually lower slope angle. He viewed eroding slopes as retreating, rather than being reduced overall, which means that some of the "initial surface" would be retained long after it would have been worn away in the Davisian concept. See Figure 16-35. Many, but not all, of Penck's ideas have been substantiated by subsequent workers.

Equilibrium Theory

In the last quarter-century or so, many geomorphologists have been paying closer attention to the physical mechanics of landform development. This approach emphasizes the delicate balance between form and process in the landscape. It is believed that the influence of crustal movement and the differing resistance of the underlying rock vary significantly from place to place, and that these variations are as significant as differences in process in determining the shape of the terrain. Thus, equilibrium theory suggests that slope forms are adjusted to geomorphic processes so that there is a balance of energy—the energy provided is just adequate for the work to be done. For example, harder rock will develop steeper slopes and higher relief, and softer rock will have gentler slopes and lower relief. The uniformity inherent in both the Davis and Penck theories is thus called into question.

A prime example of the application of equilibrium theory can be seen in a hilly area where the land is being uplifted tectonically at the same time that it is being eroded fluvially. See Figure 16-36. If the slopes are in equilibrium, they are being wasted away at the same rate as they are being regenerated by uplift. Thus the rocks are being changed through erosion from above and uplift from below, but the form of the surface remains the same; the landscape is in dynamic equilibrium. A change in either the rate of erosion or uplift will make the landscape go through a period of maladjustment until the slopes again reach a gradient at which the rate of erosion once more equals the rate of uplift.

Figure 16-36 The dynamic equilibrium concept. This schematic vertical cross section of three surfaces indicates that erosion reduces the relief of the land as rapidly as uplift raises it. The surface rocks are continually changing through uplift and erosion, but the shape of the surface remains approximately the same because removal and replacement are in balance.

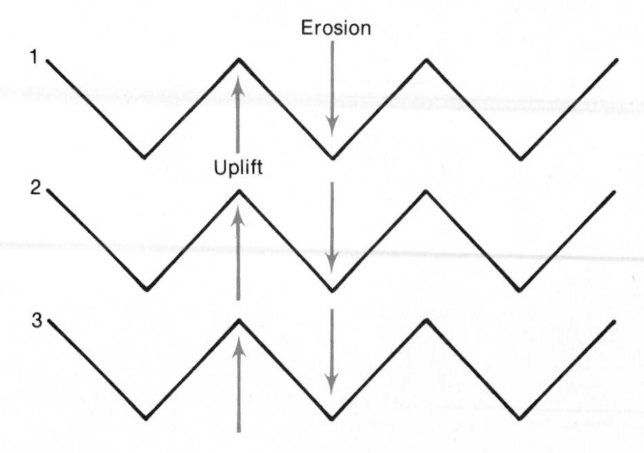

Details and implications of the equilibrium concept of landform development are still being worked out. It does not as yet provide a general theory of topographic evolution, and it has serious shortcomings in areas that are tectonically stable or have limited streamflow (as deserts). It does, however, focus more precisely on the relationship between geomorphic processes and surface forms.

Stream Rejuvenation

All parts of the continental surfaces experience episodes, sometimes frequently and sometimes rarely, when their elevation is changed with respect to sea level. This is occasionally caused by a drop in sea level (as occurred throughout the world during the various ice ages when frozen water accumulated on the land, diminishing the amount of water in the oceans), but it is much more commonly the result of tectonic uplift of the land surface. When such an event occurs, it rejuvenates the streams in the area. The increased gradient causes the streams to flow faster, which provides renewed energy for downcutting. Vertical incision, which may have been long dormant, is initiated or intensified.

If a stream, prior to rejuvenation, occupied a broad, flattish valley, the effect is to cut a steep-walled inner gorge. This means that the floodplain can no longer function as an overflow area, and instead is situated as an abandoned stretch of flat land overlooking the new gorge. This remnant of the previous valley floor is called a *stream terrace*, and it is sure evidence of rejuvenation. See Figure 16-37. Terraces often, but not always, occur in pairs, one on either side of the newly incised stream channel.

Terraces are commonplace features over the world, indicating the frequency of tectonic uplift. Indeed, in many valleys one can find not one (or one pair) but many terrace levels above the present stream channel. As many as a dozen levels are sometimes recognizable. Where more than one terrace level occurs, it is clear evidence of several successive stages of rejuvenation. The basic geomorphic history of such an area is that rejuvenative uplift caused increased downcutting for awhile, until a gentle profile was again achieved. This was followed by a period of valley widening and floodplain development due to lateral erosion. Another episode of uplift caused renewed downcutting, followed eventually by another sequence of lateral erosion. This pattern, then, was repeated a number of times, with each repetition leaving still another set of terraces on the sides of the valley.

Figure 16-37 Schematic illustration of stream terrace development, a clear indication of several instances of rejuvenation.

Under certain circumstances, rejuvenative uplift has a different and even more conspicuous topographic effect, producing *entrenched meanders*. This can come about when an area containing a meandering stream is uplifted slowly and the stream incises downward while still retaining the meandering course. In some cases, such meanders may become entrenched in narrow gorges hundreds of feet deep.

Fluvially Produced Topography

As a summation of the work of running water, we can consider the variety of topography produced by fluvial erosion and deposition. No particularly distinctive features are created on the interfluves, so our attention will be focussed on the valleys. Although the various landforms can evolve almost anywhere in the valleys, certain typical assemblages are likely to be found in the upper, middle, and lower reaches of an average valley.

Landforms of the Upper Reaches

The upper reaches of a valley normally are the steepest, and their shaping is dominated by downcutting. Typically the valley is incised into a steep-walled defile with a V-shaped profile in which the valley floor is so narrow that it is filled almost completely by the stream course. See Figure 16-38. The valley sides rise sharply from the bottom, and mass wasting can readily move material downslope into the stream. The stream channel is normally not straight but follows a zigzag path in which spurs reach down from the valley sides in interlocking fashion. The floor itself usually has an irregular gradient, primarily in response to differing resistance in the underlying rock. Thus harder rock areas are marked by knickpoints and softer rocks by a more gentle slope. Tributaries characteristically are few, short, and have very steep gradients. A valley of this type would be termed *youthful* in Davisian terminology.

Landforms of the Middle Reaches

In the midportions of a typical valley, downcutting is replaced by lateral erosion and slope wasting as the principal forms of denudation. The stream approaches an equilibrium condition, with a relatively smooth longitudinal

The Front Range—Episode Three: Erosion and More Erosion (continued from page 316)

The uplift of the intrusive batholith and its overarching sediments took place over millions of years and was accompanied throughout by significant erosion from runoff and streamflow. Some 10,000 feet (3000 m) of sediments were stripped away during this protracted event. The eventual result was the formation of a rolling plain of moderate elevation, above which rose occasional low rounded mountains or hills 1000 or 2000 feet (300 or 600 m) in height. This rolling plain represented an erosional surface that was considered by some authorities to be a peneplain with the higher rounded summits as monadnocks. Following the development of this original surface, there were from one to five other cycles of uplift and erosion, so that at various places along the Front Range there were remnants of other erosional surfaces preserved as the level summits of certain mountains and the flattish tops of many ridge crests.

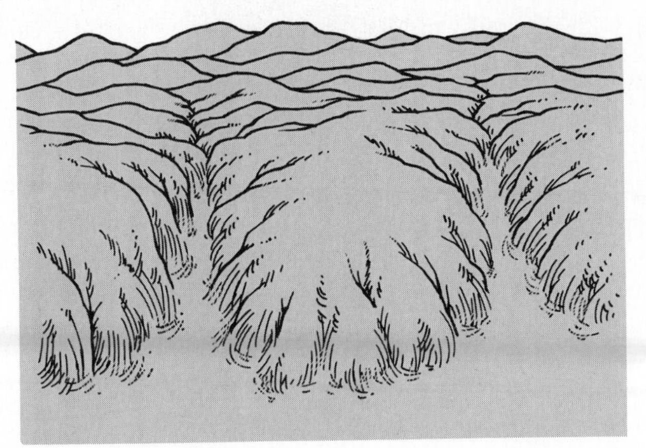

Prior to Pleistocene glaciation, the Front Range terrain consisted of a high-altitude rolling plain, carved sporadically by steep, narrow gorges. Some rounded summits (monadnocks) rose above the general level of the upland.

The most recent sequence of uplift produced a massive but gently sloping arch structure with the land raised to roughly its present height. Erosion was thus reenergized, canyons were cut, and the arch itself was locally carved into mountains. This recent so-called canyon cycle produced a number of deep gorges in the uplifted erosion surfaces, as the resistant granite was responsive to valley deepening, but not widening, by the streams that flowed across it. Although some sharp crests and rugged topography were produced in the canyon cycle, most of the main divides retained the rounded forms of subdued mountains, locally carved by deep gorges. On the eastern and western margins of the granite erosion was more effective, as the softer sedimentary beds were almost completely stripped away; only the upturned edges of a few resistant sedimentary strata remained as lengthy hogback ridges separated by longitudinal valleys.

These sequences of uplift and erosion raised the original peneplain to approximately its present height of 10,000 to 12,000 feet (3000 to 3600 m) above sea level, where it stood as a rolling surface of relatively gentle relief. A few rounded summits rose 500 to 2500 feet (150 to 760 m) above this level, as residual monadnocks, and a series of conspicuous gorges was incised into the rolling upland surface.

(To be continued. Episode Four begins on page 436.)

Figure 16-38 The upper reaches of a typical valley usually exhibit steep slopes, a narrow valley bottom, and a V-shaped profile. This scene is in southwestern Colorado. [W. Cross, U.S. Geological Survey photo.]

profile in this section. The valley floor is flattish, with at least the beginning of floodplain development. The stream course is meandering, with sinuous loops, oxbow phenomena, and meander scars. Although the valley walls are well separated by the broad valley floor, they are distinct and prominently sloping. This valley would be classed as *mature* in the Davisian system. Many tributaries with relatively steep gradients join the main valley in this section.

Landforms of the Lower Reaches

The lower course of a typical stream is dominated by depositional landforms. The valley bottom is exceedingly broad and is deeply filled with alluvium. The meandering pattern is well established and complex, with much evidence—in the form of meander scars and oxbow lakes and swamps—of frequent channel changes. See Figure 16-39. Natural levees border the stream course, and due to their growth and the filling of the channel with alluvium, a stream may be raised so that it flows in normal

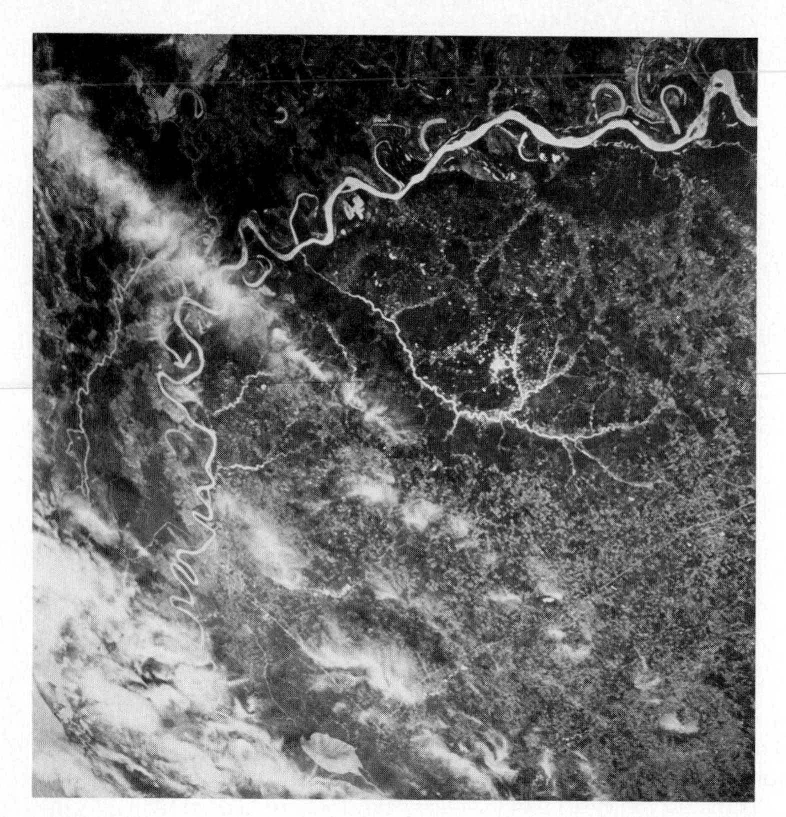

Figure 16-39 The flood plain of the lower Mississippi River is the dominant feature in this *Apollo 9* photograph. The view is southward, with the state of Mississippi to the left of the river and the state of Louisiana to the right. Numerous meander scars and oxbow lakes appear in the floodplain. [NASA photo.]

conditions well above the floodplain level. Artificial levees are sometimes constructed on top of the natural levees to prevent flooding of the alluvial lowlands. Tributaries are very few, and they often flow relatively parallel to the principal river on the floodplain for a long distance before eventually joining it. This *delayed tributary junction* is a characteristic feature of the lower course of major rivers.

In this chapter we have concentrated on the implacable force and pervasive influence of running water as a molder of topography. Overall, water is clearly the predominant external process in the shaping of the terrain of the continents. In the following chapter we will focus our attention on a somewhat more specialized environment—the dry lands of the world—where running water, although scarce, is nevertheless the principal external force involved in landform development. What we have learned about fluvial processes in this chapter, then, is also applicable, for the most part, in arid regions. Other factors, however, become influential in shaping desert terrain, and fluvial actions are somewhat modified. Thus our consideration of the evolution of terrain in dry lands still revolves around fluvial processes but with increased attention to other environmental conditions.

Review Questions

1. What is the distinction between *overland flow* and *streamflow*?
2. What determines the erosive effectiveness of streamflow?
3. How does a stream sort alluvial material?
4. Discuss some of the relationships between stream order and other aspects or characteristics of streams.
5. Is it ever possible for a stream to erode below sea level? Explain.
6. Why do most knickpoints tend to migrate upstream?
7. Explain how it is possible for a stream valley to be lengthened.
8. Describe the mechanics of *stream capture*.
9. How is it possible for stream meanders to become deeply entrenched?
10. What does the presence of stream terraces tell you about the erosional history of the stream valley?

Some Useful References

CHORLEY, R. J., ed., *Introduction to Fluvial Processes*. London: Methuen & Co. Ltd., 1971.

CRICKMAY, C. H., *The Work of the River*. London: Macmillan & Co. Ltd., 1974.

DURY, G. H., ed., *Rivers and River Terraces*. London: Macmillan & Co. Ltd., 1970.

LEOPOLD, L. B., M. G. WOLMAN, AND J. P. MILLER, *Fluvial Processes in Geomorphology*. San Francisco: W. H. Freeman & Company, Publishers, 1964.

MORISAWA, M., *Streams: Their Dynamics and Morphology*. New York: McGraw-Hill Book Company, 1968.

RICHARDS, K., *Rivers: Form and Process in Alluvial Channels*. London: Methuen & Co. Ltd., 1982.

SCHUMM, S. A., *The Fluvial System*. New York: John Wiley & Sons, Inc., 1977.

17 Topographic Development in Arid Lands

There are no clear-cut boundaries to set off arid from nonarid regions, either in terms of climate or topography. Nevertheless, the contrasts between typical terrain in arid regions and that in humid regions is distinct. In this chapter we will focus on the dry lands of the world, without attempting to establish precise definitions or borders. We will be concerned essentially with the processes that shape the topography of deserts and the landforms that result. It should be understood, however, that both the processes and the landforms under discussion are more widespread in their occurrence than the term *desert* might imply. They are clearly more prominent in, and typical of, arid lands, but they are also found in less dry regions. Thus, this chapter is about deserts, but much of what is discussed is also applicable, with generally diminishing significance, in semiarid, subhumid, and even humid regions.

Desert terrain is usually stark and abrupt in appearance. Its outline is generally unsoftened by layers of regolith, soil, and vegetation. Its rocky outcrops are exposed to view, its slopes are steep and angular, and its hues are often vivid and colorful. Its scenery is strikingly beautiful to some people and bleakly forbidding to others. But whatever emotion the desert may trigger in a person's mind, it is easy to recognize that the desert landscape is different from the "normal" landscapes of the more humid parts of the world where most people reside.

Specialized Environmental Characteristics

For the most part, the same terrain-forming processes are at work in desert areas as in humid lands. The landforms that result, however, are often conspicuously different, due to a variety of factors, most of which are related to the scarcity of rainfall, and which together constitute an environmental milieu that contrasts sharply with that of humid regions. The most important of these environmental contrasts are summarized below:

1. *Weathering*. In dry lands, mechanical weathering is dominant. Physical disintegration of rocks is much more significant than chemical decomposition (which dominates in nonarid regions). This results not only in a generally slower rate of total weathering but also in the production of more angular particles of weathered rock.
2. *Creep*. Soil creep is a relatively minor phenomenon on most desert slopes. This is due partly to the lack of soil but primarily to the lack of water to serve as lubricant for creep movement. Creep is a "smoothing" phenomenon in more humid climates, and the lack of creep accounts in part for the angularity that prevails on desert slopes.
3. *Soil and regolith*. The accumulation of unconsolidated material atop bedrock in desert areas is limited. The covering of soil and regolith is either thin or absent in most places, which exposes the bedrock more directly to erosion and contributes to the stark, rugged, rocky outline of the terrain.
4. *Impermeable surfaces*. A relatively large proportion of the desert surface is impermeable to percolating water, permitting little moisture to seep into the ground. Caprocks and hardpans of various types are relatively widespread, and what soil has formed may be thoroughly compacted, which makes for little infiltration and much runoff.
5. *Sand*. Deserts have a relative abundance of sand in comparison with other parts of the world. See Figure 17-1. This is not to say that deserts are mostly sand-covered; indeed, the notion that deserts consist of great seas of sand is quite erroneous. Nevertheless, the relatively high proportion of sand in deserts has three important influences on topographic development: (1) A sandy cover promotes infiltration of water and inhibits drainage development; (2) sand is readily moved by heavy rains; and (3) sand can be shifted and shaped by the wind.
6. *Rainfall*. A relatively high proportion of the rainfall of desert areas is intense, which, when coupled with some of the factors mentioned above, means that runoff is usually rapid. Floods, although often brief and limited, are the rule rather than the exception. Thus fluvial erosion and deposition, however sporadic and rare, are remarkably effective and conspicuous.
7. *Drainage*. Almost all streams in desert areas are inter-

Figure 17-1 Most deserts contain considerable accumulations of sand, which sometimes build up to form conspicuous dunes. This is a Death Valley (California) scene. [TLM photo.]

mittent in nature, flowing only during and immediately after a rain. Such streams may be effective agents of erosion during their brief periods of activity, shifting enormous amounts of material. However, this is mostly short-distance transportation; a large volume of unconsolidated debris will be moved to a new location not very far away, and as the stream dries up, the debris is dumped on slopes or in valleys, where it is readily available for the next rain. As a consequence, depositional features are unusually common in desert areas.

8. *Wind*. Another fallacy associated with deserts is that their landforms are produced largely by wind action. This is not true, even though a high degree of windiness is characteristic of most deserts. However, much fine-grained unconsolidated material (sand, silt) is available, and since there is a scarcity of vegetation to hold it in place, the wind action is particularly effective.
9. *Basins of interior drainage*. Desert areas contain a prevalence of basins of interior drainage, i.e., watersheds that do not drain ultimately into any ocean. For most continental surfaces, rainfall has the potential of flowing all the way into the sea. In the dry lands, however, drainage nets are frequently underdeveloped or incomplete, and the terminal of a drainage system is often a basin or valley with no external outlet. The eastern United States has a completely external drainage pattern, with very limited exceptions. Any drop of rain falling in the eastern half of the country has the potential of reaching the ocean. In the southwestern quarter of the United States, in contrast, large areas have only interior drainage. See Figure 17-2. Any rain that falls in Nevada, for example, except in the extreme southeastern and northeastern corners of the state, has no hope of reaching the sea. Certain types of landforms develop only in basins of interior drainage, and others are particularly prevalent in such areas.

Figure 17-2 The basins of interior drainage in the United States are all in the West. By far the largest area of interior drainage is in Nevada and the adjoining states.

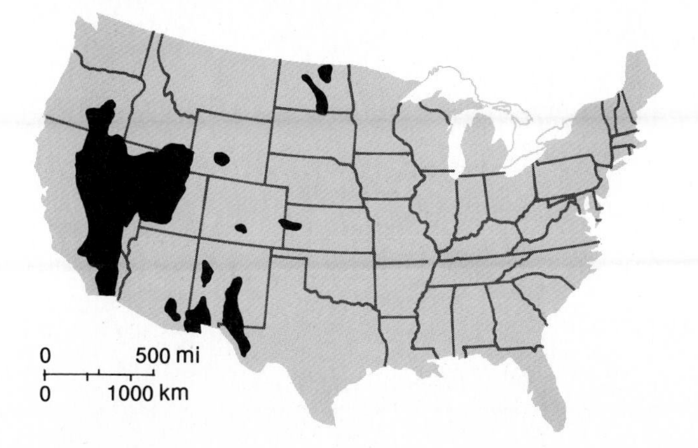

10. *Vegetation*. All the environmental characteristics listed above have important effects on topographic development, but the single most significant specialized feature of dry lands is the lack of vegetation. The absence of a continuous vegetative cover is the key factor in providing uniqueness to desert landforms. The plant cover of dry lands consists mostly of widely spaced shrubs or sparse grass, which provide little protection from the force of raindrops and function inadequately to bind the surface material with roots. Any fragment loosened by weathering is liable to rapid removal by the next rain or wind.

Unique Landscape Types of the Desert

Desert landscapes are exceedingly variable. They may encompass mountains, hills, plateaus, or plains; they may have high relief or low; they may be coastal or inland; and they occur in almost every latitudinal zone. The basic characteristics that they hold in common include a lack of surface water and a general paucity of vegetation. However, there are three particular landscape types that are found only in desert areas: The erg, the reg, and the hamada. In some locations they are very extensive, and frequently they are the dominant landscape types in the very driest portions of the dry lands.

Most notable is the *erg*, the classic case of a "sea of sand," often associated in the public mind with the term *desert*. An erg is a large area covered with loose sand, generally arranged in some sort of dune formation by the wind. The accumulation of the vast amount of sand necessary to produce an erg is not easily explained. Since desert weathering processes are very slow, it is probable that ergs have developed only where a more humid climate originally formed weathering products, products that were carried by streams into an area of accumulation; this then was followed by a change to a more arid climate in which the wind became an important agent of transportation and deposition. Several large expanses of ergs occur in the Sahara and Arabian deserts, and smaller ergs are found in most other deserts. The Australian deserts are dominated by large accumulations of sand, including extensive dunefields, but these are not true ergs because most of the sand is anchored by vegetation, and it is not free to move with the wind. Relict ergs are sometimes found in nonarid areas, indicative of a drier climate in the past. Much of western Nebraska has such relicts.

A second specialized type of desert landscape, though not as widespread as the erg, is the *reg*, a surface covering of coarse material (gravel, pebbles, boulders) from which all sand and dust have been removed by wind and water erosion. A reg, then, is a stony desert, although the surface covering of stones may be very thin (in some cases, it is literally one-pebble deep). The finer material having been removed, the surface pebbles often fit closely together, sealing whatever material is below from further erosion. For this reason, a reg is often referred to as *desert pavement* or *desert armor*. In Australia, where stony deserts are relatively widespread, they are called *gibber plains*.

A third variety of specialized desert landscape is the *hamada*, a barren surface of consolidated material. A hamada surface usually consists of exposed bedrock, but it is sometimes composed of sedimentary material that has been cemented together by salts evaporated from groundwater. In either case, fragments formed by weathering are quickly swept away by the wind so that little loose material remains.

All three of the specialized desert landscapes discussed above are limited to plains areas. Both regs and hamadas are exeedingly flat, whereas the ergs are only as high as the sand dunes built by the wind.

Although the extent of ergs, regs, and hamadas may be significant in some dry lands, the majority of the arid land in the world contains none of the three.

Running Water in Waterless Regions

Probably the most fundamental fact of desert geomorphology is that running water is by far the most important of the external agents of landform development. The erosional and depositional work of running water accounts for the shape of the terrain surface almost everywhere outside the ergs and regs. The lightly vegetated ground is defenseless to whatever rainfall may occur, and erosion by rainsplash, sheetwash, rilling, and streamflood is enormously effective. Despite the rarity of precipitation, its intense nature produces abrupt runoff, and great volumes of sediment can be moved in a very short time.

The steeper gradients of mountain streams augment their capacity for shifting large loads, but their sporadic flow regimes result in an unpredictable imbalance between erosion and deposition, so that at any given time much transportable rock debris and alluvium sit at rest in the dry steam bed, awaiting the next flow period. Loose surface material is either thin or absent on the slopes, and the underlying structure of the bedrock is often clearly exposed, with the more resistant strata standing out as caprocks and cliff faces.

Where slopes are gentle, the streams rapidly become choked with sediment as the brief floodtime subsides. Steam channels are readily subdivided by braiding, and main channels often break up into distributaries in the basins. Much fine-grained material—silt and sand—is thus left on the surface for the next flood to move, unless wind moves it first.

Desert Hydrography

Surface water in deserts is conspicuous by its absence. There are lands of sandy streams and dusty lakes, in which the presence of surface water is usually episodic and brief.

Permanent streams in the dry lands are few and far between, and with scarce exceptions, they are *exotic*, meaning that they are sustained by water that originates from outside the desert. The water comes from a more humid environment (an adjacent wetter area or a higher mountain area in the desert) and has sufficient volume of flow to survive the passage all or part way across the dry lands. In humid regions, a river becomes larger as it flows downstream, nourished by tributaries and groundwater inflow. In dry lands, however, the flow of exotic rivers diminishes downstream because of seepage of water into the riverbed, evaporation, diversion by man for irrigation, and the lack of joining tributaries. The Nile is the classic example of an exotic stream (see Figure 17-3), obtaining its water from the mountains and lakes of central Africa and Ethiopia in sufficient quantity to survive a 2000-mile (3200-km) journey across the desert without benefit of tributaries.

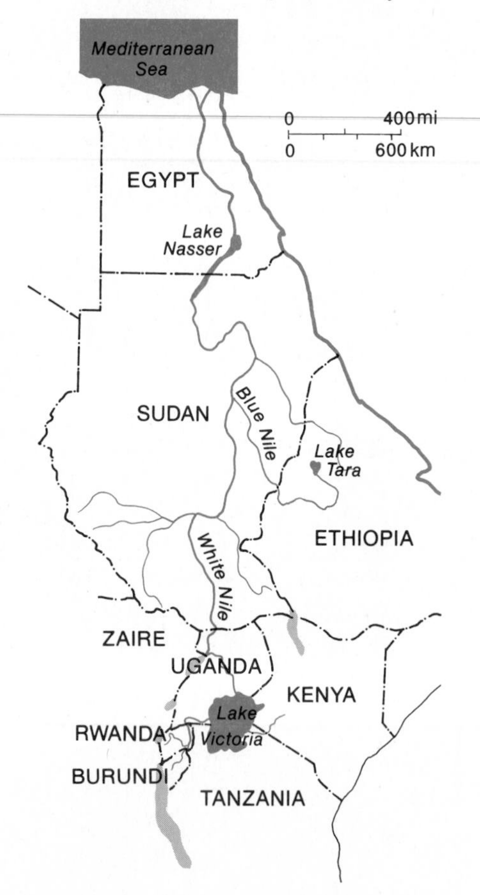

Figure 17-3 The world's preeminent example of an exotic stream is the Nile River. It flows for many hundreds of miles without being joined by a tributary.

Figure 17-4 A steep-sided arroyo in central Colorado. [W. T. Lee, U.S. Geological Survey photo.]

Although almost every desert has a few large and prominent exotic rivers, more than 99 percent of all desert streams are intermittent, flowing only in response to specific episodes of rainfall. A typical intermittent stream carries water for only a few days, or even a few hours, out of the year, and not every year. It may rise and flood with remarkable rapidity after a rainfall, but it will ebb with equal speed when the rain terminates. The brief periods of flow are times of intense activity—erosion, transportation, deposition. Most intermittent desert streamflow eventually dissipates through seepage and evaporation, although sometimes such a stream is able to reach the sea, a lake, or an exotic river before it dries up. The normally dry beds of intermittent streams typically have flat floors, sandy bottoms, and steep sides; they are referred to as *arroyos* in the United States, *barrancas* in Latin America, and *wadis* in the Sahara. See Figure 17-4.

Although lakes are uncommon in desert areas, dry lake beds are not. See Figure 17-5. We have already noted the prevalence of basins of interior drainage in dry lands; most of them have a lake bed occupying their area of lowest elevation, which functions as the local base level for that basin. These dry lake beds are called *playas* (see Figure 17-6), although the term *salina* may be used if there is an usually heavy concentration of salt in the lake-bed sediments. On rare occasions, the intermittent streams may have sufficient flow to bring water to the playa, forming a temporary lake called a *playa lake*. These are ephemeral bodies of water, destined for eventual evaporation and disappearance until the next episode of infilling.

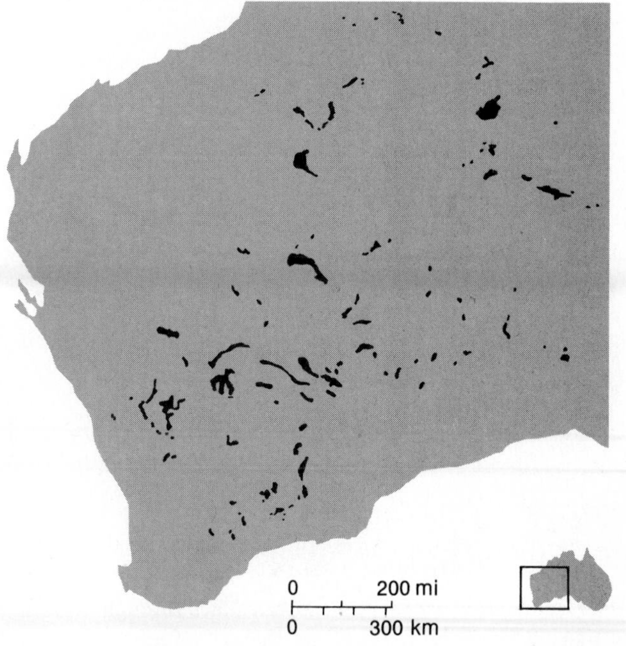

Figure 17-5 Dry lake beds often are numerous in desert areas. This map shows the principal playas and salinas in Western Australia.

A few desert lakes are permanent bodies of water. The smaller permanent lakes in the desert are nearly always the product of either subsurface structural condi-

The most prominent exotic stream in our southwestern dry lands is the Colorado River. It travels some 1450 miles (2230 km) from its source in the Rocky Mountains of north-central Colorado to its mouth at the upper end of the Gulf of California. After it leaves the headwaters area, its volume is swelled by only nine important tributaries, all of them also exotic, in 1300 miles, a fairly typical situation for exotic rivers in arid lands.

For most of its course in the state of Colorado, the river has many of the characteristics of a large mountain stream. It flows rapidly down a steep gradient, and much of its extent is through deep narrow canyons. It cuts across both structural uplifts and structural basins; its history in this area is a mixture of antecedence and superimposition. Some of the canyons hold an almost continual froth of cascading white water, the turbulence created by the steep gradient and rocky bottom. The riverbed slopes as much as 60 feet per mile (11 m/km) in some places. Numerous small tributaries join the Colorado in these upper reaches, most of them clear brooks rushing down from high-country watersheds. The Colorado River itself, however, is already opaque with suspended sediment, even in the first hundred miles (164 km) of its course.

As the river proceeds out of the mountains in western Colorado, it begins to develop a broad floodplain, more than a mile (1.6 km) in width, and several miles wide in the area around Grand Junction, where it is joined by the Gunnison River from the south. As the Colorado flows southwesterly across Utah, it is joined by only two important tributaries, the Green River from the north and the San Juan River from the east. This is "canyonlands" country; the river has incised deeply into an old plateau surface, and it occupies a steep-sided ragged canyon that twists and turns across the plateau in an intermittent series of magnificent entrenched meanders.

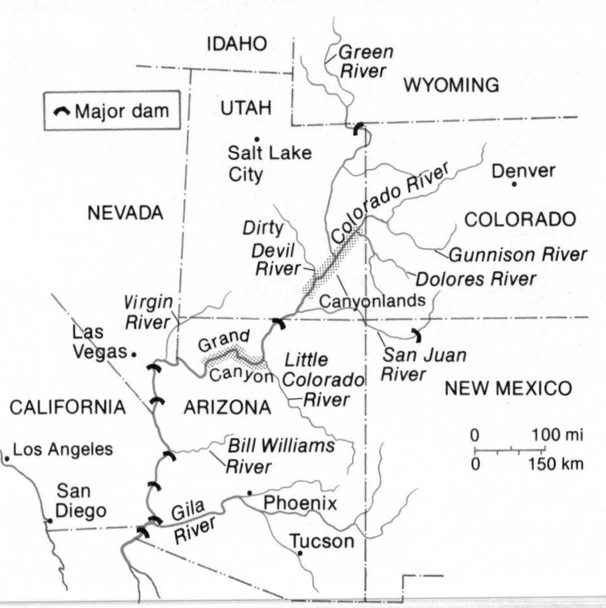

The premier exotic stream of North America is the Colorado River.

The Colorado River in eastern Utah flows through areas of limited relief, although there are many steep slopes. [E. C. La Rue, U.S. Geological Survey photo.]

tions that provide water from a permanent spring, or of exotic streams that flow from nearby mountains. The larger permanent lakes are almost all remnants of still larger bodies of water that were formed in a previously wetter climatic cycle. Utah's Great Salt Lake is the outstanding example in this country. Although it is the second largest lake wholly in the United States (after Lake Michigan), it is a mere shadow of the former Lake Bonneville, which was formed during the wetter conditions of Pleistocene time. A contemporary map of Nevada and Utah shows several other large lakes of similar origins. See Figure 17-7.

Some 600 miles (960 km) of the channel of the Colorado River are in deep canyons, but by far the most majestic portion is the 200-mile (320-km) stretch known as the Grand Canyon, which trends east-west across northern Arizona. The awesome statistics of the Grand Canyon—maximum depth of 6000 feet (1800 m), width of 5 to 18 miles (8 to 29 km), river averaging 300 feet (100 m) wide and 12 feet (4 m) deep throughout its course—are overshadowed only by its scenic grandeur. Over the years, the raging brown and red torrents of the Colorado River have carried an average of half a million tons (455,000 t) of sediment past a given point every day. The titanic canyon itself is a relatively young feature, having been carved in less than 10 million years by ordinary fluvial processes as the local area was being gently uplifted and the river continued its vertical erosion. The steplike profile of the canyon sides is festooned with projecting spurs, temples, and pyramids, many of which would be called mountains if they rose above some level surface, but here they are mere details on the chasm walls. These remarkable features expose to view bedrock outcrops from many eras of geologic time.

Downstream from the Grand Canyon, the Colorado River flows much more leisurely, with only a few relatively minor canyons interspersed among broader reaches where flat floodplains and braided, meandering channels characterize the valley. Eventually, at about the location of the international border, the valley merges with the delta, previously discussed on page 360.

In its former natural flow pattern, the Colorado River was typical of most major arid-land exotic streams in that it experienced great fluctuations in volume and carried an enormous amount of sediment. Springtime runoff in its distant Rocky Mountain headwaters produced an annual flood that moved slowly downstream to reach the delta by midsummer. Summer discharge in the lower reaches of the river was characteristically 20 to 30 times as great as the flow in January and February. Its muddiness was unchallenged among major American rivers, amounting to about 100,000 acre-feet of sediment annually.

Now, however, all of this has been significantly modified by human endeavor. The Colorado River today is the most thoroughly "controlled" of all the significant arid-land rivers of the world. Eight enormous dams have been constructed along the main stream (all but one in the lower reaches), in addition to several others on major tributaries. These damming projects, begun in the early 1930s, have greatly minimized flood problems, stabilized seasonal flow regimes, and markedly reduced the sediment load. The huge reservoirs created by the dams not only serve as sediment traps but also provide a reliable source of stored water for diversion to productive uses, often far beyond the Colorado valley itself. The most notable diversion project is the Colorado River Aqueduct, which conducts water to Los Angeles, but there are several other diversion projects, primarily for irrigation purposes. Also, hydroelectric power is generated at several of the dams.

The Colorado River, then, is an example of a major exotic stream in an arid region that has been transformed, by enormous human effort and expense, into a partially controlled and channelled waterway.

The inner gorge of the Grand Canyon of the Colorado River. [Courtesy U.S. Department of the Interior, Geological Survey.]

Fluvial Erosion in Arid Lands

Although fluvial erosion in desert areas is active during only a small proportion of any given year, its accomplishments are rapid and effective, and its results are conspicuous. In areas of any significant relief, large exposures of bedrock are common, due to the lack of soil and vegetation. The bedrock is attacked directly by mechanical weathering and by fluvial processes during the rare rains.

The result is that variations in rock type and structure produce conspicuous differences in slope angle and shape. *Differential erosion* shows up prominently. See Figure

Figure 17-6 The playa of Owens Lake in the Owens Valley of California, with the west slope of the Sierra Nevada in the background. [TLM photo.]

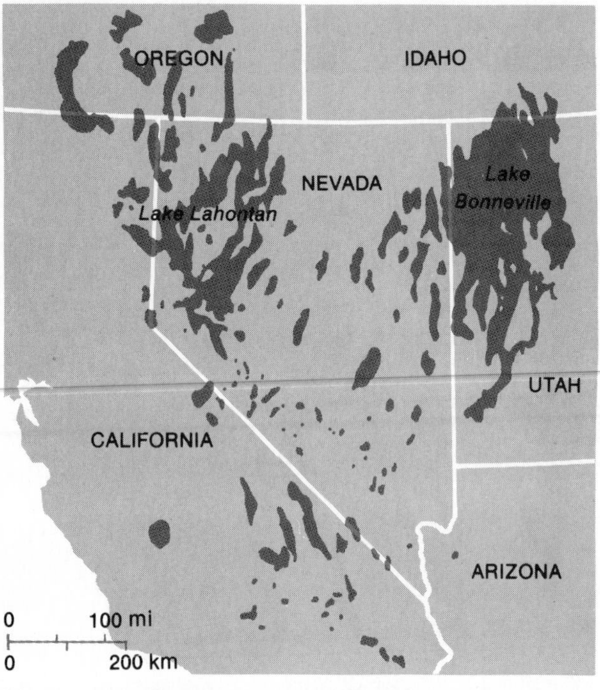

Figure 17-7 Pleistocene lakes of the intermontane region of the United States.

17-8. Every rock formation has a different resistance to the forces of weathering and erosion that act upon it; these variations are most striking in the dry lands because they are not obscured or lessened by the "insulation" of soil and vegetation. Thus the more resistant rocks stand out in bold relief, forming cliffs, pinnacles, spires, and other sharp crests. Softer rocks wear away more rapidly and produce slopes of gentler and less abrupt angles. In areas dominated by crystalline (igneous or metamorphic) bedrock, the formations tend to be massive, and differential erosion is less noticeable because the differences in rock resistance are less distinct. Where sedimentary strata predominate, differential erosion achieves maximum expression because of significant variations from layer to layer; such areas often have vertical escarpments and very abrupt changes in slope angle. Whatever the bedrock, though, fluvial erosion in desert uplands tends to produce steep, rugged, rocky surfaces.

Along the lower slopes of mountains and hills in desert areas a very different kind of surface is often developed by fluvial erosion. This is a gently inclined bedrock platform, called a *pediment*, that extends outward from

Figure 17-8 The effects of differential erosion are conspicuous on the Red Cliffs in western Colorado, near Gateway. The more resistant layers near the top form an abrupt escarpment, whereas the softer layers below are weathered and eroded into somewhat gentler slopes. [TLM photo.]

Focus on a Major Salina, Australia's Lake Eyre

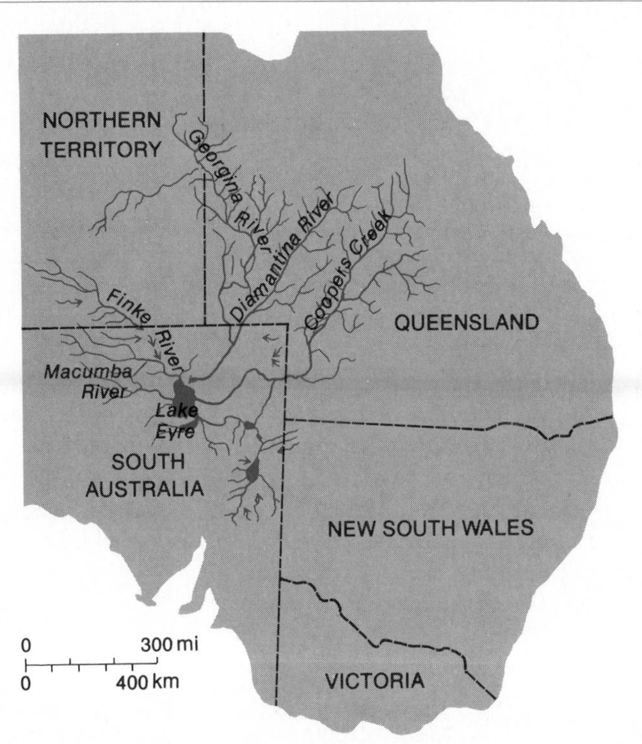

The drainage basin of Australia's Lake Eyre.

One of the largest basins of interior drainage in the world occupies the east-central portion of Australia. All the runoff from approximately 540,000 square miles (1,382,000 km^2) of the "Outback" drains toward the extensive salina of Lake Eyre, but precious little of the flow ever reaches that ultimate destination, due to evaporation and seepage along the way. The Lake Eyre basin portrays a classic example of desert hydrography on a grand scale.

The basin occupies a structural depression that has experienced a more or less continuous period of deposition of lacustrine (lake) and fluvial sediments for the last 100 million or so years. This vast internal watershed displays remarkably little relief. It has a flat to gently undulating surface that merges into a seemingly limitless horizon in all directions. The rivers that drain the basin in a generally centripetal pattern toward Lake Eyre are all intermittent. The lengthy channels of the Georgina and Diamantina rivers and Coopers Creek drain southwesterly from Queensland, the Finke/Macumba system trends southeasterly from the Northern Territory, and many shorter stream courses converge from other points of the compass.

Lake Eyre is not the only significant salina/playa in the basin, but it is the largest and the lowest, so it functions as the ultimate drainage terminal in the system. The Lake Eyre playa has an area of about 3600 square miles (9200 km^2), and its lowest point is 46 feet (14 m) below sea level. Several other large playas in the basin, particularly southeast of Lake Eyre, are interconnected by intermittent drainage channels. The total complex of ephemeral streams and salinas is a very complicated one in which all segments are rarely joined because of scarce precipitation and high rates of evaporation.

The rivers of the system rise in areas of low and erratic rainfall (annual average of 10–15 inches or 25–38 cm) and are directed toward the most arid portion of the driest continent; Lake Eyre itself is entirely within the 5-inch (12.7-cm) isohyet. In times past, the rivers probably carried much more water than they do today, but now they flow in deeply alluviated valleys only after heavy rains. The rivers have very gentle gradients; in their lower reaches, the slope is less than one foot per mile (19 cm/km). These seldom-wet streams characteristically diverge into a network of braided and interlacing channels that spread widely and are sometimes lost in windblown sand. Often a fringing gallery of riparian trees is the only clear evidence of a stream course.

Although total rainfall in this vast watershed is scanty, an occasional heavy storm sets one or more rivers into flow. Episodically, general flooding of the Queensland rivers causes a vast overflow in the southwestern portion of that state, called the "channel country." This creates what amounts to a temporary, discontinuous, shallow, moving lake over literally thousands of square miles of erstwhile dry lands.

Minor floods bring water to a portion of Lake Eyre every decade or so, but only extraordinary circumstances create major flows sufficient to fill the lake. The huge catchment basin is generally so dry that one year of heavy rains only serves to soak the ground. Two or more consecutive years of very abnormal rainfall are necessary to cover the lake bed completely. Such flows occur only two or three times per century (most recently in 1890–1891, 1949–1950, and 1973–1974). The potential evaporation rate is about 8 feet (244 cm) annually, and some $3\frac{1}{2}$ years are needed for the lake to dry completely after a full flood.

Most of the time Lake Eyre is dry, flat, and hard-surfaced, but conditions can change rapidly. For example, in 1963, Sir Donald Campbell mounted an expedition from England to attempt to establish a world land-speed record by driving a racing car over the smooth flatness of the Lake Eyre playa. His entourage came halfway around the world to Lake Eyre, a base camp was established, and the driving course was demarcated on the playa. Shortly before the record attempt, however, a rare storm brought just enough runoff to spoil the surface, and the effort had to be abandoned, despite an expenditure of close to a million dollars. The following year, Campbell repeated the endeavor and established a new world land-speed record (since broken repeatedly) on the Lake Eyre salina. Interestingly enough, a mammoth flood just a decade later (March 1974) permitted water skiers to mount a more modest expedition to Lake Eyre and to speed across the same race course in water 15 feet deep! Thus are the vicissitudes of flood and drought in the erratic regimes of desert lake basins.

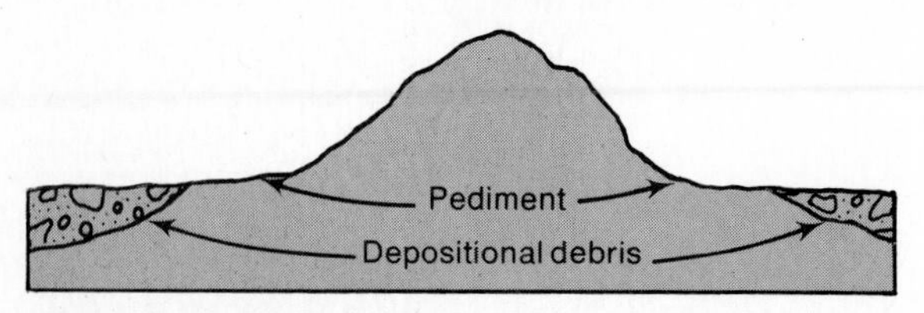

Figure 17-9 A pediment is an erosional surface usually found in the piedmont zone of a desert mountain range. It is often partly or completely covered by depositional debris.

the mountain front. See Figure 17-9. A pediment is an erosional surface produced by the entire complex of fluvial processes that wears away the slopes of the uplands, although stream dissection is not a critical agent in its formation because the pediment surface is relatively smooth and has very few lines of concentrated drainage. As the slopes of the upland are worn back by erosion, the pediment becomes more extensive. Pediments are found in all deserts and are sometimes the dominant terrain features. They are not easily recognizable, however, because almost invariably they are covered with a veneer of debris that has been deposited by water and wind, sometimes to a considerable thickness.

Stream channels in desert areas tend to be deeply incised. Both mountain canyons and flatland arroyos typically have flat, narrow bottoms and steep, often near-vertical, sides. Deep accumulations of sand and other loose debris usually cover the channel bed, although flash floods sometimes scour away all alluvial fill right down to the bedrock.

Fluvial Deposition in Arid Lands

Despite the emphasis on differential erosion and the prominence of such stark erosional features as cliffs, escarpments, and steep bedrock slopes, depositional landforms are often even more conspicuous in desert areas. This is not true in desert hills and mountains, but in most other desert landscapes, depositional features are more notable than erosional ones.

Desert uplands are dominated by residual landforms, shaped by weathering and erosion. Depositional features consist mostly of talus accumulations at the foot of steep slopes and deposits of alluvium and other fragmented debris in intermittent stream channels, which represent bedload left behind with the subsidence of the last flood.

The term *piedmont* has a generic meaning that refers to the zone at the "foot of the mountains." (It should not be confused with the term *pediment*, discussed above, which comes from the same root but refers to a specific landform.) The piedmont zone at the base of desert mountain ranges is one of the most prominent areas of active fluvial deposition. There is normally a pronounced change in the angle of slope at the mountain base (the *piedmont angle*), with a steep slope giving way abruptly to a gentle one. This is a logical area for significant accumulation of rock debris because the break in slope greatly reduces the velocity of any sheetwash or streamflow that crosses the piedmont angle, and the slowed flow has significantly diminished load capacity. Moreover, the streams issue more or less abruptly from canyons onto the more open piedmont slope, where they are freed from lateral constraint. The resulting fluvial deposition in the piedmont zone of deserts often reaches depths of hundreds of feet (dozens of meters).

The flatter portions of desert areas, particularly in basins of interior drainage, also hold a prominent accumulation of water-deposited material. Any sheetwash or streamflow that reaches into such low-lying flatlands usually had to travel a considerable distance over low-angle slopes, which means that both the volume and velocity of flow are likely to be limited. Consequently, larger rock fragments are rarely transported into the basins; instead, they are covered with fine particles of sand, silt, and clay, sometimes to a considerable depth.

The Work of the Wind

The irrepressible winds of the desert create spectacular sand and dust storms and continually reshape the minor surface configurations of depositional landforms. However, the effect of wind as an external sculptor of terrain is very limited, with the important exception of such relatively impermanent features as sand dunes. The fretting action of wind etches and pits the surface of rock outcrops, but it plays no major role in the gross configuration of the surface. Where loose sand is involved, however, wind is usually the dominant agent in shaping its surface form.

The term *wind* is theoretically restricted to horizontal air movements. Some turbulence is nearly always involved, however, so there is at least a limited vertical component of flow as well. In general, the motion of air passing over the ground is similar to that of water flowing over a stream bed. Wind moves at varying velocities at different times, places, and levels of the atmosphere. In a thin layer right at the surface, the speed of the wind is reduced to zero, but it flows progressively faster at higher elevations. The shear developed between different layers of air moving at varying speeds causes turbulence similar to that in a stream of water. Moreover, wind turbulence can be augmented by heating from below, which causes expansion and vertical motion. Thus wind has capabilities similar to that of running water for erosion, transportation, and deposition, but at a much lower order of magnitude.

Aeolian processes are those related to wind action (Aeolus was the Greek god of the winds). Aeolian processes are most pronounced, widespread, and effective in desert areas, although they are not restricted to dry lands. Wher-

Figure 17-10 The wind is sometimes prominent in shaping surfaces that are covered with loose particles, as in this west Texas scene. [USDA-Soil Conservation Service photo.]

ever fine-grained unconsolidated sedimentary material is exposed to the atmosphere, without benefit of vegetation, moisture, or some other form of protection, it is liable to wind action. In most nondesert areas, the effect of the wind is relatively insignificant, except along beaches where loose sand is frequently available in quantity.

Aeolian Erosion

The erosive effect of wind movement is simple and clear-cut; it can be divided into two categories—deflation and abrasion. *Deflation* consists of shifting loose particles by blowing them into the air or rolling them along the ground. The wind is not strong and buoyant enough, except under extraordinary circumstances, to move anything more than dust and small sand grains, so no significant landforms are created by deflation. Sometimes a *blow out* or *deflation hollow* may be formed; this is a shallow depression from which an abundance of fine material has been deflated. Most blowouts are small, but some are known that exceed a mile (1.6 km) in diameter. Deflation is also a factor in the formation of a reg surface, although fluvial erosion is contributory.

The other type of aeolian erosion is *abrasion*, which is analogous to fluvial abrasion discussed previously except that it is much less effective. Whereas deflation is accomplished entirely by air currents, abrasion requires the use of tools. The wind drives sand and dust particles against rock and soil surfaces in a form of natural sandblasting. Wind abrasion does not construct or even significantly shape a landform; it merely sculptures those already in existence. See Figure 17-10. The principal result of aeolian abrasion is the pitting, etching, faceting, and polishing of exposed rock surfaces, and the further fragmenting of rock fragments.

Aeolian Transportation

Rock materials are transported by wind in much the same fashion as they are moved by water, but less effectively. The finest particles are carried in suspension as dust. Strong turbulent winds can lift and carry thousands of tons of suspended dust; some dust storms extend for thousands of feet (hundred of meters) above the Earth's surface and may move material over more than a thousand miles (1,600 km) of horizontal distance. See Figure 17-11.

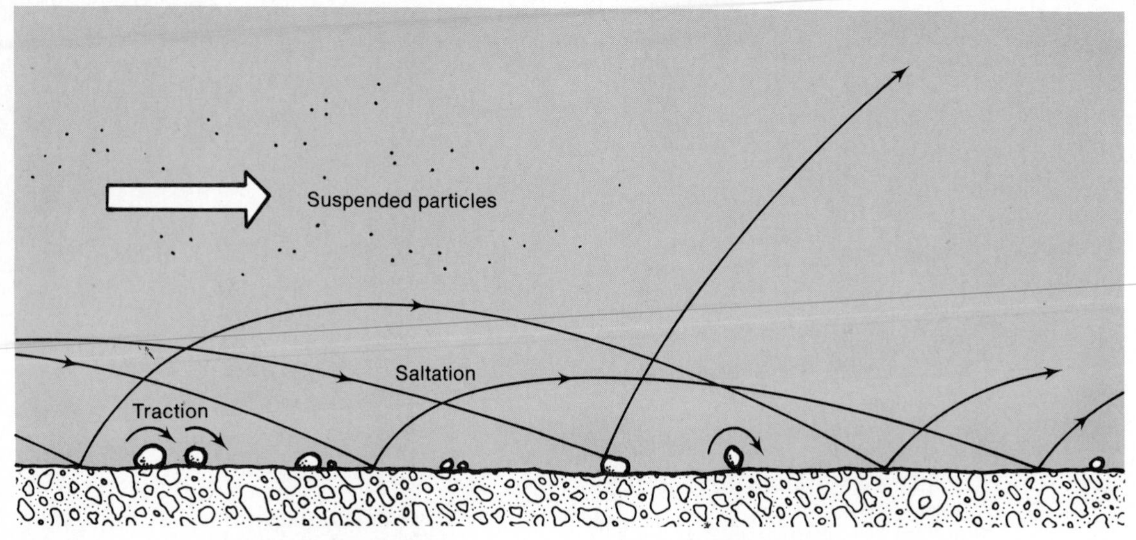

Figure 17-11 Aeolian transportation. Fine particles are carried in suspension; larger particles are moved by saltation and traction.

Larger particles are moved by wind through *saltation* and *traction*, just as in streamflow. Wind is unable to lift particles larger than the size of sand grains, and even these are likely to be carried only a foot or two above the surface. Indeed, most sand, even when propelled by a strong wind, leaps along in a low curved trajectory, striking the ground at a low angle and bouncing onward; this is called *saltation*. Larger particles are rolled or pushed along the ground by the wind, as in a stream, in a process called *traction*. At the same time, the entire surface layer of sand moves slowly downwind as a result of the frequent impact of the saltating grains; this process is called *creep*, but it should not be confused with soil creep. A true sandstorm, then, is a cloud of generally horizontally moving sand that extends for only a few inches or feet above the surface; a person standing in its path would have his legs peppered by sand grains, but his head would probably be above the level of the sand cloud. The abrasive impact of a sandstorm, while having little erosive effect on the terrain, may be quite significant for the works of man near ground level; wooden poles and posts can be rapidly cut down by the sandblasting if they are not protected artifically, and cars traveling through a sandstorm are likely to suffer etched windshields and chipped paint.

Aeolian Deposition

Sand and dust moved by the wind are eventually deposited when the wind dies down. The finer material, which may be carried long distances, is usually laid down as a thin coating of silt and has little or no landform significance. The coarser sand, however, is normally deposited locally. Sometimes it is spread across the landscape as an amorphous sheet, often called a *sandplain*, with no particular surface shape or significant relief. The most notable of all aeolian deposits, however, is the *sand dune*, in which loose windblown sand is heaped into a mound or low hill.

Desert Sand Dunes Dune topography comprises one of the world's most distinctive landscapes. It should not be thought of as a single landscape, however, for dune patterns are almost infinite in their variety. In some instances, dunefields are composed entirely of unanchored sand, comprised mostly of uniform grains of quartz (occasionally gypsum, rarely some other minerals) and usually monochromatically dun-colored, although sometimes a brilliant white.

Some dunefields are astonishingly extensive; one area of unbroken dunes in the Sahara Desert, for example, is half as large as France. In many cases, however, the expanse of live dunes is interrupted, sometimes sporadically and sometimes frequently, by such nondune features as playas, claypans, arroyos, or rocky outcrops. Another characteristic arrangement is one in which the dunes are mostly or entirely *fixed*, i.e., anchored in position and no longer shifting with the wind. Various factors can cause the fixing of dunes, but it is usually the result of anchoring by vegetation.

Dunes themselves provide little nourishment or moisture for plant growth, but desert vegetation is remarkably hardy and persistent and often is able to survive in a dune environment. Where vegetation manages to gain a foothold, it may proliferate and serve to anchor the dunes.

Live dunes normally move at the behest of local winds. They may change their position, size, and arrangement with changing wind direction and velocity. The basic mechanics of the situation are that the wind erodes the windward slope of the dune, forcing the sand grains up and over the crest of the dune to be deposited on the steeper leeward side, or *slip face*. See Figure 17-12. If the wind prevails from one direction for many hours, the dune may migrate downwind without changing its essential shape. Such migration is usually slow; but in some cases, dunes can move several hundred feet (several dozens of meters) in a year.

Several characteristic dune forms are widespread in the world's deserts; we will consider three of the most common. See also Figure 17-13.

1. Best known of all dune forms is the *barchan*, which usually occurs as an individual dune migrating across a nonsandy surface, although barchans may also be found in groups. A barchan is crescent-shaped, with the *horns* or *cusps* of the crescent pointing downwind. Sand movement in a barchan is not only over the crest, from windward side to slip face, but also around the edges of the crescent to extend the horns. Barchans form where strong winds blow consistently from one direction. They tend to be the fastest moving of all dunes and are found in all deserts except those of Australia.
2. Related to barchans are *transverse dunes*. These, too, are crescentic, but their shape is less uniform than the barchans. Transverse dunes occur where the supply of sand is much greater; normally the entire landscape is sand-covered. Like barchans, their convex sides face the prevailing direction of wind. Their crests are perpendicular to the wind vector, and they are aligned in parallel waves across the land. They migrate in the same downwind fashion as barchans, and if the sand supply decreases, they are likely to break up into barchans.
3. The *seif* or *longitudinal dune* is rare in American deserts, but it may be the most common dune form in other parts of the world. Seifs are long, narrow dunes that usually occur in multiplicity and in generally parallel arrangement. Typically they are a few dozen to a few hundred feet in height, a few tens of yards in width, and have a length that is measured in miles or even tens of miles. Their origin is still not well understood, although their lengthy, parallel orientation apparently represents an intermediate direction between two dominant wind regimes.

Figure 17-12 Loose sand, activated by the wind, frequently moves up and over an active dune, being removed from the windward side and deposited on or beyond the slip face.

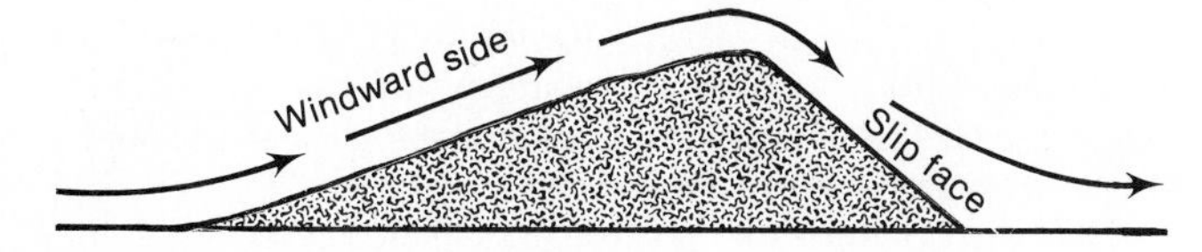

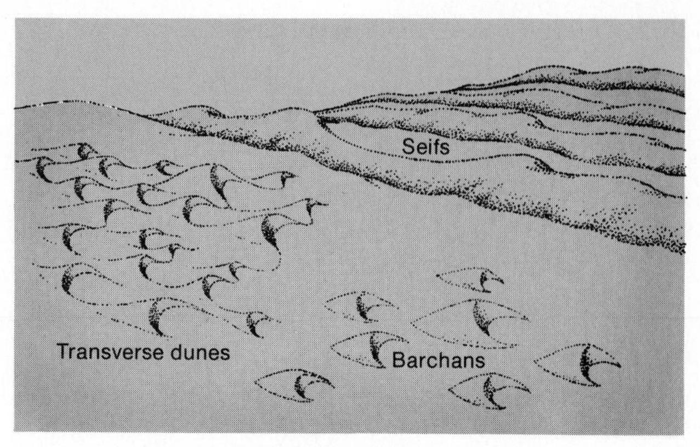

Figure 17-13 Common dune forms.

Coastal Dunes Winds are also active in dune formation along many stretches of ocean and lake coasts, whether the climate is dry or otherwise. On almost all flattish coastlines, waves deposit sand along the beach, where it is exposed except perhaps at high tide. A prominent onshore wind can blow some of the sand inland, often forming dunes. In some areas, particularly if vegetative growth is inhibited, the dunes slowly migrate inland, occasionally inundating forests, fields, roads, and even buildings. See Figure 17-14. Most coastal dune aggregations are small, but they sometimes cover extensive areas. The largest area is probably along the Atlantic coastline of southern France, where dunes extend for 150 miles (240 km) along the shore and reach inland for 2 to 6 miles (3.2 to 9.6 km).

Figure 17-14 Considerable effort has been expended along this stretch of the Maryland shore to anchor mobile sand dunes by means of fences and vegetation. [USDA-Soil Conservation Service photo.]

One of the most remarkable and extensive landform complexes in the world is represented by the distinctive dunefield deserts of Australia. More than 700,000 square miles (1,800,000 km²) of the continent, nearly one-fourth of its total area, are covered with a repetitious pattern of parallel sand ridges. The most expansive tracts of sandridge country are in the named deserts—Simpson, Tanami, Gibson, Great Sandy, and Great Victoria—but sand ridges are also widespread in other portions of central and western Australia.

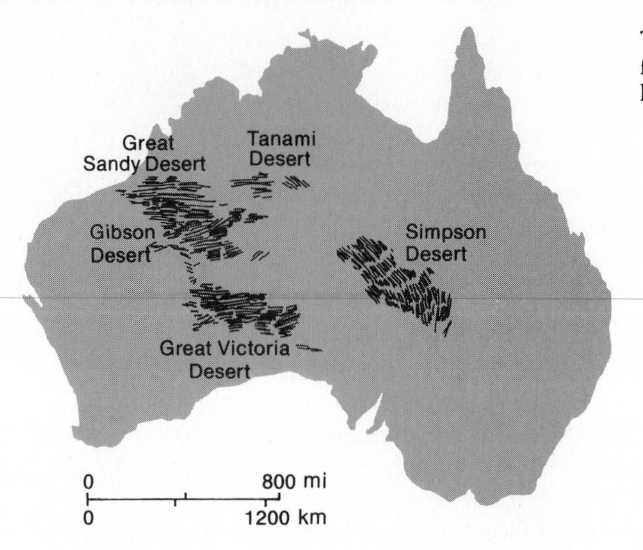

The major active dunefield deserts of Australia.

These Australian dunefields have several unusual attributes. The most striking characteristic is the seif dune form which is a relatively narrow, linear ridge of sand that is often tens of miles in length. Seifs are not uncommon in most other deserts of the world, but they are relatively much more prevalent and are developed on a greater scale in Australia than elsewhere. (The barchan, so well represented in the dry lands of other continents, is apparently absent from Australia.) In some parts of Australia, the dune pattern is complex and confused, but over the vast majority of the sandridge country, the seifs occur with incredible parallelism and regularity for thousands of square miles. For example, Charles Sturt, the first European to penetrate the Simpson Desert (in 1845), wrote:

> It was a remarkable fact that . . . after an interval of more than 50 miles, the same sandridges should occur, running in parallel lines at the same angle as before into the very heart of the interior, as if they absolutely were never to terminate.

In a similar vein, C. T. Madigan, who led the first expedition to cross the Simpson Desert (in 1939), reported:

> . . . the structure is astonishingly uniform and simple. There are no complications of pattern, no spurs or cross ridges, nothing but a succession of sandy lanes separated by low ridges of sand, ribbing the level surface in straight and parallel lines. . . . Of few places surely could it be said that there is nothing in any one particular locality to distinguish it from another a hundred miles away.

Another unusual characteristic of Australian dunefields is that very few of the dunes are unvegetated. In striking contrast to the large areas of shifting sands in most deserts on other continents, essentially all of the Australian sand ridges are fixed by vegetation; "walking" dunes are almost totally absent.

Details of the origin of the dunefields are still unclear, but apparently most were formed in a relatively recent period of somewhat greater aridity, with the sands themselves perhaps being the result of stream deposition in a still earlier, more humid period. Fossil dunes have been discovered under some of the present dunefields, and there are more northerly areas (particularly in the northwestern shoulder of the continent) where thoroughly anchored and completely vegetated dune systems occur as clear indicators of drier times in the past.

Whatever the paleoclimatic evidence, however, the general concordance of circular dune trends around the center of the continent, with prevailing atmospheric circulation around the continental anticyclone, apparently indicates that present climatic conditions are not very different from those of the dune-forming period. This conclusion is reinforced by the generally progressive increase in stabilizing dune vegetation from the arid center of the continent to the north and northwest.

Most students of dune formation agree that the longitudinal trend of the seifs does not parallel a single dominant windstream but rather approximates the vector of two strong wind directions. Despite the striking parallelism of dune trend, there is a fairly frequent joining of adjacent dunes. Where such convergence takes place, the departure from parallelism is mostly restricted to the last few hundred feet (few tens of meters) before the junction, and the convergence invariably points downwind.

Sand ridges are still being initiated in some areas, particularly in the southern portion of the Simpson Desert and adjacent sections of the Lake Eyre basin. Apparently the most characteristic pattern of development is for mounds of sand to accumulate on the leeward side of playas. As winds sweep over or around the mounds, linear dunes form in the eddy area behind. Being in accordance with prevailing winds, they may extend to great lengths and thus achieve the longitudinal seif form that is so characteristic.

The parallelism of uninterrupted seifs is not absolute. Interspersed among the dunefields are occasional playas, periodically flooded swamps, rises of higher ground, and even ephemeral stream courses. Moreover, nonlinear dune patterns sometimes intrude. Marked parallel linearity, however, is prevalent in most areas.

Spacing and height of the sand ridges are relatively uniform over broad areas. They tend to be between 30 and

The sand ridges of the Simpson Desert are spectacular for their linearity and parallelism. [TLM photo.]

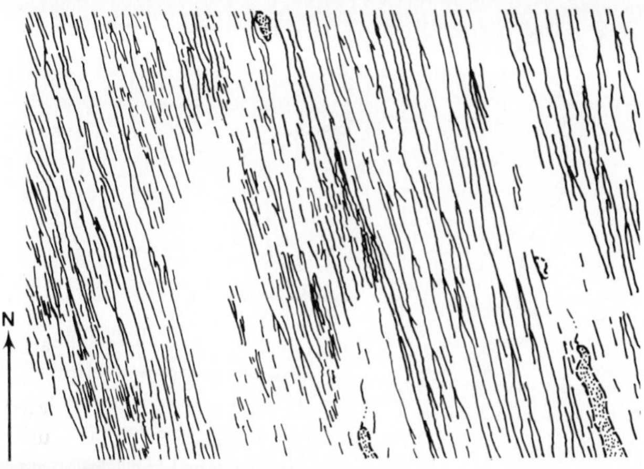

The pattern of longitudinal dunes in the Simpson Desert of South Australia, from the Poolowanna topographic sheet.

100 feet (9 and 30 m) high, and with a typical interval of from 1000 to 1500 feet (300 to 450 m) from one ridge to the next. Despite variations from these norms, the total volume of sand per unit area appears to vary little because where the dunes are higher, they tend to be more widely spaced, and vice versa. The feeling of linearity is reinforced by the extreme length of the dune ridges. Most extend for several miles or tens of miles. Some have been reported to be 200 miles (320 km) in length, but a maximum of half that figure is probably nearer reality.

From a practical standpoint, the most significant characteristic is the regularity; if one is traveling "with" the grain of a dunefield, progress is simple and easy along an inter-dune corridor with ridges paralleling the line of travel to right and left; but if one's route is across the grain, progress is slow and difficult, to say the least, as the steep sides of the dunes must be negotiated repeatedly. For example, the Madigan expedition crossed the Simpson Desert on camels from west to east in 1939, and their expedition journal describes the repetitive monotony of laborious passage over the insecure sands of 754 longitudinal dunes in 230 miles (370 km) of travel.

The character of individual sand ridges is much the same throughout Australia. The lower and middle slopes (sometimes referred to as the *plinth*) are normally smooth and stabilized by relatively close-growing grass or shrubs. The upper slopes are dotted with sparser vegetation, but the crestal zone of the ridge usually is composed of mobile sand, which shifts erratically with the wind of the day. Typically the upper slope on one side of the ridge is steeper than that on the other; for instance, in the Simpson Desert, the average east-facing upper slope angle is 25°, whereas that on the west is only 15°.

The sand itself comprises a relatively thin veneer; the deepest sand discovered in the Simpson Desert is only 120 feet (37 m). The sand on the ridges is normally fine-grained, composed mainly of quartz, and usually bright red in color. The sand in the swales between the dunes is less well sorted but tends to be finer than that of the crests.

The Australian dunefield deserts are by no means typical of the world's arid regions. They represent an extreme development of a landscape dominated by one particular type of landform, which is particularly notable because of the remarkable repetition of pattern over such a vast area.

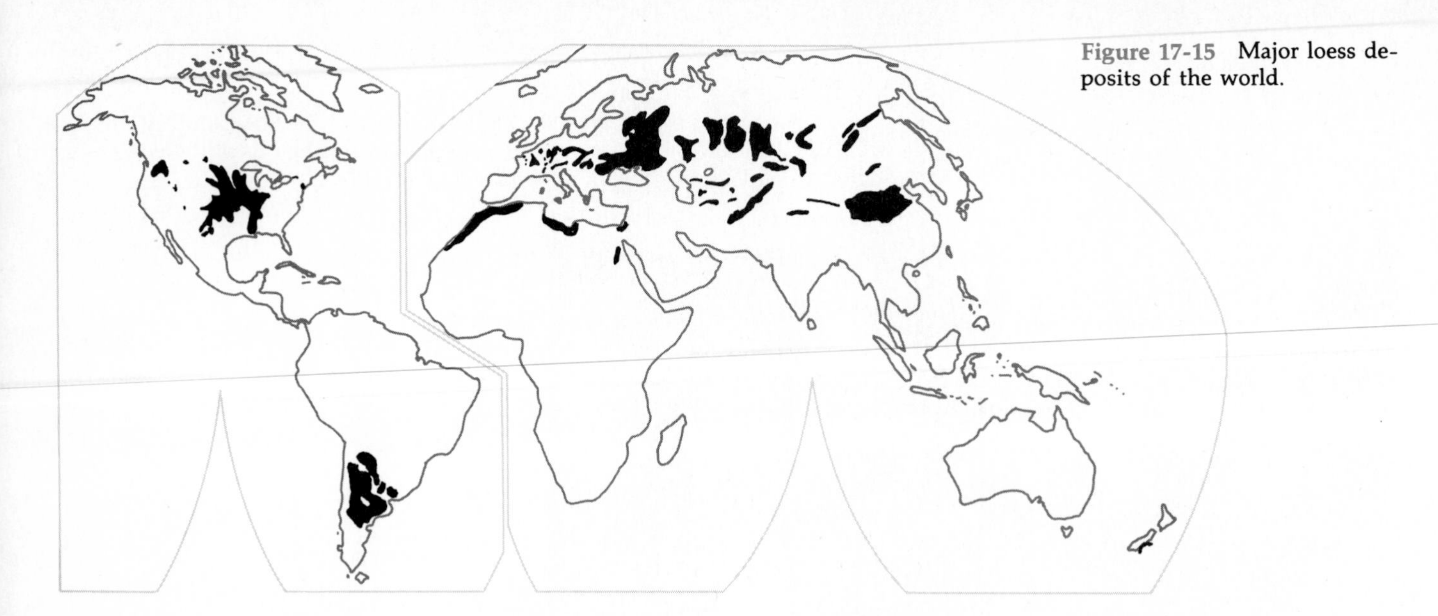

Figure 17-15 Major loess deposits of the world.

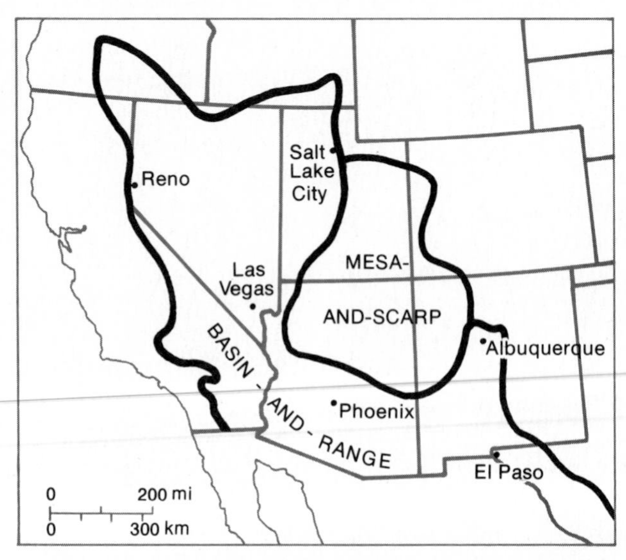

Figure 17-16 The southwestern interior of the United States contains two principal assemblages of landforms: basin and range, and mesa and scarp.

Loess A specialized form of aeolian deposit that is *not* associated with dry lands is loess. Loess (pronounced *luhss*) is a fine-grained, wind-deposited silt found in extensive deposits in the midlatitudes, particularly in the United States, the Soviet Union, China, and Argentina. See Figure 17-15. Despite its depositional origin, loess lacks horizontal stratification. Its most distinctive characteristic is its ability to stand in vertical cliffs. Loess possesses vertical jointlike cleavage planes, and although it is relatively soft and unconsolidated when exposed to erosion, it maintains almost vertical slopes.

The formational history of loess is not completely clear, although its immediate origin is obviously aeolian deposition. Much of the silt apparently was produced in association with Pleistocene glaciation. Strong winds picked up fine material from the floodplains of rivers that drained meltwater from the continental ice sheets and dropped it in some places in very thick deposits. Much loess also originated in dust storms from desert areas, especially in central Asia.

Loess deposits provide fertile possibilities for agriculture, and they comprise some of the most productive land in central United States, central Europe, northeastern China, and Argentina. Also, in China particularly, numerous cave dwellings have been excavated in loess because of its remarkable capability of standing in structurally rigid vertical walls.

Some Characteristic Desert Landform Assemblages

The dry lands of the United States have many different landscapes. Two particular assemblages of landforms, however, are by far the most common and widespread: basin-and-range terrain and mesa-and-scarp terrain. Their pattern of development is repeated time and again over thousands of square miles of the West. See Figure 17-16. An examination of these two assemblages will be useful in understanding the processes of terrain evolution in a desert environment.

Basin-and-Range Terrain

Most of the southwestern interior of the United States is characterized by basin-and-range topography. Other landscape types are encountered sporadically, but basin-and-range country predominates over almost all of Nevada, portions of adjoining states (eastern California, southeastern Oregon, southern Idaho, western Utah, and western Arizona), and in an eastward extension across southern Arizona, southern New Mexico, and western Texas.

This is a land that is largely without external drainage. Only a few exotic rivers (notably the Colorado and Rio Grande) flow through or out of the region. This proliferation of basins of internal drainage is characteristic of dry climates. (In humid regions, the basins would fill with water, overflowing eventually to create external drainage outlets.) Scattered among the basins are numerous ranges of mountains and hills that comprise the other major topographic complex of the region. See Figure 17-17. These ranges are usually long and narrow in form and typically parallel in arrangement. Over most of Nevada, for example, the pattern of parallel ranges separated by small interior basins, all with a north-south orientation, is remarkably repetitious.

Basin-and-range terrain consists of three principal complexes of features: the ranges, the piedmont zone, and the basins.

The Ranges The ranges dominate the horizon in all directions; mountains are always in the distance, although in total area covered, the basins are actually more extensive. Some of the ranges are high, whereas others are quite low, but the prevalence of steep and rocky slopes presents an aura of ruggedness. See Figure 17-18. Although their diastrophic origin is varied (most have been uplifted by faulting, but others were formed by folding, or vulcanism, or in more complex fashion), their surface features have been shaped almost entirely by weathering, mass movement, and fluvial processes.

Ridge crests and peaks are usually sharp, steep cliffs are common, and rocky outcrops protrude at all elevations. Most of the ranges are seamed by numerous gullies, gorges, and canyons, which rarely house flowing streams. These dry drainage channels are usually narrow and steep-sided and have a V-shaped profile. Typically the channel bottoms are well supplied with sand and other loose debris.

In some areas the ranges have experienced an unusually long period of erosion and have been worn down to much lower relief. Where this has happened, the slopes are gentler, the summits more rounded, and the gorges less incised. Bare rock outcrops are still prominent, however. If the range stands in isolation and the alluvial plains and basins roundabout are extensive, the term *inselberg* ("island mountain") is applied to the mountain remnant.

The Piedmont Zone At the base of the ranges there is usually a sharp break in slope (the piedmont angle) that marks the change from range to piedmont. The piedmont is a sort of transition area from the steep slopes of the ranges to the near-flatness of the basins. Slope angles vary within the piedmont itself, but they are generally gentle, with somewhat steeper components near the upper margin.

Much of the piedmont is underlain by an erosional pediment, although the pediment is rarely visible at the surface of the land. It is normally covered with several feet of unconsolidated debris because the piedmont zone is particularly well suited for deposition. During the occasional periods of rainfall, streams come roaring out of the gullies and gorges of the surrounding ranges, heavily laden with sedimentary material. As they burst out of the mouths of the confining canyons onto the piedmont, their velocity is abruptly slackened, and their load capacity is dramatically reduced. Significant deposition is the inevitable result.

One of the most prominent and widespread topographic features to be found in any desert area is the *alluvial fan*. It is particularly characteristic of the piedmont zone of basin-and-range country. As a stream leaves the narrow confines of a mountain gorge and debouches into the open land of the piedmont zone, it breaks into distributaries that wend their way down the piedmont slope, sometimes cutting shallow new channels in the loose allu-

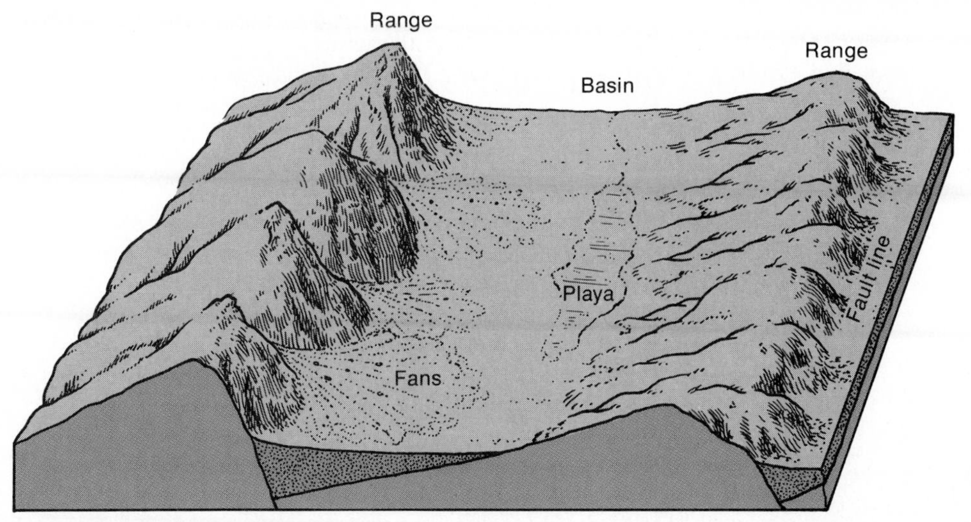

Figure 17-17 A typical basin-and-range landscape.

Figure 17-18 Slopes in a desert are typified by much rockiness and a rugged appearance, as here in the Whipple Mountains of southeastern California. [TLM photo.]

vium but frequently depositing more debris atop the old. Channels become choked and overflow, developing new ones. In this fashion a fan-shaped depositional feature is constructed below the apex at the mouth of the canyon. When one part of the fan is built up, the channelled flow shifts to another section and builds that up. See Figure 17-19. This means that the entire fan is eventually covered more or less symmetrically with sediment. As deposition continues, the fan is extended outward across the piedmont and onto the basin floor beyond.

A fan deposit has a rough and incomplete sorting of material. Larger boulders are dropped near the apex and finer material around the margins, with a considerable mixture of particle sizes throughout. The dry drainage channels across the fan surface frequently shift their positions as well as their balance between erosion and deposition.

Essentially every canyon mouth has a fan that grows down the piedmont slope. As they become larger, they often overlap with one another and are called *coalescing alluvial fans*. Continued growth and more complete overlap of the fans may eventually result in a continual alluvial surface all across the piedmont zone, slanting from the range toward the basin, in which it is difficult to distinguish between individual fans. This feature is known as a *piedmont alluvial plain* or *bajada*. See Figure 17-20. Near the mountain front, a bajada surface is undulating, with convex sections near the canyon mouths and concave surfaces in the overlap areas between the canyons. In the lower portion of the bajada, however, no undulations occur because the component fans have coalesced so thoroughly.

The Basin Floor Beyond the piedmont zone is the flattish floor of the basin, which has a very gentle slope from all sides toward some low point. Drainage channels across the basin floor are sometimes clear-cut, but more often they are shallow and ill-defined, frequently disappearing entirely before reaching their logical termination. The low

Figure 17-19 Sequential development of alluvial fans at the base of a desert mountain range.

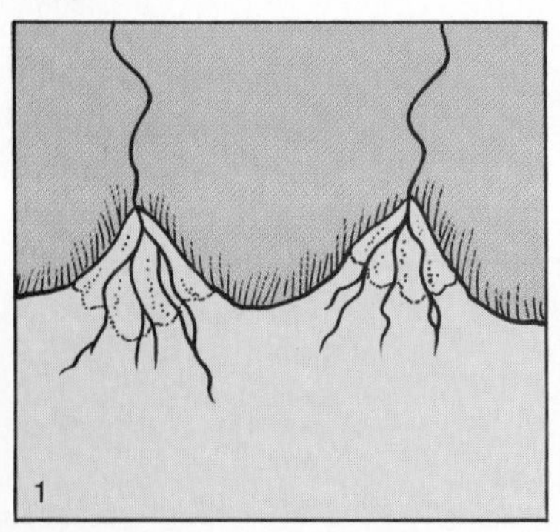

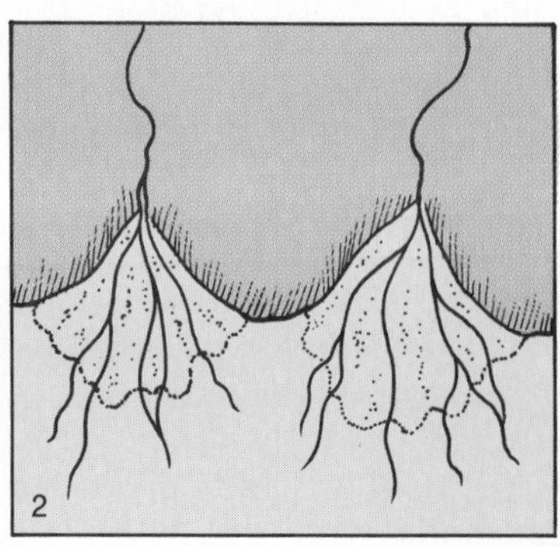

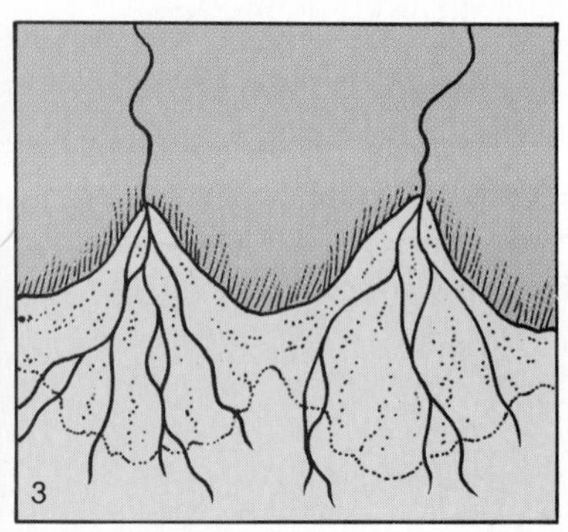

Figure 17-20 In the piedmont zone of this range, three alluvial fans have completely coalesced to form a bajada. These are the Virgin Mountains near Littlefield, Arizona. [TLM photo.]

point, then, functions as the ultimate drainage terminal for all overland and channelled flow from the near sides of the surrounding ranges, but only rarely does much water actually reach it. Most is lost by evaporation and seepage long before it can attain the center of the basin.

The focal point of the basin floor is usually in a dry lake bed or *playa*. If the playa surface is heavily impregnated with clay, it may be called a *claypan*. Frequently the playa surface has concentrations of salts, and such terms as *salina*, *saltpan*, and *alkali flat* are applied. Salt accumulations are commonplace on the floors of desert basins because soluble minerals are washed out of the surrounding watershed and brought to the drainage terminal. There the water evaporates or seeps away, but the salts cannot evaporate and are only marginally involved in seepage, so they become increasingly concentrated in the playa. The presence of the salt usually gives a brilliant whitish color to the playa surface. Many different saline compounds can be involved, and their accumulations are sometimes large enough to support a prosperous mining enterprise, which simply scrapes material off the surface.

In the rare occasions when water does flow into a playa, it becomes a *playa lake*. Such lakes may be extensive, but they are usually very shallow and normally persist for only a few days or weeks until they dry up. Saline lakes are marked by clear water and a salty froth around the edges, whereas shallow freshwater lakes have muddy water because they lack salt to make the silt settle.

The basin floor is surfaced with very fine-grained material because the contributory streams are too weak to transport large particles. Silt and sand predominate and are sometimes accumulated to remarkable depths. Indeed, the normal denudation processes in basin-and-range country tend to raise the floor of the basins. Debris is brought from the surrounding ranges and has nowhere to go but the basin of interior drainage. Thus, as the mountains are being worn down, the basin floor is gradually rising in elevation.

The fine material of basin floors is very susceptible to the wind, with the result that small concentrations of sand dunes are often found in some corner of the basin. Free-moving sand is rarely found in the center of the basin because winds push it to one side or another.

Mesa-and-Scarp Terrain

The other major landform assemblage of the desert Southwest is mesa-and-scarp terrain. It is most prominent in the "four corners country," where the states of Colorado, Utah, Arizona, and New Mexico come together, and is the dominant landscape type in the contiguous portions of those states. *Mesa* is Spanish for table, and implies a flat-topped surface. *Scarp* is short for escarpment, and pertains to steep, more or less vertical cliffs.

Mesa-and-scarp terrain is normally associated with a particular geologic structure—relatively horizontal sedimentary strata. See Figure 17-21. Such strata invariably offer differing degrees of resistance to erosion, so abrupt changes in slope angle are characteristic. The most resistant layers, typically limestone or sandstone, often play a dual role in topographic development: They can form a *caprock* or flattish erosional surface that is very extensive, and at the eroded edge of the caprock, the hard layer protects underlying strata and produces a steep vertical *cliff* or *escarpment*. Thus it is the resistant layers that are responsible for both elements of slope (*mesa* and *scarp*) that describe this terrain type.

Often the topography has a broad and irregular

California's Death Valley is justly famous for its extremes of temperature and relief. It is the locale of the highest temperature ever recorded in the Western Hemisphere, contains the lowest elevation in North America, and has one of the greatest ranges of local relief within a small area to be found in the United States. For our purposes, however, Death Valley is probably most important for the great variety of landforms that it displays. It is a vast topographic museum, a veritable primer of basin-and-range terrain.

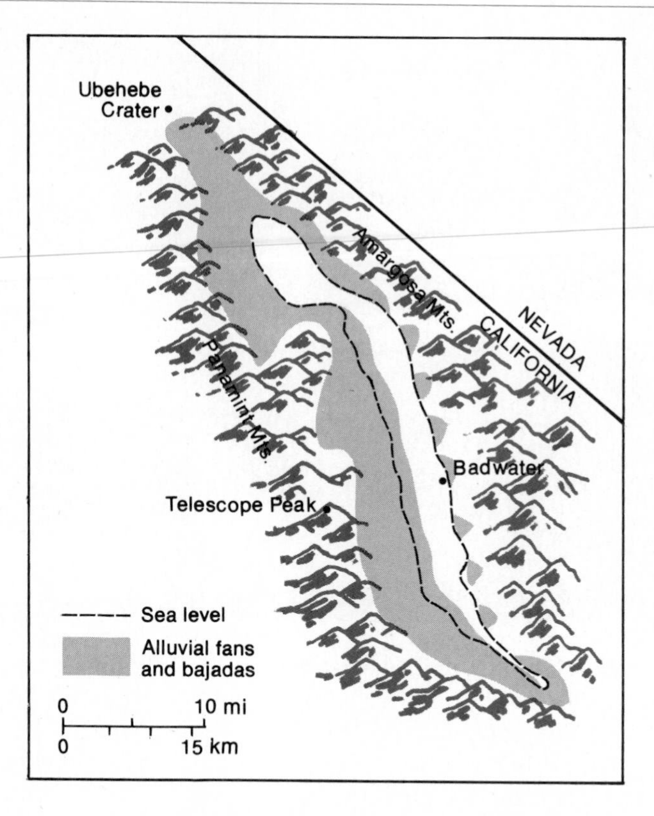

The setting of Death Valley.

The valley is located in east-central California, quite close to the Nevada border. It is a classic graben, with extensive and complex fault zones both east and west of the valley floor. The trough is about 140 miles (225 km) long, in a general northwest-southeast orientation; its width ranges from 4 to 16 miles (6 to 26 km). The downfaulting has been so pronounced that nearly 550 square miles (1425 km²) of the valley floor are below sea level, reaching an extreme downward depth of −282 feet (−86 m). Lengthy, upfaulted mountain ranges border the valley on either side. The Panamint Mountains on the west are the most prominent; their high point at Telescope Peak (11,049 feet or 3368 m above sea level) is only 18 miles (29 km) due west from the low point of the valley. The Amargosa Range on the east is equally steep and rugged but a bit lower overall (the highest peak reaches 8738 feet or 2663 m above sea level).

The most conspicuous topography associated with Death Valley is that of the surrounding mountains. The ranges exemplify all of the characteristic features of desert mountains—rugged, rocky, and generally barren. Their erosional slopes are steep, and in many places even steeper escarpments are related to faulting. The canyons that seam the ranges are invariably deep, narrow, V-shaped gorges. Many of the canyons in the Amargosa Range and some in the Panamints are of the type known as *wineglass canyons*: The cup of the "wine glass" is represented by the open area of dispersed headwater tributaries high in the range, the stem is the narrow gorge cut through the mountain front, and the base is the fan that opens out from the canyon mouth onto the piedmont.

The piedmont zone at the foot of the Panamints and the Amargosas is almost completely alluviated in one of the most extensive complexes of fans that can be imagined. Every canyon mouth is the apex of a fan (or of a fan-shaped debris flow), and the canyons are sufficiently numerous that most of the fans overlap with others to the north and south. The fans on the west side of the valley (those formed by debris from the Panamint Mountains) are much more extensive that those on the east side. For the most part, these Panamint fans are thoroughly coalesced into a conspicuous bajada that averages about 5 miles (8 km) in width, with the outer margin of the bajada as much as 2000 feet (610 m) lower than the canyon mouths at the upper edge of the fans. The Amargosa fans on the east side of Death Valley are much smaller, primarily because the fault pattern of the graben has tilted the valley eastward so that the lowest portions are nestled close to the base of the Amargosa Range, and a much longer slope extends downward from the Panamints. Thus the west-side fans have extended themselves outward onto the valley floor, whereas tilting has reduced the size of the east-side fans by creating shorter slopes and by facilitating their partial burial by valley-floor deposits. The Amargosa fans,

stair-step pattern. An extensive flat erosional platform terminates in an abrupt irregular escarpment, and a sheer rock cliff drops directly from the scarp edge, extending downward to the bottom of the resistant layer(s), from which a gentler (but commonly steeply inclined) slope continues down through softer strata. This inclined slope flattens downward to the next resistant layer, which forms either another cliff or another broad erosional platform ending in a cliff. The plains and canyons and plateaus of the "four corners country" are replete with variations of the pattern, repeated irregularly on scales large and small.

A typical Death Valley landscape includes rugged mountains, coalesced alluvial fans, and a barren basin surface. [TLM photo.]

then, are mostly short, steep, and discrete, so there is no bajada on the east side of Death Valley, although some of the fans do overlap with one another.

The floor of Death Valley is also of great topographic interest, although its flatness makes the features less easy to see and understand. The valley itself is filled with an incredible depth of alluvial material, most of which has been washed down from the surrounding mountains; in places the fill is estimated to consist of 3000 feet (915 m) of young alluvium resting atop another 6000 feet (1830 m) of relatively recent (Tertiary) sediment. The surface of the valley floor has little relief and slopes gently toward the low point near Badwater, which is a permanent salt pond located about one-third of the way north from the southern end of the valley. Drainage channels appear irregularly on the valley floor, trending toward Badwater. In some places, distinct braided channels appear; in other locations, the channels disappear in sand or playa.

Particularly in the middle of the valley, several extensive salt pans occur, deeply crusted with white salts. The thickest salt accumulations are found between Badwater and the large salt pans, with a depth of some 1200 feet (365 m) of mixed salt and sediment. Salt has been precipitated from drying water here through eons of time and is constantly being brought to the surface by capillary action from the water table below. Much of the surface in this area (called the Devil's Golf Course) is extremely rough and irregular. The recrystallizing salt expands and breaks the crust into uneven blocks. Wind-driven rain then dissolves and erodes the blocks into sharp ridges and pinnacles. Although the local relief of this jumbled surface is only a few feet, the roughness is so harsh that it is almost impassable. Several sand accumulations occur in Death Valley, with one large area of mobile dunes that covers some 14 square miles (36 km^2).

Considerable volcanic activity marks the geologic history of Death Valley, with lava outcrops showing up in many places in the mountains. The most notable volcanic landform is at the north end of the valley where Ubehebe Crater was produced by several explosive eruptions less than 2000 years ago. Volcanic ash was scattered over a radius of 3 miles (5 km) by these eruptions. More recently, perhaps only 250 years ago, eruptions created two smaller craters just south of Ubehebe.

During the most recent Ice Age (Pleistocene time), Death Valley was occupied by an immense lake. Lake Manly was more than 100 miles (160 km) long and 600 feet (960 km) deep. It was fed by three rivers that flowed into the valley from the west carrying meltwater from the Sierra Nevada glaciers. As the climate became drier and warmer, the lake eventually disappeared through evaporation and seepage, but traces of its various shoreline levels can still be seen at several places on the lower slopes; much of the salt accumulation is due to the evaporating waters of the lake.

The southwestern basin-and-range country has many landforms and many terrain complexes; Death Valley contains examples of almost all of them.

The flattish erosional platforms are sometimes very extensive; they are properly referred to as *plateaus* if they are bounded on one or more sides by a prominent escarpment. If a scarp edge is absent or relatively inconspicuous, the erosional surface is called a *stripped plain*.

The escarpment edge is worn back by ordinary fluvial erosion. The cliffs retreat, maintaining their perpendicular faces, as they are undermined by the more rapid erosion of the less resistant strata (often shale) beneath the caprock. When undermined, blocks of the caprock break off, usually along vertical joint lines. Throughout this process the harder rocks are the cliff-formers, and

The most famous mesa in the world is not a mesa at all. Mesa Verde is actually a plateau that rises abruptly and conspicuously from the surrounding flatlands of the southwestern corner of Colorado. The name was bestowed by early Spanish explorers because the broad surface of the plateau is covered with a forest of juniper and pine and thus was suggestive of a huge "green table."

The plateau is an enormous (about 250 square miles or 650 km²) erosional remnant, protected by a massive sandstone caprock from the fluvial processes that stripped away the surrounding material. The caprock formation and the underlying strata of softer sedimentary rock dip gently southward, causing the plateau surface to have a similar orientation—highest at the northern margin and inclined slightly downward toward the south. The expectable flatness of the "table top" of Mesa Verde is significantly interrupted, as the plateau surface is seamed by some two dozen long, narrow, parallel canyons that drain southward. The canyons have the typical U-shaped profile of dry-land drainage channels, with very steep sides and narrow flat bottoms. All the streams that cut the canyons are intermittent and carry flowing water only rarely. The interfluves, which occupy about three-fourths of the total surface area, retain the anticipated flatness.

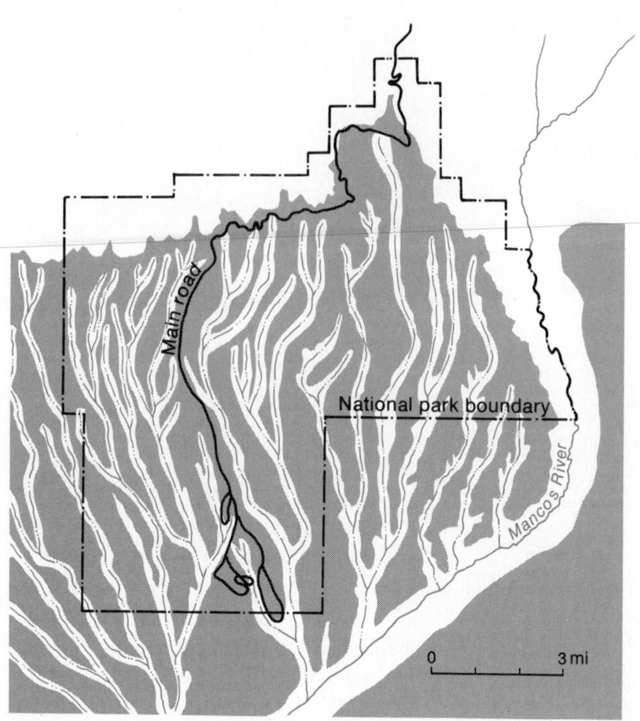

Mesa Verde is a prominent plateau that is seamed by a dendritic pattern of arroyos.

An abrupt escarpment surrounds Mesa Verde. Relief is greatest along the northern edge, where the almost unbroken crest of the scarp towers some 2000 feet (610 m) above the plains below. The

the less resistant beds develop more gently inclined slopes. Talus often accumulates at the base of the slope.

Although the term *mesa* is applied generally to many flattish surfaces in these dry environments, it properly refers to a particular landform. A *mesa* is a flat-topped, steep-sided hill with a limited summit area. It is an erosional remnant of a formerly more extensive surface, most of which has been worn away by erosion. Sometimes it stands in splendid isolation as a final remnant in an area where most of the previous surface has been removed,

Figure 17-21 Representative mesa-and-scarp terrain.

Mancos River flows along the eastern and southern sides of the plateau, its canyon walls forming a scarp almost as impressive as that on the northern side. The scarp on the western edge of Mesa Verde is very irregular and embayed and displays an almost vertical slope of more than 1000 feet (305 m) along most of its length. Offset from this western flank are a few towering buttes, residual outliers separated from the plateau by erosion. All around Mesa Verde the plateau edge has similar topographic expression. A vertical cliff drops abruptly from the crest for several hundred feet (several tens of meters). At the bottom of this cliff a high-angle erosional slope continues downward, only to be lost in talus rubble at the base.

Cliff Palace is a large cliff dwelling built under a massive alcove arch at Mesa Verde.

The topography of Mesa Verde, then, is dominated by the steep cliffs and flat tops that are representative of mesa-and-scarp terrain, both of these landform elements occurring on a massive scale. However, this prominent plateau is more renowned for its archaeology than for its topography. Preserved in the dry canyons of Mesa Verde is one of the most remarkable assemblages of aboriginal ruins to be found anywhere in the country.

A sedentary group of Indians, the Anasazi, moved onto Mesa Verde in the early years of the Christian era, and departed, for unknown reasons, about the end of the thirteenth century. For most of that time, the Anasazi resided on the flat plateau surface. However, during approximately the last century of their occupancy of Mesa Verde, the Indians moved off the mesa top into the canyons, where they constructed a magnificent series of settlements in seemingly inaccessible locations high on the vertical walls. Almost all these so-called cliff dwellings were built in naturally excavated chambers that geomorphologists call *alcove arches*. Such alcoves were formed when trickling surface water and percolating groundwater softened and dissolved cementing material of softer rocks under the more resistant overhanging cap rock. Subsequent wind and water erosion removed the loosened grains, forming shallow alcoves that slowly became enlarged.

These cliff dwellings functioned splendidly as protective residences, but they were inconvenient for most aspects of daily life. At any rate, they were totally abandoned in the late 1200s and not rediscovered for nearly 600 years.

but more commonly it occurs as an outlier not very distant from the retreating escarpment face to which it was once connected. Mesas are invariably capped by some sort of resistant material that serves to maintain its flattish summit form, even as its bulk is reduced by continuing erosion of its rimming cliffs.

A related but smaller topographic feature is a *butte*, which is an erosional remnant of very small surface area with clifflike sides; it rises conspicuously above the surroundings. Some buttes have other origins, but most are formed by the wastage of mesas until nothing is left but a final spire of resistant caprock. Buttes, like mesas, usually are found as outliers not far from some retreating escarpment face.

One of the most striking topographic features of arid and semiarid regions is the intricately rilled and barren terrain known as *badlands*. See Figure 17-22. In areas underlain by horizontal strata of shale and other clay formations that are poorly consolidated, overland flow after the occasional rains is extremely effective as an erosive agent. Innumerable tiny rills develop over the surface, which rapidly evolve into ravines and gullies that dissect

Figure 17-22 Extreme dissection, with short, very steep slopes, is characteristic of badlands terrain. This scene is in South Dakota near the town of Wall. [TLM photo.]

the land in an extraordinarily detailed manner. A maze of short but very steep slopes is etched in a filigree of rills, gullies, and gorges, with a great many ridges, ledges, and other erosional remnants scattered throughout. Erosion is too rapid to permit soil to form or plants to grow, so the badlands topography is a barren, lifeless wasteland of almost impassable terrain. Badlands are found in scattered locations (most of them mercifully small) in every western state, the most famous areas being in Bryce Canyon National Park in southern Utah and in Badlands National Monument in western South Dakota.

Mesa-and-scarp terrain is also famous for its numerous minor erosional features, most of which are produced by a combination of weathering and fluvial erosion. Probably the most spectacular are *natural bridges*, formed by stream erosion where entrenched meanders wear away the rock in a narrow neck between meander loops. *Arches*, on the other hand, are produced by undermining of a caprock ledge, with collapse of some of the material beneath the caprock crest. *Pedestals* (see Figure 17-23) and *pinnacles*, sometimes larger at the top than at the bottom, rise abruptly above the surroundings, their caps consisting of resistant material and their narrow basal stems subjected to concentrated weathering associated with rainwater trickling down the surface and dissolving the cementing material that holds the sand grains together, so that the loosened grains can easily be blown or washed away.

One other notable characteristic of mesa-and-scarp terrain is the prevalence of vivid colors in the landscape. The sedimentary outcrops and sandy debris of these regions often are resplendent in various shades of red, brown, yellow, and gray. These hues are mostly an expression of concentrations of iron compounds in the parent material.

Figure 17-23 The Goblet of Venus was a notable pedestal rock in southern Utah until it was pushed over by vandals. [TLM photo.]

Review Questions

1. List several ways in which topographic development in arid lands is distinctly different from that in humid regions.
2. Why are basins of interior drainage particularly prevalent in desert areas?
3. Characterize the hydrography of a typical desert.
4. What is the difference between a *playa* and a *salina*?
5. What is the difference between *pediment* and *piedmont*?
6. How does an *alluvial fan* differ from a *delta*?
7. How important is wind in the sculpturing of desert landforms?
8. Describe some typical landforms of basin-and-range country.
9. How does a *mesa* differ from a *butte*?
10. What is distinctive about *badlands* terrain? How did it come to be that way?

Some Useful References

Cooke, R. U., and A. Warren, *Geomorphology in Deserts*. Berkeley, Calif.: University of California Press, 1973.

Doehring, D. O., *Geomorphology in Arid Regions*. London: George Allen and Unwin Ltd., 1981.

Goudie, S. A., and A. Watson, *Desert Geomorphology*. London: Macmillan & Co. Ltd., 1981.

Mabbutt, J. A., *Desert Landforms*. Cambridge, Mass.: The M.I.T. Press, 1977.

Twidale, C. R., *Geomorphology, with Special Reference to Australia*. Sydney: Thomas Nelson (Australia) Ltd., 1968.

18 Coast and Karst Topography

In the immediately preceding chapters our attention has been focussed primarily on the terrain-sculpting activities of running water. Two other external processes also involve moving water—waves and currents. The coastlines of oceans and lakes are shaped significantly by the agitated edge of these waters, with the result that coastal terrain is often quite different from that just a short distance removed from direct contact with the water body. One other set of aqueous actions—the dissolving and deposition of soluble chemicals by underground water—also can produce distinctive terrain features.

Coastal Processes and Forms

The coastlines of the world's oceans and lakes extend for hundreds of thousands of miles. Every conceivable variety of structure, relief, and topographic expression can be found somewhere along these coasts. The distinctiveness of the coastal milieu, however, is that it is at the interface of the three major components of the Earth's environment—hydrosphere, lithosphere, and atmosphere. This interface is dynamic and highly energetic, primarily because of the restless motions of the waters. See Figure 18-1.

We have seen in Chapter 17 that wind is sometimes an important shaper of landforms in continental areas. Along the coastlines of lakes and seas the wind has an even greater influence on topographic development than in the continental areas. This is because the surface of a large body of water is subject to abrupt and rapid deformation by wind action, which engenders waves and currents that attain their maximum effectiveness on coastlines, producing certain topographic features that are found throughout the world in all latitudinal locations and in a variety of climatic regimes.

The wind is not the only generator of water movement, of course, although from the standpoint of geomorphic effects it is the most important. Oceanic coastlines also experience daily tidal fluctuations, caused by gravitational attraction of the sun and moon, which often move enormous quantities of water. Diastrophic events, particularly earthquakes, contribute to water motions, as does volcanic activity upon occasion. Even more fundamental are long-term changes in the sea level or lake level caused by tectonic and eustatic forces, and to a lesser extent by the actions of continental ice sheets.

The forces that shape the topography of coastlines of oceans and lakes are similar, with three important exceptions:

Figure 18-1 A shoreline is a place where hydrosphere, lithosphere, and atmosphere meet. It is often an interface of ceaseless movement and energy transfer. This scene is near Port Edward on the Natal coast of South Africa. [TLM photo.]

1. Tides are experienced along seacoasts but not along lakeshores.
2. The causes of sea-level fluctuations are quite different from the causes of lake-level fluctuations.
3. Reef structures are built by corals only in tropical and subtropical oceans, not in lakes.

With these exceptions, the topographic forms produced on seacoasts and lakeshores are generally similar. Even so, the larger the body of water, the more massive will be the effects of the coastal processes. Thus, topographic features developed along seacoasts are normally much larger, more conspicuous, and more distinctive than those found along lakeshores.

The Forces at Work

Coastal topography is produced in part by the "normal" geomorphic processes dealt with in previous chapters—diastrophism, vulcanism, weathering, mass wasting, running water, and wind. The seven "specialized" processes that contribute to the distinctive shaping of coastal features will be discussed below.

Long-term Changes in Water Level Sea-level changes can result from either the emergence or submergence of a landmass or from the increase or decrease of the amount of water in the oceans. During the recent history of our planet there have been various changes in sea level, sometimes of worldwide extent and sometimes relating only to one or more continents or islands. The changes of greatest magnitude and most extensive effect have been associated with the total volume of seawater before, during, and after Pleistocene glaciations, and with major tectonic movements of portions of the Earth's crust. Some coastlines show emergent characteristics, in which shoreline topography of the past is now situated well above the contemporary sea level; whereas other coasts have been submerged, drowning a portion of the previous landscape.

Most lake level changes have been less extensive and less notable. These are usually the result of the drainage or partial drainage of previous lakes, and their principal topographic expression is the exposure of ancestral beach lines and wave-cut cliffs above present lake levels.

Tidal Movements The waters of the oceans oscillate in a regular and predictable pattern due to the gravitational influence of the sun and moon. The tides rise and fall in a rhythmic cycle that takes place about every $12\frac{1}{2}$ hours, producing two high tides and two low tides within a duration of slightly more than one day on most (but not all) of the seacoasts of the Earth. The tidal range (vertical change in sea level) varies significantly from month to month and also from place to place, although it is very specific and predictable for any given locality. Despite the enormous amount of water that is moved by the tides with such frequency, the topographic effects are surprisingly small. Tides are significant agents of erosion only in narrow embayments, around the margin of shallow seas, and in passages between islands where there are strong enough currents to scour the bottom and erode cliffs and shorelines.

Waves Waves are generated largely by wind, and they represent the most important effect of wind in the shaping of landforms. Most waves in the open ocean or the interior of large lakes are no more than shapes; very little water is actually moved with the passage of a wave form, other than making a small oscillation. As the wave form reaches shallow water, however, the bottom interferes with the oscillatory motion. This causes an increase in wave height and a steepening of its slope, which quickly causes the wave to collapse or "break," sending a cascading mass of water shoreward. This surge can carry sand and rock particles onto the beach, or it can pound onto rocky headlands and sea cliffs with considerable force. When the wave is spent, the return flow sweeps seaward, carrying loose rock material with it.

Currents Currents consist of large volumes of water moving horizontally in a certain direction. Many kinds of currents flow in the oceans and lakes of the world, usually confined to surface areas, but they are sometimes deeper. Coastal topography is affected mostly by *longshore* or *littoral currents*, in which the water moves roughly parallel to the shoreline in a generally downwind direction. Such currents can be significant erosive forces on weak coastal rocks. Moreover, they are capable of transporting and depositing vast amounts of sand and other sedimentary material.

Stream Outflow The outflow of streams and rivers into oceans and lakes is included here as a coastal "force" because these outflows are such important feeders of sediments to the ocean flow system. They provide much of the sand and other sedimentary material that is moved around and deposited by coastal waters.

Ice Push The shores of bodies of water that freeze over in winter are sometimes significantly affected by *ice push*, which is usually the result of contraction and expansion of the ice due to freezing and thawing with weather changes. During expansion, the near-shore ice is shoved onto the land, where it can deform the shoreline topography by pushing against it, more or less in the fashion of a small glacial advance. Ice push is usually unimportant on seashores outside the Arctic and Antarctic, but it can be responsible for numerous minor alterations of the shorelines of high-latitude or high-altitude lakes.

Organic Secretions Several kinds of primitive aquatic animals and plants are capable of producing solid masses

Waves are rising and falling motions that are transmitted from particle to particle of a substance, carrying energy from one place to another in the process. Our interest here is in water waves, which are simply undulations in the surface layers of a water body. Although water waves appear to move water horizontally, this appearance can be misleading. Except where waves crest and break, the water itself is shifted only very slightly; rather, there is a movement of the wave form and a transmittal of energy. For example, when a pebble is dropped into a calm pool, the water surface will soon be covered with a series of concentric ripples that widen outward from the point of impact. The water appears to be diverging outward, but in fact it is simply rising and falling in a circle as the wave form moves outward and as the energy transfer from particle to particle sets up a train of diverging waves until the energy is finally dissipated.

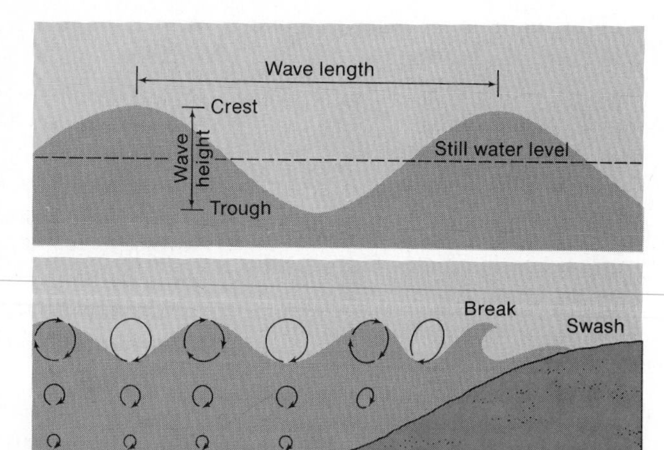

In deep water, the passage of a wave involves almost circular movements of individual water particles. Agitation diminishes rapidly with depth, as shown here by decreasing orbital diameter downward. As the wave form moves into shallow water, the orbits of the revolving particles become more elliptical, the wave length becomes shorter, and the wave becomes steeper. Eventually the wave "breaks" and dissipates its remaining energy as it washes up onto the beach.

Most waves are wind-generated; that is, they are set in motion largely by the friction of air blown across a water surface. This transfer of energy from wind to water initiates wave motion. Some water waves (called *forced waves* or simply *sea*) are generated directly by wind stress on the water surface; they can develop to considerable size if the wind is strong and turbulent, but they usually are of limited duration and do not travel long distances. Water waves become *swell* when they escape the influence of the generating wind, and they can travel enormous distances away from the source of the disturbance to impact coastlines thousands of miles away. A small number of all water waves are generated by something other than the wind, such as a tidal surge, volcanic activity, or undersea diastrophic movement.

In deep water, waves appear as undulations on the surface. The water moves up and down as each wave passes. Water particles actually make a small circular or oscillatory movement, with very little forward motion. These are called *waves of oscillation*. As the wave passes, the water moves upward, producing a *wave crest*; this is followed by a sinking of the surface that creates a *wave trough*. The combination of one crest and one trough forms a single wave. The horizontal distance from crest to crest or from trough to trough is called the *wave length*. The vertical dimension of wave development is determined by the circular orbit of the surface water particles as the wave form passes; the vertical distance from crest to trough is equivalent to the diameter of this orbit and is called the *wave height*. *Wave amplitude* is one-half the height; i.e., it is the vertical distance from the still water level either upward to the crest or downward to the trough. The passage of a wave of oscillation normally moves water only slightly in the direction of flow. Thus, an object floating on the surface simply bobs up and down without advancing, except as it may be pushed by the wind. The influence of wave movement diminishes rapidly with depth; even very high waves stir the subsurface water to a depth of only a few tens of yards or meters.

Waves often travel great distances across deep water with relatively little change in velocity or shape. As they roll into shallow water, however, a significant metamorphosis occurs. When the water depth becomes equal to about half the wave length, the wave motion begins to be affected by frictional drag on the sea bottom. The waves of oscillation then rapidly become changed into *waves of translation*, with significant horizontal movement of the surface water. Friction retards the progress of the waves so that they are slowed and bunch together, marking a decrease in wave length, while at the same time their height is increased. As the wave becomes higher and steeper, frictional drag becomes even greater, which causes the wave to tilt forward and become more and more unstable. Soon and abruptly the wave *breaks*, collapsing into white water surf or plunging forward as a breaker, or if the height is small, perhaps simply surging up the beach without cresting. The cascading forward motion of the breaking wave rushes toward shore or up the beach as *swash*. The momentum of the surging swash is soon overcome by gravity and friction, and a reverse flow, called *backwash*, drains much of the water seaward again, usually to meet the oncoming swash of the next wave.

Waves often change direction as they approach a shoreline, a phenomenon known as *wave refraction*. This is because the line of waves does not always approach exactly parallel to the shore, or because the coastline is uneven, or be-

A breaking wave. In shallow water the ocean bottom impedes the oscillation of water particles, causing the wave form to become increasingly steeper until it is so oversteepened that it collapses and tumbles forward as a breaker. The surging water then rushes up the beach as swash.

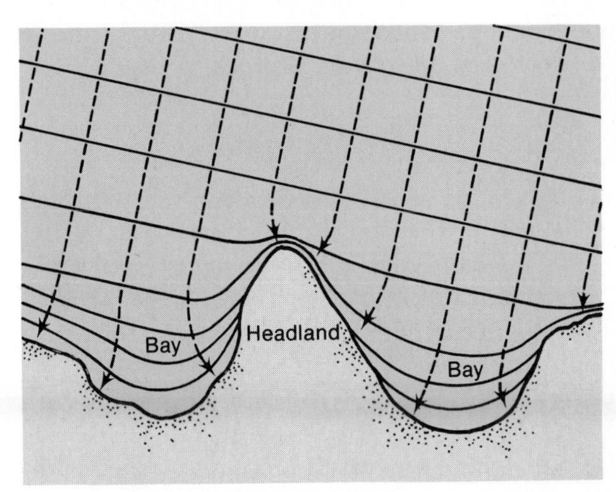

Refraction of waves on an irregular coastline. The waves approach the headland first and then pivot toward it. Thus, wave energy is concentrated on the headlands and is diminished in the bays.

cause of irregularities of water depth in the nearshore zone. For one or more of these reasons, a portion of a wave reaches shallow water sooner than other portions, and it is thus slowed down. This causes a bending or refraction of the wave line, as it pivots toward the obstructing area. The wave energy then becomes concentrated in the vicinity of the obstruction, and is diminished in other areas.

The most conspicuous geomorphic result of wave refraction is the focussing of wave action on headlands that project outward from the shoreline, subjecting them to direct onslaught of pounding waves, whereas an adjacent bay experiences much gentler, low-energy wave action. Other things being equal, the differential effect of wave refraction tends to smooth the coastal outline by wearing back the headlands and by increasing sediment accumulation in the bays.

Occasionally major oceanic wave systems are triggered by undersea tectonic or volcanic events. These waves, called *tsunamis* or *seismic sea waves* (improperly called *tidal waves*), are inconspicuous in the open sea because they are low and have enormous wave lengths (usually several tens of miles), although they can move at a speed of several hundred miles per hour. As they approach the coastlines, they build to great heights (as much as 100 feet or 30 meters) and can cause great devastation as they engulf the coastal zone. They are rare in occurrence but disastrous in effect. Happily, tsunamis can usually be anticipated long enough in advance to allow time for evacuation of the impact area, as they usually originate in deep-ocean trenches from earthquake shocks that are readily detectable by seismographs.

Whether awesome tsunamis or mild swells, the peculiar contradiction of water waves is that they normally pass harmlessly under such fragile things as boats or swimmers in open water but can wreak devastation on even the hardest rocks of a shoreline. In summary, a wave of oscillation is a relatively gentle phenomenon, but a wave of translation can be a powerful force of destruction.

of rocklike material by secreting lime. By far the most significant of these organisms is the coral polyp, which builds a hard external skeleton of calcium carbonate within which it lives. These polyps are of many species, and they cluster together in social colonies of uncounted billions of individuals. Under favorable conditions (clear, shallow, salty, warm water), the coral can accumulate into enormous masses, forming reefs, platforms, and atolls. These coral structures are commonplace features in tropical and subtropical oceans, mostly in association with island or continental coastlines.

Erosive Action

The most notable erosion, by far, along coastlines is accomplished by the action of waves. The incessant pounding of even small waves is a potent force in wearing away the shore, and the enormous power of storm waves almost defies comprehension. Waves break with abrupt and dramatic impact, hurling water and debris (and air) in a thunderous crash onto the shore. See Figure 18-2. Spray from breaking waves commonly moves as fast as 70 miles (113 km) per hour, and small jets have been measured at more than twice that velocity. The sheer weight in addition to the velocity involved in such hydraulic pounding is responsible for much coastal erosion.

Waves often carry tools that make their erosive effectiveness even greater. Along with the water, sand, pebbles, and boulders may be hurled shoreward by the breaking wave, adding an abrasive effect of pounding and grinding.

Another dimension to wave erosion is provided by the forcing of air into cracks (joints, pores, fissures) in the rock as the wave impacts the shore. The resulting compression is abruptly released as the water recedes, allowing instant expansion of the air. This pneumatic action is often very effective in loosening rock particles of various sizes.

Figure 18-2 The ceaseless pounding of waves on a rocky shore is one of the most conspicuous examples of external erosional forces at work. This is Point Lobos, near Carmel, California. [M. R. Campbell, U.S. Geological Survey photo.]

Chemical action also plays a part in wave erosion. Solution action is common in an area of continual or frequent wetness; limestone is not the only type of rock that reacts to dissolving effects, for most rocks are to some extent soluble in seawater. The crystallization of salts in crevices and pores is a further mechanism for weakening and breaking up the rock.

All these wave actions tend to wear away the exposed coastal bedrock at the foot of the steep sea cliffs. The most effective erosion is just at or slightly above sea level, so that a notch is cut in the base of the cliff. The cliff face then retreats due to collapse of the slope above the undercutting. See Figure 18-3. The resulting debris is broken, smoothed, and reduced in size by further wave action, and eventually most of it is carried seaward.

Where a shoreline is composed of sand or other unconsolidated material, currents and tides as well as the waves will cause rapid erosion. Stormy conditions greatly accelerate the erosion of sandy shores; a violent storm can remove an entire beach in just a few hours, cutting it right down to the underlying bedrock.

Transportation in Coastal Waters

Nearly all movement of rock debris along the coastlines is accomplished by wave action. The most obvious, but not always the most significant, movements involve the short-distance shifting of sand directly onshore by breaking waves and directly offshore by the immediately subsequent retreat of the water after each wave is spent. The debris (mostly sand) that advances and retreats with the repeated surging and ebbing of the waves abrades and smooths the beach, while the continual abrasion reduces the size of the particles.

Generally of greater consequence is the so-called *beach drifting* of materials along the coastline. This zigzag movement of particles results in downwind displacement parallel to the coast. See Figure 18-4. Nearly all waves approach the coast obliquely rather than directly, and the sand and other debris carried onshore by the breaking wave move up the beach at an oblique (rather than a right) angle. Some of the water soaks into the beach, but much of it returns seaward directly downslope, which is normally at a right angle to the shoreline. This return flow takes some of the debris with it, much of which is picked up by the next surging wave and carried shoreward again along an oblique path. This infinitely repetitious pattern of movement shifts the debris further and further along the coastline longitudinally. The strength, direction, and duration of the winds comprise the principal determinants of this movement.

Tides and currents play lesser roles in debris trans-

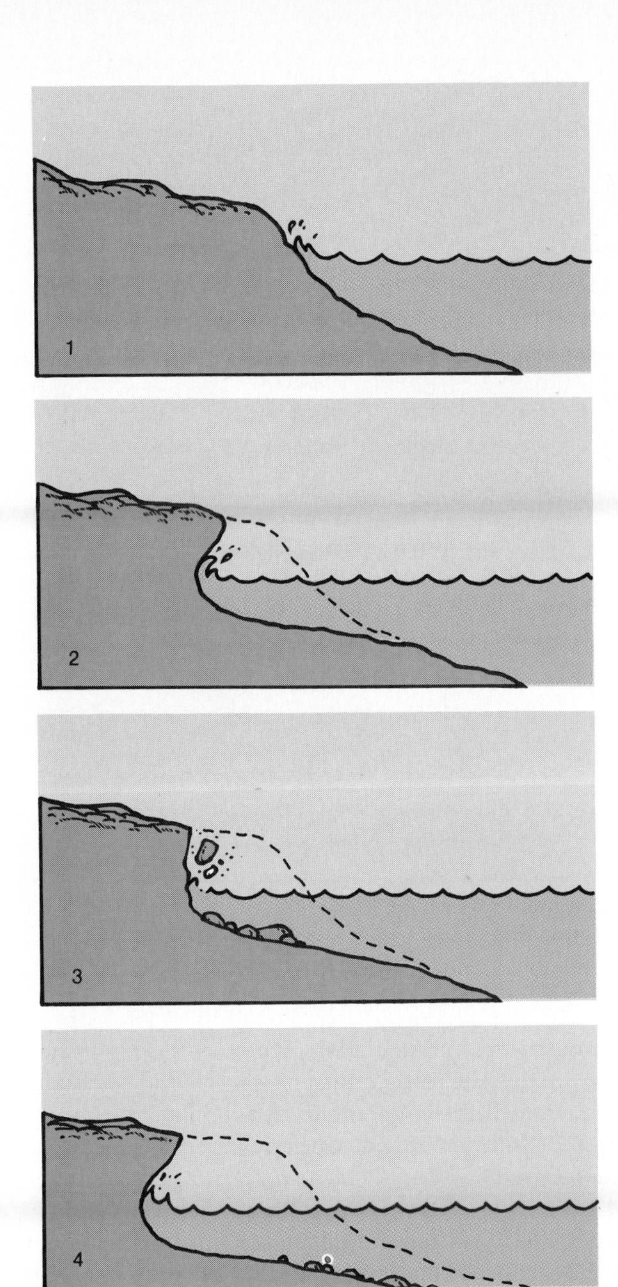

Figure 18-3 Waves pounding on exposed headlands accomplish the most effective erosion at water level, so that a notch may be cut in the base of the slope. This tends to undermine the higher portion of the headland, which may subsequently collapse, producing a steep cliff. The notching/undercutting/collapse sequence may be repeated many times, causing a retreat of the cliff face.

port. They are not very effective in moving coarser debris, but where channels are narrow and relatively shallow, they can move fine material. Most prominent is the action of longshore currents, which can transport a considerable volume of sand and silt parallel to the coastline.

One other facet of debris transport along shorelines is accomplished directly by the wind. Where sand and finer-grained particles have been carried or hurled to positions above the water level, they can be picked up by a breeze and moved overland. This frequently results in dune formation and sometimes moves sand a considerable distance inland.

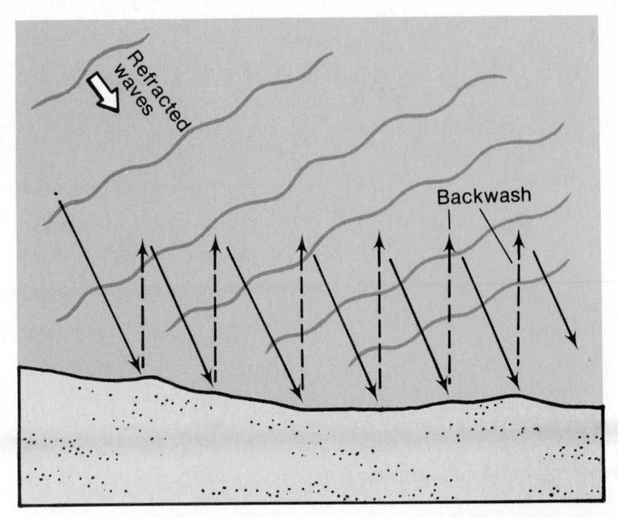

Figure 18-4 Beach drifting involves a zigzag movement in a general downwind direction along the coast. Debris is brought obliquely onto the beach by the wave, and then is returned seaward perpendicularly by the backwash. Frequent repetition of this sequence yields a net longitudinal movement of the particles.

Coastal Deposition

Although the restless waters of coastal areas accomplish notable erosion and transportation, in many cases the most conspicuous topographic features of a shoreline are formed by deposition—including beaches, offshore bars, and spits. As with other external forces, deposition occurs wherever the energy of the moving waters is diminished. Where wave and current action decreases, sediment is permitted to sink and come to rest.

Maritime depositional features along coastlines tend to be ephemeral in nature; their longevity is usually brief in comparison with noncoastal depositional forms. This is due primarily to their composition, which is typically of relatively fine particles (sand and gravel), and the fact that the sand is not stabilized by a vegetation cover. Most such features are under a constant onslaught by agitated waters, which can rapidly wash away portions of the sediment. Consequently, the sediment budget must be in some sort of balance for the feature to persist; removal of sand must be offset by addition of sand. Most marine depositional forms have a continuing sediment flux, with debris arriving at one end and departing at the other. During stormy periods the balance is often upset, and the feature is significantly reshaped or totally removed.

The most widespread marine depositional feature is the *beach*, which is an exposed deposit of loose sediment, normally composed of sand and/or gravel. See Figure 18-5. Beaches occupy the transition zone between land and water, sometimes extending well above the normal

Figure 18-5 Beaches are common shoreline features, but their distribution is quite variable. Much of the Australian coast is fringed with beaches; this example is near Port Douglas in northern Queensland. [TLM photo.]

sea level into elevations reached only by the highest storm waves. On the seaward side, beaches generally extend down to the level of the lowest tides and can often be found at still lower levels where they merge with muddy bottom deposits. Beaches sometimes extend for dozens of miles along straight coastlines, particularly if the relief of the land is slight and the bedrock is unresistant. Along irregular shorelines, beach development may be restricted largely or entirely to embayments, frequently with an alternation of rocky headlands and bayhead beaches. Beaches are not only ephemeral in longevity but very temporary in form. Their shapes may change greatly in detail from day to day and even from hour to hour. Normally they are built up during periods of quiet weather and are removed rapidly during storms. Most beaches are broader and more extensive in summer and are worn away to become much smaller in winter.

Another prominent form of coastal deposition is the *offshore bar*, also called a *barrier bar*. See Figure 18-6. This is a long, narrow sandbar built up in shallow offshore waters, sometimes only a few hundred yards from the coast but often several miles at sea. Such bars are always oriented approximately parallel to the shoreline, are extremely narrow in comparison with their length, and rise only a few feet above sea level at their highest points. They are believed to originate by the heaping up of debris where large waves (particularly storm waves) first begin to break in the shallow waters of continental shelves, although many of the larger offshore bars may have more complicated histories linked to the lowered sea level of Pleistocene time.

A third form of coastal deposition consists of spits—longitudinal features formed under certain conditions of current flow and bottom configuration. Currents moving roughly parallel to the coast carry fine-grained material into the relatively deeper waters of bays, where deposition takes place. As more sediment is brought in by the current, the embankment becomes longer, wider, and higher. It may eventually reach the surface of the water to form a low, sandy projection that grows down-current.

Significance of Recent Sea-level Fluctuations

One of the most prominent influences on the development of coastal topography is a change in sea level. Where the sea rises in relation to the land, the previous terrain is drowned and a continental environment is subjected to the actions of coastal waters. In contrast, where the land rises in relation to the sea, the previous coastal landforms are lifted to a "high and dry" location and a previously submarine landscape is exposed to coastal processes.

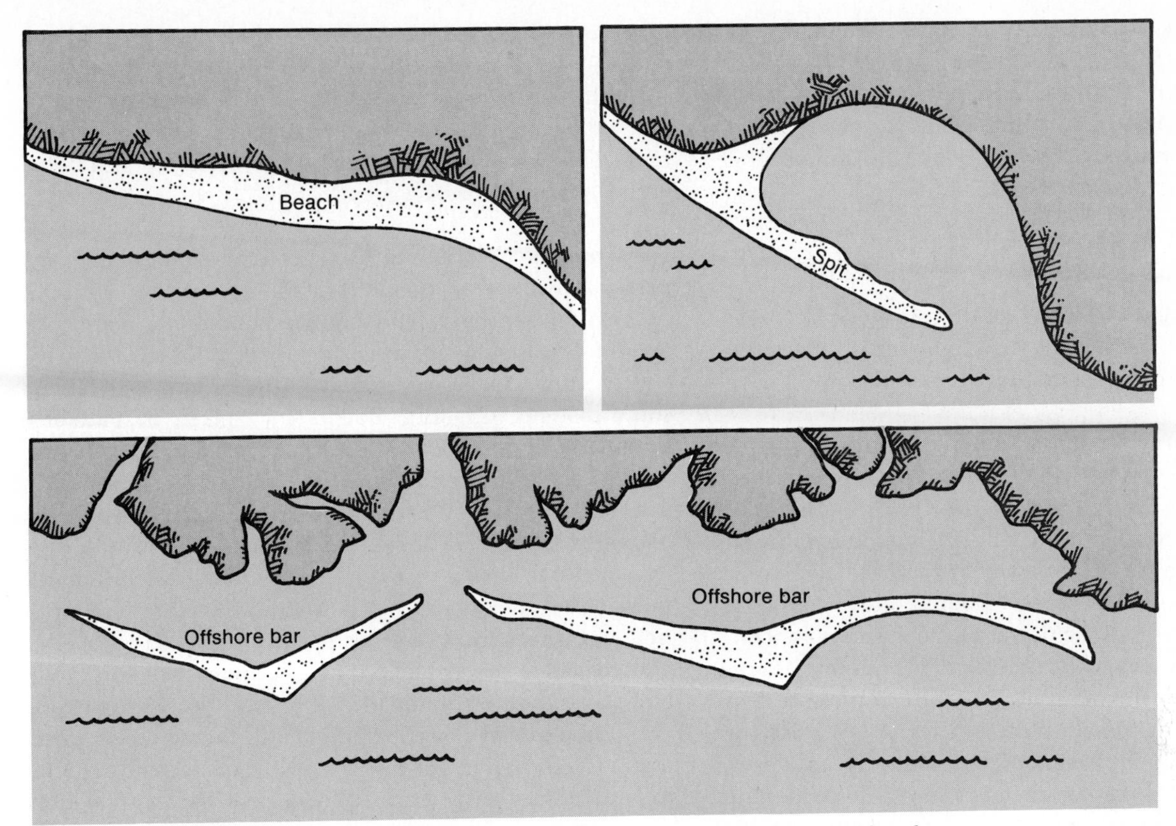

Figure 18-6 Deposition along coastlines takes many forms. Most widespread are beaches, offshore (barrier) bars, and spits.

The topographic result of such fluctuations is usually more pronounced in areas of low relief, but even where coastal mountains are involved, the modification of the previous terrain may be significant.

Shorelines of Submergence Almost all the world's oceanic coastlines show evidence of submergence during the last 15,000 years or so, due to the melting of Pleistocene ice. The most prominent result of submergence is the drowning of previous river valleys, which produces *estuaries*, or long, dendritic fingers of seawater projecting inland. A coast embayed with numerous estuaries is called a *ria* shoreline. If a hilly or mountainous coastal area is submerged, numerous offshore islets may indicate the previous location of hilltops and ridge crests.

Shorelines of Emergence The clearest topographic result of coastal emergence is the occurrence of raised shoreline features, such as cliffs, benches, terraces, and beaches, well above the present water level. Often the emerged portion of a continental shelf appears as a broad, flat coastal plain.

Compound Shorelines Many shorelines show evidence of both submergence and emergence, indicating that both events have taken place in the recent past. This is a relatively common situation because of the worldwide submergence associated with Pleistocene deglaciation and because of localized emergent conditions in many areas.

Neutral Shorelines Some shorelines display neither submergent nor emergent characteristics. These are usually associated either with large-scale deposition, such as deltas or outwash plains, or with coral-reef formation.

Human Influences on Coastal Topography

The terrain of most ocean and lake coastlines has experienced relatively little direct effect due to human activity. In some areas, however, the natural coastal processes have been significantly influenced by human endeavor, resulting in major topographic modification.

To protect coastal settlements and provide more favorable milieus for shipping, fishing, and recreational facilities, large construction projects are often initiated to deflect or redirect waves and currents. These usually consist of breakwaters, sea walls, groins, and other kinds of concrete or rock structures, set parallel or perpendicular to the coastline. Such structures usually are designed to reduce the impact of waves and/or modify the flow of currents. They are often effective in significantly reducing the magnitude of coastal processes and influencing the landforms that result.

Upstream flood-control projects on coastal rivers and streams can also be influential on coastal processes, primarily by diminishing the supply of sand and silt that is debouched from stream to ocean.

Some Prominent Landform Assemblages

Certain assemblages of coastal landforms are widespread over most of the Earth. Their development is illustrative of many of the processes discussed previously.

Cliff/Bench/Terrace One of the most common coastal landform complexes is that of the *wave-cut cliff*, *wave-cut bench*, and *wave-built terrace*. See Figure 18-7. As discussed earlier, the rocky headlands may be acted upon by pounding waves. As the waves eat away at the headland, steep cliffs are formed, which receive their greatest pounding right at the base, where the power of the waves is concentrated. The combined effects of hydraulic pounding, abrasion, pneumatic push, and solution action at the cliff base frequently cause a *notch* to be cut at water level. As the notch is enlarged, the overhang sporadically collapses, and the cliff recedes as the ocean advances. A broadening erosional platform or bench is created, usually slightly below the water level, by the continued pounding and abrasion of the waves. This produces a wave-cut configuration of the coastal profile that resembles a laid-back letter L, with the steep vertical component of the cliff descending to a notched base and the flattened horizontal component of the bench extending seaward.

The debris eroded from cliff and bench is mostly removed by the swirling waters. The larger fragments are battered into smaller and smaller pieces, until they are small enough to be transportable. Some of the sand and gravel produced in this fashion may be washed into an adjacent bay to become, at least temporarily, a part of the bayhead beach. Much of the debris, however, is shifted directly seaward, where a great deal of it is deposited just beyond the wave-cut bench as a wave-built terrace. As with all sediment transported by water, the finer particles tend to move farthest before being deposited. With the passage of time and the wearing away of the cliff by weathering and erosion, the terrace may become more extensive, the bench may be buried by sediment, and a prominent beach may form.

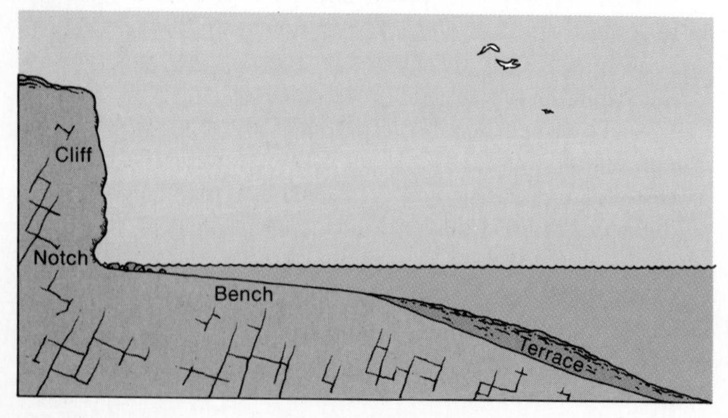

Figure 18-7 Cross section of a cliff/bench/terrace landform complex.

Bar/Lagoon A commonplace coastal landscape form, particularly where offshore waters are shallow, consists of a barrier bar separated from the shoreline by a shallow lagoon. For various reasons, such bars can be formed by the accumulation of sand heaped up by waves breaking some distance offshore. Barrier bars often become the dominant element of the coastal terrain, particularly where they extend longitudinally (parallel to the coastline). Although they usually rise at most only a few feet or meters above sea level, and are typically only a few hundred yards (meters) wide, they may extend many miles or kilometers in length. Most of the eastern (Atlantic) and southern (Gulf of Mexico) coastlines of the United States are paralleled by lengthy barrier bars, several more than 30 miles (48 km) long. See Figure 18-8.

An extensive barrier bar isolates the water between it and the shoreline, forming a body of quiet salt or brackish water called a *lagoon*. See Figure 18-9. Its very isolation

Figure 18-8 Most of the Atlantic and Gulf coasts of the United States are fringed by barrier bars and lagoons. This map depicts the situation along the Texas coastline.

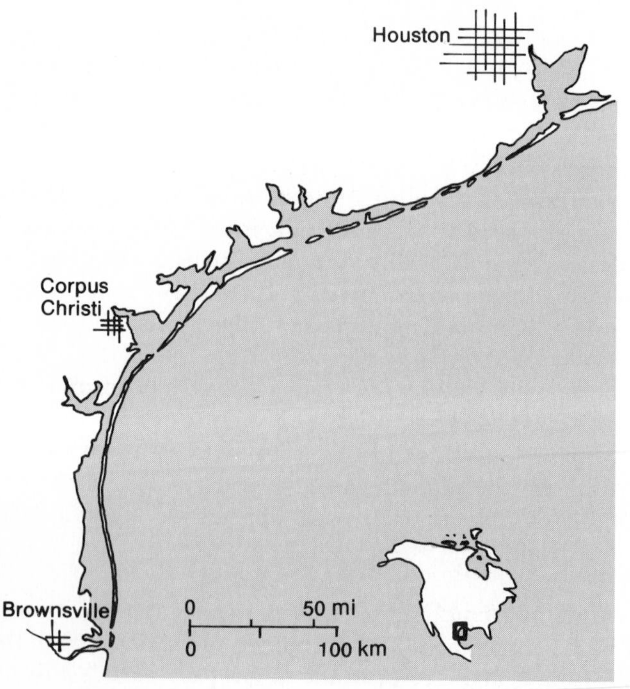

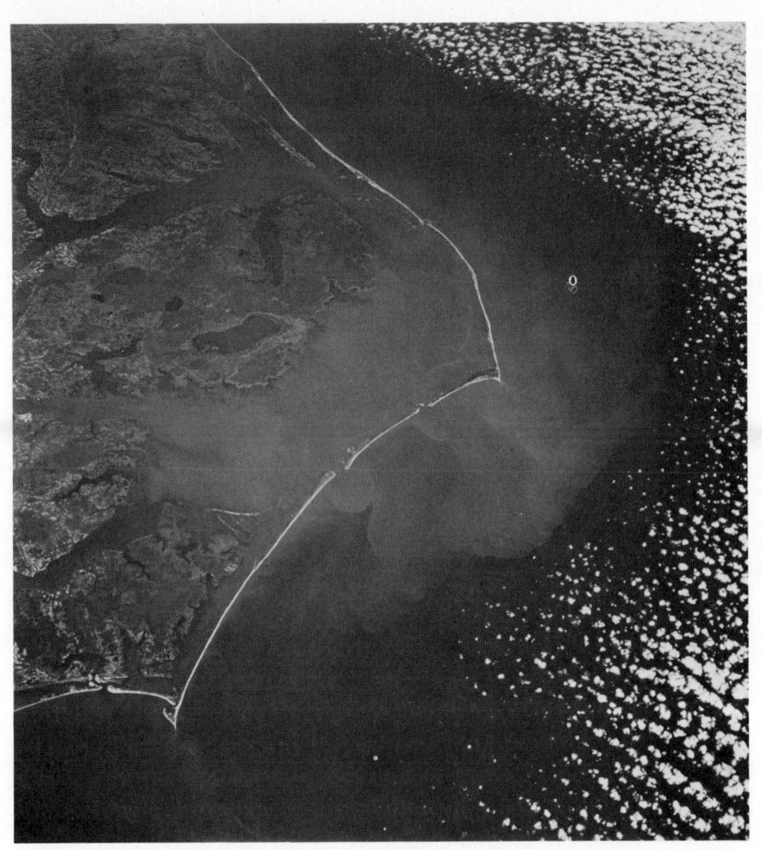

Figure 18-9 Very prominent along the Atlantic coastline are the barrier bars of North Carolina. The prominent "points" of the bars are Cape Hatteras in the center and Cape Lookout near the bottom. Extensive lagoons (called Pamlico Sound in the center and Albermarle Sound in the north) are separated from the Atlantic Ocean by the bars. [*Apollo 9* photo; courtesy NASA.]

contributes to its eventual disappearance, as it becomes increasingly filled with water-deposited sediment from coastal streams and wind-deposited sand from the bar. The ultimate destiny of most such coastal lagoons, unless inlets across the bar permit vigorous tides or currents to carry lagoon debris seaward, is to slowly be transformed into marshes and then into meadows by the long-term accumulation of sediment and by the growth of vegetation.

After barrier bars have attained considerable size, they often begin to migrate slowly shoreward, due to attrition by waves on the seaward side. The eventual result, if the pattern is not interrupted by such things as changing sea level, is for the bar and the shore to merge, through the infilling of the lagoon and the landward retreat of the bar.

Spits The nearshore coastal zone normally contains a considerable amount of relatively fine-grained sediment that is shifted about by wave, current, and tidal action. Most of this material has been derived through erosion by waves or transport of land sediments by stream outflow. As pointed out earlier, the major components of coastal sediment movement are beach drifting and longshore transport along the coastline. Irregularities in the coastal configuration can cause littoral currents to shift debris into the mouths of bays and other deeper water areas, rather than following all the indentations of the coastline. Here, whenever the current enters deeper water, it loses some of its velocity and therefore drops some of its sediment load. The deposit normally grows in a linear direction as more deposition takes place. Incoming waves wash material up the seaward slope and, aided by wind, may build the deposit above water level. The growing embankment guides the current further into the deep water, where still more material is deposited. Any such linear deposit attached to the land is called a *spit*.

Although most spits are straight, linear features projecting as sandy peninsulas out into an embayment or other coastal indentation, the vicissitudes of currents, winds, and waves often give them other configurations. See Figure 18-10. In some cases, the spit becomes extended clear across the mouth of a bay to connect with a headland on the other side, producing a *connecting bar* or *baymouth bar*, and transforming the bay into a lagoon. Another common modification of spit shape is caused by conflicting water movements in the bay, which can guide the deposition in a recurved fashion toward the coast, forming a curving sandbar, called a *hook*, at the outer end of the spit. A less common but no less distinctive development is a *tombolo*; this is a spit formed by sand deposition of waves converging in two directions on the landward side of a nearshore island, so that the bar connects the island to the land.

Fjorded Coasts The most spectacular of all coastlines occur where coastal terrain of relatively high relief has undergone extensive glaciation. Glacial troughs in hilly

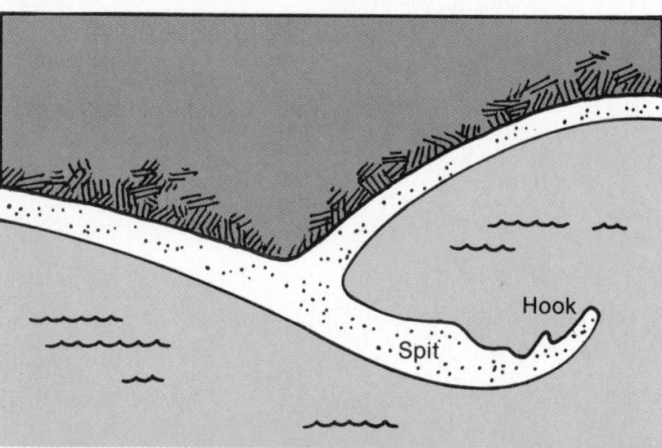

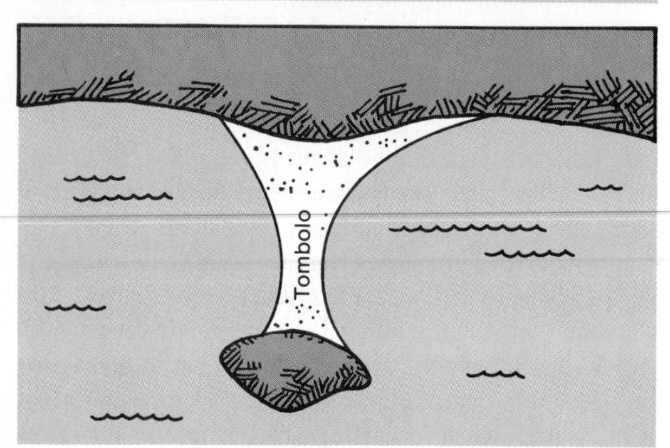

Figure 18-10 Some forms of coastal deposition.

coastal country, gouged out either by valley glaciers or the edges of continental ice sheets, can be cut far below the present sea level by submarine erosion by the ice, or by subsequent drowning of subaerial valleys. In either case, once the ice has melted, a fjord is produced. In some localities these deep, sheer-walled coastal indentations occur in multiplicity, creating an extraordinarily irregular coastline, often with long narrow fingers of salt water reaching more than 100 miles (160 km) inland. The most extensive and spectacular fjorded coasts are found in Norway, western Canada, Alaska, southern Chile, the South Island of New Zealand, Greenland, and Antarctica.

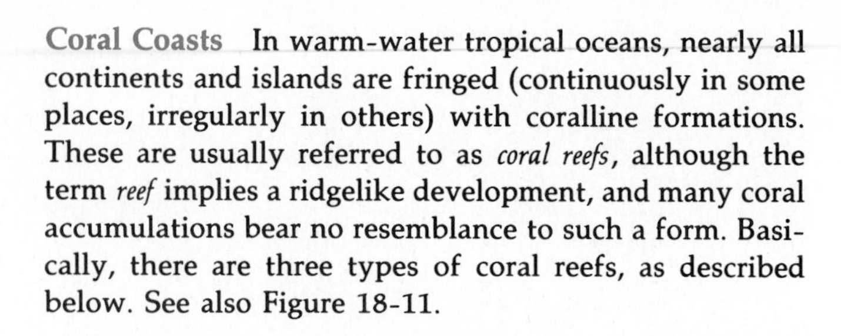

Coral Coasts In warm-water tropical oceans, nearly all continents and islands are fringed (continuously in some places, irregularly in others) with coralline formations. These are usually referred to as *coral reefs*, although the term *reef* implies a ridgelike development, and many coral accumulations bear no resemblance to such a form. Basically, there are three types of coral reefs, as described below. See also Figure 18-11.

Figure 18-11 The three basic types of coral reefs: (A) fringing reef, (B) barrier reef, (C) atoll.

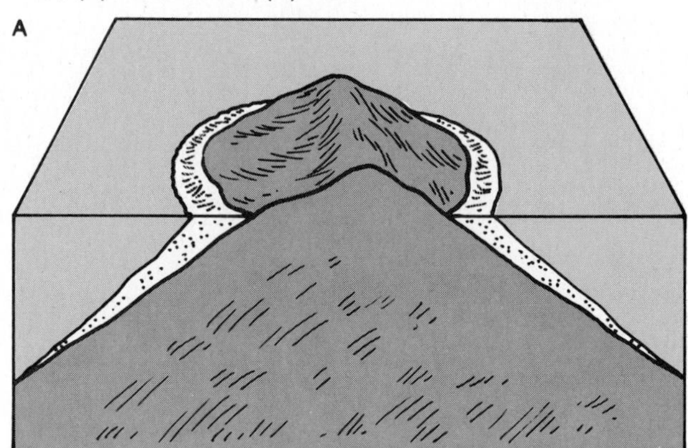

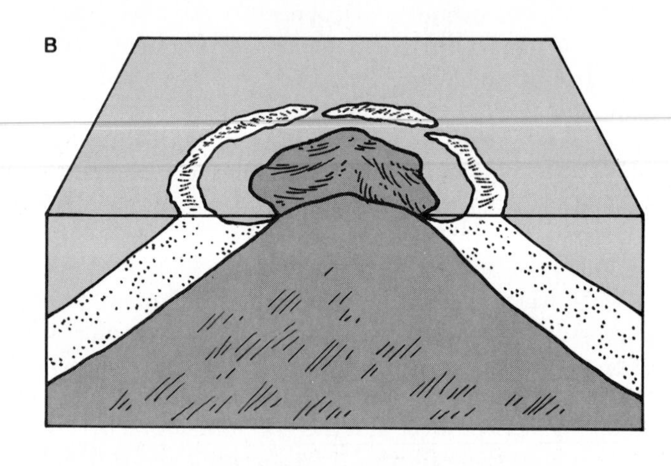

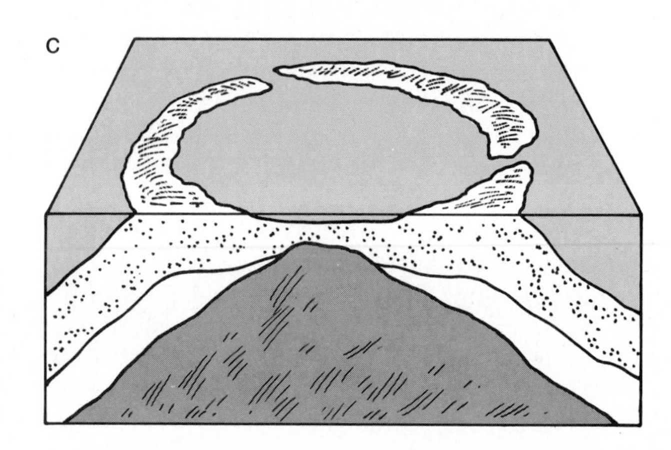

Figure 18-12 A barrier reef off the coast of Matuku Island in Fiji. The open sea is to the left, and the calm waters of a shallow lagoon separate the reef from the island. [TLM photo.]

1. *Fringing reefs* are built out laterally from the shore, forming a broad bench that is only slightly below sea level, often with the tops of individual coral "heads" exposed to the open air at low tide.
2. *Barrier reefs* consist of prominent ridges of coral that roughly parallel the coastline; they lie from a few yards to a few miles offshore, with a shallow lagoon between the reef and the coast. See Figure 18-12. Characteristically, the lagoon is also dotted with coral heads, often in masses. The surface of a barrier reef is usually right at sea level, with some portions projecting upwards into the open air. The outer edge of a barrier reef is an area of very agitated water, with waves crashing onto it most of the time; this contrasts sharply with the placid waters of the lagoon, which is totally sheltered from the open sea except for narrow gaps in the reef. However, the largest and most famous of all reefs, the Great Barrier Reef off the northeastern coast of Australia, is not a reef at all, but rather an immense shallow-water platform largely but not entirely covered with coralline features. Its enormously complex structure includes many individual reefs, irregular coral masses, and a number of islands. For much of its extent the Great Barrier Reef is more than 50 miles (80 km) distant from the Australian coastline.
3. *Atolls* are irregularly shaped circular reefs that enclose a lagoon within which no land is at or above sea level. See Figure 18-13. Atolls are believed to have originated

Figure 18-13 Atolls aplenty. This *Apollo 7* photo shows a portion of the Tuamoto Archipelago in French Polynesia in the South Pacific. [Courtesy NASA.]

Coralline structures are built by a complicated series of events that involves animals, plants, and various physical and chemical processes. The critical element in their development is a group of anthozoan animals (members of the class *Anthozoa* that are closely related to jellyfish and sea anemones) called *stony corals*. These tiny creatures (most are only a fraction of an inch long) live in colonies of countless individuals, attaching themselves to one another both with living tissue and with their external skeletons. Each individual *polyp* extracts the calcium from seawater and secretes a limy skeleton around the lower half of its body. Most polyps withdraw into their skeletal cups during the day and extend their sacklike bodies at night to feed. At the top of the body is a mouth surrounded by rings of tentacles, which gives them a blossomlike appearance that for centuries caused biologists to believe that they were plants rather than animals. They feed on minute animal plankton.

Their initial reproduction is sexual. Male spermatozoa are released to float around the colony, and some are drawn into other polyps, where eggs are fertilized and larvae begin to develop. These larvae are duly released to float around in search of a new home. Those fortunate enough to escape predation while thus exposed eventually settle on some hard surface, where they mature and secrete their external skeletons. Once established, they reproduce rapidly in asexual fashion by budding, thus building a new colony. The colonies build upward and

Coral takes many forms. One of the most prominent of the reef-building corals is called "brain coral," which is pictured here in the Caribbean. [Courtesy Florida News Bureau, Department of Commerce.]

as fringing or barrier reefs around volcanic islands that have subsequently subsided beneath the sea. As the islands slowly sank, the coral was gradually built up by the polyps (which can live only in shallow water). Thus the reefs persist, although the islands have long since disappeared. The term *atoll* implies a ring-shaped structure. In actuality, however, the ring is rarely a complete enclosure; rather it consists of a string of closely spaced coral islets, called *motus*, separated by narrow channels of water.

Solution Processes and Landforms

In our consideration of topographic development we have paid a great deal of attention to the role of water. We have noted that running water on the land is the most significant of the external shapers of terrain, and that coastal waters produce very distinctive landforms around the margins of oceans and lakes. In both of these cases there is an emphasis on relatively rapid water movement and the expenditure of much energy in erosion, transportation, and deposition.

In the final few pages of this chapter, we will focus our interest in water in a different sort of environment—underground. Underground water functions in a much more restricted fashion than surface water; it is confined, it is largely unchannelled and therefore generally diffused, and it moves very slowly for the most part. Consequently, it is almost totally ineffective in terms of hydraulic power, corrasion, and other kinds of mechanical erosion. Indeed, the mechanical effects of underground water on topographic development are limited almost entirely to sub-

outward, forming a fringe of living coral on the periphery of a great mass of limy skeletons of their departed forebears.

The ubiquity of coral reefs in shallow tropical waters is a tribute to the remarkable fecundity of the polyps because they are not actually very hardy creatures. They cannot survive in water that is very cool or very fresh or very dirty. Moreover, they require considerable light, so they cannot live more than a few tens of feet below the surface of the ocean.

Not all corals build reefs. Some species are "soft" corals, which do not secrete external skeletons. Even stony corals sometimes grow too slowly to resist the battering of tide, surf, and currents. Such retarded growth is usually due to the absence of a particular variety of microscopic algae that normally exist in vast numbers (thousands in an individual polyp) in the tissues of stony coral. In addition, some of the fastest growing of the stony corals produce porous, fragile skeletons that are easily broken or damaged. Furthermore, all corals are susceptible to a multitude of boring organisms, from sponges to clams, which weaken their skeletal structure.

Under the best of circumstances, a coral colony cannot build a reef unaided. The many openings in and among the skeletons must be filled in and cemented together by other agencies. Much of the infilling material consists of fine sediment that is provided by other animals through their borings and waste products, and a host of tiny animals and plants contribute their skeletons and shells when they die. Indispensible to the reef-forming process are cement and mortar provided by algae and other tiny forms of "seaweed."

The final stage in reef formation is further compaction and cementation, provided by the soluble nature of the limy material. The relatively unstable crystals of corals and shells gradually break down to become rearranged as a more stable form of limestone.

The coralline structures created by these complex processes are extremely variable in form and characteristics. All are notable, however, for the extraordinary variety of marine life, animal and plant, living in association.

From a topographic standpoint, the significance of coral reefs is at least fourfold:

1. The features themselves—the reefs and other coralline forms—comprise an important element in the landscape of coastal and nearshore locations in tropical portions of the world. They are most widespread in the Pacific, reflecting the vast number of islands in low-latitude portions of that ocean, but are also notable in parts of the Indian and Atlantic oceans.
2. Fringing reefs and especially barrier reefs provide an important element of stability for many tropical coastlines. These reefs absorb most or all of the energy of the pounding seas, significantly insulating the shoreline from coastal erosion.
3. The presence of coral formations, living and dead, provides evidence of past geomorphic conditions. We are learning more and more of the subtleties of such evidence. The most obvious indicators are shown by atolls. Most present atolls are on former volcanic islands that are now worn away and/or submerged.
4. The coral milieu is very different from that of areas where coral is absent. This is shown most strikingly, perhaps, by atolls, which are "special" islands that engender "special" feelings in people familiar with them. As an illustration, we might turn to the colorful prose of James Michener in *Return to Paradise* (New York: Random House, Inc., 1951, pp. 6–7):

> A coral atoll, circular in form, subtended a shallow lagoon. On the outer edge giant green combers of the Pacific thundered in majestic fury. Inside, the water was blue and calm. Along the shore of the lagoon palm trees bent their towering heads as the wind directed. . . . The world contains certain patterns of beauty that impress the mind forever. . . . The list need not be long, but to be inclusive it must contain a coral atoll with its placid lagoon, the terrifyingly brilliant sands, and the outer reef shooting great spires of spindrift a hundred feet into the air. Such a sight is one of the incomparable visual images of the world.

surface weathering and to the encouragement of certain forms of mass wasting (such as earthflows and slumps) through lubrication of loose materials.

Through chemical action, underground water assumes some prominence as a shaper of topography. As a solvent, it dissolves certain rock-forming chemicals, carrying them away in solution and depositing them elsewhere. Under certain circumstances the topographic results of this solution process are widespread and distinctive.

The Actions Involved: Solution and Precipitation

The chemical reactions involved in the work of underground water are relatively simple. Although pure water is a poor solvent, almost all underground water is laced with enough chemical impurities to become an active dissolver of a few common minerals. Basically, underground water functions as a weak solution of carbonic acid because it accumulates carbon dioxide. Limestone, a common sedimentary rock, is composed largely of calcium carbonate, which reacts strongly with carbonic acid to yield calcium bicarbonate, a compound that is very soluble in water. Certain other limy rocks, such as dolomite, gypsum, and chalk, experience similar reactions.

The pertinent chemical equation is:

$$\underset{\text{lime}}{CaCO_3} + \underset{\text{water}}{H_2O} + \underset{\text{carbon dioxide}}{CO_2} \longrightarrow \underset{\text{calcium bicarbonate}}{Ca(HCO_3)_2}$$

This represents the most notable of the solution processes. Water percolating down into a limy bedrock dissolves and carries away a part of the rock mass. Since limestone and related rocks are composed largely of soluble minerals, great volumes are sometimes taken into solution and removed, leaving conspicuous voids in the bedrock. This action occurs more rapidly and on a vaster scale in a humid climate, where abundant precipitation provides plenty of the aqueous medium for solution, than in dryer climates. Solution action is unusual in arid regions except for relict features dating from a more humid period in the past.

The structure of the bedrock is also important to the success of the solution process. A profusion of joints and bedding planes will permit the ground water to penetrate the rock readily. Active movement of the water also helps, so that as the water becomes saturated with dissolved calcium bicarbonate, it can drain away and be replaced by "fresh" unsaturated water for more dissolving. Such drainage is enhanced by some outlet at a lower level, such as a deeply entrenched surface stream.

Most limestone is physically resistant to surface erosion and often produces rugged topography. Thus its ready solubility contrasts notably with its physical durability—a vulnerable interior beneath a durable surface.

Complementary to the removal of lime in solution is the precipitation of lime from solution. Mineralized water may trickle in along the roof or wall. The reduced pressure of the open cavern, sometimes supported by evaporation, induces precipitation of whatever calcium minerals the water is carrying.

One other type of precipitation deposit is worth mentioning, despite its scarcity, because of its dramatic distinctiveness. Hot springs and geysers nearly always provide an accumulation of precipitated minerals, frequently brilliant white in color but sometimes colorfully hued in orange or green or some other color due to the associated algae. Where underground water comes in contact with subsurface magma and becomes heated, it sometimes finds its way back to the surface through a natural opening so rapidly that it is still hot when it reaches the open air. Hot water is a much more potent solvent than cold, so a hot spring or geyser usually contains a significant quantity of dissolved minerals. When exposed to the open air, the hot water normally precipitates much of its mineral content due to the reduction in temperature and pressure, the dissipation of gases in the water that helped to keep the minerals in solution, and the presence of algae and other organisms that live in the hot water and secrete mineral matter. These deposits take the form of mounds, terraces, walls, and peripheral rims.

Caverns and Related Features

Some of the most spectacular landforms produced by solution action are not visible at the Earth's surface. Solution action along joints and bedding planes in limestone beneath the surface often creates holes, sometimes large and extensive, in the form of vertical shafts or caverns or cave systems. The larger openings are usually more expansive horizontally than vertically, indicating a development along bedding planes. In many cases, however, the cave pattern has a rectangularity that demonstrates a relationship to the joint system.

Cavern development is very widespread in suitable rocks. Caves are found in most areas of the world where there is a fairly massive limestone deposit at or near the surface. Some caverns are very extensive, with an elaborate system of galleries and passageways, usually very irregular in shape, and sometimes including massive openings ("rooms") scattered here and there along the galleries. A stream may flow along the floor of a large cavern, adding another dimension to the erosional and depositional processes.

Almost all limestone caverns are decorated with a wondrous variety of *dripstone* features formed by precipitated deposits of minerals. See Figures 18-14 and 18-15. These are formed when water leaves behind the minerals it was carrying in solution, due to pressure reduction and

Figure 18-14 The interior of most caverns is heavily ornamented with dripstone features. This scene is in Oregon Caves. [Courtesy National Park Service; photo by George A. Grant.]

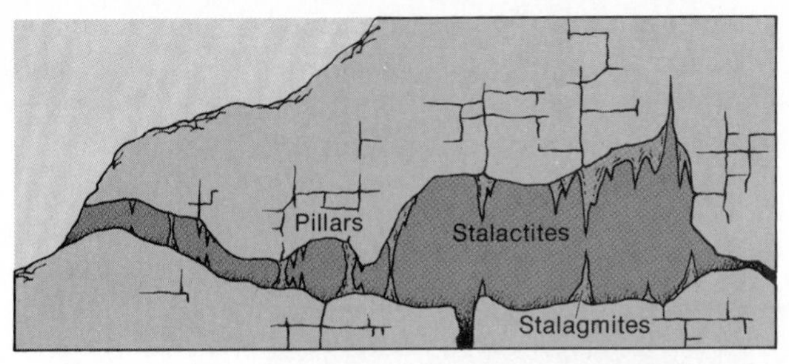

Figure 18-15 A cavern with various dripstone features.

evaporation. Much of the precipitation occurs on the sides of the cavern, but the most striking features are formed on the roof and floor. Where water drips from the roof, a pendant structure may grow slowly downward like an icicle (*stalactite*). Where the drip hits the floor, a companion feature (*stalagmite*) grows upward, usually in more massive form. Stalactites and stalagmites may be extended until they meet, forming a *column* or *pillar*.

Karst Topography

In many areas of limestone bedrock, the solution process has been so widespread and effective that a distinctive landform assemblage has developed at the surface, in addition to whatever caves may exist underground. The term *karst* is applied to such solution topography, wherever it may be, despite the great variety in the terrains of different areas. The name derives from the Karst (meaning "barren land") region of western Yugoslavia, a rugged hilly area that has been shaped almost entirely by solution action in massive limestone formations.

The most common surface features of karst landscapes are the *sinkholes* or *dolines*; these occur by the hundreds and sometimes by the thousands. See Figure 18-16. They are usually relatively small, rounded depressions formed by the dissolution of surface limestone, typically at joint intersections. Sinks that result from the collapse of the roof of a subsurface cavern, are called *collapse dolines*. In some karst regions, these various solution depressions are so common as to occupy most of the surface area.

Figure 18-16 A karst landscape dominated by sinkholes and collapse dolines.

Apart from the ubiquity of sinkholes, karst areas show considerable topographic diversity. Where the relief is slight, as in central Florida, sinkholes are the dominant features. Where the relief is greater, cliffs and steep slopes alternate with flat-floored, streamless valleys. Limestone bedrock exposed at the surface tends to become pitted, grooved, etched, and fluted, with a great intricacy of erosive detail.

In many ways the most notable feature of karst regions is what is missing—surface drainage. Most rainfall and snowmelt seep downward along joints and bedding planes, enlarging them by solution in the process. Surface runoff that does become channelled usually does not go far before it disappears into a sink or an enlarged joint crack. The water that collects in sinkholes generally percolates downward in gradual fashion, but some sinkholes have distinct openings at their bottom (called *swallow holes*), through which surface drainage can pour directly into an underground channel, often to reappear at the surface through another hole some distance away. Where solution action has been effective for a long time, there may be a complex underground drainage system that has superseded any sort of surface drainage net. An appropriate generalization concerning surface drainage in karst regions is that valleys are relatively scarce and mostly dry.

Hydrothermal Features

In many areas in the world, mostly of very limited extent, hot water comes to the surface through natural openings. The outpouring or ejection of hot water, often accompanied by steam, is a *hydrothermal* activity and usually takes the form either of a hot spring or a geyser.

Hot Springs The appearance of hot water at the surface of the Earth usually indicates that the underground water has come in contact with heated rocks or magma beneath the surface and has been forced upward through a fissure or crack by the pressures that develop when water is heated anywhere. See Figure 18-17. The usual result at the surface is the formation of a hot spring, with water bubbling out either continuously or intermittently. The hot water is invariably charged with a large amount of dissolved mineral matter, and a considerable proportion of this load is precipitated out as a distinctive deposit as soon as the water encounters the reduced temperature and pressure conditions of the surface environment.

The deposits around and downslope from hot springs can take many forms. If the opening is on sloping land, terraces are usually formed. Where the springs emerge onto flat land, there may be cones or domes or irregularly concentric deposits. Since lime is so readily soluble, the deposits of most springs are composed largely of massive (*travertine*) or porous (*tufa* or *sinter*) accumula-

Hydrothermal features are relatively common in volcanic areas throughout the world. They are particularly notable in New Zealand, Iceland, Japan, Java, and on Siberia's Kamchatka Peninsula. By far the largest concentration, however, occurs in Yellowstone National Park in northwestern Wyoming, which has probably more hydrothermal phenomena than in the rest of the world combined.

The Yellowstone Park area consists essentially of a broad, flattish plateau bordered by extensive mountains (the Absaroka Range) on the east and more limited highlands (particularly the Gallatin Mountains) on the west. The bedrock surface of the plateau is composed almost entirely of volcanic materials—extrusive tuffs and breccias and intrusive rhyolites and basalts. Although no volcanic cones are in evidence, this clearly is a region of extensive recent vulcanism.

The uniqueness of Yellowstone's geologic setting, however, stems from the presence of a large magmatic chamber at a relatively shallow depth beneath the plateau. Test holes bored into the plateau surface reveal an abnormally high thermal gradient, in which the temperature increases with depth at a rate of about 36° F per 100 feet (20° C/30 m), indicating molten material at a depth of less than one mile (1.6 km) below the surface. This provides a remarkable heat source—the most important of the three necessary conditions for the development of hydrothermal features.

The second requisite is an abundance of water to seep downward and become heated, subsequently to rise to the surface in the form of hot water and steam. Yellowstone receives relatively copious summer rain and a deep winter snowpack (averaging more than 100 inches or 254 cm throughout the park), thus fulfilling the second condition.

The third necessity for hydrothermal development is weak or broken surficial material to allow the water ease of movement upward and downward. Yellowstone provides this requirement too. The Earth's crustal surface in the vicinity of Yellowstone is very unstable and is subject to frequent earthquakes and faulting in addition to volcanic activity. Consequently, many fractures and zones of weakness provide easy avenues of access for vertical water movements.

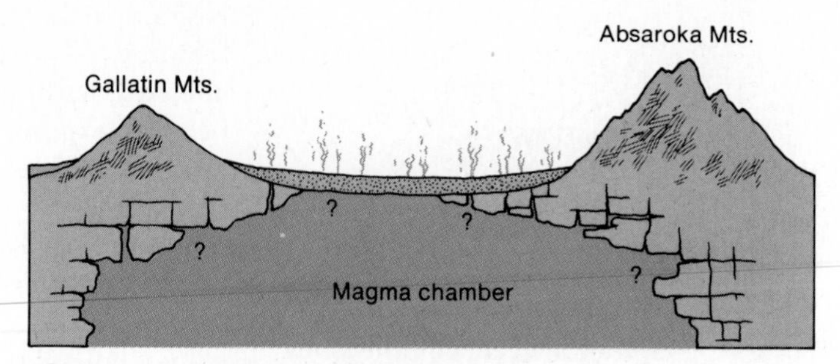

Schematic west-east cross section through the Yellowstone Plateau.

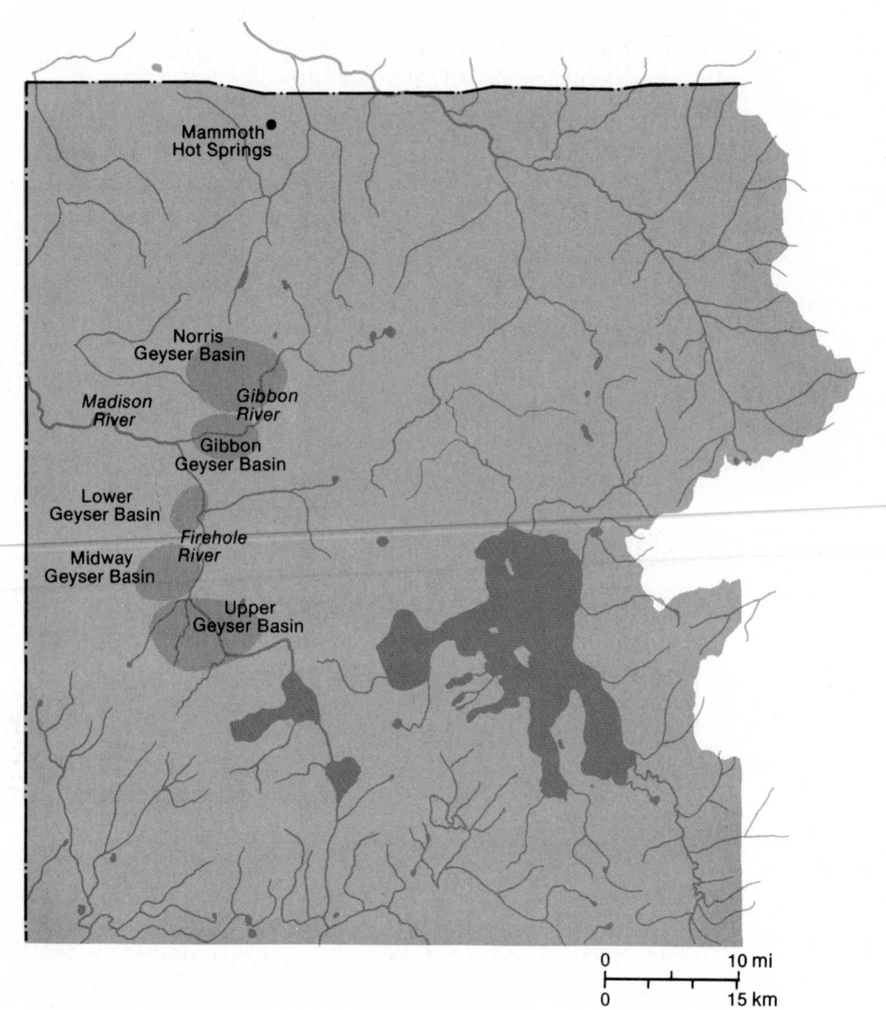

Yellowstone Park and its major geyser basins.

Yellowstone Park is the supreme thermal area of the world. It contains about 250 geysers, more than 3,000 hot springs, and some 7,000 other thermal features (fumaroles, steam vents, hot water terraces, hot mud cauldrons, etc.). There are five major geyser basins, a half dozen minor ones, and an extensive scattering of individual or small groups of thermal features.

The principal geyser basins are all

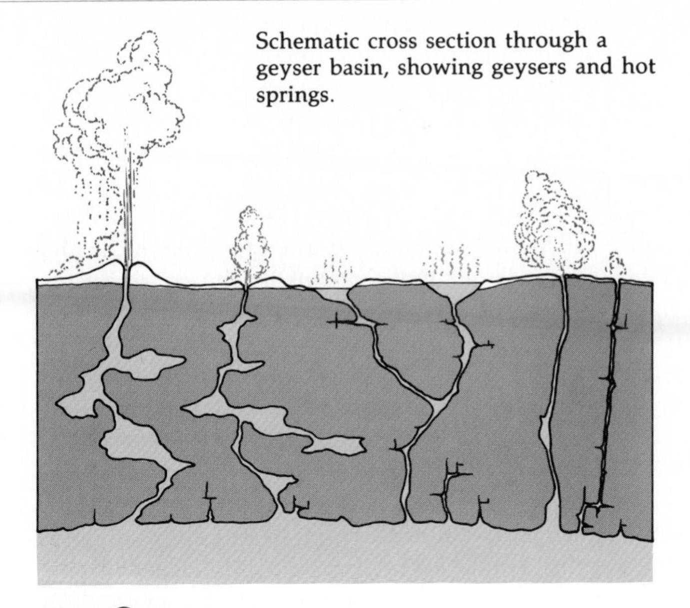

Schematic cross section through a geyser basin, showing geysers and hot springs.

A small portion of the travertine terraces at Mammoth Hot Springs in Yellowstone Park. [TLM photo.]

Schematic diagram of Jupiter Terrace at Mammoth Hot Springs, Yellowstone Park. Water from rain and snowmelt percolates down into the underlying limestone, where it is heated from below and seeps downslope. Some of the hot water issues onto the surface and precipitates travertine deposits when exposed to the air.

in the same watershed on the west side of the park. The Gibbon River from the north and the Firehole River from the south unite to form the Madison River, which flows westerly into Montana, eventually to join two other rivers in forming the Missouri. The Gibbon River drains the Norris and Gibbon geyser basins, whereas the Firehole drains the Upper, Midway, and Lower basins. The Firehole River derives its name from the great quantity of hot water that is fed into it from the hot springs and geysers along its way. Approximately two-thirds of the hydrothermal features of Yellowstone Park are in the drainage area of the Firehole River.

All the major geyser basins are similar in appearance. They consist of gently undulating plains or valleys covered mostly with glacial sediments and large expanses of whitish-colored siliceous sinter, which is usually called *geyserite*. Each basin contains from a few to several dozen geysers, some of which are simply inconspicuous holes in the geyserite, whereas others have built up fairly impressive cones that rise a few feet above the general level of the basin. In addition, each basin contains a number of hot springs and fumaroles; the former have a nearly continuous overflow of scalding water, sometimes in very small quantities but occasionally involving several cubic feet per second; the latter are marked by continual or sporadic issuance of clouds of steam.

Yellowstone's geysers exhibit an extraordinary range of behavior. Some erupt continually; others have experienced only a single eruption in all history. Most, however, erupt irregularly several times a day or week. Some shoot their hot water only a few inches into the air, but the largest (Steamboat, Excelsior) erupt to heights of 300 feet (100 m), with clouds of steam rising much higher.

In the northeastern portion of the park is the most remarkable aggregation of hot-water terraces in the world, the Mammoth Hot Springs Terraces. There, groundwater percolates down from surrounding hills into thick layers of limestone. Hot water, carbon dioxide, and other gases rise from the heated magma below to mingle with the groundwater and produce a mild carbonic acid that rapidly dissolves great quantities of the limestone. Saturated with lime, the temporarily "carbonated" water seeps downslope until it gushes forth near the base of the hills as the Mammoth Hot Springs. In the open air the carbon dioxide escapes, and the lime is precipitated as massive deposits of travertine in the form of flat-topped, steep-sided terraces, as the hot water continually trickles across and down, building the terraces ever larger.

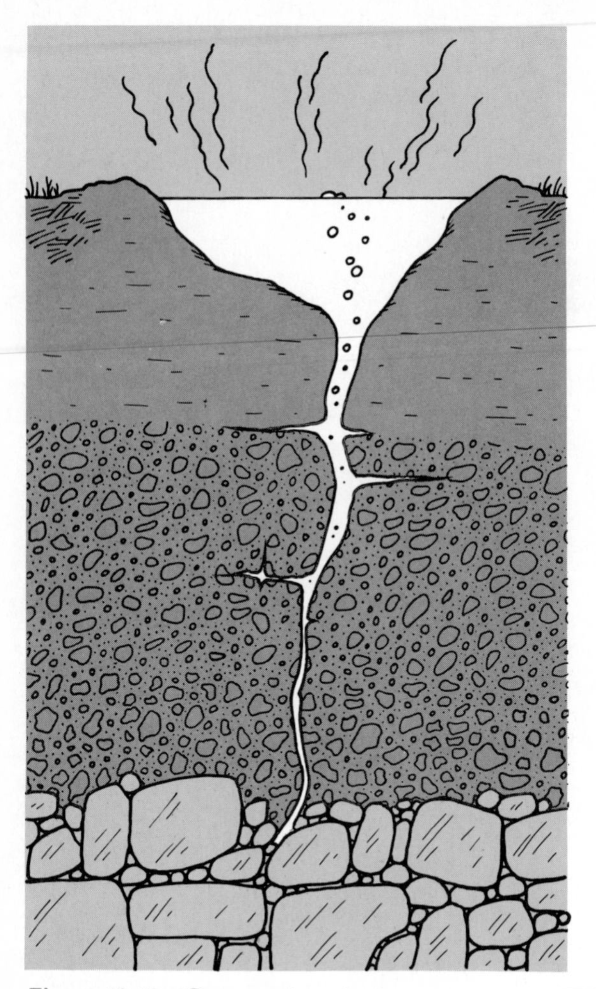

Figure 18-17 Cross section of a hot spring.

tions of calcium carbonate. Various other minerals are also encompassed in the deposits on occasion, especially silica, but these substances are much less common than the calcium compounds.

Sometimes the water bubbling out of a hot spring builds a continually enlarging mound or terrace. As the structure is built higher by the precipitation of mineral matter, the opening through which the hot water comes to the surface also rises so that the water is always emerging above the highest point. As the water flows down the sides of the structure, more deposition takes place there, thus broadening the structure as well, often with brilliantly colored algae, which add to the striking appearance as well as contributing to the deposit through mineral secretions.

Geysers A specialized form of intermittent hot spring is the *geyser*. Hot water usually issues from a geyser only sporadically, and most or all of the flow is a temporary ejection (called an *eruption*) in which hot water and steam are spouted upward for some distance. Then the geyser subsides into apparent inactivity until the next eruption.

The basic principle of geyser activity involves the building up of critical steam pressure within a restricted subterranean tube until the pressure is relieved by an eruption. Underground water seeps into subterranean openings that are connected in a series of narrow caverns and shafts. Heated rocks and/or magma are close enough to these storage reservoirs to provide a constant source of heat. As the water accumulates in the reservoirs, it is heated to very high temperatures and much of it becomes steam. The accumulation of steam deep in the tube and of boiling water above eventually causes a great upward surge, in almost explosive fashion, which sends water and steam showering out of the geyser vent. This releases the pressure, and when the eruption subsides, underground water again begins to collect in the reservoirs in preparation for a repetition of the process.

Some geysers erupt continuously, indicating that they are really hot springs with a constant supply of water through which steam is escaping. Most geysers are only sporadically active, however, apparently depending on the accumulation of sufficient water to force an eruption. Some eruptions are very brief, whereas others may continue for many minutes. The interval between eruptions for most geysers is variable. Most erupt at intervals of a few hours or a few days, but some wait for years or even decades between eruptions. The most famous of all geysers is Old Faithful in Yellowstone Park (see Figure 18-18); its reputation is based in part on the size of its

Figure 18-18 Yellowstone's Old Faithful is by far the most famous geyser in the world. [Courtesy Wyoming Travel Commission.]

eruptions (the eruptions send a column of water more than 100 feet or 30 meters into the air) but primarily on its regularity (its average interval of 65 minutes between eruptions is remarkably constant, day and night, winter and summer).

The depositional results of geyser activity are usually much less notable than those associated with hot springs, although this is dependent in part on the composition of the bedrock. Some geysers erupt from open pools of hot water, throwing tremendous sheets of water and steam into the air, but usually producing relatively minor depositional features. Other geysers are of the "nozzle" type, which build up a depositional cone and erupt through a small opening in it. Most deposits resulting from geyser activity are simply sheets of precipitated mineral matter spread irregularly over the ground.

Fumaroles A third hydrothermal feature closely related to hot springs and geysers is the fumarole. It consists of a surface crack that is directly connected with a deep-seated source of heat. For some reason, very little water drains into the tube of the fumarole. The water that does drain in, is instantly converted to steam by the heat and gases, and a cloud of steam is then expelled from the opening. Thus a fumarole is marked by the continual or sporadic issuance of steam from a surface vent; in essence a fumarole is simply a hot spring that lacks water.

Review Questions

1. What effect did Pleistocene glaciation have on sea-level fluctuations?
2. What is the difference between *waves of oscillation* and *waves of translation*? How does one develop from the other?
3. Explain the roles of *beach drifting*, *tides*, and *currents* in sand transportation.
4. Distinguish between the topography of shorelines of *submergence* and that of shorelines of *emergence*.
5. Explain the mechanics of formation of a wave-cut cliff.
6. Discuss the effect of human activities on natural beach development.
7. What is the effect of barrier coral reefs on coastal topography?
8. How is it possible for percolating groundwater to both erode and deposit?
9. Explain how a *sinkhole* is formed.
10. What is the difference between a *hot spring* and a *geyser*? What causes this difference?

Some Useful References

BIRD, E. C. F., *Coasts*. Cambridge, Mass.: The M.I.T. Press, 1969.
DAVIES, J. L., *Geographical Variation in Coastal Development*. New York: Longman, Inc., 1977.
HERAK, M., AND V. T. STRINGFIELD, eds., *Karst*. Amsterdam, N.Y.: American Elsevier Publishing Co., 1972.
JENNINGS, J. N., *Karst*. Cambridge, Mass.: The M.I.T. Press, 1971.
KING, C. A. M., *Beaches and Coasts*, 2nd ed. New York: St. Martin's Press, 1972.
SHEPARD, F. P., AND H. R. WANLESS, *Our Changing Coastlines*. New York: McGraw-Hill Book Company, 1971.

19 Glacial Modification of Terrain

Various aspects of climate change with time, and such changes may have significant effects on geomorphic processes. Most prominent among climatic fluctuations are trends toward colder, wetter conditions that can bring about an "ice age." During such intervals, large areas of the Earth's surface are covered with glacial ice, sea levels are lowered in all oceans, and increased precipitation often occurs in arid lands.

In the long history of our planet, ice ages have occurred an unknown number of times. However, all, or nearly all, evidence of past glacial periods, with one outstanding exception, has been eradicated by other kinds of geomorphic events that have taken place subsequently. Only the last of the ice ages was recent enough to have created a significant impact on contemporary topography. Consequently, when referring to the Ice Age, we usually mean "the most recent ice age," which distinguishes the epoch known as Pleistocene. The Pleistocene epoch was only "yesterday" in terms of geologic time, a period that began at least 1.5 million years ago and ended less than 10,000 years ago, if indeed it has yet ended.

In this chapter we will be concerned particularly with the events of Pleistocene time both because they produced such significant modifications of pre-Pleistocene topography and because their results are so thoroughly imprinted on many parts of the continental terrain today. Glacial processes are still at work, to be sure, but their importance is much less now than it was just a few thousand years ago simply because so much less glacial ice is present today.

Glaciations Past and Present

The amount of ice that has built up on the surface of the Earth has varied remarkably over the last few million years of our planet's history. Periods of accumulation have been interspersed with periods of melting. Times of ice advance have alternated with times of ice retreat. These have been conspicuous and clear-cut events, although they did not leave precise evidence to chronicle their fluctuations, and we can discuss most of them only in general terms. Even so, a great deal of secondary evidence was left behind by the moving and melting ice, and scientists have been remarkably perspicacious in piecing together the chronology of past glaciations. Nevertheless, the record is incomplete and often approximate. As is to be expected, the more recent events are best documented; the further one delves into the past, the murkier the evidence becomes.

Pleistocene Glaciation

The Pleistocene epoch is of indeterminate extent because its precise boundaries are unknown at both ends of the time scale. It began about 1.5 million years B.P. (before the present), but evidence of earlier glaciation is discovered almost every year, and geochronologists keep pushing the starting date further and further back. Some parts of the Northern Hemisphere experienced glaciation as much as 2.5 million years B.P. The unearthing of new evidence has also brought forward the "close" of the Pleistocene; it is now considered that the last major ice retreat took place only about 9000 years ago. And even this may not be an adequate representation of the actual situation because the Ice Age may not yet have ended at all; we will consider this possibility later in the chapter. To the best of present knowledge, then, the Pleistocene epoch has occupied all or almost all of the most recent couple of million years of the Earth's history.

The dominant environmental characteristic of the Pleistocene was the refrigeration of high-latitude and high-altitude portions of the Earth so that a vast amount of ice accumulated in many places. However, the epoch was by no means universally icy. During several lengthy periods, most or all of the ice melted, only to be followed subsequently by other intervals of accumulation. In broad terms, the Pleistocene consisted of an alternation of glacial (times of ice accumulation) and interglacial (times of ice retreat) periods. It is known that there were at least five major glacial episodes, and there may have been as many as two dozen. The accompanying chart (see Figure 19-1) displays the chronology of the five recognized glacial stages and their accompanying interglacials, for North America and the Alpine portion of Europe. It should be understood, however, that there were lesser permutations within this broad pattern. Each of these stages experienced several briefer intervals of ice advance and retreat, presumably reflecting minor climatic changes.

At its maximum Pleistocene extent, ice covered just about one-third of the total land area of the Earth—nearly 19,000,000 square miles or 47,000,000 km^2. See Figure 19-2. The thickness of this coverage was variable and can be estimated only roughly. In some areas, the ice reached a depth of several thousands of feet.

North America experienced the greatest total area of ice coverage. The Laurentide ice sheet, which covered

Figure 19-1 The major recognized glacial stages of the Pleistocene epoch in North America and their equivalent names in the European alpine areas.

(Younger ↑ / Older ↓)	North America	European Alps
Younger	Wisconsin	Würm
	Illinoian	Riss
	Kansas	Mindel
	Nebraskan	Günz
Older	Pre-Nebraskan	Donau

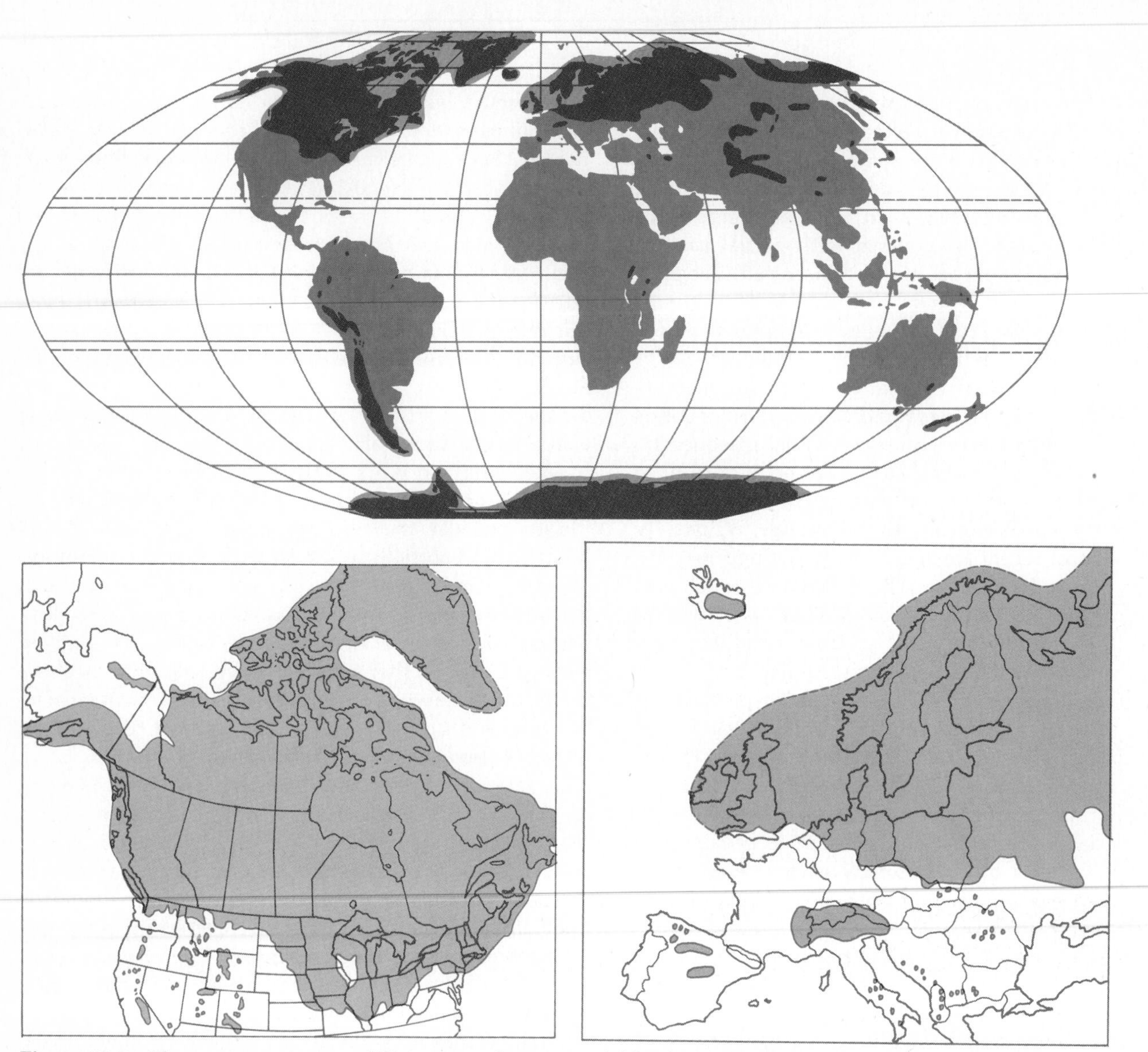

Figure 19-2 The maximum extent of Pleistocene glaciation: worldwide, North America, and Europe.

most of Canada and a considerable portion of northeastern United States, was the most extensive of all Pleistocene ice masses; its area was somewhat greater than that of the present ice cap covering Antarctica. It extended southward into the United States to a position approximating the present location of Long Island, the Ohio River, and the Missouri River. Most of western Canada and much of Alaska were covered by an interconnecting mass of smaller ice sheets, piedmont glaciers, and mountain glaciers.

However, there was a small area in northwestern Canada as well as much more extensive portions of northern and western Alaska that were never glaciated during the Pleistocene. Moreover, a small area (some 11,200 square miles or 29,000 km^2) in southwestern Wisconsin and parts of three adjoining states was also avoided by ice—this is referred to as the Driftless Area. See Figure 19-3.

Elsewhere there was a continuous ice cover over about the northern one-third of the state of Washington, which was connected with the ice complex of western Canada. In addition, there were about 75 separate areas of highland ice accumulation in the western states, many of which extended outward into surrounding lowlands; the largest single accumulation in the West was in and around Yellowstone Park. The southernmost areas of Pleistocene glaciation in the United States were in northern New Mexico, northern Arizona, and the San Bernardino Mountains of southern California.

More than half of Europe was overlain by ice during the Pleistocene. The Scandinavian ice sheet covered all of Scandinavia and Finland, much of European Russia,

Figure 19-3 The Driftless Area of the United States.

and extended as far south as central Germany and Poland. The British Isles were almost entirely covered by several small coalescing ice sheets, which also spread eastward to meet the Scandinavian ice sheet in the North Sea. Most of the mountains of southern Europe, particularly the Alps, Pyrenees, and Caucasus, were inundated by mountain icefields and glaciers.

Asia was less extensively covered, presumably because much of its subarctic portion received inadequate precipitation for ice to persist. Nevertheless, some medium-sized ice sheets occupied Siberia, and extensive glaciation occurred in most Asian mountain ranges.

In the Southern Hemisphere, the Antarctic ice cap was only slightly larger than it is today, a large ice complex covered southernmost South America, and the South Island of New Zealand was largely covered with an ice mass. Other high mountain areas all over the world—in central Africa, New Guinea, Hawaii—experienced more limited glaciation.

The accumulation of ice and the movement and melting of the resulting glaciers exerted enormous effects on preexisting topography and drainage, which will be discussed later in this chapter. Several other indirect effects of Pleistocene glaciation also had significant influence on topographic development, as discussed below:

1. *Periglacial processes.* Beyond the outermost extent of ice advance is an area of indefinite size called the *periglacial zone*, which was never touched by glacial ice but where indirect influence of the ice was felt. The most important periglacial process was the erosional and depositional work of the prodigious amounts of meltwater released as the glaciers waned. Also important was frost weathering engendered by the coldness of the periglacial zone, and the associated solifluction in areas of frozen subsoil.
2. *Sea level changes.* The buildup of ice on the continents meant that less water was available to drain into the oceans, which resulted in a worldwide lowering of sea level during every episode of glacial advance. This caused a significant difference in drainage patterns and topographic development on both seashores and coastal plains.
3. *Crustal depression.* The enormous weight of accumulated ice on the continents caused the sinking of portions of the Earth's crust, in some cases by as much as 4000 feet (1200 m). After the ice melted, the crust slowly rebounded in isostatic adjustment to the reduced load. The adjustment has not yet been completed, and some portions of Canada and northern Europe are still rising as much as 8 inches per decade (2 cm per year).
4. *Pluvial developments.* Concurrent with Pleistocene glaciation came a considerable increase in available moisture on all continents, partly due to meltwater runoff and partly due to decreased evaporation. A prominent result was the creation of many lakes, some of them enormous, in areas where none had previously been. Most of these lakes have subsequently been drained or have been significantly reduced in size, but they have left lasting imprints on the landscape. Pleistocene lakes in the western part of the United States have already been discussed in Chapter 17.

Contemporary Glaciation

In marked contrast to maximum Pleistocene glaciation, the extent of ice covering the continental surfaces today is very limited, with the notable exceptions of Antarctica and Greenland. About 10 percent of the Earth's land surface—amounting to some 6,000,000 square miles or 15,000,000 km^2—today is covered with ice, but more than 96 percent of that total is included in the Antarctic and Greenland ice caps. Something more than two-thirds of all the world's freshwater is frozen into glacial ice.

The Antarctic ice cap is by far the largest in existence. It encompasses about 85 percent of the world's land ice. In some places it is more than 13,000 feet (4000 m) thick, and over most of the Antarctic continent the ice is more than a mile (1.6 km) deep.

The Greenland ice sheet is much smaller (670,000 square miles or 1,740,000 km^2) but still of impressive size. Elsewhere there are only relatively small ice caps on certain islands in the Canadian Arctic, Iceland, and some of the islands north of Europe.

Many of the higher mountains of the world contain glaciers, although most of these are quite small. The distribution of ice sheets and glaciers in the Northern Hemisphere is shown in Figure 19-4. The principal concentrations are in the coastal ranges of Alaska/Yukon/British Columbia, the mountain complexes of central Asia, and the South Island of New Zealand. In the conterminous

Antarctica is a unique continent in many ways. It is the only polar-oriented continent; it is by far the remotest of the world's major landmasses; and it is the only continent totally without permanent human habitation. However, it is the ice cover that lends this continent its particular distinctiveness, as more than 95 percent of its surface is covered with glacial ice.

The bedrock floor beneath the Antarctic ice sheet is irregular in profile and deeply buried by the ice. In places, the subglacial surface is actually below sea level. *Nunataks*, the uppermost rocky pinnacles of high mountains, protrude through the ice in a number of locations, but only around the margin of the continent is much bare rock exposed, primarily in the form of rimming mountains.

Physically the continent can be thought of as consisting of two unequal sections separated by the wide upland belt of the Transantarctic Mountains, which extends for some 2500 miles (4000 km). West Antarctica, the smaller of the two sections, is generally mountainous. It contains, however, a few interior valleys that are curiously ice-free; these are thought to be the result of the retreat of shallow individual glaciers that developed within the ice sheet. If West Antarctica were to lose its ice, it would probably appear as a considerable number of scattered islands. East Antarctica is more extensive, and its subglacial relief is less varied, appearing to be largely a broad plateau with scattered mountains. The ice is considerably deeper in West Antarctica, and the surface of the ice is generally at a greater altitude than in the eastern section. Most of the surface of West Antarctica is more than 8000 feet (2.4 km) above sea level, and a considerable portion exceeds 10,000 feet (3 km) in altitude.

The Antarctic ice sheet continues its outward advance of ice in all direc-

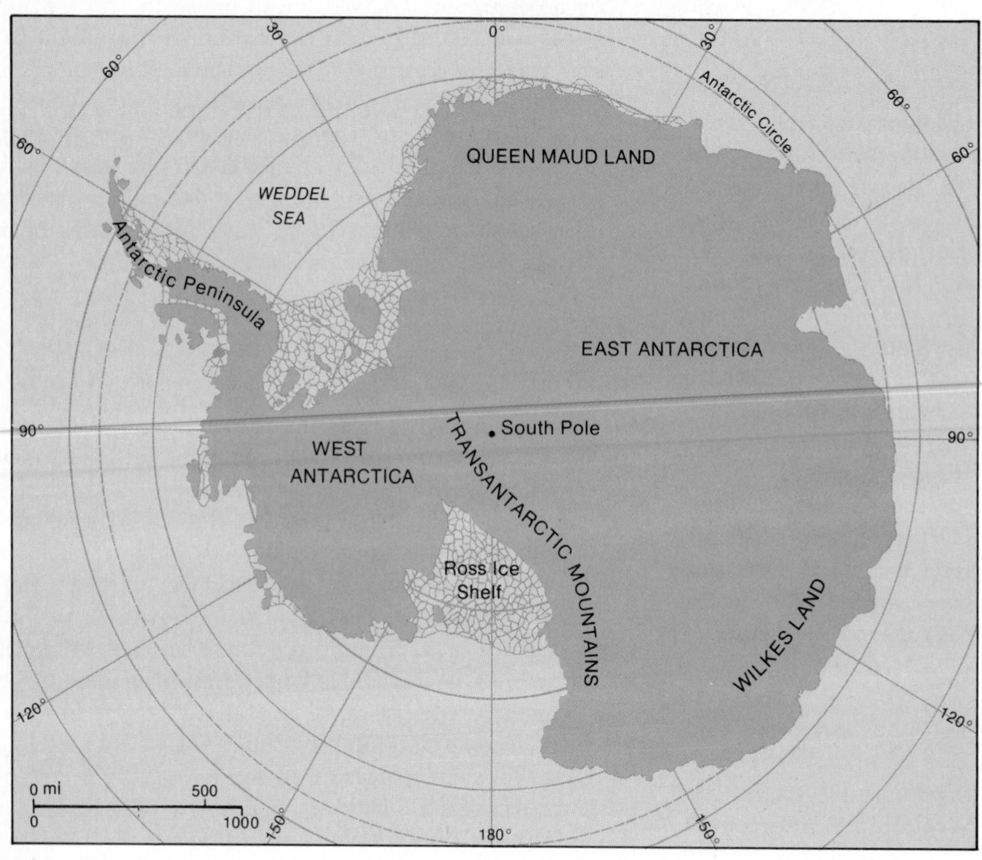

The Antarctic continent.

United States, most glaciers are in the Pacific Northwest, and more than half of these are in the North Cascade Mountains of Washington. The southernmost U.S. glaciers are in southern Colorado and the southern Sierra Nevada of California, but they are so small that their size is measured in acres rather than in miles.

Types of Glaciers

Although glacial ice behaves in similar fashion wherever it accumulates, its pattern of movement and its effect on topographic shaping can vary considerably depending on its quantity and particularly its environmental setting.

tions, which means that icebergs are being calved (broken off) into the sea more or less continuously around the perimeter of the continent. Some of these icebergs originate through outlet glaciers, but many are broken off from ice shelves. There are several of these great plates of floating ice, particularly in West Antarctica, with the Ross Ice Shelf being the largest (200,000 square miles or 520,000 km^2).

Despite its remoteness, Antarctica exerts a prominent influence on the world's environment, particularly in terms of sea level, oceanic temperature and circulation patterns, nutrient content of the oceans, and atmospheric circulation. The volume of ice encompassed in the Antarctic ice sheet amounts to about 7,200,000 cubic miles (30,000,000 km^3), which could be very significant to the Earth's heat and water budgets if the ice sheet melted. If all the Antarctic ice were to melt, the sea level around the world would rise by about 250 feet (80 m)—up to the nose of the Statue of Liberty! This is an unlikely eventuality, however, since from all indications, both precipitation and wastage have been slight in Antarctica for a very long time, and the regimen appears to be nearly in equilibrium.

A view across the Antarctic ice cap, looking westward over the Daniell Peninsula. Most of the world's ice is encompassed in this contiguous sheet. [W. B. Hamilton, U.S. Geological Survey photo.]

These variations are best understood by first considering the different forms or types of glaciers.

Continental Ice Sheets By far the most significant, because of their immense size, are the continental ice sheets. Only two true ice sheets exist today, those of Antarctica and Greenland. These are vast blankets of ice that completely inundate the underlying terrain to depths of hundreds or thousands of feet. The ice accumulates to great depths in the interior of the sheets and flows outward toward the perimeter, where it is much thinner. Around the margin of the sheet, some long tongues of ice, called

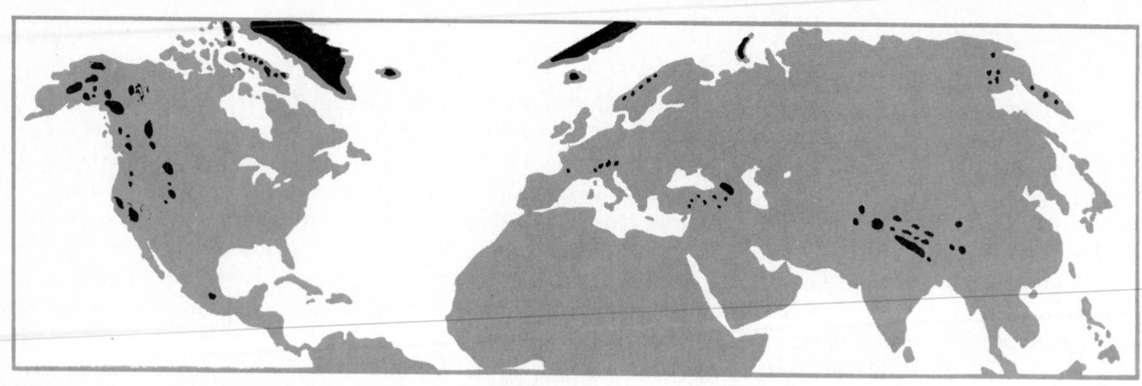

Figure 19-4 The distribution of contemporary ice sheets and glaciers in the Northern Hemisphere. Many small glaciers are not shown because of the small scale of the map.

outlet glaciers, extend between rimming hills to the sea. In other places, the ice reaches the ocean along a massive front, where it sometimes projects out over the sea as an *ice shelf*. Great chunks of ice frequently break off, both from the ice shelves and from the ends of outlet glaciers, to produce floating tabular *icebergs*.

Highland Icefields In a few high mountain areas, ice accumulates in a relatively unconfined sheet that may cover a few hundred or a few thousand square miles, submerging all the underlying topography except perhaps for some rocky pinnacles (called *nunataks*) that protrude above the icefields. Such icefields are notable in parts of the high country of western Canada and southern Alaska and on various Arctic islands (particularly Iceland). See Figures 19-5 and 19-6. Their outlets are often in the form of valley glaciers that descend from the mountain fastness for several miles.

Alpine Glaciers Alpine glaciers are individual mountain glaciers that develop at the heads of valleys and normally move down-valley for greater or lesser distances. See Figure 19-7. Very small glaciers confined to the basins where they originate are called *cirque glaciers*. Normally, however, glaciers spill out of their originating basins and flow down-valley as long, narrow features that resemble rivers of ice; these are known as *valley glaciers*. Occasionally they extend to the mouth of the valley, where the ice then spreads out broadly over flat land, forming a *piedmont glacier*. See Figure 19-8 for the three types of alpine glaciers.

Regimen of Glaciers

The snow falls and ice accumulates in many parts of the world, but glaciers do not always develop from these events. Glaciers require certain circumstances to form.

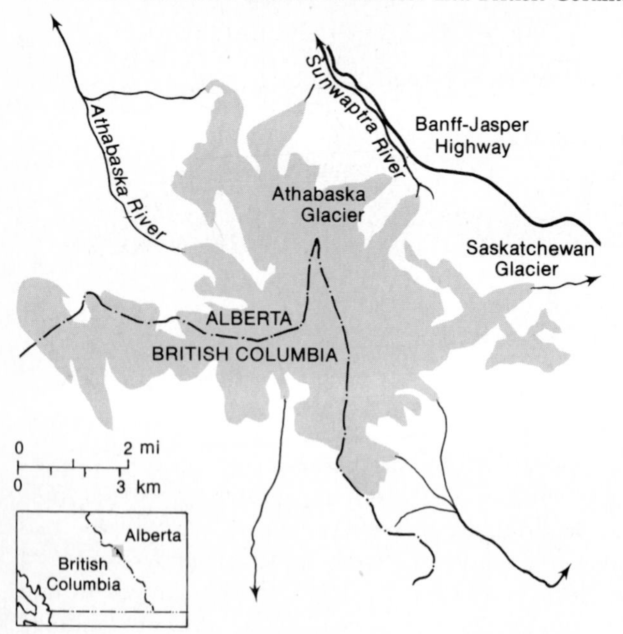

Figure 19-5 The largest highland icefield in the Rocky Mountains is the Columbia icefield in Alberta and British Columbia.

Figure 19-6 Alaska's Juneau icefield covers an area of some 4000 square miles (10,000 km²). Many nunataks rise above the general level of the ice. [TLM photo.]

Figure 19-7 An alpine glacier in southeastern Alaska. Two rivers of ice flow out of the high country to unite, forming the North Sawyer Glacier, which extends down to tidewater. [Courtesy U.S. Forest Service.]

Figure 19-8 The three basic types of alpine glaciers: (A) cirque glacier, (B) valley glacier, and (C) piedmont glacier.

They are often massive, monolithic features that are capable of thoroughly reshaping the terrain. Yet they are also temporary and fragile, depending on just the right combination of temperature and moisture to survive; a slight warming or drying trend for a few decades can cause even the most extensive ice sheet to disappear. The persistence of any glacier depends on the balance between *accumulation* (addition of ice by incorporation of snow) and *ablation* (wastage of ice through melting and sublimation).

Formation of a Glacier

Glaciers can develop wherever more snow collects in winter than melts in summer, over a period of time. This occurs in certain high-latitude or high-altitude areas of the world, but in an irregular and not always predictable pattern. The concept of *snowline*, the elevation above which some winter snow is able to persist throughout the year, is useful in understanding where glaciers form. In equatorial regions, the snowline is usually above 20,000 feet (6,100 m), and it descends in generally progressive fashion in a poleward direction until it reaches sea level in subpolar latitudes. However, at any latitude, the snowline is often zigzag and indefinite, depending on exposure, slope, vegetative cover, prevailing wind patterns, and other environmental factors.

Snow is not frozen water; rather, it has crystallized directly from water vapor in the atmosphere and floats to earth as lacy, hexagonal ice crystals that are only about one-tenth as dense as water. Sooner or later (within a few hours if the temperature is near freezing, but only over a period of years in very cold situations), the new-fallen crystalline snow is compressed into granular form with a density that is approximately doubled. With more time and further compression, the granules are packed more closely and begin to coalesce, achieving a density about half as great as that of water. This material is called *névé* or *firn*. As time passes, the metamorphosis from snow to ice continues. The pore spaces with their trapped air among the whitish firn crystals gradually diminish as the air is squeezed out, the density approaches 90 percent of that of water, and the material takes on the bluish tinge of glacial ice. See Figure 19-9. The "solid ice" that results usually continues to change, though very slowly, with more air being forced out, the density increasing slightly, and the ice crystals increasing in size.

Figure 19-9 A close view of ice on the margin of the Mendenhall Glacier in southeastern Alaska. The granular nature of the ice is conspicuous. Meltwater percolating among the granules gives the appearance of a wrinkled surface. [L. L. Ray, U.S. Geological Survey photo.]

Glacial Movement

Glaciers are different from all other types of ice in that they flow outward or downslope from their area of accumulation. In point of fact, glaciers are rarely static; they are very sensitive to minor climatic variations, as shown by their movements in response to these changes. Glaciers are often likened to "rivers of ice" that flow through their valleys, but there is actually very little similarity between liquid flow and glacial movement. The "flow" of a glacier involves an orderly molecular rearrangement, which is quite unlike the disordered, pell-mell movement of molecules in flowing water.

We usually think of ice as being a brittle substance that breaks rather than bending and that resists any sort of deformation. This is generally true for surface ice, as evidenced by the cracks and crevasses that often appear at the surface of a glacier. However, ice under considerable confining pressure, as below the surface of a glacier, behaves quite differently. It experiences internal deformation both within the individual ice crystals and between adjacent crystals. Moreover, partial melting, due to the stresses within, and the pressure at the bottom of the glacier, provides lubrication for movement.

When a mass of ice attains a thickness of several dozens of yards (meters), it begins to flow in response to the overlying weight. The entire mass does not move; rather there is an oozing outward from around the edge of an ice sheet or down-valley from the toe of an alpine glacier. There is also *laminar flow* within the ice, which causes different portions of the glacier to move with different velocities, and *basal slip* at the bottom of the glacier, in which the entire mass slides over its bed on a lubricating film of water. The glacier more or less molds itself to the shape of the terrain over which it is riding, although it simultaneously reshapes the terrain significantly by erosion.

The rate of movement of a glacier is normally quite slow; indeed, *glacial* is sometimes used as an adjective synonymous with *exceedingly slow*. The movement of most glaciers can be measured in fractions of an inch per day, although an advance of several feet per day would not be unusual, and extreme examples of movement of nearly

100 feet (30 m) in a 24-hour period have been recorded. Also, the flow is often erratic, with irregular pulsations and surges over a short span of time. As might be expected, all parts of a glacier do not move at the same rate. The fastest movement is experienced at and near the surface, and if the glacier is confined, such as a valley glacier, the center of the surface ice will move faster than the sides.

One of the most important principles of glacier movement is to recognize that as long as a glacier exists, it continues to move forward. This does not necessarily mean that the outer edge of the ice is advancing, because a distinction must be made between glacier flow and glacier advance. The generalization is that ice at any point on the glacier will continually move forward (outward or downhill), but the outer margin of the glacier may or may not be advancing, depending on the balance between accumulation and wastage of the ice. See Figure 19-10. Thus a glacier can be thought of as a sort of conveyor belt that is always moving forward, but the movement does not affect the position of the end of the conveyor belt. Even a retreating glacier (i.e., one whose outer margin is retracting toward its point of origin due to heavy wastage) is likely to be advancing throughout; it is retreating simply because its rate of wastage at the outer margin is more rapid than the rate of advance.

Figure 19-10 A flowing glacier is not necessarily an advancing glacier. In this sequential illustration, the front of the glacier is clearly retreating, but the ice itself continues to flow forward like a conveyor belt. The boulder marked by an arrow within the ice illustrates the principle.

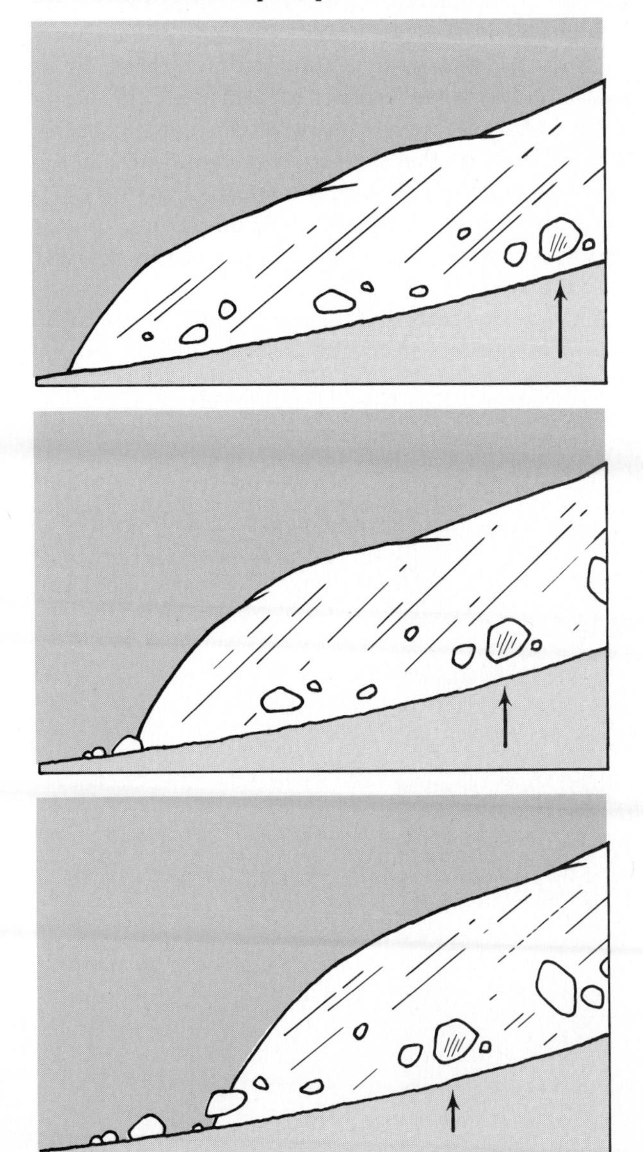

Every glacier can be divided into two portions on the basis of the balance between accumulation and ablation. The upper portion is called the *accumulation zone*, in which the annual increment of ice exceeds wastage. The lower portion is the *ablation zone*, in which there is a net loss of ice on an annual basis due to melting and sublimation. Separating the two zones is a theoretical *equilibrium line*, along which accumulation exactly balances ablation. See Figure 19-11. Whether a glacier is expanding or contracting depends on the relationship between accumulation and ablation. Most of the world's glaciers have shown net retreat during the twentieth century, a clear indication of an excess of wastage over accumulation.

Glacial Erosion

Despite their spectacular nature, glaciers do not create landscapes; they merely modify existing ones. Like the other external agents, this is accomplished by erosion and deposition. The very nature of a large moving mass of ice, however, is spectacular, and the amount of work that it can accomplish in a geologically short time is impressive.

Figure 19-11 Schematic longitudinal cross section of an alpine glacier. The upper portion is an area of net accumulation of ice. Below the equilibrium line there is more ablation (wastage) than accumulation.

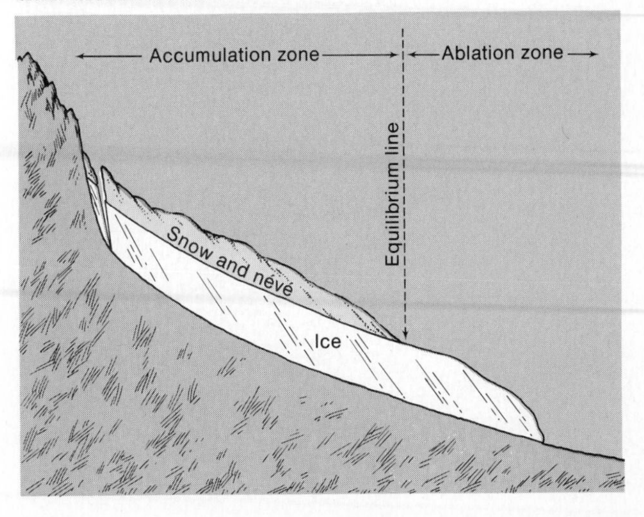

As with streams, volume and velocity are the important determinants of the effectiveness of glacial erosion. The amount of erosion is roughly proportional to the thickness of the ice and its rate of flow. The depth of the erosion is limited in part by the structure and texture of the bedrock and in part by the relief of the terrain. Glacial erosion is inhibited by low relief and is enhanced by high relief. On flattish land, the induced changes are relatively minor; but in mountainous areas, the modifications may be strikingly conspicuous.

The direct erosive power of moving ice is greater than that of flowing water, but not remarkably so. As the slowly moving ice impacts against bedrock, melting is induced by the frictional contact, which reduces the pressure on the rock. Melted ice, however, can refreeze around rocky protrusions and can exert a significant plucking force as it is pushed by the ice behind it. Probably the most significant erosive work of glaciers is accomplished by this *plucking* or *quarrying*. Rock particles beneath the ice are grasped by the freezing of meltwater in joints and fractures. As the ice moves along, these particles are pried out and dragged along in the general flow. This action is particularly effective on leeward slopes.

Glaciers also erode by *abrasion*, in which the bedrock is scoured or worn down by the rock debris dragged along in the moving ice. As with running water, moving ice carries abrasive tools along with it, including much larger chunks of rock than streams can move. Thus the sandpaper effect of a glacier is more effective than that of a river. Abrasion mostly produces minor features, such as polished bedrock surfaces, striations (fine parallel indentations), and glacial grooves (deeper and larger indentations). Whereas plucking tends to roughen the underlying surface and provide the glacier with abrasive tools, abrasion tends to grossly smooth the subglacial surface and dig striations and grooves.

In plains areas, the topography produced by glacial erosion is relatively inconspicuous. Prominences are smoothed and small hollows may be excavated, but the general appearance of the terrain changes little. In hilly areas, however, the effects of glacial erosion are much more notable. Mountains and ridges are sharpened, valleys are deepened and steepened and made more linear, and the entire landscape becomes more angular and rugged.

Transportation by Glaciers

Although their movement is slow, glaciers are extremely competent, as well as indiscriminate, in their ability to transport rock debris. Because they are solid bodies, they are able to move immense blocks of rock, literally the size of houses. Moreover, they may transport these gigantic pieces for dozens or even hundreds of miles. Most of a glacier's load, however, does not consist of such huge blocks, but rather is a heterogeneous collection of particles of all sizes. Perhaps the most typical component of the load is *glacial flour*, which is rock material that has been ground to the texture of very fine talcum powder; a large proportion of the sediment carried by glacial meltwater consists of glacial flour. See Figure 19-12.

Most of the material transported by a glacier is plucked or abraded from the underlying surface, so it is

Figure 19-12 A meltwater stream issuing from a subglacial tunnel at the front of Valdez Glacier in southern Alaska. The murky appearance of the water is due to its heavy load of glacial flour. [F. C. Schrader, U.S. Geological Survey photo.]

carried along primarily at the base of the ice. Thus there is a narrow zone at the bottom of the glacier that is likely to be well armored by miscellaneous rock debris frozen into it, whereas most of the rest of the glacial ice is relatively free of rock fragments. In alpine glaciers, some material is also transported on top of the ice, as debris falls down from the surrounding slopes from mass wasting or surface erosion. Also, a certain amount of debris is pushed along in front of the advancing terminus of the glacier, in a sort of bulldozer effect.

Glacial transportation moves the load outward or down-valley at a variable speed. The rate of flow usually increases in summer and slows in winter, but this is also dependent on variations in ice accumulation and in the gradient of the underlying slopes.

One other important aspect of glacial transportation is the role of flowing water on, in, and under the ice. During the warmer months, streams of meltwater normally flow along with the moving ice. See Figure 19-13. Such streams may run along on the surface of the glacier until they find cracks or crevasses into which to plunge, continuing their flow as subglacial streams either within the ice or along the interface between the glacier and the underlying bedrock. Wherever such streams flow, they transport rock debris, particularly smaller particles, providing an effective mechanism for shifting debris from the surface of the ice to a position within or at the bottom of the glacier.

The "conveyor belt" action of a glacier continues throughout its existence. Even if the ice margin is retreating because of excessive wastage, there is still forward transport of debris by the advancing ice throughout the general extent of the glacier. The transport function of a glacier persists indefinitely unless and until the ice becomes so thin that subglacial obstacles, like a hill, prevent further flow.

Glacial Deposition

In addition to erosion and transportation, the depositional work of glaciers is also impressive. Probably their major role in landscape modification is to remove lithospheric material from one area and take it to some distant region where it is left in a fragmented and vastly changed form. This is clearly displayed in North America where an extensive portion of central Canada has been glacially scoured of its soil, regolith, and much of its surface bedrock, leaving a relatively barren, rocky, gently undulating surface dotted with bodies of water. Much of the material that was removed was taken southward and subsequently deposited in the midwestern states, producing an extensive plains area of remarkably fertile soil. Thus the legacy of Pleistocene ice sheets for the Midwest was the evolution of one of the largest areas of productive soils ever known, at the expense of central Canada, which was left impoverished by those same glaciers.

The debris that is deposited by glaciers occurs in all sizes, shapes, and degree of agglomeration. The general term that is applied to all such material moved by glaciers is *drift*, a misnomer coined in the eighteenth century when it was believed that the vast debris deposits of the Northern Hemisphere were leftovers from Biblical floods. De-

Figure 19-13 A meltwater stream flowing down the surface of Athabaska Glacier. [TLM photo.]

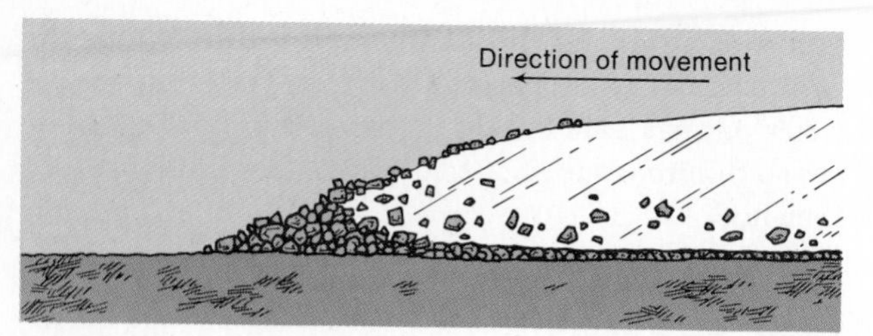

Figure 19-14 Till accumulates at the toe of a glacier, especially during periods of stagnation or retreat of the ice.

spite the erroneous concept, the term is fully accepted today as applicable to all materials carried by glaciers, even for debris actually deposited by meltwater or subsequently moved and redeposited by meltwater.

Rock debris deposited directly by moving or melting ice, with no meltwater flow or redeposition involved, is given the more distinctive name of *till*. See Figure 19-14. Direct deposition by ice usually is the result of melting around the margin of an ice sheet or near the lower end of an alpine glacier, but it is also accomplished by the plastering of debris on the subglacial surface beneath the moving ice. In either case, the result is an unsorted and unstratified agglomeration of fragmented rock material of all sizes. Most of the fragments are angular or subangular in form, as they have been held in a relatively fixed position while carried in the ice and have had little opportunity to become rounded by frequent impact, as happens to pebbles in a stream. Sometimes outsize boulders are included in the glacial till; such enormous fragments, which may be very different from the local bedrock, are called glacial *erratics*. See Figure 19-15.

Glaciofluvial Deposition

Much of the debris carried along by glaciers eventually is deposited or redeposited by glacial meltwater. In some cases, this is accomplished by subglacial streams issuing directly from the ice and carrying sedimentary material washed from positions in, on, or beneath the glacier. Much meltwater deposition, however, involves debris that was originally deposited by ice and subsequently picked up and redeposited by the meltwater well beyond the outer margin of the ice. Such *glaciofluvial deposition* occurs around the margins of all glaciers, as well as far out in some periglacial zones.

Continental Ice Sheets

Apart from the oceans and the continents themselves, continental ice sheets are the most extensive features ever to appear on the face of the Earth. Their actions during Pleistocene time significantly reshaped both the terrain and the drainage of nearly one-fifth of the total surface area of the continents.

Figure 19-15 A glacial erratic perched precariously on a scoured surface in northern Maine. This particular boulder was carried at least a dozen miles by a Pleistocene ice sheet. [E. S. Bastin, U.S. Geological Survey photo.]

Development and Flow

Pleistocene ice sheets, with the exception of the one covering Antarctica, did not originate in the polar regions. Rather, they developed in subpolar and midlatitude locations, from which they spread outward in all directions, including poleward. Several (perhaps several dozen) centers of original ice accumulation were instigated when greater quantities of snow built up than melted. The accumulation of an increasing mass of snow/névé/ice eventually produced such a heavy weight that the ice began to flow outward in a more or less radial pattern from each center of accumulation.

The initial flow was channelled by the preexisting terrain along valleys and other low-lying areas, but in time the ice developed to such depths that it overrode all, or almost all, preglacial topography. In many places, it submerged even the highest parts of the landscape under thousands of feet of ice. Eventually the various ice sheets coalesced into only one or two or three massive sheets on each continent. These vast ice sheets flowed and ebbed as the climate changed, advancing and retreating, and always modifying the landscape with their enormous erosive power and the great masses of debris they deposited. The elaborate result was nothing less than a total reshaping of the land surface and a total rearrangement of the drainage pattern.

Figure 19-16 Development of a *roche moutonnée.* The moving ice grinds along the stoss side of the hill, shaping it largely by abrasion. As the ice moves down the lee side, plucking action is prominent.

Erosion by Ice Sheets

The inexorable force of a monstrous mass of moving ice overrides almost all preexisting topography; grinding, gouging, and plucking as it goes. It scrapes off the soil and regolith, knocks over hills, removes steep slopes, generally flattens the terrain, scours out valleys, and abrades exposed bedrock surfaces. The principal topographic expression resulting from ice sheet erosion is a gently undulating surface, except in mountainous areas of great initial relief.

The most conspicuous topographic features carved by the direct erosion of ice sheets are excavations gouged and deepened by the moving ice in valley bottoms. Such U-shaped troughs are deepest where the preglacial valleys were oriented parallel to the direction of ice movement, particularly in areas of softer bedrock. A prime example of such development is the Finger Lakes District of central New York, where a set of parallel stream valleys was reshaped by glaciation into a group of long, narrow, relatively deep lakes. However, even where valley orientation does not have such configuration, glacial gouging and scooping normally produce a large number of shallow excavations that become lakes after the ice disappears. Indeed, the postglacial landscape in areas of ice sheet erosion is notable for its profusion of lakes.

Hills are generally sheared off and rounded by the moving ice. A characteristic shape is that of the *roche moutonnée* (a French term meaning "sheep's back"), which is often produced when a bedrock hill or knob is overridden by moving ice. See Figure 19-16. The *stoss* side (facing in the direction from which the ice came) of a roche moutonnée is smoothly rounded and streamlined by grinding abrasion as the ice rides up the slope, but the *lee* side (facing away from the ice) is shaped largely by plucking, which produces a steeper and more irregular slope.

Most bedrock surfaces exposed to direct contact with the ice and its load of abrasive debris are marked in various ways by the impact. Such abrasive marks are commonly in the form of *grooves* and *striations* that parallel the direction of ice movement, although other kinds of marks are also found in some cases. Another indication of glacial abrasion is a distinctly polished bedrock surface, usually associated with very resistant rocks, caused by the persistent rubbing of glacial flour as the base of the glacier slides across the rock surface.

The postglacial landscape produced by ice-sheet erosion is one of relatively low relief but not absolute flatness. The principal terrain elements are ice-scoured rocky knobs and scooped-out depressions. Soil and weathered materials are largely absent, with bare rock and lakes dominating the surface. Stream patterns are erratic and inadequately developed because the preglacial drainage net has been deranged by ice erosion. Most of this erosional landscape, however, is subjected to significant further modification, either simultaneously or subsequently, by glacial deposition. Some areas that have been scoured by ice sheets experience little deposition, but most glaciated regions are subjected to both erosion and deposition. Thus the starkness of the erosional landscape may be modified in short order by the superposition of depositional debris.

Deposition by Ice Sheets

The unsorted debris transported by ice sheets is eventually deposited. If the deposition is directly from the ice, without reworking by stream or meltwater, the debris is called *till*. In some cases, the till is deposited heterogeneously and extensively, without forming any particular identifiable topographic features; a veneer of unsorted debris is simply laid down over the preexisting terrain. This veneer is sometimes quite shallow and does not mask the original terrain configuration. In other cases, till is sometimes deposited to a depth of several hundred feet, completely obliterating the shape of the preglacial landscape. In either case, deposition tends to be rather uneven, producing an irregularly undulating surface of broad, low rises and shallow depressions. Such a surface is referred to as a *till plain*.

In many instances, glacial sediments are laid down in somewhat more precise patterns, creating characteristic and identifiable landforms. See Figure 19-17. The largest

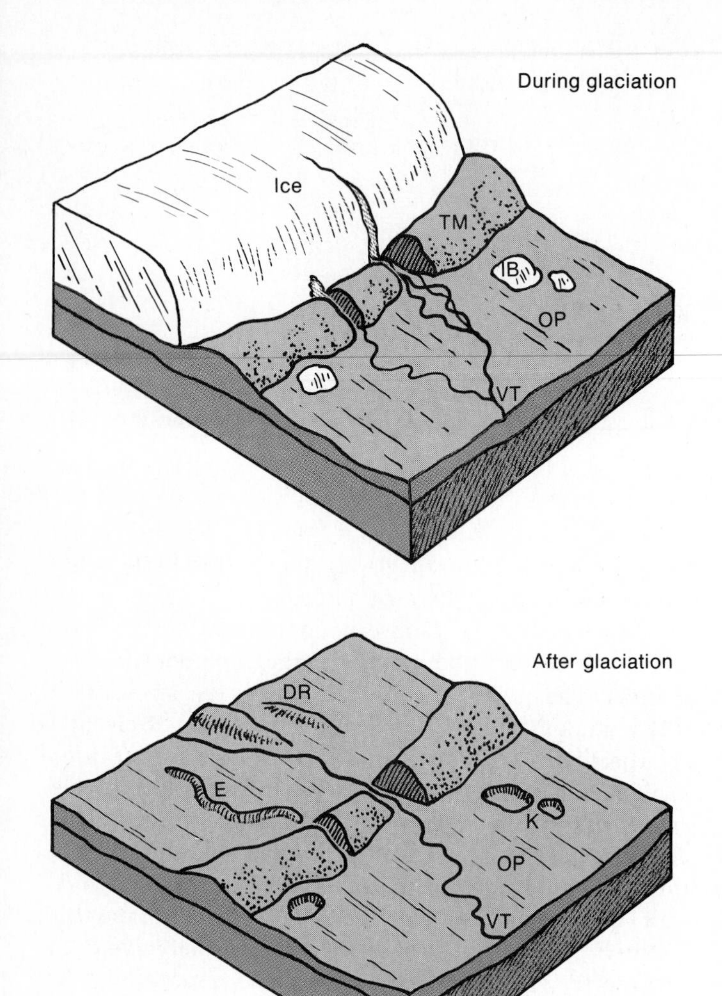

Figure 19-17 Common depositional features associated with the margin of an ice sheet. Key: TM—terminal moraine; OP—outwash plain; E—esker; VT—valley train; IB—ice block; K—kettle; DR—drumlin.

and generally most conspicuous landform features produced by glacial deposition are *moraines*, which are irregular rolling topography rising somewhat above the level of the surrounding terrain. Moraines are usually much longer than they are wide, although the width can vary from a few tens of feet to as much as several miles. Some moraines are distinctly ridgelike, whereas others are much more irregular in shape. Their relief is not great, varying from a few feet to a few hundred feet. When originally formed, moraines tend to have relatively smooth and gentle slopes, which become more uneven with the passage of time, as the blocks of stagnant ice, both large and small, included within the till, eventually melt, creating irregular depressions, known as *kettles*, in the morainal surface.

Two types of moraines are particularly associated with deposition from continental ice sheets—terminal moraines and recessional moraines. A *terminal moraine* builds up at the outermost extent of ice sheet advance, whereas a *recessional moraine* is formed during a lengthy pause in the retreat of the ice margin. See Figure 19-18. Both features are fed by debris brought forward by the conveyor belt movement of the ice, which, as we have seen, continues throughout the life of a glacier. Both terminal and recessional moraines normally occur in the form of concave arcs that bulge outward in the direction of ice movement, indicating that the ice sheets did not advance along an even line but rather as a connecting series of great lobes or tongues of ice, each with a curved front.

Another prominent feature formed by ice sheet deposition is a low, elongated hill called a *drumlin*. A drumlin is a much smaller landform than a moraine, but it is composed of similarly unsorted till. The long axis of the drumlin is aligned parallel with the direction of ice movement; the end of the drumlin facing the direction from which

Figure 19-18 End (terminal and recessional) moraines in the United States resulting from the Wisconsin stage of glaciation.

the ice came is somewhat blunt and slightly steeper than the narrower and more gently sloping opposite end. Thus the configuration is the reverse of that of a roche moutonnée. The origin of drumlins is not completely understood, but most of them apparently are the result of ice readvance into an area of previous glacial deposition; in other words, they are depositional features subsequently shaped by erosion. Drumlins usually occur in groups, sometimes numbering in hundreds, with each drumlin oriented parallel to the others in the group. See Figure 19-19. The greatest concentrations of drumlins in the United States are found in central New York and in eastern Wisconsin.

Glaciofluvial Features

The deposition or redeposition of debris by glacial meltwater produces certain glaciofluvial features in areas of ice sheet activity and in the periglacial region beyond. These features are composed of *stratified drift*, which means that there has been some sorting or stratification of the debris as it was carried along by the flow of the meltwater. *Glaciofluvial* features, then, are composed largely or entirely of smaller rock particles—mostly gravel, sand, and silt—because the meltwater is incapable of moving larger material.

The most extensive glaciofluvial features are *outwash plains*, which are relatively smooth, flattish alluvial aprons deposited beyond the recessional or terminal moraines by streams issuing from the ice. Such outwash deposits sometimes cover many hundreds of square miles. They are occasionally pitted by *kettle holes*, formed by the slow melting of blocks of stagnant ice in similar fashion to that for morainal kettles, and which often become ponds or small lakes. Beyond the outwash plain there is sometimes a lengthy deposit of glaciofluvial alluvium confined to a valley bottom; such a deposit is termed a *valley train*.

Less common than outwash plains but more conspicuous are long sinuous ridges of stratified drift called *eskers*. These are composed largely of glaciofluvial gravel and are thought to originate by the choking of subglacial streams during a time of glacial stagnation. Streams flowing within or beneath a stagnating ice sheet often carry a great deal of debris, and as the ice continues to melt, the stream may deposit much of its load in the subglacial tunnel, later to be exposed by subsequent melting of the overlying ice. Eskers are usually a few dozen feet high, a few dozen feet wide, and may be a few dozen or even a few hundred miles long.

Relatively small, steep-sided mounds or conical hills of stratified drift are found sporadically in areas of ice sheet deposition. These *kames* appear to be of diverse origin, but they are clearly associated with meltwater deposition in close association with stagnant ice. See Figure 19-20. In fact, many seem to have been built like steep fans or deltas against the edge of the ice, with partial collapse of the mound resulting when the ice melted. Morainal surfaces with a number of mounds and depressions may be referred to as *kame and kettle topography*.

Figure 19-19 A drumlin field in Michigan.

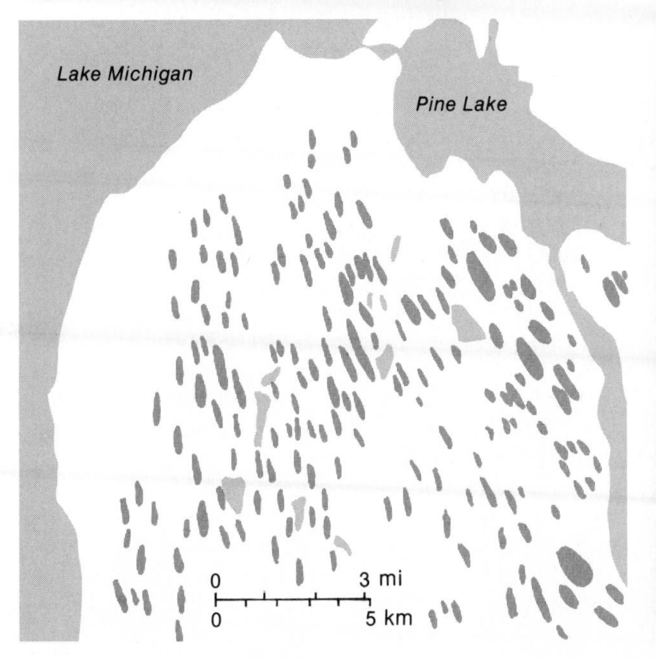

Figure 19-20 A kame in southeastern Wisconsin, near Dundee. [TLM photo.]

Focus on Proglacial Lakes

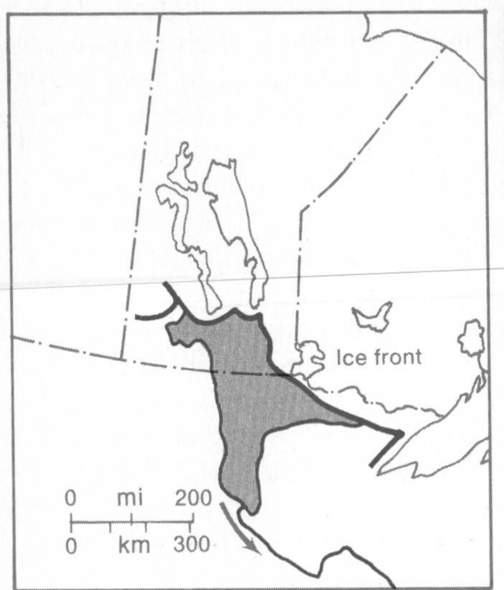

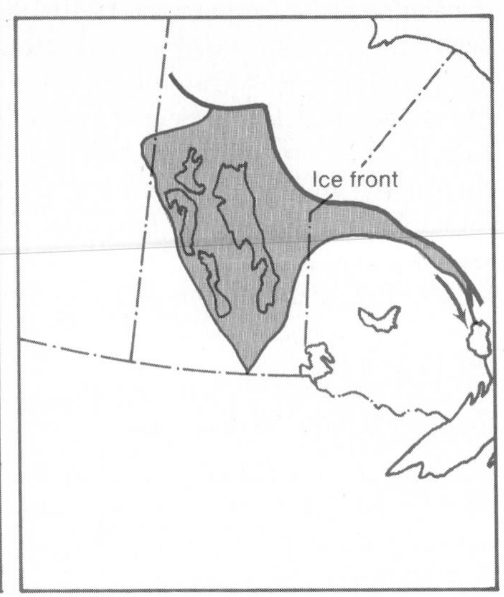

The presumed sequential development of Lake Agassiz as the ice sheet melted.

Where ice flows across or against the general slope of the land, the natural drainage is impeded or completely blocked, and meltwater from the ice can become impounded against the ice front, forming a marginal or *proglacial lake*. This happens sometimes in alpine glaciation, but is much more common along the margin of continental ice sheets, particularly when the ice stagnates. Most proglacial lakes are small and quite temporary because subsequent ice movements cause drainage changes or because normal fluvial processes, accelerated by the growing accumulation of meltwater in the lake, cut spillways or channels to drain the impounded waters.

Sometimes, however, proglacial lakes are large and relatively long-lived. Such major lakes are characterized by considerable fluctuations in size, due to the changing location of the receding or advancing ice front. Several huge proglacial lakes were impounded along the margins of the ice sheets as they advanced and retreated during the Pleistocene epoch in North America, Europe, and Siberia.

By far the largest of the proglacial lakes in North America was Lake Agassiz, named after the Swiss scientist who first developed the theory of an "ice age." Lake Agassiz consisted of meltwater impounded between the great ice sheet on the north and various ridges (especially the Manitoba Escarpment) and moraines on the south. The lake was first formed about 13,000 B.P. (before the present), and existed for some 5000 years before its final drainage. During its life it covered a total area of about 180,000 square miles (466,000 km^2), but only about half of this area was occupied by the lake at its maximum extent, which was a more extensive water surface than that of all five Great Lakes combined today. Lake Agassiz initially drained southward into the Mississippi River system, but as the ice sheet retreated northward, it found a new drainage outlet into Lake Superior and eventually drained into Hudson Bay.

One of the significant results of the formation of proglacial lakes was the development of *lacustrine* (lake-built) topographic features. Beaches and terraces were built by erosion and deposition along the lakeshores, many of which are still found on the land today, well removed from any present-day lakes. More important from an economic standpoint was the accumulation of fine sediments in the old lake beds. The plain of Lake Agassiz, for example, is a remarkably flat surface in parts of two states (Minnesota and North Dakota) and two provinces (Manitoba and Saskatchewan), underlain by very productive soils derived from the lacustrine sediments.

The five Great Lakes that occur along the United States-Canada border are essentially proglacial lakes, although their formation involved a much more complex series of events, still not completely understood. The lake basins probably originated as broad lowlands that were developed on weak sedimentary rocks by ordinary fluvial processes, presumably draining into the Mississippi River system. The early advances of the ice sheet over this region deepened the basins by glacial erosion.

Subsequent retreat of the ice produced the first large proglacial lakes—Chicago and Maumee—whose drainage was southward. Further minor advances and retreats of ice lobes broke up Lake

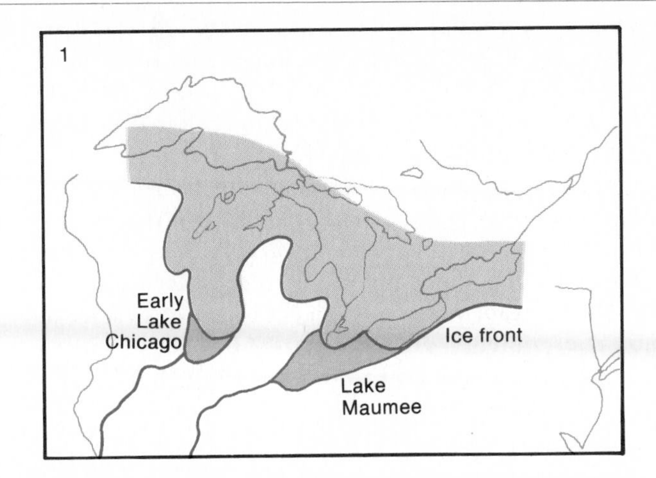

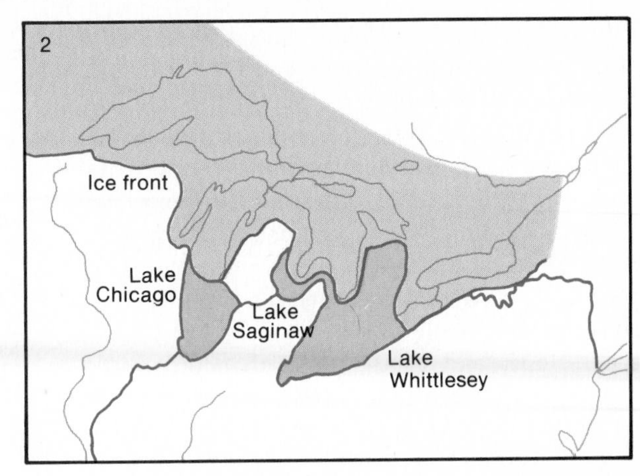

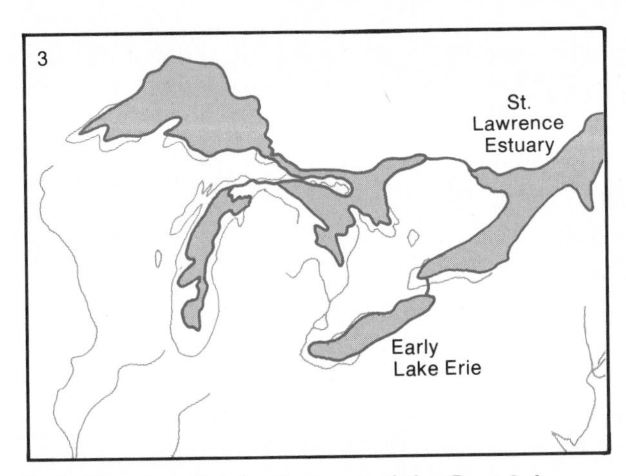

Probable sequential development of the Great Lakes.

Maumee and formed the smaller Lake Saginaw to the west and the huge Lake Whittlesey to the east, as well as establishing a new drainage outlet eastward along the ice front into what is now the Mohawk Valley of New York. The final retreating phase of the ice sheet opened up Lake Superior for the first time, and established a drainage outlet down the valley of the Ottawa River into a long arm of the Atlantic Ocean that comprised the St. Lawrence estuary.

The immediate postglacial stage of the Great Lakes was the maximum extent of their development, with the three upper lakes being joined into a single enormous Lake Nipissing, which had three different drainage outlets. The final form of the lakes came when isostatic rebound (due to the unloading of the great weight of ice) uplifted the shield area to the north, draining Lake Nipissing sufficiently to create the three separate upper lakes and shutting off the Ottawa River drainage outlet. The Great Lakes were then in their contemporary form, but they still had two drainage outlets; one via the Illinois River into the Mississippi and the other down the lakes into the St. Lawrence River. A final lowering of the lake levels caused the abandonment of the Illinois River outlet, and the entire Great Lakes system now drains into the Atlantic via the St. Lawrence River.

The Great Lakes are unmatched in magnitude by any other group of glacially formed lakes in the world and constitute the most conspicuous result of Pleistocene glaciation in the landscape of North America.

Glaciation in Mountainous Areas

Most of the world's high mountain regions experienced massive Pleistocene glaciation, and many of them are still undergoing active glaciation, on a somewhat reduced scale, at present. The reshaping of terrain by mountain glaciers is usually not as complete as that by continental ice sheets, partly because some portions of the mountains protrude above the ice and partly because the movement of mountain glaciers is significantly channelled by preexisting topography. However, the effect of glacial action, particularly erosion, on mountainous topography is to create steeper slopes and greater relief than previously occurred, in contrast to ice sheet action, which tends to smooth and round the terrain.

Formation and Movement

As with continental ice sheets, glaciers are initiated in mountainous country by a net annual accumulation of snow over a considerable period of time. A vast accumulation of snow/névé/ice can produce an extensive highland icefield that extends broadly across the high country, submerging all but the uppermost peaks, and finding its outlet in a series of valley glaciers that issue from the icefield down adjacent drainage channels. Much more common is the initiation of ice accumulation in sheltered depressions near the heads of stream valleys (but often many hundreds of feet below the level of the peaks or ridge crests) in mountainous areas. Glaciers from either source advance downslope, following the urging of gravity and normally finding the path of least resistance along a preexisting stream valley. See Figure 19-21. A system of major and tributary glaciers usually develops, with a trunk glacier in the main valley joined by tributary glaciers from smaller valleys.

Erosion by Mountain Glaciers

Erosion by highland ice fields and alpine glaciers reshapes the topography in dramatic fashion. It largely or totally remodels the peaks and ridges of the summit areas and

Figure 19-21 Sequential shaping of terrain by glaciation in a mountainous area: (1) First is the preglacial landscape; (2) then long-term net accumulation of snow in the higher valleys compresses into ice, and glaciers begin to develop; (3) at the time of maximum glacial advance, the main valleys are modified into deep glacial troughs, and tributary valleys are also severely eroded; (4) the postglacial landscape is composed of jagged ridges, numerous cirques, and deep U-shaped valleys.

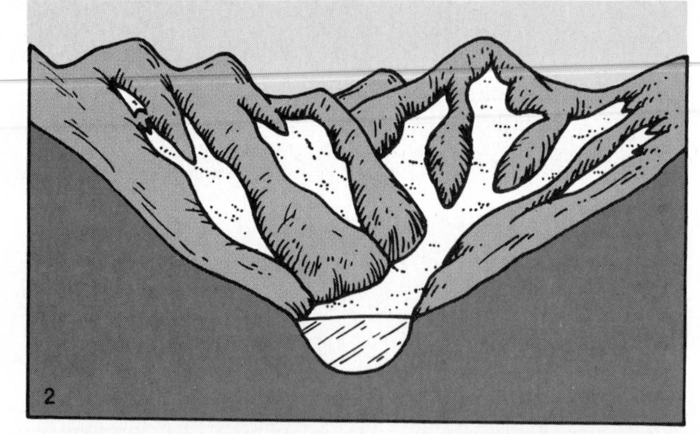

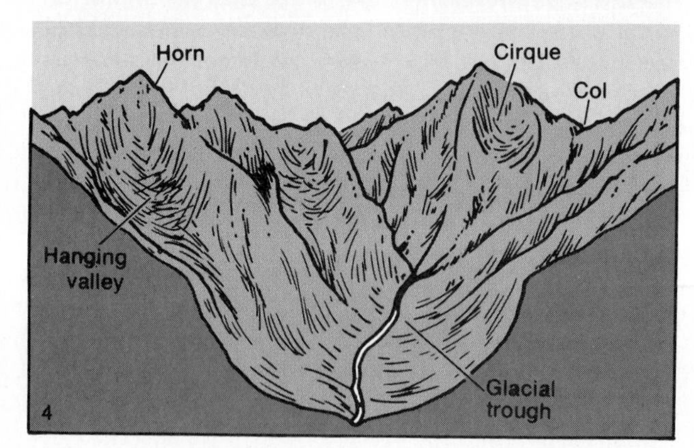

Figure 19-22 Three little cirques lined up in a row on Mt. Nebo in central Utah. This November scene shows snowfields in the cirques, but no glaciers. From the shape of the valleys below the cirques, it appears that the original cirque glaciers never moved out to become valley glaciers. [TLM photo.]

thoroughly transforms the valleys leading down from the high country.

In the High Country The basic landform feature in glaciated mountains is the *cirque*, a broad amphitheater hollowed out at the head of a glacial valley. It has very steep, often perpendicular, head and side walls, and a floor that is flattish or gently sloping, or even gouged enough to form a saucerlike basin. A cirque is the place of origin of an alpine glacier. It is the first landform feature produced by alpine glaciation, although the precise mechanics of its formation are hazy. See Figure 19-22. Essentially, a cirque is quarried out of the mountainside by the plucking action of the ice, abetted by mass wasting and frost wedging. As the glacier grows, its erosive effectiveness within the cirque increases, and when the glacier begins to extend itself out of the cirque down-valley, quarried fragments from the cirque are carried away with the flowing ice. Cirques vary considerably in size, ranging from a few acres to a few square miles in extent.

Enlargement of a cirque is accomplished largely by ice plucking that eats back into the head and side walls. Where cirques are relatively close together and glaciation is active, adjacent cirques may be expanded toward one another so that the upland interfluve between them is reduced to little more than a steep rock wall. Where several cirques have been cut back into an interfluve from opposite sides of a divide, a narrow, jagged, serrated spine of rock may be all that is left of the ridge crest; this is called an *arête*. See Figure 19-23. If two adjacent cirques on opposite sides of a divide are cut back enough to remove part of the arête between them, the sharp-edged pass or saddle through the ridge is referred to as a *col*. An even more prominent feature of glaciated highland summits is a *horn*, a steep-sided, pyramidal rock pinnacle formed by expansive quarrying of the headwalls where

Figure 19-23 The Grand Teton Mountains in Wyoming constitute a massive and spectacular arête, with several horn peaks protruding skyward. [TLM photo.]

three or more cirques intersect; the name is derived from Switzerland's Matterhorn, the most famous example of such a glaciated spire.

When a cirque is deglaciated, i.e., when the glacial ice has melted away, there is sometimes enough of a depression formed to hold water in the form of a small lake or pond. Such a cirque lake is called a *tarn*.

In the Valleys Some alpine glaciers never leave their cirques, presumably because of insufficient accumulation of ice to force a down-valley movement. The valleys below such *cirque glaciers* are unmodified by glacial erosion. Most alpine glaciers, however, as well as those issuing from highland icefields, flow down preexisting valleys and remodel them completely.

A glacier moves down a mountain valley with much greater erosive effectiveness than a stream. It is denser, carries more abrasive tools, and has an enormously greater volume than a stream. It erodes by both abrasion and plucking. The lower layers of the ice can even flow uphill for some distance if blocked by resistant rock on the valley floor, permitting the dragging of rock fragments out of depressions below the general valley level.

The principal erosive work of a valley glacier is to deepen, steepen, and widen its valley. Abrasion and plucking take place not only on the valley floor but along the sides as well. Thus the valley floor is made deeper, the valley walls are made further apart, and the valley sides are oversteepened. The cross-sectional profile is changed from its stream-cut V-shape to an ice-eroded U-shape that is somewhat flared at the top. Moreover, the general course of the valley is significantly straightened because the ice does not meander like a stream; rather it tends to grind away the protruding spurs that separate side canyons, creating what are called *truncated spurs*, and thereby replacing the sinuous course of the stream with a straight *glacial trough*. See Figure 19-24.

The grinding of a glacier along the floor of a glacial trough does not produce a very smooth surface, as might be expected. Valley glaciers do not erode a continuously sloping channel; the principle of differential erosion works with ice as well as with water. See Figure 19-25. The more highly resistant rock on the valley floor is gouged less deeply, whereas weaker or more fractured rock may be scoured out to some depth. The long profile of the glaciated valley floor is irregular, with parts that are gently sloping, flat, or steep, and with some excavated depressions alternating in erratic sequence.

The resulting landscape of the glacial trough, after the ice has melted away, usually shows an irregular series

Figure 19-24 Looking west down Little Cottonwood Canyon, a U-shaped glacial trough near Salt Lake City, Utah. [TLM photo.]

Figure 19-25 The classic glacial landscape of Yosemite, looking past Half Dome up the Tenaya Valley. [TLM photo.]

The Front Range—Episode Four: Pleistocene Glaciation (continued from page 367)

Although fluvial erosion in the "canyon cycle" produced some sharp crests and rugged terrain, most of the main divides rising above the uplifted peneplain surface retained the rounded form of subdued mountains. The peneplain remnants themselves were only gently rolling, although stream-carved canyons cut deeply into the uplands. During the Pleistocene epoch, these lofty but unspectacular mountains received abundant snowfall and became the setting for the development of scores of alpine glaciers.

The Front Range was seamed with a great many stream valleys, some deep with gentle gradients and some shallow with steep gradients. Most of these valleys extended directly down from the high country, trending either easterly toward the plains or westerly toward the jumbled terrain of central Colorado. At the head of every main valley, as well as at the heads of tributary valleys, snow and ice accumulated, and glaciers were formed. Cirques were plucked out of the heads of the valleys, generally between 10,000 and 12,000 feet (3000 and 3600 m) above sea level. As the cirque glaciers continued to grow, they spilled over the edges of the cirques and moved down the valleys, grinding and gouging and plucking as they went.

The valley glaciers reshaped the stream-cut gorges in classic fashion, deepening and broadening them and cutting U-shaped profiles. Most of the glaciers extended several miles down-valley at their maximum development. The longest achieved a length of some 15 miles (24 km) and were at least 1000 feet (300 m) thick in their upper reaches. They did not, however, extend to the base of the mountains, and none ever reached the plains beyond. Indeed, very few of the Front Range glaciers descended much below 8000 feet (2400 m).

During the height of Pleistocene glaciation, extensive alpine glaciers filled all the high country valleys of the Front Range. The glaciers extended toward the foothills but did not reach the plains. Notable lateral and terminal moraines were built by the glacial action.

Nearly all of the glaciers built modest terminal moraines at their lower ends, and many of them produced a series of small recessional moraines as they retreated erratically during deglaciation. However, there were some lateral moraines of very impressive size (up to 1000 feet high and several miles long) that were heaped up during the repeated glacial advances.

The crest of the mountains was also affected significantly by the work of Pleistocene ice. For the most part, the crestal drainage divide of the Front Range is just one long north-south arête, joined by numerous smaller east-west trending side ridges, also often highly serrated. The glaciers cut their cirques back into the divide from both east and west, producing an abundance of saw-tooth spines and steep cliffs but only a few prominent horn peaks.

Today less than two dozen glaciers remain in the entire Front Range, and they are all mere remnants, largely confined to their original cirques. The largest, Arapaho Glacier, is less than a mile long. But the spectacular scenery of the range—peaks, cliffs, waterfalls, and lakes—is primarily a legacy of this glacial epoch.

(*To be continued. Episode five begins on page 439*).

of rock steps or benches, with steep (though usually short) cliffs on the down-valley side and small lakes in the shallow excavated depressions of the benches. This landform expression is called *cyclopean stairs*. The postglacial stream that flows down-valley out of the cirque has a relatively straight course but a fluctuating gradient. Rapids and waterfalls are common, particularly on the cliffs below the benches. The various shallow lakes occur in a sequence called *paternoster lakes*, after a fancied resemblance to beads on a rosary.

Some of the most spectacular glacial troughs occur along coastlines where valleys have been partly drowned by the sea. These spectacular valleys, called *fjords*, are particularly noteworthy in parts of Alaska, Canada, Norway, Scotland, Chile, and New Zealand. See Figure 19-26. Many fjords were gouged out while the land was above sea level and were subsequently flooded, but others were actually carved in part by glaciers whose bottoms were below sea level. Some fjords are well over 100 miles (160 km) long.

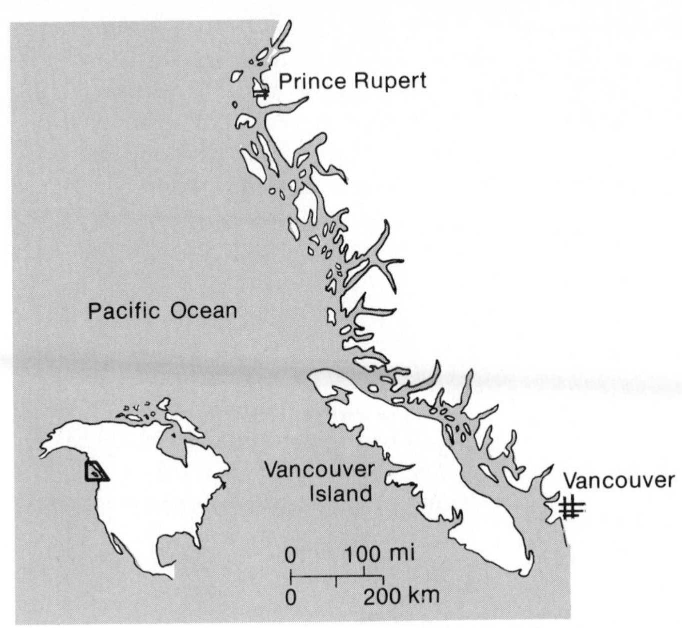

Figure 19-26 The fjorded coastline of British Columbia.

Our interest in the last few paragraphs has been on major valleys occupied by trunk glaciers. The same processes are at work and the same features are produced in smaller tributary valleys, although usually on a lesser scale. However, one important distinction between main valleys and tributary valleys is in terms of the amount of glacial erosion. Erosive effectiveness is determined largely by the amount of ice that is brought to bear on the valley; thus the smaller ice streams of tributary valleys cannot widen and deepen as much as the main valley glaciers. When occupied by glaciers, both main and tributary valleys may appear equal in depth because the top of the ice joins relatively smoothly at the same level. When deglaciated, however, the discordance of the valley floors is conspicuous; the mouths of the tributary valleys are characteristically perched high along the sides of the major troughs, forming *hanging valleys*. Typically, streams that drain the tributary valleys must plunge over waterfalls to reach the floor of the main trough. Several of the world-famous falls of Yosemite National Park are of this type. See Figure 19-27.

Figure 19-27 The waters of Upper Yosemite Falls plummet down from the hanging valley of Yosemite Creek into the glacial trough of the Merced River in Yosemite National Park. [TLM photo.]

Erosion by mountain glaciers, then, produces considerable landform contrast. The high country is made more rugged and jagged, more saw-toothed and spectacular than before. The valleys are enlarged, deepened, and widened, but their lines are less stark and abrupt than in the high country.

Deposition by Mountain Glaciers

Depositional features are relatively much less significant in areas of mountain glaciation than where continental ice sheets have been at work. The high country is almost totally devoid of drift; only in the middle and lower courses of glacial valleys can much deposition be found.

The principal depositional landforms associated with mountain glaciation are moraines. *Terminal moraines* and *recessional moraines* form in the same fashion as previously discussed with ice sheets. They are, however, much smaller and less conspicuous because they are restricted to the narrow confines of the glacial troughs.

The largest depositional features produced by mountain glaciation are *lateral moraines*; these are well-defined ridges of unsorted debris built up along the sides of valley glaciers, parallel to the valley walls. See Figure 19-28. The debris is accumulated in part by the glacier as it moves along and in part by rock material that falls or is washed down the valley walls.

Where a tributary glacier joins a trunk glacier, their lateral moraines become united at the intersection and often continue together down the middle of the combined glacier as a dark band of rocky debris known as a *medial moraine*. Medial moraines are sometimes found in multiplicity, indicating that several glaciers have joined; this produces a candy-cane effect of black (moraine) and white (ice) bands extending longitudinally down the valley.

Figure 19-28 A recently formed lateral moraine alongside Athabaska Glacier in the Canadian Rockies of Jasper National Park. [TLM photo.]

The Front Range—Episode Five: Resumed Fluvial Action (continued from page 436)

In the brief span of 80 or 90 centuries that has transpired since the "end" of the last glacial stage in the Front Range, there has been relatively little change in its topography. The melting ice revealed a landscape that has been little altered by postglacial processes.

The high country retains its glacially produced starkness. Steep cliffs, sharp spires, and abrupt slopes mark the horizon everywhere. Broken fragments, large and small, of frost-riven rock cover the rolling upland remnants of the old peneplain surfaces. The upper ends of the valleys are dominated by the barren amphitheaters of their cirques, most of which now contain neither glacier nor permanent snowfield, although some hold small, shallow tarns.

The contemporary Front Range has a rolling crestline that is a relic of the assumed peneplaination. Some jagged, glaciated peaks rise above the general crest, and many U-shaped glacial gorges are cut deep into it.

The steep-sided glacial troughs all contain small flowing streams in their bottoms, but these streams have managed to accomplish precious little in the way of erosion during the brief postglacial era. Most of them flow swiftly, with many cascades and small waterfalls. Little lakes, set in paternoster pattern, dot the floors of many of the troughs. Most of these lakes are shallow and occupy glacially gouged depressions, but some were formed behind recessional or terminal moraines that have not yet been carved through by fluvial action. Further down the valleys are occasional broad, flattish meadows (called "parks" in Colorado), most of which represent the beds of somewhat larger postglacial lakes that were impounded for a short time behind end moraines, before eventually being drained by the downcutting of the outlet stream through the morainal dam.

The foothill topography looks much as it has for the last million years or so. Erosion by running water has been the principal process at work in this zone during all that time, slowly reducing the level of the land while retaining the V-shaped nature of the valley profiles.

(*The end.*)

Glaciofluvial deposition below mountain glaciers is also similar to that bordering ice sheets because similar outwash is produced. *Outwash plains* occur down-valley from terminal and recessional moraines, and lengthy *valley trains* develop beyond, with the finest materials being carried farthest from the source of sediment.

Are We Still in an Ice Age?

We have seen that glaciation is a remarkable modifier of the landscape. In lowlands it smooths the terrain, deranges the drainage patterns, impounds a great many standing bodies of water, and either removes the soil (through erosion) or significantly adds to the amount of soil-forming material (through deposition). In highlands it serrates the interfluves, enlarges the valleys, and produces spectacular scenery.

Ice ages in which these events took place have occurred several, perhaps many, times in the history of the Earth. They have come and gone, leaving prominent evidence of their passing, evidence that was eventually obliterated through time, except for that of the Pleistocene epoch, which is too recent to have been erased.

Ice ages are fascinating not only because of the landscape changes they bring about but also because of the mystery they manifest. What initiates massive accumulations of ice on the continental surfaces? What stimulates the advances and retreats of these ice masses? What causes their disappearance? Scientists and other scholars have pondered these questions for decades. They have developed many scenarios, postulated many theories, and constructed many models in attempts to explain the sporadic glaciation and deglaciation of our planet.

Any satisfactory theory must be able to account for:

1. The accumulation of ice masses more or less simultaneously at various latitudes in both Northern and Southern Hemispheres, but without uniformity (for example—much less in Siberia and Alaska than in similar latitudes in Canada and Scandinavia).
2. The apparently concurrent development of pluvial (wetter) conditions in dry-land areas.
3. Multiple cycles of ice advance and retreat, encompassing minor fluctuations over decades and centuries as well as major stages of glaciation and deglaciation over intervals of tens of thousands of years.
4. Eventual total deglaciation, either actual (in past geologic eras) or potential (for the Pleistocene epoch).

It is easy enough to state that glaciers develop when there is a net accumulation of snow on an annual basis over a period of time, and that glaciers will waste away when summer melting exceeds winter snowfall over a period of time. Beyond that simplistic statement, however, theorists are in dispute and cannot even agree whether a colder climatic regime would be more conducive to glaciation than a warmer one! Although colder conditions would inhibit summer wastage and thus enhance the longevity of the winter accumulation, cold air cannot hold much water vapor. Hence, warmer winters would favor increased snowfall, whereas cooler summers are needed for decreased melting. Even a theory that accommodated either significantly increased snowfall or significantly decreased melting, or a combination of both, would still have to take into account the waxing and waning, the advance and retreat of the ice.

Various indeed are the theories that have been propounded as explanations of Pleistocene climatic changes (and, by extension, of climatic fluctuations during previous ice ages). Some theories are based on variations in the intensity of solar radiation received by the Earth; others are founded on such astronomical cycles as the shifting of the Earth's axis or variations in the eccentricity of the Earth's orbit around the sun; still others are focussed on changes in the amount of carbon dioxide in the atmosphere; some are founded on changes in the position of the continents and ocean circulation patterns; some are rooted in the increased altitude of continental masses after a period of tectonic upheaval; and some incorporate elements of more than one of the above. We will make no attempt here to summarize this multitude of theories, primarily because none of them is widely accepted; the search for a convincing explanation continues.

One significant outgrowth of our inadequate understanding of climatic controls during the Pleistocene epoch is the further puzzle as to whether or not the Pleistocene has ended. Are we now, in the twentieth century A.D., living in a postglacial period or in an interglacial stage? Has the Pleistocene epoch closed? Are we now in the early centuries of a new geologic era? Or have the glaciers merely "gone back to get some more rocks"? We are, of course, much too close to the event (in a temporal sense) to obtain a definitive answer.

However, some hints may be straws in the wind, or they may be only minor aberrations in a long-term pattern. For most of the last 10,000 years or so, there is evidence of a general deglaciation of the Earth—a general retreat of almost all glaciers. Over the last quarter of a century, however, this trend has been largely, but not entirely, either slowed down or reversed. Many glaciers with a history of nothing but retreat as long as they have been known by mankind, began—in the 1960s and 1970s—to readvance. The world has been experiencing several years in which the weather has been more favorable for the nourishment of glaciers. Are these temporary conditions, or does this mark the onset of another glacial stage—and the continuation of the Pleistocene?

Review Questions

1. Why is the Pleistocene epoch so important to physical geography, whereas other ice ages are not?
2. Describe the extent of glaciation during the Pleistocene epoch.
3. Discuss *periglacial* and *pluvial* developments that were associated with Pleistocene glaciation.
4. List and briefly describe the three types of alpine glaciers.
5. Why are mountain areas that have experienced glaciation usually quite rugged?
6. Why are regions that have experienced continental glaciation usually poorly drained?
7. How is it possible for glacial ice to be advancing while the margin of the glacier is retreating?
8. How does *glaciofluvial deposition* differ from *glacial deposition*?
9. Why are *recessional moraines* less prominent than *terminal moraines*?
10. Are we still in an ice age?

Some Useful References

EMBLETON, C., AND C. A. M. KING, *Glacial Geomorphology*. New York: John Wiley & Sons, Inc., 1975.

FLINT, R. F., *Glacial and Quaternary Geology*. New York: John Wiley & Sons, Inc., 1971.

IMBRIE, J., AND K. P. IMBRIE, *Ice Ages: Solving the Mystery*. Short Hills, N.J.: Enslow Publishers, 1979.

PRICE, L. W., *The Periglacial Environment, Permafrost, and Man*. Washington, D.C.: Commission on College Geography Resource Paper No. 14, Association of American Geographers, 1972.

SCHULTZ G. M., *Ice Age Lost*. Garden City, N.Y.: Anchor Books/Doubleday & Co., Inc., 1974.

SUGDEN, D. E., AND B. S. JOHN, *Glaciers and Landscape: A Geomorphological Approach*. London: Edward Arnold (Publishers) Ltd., 1976.

WRIGHT, H. E., AND D. G. FREY, EDS., *The Quaternary of the United States*. Princeton, N.J.: Princeton University Press, 1965.

Appendix I

Color Maps

Plate 6
Major Landform Assemblages of the World

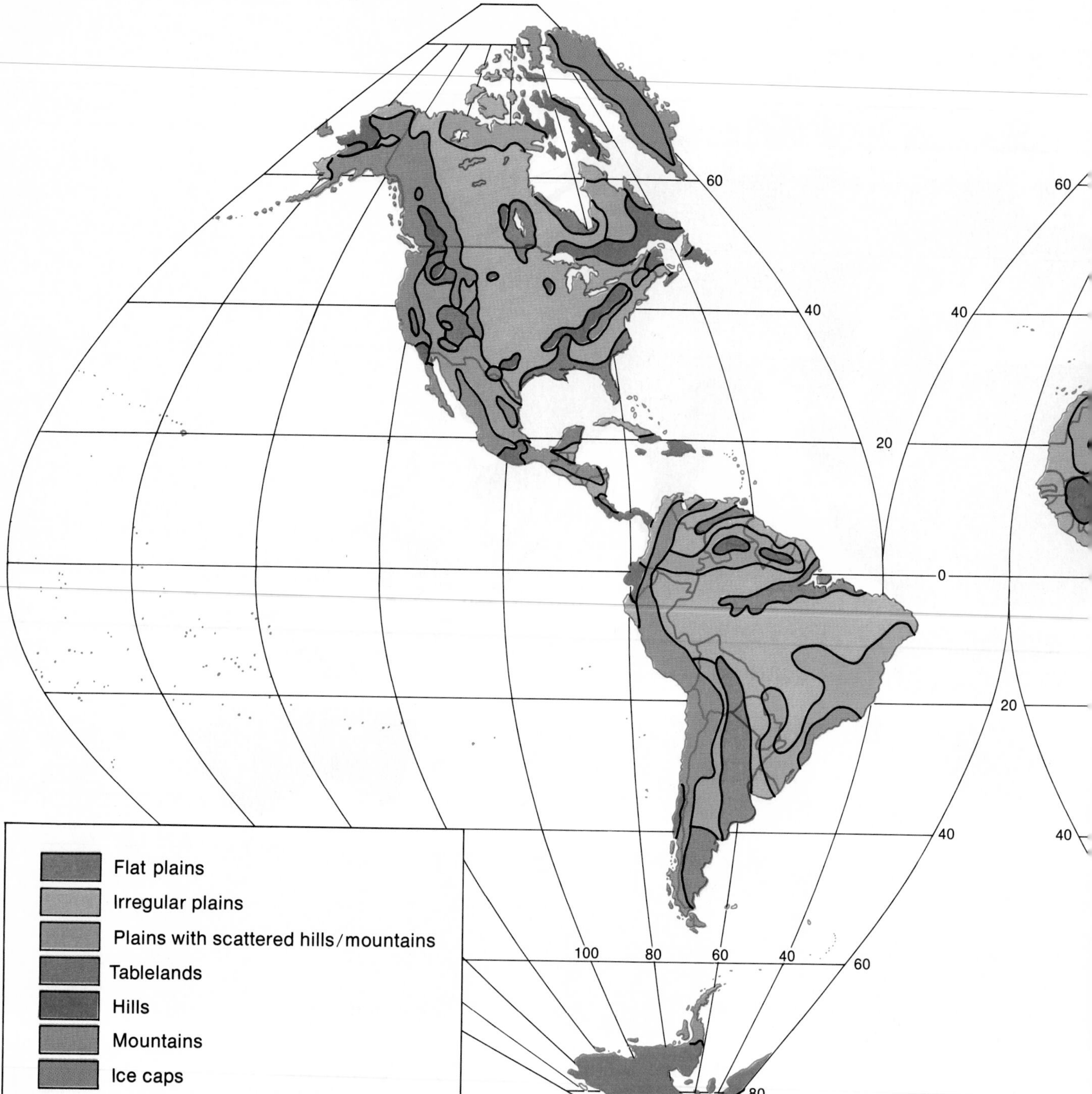

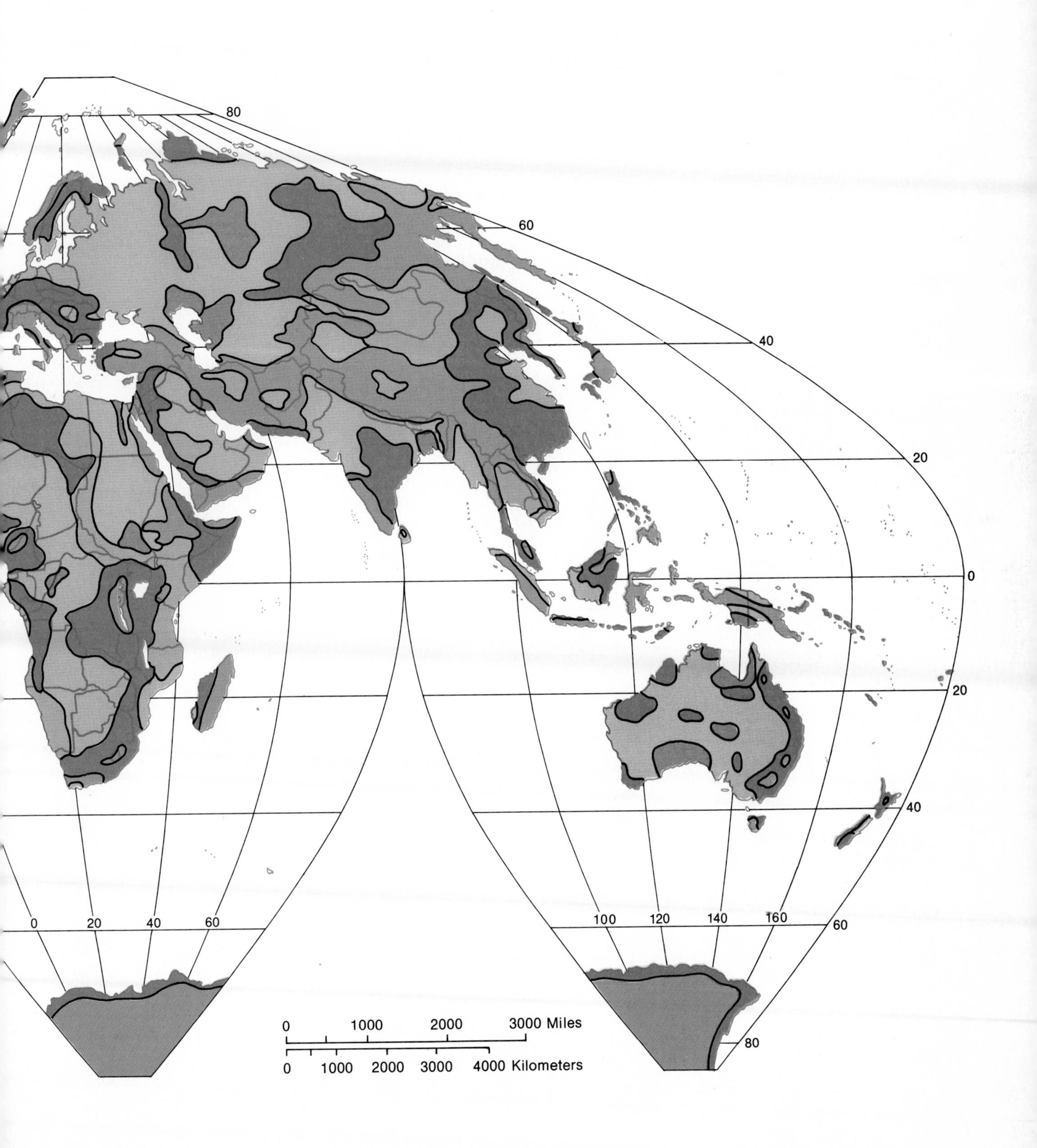

80
60
40
20
0
20
40
60
80
0
20
40
60
100
120
140
160
0 1000 2000 3000 Miles
0 1000 2000 3000 4000 Kilometers

Appendix II

The International System of Units—SI

Three aspects of the modernized metric system are presented here, based on the international system of units—SI.

A. The base and supplementary units of the system.

B. Some common multiples and prefixes that are applicable to all SI units.

C. Conversion tables for U.S. customary units and SI units in describing length, area, volume, and mass.

TABLE A SI Units

Quantity	Unit	Symbol
Base units		
Length	Meter	m
Mass	Kilogram	kg
Time	Second	s
Electric current	Ampere	A
Temperature	Kelvin	K
Amount of substance	Mole	mol
Luminous intensity	Candela	cd
Supplementary units		
Plane angle	Radian	rad
Solid angle	Steradian	sr

TABLE B Common Multiples and Prefixes (applicable to all SI units)

Multiple	Value	Prefix	Symbol
1,000,000,000,000	10^{12}	tera	T
1,000,000,000	10^{9}	giga	G
1,000,000	10^{6}	mega	M
1000	10^{3}	kilo	k
100	10^{2}	hecto	h
10	10^{1}	deka	da
0.1	10^{-1}	deci	d
0.01	10^{-2}	centi	c
0.001	10^{-3}	milli	m
0.000001	10^{-6}	micro	μ
0.000000001	10^{-9}	nano	n
0.000000000001	10^{-12}	pico	p

TABLE C Conversion Tables

Length

Symbol	Known Unit	Multiply by	Unknown Unit	Symbol
in.	Inches	2.54	Centimeters	cm
ft	Feet	0.3048	Meters	m
yd	Yards	0.9144	Meters	m
mi	Miles	1.6093	Kilometers	km
mm	Millimeters	0.039	Inches	in.
cm	Centimeters	0.39	Inches	in.
m	Meters	3.281	Feet	ft
km	Kilometers	0.621	Miles	mi

Area

Symbol	Known Unit	Multiply by	Unknown Unit	Symbol
$in.^2$	Square inches	6.452	Square centimeters	cm^2
ft^2	Square feet	0.0929	Square meters	m^2
yd^2	Square yards	0.836	Square meters	m^2
mi^2	Square miles	2.59	Square kilometers	km^2
—	Acres	0.4	Hectares	ha
cm^2	Square centimeters	0.155	Square inches	$in.^2$
m^2	Square meters	10.764	Square feet	ft^2
m^2	Square meters	1.196	Square yards	yd^2
km^2	Square kilometers	0.386	Square miles	mi^2
ha	Hectares	2.471	Acres	—

TABLE C—Conversion Tables (*continued*)

Volume

Symbol	Known Unit	Multiply by	Unknown Unit	Symbol
in.^{3}	Cubic inches	16.387	Cubic centimeters	cm^3
ft^3	Cubic feet	0.028	Cubic meters	m^3
yd^3	Cubic yards	0.7646	Cubic meters	m^3
fl oz	Fluid ounces	29.57	Milliliters	ml
pt	Pints	0.47	Liters	l
qt	Quarts	0.946	Liters	l
gal	Gallons	3.785	Liters	l
cm^3	Cubic centimeters	0.061	Cubic inches	in.^{3}
m^3	Cubic meters	35.3	Cubic feet	ft^3
m^3	Cubic meters	1.3	Cubic yards	yd^3
ml	Milliliters	0.034	Fluid ounces	fl oz
l	Liters	1.0567	Quarts	qt
l	Liters	0.264	Gallons	gal

Mass (weight)

Symbol	Known Unit	Multiply by	Unknown Unit	Symbol
oz	Ounces	28.35	Grams	g
lb	Pounds	0.4536	Kilograms	kg
—	Tons (2000 lb)	907.18	Kilograms	kg
—	Tons (2000 lb)	0.90718	Tonnes	t
g	Grams	0.035	Ounces	oz
kg	Kilograms	2.2046	Pounds	lb
kg	Kilograms	0.0011	Tons (2000 lb)	—
t	Tonnes	1.1023	Tons (2000 lb)	—

Glossary

Ablation Wastage of glacial ice through melting and sublimation.

Ablation zone The lower portion of a glacier where there is a net annual loss of ice due to melting and sublimation.

Abrasion The chipping and grinding effect of rock fragments as they are swirled or bounced or rolled downstream by moving water.

Absolute humidity A direct measure of the water vapor content of air, expressed as the weight of water vapor in a given volume of air, usually as grams of water per cubic meter of air.

Absorption The ability of an object to assimilate energy from electromagnetic waves that strike it.

Accelerated erosion Soil erosion occurring at a rate faster than soil horizons can be formed from the parent regolith.

Accumulation Addition of ice into a glacier by incorporation of snow.

Accumulation zone The upper portion of a glacier where there is a greater annual increment of ice than there is wastage.

Adiabat (dry adiabatic rate) The relatively steady rate of $5\frac{1}{2}$°F per 1000 feet (10°C per km) at which a parcel of air cools as it rises.

Adiabatic cooling Cooling by expansion in rising air.

Adiabatic warming Warming by contraction in descending air.

Adret slope A slope oriented so that the sun's rays arrive at a relatively direct angle. Such a slope is relatively hot and dry, and its vegetation will not only be sparser and smaller but is also likely to have a different species composition from adjacent slopes with different exposures.

Advection Horizontal movement of air across the Earth's surface.

Advectional inversion An inverted temperature gradient caused by a horizontal inflow of colder air into an area—usually colder air "draining" downslope into a valley—or by cool maritime air blowing into a coastal locale. Advectional inversions are usually short-lived and shallow and are more common in winter than summer.

Advection fog The condensation that results when warm, moist air moves horizontally over a cold surface.

Aeolian processes Processes related to wind action that are most pronounced, widespread, and effective in dry lands.

Aerial photograph A photograph taken from an elevated "platform," such as a balloon, airplane, rocket, or satellite.

Aggradation The process in which a stream bed is raised as a result of the deposition of sediment.

Agonic line The isogonic line on which there is no magnetic declination because of the exact alignment of true north and magnetic north.

A horizon Upper soil layer in which humus and other organic materials are mixed with mineral particles.

Air drainage The sliding of cold air downslope to collect in the lowest spots, usually at night.

Air mass An extensive body of air that has relatively uniform properties in the horizontal dimension and moves as an entity.

Albedo The fraction of total solar radiation that is reflected back, unchanged, into space.

Alcove arches Naturally excavated chambers that formed when trickling surface water and percolating groundwater softened and dissolved the underlying rock, cementing the material of this softer rock under more resistant overhanging caprock. The loosened grains were removed by water and wind erosion.

Alfisols A widely distributed soil order distinguished by a subsurface clay horizon and a medium-to-generous supply of plant nutrients and water.

Alluvial fan A fan-shaped deposition feature laid down by a stream issuing from a mountain canyon.

Alluvium Any stream-deposited sedimentary material.

Alpine glacier Individual glacier that develops near a mountain crest line and normally moves down valley for some distance.

Altocumulus Middle-level clouds, between about 6500 and 20,000 feet (2 and 6 km), which are puffy in form and are composed of liquid water.

Altostratus Middle-level clouds, between 6500 and 20,000 feet (2 and 6 km), which are layered and are composed of liquid water.

Amphibians Semiaquatic vertebrate animals. In the larval stage they are fully aquatic and breathe through gills; as adults they are air-breathers by means of lungs and through glandular skin.

Anadromous fish Fish that spawn in freshwater streams but live most of their lives in the ocean.

Andesite A volcanic rock composed largely of a distinctive mineral association of plagioclase feldspar.

Anemometer An instrument that measures wind speed.

Angiosperms Plants that have seeds encased in some sort of protective body, such as a fruit, a nut, or a seedpod.

Angle of incidence The angle at which the sun's rays strike the Earth's surface.

Angle of repose Steepest angle that can be assumed by loose fragments on a slope without downslope movement.

Animalia The kingdom of organisms that consists of all multicellular animals.

Annual plants Plants that perish during times of climatic stress but leave behind a reservoir of seeds to germinate during the next favorable period.

Annular drainage pattern A network in which the major streams are arranged in a ringlike, concentric pattern in response to a structural dome or basin.

Antarctic Circle The parallel of $66\frac{1}{2}$° south latitude.

Antecedent stream A stream that antedates the existence of an uplift.

Antelope A collective term that refers to several dozen species of hoofed animals, nearly all of which have permanent, unbranched horns.

Anticline A simple symmetrical upfold.

Anticyclone A high-pressure center.

Aphelion The point in the Earth's elliptical orbit at which the Earth is farthest from the sun (94,555,000 miles or 152,171,500 km).

Apogee The point at which the moon is farthest from the Earth in its elliptical orbit.

Aquale A suborder of Alfisol soils that has characteristics that are associated with wetness.

Aquent One of the suborders of Entisol soils that occupies wet environments where the soil is more or less continuously saturated with water.

Aquiclude An impermeable rock layer that is so dense as to exclude water.

Aquifer A permeable subsurface rock layer that can store, transmit, and supply water.

Arboreal Tree-dwelling.

Arctic Circle The parallel of $66\frac{1}{2}°$ north latitude.

Arcuate delta A delta with its shoreline curved convexly outward from the land.

Arent A suborder of Entisol soils that lacks horizons because of human interference.

Arête A narrow, jagged, serrated spine of rock; remainder of a ridge crest after several cirques have been cut back into an interfluve from opposite sides of a divide.

Argid A suborder of Aridisol soils that has a distinctive subsurface horizon with clay accumulation.

Aridisol A soil order occupying dry environments that do not have enough water to remove soluble minerals from the soil; typified by a thin profile that is greatly lacking in organic matter and a sandy texture.

Arroyo The term used in the United States for the normally dry bed of an intermittent stream.

Artesian well The free flow that results when a well is drilled from the surface down into the aquifer and the confining pressure is sufficient to force the water to the surface without artificial pumping.

Asthenosphere Somewhat plastic layer of undetermined extent that underlies the embedded mosaic of rigid plates in the Earth's crust.

Astronomical unit (AU) The average distance from the Earth to the sun (92,955,806 miles ± 3 m) (149,597,892 km ± 5 km).

Atmosphere The gaseous envelope surrounding the Earth.

Atmospheric pressure The weight of the air.

Atoll Coral reef in the general shape of a ring or partial ring that encloses a lagoon.

Autumnal equinox Equinox that occurs about September 23rd in the Northern Hemisphere and March 21st in the Southern Hemisphere.

Axis (Earth's axis) The diameter line that connects the points of maximum flattening on the Earth's surface.

Azimuthal projections A family of maps derived by the perspective extension of the geographic grid from a globe to a plane that is tangent to the globe at some point.

Backswamp Area of low, swampy ground on a floodplain between the natural levee and the bluffs.

Backwash Water moving seaward after the momentum of the wave swash is overcome by gravity and friction.

Badlands Intricately rilled and barren terrain of arid and semiarid regions, characterized by a multiplicity of short, steep slopes.

Baguio The term used for a tropical cyclone affecting the Philippines.

Bajada (piedmont alluvial plain) A continual alluvial surface that extends across the piedmont zone, slanting from the range toward the basin, in which it is difficult to distinguish between individual fans.

Bar Ridge of sand or gravel deposited offshore or in a river.

Barchan A crescent-shaped sand dune with cusps of the crescent pointing downwind.

Barometer An instrument that measure atmospheric pressure.

Barrier reefs Prominent ridges of coral that roughly parallel the coastline, but lie offshore, with a shallow lagoon between the reefs and the coast.

Basal slip The term used to describe when the entire mass at the bottom of a glacier slides over its bed on a lubricating film of water.

Basalt Fine-grained speckled extrusive rock.

Base level An imaginary surface extending underneath the continents from sea level at the coasts, and indicating the lowest level to which land can be eroded.

Batholith The largest and most amorphous of igneous intrusions.

Baymouth bar A spit that has become extended across the mouth of a bay to connect with a headland on the other side, transforming the bay into a lagoon.

Beach An exposed deposit of loose sediment, normally composed of sand and/or gravel, and occupying the coastal transition zone between land and water.

Bedload Sand, gravel, and larger rock fragments moving in a stream by saltation and traction.

Bedrock Residual rock that has not experienced erosion.

Benthos Animals and plants that live at the ocean bottom.

B Horizon Mineral soil horizon located beneath the A horizon.

Biomass The total weight of all living organisms.

Biome A large, recognizable assemblage of plants and animals in functional interaction with its environment.

Biosphere The living organisms of the Earth.

Biota The total complex of plant and animal life.

Bird's-foot delta A delta that has long, projecting distributary channels branching outward like the toes or claws of a bird.

Blackbody A body that emits the maximum amount of radiation possible, at every wavelength, for its temperature.

Blowout (deflation hollow) A shallow depression from which an abundance of fine material has been deflated.

Bluff A relatively steep slope at the outer edge of a flood plain.

Boralf A suborder of Alfisol soils associated with wetness.

Boreal forest (taiga) An extensive needle-leaf forest in the subarctic regions of North America and Eurasia.

Braided streams Streams that consist of a multiplicity of interwoven and interconnected shallow channels separated by low islands of sand, gravel, and other loose debris.

Broadleaf trees Trees that have flat and expansive leaves.

Bryophytes Mosses and liverworts.

Butte An erosional remnant of very small surface area and clifflike sides that rises conspicuously above the surroundings.

Calcification One of the dominant pedogenic regimes in areas where the principal soil moisture movement is upward because of a moisture deficit. This regime is characterized by a concentration of calcium carbonate ($CaCo_3$) in the B horizon forming a hardpan, an upward movement of $CaCo_3$ by capillary water by grass roots, and a return of $CaCO_3$ when grass dies.

Caldera Large, steep-sided, roughly circular depression resulting from the explosion and subsidence of a large volcano.

Capillarity The action by which water can climb upward in restricted confinement as a result of its high surface tension, and thus the ability of its molecules to stick closely together.

Capillary water Moisture held at the surface of soil particles by surface tension.

Caprock A flattish erosional surface formed of the most resistant layers of horizontal sedimentary or volcanic strata.

Carbonates Minerals that are carbonate compounds of calcium or magnesium.

Carbonation A process in which carbon dioxide in water reacts with carbonate rocks to produce a very soluble product (calcium bicarbonate), which can readily be removed by runoff or percolation, and which can also be deposited in crystalline form if the water is evaporated.

Carbon cycle The change from carbon dioxide to living matter and back to carbon dioxide.

Carnivore A flesh-eating animal.

Centripetal drainage pattern A basin structure in which the streams converge toward the center.

Cetaceans Members of an order of aquatic, largely marine mammals, such as whales and dolphins.

Chaparral Shrubby vegetation of the mediterranean climatic region of North America.

Chemical weathering The decomposition of rock by the alteration of rock-forming minerals.

Chinook A localized downslope wind of relatively dry and warm air, which is further warmed adiabatically as it moves down the leeward slope of the Rocky Mountains.

Chordata The phylum within the animal kingdom that includes all animals with backbones.

C horizon Lower soil layer composed of weathered parent material that has not been significantly affected by translocation or leaching.

Cinder cone Small, common volcano that is composed primarily of pyroclastic material blasted out from a vent in small but intense explosions. The structure of the volcano is usually a conical hill of loose material.

Circle of illumination The edge of the sunlit hemisphere that is a great circle separating the Earth into a light half and a dark half.

Cirque A broad amphitheater hollowed out at the head of a glacial valley by ice erosion.

Cirque glacier A small glacier confined to its cirque and not moving down-valley.

Cirriform clouds Clouds that are thin and wispy, composed of ice crystals rather than water particles, and found at high elevations.

Cirrocumulus High cirriform clouds arranged in a patchy pattern.

Cirrostratus High cirriform clouds that appear as whitish, translucent veils.

Cirrus High cirriform clouds of feathery appearance.

Clay Very small inorganic particles produced by chemical alteration of silicate minerals.

Clay pan A playa surface that is heavily impregnated with clay.

Cliff Sheer residual rock face.

Climate An aggregate of day-to-day weather conditions over a long period of time.

Climax vegetation A stable plant association of relatively constant composition that develops at the end of a long succession of changes.

Col A pass or saddle through a ridge produced when two adjacent cirques on opposite sides of a divide are cut back enough to remove part of the arête between them.

Cold front The leading edge of a cool air mass actively displacing warm air.

Collapse doline A sinkhole produced by the collapse of the roof of a subsurface cavern.

Colloids Organic and inorganic microscopic particles of soil that represent the chemically active portion of particles in the soil.

Combined water Water held in chemical combination with various soil minerals.

Composite volcanoes Volcanoes with the classic symmetrical cone-shaped peak, produced by a mixture of lava outpouring and pyroclastic explosion.

Condensation Process by which water vapor is converted to liquid water.

Condensation nuclei Tiny atmospheric particles of dust, smoke, and salt that serve as collection centers for water molecules.

Conduction The movement of energy from one molecule to another without changing the relative positions of the molecules. This enables the transfer of heat between different parts of a stationary body.

Cone of depression The phenomenon whereby the water table has sunk into the approximate shape of an inverted cone in the immediate vicinity of a well as the result of the removal of a considerable amount of the water.

Conglomerate Sedimentary rock composed of pebbles or larger particles in a matrix of finer material.

Conic projections A family of maps in which one or more cones is set tangent to, or intersecting, a portion of the globe and the geographic grid is projected onto the cone(s).

Consequent Stream A stream that follows the initial slope of the land.

Continental Drift Theory that proposes that the present continents were originally connected as one or two large landmasses that have broken up and literally drifted apart over the last several million years.

Continental shelf Submerged margin of continents.

Contour line A line joining points of equal elevation.

Convection Vertical movements of parcels of air due to density differences.

Convergent precipitation Showery precipitation that occurs as a result of the forced uplift of air due to crowding in areas of air convergence.

Coral reefs Coralline formations that fringe continents and islands in warm-water tropical oceans.

Core Spherical central portion of the Earth; divided into a molten outer layer and a solid inner layer.

Coriolis effect The apparent deflection of free-moving objects to the right in the Northern Hemisphere and to the left in the Southern Hemisphere, in response to the rotation of the Earth.

Corrosion Chemical reactions (especially hydrolysis and solution action) associated with streamflow that assist in erosion.

Creep The slowest and least perceptible form of mass wasting, which consists of a very gradual downhill movement of soil and regolith.

Crust The outermost solid layer of the Earth.

Cuesta A low asymmetrical ridge formed by a resistant, gently dipping sedimentary layer.

Cumuliform clouds Clouds that are massive and rounded, usually with a flat base and limited horizontal extent, but often billowing upward to great heights.

Cumulonimbus Cumuliform cloud of great vertical development, often associated with a thunderstorm.

Cumulus Puffy white cloud that forms from rising columns of air.

Cumulus stage The early stage of thunderstorm formation in which updrafts prevail and the cloud continues to grow.

Cutoff meander A sweeping stream channel curve that is isolated from streamflow because the narrow meander neck has been cut through by stream erosion.

Cycle of erosion *See* Geomorphic cycle.

Cyclone Low-pressure center.

Cyclopean stairs The resulting landscape of a glacial trough after the ice has melted away, which shows an irregular series of rock steps or benches, with steep (though usually short) cliffs on the down-valley side and small lakes in the shallow excavated depressions of the benches.

Cylindrical projections A family of maps derived from the concept of projection onto a paper cylinder that is tangential to, or intersecting with, a globe.

Debris flow Streamlike flow of muddy water heavily laden with sediments of various sizes.

Deciduous tree A tree that experiences an annual period in which all leaves die and usually fall from the tree, due either to a cold season or a dry season.

Deflation The shifting of loose particles by wind blowing them into the air or rolling them along the ground.

Delta A landform at the mouth of a river produced by the sudden dissipation of a stream's velocity and the resulting deposition of the stream's load.

Dendritic drainage pattern A treelike, branching pattern that consists of a random merging of streams, with tributaries joining larger streams irregularly but always at acute angles.

Dentrification The conversion of nitrates into free nitrogen in the air.

Denudation The total effect of all actions that lower the surface of the continents.

Dew The condensation of beads of water on relatively cold surfaces.

Dew point The critical air temperature at which saturation is reached.

Diastrophism A general term that refers to the deformation of the Earth's crust.

Dike A vertical or nearly vertical sheet of magma that is thrust upward into preexisting rock.

Discharge Volume of flow of a stream.

Dissolved load The minerals, largely salts, that are dissolved in water and carried invisibly in solution.

Distributaries Branching stream channels that cross a delta.

Doldrums Belt of calm air associated with the region between the trade winds of the Northern and Southern hemispheres; generally in the vicinity of the equator.

Drainage basin An area that contributes overland flow and ground water to a specific stream. (Also called a *watershed* or *catchment*.)

Drainage divide The line of separation between runoff that descends into two different drainage basins.

Drainage net The complex of streams within a drainage basin.

Drift All material carried and deposited by glaciers.

Dripstone Features formed by precipitated deposits of minerals that decorate the walls, floor, and roof of a cavern.

Drumlin A low, elongated hill formed by ice sheet deposition. The long axis is aligned parallel with the direction of ice movements, and the end of the drumlin that faces the direction from which the ice came is somewhat blunt and slightly steeper than the narrower and more gently sloping end that faces in the opposite direction.

Dry adiabatic rate (adiabat) The relatively steady rate at which a parcel of air cools as it rises ($5\frac{1}{2}$° F per 1000 feet) (10° C/km).

Dyne The force needed to accelerate one gram of mass one centimeter per second.

Earthquake Abrupt movement of the Earth's crust.

Easterly wave A long but weak migratory low-pressure trough in the tropics.

Ebb tide A periodic falling of sea level.

Ecosystem The totality of interactions among organisms and the environment in the area of consideration.

Ecotone The transition zone between biotic communities in which the typical species of one community intermingle or interdigitate with those of another.

Edaphic Having to do with soil.

Electromagnetic spectrum Electromagnetic radiation, arranged according to wavelength.

Elliptical projections A family of map projections in which the entire world is displayed in an oval shape.

Eluviation The process by which gravitational water picks up fine particles of soil from the upper layers and carries them downward.

Energy The capacity to do work.

Entisols The least developed of all soil orders, with little mineral alteration and no pedogenic horizons. These soils are commonly thin and/or sandy and have limited productivity, although those developed on recent alluvial deposits tend to be quite fertile.

Entrenched meanders A winding, sinuous stream valley with abrupt sides.

Ephemeral streams Streams that carry water only during the "wet season" or during and immediately after rains.

Epiphytes Plants that live above ground level out of contact with the soil, usually growing on trees or shrubs.

Equator The parallel of 0° latitude.

Equatorial counter current An east-moving ocean current, between the two equatorial currents.

Equilibrium line A theoretical line separating the ablation zone and accumulation zone of a glacier along which accumulation exactly balances ablation.

Equilibrium theory Idea that slope forms are adjusted to geomorphic processes so that there is a balance of energy—the energy provided is just adequate for the work to be done.

Equinox The time of the year when the perpendicular rays of the sun strike the equator, the circle of illumination just touches both poles, and the periods of daylight and darkness are each 12 hours long all over the Earth.

Erg "Sea of sand." A large area covered with loose sand, generally arranged in some sort of dune formation by the wind.

Erosion Detachment and removal of fragmented rock material.

Eskers Long, sinuous ridges of stratified drift composed largely of glaciofluvial gravel and formed by the choking of subglacial streams during a time of glacial stagnation.

Estivation The act of spending a dry/hot period in a torpid state.

Estuaries Fingers of the sea projecting inland along drowned river valleys.

Evaporation Process by which liquid water is converted to gaseous water vapor.

Evaporation fog The condensation that results from the addition of water vapor to cold air that is already near saturation.

Evapotranspiration The transfer of moisture to the atmosphere by transpiration from plants and evaporation from soil and plants.

Evergreen A tree or shrub that sheds its leaves on a sporadic or successive basis, but at any given time appears to be fully leaved.

Exfoliation Weathering process in which curved layers peel off bedrock in sheets. This process apparently occurs only in granite and related intrusive rocks.

Exfoliation dome A large rock mass with a surface configuration that consists of imperfect curves punctuated by several partially fractured shells of the surface layers.

Exotics Organisms that are introduced into "new" habitats in which they did not naturally occur.

Exotic streams Streams that flow into a dry region, bringing their water from somewhere else.

Extinction The dying out of the entire population of a taxa.

Extratropical anticyclone An extensive, migratory high-pressure cell of the midlatitudes that moves generally with the westerlies.

Extratropical cyclone Large migratory low-pressure system that occurs within the middle latitudes and moves generally with the westerlies.

Extrusive rock Rock ejected onto the Earth's surface while still in a molten state, solidifying quickly in the open air.

Eye The nonstormy center of a tropical cyclone, which has a diameter of 10 to 25 miles (16 to 40 km) and is a singular area of calmness in the maelstrom that whirls around it.

Eye wall Peripheral zone at the edge of the eye where winds reach their highest speed.

Fault A fracture or zone of fracture where the rock is forcefully broken with an accompanying displacement; i.e., an actual movement of the crust on one or both sides of the break. The movement can be horizontal or vertical, or a combination of both.

Fault-block mountains Mountains formed under certain conditions of crustal stress, whereby a surface block may be severely faulted and upthrown on one side without any faulting or uplift on the other side. The block is tilted asymmetrically, producing a steep slope along the fault scarp and a relatively gentle slope on the other side of the block.

Fault line The intersection of a fault zone with the Earth's surface.

Fault plane Interface along which movement occurs in faulting.

Fault scarp Cliff formed by faulting.

Fault zone Zone of weakness in the crust where faulting may take place.

Fauna Animals.

Field capacity The maximum amount of water that can be retained in the soil after the gravitational water has drained away.

Firn (névé) Snow granules that have become packed and begin to coalesce due to compression, achieving a density about half as great as that of water.

First-order relief The largest scale of relief features, consisting of continental platforms and ocean basins.

Fjords Glacial troughs that have been partly drowned by the sea.

Flood basalt A large-scale outpouring of basaltic lava that may cover an extensive area of the Earth's surface.

Floodplain A flattish valley floor covered with stream-deposited sediments and subject to periodic or episodic inundation by overflow from the stream.

Flood tide The movement of ocean water toward the coast—from the ocean's lowest surface level the water rises gradually for about 6 hours and 13 minutes.

Flora Plants.

Flow A gentle, smooth motion as when a sector of a slope becomes unstable, normally due to an addition of water, and slips gently downhill.

Fluvent A suborder of Entisol soils that form on recent water-deposited sediments that have satisfactory drainage.

Fluvial Running water—including both overland flow and streamflow.

Foehn *See* Chinook. The word *foehn* is used particularly in Europe.

Fog A cloud whose base is at or very near ground level.

Folding The bending of crustal rocks by compression and/or uplift.

Foliation Bending that gives a wavy layered appearance to rock.

Food chain Sequential predation in which organisms feed upon one another, with organisms at one level providing food for organisms at the next level, etc. Energy is thus transferred through the ecosystem.

Food pyramid *See also* Food chain. Another conceptualization of energy transfer through the ecosystem from large numbers of "lower" forms of life through succeedingly smaller numbers of "higher" forms, as the organisms at one level are eaten by the organisms at the next higher level.

Forbs Broadleaf herbaceous plants.

Forest An assemblage of trees growing closely together so that their individual leaf canopies generally overlap.

Fractional scale Ratio of distance between points on a map and the same points on the Earth's surface; expressed as a ratio or fraction.

Freezing Change from liquid to solid state.

Fringing reef A coral reef built out laterally from the shore, forming a broad bench that is only slightly below sea level, often with the tops of individual coral "heads" exposed to the open air at low tide.

Front A zone of discontinuity between unlike air masses.

Frost wedging Fragmentation of rock due to expansion of water that freezes in rock openings.

Fumarole A hydrothermal feature consisting of a surface crack that is directly connected with a deep-seated source of heat. The little water that drains into this tube is instantly converted to steam by heat and gases, and a cloud of steam is then expelled from the opening.

Geomorphic cycle (cycle of erosion) A conceptual model of landscape development that was propounded by William Morris Davis. Davis envisioned a circular sequence of terrain evolution in which a relatively flat surface was uplifted from a lower to a higher elevation, where it was incised by fluvial erosion into a landscape of slopes and valleys, and then thoroughly denuded until it became a flat surface at low elevation. He recognized three stages—"youth," "maturity," and "old age."

Geomorphology The study of the characteristics, origin, and development of landforms.

Geostrophic wind A wind that moves parallel to the isobars as a result of the balance between the pressure gradient force and the Coriolis effect.

Geyser A specialized form of intermittent hot spring with water issuing only sporadically as a temporary ejection, in which hot water and steam are spouted upward for some distance.

Geyserite Whitish-colored siliceous sinter that often covers a geyser basin.

Gibber plain A gravel-covered desert in Australia.

Glacial erratic Outside boulder included in the glacial till, which may be very different from the local bedrock.

Glacial flour Rock material that has been ground to the texture of very fine talcum powder by glacial action.

Glacial trough A valley reshaped by an alpine glacier.

Glacier A large natural accumulation of land ice that flows either downslope or outward from its center of accumulation.

Glaciofluvial deposition The action whereby much of the debris that is carried along by glaciers is eventually deposited or redeposited by glacial meltwater.

Gleization The dominant pedogenic regime in areas where the soil is saturated most of the time due to poor drainage.

Gley soils The general term for soils produced by gleization.

Gneiss One of the most common metamorphic rocks; it is characterized by broad foliations.

Gondwanaland Presumed Southern Hemisphere supercontinent resulting from the initial break-up of Pangaea.

Graben A block of land bounded by parallel faults in which the block has been downthrown, producing a distinctive structural valley with a straight, steep-sided fault scarp on either side.

Gradient Horizontal rate of variation of some quantity, such as atmospheric pressure or elevation of streambed.

Granite The most common and well-known intrusive rock.

Graphic scale The use of a line marked off in graduated distances as a map scale.

Grassland Plant association dominated by grasses and forbs.

Gravitational water Soil water that is temporary in occurrence in that it results from prolonged infiltration from above (usually due to prolonged precipitation) and is pulled downward through the interstices toward the groundwater zone below by gravitational attraction.

Gravity (surface gravity of the Earth) The force attracting bodies toward the center of the Earth.

Great Artesian Basin The most notable of the world's artesian basins, which underlies some 670,000 square miles (1,750,000 km^2) of east-central Australia.

Greenhouse effect The trapping of heat in the lower troposphere because of differential transmissivity for short and long waves—the atmosphere lets shortwave radiation in but doesn't let long-wave radiation out.

Ground ice Ice located below the surface of the land.

Groundwater Water found in the vadose zone.

Gully erosion Overland flow that erodes conspicuous channels in the soil.

Gymnosperms ("naked seeds") Seed-reproducing plants that carry their seeds in cones.

Hadley cells Two complete vertical circulation cells between the equator, where warm air rises, and 25° to 30° of latitude, where much of the air subsides.

Hail Rounded or irregular pellets or lumps of ice produced in cumulonimbus clouds as a result of active turbulence and vertical air currents. Small ice particles grow by collecting moisture from supercooled cloud droplets.

Halley, Edmund An English astronomer and cartographer who, in his published map of 1700, was the first person to use isolines that appeared in print. His map showed isogonic lines in the Atlantic Ocean.

Hamada A barren desert surface of consolidated material that usually consists of exposed bedrock but is sometimes composed of sedimentary material that has been cemented together by salts evaporated from groundwater.

Hanging valley A tributary glacial trough, the bottom of which is considerably higher than the bottom of the principal trough that it joins.

Hardwoods Angiosperm trees that are usually broad-leaved and deciduous. Their wood has a relatively complicated structure, but it is not always hard.

Heat A form of energy associated with the random motion of molecules. Things are made hotter by the collision of the moving molecules.

Herbaceous plants Plants that have soft stems—mostly grasses, forbs, and lichens.

Heterosphere That portion of the atmosphere above the homosphere where there is an irregular layering of gases in accordance with their molecular weights.

Hibernation The act of spending winter in a dormant state.

Histosol A soil order characterized by organic, rather than mineral, soils, which is invariably saturated with water all or most of the time.

Homosphere A zone of homogenous composition comprising the lowest 50 miles (80 km) of the atmosphere, where the principal gases have a uniform pattern of vertical distribution.

Hook A curving sandbar at the outer end of a spit, produced by conflicting water movements in a bay that guide deposition in a curved fashion toward the coast.

Horizons The more or less distinctly recognizable layers of soil, distinguished from one another by their differing characteristics and forming a vertical zonation of the soil.

Horn A steep-sided, pyramidal, rock pinnacle formed by expansive quarrying of the headwalls where three or more cirques intersect.

Horse latitudes Areas in the Subtropical Highs characterized by warm, tropical sunshine and an absence of wind.

Horst An uplifted block of land between two parallel faults.

Hot spots Places where magma is expelled through cracks in a crustal plate.

Hot spring Hot water at the Earth's surface that has been forced upward through fissures or cracks by the pressures that develop when underground water has come in contact with heated rocks or magma beneath the surface.

Humidity Water vapor in the air.

Humus A dark-colored gelatinous, chemically stable fraction of organic matter on or in the soil.

Hurricane A tropical cyclone affecting North or Central America.

Hydration The process whereby water molecules become attached to other substances and cause a wetting and swelling of the original substance without any change in its chemical composition.

Hydrologic cycle A series of storage areas interconnected by various transfer processes, in which there is a ceaseless interchange of moisture in terms of its geographical location and its physical state.

Hydrology Study of water on or below the surface of the Earth.

Hydrolysis A chemical union of water with another substance to produce a new compound that is nearly always softer and weaker than the original.

Hydrophytic (water-loving) The term used to describe those plants particularly suited to growing in a wet terrestrial environment.

Hydrosphere Total water realm of the Earth, including the oceans, surface waters of the lands, groundwater, and water held in the atmosphere.

Hydrothermal activity The outpouring or ejection of hot water, often accompanied by steam, which usually takes the form of either a hot spring or a geyser.

Hygrometer Any instrument for measuring humidity.

Hygrophytic See Hydrophytic.

Hygroscopic water A microscopically thin film of moisture that is bound rigidly to soil particles by adhesion.

Icebergs Great chunks of floating ice that break off an ice shelf or the end of an outlet glacier.

Ice cap A small ice sheet, normally found in the summit area of high mountains.

Ice floes Masses of ice that break off from larger ice bodies (ice sheets, glaciers, ice packs, and ice shelves) and float independently in the sea. This term is generally used with large, flattish, tabular masses.

Ice pack The extensive and cohesive mass of floating ice that is found in the Arctic and Antarctic oceans.

Ice sheet A vast blanket of ice that completely inundates the underlying terrain to depths of hundreds or thousands of feet.

Ice shelf A massive portion of an ice sheet that projects out over the sea.

Igneous rock Rock formed by solidification of molten magma.

Illuviation The process by which fine particles of soil from the upper layers are deposited at a lower level.

Inceptisol An immature order of soils that has relatively faint characteristics; not yet prominent enough to produce diagnostic horizons.

Infiltration Downward movement of water into the soil and regolith.

Infrared radiation Electromagnetic radiation in the wavelength range of about 0.7 to 1000 micrometers.

Inselberg ("island mountain") Isolated summit rising abruptly from a low-relief surface.

Insolation Incoming solar radiation.

Interfluve The higher land above the valley sides that separates adjacent valleys.

Intermittent stream A stream that carries water only part of the time, during the "wet season" or during and immediately after rains.

International date line The line marking a time difference of an entire day from one side of the line to the other. Generally, this line falls on the 180th meridian except where it deviates to avoid separating an island group.

Interstices The pore spaces; a labyrinth of interconnecting passageways among the soil particles that makes up nearly half the volume of an average soil.

Intertropical convergence zone The region where the northeast trades and the southeast trades converge.

Interval The numerical difference between one isoline and the next.

Intrusive igneous rock Rocks that cool and solidify beneath the Earth's surface.

Invertebrates Animals without backbones.

Ionosphere A deep atmospheric layer containing electrically charged molecules and atoms in the upper mesosphere and lower thermosphere.

Isobar A line joining points of equal atmospheric pressure.

Isogonic line A line joining points of equal magnetic declination.

Isohyet A line joining points of equal numeric value of precipitation.

Isoline A line on a map connecting points that have the same quality or intensity of a given phenomenon.

Isostasy Maintenance of the hydrostatic equilibrium of the Earth's crust.

Isotherm A line joining points of equal temperature.

Jet stream A rapidly moving current of air concentrated along a quasihorizontal axis in the upper troposphere or in the stratosphere, characterized by strong vertical and lateral wind shears and featuring one or more velocity maxima.

Joints Cracks that develop in bedrock due to stress, but in which there is no appreciable movement parallel to the walls of the joint.

Kame A relatively steep-sided mound or conical hill composed of stratified drift found in areas of ice sheet deposition and associated with meltwater deposition in close association with stagnant ice.

Karst Topography developed as a consequence of subsurface solution.

Katabatic winds Winds that originate in cold upland areas and cascade toward lower elevations under the influence of gravity.

Kettles Irregular depression in a morainal surface created when blocks of stagnant ice eventually melt.

Kilopascal Equivalent to 10 millibars.

Knickpoint A sharp irregularity (such as a waterfall, rapid, or cascade) in a stream channel profile.

Köppen system A climatic classification of the world devised by Wladimir Köppen.

Laccolith An igneous intrusion produced when slow-moving viscous magma is forced between horizontal layers of preexisting rock. The magma resists flowing and builds up into a mushroom-shaped mass that domes the overlying strata. If near enough to the Earth's surface, a rounded hill will rise above the surrounding area.

Lagomorphs Rabbits and hares.

Lagoon A body of quiet salt or brackish water in an area between a barrier island or a barrier reef and the mainland.

Lake A body of water surrounded by land.

Land breeze Local wind blowing from land to water, usually at night.

Landform An individual topographic feature.

Landforms Topography.

Landsat A series of unmanned satellites that orbit the Earth at an altitude of 570 miles (915 km) and are capable of imaging all parts of the Earth, except the polar regions, every nine days.

Landslide A large mass of rock and earth slides bodily downslope, abruptly and often catastrophically, in an event that lasts for only a few seconds or minutes. An instantaneous collapse of a slope.

Langley (ly) The unit of measure of radiation intensity that is one calorie per square centimeter.

Lapse rate The rate of temperature decrease with height in the troposphere. The average lapse rate has been calculated at about 3.6°F per 1000 feet (6.5°C/km).

Large-scale map A map with a scale that is a relatively large representative fraction and therefore portrays only a small portion of the Earth's surface, but in considerable detail.

Latent heat Energy stored or released when a substance changes state.

Latent heat of vaporization Stored energy absorbed by escaping molecules during evaporation.

Lateral erosion Erosion that occurs when the principal current of a stream swings laterally from one bank to the other, eroding where the velocity is greatest and depositing where it is least.

Lateral moraine Well-defined ridge of unsorted debris built up along the sides of valley glaciers, parallel to the valley walls.

Laterization The dominant pedogenic regime in areas where temperatures are relatively high throughout the year and which is characterized by rapid weathering of parent material, dissolution of nearly all minerals, and the speedy decomposition of organic matter.

Latitude Distance measured north and south of the equator.

Latosols The general term applied to soils produced by laterization.

Laurasia Presumed Northern Hemisphere supercontinent resulting from the initial break-up of Pangaea.

Lava Molten magma that is extruded onto the surface of the Earth, where it cools and solidifies.

Lava vesicles Holes of various sizes, usually small, that develop in cooling lava when gas is unable to escape during solidification.

Leaching The process in which gravitational water dissolves soluble materials and carries them downward in solution to be redeposited at lower levels.

Lifting condensation level The altitude at which rising air cools sufficiently to reach 100 percent relative humidity at the dew point temperature, and condensation begins.

Lightning A luminous electric discharge in the atmosphere caused by the separation of positive and negative charges associated with cumulonimbus clouds.

Limestone A common, chemically accumulated sedimentary rock composed largely of calcium carbonate.

Linnaen system A biological classification focussed on the morphology of the organisms and grouping them on the basis of structural similarity.

Lithosphere The solid, inorganic portion of the earthly fundament comprised of the rocks of the Earth's crust as well as of the broken and unconsolidated particles of mineral matter that overlie the unfragmented bedrock.

Litter The collection of dead plant parts that accumulate at the surface of the soil.

Loam A soil texture in which none of the three principal soil separates—sand, silt, and clay—dominates the other two.

Loess A fine-grained, wind-deposited silt. Loess lacks horizontal stratification, and its most distinctive characteristic is its ability to stand in vertical cliffs.

Longitude Distance measured in degrees, minutes, and seconds, east and west from the prime meridian on the Earth's surface.

Longshore currents (littoral currents) Currents in which water moves roughly parallel to the shoreline in a generally downwind direction.

Longshore drift The zigzag movement of particles in which the net result is a displacement parallel to the coast in a downwind direction.

Loxodrome (rhumb line) A true compass heading; a line of constant compass direction.

Magma Molten material in the Earth's interior.

Magnetic declination The angular difference between a magnetic north line and a true north line, expressed as degrees east or west of the meridian in question.

Magnetic north The direction that a magnetic compass needle points.

Mammals The highest form of animal life, distinguished from all other animals by several internal characteristics and two prominent external features—the production of milk to feed their young and the possession of true hair.

Mantle That portion of the Earth beneath the crust and surrounding the core.

Map projection A systematic representation of all or part of the three-dimensional Earth surface on a two-dimensional flat surface.

Marble Metamorphosed limestone.

Marine terrace A platform of marine erosion that has been uplifted above sea level.

Marsh Flattish surface area that is submerged in water at least part of the time, but is shallow enough to permit the growth of water-tolerant plants—primarily grasses and sedges.

Marsupials A small group of mammals whose females have pouches in which the young, which are born in a very undeveloped condition, live for several weeks or months after birth.

Mass wasting The downslope movement of broken rock material by gravity, sometimes lubricated by the presence of water.

Maturity (mature stage) The second stage in Davis's Geomorphic Cycle, which is dominated by slope land and marked by the absence of the vast area of initial surface.

Meandering stream A stream with a sinuous course of elaborate, smooth curves.

Meander scar A former stream meander through which the stream no longer flows.

Medial moraine A dark band of rocky debris down the middle of a glacier created by the union of the lateral moraines of two adjacent glaciers.

Mercator projection A cylindrical projection mathematically adjusted to attain complete conformality, and which has a rapidly increasing scale with increasing latitude.

Meridians Imaginary lines of longitude extending from pole to pole, crossing all parallels at right angles, and being aligned in true north-south directions.

Mesa A flat-topped, steep-sided hill with a limited summit area.

Mesopause Transition zone at the top of the mesosphere.

Mesosphere Atmospheric layer above the stratopause where temperature decreases with height.

Mesozoic era The second youngest geologic age; the name refers to "middle life."

Metamorphic rocks Rocks that were originally something else but have been drastically changed by massive forces of heat and/or pressure working on them from within the Earth.

Midocean ridge A lengthy system of deep-sea mountain ranges, generally located at some distance from any continent.

Millibar An "absolute" measure of pressure, consisting of one-thousandth part of a bar, or 1000 dynes per square centimeter.

Mineral A naturally formed inorganic substance that has an unvarying chemical composition.

Mistral A cold, high-velocity wind that sometimes surges down France's Rhône Valley, from the Alps to the Mediterranean Sea.

Mollisol A soil order characterized by the presence of a mollic epipedon, which is a mineral surface horizon that is dark, thick, contains abundant humus and base nutrients, and retains a soft character when it dries out.

Monadnock Erosional remnant of resistant rock rising slightly above a landscape of limited relief.

Monera The kingdom of organisms that comprises the simplest known organisms—one-celled bacteria and blue-green algae.

Monocline A one-limbed fold connecting horizontal or gently inclined strata.

Monotremes Egg-laying mammals.

Monsoon A seasonal reversal of winds; a general onshore movement in summer and a general offshore flow in winter, with a very distinctive seasonal precipitation regime.

Moraines The largest and generally most conspicuous landform features produced by glacial deposition, which consist of irregular rolling topography that rises somewhat above the level of the surrounding terrain.

Motus Coral islets that are closely spaced and separated by narrow channels of water and that together form a ring-shaped atoll.

Mudflow Downslope movement of a thick mixture of soil and water.

Natural levee An embankment of slightly higher ground fringing a stream channel in a floodplain; formed by deposition during floodtime.

Natural Selection The Darwinian theory of "the survival of the fittest," which explains the origin of any species as a normal process of descent, with variation, from parent forms.

Neap tides The lower than normal tidal variations that occur twice a year as the result of the alignment of the sun and moon at a right angle to one another.

Nebula A cloud of interstellar gas and dust, perhaps one light-year in diameter.

Needle-leaf trees Trees adorned with thin slivers of tough, leathery, waxy needles rather than typical leaves.

Nekton The term applied to animals that swim freely in the oceans.

Névé *See* Firn.

Newton The force that must be exerted on a mass of 1 kg in order to accelerate it at a rate of one meter per second per second.

Nimbostratus A low, dark cloud, often occurring as widespread overcast and normally producing precipitation.

Nitrogen cycle An endless series of processes in which nitrogen moves through the environment.

Nitrogen fixation Conversion of gaseous nitrogen into forms that can be used by plant life.

Normal fault The result of tension producing a steeply inclined fault plain, with the block of land on one side being pushed up, or upthrown, in relation to the block on the other side, which is downthrown.

Nunatak A rocky pinnacle protruding above an ice field.

Obsequent streams Streams that commonly flow in the opposite direction of consequents and are often short tributaries of subsequents.

Occluded front A complex front formed when a cold front overtakes a warm front.

Oceanic trench Deep linear depression in the ocean floor where subduction is taking place.

Offshore bar (barrier bar) Long, narrow sandbar built up in shallow offshore waters.

Old age Third stage in Davis's Geomorphic Cycle in which the entire landscape has been eroded down to a plain of low relief.

Orographic precipitation Precipitation that occurs when air, forced to ascend over topographic barriers, cools to the dew point.

Orthent A suborder of Entisol soils that develops on recent erosional surfaces.

Orthid A suborder of Aridisol soils that does not have a distinctive subsurface horizon.

Outcrop Surface exposure of bedrock.

Outlet glacier A tongue of ice around the margin of an ice sheet that extends between rimming hills to the sea.

Outwash plain Extensive glaciofluvial feature that is a relatively smooth, flattish alluvial apron deposited beyond recessional or terminal moraines by streams issuing from ice.

Overland flow The general movement of surface water down the slope of the land surface.

Overthrust fault A fault created by compression forcing the upthrown block to override the downthrown block at a relatively low angle.

Overthrust fold A fold in which the pressure was great enough to break the oversteepened limb and cause a shearing movement.

Overturned fold An upfold that has been pushed so vigorously from one side that it becomes oversteepened enough to have a reverse orientation on the other side.

Oxbow lake A cutoff meander that initially holds water.

Oxbow swamp An oxbow lake that has been at least partly filled with sediment and vegetation.

Oxidation The chemical union of oxygen atoms with atoms from various metallic elements to form new products, which are usually more voluminous, softer, and more easily eroded than the original compounds.

Oxisol The most thoroughly weathered and leached of all soils. This soil order invariably displays a high degree of mineral alteration and profile development.

Oxygen cycle The movement of oxygen by various processes through the environment.

Ozone A gas composed of molecules consisting of three atoms of oxygen, O_3.

Ozone layer A layer of ozone between 10 and 25 miles (16 and 40 km) high, which absorbs ultraviolet solar radiation.

Ozonosphere Zone of relatively rich concentration of ozone in the atmosphere.

Paleomagnetism Past magnetic orientation.

Pangaea The massive supercontinent that Alfred Wegener postulated to have existed about 200 million years ago. He visualized Pangaea as breaking up into several large sections that have continually moved away from one another and that now comprise the present continents.

Parallel A small circle resulting from an isoline connecting all points of equal latitude.

Parallel drainage pattern A drainage pattern that can emerge in areas of pronounced regional slope, particularly if the gradient is gentle, with long consequent streams flowing parallel to one another.

Parasites Organisms of one species that infest the body of a creature of another species, obtaining their nutriment from the host, which is almost invariably weakened and sometimes killed by the actions of the parasite.

Parent material The source of the weathered fragments of rock from which soil is made—solid bedrock or loose sediments that have been transported from elsewhere by the action of water, wind, or ice.

Pascal (Pa) A pressure of one newton per square meter.

Paternoster lakes A sequence of small lakes found in the shallow excavated depressions of a glacial trough.

Pediment A gently inclined bedrock platform that extends outward from a mountain front, usually in an arid region.

Pedogenic regimes Soil-forming regimes that can be thought of as environmental settings in which certain physical/chemical/biological processes prevail.

Peds The larger masses or clumps that individual soil particles tend to aggregate into and that determine the structure of the soil.

Peneplain A flat and relatively featureless landscape with minimal relief; considered to be the end product of the geomorphic cycle.

Perennials Plants that can live more than a single year despite seasonal climatic variations.

Perennial stream A permanent stream that contains water the year-round.

Perigee The point at which the moon is closest to the Earth in its elliptical orbit—231,200 miles (370,000 km).

Periglacial zone An area of indefinite size beyond the outermost extent of ice advance that was indirectly influenced by glaciation.

Perihelion The point in its orbit at which a planet is nearest the sun.

Permafrost Permanent ground ice or permanently frozen subsoil.

Permeability A soil or rock characteristic in which there are interconnected pore spaces through which water can move.

Photogrammetry The science of obtaining reliable measurements from photographs and, by extension, mapping from aerial photos.

Photoperiodism The response of an organism to the length of exposure to light in a 24-hour period.

Photosynthesis The basic process whereby plants produce stored chemical energy from water and carbon dioxide and which is activated by sunlight.

Phreatic zone (zone of saturation) The second hydrologic zone below the surface of the ground, whose uppermost boundary is the water table. The pore spaces and cracks in the bedrock and regolith of this zone are fully saturated.

Physical weathering The mechanical disintegration of rock material without any change in its chemical composition.

Piedmont Zone at the "foot of the mountains."

Piedmont angle The pronounced change in the angle of slope at a mountain base, with a steep slope giving way abruptly to a gentle one.

Piedmont glacier A valley glacier that extends to the mouth of the valley and spreads out broadly over the flat land beyond.

Placental mammals Mammals whose young grow and develop in the mother's body, nourished by an organ known as the *placenta*, which forms a vital connecting link with the mother's bloodstream.

Plane of the ecliptic The imaginary plane that passes through the sun and through the Earth at every position in its orbit around the sun.

Plankton Plants and animals that float about, drifting with the currents and tides—mostly microscopic in size.

Plantae The kingdom of organisms that includes the green plants and higher algae.

Plant succession The process whereby one type of vegetation is replaced naturally by another.

Plateau Flattish erosional platform bounded on at least one side by a prominent escarpment.

Plate tectonics A coherent theory of massive crustal rearrangement based on the movement of continent-sized crustal plates.

Playa Dry lake bed in a basin of interior drainage.

Playa lake Shallow and short-lived lake formed when water flows into a playa.

Plucking (quarrying) Action in which rock particles beneath the ice are grasped by the freezing of meltwater in joints and fractures and pried out and dragged along in the general flow of a glacier.

Plutonic activity Activity of deep-seated magma that solidifies far below the surface.

Podzolization The dominant pedogenic regime in areas where winters are long and cold, and which is characterized by slow chemical weathering of soils and rapid mechanical weathering from frost action, resulting in soils that are shallow, acidic, and with a fairly distinctive profile.

Podzols General term for soils formed by podzolization.

Polar easterlies A global wind system that occupies most of the area between the Polar Highs and about 60° of latitude. The winds move generally from east to west and are typically cold and dry.

Polar front The contact between unlike air masses in the subpolar low-pressure zone.

Polar Highs High-pressure cells situated over both polar regions.

Pond A lake of very small size.

Porosity The amount of pore space between the soil particles and between the peds, which is a measure of the capacity of the soil to hold water and air.

Potential evapotranspiration The maximum amount of moisture that could be lost from soil and vegetation if the water were available.

Prairie A tall grassland in the midlatitudes.

Primates Humans and such humanlike mammals as apes and monkeys.

Prime meridian The meridian passing through the Royal Observatory at Greenwich (England), just east of London, and from which longitude is measured.

Proglacial lake A lake formed when ice flows across or against the general slope of the land and the natural drainage is impeded or completely blocked so that meltwater from the ice becomes impounded against the ice front.

Protista The kingdom of organisms that consists of one-celled organisms outside the Monera Kingdom, and some simple multicelled algae.

Psamment A suborder of Entisol soils that occurs in sandy situations.

Pseudoadiabat *See* Wet adiabatic rate.

Pteridophytes Spore-bearing plants such as ferns, horsetails, and clubmosses.

Pyroclastic material Rock fragments thrown into the air by volcanic explosions.

Quartz A mineral composed of silicon dioxide.

Quartzite A metamorphosed rock derived from sandstone or another rock composed largely of quartz.

Quick clays Clay formations that spontaneously change from a relatively solid mass into a near-liquid condition as the result of a sudden disturbance or shock.

Radial drainage pattern Drainage pattern in which consequent streams flow outward in all directions from a central dome or peak.

Radiation The process by which energy is emitted from a body.

Radiational inversion Surface inversion that results from rapid radiational cooling of lower air, typically on cold winter nights.

Radiation fog A fog produced by condensation near the ground, where air is cooled to the dew point by contact with the colder ground.

Rain The most common and widespread form of precipitation, consisting of drops of liquid water.

Rain gauge An instrument used to measure the amount of rain that has fallen.

Rain shadow Area of low rainfall on the leeward side of a topographic barrier.

Recessional moraine A glacial deposit formed during a pause in the retreat of the ice margin.

Rectangular drainage pattern Pattern where streams are essentially all subsequents that follow sets of faults and/or joints, with prominent right-angled relationships.

Reflection The ability of an object to repel waves without altering either the object or the waves.

Reg A desert surface of coarse material from which all sand and dust have been removed by wind and water erosion. Often referred to as *desert pavement* or *desert armor*.

Regolith A layer of broken and partly decomposed rock particles that covers bedrock.

Rejuvenation Concept in which regional uplift could raise the land and interrupt the geomorphic cycle at any stage by re-energizing the degradational processes.

Relative humidity An expression of the amount of water vapor in the air in comparison with the total amount that could be there if the air were saturated. This is a ratio that is expressed as a percentage.

Relief Refers to the difference in elevation between the highest and lowest points in an area; the vertical variation from mountain top to valley bottom.

Remote sensing Study of an object or surface from a distance by using various instruments.

Representative fraction (r.f.) The ratio that is an expression of a fractional scale that compares map distance with ground distance.

Reptiles Cold-blooded vertebrates of which most are land-based.

Resequent streams Streams that flow in the same direction as consequents but develop on slopes that were formed later. They are found typically on the upslope sides of subsequent valleys.

Reverse fault A fault produced from compression, with the upthrown block rising steeply above the downthrown block, so that the fault scarp would be severely oversteepened if erosion did not act to smooth the slope.

Ria shoreline An embayed coast with numerous estuaries.

Richter scale A scale of earthquake magnitudes, devised by California seismologist Charles F. Richter in 1935, to describe the amount of energy released in a single earthquake.

Rift valleys Downfaulted graben structures extended for extraordinary distances as linear structural valleys enclosed between typically steep fault scarps.

Rill erosion A more concentrated flow than that of sheet erosion, which loosens additional material and scores the slope with numerous parallel seams.

Rills Tiny drainage channels.

Riparian vegetation Anomalous stream-side growth, particularly prominent in relatively dry regions, where stream courses may be lined with trees, although no other trees are to be found in the landscape.

Roche moutonnée A characteristic landform produced when a bedrock hill or knob is overridden by moving ice. The stoss side is smoothly rounded and streamlined by grinding abrasion as the ice rides up the slope, but the lee side is shaped largely by plucking, which produces a steeper and more irregular slope.

Rock Solid material composed of aggregated mineral particles.

Rock glaciers An accumulated talus mass that moves slowly but distinctly downslope under its own weight.

Rodents Gnawing mammals.

Rossby waves Very large north-south undulations of the upper-air westerlies.

Runoff Flow of water from land to oceans by overland flow, streamflow, and groundwater flow.

Rusting Production of reddish/yellowish/brownish iron oxide minerals by oxidation.

Sag pond A pond caused by the collection of water from springs and/or runoff into sunken ground, resulting from the jostling of the Earth in the area of fault movement.

Salina Dry lake bed that contains an unusually heavy concentration of salt in the lake-bed sediment.

Salinity A measure of the concentration of dissolved salts.

Salinization One of the dominant pedogenic regimes in areas where principal soil moisture movement is upward because of a moisture deficit.

Saltation Process in which small particles are moved along by streamflow or wind in a series of jumps or bounces.

Salt wedging Rock disintegration caused by the crystallization of salts from evaporating water.

Sand dune A mound, ridge, or low hill of loose windblown sand.

Sandplain An amorphous sheet of coarse sand spread across the landscape with no particular surface shape or significant relief.

Sandstone A common mechanically accumulated sedimentary rock composed largely of sand grains.

Sand storm A cloud of generally horizontally moving sand that extends for only a few inches or feet above the surface.

Savanna A low-latitude grassland characterized by tall forms.

Scattering A change in direction, but not in wavelength, of light waves.

Schist One of the most common metamorphic rocks in which the foliations are very narrow.

Scree *See* Talus.

Sea breezes Wind that blows from the sea toward the land, usually during the day.

Sea-floor spreading The pulling apart of crustal plates to permit the rise of deep-seated magma to the Earth's surface in midocean areas.

Second-order relief Major mountain systems and other extensive surface formations of subcontinental extent.

Sediment Small particles of rock debris or organic material deposited by water, wind, or ice.

Sedimentary rock Rock formed of sediment that is consolidated by the combination of pressure and cementation.

Seif (longitudinal dune) Long, narrow desert dunes that usually occur in multiplicity and in parallel arrangement.

Selva (tropical rainforest) A distinctive assemblage of tropical vegetation that is dominated by a great variety of tall, high-crowned trees.

Sensible temperature A concept of the relative temperature that is sensed by a person's body.

Separates The size groups within the standard classification of soil particle sizes.

Seventh Approximation The U.S. Comprehensive Soil Classification System, which is generic in nature and focusses on the existing properties of the soil rather than on environment, genesis, or the properties it would possess under virgin conditions.

Shale A common mechanically accumulated sedimentary rock composed of silt and clay.

Sheet erosion Water flows across the surface as a thin sheet, transporting material already loosened by splash erosion.

Shield volcanoes Volcanoes built up in a lengthy outpouring of very fluid basaltic lava.

Shrub Woody, low-growing perennial plant.

Shrubland Plant association dominated by relatively short woody plants.

Sial The upper layer of two igneous rock layers, which is discontinuous, apparently underlying only the continental masses, where it sits as immense bodies of rock embedded in the sima beneath. "Sial" is named for its common constituents of *si*lica and *al*uminum.

Sidereal day A complete rotation of the Earth with respect to the stars (23 hours, 56 minutes, and 4.099 seconds).

Silica Silicon dioxide in any of several mineral forms.

Sill A long, thin intrusive body that is formed when magma is forced between layers of preexisting rock to solidify eventually in a sheet.

Sima The lower of two igneous rock layers that is continuous and underlies both the ocean basins and the continents. "Sima" is named for its two most prominent mineral compounds, *si*lica and *ma*gnesium.

Sinkhole (doline) A small, rounded depression that is formed by the dissolution of surface limestone, typically at joint intersections.

Slate Metamorphosed shale.

Sleet Small raindrops that freeze during descent.

Slip face Steeper leeward side of a sand dune.

Slump A slope collapse with a backward rotation.

Small-scale map A map whose scale is a relatively small representative fraction and therefore shows a large portion of the Earth's surface in limited detail.

Snow Solid precipitation in the form of ice crystals, small pellets, or flakes, which is formed by the direct conversion of water vapor into ice.

Snowline The elevation above which some winter snow is able to persist throughout the year.

Softwoods Gymnosperm trees—nearly all are needle-leafed evergreens—with wood of simple cellular structure but not always soft.

Soil An infinitely varying mixture of weathered mineral particles, decaying organic matter, living organisms, gases, and liquid solutions. Soil is that part of the outer skin of the Earth occupied by plant roots.

Soil profile A vertical cross section from the Earth's surface down through the soil layers into the parent material beneath.

Soil water Water found in the phreatic zone.

Soil-water balance The relationship between gain, loss, and storage of soil water.

Soil-water budget An accounting that demonstrates the variation of the soil-water balance over a period of time.

Solar constant The fairly constant amount of solar insolation received at the top of the atmosphere—slightly less than 2 calories per square centimeter per minute or 2 langleys per minute.

Solar day A complete rotation of the Earth with respect to the sun (24 hours).

Solifluction A special form of creep in tundra areas that produces a distinctive surface appearance. During the summer the near-surface portion of the ground thaws, but the meltwater cannot percolate deeper because of the permafrost below. The spaces between the soil particles become saturated, and the heavy surface material sags slowly downslope.

Solstices Those two times of the year in which the sun's perpendicular rays hit the northernmost or southernmost latitudes during the Earth's cycle of revolution.

Solum The true soil that includes only the top three horizons: O, the organic surface layer; A, the topsoil; and B, the subsoil.

Source regions Parts of the Earth's surface that are particularly suited to generate air masses.

Specific heat The amount of energy required to raise the temperature of a unit mass of a substance by 1° C.

Specific humidity A direct measure of water vapor content expressed as the mass of water vapor in a given mass of air (grams of vapor/kg of air).

Spit A linear deposit of marine sediment that is attached to the land at one or both ends.

Splash erosion The direct collision of a raindrop with the ground, which blasts fine particles upward and outward, shifting them a few millimeters laterally.

Spodosols A soil order characterized by the occurrence of a spodic subsurface horizon, which is an illuvial layer where organic matter and aluminum accumulate, and which has a dark, sometimes reddish, color.

Spring tide A time of maximum tide that occurs as a result of the alignment of sun, moon, and Earth in a straight line.

Squall line A line of intense thunderstorms.

Stalactite A pendant structure hanging downward from a cavern's roof.

Stalagmite A pendant structure growing upward from a cavern's floor.

Stationary front The common "boundary" between two air masses in a situation in which neither air mass displaces the other.

Steppe A plant association dominated by short grasses and bunchgrasses of the midlatitudes.

Stock A small body of igneous rock intruded into older rock; amorphous in shape and indefinite in depth.

Storm surge A surge of wind-driven water as much as 25 feet ($7\frac{1}{2}$ m) above normal tide level, which occurs when a hurricane pounds into a shoreline.

Strata Distinct layers of sediment.

Stratified drift Drift that was sorted as it was carried along by the flowing glacial meltwater.

Stratiform clouds A cloud form characterized by clouds that appear as grayish sheets or layers that cover most or all of the sky, rarely being broken into individual cloud units.

Stratocumulus Low clouds, usually below 6500 feet (2 km), which sometimes occur as individual clouds, but more often appear as a general overcast.

Stratopause The top of the stratosphere [elevation about 30 miles (48 km)] where maximum temperature is reached.

Stratosphere Atmospheric layer directly above the troposphere.

Stratus Low clouds, usually below 6500 feet (2 km), which sometimes occur as individual clouds, but more often appear as a general overcast.

Stream capture (stream piracy) An event where a portion of the flow of one stream is diverted into that of another by natural processes.

Streamflow Channelled movement of water along a valley bottom.

Stream load Solid matter carried by a stream.

Stream order Concept that describes the hierarchy of a drainage net.

Stream terrace Remnant of a previous valley floodplain of a rejuvenated stream.

Striations Marks in a bedrock surface produced by the direct impact of glacial ice and its load of abrasive debris, which parallel the direction of ice movement.

Stripped plain A flattish erosional platform where the scarp edge is absent or relatively inconspicuous.

Structure Nature, arrangement, and orientation of the materials.

Subartesian well The free flow that results when a well is drilled from the surface down into a confined aquifer and which requires artificial pumping to raise the water to the surface because the confining pressure only forces the water partway up the well shaft.

Subduction Descent of the edge of a crustal plate under the edge of an adjoining plate, presumably involving melting of the subducted material.

Sublimation The process by which water vapor is converted directly to ice, or vice versa.

Subpolar low A zone of low pressure that is situated at about 50° to 60° of latitude in both Northern and Southern Hemispheres (also referred to as the *polar front*).

Subsequent streams Streams that develop along zones of structural weakness and often trend at right angles to other drainage channels.

Subsidence inversions Temperature inversions that occur well above the Earth's surface as a result of air sinking from above.

Subtropical high pressure Large semipermanent high-pressure cells centered at about 30° latitude, which have average diameters of 2000 miles (3200 km) and are usually elongated east-west.

Succulents Plants that have fleshy stems that store water.

Summer solstice The dates that represent the most poleward extent of the perpendicular rays of the sun. (Northern Hemisphere—23 $\frac{1}{2}$° N on about June 22nd; Southern Hemisphere—$23\frac{1}{2}$° S on about December 22nd).

Supercooled water Water that persists in liquid form at temperatures below freezing.

Superposed stream A stream that was originally established upon a higher sequence of land that has been entirely or largely eroded away, so that the original drainage pattern becomes incised into an underlying sequence of rocks of quite different structure.

Suspended load The very fine particles of clay and silt that are in suspension and move along with the flow of water without ever touching the stream bed.

Swallow holes The distinct openings at the bottom of some sinks through which surface drainage can pour directly into an underground channel.

Swamp A flattish surface area that is submerged in water at least part of the time, but is shallow enough to permit the growth of water-tolerant plants—predominantly trees.

Swash The cascading forward motion of a breaking wave that rushes up the beach.

Swells Water waves, usually produced by stormy conditions, that can travel enormous distances away from the source of the disturbance.

Symbiosis A mutually beneficial relationship between two organisms.

Syncline A simple downfold.

Taiga (boreal forest) The great northern coniferous forest.

Talus (scree) Pieces of rock, of whatever size, that fall directly downslope.

Talus cone Sloping, cone-shaped heaps of dislodged talus.

Talus slope (talus apron) The fragments of rocks (talus) that accumulate relatively uniformly along the base of the slope.

Tarn Small lake in the shallow excavated depression of benches of a glacial trough or cirque.

Taxonomy The science of classification.

Tectonic activity Crustal movements of various kinds.

Temperature inversion A situation in which temperature increases upward, and the normal condition is inverted.

Terminal moraine A glacial deposit that builds up at the outermost extent of ice advance.

Terracettes A complicated terracing effect, resembling a network of faint trails, which is produced by a creep, usually on steep grassy slopes.

Thermometer An instrument designed for the measurement of temperature.

Thermosphere The highest recognized thermal layer in the atmosphere, above the mesopause, where temperature remains relatively uniform for several miles and then increases continually with height.

Third-order relief Specific landform complexes of lesser extent and generally of smaller size than those of the second order.

Thunder The abrupt heating occasioned by a lightning bolt produces instantaneous expansion of the air, which creates a shock wave that becomes a sound wave.

Thunderstorm A relatively violent convective storm accompanied by thunder and lightning.

Tidal bore A wall of seawater several inches to several feet in height that rushes up a river as the result of enormous tidal inflow.

Tidal range The vertical difference in elevation between high and low tide.

Tides The rise and fall of the coastal water levels caused by the increasing and then decreasing gravitational pull of the moon and the sun on varying parts of the Earth's surface.

Till Rock debris that is deposited directly by moving or melting ice, with no meltwater flow or redeposition involved.

Till plain An irregularly undulating surface of broad, low rises and shallow depressions produced by the uneven deposition of glacial till.

Tombolo A spit formed by sand deposition of waves converging in two directions on the landward side of a nearshore island, so that a spit connects the island to the land.

Topography Surface configuration of the Earth.

Tornado A localized cyclonic low-pressure cell surrounded by a whirling cylinder of wind spinning so violently that centrifugal force creates partial vacuum within the funnel.

Torrert A suborder of Vertisol soils found in arid regions where cracks remain open most of the time.

Tracheophyta A division in the plant kingdom that consists of vascular plants that have efficient internal systems for transporting water and sugars and a complex differentiation of organs into leaves, stem, and roots.

Traction Process in which coarse particles are rolled or slid along the stream bed.

Trade winds The major wind system of the tropics, issuing from the equatorward sides of the Subtropical Highs and diverging toward the west and toward the equator.

Transcurrent boundary Two plates slipping past one another laterally in a typical fault structure.

Transcurrent fault A fault produced by shearing, with adjacent blocks being displaced laterally with respect to one another. The movement is entirely horizontal.

Transferscope An instrument that allows the operator to bring a map and photo, or photo and other image, to a common scale for comparison.

Transmission The ability of a medium to allow rays to pass through it.

Transpiration The transfer of moisture from plant leaves to the atmosphere.

Transverse dunes Crescent-shaped dunes that have convex sides facing the prevailing direction of wind and occur where the supply of sand is great. Their crests are perpendicular to the wind vector, and they are aligned in parallel waves across the land.

Travertine Massive accumulation of calcium carbonate.

Trellis drainage pattern A drainage pattern that is usually developed on alternating bands of hard and soft strata, with long parallel subsequent streams linked by short right-angled segments and joined by short obsequent and resequent tributaries.

Tropical cyclones The storms most significantly affecting the tropics and subtropics, which are intense, revolving, rain-drenched, migratory, destructive, and erratic. These storm systems consist of prominent low-pressure centers that are essentially circular in shape and have a steep pressure gradient outward from the center.

Tropical rainforest *See* Selva.

Tropical year The amount of time it takes for the Earth to revolve completely around the sun (365 $\frac{1}{4}$ days).

Tropic of Cancer The parallel of 23$\frac{1}{2}$° north latitude, which marks the northernmost location reached by the vertical rays of the sun in the annual cycle of the Earth's revolution.

Tropic of Capricorn The parallel of 23$\frac{1}{2}$° south latitude, which marks the southernmost location reached by the vertical rays of the sun in the annual cycle of the Earth's revolution.

Tropopause A transition zone at the top of the troposphere, where temperature ceases to decrease with height.

Troposphere The lowest thermal layer of the atmosphere, in which temperature decreases with height.

True (geographic) north The actual direction toward the North Pole from any point, measured along the meridian that passes through that point.

Tsunami Very long sea wave generated by submarine earthquake or volcanic eruption.

Tufa (sinter) Porous accumulations of calcium carbonate.

Tundra A complex mix of very low-growing plants, including grasses, forbs, dwarf shrubs, mosses, and lichens, but no trees. Tundra occurs only in the perennially cold climates of high latitudes or high altitudes.

Turbulent flow The general downstream movement is interrupted by continuous irregularities in direction and speed, producing momentary currents that can move in any direction, including upward.

Typhoons The term used for tropical cyclones affecting the western North Pacific region.

Ubac slope A slope oriented so that sunlight strikes it at a low angle and is much less effective in heating and evaporating than on the adret slope, thus producing more luxuriant vegetation of a richer diversity.

Udalf A suborder of Alfisol soils, characterized by brownish or reddish soils of moist midlatitude regions.

Udert A suborder of Vertisol soils, found in humid areas where cracking is irregular.

Ultisols A soil order similar to Alfisols, except that they are more thoroughly weathered and more completely leached of bases.

Ultraviolet waves Waves in the electromagnetic spectrum between 0.4 and 0.1 micrometers in length.

Ungulates Hoofed mammals.

Upslope fog Condensation that occurs when humid air is caused to ascend a topographic slope and consequently cools adiabatically.

Ustalf A suborder of Alfisol soils, characterized by brownish or reddish soils of subtropical regions and a hard surface layer in the dry season.

Ustert A suborder of Vertisol soils, associated with monsoonal climates and with a complicated cracking pattern.

Vadose zone (zone of aeration) The topmost hydrologic zone within the ground, which contains a fluctuating amount of moisture (soil water) in the pore spaces of the soil (or soil and rock).

Valley That portion of the total terrain in which a drainage system is clearly established.

Valley glaciers Long, narrow features resembling rivers of ice, which spill out of their originating basins and flow downvalley.

Valley train A lengthy deposit of glaciofluvial alluvium confined to a valley bottom beyond the outwash plain.

Vein Small igneous intrusions, usually with vertical orientation.

Vernal equinox The equinox that occurs about March 21st in the Northern Hemisphere and September 23rd in the Southern Hemisphere.

Vertebrates Animals that have a backbone that protects their spinal cord—fishes, amphibians, reptiles, birds, and mammals.

Vertisols A soil order comprising a specialized type of soil that contains a large quantity of clay and has an exceptional capacity for absorbing water. An alternation of wetting and drying, expansion and contraction, produces a churning effect that mixes the soil constituents, inhibits the development of horizons, and may even cause minor irregularities in the surface of the land.

Visible light Waves in the electromagnetic spectrum in the narrow band between about 0.4 and 0.7 micrometers in length.

Volcanic ash Fine particles of extrusive igneous rock blown out of a volcanic vent.

Volcanic neck Small, sharp spire that rises abruptly above the surrounding land. It represents the pipe or throat of an old volcano, filled with solidified lava after its final eruption. The less resistant material that makes up the cone is eroded, leaving the harder, lava-choked neck as a residual remnant.

Volcano A conical mountain or hill from which extrusive material is ejected.

Vulcanism General term that refers to movement of magma from the interior of the Earth to or near the surface.

Wadi The normally dry beds of an intermittent stream. This term is used in the Sahara Desert region.

Warm front The leading edge of an advancing warm air mass.

Waterfall Abrupt descent of a stream over a prominent knickpoint.

Water gap A narrow notch in which a stream flows through a ridge.

Waterless zone The fifth and lowermost hydrologic zone that generally begins several miles or kilometers beneath the land surface and is characterized by the lack of water in pore spaces due to the great pressure and density of the rock.

Water table The top of the saturated zone within the ground.

Wave amplitude One-half the wave height; i.e. it is the vertical distance from still water level, either upward to the crest or downward to the trough.

Wave-built terrace Submarine deposit of sand at the outer margin of an erosional platform or bench.

Wave crest Highest point of a wave.

Wave-cut notch An indentation in a sea cliff cut at water level by the combined effects of hydraulic pounding, abrasion, pneumatic push, and solution action.

Wave height The vertical distance from crest to trough.

Wave length The horizontal distance from crest to crest or from trough to trough.

Wave refraction Phenomenon whereby waves change their directional trend as they approach a shoreline.

Wave trough Lowest part of a wave.

Weather A temporary concept that refers to short-term atmospheric conditions for a given time and a specific area.

Weathering The physical and chemical disintegration of rock that is exposed to the weather.

Westerlies The great wind system of the midlatitudes that flows basically from west to east around the world in the latitudinal zone between about 30° and 60° both north and south of the equator.

Wet adiabatic rate (pseudoadiabat) The diminished rate of cooling, averaging about 2°F per 1000 feet (6.5°C/km), of rising air above the lifting condensation level.

Wetlands Landscape characterized by shallow standing water all or most of the year, with vegetation rising above the water level.

Wilting point The point at which plants are no longer able to extract moisture from the soil because the capillary water is all used up or evaporated.

Wind gap An abandoned water gap.

Winter solstice The dates at which the most poleward extent of the perpendicular rays of the sun occur in the opposite hemisphere.

Woodlands Tree-dominated plant associations in which the trees are spaced more widely apart than those of forests and do not have interlacing canopies.

Woody plants Plants that have stems composed of hard fibrous material—mostly trees and shrubs.

Xeralf A suborder of Alfisol soils found in regions of mediterranean climate and characterized by a massive hard surface horizon in the dry season.

Xerert A suborder of Vertisol soils found in mediterranean climates and characterized by cracks that open and close regularly once each year.

Xerophytic The descriptive term that is applied to plants that are structurally adapted to withstand protracted dry conditions.

Youth (youthful stage) The initial, down-cutting stage in Davis's Geomorphic Cycle, in which the consequent streams are established and a drainage net begins to take shape.

Zone of aeration (vadose zone) The topmost hydrologic zone within the ground, which contains a fluctuating amount of moisture (soil water) in the pore spaces of the soil (or soil and rock).

Zone of confined water The third hydrologic zone below the surface of the ground, which contains one or more permeable rock layers (aquifers) into which water can infiltrate and is separated from the zone of saturation by impermeable layers.

Zone of saturation (phreatic zone) The second hydrologic zone below the surface of the ground, whose uppermost boundary is the water table. The pore spaces and cracks in the bedrock and the regolith of this zone are fully saturated.

Index

A

Note: Page numbers in *italics* refer to illustrations.

B

C

D

E

F

G

H

I

J

K

L

M

N

O

P

Q

R

T

U

V

W

X

Y

Z